World Sustainability Series

Series editor

Walter Leal Filho, Hamburg, Germany

More information about this series at http://www.springer.com/series/13384

Walter Leal Filho · Luciana Brandli
Olga Kuznetsova
Arminda Maria Finisterra do Paço
Editors

Integrative Approaches to Sustainable Development at University Level

Making the Links

Springer

Editors
Walter Leal Filho
HAW Hamburg
Hamburg
Germany

and

Manchester Metropolitan University
Manchester
UK

Luciana Brandli
Faculty of Engineering and Architecture
University of Passo Fundo
Passo Fundo
Rio Grande do Sul
Brazil

Olga Kuznetsova
Business School
Manchester Metropolitan University
Manchester
UK

Arminda Maria Finisterra do Paço
Department of Business and Economics
University of Beira Interior
Covilhã
Portugal

ISSN 2199-7373 ISSN 2199-7381 (electronic)
ISBN 978-3-319-10689-2 ISBN 978-3-319-10690-8 (eBook)
DOI 10.1007/978-3-319-10690-8

Library of Congress Control Number: 2014951054

Springer Cham Heidelberg New York Dordrecht London

Printed on acid-free paper

Springer is part of Springer Science+Business Media (www.springer.com)

Preface

Due to its scope and spectrum, sustainable development is a theme which is both interdisciplinary in nature, and flexible in respect of its usefulness and applications. An integrative approach to sustainability, i.e., an approach that combines aspects of several schools of thought, several methods and tools, is therefore very useful not only in respect of teaching but also in research as well, yielding results that monolitical approaches could not possibly achieve. The potential of integrative approaches is threefold:

- First, it caters for a wide range of techniques which foster learning, not only in respect of environmental elements, but also in terms of social issues, economic matters and political influences, which are known to permeat the sustainability debate.
- Second, integrative approaches entail the use of various techniques, from usual classroom discussions, but also fieldwork and the use of e-learning, hence fostering a broaders understanding of what sustainable development is and means.
- Third, integrative approaches take into consideration the needs and requirements of various stakeholders, being in a position to lead to action which do not ignore people's backgrounds, abilities and needs.

But even though integrative approaches to sustainability are characterized by a wide range of applications, their use is at present is rather limited. Part of the problem lies on the fact that most publications on sustainability tend to have a focus on individual matters (e.g. curriculum improvements, campus greening, etc.) and less on cross-cutting issues. Therefore, this publication has been conceived. It is an attempt to address a gap in the literature, at the same that that it shows, in practice, how integrative approaches work.

The book "Integrative Approaches to Sustainable Development at University Level: Making the Links" showcases different means, tools and initiatives, which illustrate how integrative approaches to sustainable development may help to bring the main messages across. Consisting of case studies, descriptions of practical experiences and empirical analyses, the book demonstrates how efficient integrative approaches to sustainability can be, at the same time that it documents and promotes some of the excellent work being undertaken, around the world, on this emerging area. The book is structured around two main parts:

Part I is about *Contextualising Integration*, which includes teaching and learning approaches aimed at the successful integration of sustainability issues in higher education systems, referring to the use of e-learning, campus engagement, sustainability delivery and reporting. It also includes matters related to indicators, and elements of transition and curriculum integration.

Part II of the book refers to *Convergent Approaches*, outlining examples of training programme, curriculum innovation and organisational changes, and including simulations and appraisals of problems and barriers, with examples of good practice.

This book is a further volume of the new "World Sustainable Development Book Series" and contains some of the papers presented at the "World Symposium on Sustainability in Higher Education" (WSSD-U-2014) held in Manchester, UK, on 3–5 September 2014. We want to thank the authors for sharing their work, their know-how and their expertise to a world audience, and hope the many experiences amassed in this book will help to support the work of sustainability lecturers, researchers and students working in this very important field.

Autumn 2014

Walter Leal Filho
Luciana Brandli
Olga Kuznetsova
Arminda Maria Finisterra do Paço

Contents

Part I
Contextualising Integration

Digital Learning for Sustainability: An Interactive Magazine for Students, Academics and Expert Practitioners

Antonius Raghubansie, Wendy Corbett, Katy Boom and Lorraine Weaver

Abstract

The complexity which underpins the evolving concept of learning for sustainability requires effective inter-disciplinary collaboration and networking to succeed. Developing a shared understanding and route maps for sustainability would not be feasible without the rapid advances in information technology, global access to information and the evolution of self-organised socio-technical networks. The University of Worcester in the UK has developed a learning elective programme (cross-faculty) for sustainability, which aims to build skills to enhance employability using innovative teaching and learning techniques. A new dimension to this innovative pedagogy is the development of an online magazine and virtual learning environment called susthingsout to promote and facilitate learning for sustainability in the wider undergraduate curriculum, including activities on campus and in the wider community. This paper outlines how this new platform has been co-created by students, academics and expert practitioners and how it supports the application of theoretical knowledge and learning to consolidate and enhance students' employment prospects.

A. Raghubansie (✉) · W. Corbett · K. Boom · L. Weaver
Institute for Science and the Environment, Worcester Business School,
Institute for Humanities and Creative Arts, Sustainability Department,
University of Worcester, Henwick Grove WR2 6AJ, UK
e-mail: a.raghubansie@worc.ac.uk

W. Corbett
e-mail: w.corbett@worc.ac.uk

K. Boom
e-mail: k.boom@worc.ac.uk

L. Weaver
e-mail: l.weaver@worc.ac.uk

W. Leal Filho et al. (eds.), *Integrative Approaches to Sustainable Development at University Level*, World Sustainability Series, DOI 10.1007/978-3-319-10690-8_1

The main findings are that the model can be effective but it has to draw on a range of skills beyond disciplines, the institution and the classroom, a major challenge to established academic practice. Students developed a much wider understanding of sustainability, saw the issues come to life and felt that the experts who contributed was one of the strongest element. The requirement to work in teams for a live project was the most daunting but ultimately rewarding experience. Major issues to be addressed include steep learning curves (for all), cross-functional planning, managing issues beyond sustainability, marketing programme for external followers, providing safe spaces for students to make mistakes and better linkages to assessments and inputs.

Keywords
Education for sustainable development (ESD) · Employability · Innovative pedagogy · Digital learning · Trans-disciplinarity

1 Introduction: The Worcester Sustainability Elective

The Sustainability *Elective* is a curriculum innovation which sits alongside the typical degree programme. It is trans-disciplinary, made of year-long, 30 credit modules, and open to students (level 4 and 5) from across the University. It has a distinct timetable slot across all faculties to assist choice. It is distinct in the UK HE sector in these regards, as the elegant solutions made education for sustainable development (ESD) available to the whole undergraduate community. Its application of digital technology is innovative but also its pedagogic structure—experts, practitioners and students engaging in the same space. The virtual learning environment (VLE) susthingsout.com is the focus of this study. It aims to be a developmental learning platform, source of information creation and dissemination and engages multiple stakeholders in on-going dialogue and networking. Our proposition is that greater impact will be achieved if all these stakeholders are linked. Creative digital technologies and online learning and teaching strategies are applied to deliver academic and university strategic objectives (see Fig. 1). The inclusion of the community, campus, curriculum and students in one space is unique in this VLE which has a public and private side.

In the English counties of Worcestershire and Herefordshire the University of Worcester (UoW) is the sole institution with about 10,000 students. One of the core values is sustainable development through active community engagement (UoW 2013), evident in its top five ranking for the People and Planet Green League and achieving the first EcoCampus Platinum rating for the whole organisation in England.

This paper sets out the VLE model, the journey of the team, the impact and major lessons learned from a complex and ambitious endeavour. The research questions:

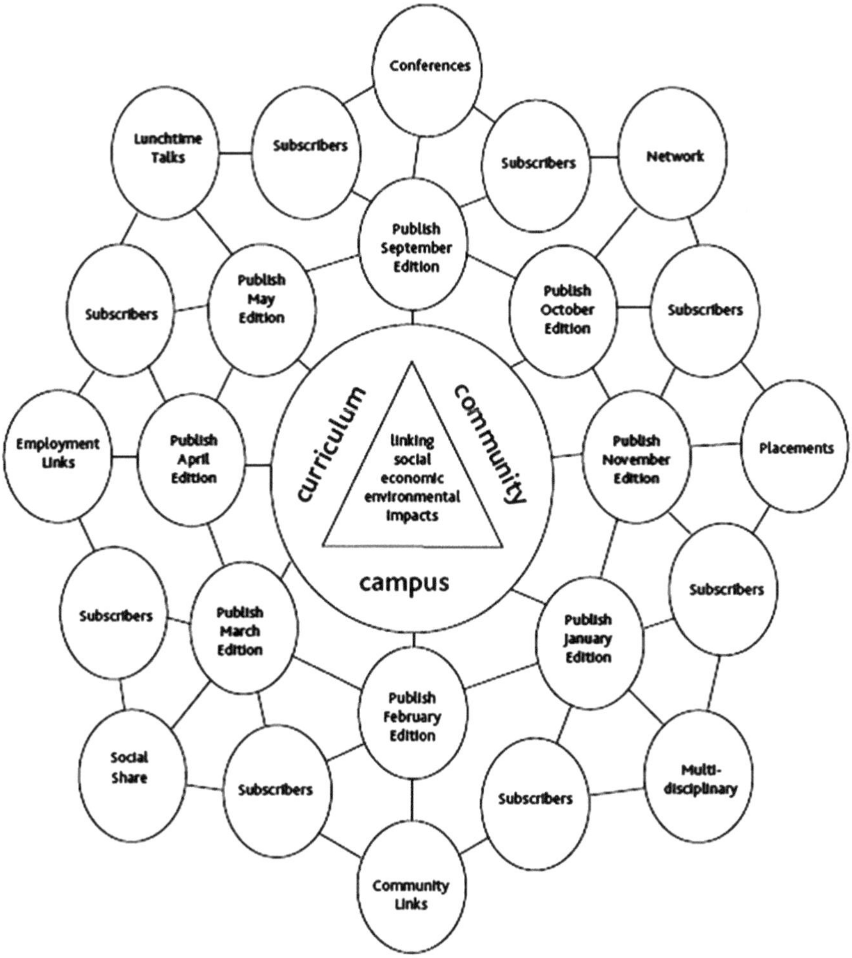

Fig. 1 Linking students, curriculum, campus, community: co-creating knowledge through creative digital technologies and online learning and teaching strategies

- What are students, 'academics and experts' views on a digital collaborative learning model?
- What good practice has been identified?
- What were the key lessons?/What areas to be developed?

2 VLE Model

Susthingsout.com is a digital magazine hosted by the University of Worcester and published eight times a year, offering free subscription. Its tag line is *change today, protect tomorrow*. It has an external facing side (as a publishing platform) and an internal learning side, which is private to the students and academics. Its impact is in linkages to sustainability projects (Academic, Social, Economic and Environmental) introduced into the curriculum by live internal and external events, student-led projects, talks, conferences, special features, guest contributors and case histories so students can see that they can change today to protect tomorrow. There are links to the rest of the UK and the world as a number of the experts and practitioners are world experts in their fields.

2.1 Initial Journey

Stage 1—Basic Concept Discussions
During a peer observation in which a website was being used to facilitate a year 1 UG class, a question was posed as to how this strategy might work for the Worcester Elective. This was the start of weeks of discussions with colleagues.

Objectives:

- Create a secure digital platform to facilitate a multi-disciplinary, online learning and teaching content strategy, for use by multiple authors with the functionality to facilitate sharing and knowledge co-creation.
- To include a private, secure, online collaborative space for use by tutors and students including functionality for e-portfolios, feedback mechanisms and assessment as part of a blended learning strategy.
- Develop student employability skills, sharing volunteer and paid opportunities, and participate in an online multi-disciplinary network of potential employers, academics and the wider community.

Stage 2—Project Scoping

- Interactive digital magazine using web 2.0 technologies with VLE capabilities. Wordpress was chosen as the system.
- A content strategy linked to the curriculum and facilitated by a multi-disciplinary editorial team made up of academics, experts and students.
- A marketing strategy to build an online community.

Stage 3—Final Decisions

- susthingsout was approved including full social media embedded (Twitter, comments, blogs, videos etc.) to launch it in 2013–2014

- To set up the editorial team to implement
- Re-launch in September 2014 with improvements from the initial run

3 Literature Review

The role of universities in providing skills for ESD is well established as an area of study (Martin 2008; Martin et al. 2013). Much of this has been aimed at environmental sciences in the first place (Roberts and Roberts 2008), then towards broader institutional change (Beringer et al. 2008; Junyent and de Ciurana 2008). If HE is to fulfil its role, traditional disciplinary silos need to be broken down (Tilbury 2011). In the UK academic community, networking and communication groups (via EAUC, JISC SHED-SHARE) have been established to address the broader challenge of breaking down silos. However, there have been few which bring together the academic, the student and the broader society, which is what the Worcester online resource provides. This section seeks to identify a theoretical framework and set UoW's sustainability e-zine into this context.

3.1 Digital Resources and Sustainability

Technology has made it much easier for interested parties from all over the country and the world to access resources. Technology enhanced learning (TEL) has been growing apace (Jenkins et al. 2011). For the first time, we also have the so-called "Google generation" at University, for the vast majority of whom self-organising social networks (SN) have been a part of their everyday activity. Li and Pak (2010) summarise the following criteria for determination of web quality—ease of use, usefulness, information content, security, responsiveness and personalization.

3.2 Education for Sustainability

ESD has a goal to encourage the adoption of sustainable habits by learners and those they influence (Ballantyne and Packer 2005). However, Bonnett (2002) argues that there is too much reliance on a "right way" which is eco-dominant, that education needs to move to a much broader discussion including change at an individual, metaphysical level, which poses a challenge for western education. The dominant environmental aspect is evident in a survey to capture students' views and attitudes towards sustainability issues (Kagawa 2007).

This study is not quantitative, nor at the level of the full student body but a smaller group as we are examining motivations and personal views rather than breadth of opinion, as the case study investigates a model that is deliberately different. Lange (2004) agrees with Bonnett in finding that a more dialectic approach produces deep learning in a US study among adult learners on an active citizenship course (also Martin et al. 2006; Carlos and Capriotti 2009). The Worcester e-zine is aimed at young people within the context of a taught full module over an entire year.

Lugg (2007) discusses the need to develop a more holistic approach proposing using outdoor learning for sustainability, allowing for experiential integration of knowledge from a range of disciplines. Warburton (2003) also argues for a deeper learning on a similar level examining the impact of the students' background and their abilities to engage in a classroom. In an idealistic paper, Ballantyne and Packer (2005) go so far as to suggest that Education for Sustainability (EfS) should be free choice, with an attempt to capture informal settings and their effectiveness. Shephard (2008) explored the relationship between curriculum and the values, attitudes and actions of learners and how these might be applied to sustainability. Kurland et al. (2010) attempt to take this on with an undergraduate sustainability module in the US studied from seven different disciplines. This was a pilot as part of a larger institutional sustainability curriculum initiative. The Worcester case study is consistent with this context but within a new, UK institution.

We found one Canadian study of EfS with first year UG students but within the context of public sphere models to encourage interactions between class and community (Van Wynsberghe and Andruske 2007). Sibbel (2009) observes that the focus of EfS has been on consumer behaviour to influence consumption habits, arguing that you have to develop self-efficacy, advocacy skills and a holistic view in a new way of considering professional development. Hence, we deliberately involve professionals, external activities, project work and individual work hosted on the e-zine.

3.3 Studies in the Context of Blended Learning

McGuire and Castle (2010) in a US study explored students' views of blended learning comparing face to face, online and hybrid, and how students believe these help them to learn but generally rather than in the context of sustainability education. Conole (2010) finds that despite the desire for help, the access to resources and the communication of these via Web 2.0, adoption has been slow. Kassens-Noor (2012) looks at informal use of Twitter finding that despite Web 2.0 being more flexible there are also pitfalls. Shepherd and Skrabut (2011) examine electronic portfolios generically, noting they can deepen learning, produce materials, develop learning abilities and provide a basis for communication with third parties. However, there are many limitations. A recent study of Finnish and US tourism module's use of Facebook employs a case study, as we do here. Overall, students gave positive feedback on the informal nature of the learning tool but there were barriers to group working and there were some individual confidence issues (Isacsson and Gretzel 2011). The Worcester case is of a specific UK university but also across disciplines.

3.4 Digital Learning and Sustainability

Dengler (2008) studies postgraduate geography students showing that an online forum increased participation and reduced inhibitions (also Jenkins et al. 2011). The Worcester case looks at level 4 students who are less confident and have less experience than PG students. We have not located a UK study that has students from many different disciplines such as this. Wiek et al. (2013) in a US/German online module do focus on sustainability but within the cross-cultural, national, and geographical boundaries to understand the historical, epistemological and ethical underpinnings of diverse viewpoints.

3.5 Propositions

Following McGuire and Castle (2010) to support the central theme of this paper we propose a triangle—(1) course content, and (2) instructor competence result in higher levels of student self-assessment when (3) TEL is employed. However, if (1) and (2) are not in place then (3) would not help.

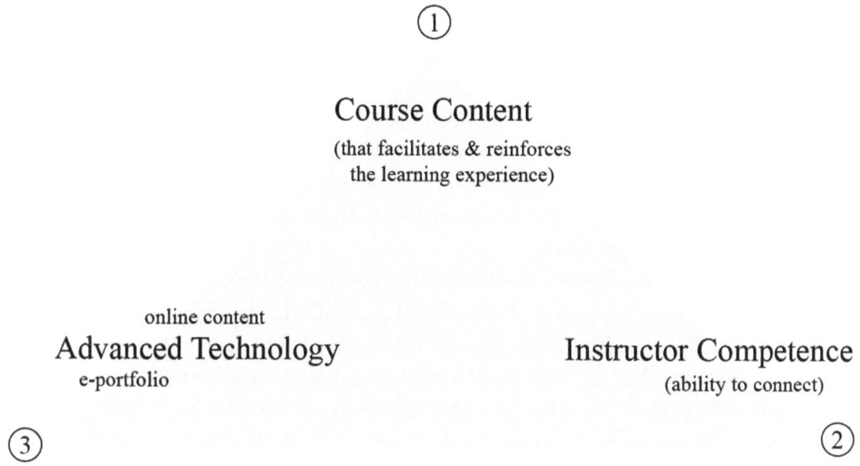

Student Self Assessment & Learning Triangle

An analysis of student self-assessment of online, blended, and face-to-face learning environments: Implications for sustainable education delivery (Castle & McGuire, 2010).

This model sets out a proposition that facilitating online student response to course content requires careful planning—introducing tasks for completion online, in teams and as an individual—and embedding these tasks into the course curriculum. One way to do this would be to explore questions against each of the three parts of the triangle (with part three split (a) e-portfolio and (b) online content, then a general conclusion from students on their experience summarised below.

- Key positives from the student perspective
- Key areas that require greater support
- Taking the whole picture in consideration preferences for this model of delivery (internal/external, face to face and online)

4 Methodology

This paper takes a constructivist view to reflect the dynamic, changing, human nature of the curriculum development and implementation under study (Stake 1995). This philosophy reflects the changing nature of the subject and the context as seen by the participants. It also allows deep connections between the researchers and the people involved (Johnson and Onwuegbuzie 2004) most often applying qualitative data collection tools. Enonbun (2010) argues that the advent of Web 2.0 means that online research has become more constructivist, as is the case here.

4.1 Case Study

Yin (2003) argues that case study is appropriate where the researcher wants to delve deeply, where the context is as important as the event being investigated and where it is difficult to separate the two. The decision to introduce an online resource could not be done without the University commitment to an external engagement agenda (Baxter and Jack 2008). It would be very difficult to understand the reasons for engaging or not engaging with the resource, without considering the influence of the broader student experience.

Yin (2003) and Stake (1995) observed that delimiting the borders of the case can enhance the focus of the study. Following Creswell (1998), this study looks over academic year 2013–2014 within one UK University and within a particular module (Stake 1995) and within the context of a pilot programme (after Miles and Huberman 1994). Therefore, here we analyse the process of developing the online learning magazine, of the on-going programme of content management as it was designed and how it evolved over the period. In this regard, the study does not attempt to examine causal relations but to explore and to describe the case as a single whole (Yin 2003).

4.2 Data Sources

Consistent with case studies, we use a number of data sources to develop a fuller, more credible understanding of the stakeholders' views of the online magazine as a learning tool (Yin 2003). We used formal module documents, planning meeting records, three semi-structured depth interviews with student editors, the online website and materials themselves, direct observations, and a focus group of six students. They are treated as part of the picture rather than as separate items for analysis, contributing to a fuller view.

4.3 Conceptual Framework and Analysis

Following Miles and Huberman (1994), we employ a broad conceptual framework developed from McGuire and Castle (2010)—the student self-assessment and learning triangle. This has helped us to set out the boundaries of the study and to focus the analysis and interpretation of the case study (Thomas 2011). We can focus on learners, academics, experts and editorial team.

Yin (2003) develops five techniques to analyse case data but here we will primarily employ, pattern matching, explanation building and logic models. The key test is to focus on the evidence and to relate these to the research propositions/questions being addressed here; the major elements of the case study (Baxter and Jack 2008).

5 Analysis and Discussion

We propose to use the logic model below to underpin the analysis (Fig. 2).

5.1 Inflow Stage

Institutional and pedagogic direction was already established. The elective structure had been approved a year and half prior to the launch of susthingsout. At the Inflow stage of the logic model applied here, there were no resources specifically dedicated to the website. However, staff time was allocated by the different academic departments to the teaching staff. One of the original members of the elective

INFLOW	OUTFLOW		OUTCOMES		
	Actions	Involvement	Learning	Behaviour	Results
Resources?	What can be done?	Stakeholders?	Learning outcomes	What will change during the year?	Tangible acts
Staff Helpers Time ££ Tools Technologies Relationships	Curriculum Meetings Workshops Presentations Publications Writing Editing Commenting Sharing	Students External experts Management Supportive? Opposing?	Employability skills and LOs	Actions Behaviour Decisions Involvement Feedback	Use of the e-zine Active in discussions Work well in teams Reflect

Fig. 2 A logic model for design and evaluation

development group was a non-academic and undertook to explore possibilities to set aside some funds to bring in an expert to support the website, illustrating like Kurland that personal motivation is an important driver for EfS initiatives. It is to the creativity and interest of the planning team that the resources were found from various sources, consistent with the work of Lugg (2007).

Personal relationships were essential in bringing colleagues to support content, editing, consultancy, guest speaking, videos, interviews etc. (evident also in Wiek et al. 2013). The travel costs of external speakers were paid by the university. A successful application was made to a university 'scaffolding' scheme for embedding sustainability in the curriculum which was used to bring expert practitioner advice to the editorial team. The setting up of a student editorial team alongside the professional team was a decisive and valuable initiative.

Technological platforms were also discussed with the media expert on the planning team. Computer tablets were purchased for the student team and for use in the classrooms from budgets established to facilitate embedding sustainability within the curriculum. Office space for the student editors was provided by the university.

5.2 Outflow Stage

The routine, planning activities were well established in the pilot. However, the more interesting area for other HEIs is the sub-section—Involvement. Resulting from the inter-departmental nature of the group, the team could influence senior decision-makers across the university from different directions essential to success as demonstrated in the various studies of Martin et al. (2013). The team also worked to secure senior management sponsors, primarily through inter-personal communication with formal communication in parallel or following on. These were immensely helpful in addressing potential bottlenecks and heading off conflict areas.

For the academic team, it was a steep learning curve with the technology. There was some skepticism to start with. Training had to be done to address these (Jenkins et al. 2011). Members of the team clearly all had other responsibilities and this meant that on occasions, decisions were delayed; some compromises had to be made (Isacsson and Gretzel 2011). As a result, meetings started to almost always have a digital element (skype or conference call). This went alongside the requirement to engage in virtual, public and private collaboration among team members.

> All in all, susthingsout is an engaging, thought provoking project that not only increased our employability skills, but also increased our knowledge and passion for sustainability.
>
> —Student editor

5.3 Outcomes Stage

These will be addressed using the Student Self-Assessment and Learning Triangle (Table 1).

Table 1 Student self-assessment and learning triangle

Positives	Challenges
Overall	
"We believe working on susthingsout has helped us improve our existing skills, such as working under pressure and towards tight deadlines, in a close team" —Student (Focus Group)	
Course Content	
Students have developed cross disciplinary capability from internal/external speakers and in developing ideas for the e-zine (Isacsson and Gretzel 2011) They feel more confident in their understanding of issues	The major issue is in understanding the role of susthingsout—an academic resource, or a public information guide consistent with Li and Pak (2010) Aligned to this is a fear of "looking stupid" in commenting in public (Isacsson and Gretzel 2011)
The assessment (group work) on a live university (Go Green Week) was a positive experience because of the scale, the responsibilities and the opportunity to engage with internal/external organisations involved with sustainability (Isacsson and Gretzel 2011). This is particularly relevant as group work has been highlighted in prior studies	Set group tasks in the class, and link as academic reference or case history (i.e. the usefulness criteria) Better clarity of links between guest speakers and session topics and assessment
e-portfolio—students feel that there is a positive impact on their learning (Shepherd and Skrabut 2011) but some of the quality and quantity of work could be better (Kassens-Noor 2012)	"I would be more likely to publish to susthingsout as a group. There will be more to say too. And more fun!"—Student (Focus Group)
Instructors	
Building on tutor knowledge	Relate/link tasks to content on susthingsout
Students felt supported with conceptual and practical guidance but that it was a "rollercoaster journey"	Build student confidence through online group tasks in the class sessions
The editorial students felt as they wrote more articles, attended editorial meetings and involved in the strategy and positioning, they developed a far wider capability than their core degree (Bonnett 2002; Lange 2004; Carlos and Capriotti 2009)	Students to choose own learning groups as a connection through similar goals builds trust and confidence. Multi-disciplinarity was the additional dimension to overcome
"It saves me lots of admin and facilitator time and when I ask the students if they want to use this method or another they always say this has helped their learning" —*Creative Academic*	Tutor help to facilitate the interdisciplinary conversations among students/lecturer to lead discussions (off and online too)
	Research skills were highlighted as a challenge in extracting the most relevant content from video interviews with external speakers

(continued)

Table 1 (continued)

Positives	Challenges
Experts	
Employability skills were enhanced by the discussions but also in students taking opportunities offered by speakers	People are very busy and so we have to consider how to make it easy for them to contribute
Presenting in several modes—in person, video interview, lunch-time cross-university public talk and on susthingsout.com	"I have had positive feedback from a couple, but it hasn't resulted in posts"—Expert witness
More than half of the speakers are returning. They have posted and made comments	
"I see this as giving back and I am very excited"—Sustainability recruitment agency	
Technology	
Li and Pak (2010) identify security as an issue but there is little risk in this case. Simple to post and comment	Ease of use could be better (Li and Pak 2010) mainly because the students are using a competing VLE in other modules
Similarly, on responsiveness susthingsout was updated regularly and the student editors brought a different perspective. Simple to post and comment	Would like to see the internal and external face look similar
Personalization was largely not an issue either (except 1 question about anonymity) as the students had their own zone, the external, its own look and then content presented in multiple formats aimed at different audiences	e-portfolio—some students wanted to have private groups/posts as they felt pressure if knowledge is lacking or they get it wrong

5.4 Transdisciplinary Observations

During the analysis of the case both from secondary and primary data, there arose a number of issues which could not be captured in the analytical framework set out above. We briefly point these out here.

- Student editors were especially interested in understanding target audiences (more in the domain of marketing communications) and of having different media to use where suited.
- Time pressures were stressful but also exciting enjoying the real pressure of working in publishing. And meeting creative professionals/academics they would not have met otherwise.
- Skills developed which were well outside their expectations—journalism, magazine marketing and writing for search engine optimisation.
- Awareness of "*the bigger picture*", that they were involved in a process which is about leading for tomorrow.

6 Conclusions

The susthingsout.com model works but it has to be very carefully planned and must draw from a range of skills, not usually available in the typical curriculum. Students from both editorial and class groups have shown that susthingsout.com has widened their view of sustainability. Academics acknowledge that learning together has helped to make the VLE potentially even more powerful, recognising that the external face is still to be promoted. Experts' involvement has had the most positive feedback. Their involvement was enthusiastic and made the issues come to life.

- Constant reflection, change and evaluation are required
- Susthingout.com provides educational, publishing and training opportunities
- Virtual collaboration skills for academics, students and practitioners
- Students were mobilised to engage in major (assessed) live projects which justified their choice of the elective

  Future prospects:

- Develop a database for the external followers and plan for a full public launch.
- Providing "safe" spaces (virtual and real) for students to make mistakes
- In class tutor-led discussions on speaker sessions and topics
- Better linkages of topics and assessments/online learning tasks
- Consider skills which are "outside" of the subject curriculum

References

Ballantyne R, Packer J (2005) Promoting environmentally sustainable attitudes and behaviour through free-choice learning experiences: what is the state of the game? Environ Educ Res 11 (3):281–295

Baxter P, Jack S (2008) Qualitative case study methodology: study design and implementation for novice researchers. Qual Rep 13(4):544–559

Beringer A, Wright T, Malone L (2008) Sustainability in higher education in Atlantic Canada. Int J Sustain High Educ 9(1):48–67

Bonnett M (2002) Education for sustainability as a frame of mind. Environ Educ Res 8(1):9–20

Conole G (2010) Facilitating new forms of discourse for learning and teaching: harnessing the power of Web 2.0 practices. Open. Learning 25(2):141–151

Carlos RJ, Capriotti P (2009) Communicating CSR, citizenship and sustainability on the web. J Commun Manag 13(2):157–175

Creswell J (1998) Research design: qualitative, quantitative, and mixed methods approaches, 2nd edn. Sage, Thousand Oaks

Dengler M (2008) Classroom active learning complemented by an online discussion forum to teach. Sustain J Geogr High Educ 32(3):481–494

Enonbun A (2010) Constructivism and Web 2.0 in the emerging learning era: a global perspective. J Strateg Innov Sustain 6(4):17–27

Isacsson A, Gretzel U (2011) Facebook as an edutainment medium to engage students in sustainability and tourism. J Hosp Tour Technol 2(1):81–90

Jenkins M, Browne T, Walker R, Hewitt R (2011) The development of technology enhanced learning: findings from a 2008 survey of UK higher education institutions. Interact Learn Environ 19(5):447–465
Johnson RB, Onwuegbuzie AJ (2004) Mixed methods research: a research paradigm whose time has come source. Educ Res 33(7):14–26
Junyent M, de Ciurana AMG (2008) Education for sustainability in university studies: a model for reorienting the curriculum. Br Educ Res J 34(6):763–782
Kagawa F (2007) Dissonance in students' perceptions of sustainable development and sustainability: implications for curriculum change. Int J Sustain High Educ 8(3):317–338
Kassens-Noor E (2012) Twitter as a teaching practice to enhance active and informal learning in higher education: the case of sustainable tweets. Act Learn High Educ 13(1):9–21
Kurland NB, Michaud KEH, Best M, Wohldmann E, Cox H, Pontikis K, Vasishth A (2010) Overcoming silos: the role of an interdisciplinary course in shaping a sustainability network. Acad Manag Learn Educ 9(3):457–476
Lange EA (2004) Transformative and restorative learning: a vital dialectic for sustainable societies. Adult Educ Q 54(2):121–139
Li RYM, Pak DHA (2010) Resistance and motivation to share sustainable development knowledge by Web 2.0. J Inf Knowl Manag 9(3):251–262
Lugg A (2007) Developing sustainability-literate citizens through outdoor learning: possibilities for outdoor education in higher education. J Adventure Educ Outdoor Learn 7(2):97–112
Martin S, Dillon J, Higgins P, Peters C, Scott W (2013) Divergent evolution in education for sustainable development policy in the United Kingdom: current status, best practice, and opportunities for the future. Sustainability 5(4):1522–1544
Martin S, Dawe G, Jucker R (2006) Embedding education for sustainable development in higher education in the UK. Drivers and barriers for implementing sustainable development in higher education. UNESCO, Paris, pp 61–67
Martin S (2008) Sustainable development, systems thinking and professional practice. J Educ Sustain Dev 2(1):31–40
McGuire CJ, Castle SR (2010) An analysis of student self-assessment of online, blended, and face-to-face learning environments: implications for sustainable education delivery. Int Educ Stud 3 (3):36–40
Miles MB, Huberman AM (1994) Qualitative data analysis: an expanded sourcebook, 2nd edn. Sage, Thousand Oaks
Roberts C, Roberts J (2008) Starting with the staff: how swapshops can develop ESD and empower practitioners. Environ Educ Res 14(4):423–434
Shephard K (2008) Higher education for sustainability: seeking affective learning outcomes. Int J Sustain High Educ 9(1):87–98
Shepherd CE, Skrabut S (2011) Rethinking electronic portfolios to promote sustainability among teachers. TechTrends 55(5):31–38
Sibbel A (2009) Pathways towards sustainability through higher education. Int J Sustain High Educ 10(1):68–82
Stake RE (1995) The art of case study research. Sage, Thousand Oaks
Thomas G (2011) A typology for the case study in social science following a review of definition, discourse, and structure. Qual Inq 17(6):511–521
Tilbury D (2011) Higher education for sustainability: a global overview of commitment and progress. High Educ World 4:18–28
University of Worcester (2013) Strategic Plan 2013–2018. http://www.worcester.ac.uk/documents/university-worcester-strat-plan-2013-18.pdf. Last Accessed 29 March 2014
Van Wynsberghe R, Andruske CL (2007) Research in the service of co-learning: sustainability and community engagement. Can J Educ 1(1):349–376
Warburton K (2003) Deep learning and education for sustainability. Int J Sustain High Educ 4 (1):44–56

Wiek A, Bernstein MJ, Laubichler M, Caniglia G, Minteer B, Lang DJ (2013) A global classroom for international sustainability education. Creative Educ 4(4A):19–28

Yin RK (2003) Case study research: design and methods, 3rd ed. edn. Sage, Thousand Oaks

Authors Biography

Antonius Raghubansie is Associate Head and Head of Marketing Subject Group, Worcester Business School. Antonius led the University of Worcester Green Academy team as part of the HEA pilot with eight UK HEIs exploring how to embed sustainability into the curriculum. Antonius is also a member of the university Learning and Teaching and Environment Committees.

Wendy Corbett is a visiting lecturer in Creative Digital Media with over 15 years expertise delivering digital publishing projects. For 6 years she was Director of the International Herald Tribune and 5 years as Senior Multimedia Designer developing digital strategies for government sustainability projects.

Katy Boom is Director of Environmental Sustainability. She has led the development of wide ranging sustainability initiatives on campus, community and in the curriculum. The University consistently tops the national sustainability leagues and became the first English University to be awarded platinum EcoCampus status. For overseeing students as partners on major sustainability change programmes and building sustainability skills with students and student volunteers, Katy's work has won a number of national awards.

Lorraine Weaver is Head of the Biological Sciences and is lead academic for the sustainability elective modules. She is a principal lecturer and UoW Teaching Fellow. Along with being a Fellow of the Society of Biology and a Chartered Biologist, Lorraine and is also chair of the university Strategic Biodiversity Management Group.

The Relevance of Transdisciplinary Teaching and Learning for the Successful Integration of Sustainability Issues into Higher Education Development

Johannes Merck and Marina Beermann

Abstract

The idea of transdisciplinary research is understood as a promising teaching method in the context of sustainability. Against the background of motivational research it will be argued that transdisciplinary teaching can act as a relevant motivational element for the sustainability learning processes.

Keywords

Transdisciplinary teaching · Higher education · Sustainability issues · Motivational research

1 Introduction

The relevance of transdisciplinary research within sustainability science (in addition to interdisciplinary research) is almost mainstream nowadays. In contrast to this, transdisciplinarity receives rather limited consideration within academic teaching and learning concepts. Accordingly, this paper intends to explore the chances and advantages of transdisciplinary teaching and learning within the field of sustainability programs. Transdisciplinarity includes the participation of society as a whole, whereupon in this article transdisciplinary teaching focuses on the participation of practitioner, e.g. experienced CSR manager.

J. Merck (✉)
Otto Group (VV-CR), Otto GmbH & Co KG, Wandsbeker Straße 3-7, 22172 Hamburg, Germany
e-mail: Johannes.Merck@ottogroup.com

M. Beermann
Systain Consulting GmbH, Spaldingstrasse 218, 20097 Hamburg, Germany
e-mail: marina.beermann@systain.com

W. Leal Filho et al. (eds.), *Integrative Approaches to Sustainable Development at University Level*, World Sustainability Series, DOI 10.1007/978-3-319-10690-8_2

2 Specific Demands Due to Sustainable Development

Despite the huge variety of sustainability concepts and the ongoing debates on possible scopes of sustainability on a civil, legal, corporate and individual level, sustainability as a guiding principle is broadly established. Nevertheless, there are still many challenges regarding the implementation of sustainability. This is due to the nature of sustainability. Sustainability normally is linked to multidimensional, complex, inter- and transdisciplinary, ambivalent challenges with different time horizons (cf. Beermann et al. 2014). Consequently, teaching and learning in the context of sustainability necessitates specific competencies—as lecturer but also from the perspective of future sustainability managers (cf. Kopp and Martinuzzi 2013; Merck 2013; Nölting et al. 2013). Hence, specific demands result for teaching sustainability issues in higher education. Furthermore, the question arises, what kind of skills and competencies need to be acquired for a successful education of future sustainability managers.

2.1 The Imparting of Sustainability Specific Skills and Competencies

Looking back at the landscape of sustainability related research and science in the context of higher education certain shifts can be identified. Tilbury (2011) clearly shows that nowadays new demands towards sustainability related research arise (Table 1):

This shows that new claims concerning the role of sustainability related research have been raised. The self-image of sustainability related research as passive and mainly informing actor changed towards a new role as "change agent". Research and science shall nudge changes towards sustainability by itself. This trend can be seen in research but also in the area of teaching and learning.

Table 1 Shifts in the sustainability related research landscape in the context of higher education

Shifts from...		...to be more inclusive of...
Research which is discipline focused	➡	Research which is inter- and multidisciplinary
Research that has academic impacts		Research which has social impact
Research that informs	➡	Research that transforms
Research on technological and behaviour change		Research that focuses on social and structural change
Researcher as expert	➡	Researcher as partner
Research on people		Research with people

Source Tilbury (2011, p. 6)

According to the UNESCO report on education and skills for inclusive and sustainable development beyond 2015, one eminent trend is to recognize (...) "that learning is increasingly happening individually beyond formal educational settings, the role of teachers will have to evolve from dispensers of information and knowledge to facilitators and enablers of learning" (UNESCO 2012a).

Amongst others, the Bologna reform has induced such changes since 1999 due to the enhancement of competence development. It acts as a frame for competence based learning and initiates a shift from defining aims and objectives into learning outcomes. Although Schaeper (2008) clarifies that key competencies are not explicitly mentioned in the official statements on the Bologna process, competencies are closely related to an explicit objective of the main reform efforts: enhancing the employability of higher education graduates. Therefore, Schaeper (2008) states that key competencies are more an implicit or rather derived objective of the Bologna reform.

The indicated entitlement to a stronger focus of academic education on the labour market shows that also sustainability related graduation programs have to consider these requirements. Against this background a study carried out by Kopp and Martinuzzi (2013, p. 195) is of high interest. Based on analyses of current publications, which are based on interviews of leaders in middle and top-positions and their view on necessary sustainability competences, Kopp and Martinuzzi show the multifaceted profile of requirements for sustainability managers/coordinators, especially in business organizations (Table 2).

Subsequently, following questions arise for the implementation of sustainability into higher education: How can students be trained as promising future sustainability managers? Which barriers need to be taken into consideration? What are possible success factors taking into account the Bologna reform's aim of enhancing employability? And which teaching and learning methods can be used?

According to the UNESCO (2012b, p. 25) nine forms of learning can be differentiated in the context of sustainable development. These had been identified as part of a global evaluation (GMES) (Table 3):

Table 2 Sustainability competences for sustainability manager/coordinators

Advanced comprehension of sustainable development	Handling of complexity	Balance between local and global perspectives	Diversity, respect for diverging perspectives
Comprehension of effects, risks and opportunities	Integrative approach	Innovation and creativity	Emotional intelligence and self-perception
Vision, power of persuasion, organisation of transformation	Understanding of interdisciplinary connections	Handling of insecurities, ambiguities and dilemmas	Perception of "shadow issues"
Systematic/holistic thinking	Long-term perspectives	Support in decision-making through dialogue and intuition	Learning and development

Source Based on Kopp and Martinuzzi (2013, p. 95)

Table 3 Various forms of ESD-related forms of learning (UNESCO 2012b)

ESD-related forms of learning (UNESCO 2012b)	Short description
1. Discovery learning	Learners are immersed in a rich context where they encounter some element of mystery; they become curious and begin to make sense of their experience through their own exploration
2. Transmissive learning	Using didactic skills (e.g. presenting, lecturing, story-telling) and supporting materials (e.g. workbooks, instruction forms, visuals) a body of knowledge, set of rules or code of conduct is transferred to the learners
3. Participatory/collaborative learning	Although not identical, both emphasize working together with others and active, not passive, participation in the learning process, which tends to focus on resolving a joint issue or task
4. Problem-based learning	Focused on solving real or simulated problems, to better understand the issue or find ways to make real-life improvements. Issues are either identified by the learners, or pre-determined (e.g. by teachers, experts, commissioning bodies)
5. Disciplinary learning	Taking questions of a disciplinary nature (e.g., geographical and biological) as a starting point, to better understand underlying principles and expand the knowledge base of that discipline
6. Interdisciplinary learning	Taking issues or problems as a starting point, then exploring them from different disciplinary angles to arrive at an integrative perspective on possible solutions or improvement
7. Multi-stakeholder social learning	Bringing together people with different backgrounds, values, perspectives, knowledge and experience, from both inside and outside the group initiating the learning process, to set out on a creative quest to solve problems that have no ready-made solutions
8. Critical thinking-based learning	Exposing the assumptions and values people, organizations and communities live by and challenging their merit from a normative point of view (e.g. animal well-being, eco-centrism, human dignity, sustainability) to encourage reflection, debate and rethinking
9. Systems thinking-based learning	Looking for connections, relationships and interdependencies to see the whole system and recognize it as more than the sum of its parts; and to understand an intervention in one part affects other parts and the entire system

Source Based on UNESCO (2012b, p. 25)

While some forms of learning can be considered as conventional (transmissive learning, disciplinary learning), others are much more innovative and novel such like multi-stakeholder social learning or systems thinking-based learning. It likewise makes the point quite clearly that there is no explicit relationship to transdisciplinary forms of teaching and learning. The specific relevance of this form of teaching and learning will therefore be examined below.

3 The Role of Transdisciplinary Teaching and Learning Within Sustainability Programs

Teaching sustainability issues necessitates sensitizing students for a realistic view on sustainability challenges as well as awareness-raising for potential discrepancies between theory and practice. The logic of science (objective establishment of the truth) on the one hand and the logic of economy (appropriateness) on the other hand can lead to imbalances and frustration for graduates. One challenge therefore is to actively reflect and discuss the necessary balancing act of the scientific preoccupation with sustainability issues and the economic reality. Against this background, the risk of a lack of motivation arises immediately. This is due to the fact that the examination of sustainability issues is often attended by complex deficits and rarely with positive success stories. Enhancing motivation instead needs to show that changes towards sustainability are possible. This is confirmed by motivational research (Deci and Ryan 1993; Bandura 1977). Transdisciplinary teaching is seen as a crucial success factor doing so.

3.1 Transdisciplinary Teaching Against the Background of Motivational Research

Motivational research has shown that physiological needs, emotions and psychological needs are essential drivers of human motivation. According to the self-determination theory by Deci and Ryan (1993), psychological needs are of extraordinary relevance for motivation. The authors thereby differentiate between three basic psychological needs for autonomy, competence and relatedness (BPNT —Basic Psychological Needs Theory). In this context Bandura's social cognitive theory, in particular his model of self-efficacy, is of high interest. According to Bandura, self-efficacy is "the belief in one's capabilities to organize and execute the courses of action required to manage prospective situations". Self-efficacy thereby is influenced by primarily four sources: 1. Mastery experiences, 2. Vicarious experiences, 3. Social persuasions, 4. Physiological states (Bandura 1997).

In the following, the focus will be on the source of vicarious experiences, which Bandura describes as follows: "Seeing people similar to oneself succeed by sustained effort raises observers' beliefs that they too possess the capabilities master comparable activities to succeed" (Bandura 1994, p. 72).

The authors argue that transdisciplinary teaching can alter the level and strength of self-efficacy due to a specific activation of the source "vicarious experiences". Whereas "conventional" teaching methods are hardly able to alter self-efficacy of the three other sources, transdisciplinary teaching can have a positive influence due to its inherent nature of vicarious experiences. This is of extraordinary importance in the context of sustainability programs with regard to the following circumstances:

- High level of discrepancies between logic of science and the logic of economy - due to a high (scientific) demand on sustainability approaches and understanding on the one side and a rather pragmatic and often limited and slow implementation process of sustainability in daily work on the other side
- High level of difficult tasks due to a high complexity, longsome circumstances, interdisciplinary problems and missing experience
- A resulting high potential for frustration and lack of motivation

4 Conclusion

Transdisciplinary teaching can have a positive influence on the motivation of graduates, especially in sustainability programs, and allows an active imparting of practice-based knowledge.

References

Bandura A (1977) Self-efficacy: toward a unifying theory of behavioral change. Psychol Rev 84:191–215

Bandura A (1994) Self-efficacy. In: Ramachaudran VS (ed) Encyclopedia of human behavior, vol 4. Academic Press, New York, pp 71–81

Bandura A (1997) Self-efficacy: the exercise of control. Freeman, New York

Beermann M, Schattke H, Günther S (2014) Heterogenität in Lehr-Lern-Settings. Konzepte zum Umgang mit Heterogenität in der Lehre, HDS Journal (1/2014) Perspektiven guter Lehre, HDS Tagungsedition

Deci EL, Ryan R (1993) Die Selbstbestimmungstheorie der Motivation und ihre Bedeutung für die Pädagogik. Zeitschrift für Pädagogik, 39. Jahrgang, Nr. 2, 223–238

Kopp U, Martinuzzi A (2013) Teaching sustainability leaders in systems thinking. Bus Syst Rev 2 (2):191–215

Merck J (2013) Herausforderungen und Kompetenzen für unternehmerische Nachhaltigkeitsstrategien. Interview mit Johannes Merck zum Beitrag "Nachhaltigkeitskompetenzen in Unternehmen und Organisationen". Die Unternehmung, 67. Jahrgang, 2/2013, 191–193

Nölting B, Schäpke N, Pape J (2013) Nachhaltigkeitskompetenzen in Unternehmen und Organisationen. Die Unternehmung, 67. Jahrgang, 2/2013, 175–191

Schaeper H (2008) The role of key competencies in the Bologna process: rhetoric and reality. http://www.dzhw.eu/pdf/pub_vt/22/2008-07-26_Vortrag_Schaeper_ICP.pdf. Last Accessed 2 Sept 2014

Tilbury D (2011) Assessing ESD experiences during the DESD: an expert review on processes and learning for ESD. UNESCO, Paris. http://insight.glos.ac.uk/sustainability/Education/Documents/GUNI%20HE%20in%20the%20World%204%20HE%27s%20Committment%20to%20Sus.pdf. Last Accessed 3 May 2014

UNESCO (2012a) Education and skills for inclusive and sustainable development beyond 2015, UN system task team on the post-2015 development agenda. http://www.un.org/en/development/desa/policy/untaskteam_undf/thinkpieces/4_education.pdf. Last Accessed 2 Sept 2014

UNESCO (2012b) Shaping the education of tomorrow. 2012 report on the UN decade of education for sustainable development. http://unesdoc.unesco.org/images/0021/002166/216606e.pdf. Last Accessed 3 May 2014

Authors Biography

Dr. Johannes Merck is Director of Corporate Responsibility at the Otto Group (VV-CR) in Hamburg, Germany.

Dr. Marina Beermann is a senior consultant at Systain Consulting GmbH, in Germany.

Making Sustainability Part of Every Student's Curriculum

Alison J. Greig

Abstract

Anglia Ruskin University (ARU) has made a corporate commitment to ensure that "sustainability will be a feature of all our students' experience". In order to address this goal ARU has sought to ensure that sustainability is embedded within every taught course across each of our four faculties (Science and Technology, Arts, Law and Social Sciences, Health Social Care and Education and the Lord Ashcroft International Business School). This paper provides a critical review of the individual, discipline specific and institutional challenges encountered and the how these have been addressed. Special mention is made of the significance of strategic level interventions and the support provided by the Higher Education Academy's Green Academy programme in levering such interventions. The crucial importance of developing a definition of sustainability which is meaningful, non-threatening and encourages engagement, both from academic staff and staff supporting academic activities will be stressed. This paper will be useful to anyone involved in embedding sustainability into HE curricula.

Keywords

Education for Sustainable Development (ESD) · Education for Sustainability (EfS) · Curriculum · Change

A.J. Greig (✉)
Global Sustainability Institute, Anglia Ruskin University, East Road, Cambridge CB1 1PT, England, UK
e-mail: Alison.greig@anglia.ac.uk

W. Leal Filho et al. (eds.), *Integrative Approaches to Sustainable Development at University Level*, World Sustainability Series, DOI 10.1007/978-3-319-10690-8_3

1 Introduction

Education is an indispensable component of our aim to achieve sustainability (UNESCO 2005a; Tilbury 2014) as it has the potential to influence many minds and lives. Higher Education Institutions (HEIs), which have access to many of the world's future leaders and decision makers during a formative part of their lives, therefore hold a unique opportunity to determine the direction of our future (Orr 2004; Stern 2009). Education for Sustainable Development (ESD) or Education for Sustainability (EfS) throughout the world has a common goal, that is, to empower people to assume responsibility for ensuring that future is sustainable. Sterling (2004), argues this will only happen through a transformation of education, including its pedagogy, curricula, policy and institutional structures. Much of the published literature on EfS focuses on pedagogy and curricula, typically rehearsing the imperative for change and then offering suggestions of good practice, sometimes on a discipline by discipline basis (e.g. Cortese 1999; Corcoran and Wals 2004; Blewitt and Cullingford 2004; Jones et al. 2010). Progress towards systemic educational change, or indeed the prospect of sustainability, has however been slow, and Corcoran (2010) notes that this is due largely to our inability to understand how institutions overcome institutional inertia and disciplinary traditions.

This paper, through the use of a case study, directs its attention to the practice of change. It focuses on policy and institutional structures and highlights the importance of working within existing structures and systems, at a personal and organisational level, in order to lever change. Concepts, examples and techniques are borrowed from psychology, the world of business and the narrative of climate change in order to overcome institutional and individual barriers and inertia.

2 UK Higher Education Context

In the UK Higher Education Institutions and the academic staff they employ enjoy and warily protect a considerable autonomy, including with regards their curriculum and course provision (Sterling and Scott 2008). HEIs tend to resist direction from central government and in the same way academic staff resist control by university management. Each HEI is a unique and complex organisation with a distinct organisational and academic structure and ethos and a unique set of priorities and policies. Sterling and Scott (2008, p. 387) describe the consequences of this for EfS, noting that the embedding of sustainability in the UK exists as "a complex, largely decentralised, and multi-stranded process undertaken by disparate groups of academics variously involved in raising the debate, developing policy and theoretical frameworks, networking, influencing peers, using existing funding streams, researching, disseminating, working with professional bodies, etc.".

This paper reflects on the multi-stranded process of embedding sustainability at Anglia Ruskin University. Whilst it is unlikely to provide solutions for other institutions who have embarked on a similar journey the hope is that it will provide some lessons which can be adapted and reused.

3 Anglia Ruskin University

Anglia Ruskin University educates around 30,000 students from three main campuses in Cambridge, Chelmsford and Peterborough in the UK and a number of partner colleges overseas. It was designated as a university in 1992 but can trace its origins back over one hundred and fifty years to the opening, in 1858, of the Cambridge School of Art, by John Ruskin.

Ruskin (1819–1900), widely considered as the leading art critic of the 19th century, was also a hugely influential social entrepreneur and his work is recognized today as fundamentally relating to sustainability, particularly in terms of fairness and equality, social justice and the protection of the environment. For example, his essay *Unto this Last* (Ruskin 1862), which he considered the central work of his life (Cook 2011), attacks 18 and 19th century economists for their 'dehumanising' capitalist ideas and argues strongly for the need to consider a politics of social justice. Ruskin also recognised, and is highly critical of the destructive effects of 19th century industrialization upon the natural world and of the social and economic implications of this.

This context is important, as it has determined ARUs point of engagement with the modern education for sustainability agenda, that is, as an agenda which fundamentally relates to social reform. Ruskin's legacy is visible in Anglia Ruskin's present day vision; that "our key contribution is to the enhancement of social, cultural and economic well-being" (ARU 2011a). Like all 21st century universities however, this purposive goal has to survive alongside what it sees as its other major functions; the generation of knowledge, the creation of future decision makers and leaders and the preparation of a skilled labour force (Sterling 2013). As a 'post-1992' university (Further and Higher Education Act 1992) whose core business relates to widening participation and delivering high quality mass education (Scott 1995), it is this latter function which normally assumes the greatest emphasis. There is also a point of potential conflict which needs to be considered when framing the place of sustainability within ARU, that is, between what Sterling (2013) describes as Universities purposive and operational goals. At ARU operational concerns drive most university policy. Attaining financial security through a balance of student fee, research and other external income, and reputational issues particularly related to student feedback on the quality of learning and teaching dominate. Effective recruitment and retention policies, the employability of its students and improving research outputs are also high on the agenda for senior managers.

4 Corporate Engagement with the Sustainability Agenda

In common with many HEIs in the UK and overseas ARUs journey towards embedding sustainability began outside the curriculum and was related to 'greening' the University estate (Tilbury 2014). It was the promise of operational cost savings from reducing energy use and waste that attracted corporate attention in the mid 2000s. In 2009 the University's environmental management system became certified to the international standard ISO 14001 making ARU one of the first UK Universities to gain whole campus certification. Although primarily focusing on reducing the environmental impact of the university estate the environmental policy and strategy created to comply with ISO14001 certification included a section on engagement with staff and students and a section relating to education for sustainability which assumed a link to the curriculum. Education for sustainability thereby became a requirement of retaining certification, and was measured, monitored and reported through University committee structures and thereby gained management attention. ISO14001 then also became an effective lever through which it was possible to argue for EfS to become an integral part of the ARU Corporate Plan. Effective lobbying by a small number of determined individuals ensured the Corporate Plan included a commitment to:

- Strive to exceed national and sector benchmarks for the sustainability of our buildings and processes;
- A commitment to making sustainability a part of every student's experience, and;
- Taking advantage of the University's decision to establish a research institute in each of its academic faculties; a commitment to internationally recognised research in sustainability.

Sustainability became one of ARU's guiding principles and an element in 4 of its 12 specific corporate aims (Fig. 1).

5 Integrating Sustainability

In order to deliver on the commitments of ARUs Corporate Plan specific parts of the University assumed responsibility for particular sustainability goals. For example, a new post, Director of EfS, was given responsibility for embedding sustainability across the formal curriculum and making links with the informal curriculum explicit. The Quality Assurance office held overall responsibility for measuring and monitoring the embedding of sustainability into the curriculum. The Global Sustainability Institute became the focus and catalyst for sustainability related research and the Environment Team within Estates and Facilities gained responsibility for delivering our emission reduction and environmental strategies (against an agenda for growth) including campus energy, transport and wider staff engagement. Encouraging and supporting students to engage with sustainability through work placements, volunteering and various employability activities became the responsibility of the University Student Services team while the Student Union led on co-ordination of the National Union of Students 'Green Impact' activities (Acevedo and Johnson 2013), and promoting student led environmental activities and societies. Finally the University's People Strategy (Human Resources) was given responsibility for overseeing the provision of support for academic and support staff in how to incorporate sustainability into their work and teaching.

<table>
<tr><th colspan="3">Anglia Ruskin's Corporate Plan 2012-14: Commitments to Sustainability</th></tr>
<tr><td>University Vision</td><td>P4</td><td>Our key contribution is to the enhancement of social, cultural and economic well-being.</td></tr>
<tr><td>Values</td><td>P4</td><td>Concern for the environment

We want our concern for a sustainable environment to inform every aspect of what we do.</td></tr>
<tr><td>Goal 4: Sustainability</td><td>P12, P14</td><td>Make sustainability a theme that runs through much of our research work.</td></tr>
<tr><td rowspan="2">Research</td><td>P3</td><td>We will build further on our world-leading research and will be held in high regard by a broad range of stakeholders for our success in research, professional practice, sustainability and social, cultural and economic development.</td></tr>
<tr><td>P5</td><td>The Global Sustainability Institute will have national recognition.</td></tr>
<tr><td>Goal 6: Staff Development</td><td>P15</td><td>...targeted staff development aligned with corporate priorities (including on such institution-wide issues as Customer Care and our 'green' agenda – particularly carbon reduction and a sustainable energy initiatives)</td></tr>
<tr><td>Goal 7: Student experience</td><td>P16, P17</td><td>...sustainability will be a feature of their experience. We will: incorporate sustainability across the curriculum and embed generally in student life and activities, achieve a place in the top ten of at least one environmental benchmarking scheme.

<table><tr><th>Milestones</th><th>Target 2012</th><th>Target 2013</th><th>Target 2014</th></tr><tr><td>Percentage of students who say that sustainability was a feature of their experience</td><td>10%</td><td>20%</td><td>50%</td></tr></table></td></tr>
<tr><td>Goal 8: University Estate</td><td>P19</td><td>We shall strive to exceed national and sector benchmarks for the sustainability of our buildings and processes. Use data to inform staff and students about sustainability issues and to improve active participation in green initiatives.</td></tr>
</table>

Fig. 1 References to sustainability in Anglia Ruskin's Corporate Plan

The Higher Education Academy's (HEA) Change Programme, the Green Academy II provided a timely and supportive vehicle to help facilitate the integration of these various responsibilities. The HEA's first Green Academy ran from 2011 to 2012 and included 8 UK HEIs which it helped support develop curricula for sustainability. The second Green Academy (2012–2013) was larger (10 HEIs), and able to draw on a range of evidence and sector-leading expertise from Green Academy I. It was also broader, aiming to support change at an institutional level, by enabling HEIs to embed sustainability in the overall student experience in more strategic and holistic ways. The Anglia Ruskin Green Academy Team includes a senior manager, the Pro-Vice Chancellor with responsibility for sustainability, the curriculum lead (Director of EfS), Director of the Global Sustainability Research Institute, Head of Environmental Management and Student Union President.

6 Sustainability in the Formal Curriculum

Much of what a university does, particularly in a post-1992 UK university, is in support of student learning. Successfully embedding sustainability across the formal curriculum must, therefore, be a priority. The driver for action related to embedding sustainability within the formal curriculum at Anglia Ruskin has been shaped by the University's Learning, Teaching and Assessment Strategy (ARU 2011b). This strategy relates specifically to the formal curriculum but extends to also cover aspects of students' informal learning. It is intended to underpin the vision and goals contained in the Corporate Plan. It therefore followed that this should also contain a number of statements and targets relating to sustainability and a specific measurable target related to embedding EfS within the formal curriculum (Fig. 2).

These goals require that sustainability becomes part of all courses in all four faculties and encompass disciplines which range from performing arts, through international business and allied health to fields which are highly technical and scientific and heavily regulated by professional bodies. Research undertaken at Plymouth University however found "no obvious correlation between subject area and the belief that sustainability is relevant to their discipline" (Cotton and Winter 2010, p. 43), suggesting that that this goal is potentially achievable. The challenge then was one of realising this potential.

7 Sustainability as Part of Anglia Ruskin's Academic Regulations

All taught courses at Anglia Ruskin University are required to adhere to its Academic Regulations. These establish the internal regulatory framework required for maintaining the academic standard of taught courses and take account of the Quality Assurance Agency UK Quality Code (QAA 2014). The inclusion of a regulation relating specifically to a consideration of sustainability as a generic learning outcomes, would therefore apply to all new taught courses, and be monitored and measured at the 3 or 5 year review of existing courses. Changing academic regulations is, however, a bureaucratic process involving large and conservative committee structures. Resistance to a new sustainability specific regulation was considerable, despite the corporate imperative, and based on three major concerns:

- Sustainability is an ephemeral addition to the higher education agenda and should therefore not be responsible for a change in core (and largely operational) educational policy;
- Sustainability is already covered in the regulations by a reference to ethics;

<table>
<tr><th colspan="3">Anglia Ruskin University Learning and Teaching Strategy 2011 - Commitments to Sustainability</th></tr>
<tr><td>Guiding principles</td><td>P1</td><td>Students and staff engage with education for sustainable development.</td></tr>
<tr><td>Ambition</td><td>P3</td><td>An educational process that integrates the principles, values and practices of sustainable development.</td></tr>
<tr><td>Theme 11</td><td>P5</td><td>Students and staff engage with education for sustainable development.

We shall help students 'develop the attitudes, skills and knowledge to make informed decisions for the benefit of themselves and others, now and in the future, and to act upon these decisions.' (UNESCO, 2010) This is a key preparation of our students for a global outlook, including their employability and ability to cope with the challenges of the future. Education for Sustainable Development (ESD) should permeate the whole student experience.</td></tr>
<tr><td>Aim 2: Improving Assessment and Feedback Practice</td><td>P8</td><td>Implement online submission processes which enable students to submit written coursework either on- or off-campus and contribute to our education for sustainability targets.</td></tr>
<tr><td>Aim 3: Supporting and Engaging Students</td><td>P10</td><td>Embed an ethos of education for sustainability, which will feature throughout students study and support processes.</td></tr>
<tr><td>Aim 4: Using Technology Enhanced Learning</td><td>P12</td><td>Digital technologies will enhance and create innovation in the student experience support our sustainability agenda.</td></tr>
<tr><td>Aim 6: Curriculum Design and Enhancement</td><td>P16</td><td>Ensure that the curriculum is embedded with learning and teaching materials and our assessment practices address the issues of sustainable development.
Milestones:
<table>
<tr><th>2011-12</th><th>2012-13</th><th>2013-14</th></tr>
<tr><td>A vision for the nature of EfS and how its quality and success can be assessed is established and communicated.</td><td>50% of courses include EfS in course review and approval.</td><td>100% of courses include EfS in course review and approval.</td></tr>
</table></td></tr>
</table>

Fig. 2 Sustainability in Anglia Ruskin University's Learning, Teaching and Assessment Strategy

- It would be very difficult to measure and monitor the embedding of sustainability in courses through existing QA processes.

Supplying evidence which responded to these concerns was met with further debate and argument (Garrett and Davies 2010). However progress was made by instead emphasising the positive benefits (in particular to the University's operational requisites) of a new sustainability related regulation. This included emphasising ARU's growing reputation as a leader in EfS, including through evidence of engagement with the HEA's Green Academy and recent sustainability awards, at a time when the importance of sustainability as a future requirement of higher education was being reinforced by QAA and HEFCE. The third concern was successfully addressed by working with, but crucially transferring responsibility for, drafting a set of words acceptable to the Head of Quality Assurance. QA staff and review panel chairs were also offered guidance and training to support them in their regulatory roles.

The approved wording, which was accepted in September 2013 as a generic learning outcome which all courses must address, is:

Knowledge and understanding

(Level 6) Sustainability: The learner has the awareness and ability to apply their knowledge and understanding and work with others to take action which promotes the principles of sustainability.

(Level 7) Sustainability: The learner has the awareness and ability to apply critically their knowledge and understanding and work with others to take proactive action which promotes the principles of sustainability.

Affective and transferable skills (generic)

(Level 6) Sustainability: The learner has developed the attitudes and skills to make informed decisions that reflect care, concern and responsibility for themselves, for others and the environment, now and in the future.

(Level 7) Sustainability: The learner has developed the attitudes and skills and is able to apply their knowledge to make informed decisions and take actions that reflect care, concern and responsibility for themselves, for others and the environment, now and in the future.

8 Embedding Sustainability in the Curriculum: Influence Versus Authority

Despite persuasive calls for fundamental changes in higher education for at least a decade (Sterling 2004) we have seen only modest progress in sustainability education becoming a priority for most universities (Richmond 2010). As Cotton and Winter (2010) observe, whilst there is potential to embed sustainability in all curriculum areas, this will only be realised if lecturers feel that it is important to do so. Jones, Selby and Sterling (2010, p. 9), describe the principle inhibitor in embedding EfS in University curricula as "academic staff, jealously guarding their academic freedom"...who see EfS as "an imposition, not commensurate with their discipline or student expectations of their discipline"... and who are "uncomfortable about the interdisciplinary teaching..."and consider EfS curricula innovation as providing "no prospect of reward or career advancement". Unfortunately they offer no suggestions on how to overcome these barriers. These inhibitors have also applied to other higher education agendas including internationalism and employability (Stoner and Milner 2010). Garrett and Davies (2010) describe the cultural landscape of higher education institutions in the UK as having a strong historical legacy, characterised by individuality and a unwillingness to conform and as a place where change is obstructed by dissent and argument. Gilley et al. (2009) go further, describing academics as part of a system which strongly defends the status quo and resists change and which can become adversarial and hostile when provoked. These characteristics are indicative of a particular cultural landscape which cannot be ignored if EfS, or indeed any other change agenda is to be successful.

It is perhaps not surprising therefore that efforts to persuade academic staff that it is important for them to consider the EfS (or any other non-discipline specific) agenda in their already overstretched curriculum has met with limited success. This traditional approach to implementing change is known as the information deficit model (Darnton 2008). In relation to the EfS agenda this approach might include for example, the provision of key statistics relating to business needs (e.g. Nidumolu et al. 2009; BIS 2011) and student demand (e.g. Bone and Agombar 2011; Drayson et al. 2012, 2013) for EfS skills. The information deficit model is based on the assumption that change occurs through communicating information in an objective and analytical format, and is based on the assumption that people process information largely in an analytical matter. In this model, university staff, who have high intellectual ability and whose core abilities relate to thinking and acting creatively, might therefore be expected to therefore accept change (e.g. to their curriculum) once significant evidence has been presented. However, this approach ignores the fact that this change requires academics, who are expert in argument and counter argument acknowledging their current position is deficient and switching to a new 'enlightened' position.

Garrett and Davies (2010) suggest an alternative approach, which involves accepting and gaining an understanding of the prevailing culture. Working from this starting point it may be possible to undertake change progressively in a way which both disarms defence and enhances and rewards change skills (Gilley et al. 2009). Behavioural and cognitive psychologists discuss similar outcomes in terms of making communication persuasive through taking into account the interrelation between cognitive, experiential and normative influences on behaviour. Van der Linden (2014) provides an accessible non-technical summary of these approaches and their application to another sustainability related change agenda, that of climate change.

9 Defining Sustainability

The terms sustainability and sustainable development are broad and contested terms and their lack of clarity and complexity can be a major challenge to their inclusion in the curriculum. (Cotton and Winter 2010). For many staff at ARU, both terms

- Respect forall people throughout the world and a commitment to social and economic justice for all;
- Respect for the human rights of future generations and a commitment to intergenerational responsibility;
- Respect and care for the greater community of life in all its diversity which involves the protection and restoration of the Earth's ecosystems;
- Respect for cultural diversity and a commitment to build locally and globally a culture of tolerance, non-violence and peace.

Fig. 3 UNESCOs (2005b) explanation of sustainability

were strongly associated with the physical environment. Being responsible for a curriculum with little relevance to the environment was a common justification given for non-engagement in the EfS agenda. For some staff the term sustainability was interpreted literally as ensuring longevity, but with little regard of the consequences to people or planet. Offering a definition which provided sufficient scope and clarity to be recognised and seen as relevant and normative across a broad range of disciplines was a key milestone at Anglia Ruskin. Like several other Universities (including Bristol, Plymouth and Bradford) Anglia Ruskin uses UNESCO's (2005b) values based definition (Fig. 3). Whilst it is possible to identify shortcomings in this definition, (including its not requiring any action), it is essentially about values which extend easily across discipline boundaries and are generally accepted as an axiom for 'good'. It is commonly received as non-threatening and familiar and provides a strong normative position from which it is very difficult to argue against.

10 Audits of Sustainability

As part of their journey to embed sustainability into curricula a number of UK HEIs including members of the First Green Academy Change programme have undertaken reviews and inventories of where sustainability is already present within the curriculum. Most commonly this has involved a desk-top survey of official course descriptions undertaken by a researcher although some (e.g. Trinity St David's) have involved a questionnaire survey of staff. Whilst these may appear to be a practical and thorough way to benchmark and monitor progress, a review of methodologies and results, undertaken as part of the second Green Academy workshop identified a number of significant practical difficulties and methodological shortcomings including;

- The large number of documents to be read and audited,
- The extent to which the official documents represent the content or style of teaching/learning delivered,
- The broad range of criteria used to identify 'sustainability' by staff,
- Researcher bias—researchers tend to 'find' more sustainability in disciplines they are familiar with.

The recommendation of the Green Academy Workshop was that curriculum reviews are most useful as a tool for engaging staff in a dialogue about education for sustainability. At Anglia Ruskin this was undertaken by holding focus group meetings with course leaders where the meaning of EfS was considered [using the UNESCO (2005b) explanation as a starting point] and discipline specific examples and case studies identified and explored. Course leaders were asked to reflect on these conversations and to identify what sustainability relevant content or pedagogy was currently being delivered. This information could also be used as evidence of compliance with the new sustainability regulation. This approach not only marks where sustainability relevant material is already present but also helps raise staff awareness and understanding of the agenda. It is however only a first step towards meeting ARUs corporate commitment that sustainability is not only embedded but recognised as being part of their experience by students.

11 Sustainability Lenses at Anglia Ruskin

The published and non-published literature contains an ever increasing number of examples and case studies of how specific disciplines relate to sustainability (Stewart 2014), which can form the basis of a conversation with course staff. The experience at ARU has been that course leaders will often find their own unique intersection points between sustainability and their curriculum. The following paragraph provide a few examples, taken from what might be considered 'hard to reach' disciplines.

Psychology: With its emphasis on human behaviour psychology has strong links with sustainability. These are most often expressed as links between human behaviour and the environment; ecophyschology, conservation psychology, consumer psychology etc. (Scott and Koger 2005). At Anglia Ruskin staff most strongly identified with sustainability as an issue central to social and health psychology and throughout their degree students are expected to respect different perspectives and challenge their own assumptions, for example about a range of groups. Social Psychology modules require students to learn about and respect different perspectives (e.g. on knowledge) and explore theories of groups and intergroup relations. They cover topics such as racism, prejudice and conflict between groups. In their module Abnormal and Health Psychology students critique the cultural (and historical) influences on societies' definitions of normal and abnormal behaviour and experience, and the benefits of diversity in human behaviour. In the module entitled Sex, Sexuality and Gender they explore issues related to sexual orientation and gender identities including perspectives relating to sex work, in particular pornography and prostitution, considering the views of those who would maintain the illegitimacy of such roles as work well as counter-arguments for the acceptance of these as forms of employment. In the final year module Critical Issues in Health Psychology students evaluate the role of poverty, culture, gender and class in the experience and prevalence of illness and disability.

Optometry and Ophthalmic Dispensing: The curriculum of these degrees is accredited by the General Optical Council who specify much of the largely competence curriculum and maintain the register of those suitably qualified and assessed as fit to practice. A consideration of sustainability within this prescriptive curriculum is achieved by giving students an opportunity to 'step back' from the technical detail and consider the implications of their work. For example, they are given the opportunity to reflect on how, through gaining knowledge on how to optimise the residual vision of their patients with impaired vision, they are able to assist those with visual impairment retain independence and dignity. The curriculum also includes discussion of how to look after those with various disabilities and of the need to respect different cultures.

Computer Science: This is another very technical subject which nevertheless has several strong links to sustainability. The most obvious perhaps, relates to the application of computer science knowledge and skills to green/clean technologies and energy efficiency, including of computer electronics. Through conversation staff were able to identify other intersection points. These included the link between sustainability and innovation, though, for example the designing out of actual and embedded energy. All students are also required to reflect on the application of computers and technology in the modern world and the advances in, for example, healthcare and education that it supports.

12 Conclusion

Higher education plays an essential role in shaping the leaders and decision makers of tomorrow and is therefore crucial that it should equip its graduates with the capabilities and motivation to shape and create a sustainable future (Thomas 2014). This will require that students are able to view the world as complex, interconnected, finite, ecological-social-psychological-economic systems (Meadows 2008) rather than discipline specific sets of knowledge or problems. The frequent calls, over the past two decades, for a transformation in education to address this issue (Blake, Sterling and Goodson 2013) continue to relay this imperative but few focus on understanding what is required within HEIs, to bring about this change. It is perhaps not surprising, therefore, that the rate of change has been frustratingly slow.

The aim of this paper has been to explore the *practice* of embedding sustainability within a UK University. It provides practical examples based on the work being conducted at Anglia Ruskin University to ensure that sustainability is a feature of all its students' higher education experience. The starting point at ARU was gaining corporate recognition of the EfS agenda. This was achieved by reference to the University's purposive principles. This in turn provided a lever with which to create new institutional requirements (e.g. a new academic regulation) which was attractive to managers as it also had the potential to deliver reputational benefits. A very different approach was used to engage individual academics and this borrowed from psychology and the experiences of researchers working with climate change sceptics to encourage and support staff in making education for sustainability a positive and normative experience.

Overall the journey to embed EfS at Anglia Ruskin has focused on acknowledging and working alongside existing structural and cultural norms and emphasising the positive outcomes of engaging with an EfS agenda. It is a journey of reformation rather than transformation. Higher education institutions are complex, interconnected, ecological-social-psychological-economic systems and must be acknowledged as such if EfS leads, like the students they are responsible for, are to bring about significant and lasting change in higher education.

References

Acevedo B, Johnson S (2013) Sustainability in practice: action learning and the green impact initiative. In: Atfield R, Kemp P (eds) Enhancing education for sustainable development case study publication. Higher Education Academy

ARU (2011a) Anglia Ruskin University Corporate Plan, 2012–14. Chelmsford: Anglia Ruskin University. http://www.anglia.ac.uk/ruskin/en/home/your_university/about_anglia_ruskin/corporate-plan.Maincontent.0001.file.tmp/corporate_plan_2012-14.pdf. Accessed 30 Apr 2014

ARU (2011b) Learning, Teaching and Assessment Strategy, 2012–14. Chelmsford: Anglia Ruskin University. http://www.lta.anglia.ac.uk/cmsAdmin/uploads/LTAStrategy2011_001.pdf. Accessed 30 Apr 2014

BIS (2011) Skills for a green economy: a report on the evidence. Department for Business Innovation and Skills. HMRC, London, pp 36

Blake J, Sterling S, Goodson I (2013) Transformative learning for a sustainable future: an exploration of pedagogies for change at an alternative college. Sustainability 5(12):5347–5372

Blewitt J, Cullingford CC (2004) The sustainability curriculum: the challenge for higher education. Earthscan, London, pp 258

Bone E, Agombar J (2011) First-year attitudes towards, and skills in, sustainable development. A report for the Higher Education Academy

Cook ET (2011) The life of John Ruskin, vol 2. George, Allen and Company, London, p 2

Corcoran PB (2010) Sustainability education in higher education: perspectives and practices across curriculum. In: Jones P, Selby D, Sterling S (eds) Sustainability education: perspectives and practices across higher education, xiii–xv. Earthscan, London

Corcoran PB, Wals AEJ (2004) Higher education and the challenge of sustainability: problematics, promise, and practice. Springer, Netherlands, pp 262

Cortese AD (1999) Education for sustainability: the need for a new human perspective. Second Nature Inc., Boston, pp 10

Cotton D, Winter J (2010) 'It's not just bits of paper and light bulbs': a review of sustainability pedagogies and their potential for use in higher education. In: Jones P, Selby D, Sterling S (eds) Sustainability education perspectives and practices across higher education. Earthscan, London, pp 39–54

Darnton A (2008) Reference report: an overview of behavioural change models and their uses. Government Social Research (GSR) Behaviour Change Knowledge Review. Routledge, London

Drayson R, Bone E, Agombar J (2012) Student attitudes towards and skills for sustainable development. A report for the Higher Education Academy

Drayson R, Bone E, Agombar J, Kemp S (2013) Student attitudes towards and skills for sustainable development. A report for the Higher Education Academy

Further and Higher Education Act (1992) England and Wales. Chapter 13. HMSO, London

Garrett G, Davies G (2010) Herding cats: being advice to aspiring academic and research leaders. Triarchy Press, Devon
Gilley A, Godeck M, Gilley JW (2009) The university immune system: higher education. Overcoming resistance to change. Contemporary issues in education research. Third Quart 2 (3):1–6
Jones P, Selby D, Sterling S (eds) (2010) Sustainability education perspectives and practices across higher education. Earthscan, London, pp 1–16
Meadows DH (2008) Thinking in systems: a primer. Earthscan, London, pp 218
Nidumolu R, Prahalad CK, Rangaswami MR (2009) Why sustainability is now the key driver of innovation. Harvard Business Review. Reprint R0909E
Orr DW (2004) Earth in mind: on education, environment, and the human prospect. Island Press, Washington
QAA (2014) The UK quality code for higher education. The Quality Assurance Agency for Higher Education
Richmond R (2010) Envisioning, coordinating and implementing the UN decade of education for sustainable development. In: Witthaus M, McCandless K, Lambert R (eds) Tomorrow today. Tudor Rose on Behalf of UNESCO, Leicester. http://unesdoc.unesco.org/images/0018/001898/189880e.pdf. Accessed 30 Apr 2014
Ruskin J (1862) Unto this last: four essays on the first principles of political economy. Smith Elder and Company, London, p 174
Scott P (1995) The meanings of mass higher education. Open University Press, Buckingham, p 198
Scott BA, Koger SM (2005) Teaching psychology for sustainability. http://www.teachgreenpsych.com/. Accessed 01 May 2014
Sterling S (2004) Higher education, sustainability and the role of systemic learning. In: Concoran PB, Wals AEJ (eds) Higher education and the challenge of sustainability: problematics, promise and practice. Kluwer Academic Publishers, Netherlands, pp 47–70
Sterling S (2013) The sustainable university: challenge and response. In: Parkin S, Sterling S, Maxey L, Luna H (eds) The sustainable university: progress and prospects. Routledge, London, pp 17–50
Sterling S, Scott W (2008) Higher education and ESD in England: a critical commentary on recent initiatives. Environ Educ Res 14(4):399–412
Stern N (2009) Foreword. In: HEFCE, sustainable development in higher education 2008 update to strategic statement and action plan. HEFCE, Bristol
Stewart AW (2014) Integrating sustainability within higher education. In: Thomas KD, Muga HE (eds) Handbook of research on pedagogical innovations for sustainable development. IGI Global, Hershey
Stoner G, Milner M (2010) Embedding generic employability skills in an accounting degree: development and impediments. Account Educ: Int J 19(1–2):123–138
Thomas I (2014) Editorial: Special issue—pedagogy for education for sustainability in higher education (HE). Sustainability (6):1705–1708
Tilbury D (2014) Education for sustainability in higher education. Report commissioned by UNESCO ESD secretariat to inform the 3rd Global DESD report and global action programme (In Prep)
UNESCO (2005a) UNESCO and sustainable development. UNESCO, pp 41. http://unesdoc.unesco.org/images/0013/001393/139369e.pdf. Accessed 30 Apr 2014
UNESCO (2005b) The UN decade for education for sustainable development (DESD 2005–2014) international implementation scheme. UNESCO, pp 19
Van der Linden S (2014) Towards a new model for communicating climate change. In: Cohen S, Higham J, Peeters P, Gössling S (eds) Understanding and governing sustainable tourism mobility: psychological and behavioural approaches. Routledge, Oxford, pp 243–275

Author Biography

Alison J. Greig is Director of Education for Sustainability at Anglia Ruskin University. She is responsible for embedding sustainability across the curriculum and for the EfS research theme within the Global Sustainability Institute.

Ecological Footprinting as a Top-Down and Bottom-Up Approach to Complete Campus Engagement and Transformation Towards the One Planet Goal

Sara Rickards, Richie Howitt and Sandie Suchet-Pearson

Abstract
Enabling students to apply well-informed and skill-based understandings to their own decision-making about the dynamic relationships between individuals, societies, environments and economies in a resource constrained world is a central sustainability goal in the tertiary sector. Ecological Footprinting (EF) measures how much people have, how much they use and identifies who uses what. At Macquarie University, in Sydney Australia, EF has been applied at the campus, faculty and building scales and integrated into student learning through the undergraduate Environmental Management curriculum, where students consider their own footprints (personal and household) and investigate EF at the faculty scale. At the building scale, the University's Property Department calculates the EF of individual buildings (new and existing) to ensure as the University grows, it reaches a One Planet campus target by 2030. The students' research projects utilise the Property Department's EF tool. This has generated a cross-campus partnership across academic-non-academic, disciplinary and department structures towards institutional goals. This integrated, top-down and bottom-up approach fosters transformative engagement across the entire campus.

S. Rickards (✉) · R. Howitt · S. Suchet-Pearson
Department of Environment and Geography, Macquarie University,
Balaclava Rd, North Ryde, NSW 2109, Australia
e-mail: sara.rickards@mq.edu.au

R. Howitt
e-mail: richie.howitt@mq.edu.au

S. Suchet-Pearson
e-mail: sandie.suchet@mq.edu.au

W. Leal Filho et al. (eds.), *Integrative Approaches to Sustainable Development at University Level*, World Sustainability Series, DOI 10.1007/978-3-319-10690-8_4

This paper presents the strategic and methodological approach of the integrated sustainability strategy, reviews early stage results and next steps towards a One Planet campus and society.

Keywords
Sustainable development · Curriculum · Ecological footprint

1 Introduction and Background

Enabling students to apply well-informed and skill-based understandings to their own decision-making about the dynamic relationships between individuals, societies, environments and economies in a resource constrained world is a central goal of education for sustainability in the tertiary sector (Cooper and Scott 2009). As no single discipline has a monopoly on insight and understanding of these issues, cross-disciplinary collaboration is essential (Cortese 2003). At Sydney's Macquarie University, complete campus engagement for the common goal of a "One Planet" campus has been proposed as an appropriate framework that will break down the barriers created by both the disciplinary and administrative structures of more traditional universities.

For centuries universities have been incubators for future leaders, entrepreneurs and decision makers (Lozano et al. 2013). However, this has not been without significant environmental cost and ongoing environmental risk:

> The depletion and pollution of the planet is not the work of ignorant people. Rather it is largely the result of work by people with BAs, BSs, LLBs, MBAs and PhDs …The truth is that without significant precautions education can equip people merely to be more effective vandals of the earth. If one listens carefully, it may even be possible to hear the creation groan every year when another batch of smart degree-holding, but ecologically illiterate, *Homo sapiens* who are eager to succeed are launched into the biosphere (Orr 1994, p. 7)

It is widely acknowledged that behavior is a repercussion of intention (Austin and Vancouver 1996; Eagly and Chaiken 1993; Gollwitzer and Moskowitz 1996; Maddux 1999; Conner and Norman 2005; Webb and Sheeran 2006). Therefore, one of the important questions facing education for sustainability is: "Does tertiary environmental education transform the behavior of students?" Faced with this question the Macquarie University's Property Department and the Department of Environment and Geography developed a holistic approach, which utilises Ecological Footprinting (EF) in the curriculum, research and operations. This approach has created a campus wide living laboratory and allows for cross-disciplinary collaboration. It brings together teaching, research and facilities management staff to consider both student behaviours and the EF of the campus community as a whole.

Segalas et al. (2010) recommend that universities should be used as experimental laboratories to create ecological learning environments, while simultaneously monitoring the effectiveness of universities in terms of Ecological Sustainable

Development. At the 1972 Stockholm Conference, education was recognized at an international scale for fostering environmental protection and conservation (Lozano 2003). In addition, The 2001 Universities of Australian Ecological Development Charter declares that universities are responsible to act as role models for the broader community. Therefore universities should be walking the walk when they educate future generations about sustainable practice, by incorporating these practices into their own operations (National Union of Students 2001; McNamara 2010; Collins 2012).

It is commonly recognized that the environmental impact of human behaviors isn't sustainable (Rees 1996; Haberl et al. 2004; Kitzes et al. 2007; McNichol et al. 2011). The regenerative and absorptive capacity of the biosphere is being exceeded as a repercussion of humanity's demand on nature and evident through species extinction, deforestation and global warming (Borucke et al. 2013). To begin to combat overconsumption, a method of measuring the impact of consumption can assist in defining what level of consumption could be considered sustainable and what can be done to reduce overconsumption (Pearce and Uhl 2003).

EF measures how much people have, how much they use, identifies who uses what and has been defined as the world's premier measure of humanity's demand on nature (Chikoti 2012; GFN 2013). Using the National Footprint Accounts (NFA) humanity's demand on the biospheres supply was 1.5 planets in 2008 (Borucke et al. 2013). EF is an ecological camera providing an annual snap shot of how much biocapacity is consumed with respect to how much is available (Chi and Stone 2005) That is, how much humanity is consuming compared to the biosphere's ability to regenerate (Wackernagel and Rees 1996). The difference is called ecological overshoot and can be liken to be living on your credit card (Borucke et al. 2013). If you spend more than 150 % of your earnings per annum you aren't living within your means and won't be able to sustain the same level of consumption for a long period of time.

Assessment of campus EF's has been applied in the United States (Venetoulis 2001; Conway et al. 2008; Klein-Banai and Theis 2011; Bekmann et al. 2013), Australia (Flint 2001; Bekmann et al. 2013) and China (Li et al. 2008; Gu et al. 2005). Ongoing and detailed assessments have not been common to date, which is possibly due to the time consuming nature of reporting and limitations of the method, such as not considering all greenhouse gases or non-renewable resources (Graymore et al. 2008). The Macquarie University Property Department was tasked by the University's executive with decoupling campus growth and increased ecological consumption. It recently commissioned The Footprint Company™ to create an appropriate campus specific calculator to reduce the cost and complexity of sustainability analysis. This has enabled Macquarie to utilise EF in operations and facilities management. Staff in the Department of Environment and Geography secured a small teaching development grant to explore how this capacity might also be drawn into the University's learning and teaching activities.

Universities are in a unique position in that they act as prototypes of mini cities. With plans for campus growth Macquarie University wanted to ensure they weren't consuming more than their fair share of resources, while attempting to decouple growth from increased emissions and consumption. However, with competing interests and limited budget, it was essential to create transparency around the key drivers of Macquarie's environmental impact, in order to create strategic sustainability strategies that target the aspects of greatest impact. This is achievable utilising EF because the method allows for comparison of components (energy, water, transport, buildings, operational expenditure etc.) in a common metric—the global hectare (Bekmann et al. 2013).

The initial determination of the campus EF for the baseline year (2011) identified sustainability strategy hotspots, allowed for projections of strategies to be carried out, and generated benchmarks for new buildings and refurbishments, based on the assessment of the existing building stock. The process also allowed Macquarie University Property to identify a proposed One Planet pathway and to begin to develop strategies to target areas of greatest impact at a variety of scales (campus, faculty[1] and building). Ultimately, embedding the footprinting process has begun across multiple aspects of the institution in a synergistic way.

Employing a scaffolded approach, EF was embedded into the undergraduate Environmental Management major (available in the B.A, B.Sc and B.Env) at first, second and third year. Students measure their personal EF in first and third year and investigate EFs of the university's four academic faculties (Faculty of Arts, Faculty of Business and Economic, Faculty of Human Sciences and Faculty of Science) in the second year. The curriculum project thus generates longitudinal data of student's personal footprints in addition to analyzing EFs at Faculty and Campus scales. Over time, this approach will contribute to developing an understanding of EFs at individual, building and faculty scales in addition to the Property Department's analyses at building and campus scales. The university and curriculum is partially responsible for the shaping of students' personal and business selves (Huyuan and Yang 2012). In the future, the project will invite environmental management alumni to revisit the exercise 3 and 5 years after graduation to assess if students' intention to reduce EFs at a variety of scales has been maintained over time.

Targeting students, faculties and buildings through facilities management as well as learning and teaching generates a simultaneous top-down bottom-up One Planet sustainability strategy. The strategy actively engages with students, staff, departments, faculties and the campus with the hope of creating a working example for the broader community by demonstrating and promoting "One Planet" consumption.

2 Methodology

2.1 The Study Site

Macquarie University Campus covers 113 ha and is located in North Ryde, Sydney. In the 2011 baseline year, there were 3,488 full time equivalent (FTE) academic and support staff and 27,084 FTE students. The University has four academic faculties (Arts, Science, Human Sciences, Business and Economics (B&E)). The campus has

[1] Faculty: A group of university departments concerned with a major division of knowledge.

implemented a One Planet 2030 strategy. This initiative aims to see Macquarie University consume the equivalent of One Planet (or less) worth of resources by 2030 (hence seeing itself as sustainable). This has fueled the incorporation of EF into curriculum, future property development and refurbishments of existing infrastructure, in order to ensure complete campus engagement of the project.

2.2 One Planet 2030 Strategy

At present 30 % of Macquarie University is developed land which includes buildings and external hard surfaces, with plans for growth and development to accommodate more students, as well as research and development partnerships. In order to achieve this growth and ensure that the impact on the planet is kept to a minimum, Macquarie University commissioned the The Footprint Company™ to utilise their existing building calculators and created a campus specific calculator. Together these calculators allow Macquarie to measure its whole-of institution impact upon the environment and ensure future growth decouples increased built form density from increased ecological consumption.

The initial stages of the project involved calculation of the campus EF and the individual primary buildings and faculties. This progressed to the calculation of estimated EFs of new buildings and refurbishments in the design pipeline, in order to make recommendations for future management and development. All future campus property developments will undergo EF assessments, which will be coupled to One Planet equivalent benchmarks to ensure new buildings at least meet, if not exceed, this standard.

While the design and construction of new and refurbished buildings is largely a concern for the University's Property Department, the day-to-day operations of learning and teaching are the responsibility of departments based within faculties. As such, a key target of the 2030 strategy roll out is ongoing engagement with departments and faculties with regard to operational activities and consumption. The One Planet strategy has three primary points of intervention campus, faculties and buildings (Figs. 1 and 2).

2.3 Pedagogical Approach

In order to investigate environmental educational outcomes, a four stage (Year 1, 2, 3 and alumni) two-tiered learning strategy has been developed for incorporation into the Bachelor of Environment program, specifically targeting students majoring in Environmental Management (Table 1). The Bachelor of Environment and Environmental Management major programs are administered by the Department of Environment and Geography within the Faculty of Science. The development and incorporation of EF modules into the compulsory units of study has established a common thread through the Environmental Management major curriculum. For detailed discussion of the methodology see Howitt and Rickards (2013).

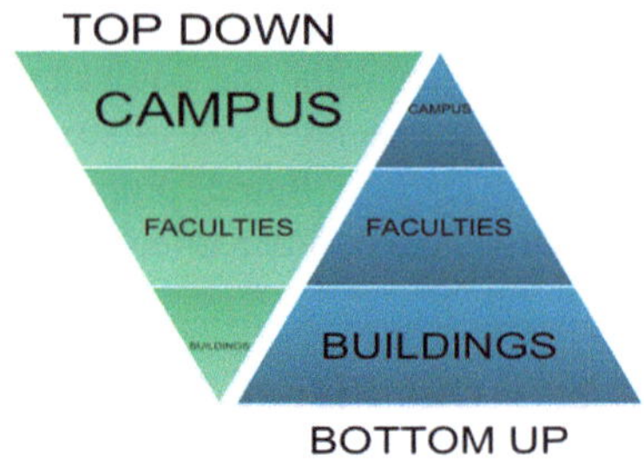

Fig. 1 Ecological footprinting as a top-down and bottom-up approach

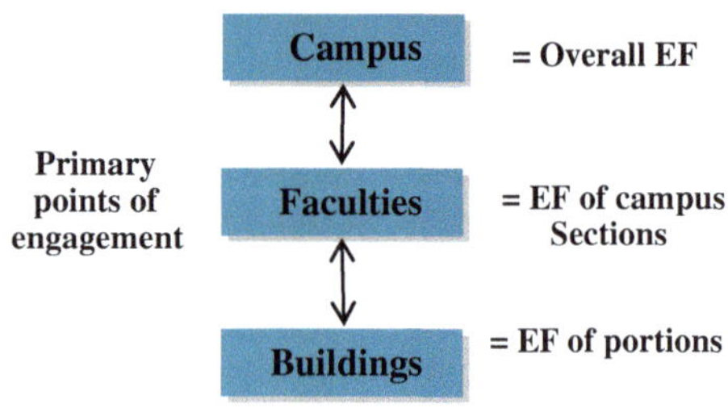

Fig. 2 Conceptual image of implementation of One Planet 2030 Master plan

Table 1 Summary of embedding eco-footprinting into curriculum

Year	Pedagogical strategy	Tier	Focus
Year 1	*Lecture* Background development, global and Australian contextualisation	1	Personal
	Tutorial Personal EF calculation and questionnaire		
Year 2	*Lecture* Campus contextualisation	2	Cross disciplinary collaboration and faculty business context
	Tutorial Software demonstration		
	Research assignment Environmental report and presentation		
Year 3	*Lecture* Investigation of EF system, limitations and assumptions	1	Personal
	Tutorial Personal EF calculation and questionnaire		
Alumni (Currently proposed for calculation at 3 and 5 years)	*Web forum* Personal EF calculation and questionnaire	1 & 2	Personal and business self

2.4 Tier One: Personal EF Calculation

The first tier engages students in a personal EF assessment at commencement of their undergraduate studies (Year 1), which is repeated again during their third year of study (Year 3) and into Alumni.

At these designated times, students perform two tasks, which allow for quantitative and qualitative data collection. The first task involves using the Global Footprint Networks (GFN) personal footprint calculator to determine the student's current EF. The second task is a questionnaire, which allowed for data tracking of the student's own reported behaviours and contextual changes that might affect their EF. This approach was chosen as raw data driven approaches without additional qualitative insights have the potential to be less instructive of the situation (Eckel et al. 1999). It should be noted that the study population changes from year to year due to enrolment and exchange programs etc. Therefore, the students are tracked using their student number, to allow for the comparison of their personal EF and behavioural reflections.

2.4.1 The Global Footprint Network Personal Calculator

The GFN personal EF calculator utilises data from the NFAs and the calculator matrix methodology and is in accordance with the international EF standards (Borucke et al. 2013). In order to determine a personal EF, different categories (food, shelter, mobility, goods and services) are allocated to land types (forest, crop, grazing, energy, and fishing), resulting in a matrix, which utilises a country's average consumption profile. The personal EF calculator increases or decreases the personal EF matrix, in response to answering questions related to the stated categories, relative to the national average behaviour (GFN 2013). For in depth details on EF methodology and calculations see Borucke et al. (2013).

2.5 Tier Two: Faculty EF Calculation

The second tier engages students with a faculty EF to investigate the key drivers of each faculty's footprint. This allows students to recommend and model faculty specific sustainability strategies while researching innovative solutions and best practice within industry and other universities. Students with exceptional understanding are then recruited to internships within the Property Department for further investigation and justification of identified interventions.

2.5.1 Faculty and Campus EF Calculation

The campus and faculty EF's were carried out using the same methodology unless otherwise stated (Table 4) (campus specific assessments discussed in Bekmann et al. (2013).

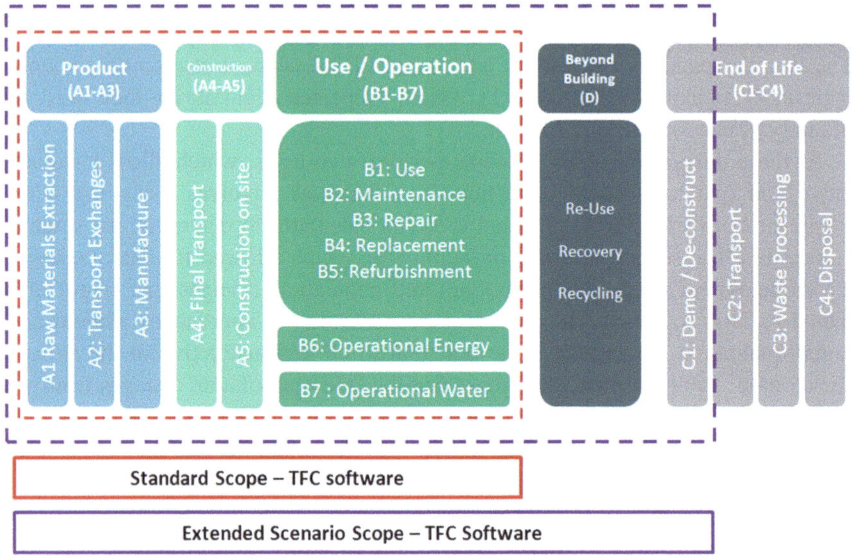

Fig. 3 The Footprint Company™ calculator's system boundary

The system boundary of the faculties is represented by the physical site boundary of faculty-designated buildings. The system boundary of the campus is represented by the physical site boundary of the campus and the facilities that Macquarie has operational control over.

These assessments consider the capital formation (the base building materials and infrastructure) of the Faculty/Campus, assuming a 25 year life-cycle. In addition, building fit-outs are also assessed, assuming a 10-year life cycle. Operational impacts per annum associated with energy, water, transport, tenancy and operational consumption are also incorporated into the assessments.

Figure 3 illustrates the system boundaries as used by the EF calculator. The aspects and activities within the dotted red line define the system boundary considered within the calculation methodology. The primary data coverage and boundary is highlighted in Table 2.

2.5.2 The Footprint Company™ Campus Calculator

In accordance with the Global footprint Standards, The Footprint Company™ (TFC™) has developed a campus EF calculator, which is used in this project as both a teaching tool and as a way of measuring the institutional footprint of the campus and its faculties.

Within the calculators a hybrid life cycle assessment (HLCA) method is employed, which combines lifecycle analysis (LCA) and input-output analysis, increasing study completeness (Crawford 2008). This is in comparison to standard LCA methodology, which is typically imperfect due to incomplete or unreliable data sources (Kim et al. 2008). In addition, input-output models in isolation do not

Table 2 Primary data coverage and boundary

Footprint aspect	Coverage and meaning	Boundary and inclusions
Developed area (Site area and usable floor area)	The ecological value of land removed from service through the construction of the building net of any "credit" for on-site ecological additions or improvements	The ground plane footprint area of the proposed building(s)
Energy	The ecological value of carbon emissions arising for stationary energy	Base building energy requirements as entered and covering scope 1, 2 and 3 emissions
Water	The ecological and carbon footprint of water consumption by source, arising from operational requirements	Incorporates all tenant/occupant use
Buildings	The ecological impact of and carbon footprint of all materials required to create the building as measured finished in place	The ecological and embodied carbon impacts from the site all the way upstream to the point of extraction of natural resources including all intermediate processing and transport. "Approximate" buildings are based on Australian national averages
Construction materials[a]	The ecological and carbon footprint of all materials required to create the building as measured finished in place	The ecological and embodied carbon impacts from the site all the way upstream to the point of extraction of natural resources including all intermediate processing and transport
Other capital items (Non-building items)	The ecological impact of and carbon footprint of all materials required to create the other capital items as measured finished in place	The ecological and embodied carbon impacts from the site all the way upstream to the point of extraction of natural resources including all intermediate processing and transport
Operational items	The ecological impact of annual operational consumption	Operational items such as IT and electronics, paper and print services, consumables, food and drink and cleaning
Transport[b]	The ecological impact of all occupant and user related transport to access the site	All tenants/occupants on the basis of trip mode and vehicle kilometers travelled
Offsets (Biocapacity)	The ecological value of any activity specific to restoring, protecting or enhancing biological capacity whether on-site or off-site	Off-site ecological protection or restoration activities are allocated to offsets. Energy offsets purchased from 3rd parties are noted in offsets

[a]Not directly measured in the campus/faculty calculation. Construction materials were assessed in the pre-assessment of campus buildings using The Footprint Company™ multi-use calculator
[b]Data was only available for the campus, in following years differences between faculties will be investigated

Table 3 The Footprint Company LCA methodology summary

LCA method element	Valid for this report	Comment
Principle methodology	*Hybrid LCA*	
Goal	Quantify anticipated EF of the registered project	Capital and one year of operations
Scope	Anticipated EF as represented in the input parameters for the registered project	TFC™ can currently investigate individual buildings and campuses as a whole (or area of a campus). Table 2 defines scope
Scope exclusions	Waste generated on site during construction and operations	Not covered by footprint methodology
Boundary	Site boundary and all materials as finished in place for building and fit-out in total terms upstream to the point of extraction—unbounded	Direct and indirect covered
Functional unit	Principle functional units include;	UFA as define by Australian Institute of Quantity Surveyors (AIQS's)
	Global hectares absolute (gha)	
	Usable floor area (UFA)	
	Global meters square absolute (gm^2)	
	Gha and gm^2 per meter square of lettable area (gha/m^2) or (gm^2/m^2)	
	Planet equivalent	
Geographic and temporal relevance	EF national footprint accounts, 2011 (GFN)	
	GHG AGO workbooks, 2011/2012	
	Materials sources including; CSIRO, 2005 (Balancing Act); BPIC database; CSIRO; TFC custom datasets	

have the ability to demonstrate the benefit of more sustainable (which can also mean more expensive) materials (Majeau-Bettez et al. 2011). The HLCA used in this study, uses actual life cycle cost quantities and where such quantities cannot be identified, the monetary value of an item is used with a gross national carbon intensity of economic end-use substituted (York et al. 2004).

The methodology considers all impacts from cradle to gate,[2] as well as incorporating all service and consumption inputs through monetary consideration. The key limitation of this technique derives from the use of the input-output model, which is where the monetary value of an item is used with a gross national carbon intensity of economic end-use. For TFP™ methodology summary see Tables 3 and 4.

[2] Cradle-to-gate is an assessment of a *partial* product life cycle from resource extraction (*cradle*) to the factory gate (i.e. before it is transported to the consumer).

Table 4 Primary assumption differences faculty versus campus

Aspect	Faculty	Campus
Space (Building footprint and UFA)	Physical building footprint and actual occupied space of faculty buildings	Physical campus boundary, excluding buildings not within operational control
Biological capacity	Assumed all built on area	Determined from complete site area
Energy	Actual building consumption	Actual campus consumption
Water	Pro-rated campus water consumption per m^2	Actual campus consumption
Non building items	Assumed nil and not within building boundary	Actual campus non-building items audited
Operational items	Actual operational expenditure per faculty	Actual campus operational consumption
Transport	Transport mode shift and average one way distance assumed equal to campus. Actual FTE of faculty used	Actual campus data

2.6 2030 Projections

Four alternative pathway models were developed. Two main models are based on the low and high growth models defined by the Macquarie University Property Department. Two of the pathways investigated the two models under a Business as usual (BAU) approach (current sustainability performance). The remaining two pathways were modelled under a high growth scenario and investigated an innovative and proposed One Planet pathway (Table 5).

3 Results

The current status of Macquarie University's One Planet project has preliminary results available from various elements of the project. These results are seen as indicative of the value and workability of the approach.

3.1 Embedding EF Into the Curriculum

3.1.1 Tier 1: Personal EF Calculation Preliminary Findings

It should be noted that presented data is representative of early stage results from the Year 3, 2013 module trial. Personal EF calculations were determined for a 39-student study group. The class average EF was 5.7 gha, which is lower than the Australian average of 6.68 gha (WWF 2012). Assuming the current global population is 7 billion people, and the biocapacity is 13.3 billion gha, if everyone lived like the third year Environmental management class, 3 planets would be required to sustain their current lifestyle.

Table 5 Primary model parameters for 2030 projections

Aspect	Units	2011 Baseline	2030 High growth proposed innovation	2030 High growth proposed One Planet pathway	2030 High growth BAU	2030 Low growth BAU
Biological capacity						
Site area	Ha	2011 physical and operational boundary	Remains at 2011 area	Remains at 2011 area	Remains at 2011 area	Remains at 2011 area
Building % of total site area	%	9	12	12	12	9
External hard surfaces	%	21	21	21	21	21
Energy						
Electricity/gas	kWh	2011 Current efficiency level	4.5 NABERS[3] base building only[b]	40 % Reduction base building and tenants[a]	4.5 NABERS base building only	4.5 NABERS base building only
Water						
Portable	kL	2011 Current efficiency level	4.5 NABERS base building only	40 % Reduction base building and tenants[a]	4.5 NABERS base building only	4.5 NABERS base building only
Transport						
Train mode	%	24	30	40	24	24
Bike mode	%	1	3	3	3	1
Operational items						
Total operational expenditure	$/FTE	2011 levels	No net change	See below	No net change	No net change
Purchased print items	$/FTE	2011 levels	No net change	40 % Reduction[a]	No net change	No net change
Waste disposal	$/FTE	2011 levels	No net change	50 % Reduction[a]	No net change	No net change
Buildings						
New academic	Gm^2/m^2	Baseline	−37 % 432[a]	−37 % 432[a]	BAU 685	BAU 685
New commercial	Gm^2/m^2	Baseline	432[a]	432[a]	685	685

[a]Method to be determined
[b]The National Australian Built Environment Ratings System (NABERS) is a performance-based environmental impact rating system for existing buildings, measuring the performance of Australian buildings. http://ee.ret.gov.au/national-australian-built-environment-rating-system#sthash.YaBkmcvz.dpuf

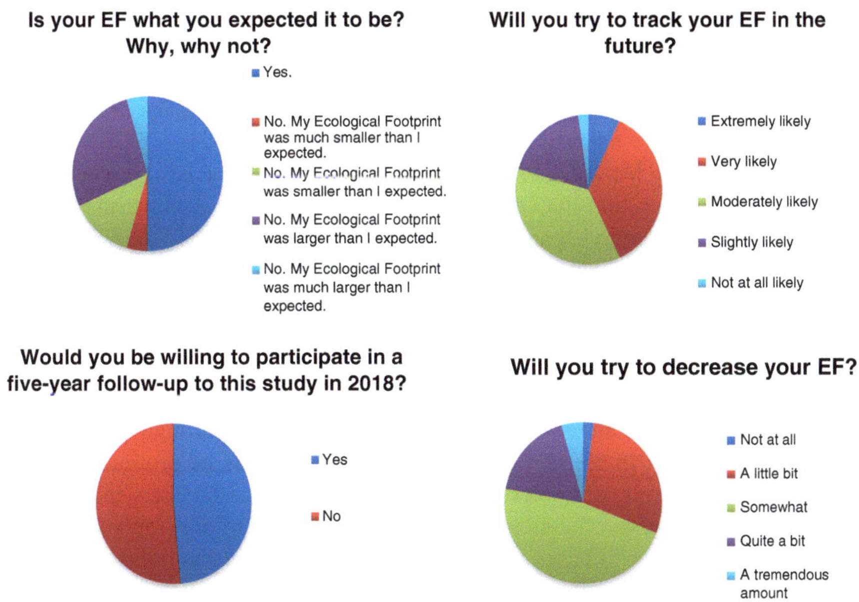

Fig. 4 Summary of key findings from EF student survey 2013 pilot

Summary of key findings:

- 81 % of students had calculated their EF before (Fig. 4)
- 50 % of students expected their EF to be equal to what the GFN calculator calculated (Fig. 4)
- 32 % of students found their EF to be larger than they anticipated (Fig. 4)
- 69 % of students said they will try to decrease their EF (more than a little bit) (Fig. 4)
- 93 % Students found the exercise interesting (more than a little bit) (Fig. 5)
- 49 % would be willing to participate in the 5-year follow up study (Fig. 4)
- Greatest impact categories were food and mobility.

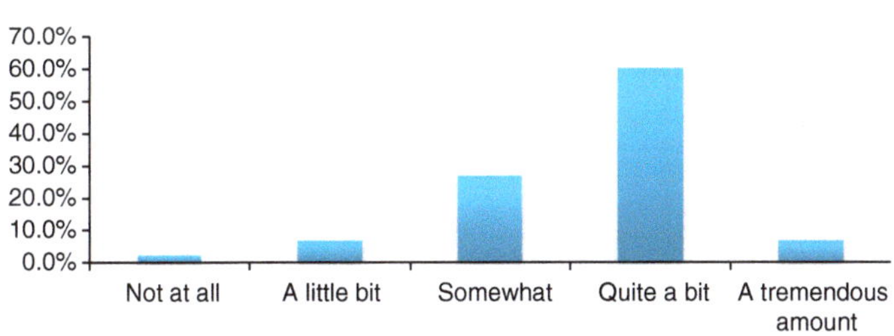

Fig. 5 Student survey question “Was this exercise interesting?”

3.1.2 Tier 2: Preliminary Faculty EF's

While campus, building and individual scale footprinting offers valuable insights into environmental burdens of campus activity, the institutional structures that dominate decision-making are also important. Second-year students are therefore asked to undertake an analysis framing the available data in terms of the University's four academic faculties. The initial investigation of the faculties, undertaken prior to incorporation into the curriculum, revealed that in absolute terms the faculties do not have a large EF and that the ancillary component of the University is the primary biocapacity consumer (as discussed further below) (Fig. 6). However, the faculties remain a primary point of intervention, due to 87 % of the campus population consisting of academic staff and students who are primarily based in faculties.

The ancillary components include faculty support buildings such as the chancellery, library and human resources, which will be addressed through the top-down bottom-up approach. In essence, these ancillary areas should be pro-rated into the faculty EF's. However, our work to date has examined the faculties in isolation. The resultant faculty-specific case studies have allowed us to develop an approach that can be adapted for different university administrative units and offices, and will support future collaborations with other universities for "best practice" guideline benchmark development.

Preliminary findings, based on data collected from 2011, demonstrate that the Science Faulty has the largest impact, in terms of absolute EF and global meters squared per square meter of building (Fig. 7 and Table 6). The Science Faculty has the largest staff numbers (33 %), but only 17 % of FTE students. In contrast, the Faculty of B&E with 42 % of FTE students hosts only 17 % of FTE staff (Table 7). In addition, the energy usage with respect to FTE population was largest for the Science Faculty, followed by Human Sciences, Arts and B&E, respectively (Fig. 8 and Table 7). It should be pointed out that the Science Faculty used the least amount of Energy with respect to UFA (Table 7). The Science Faculty is a large

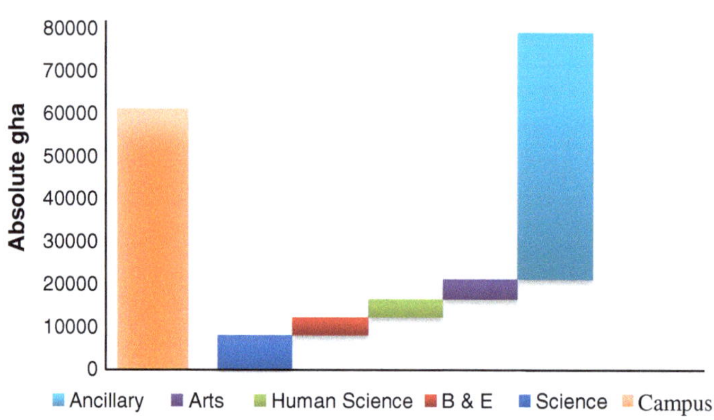

Fig. 6 Absolute ecological footprint by component

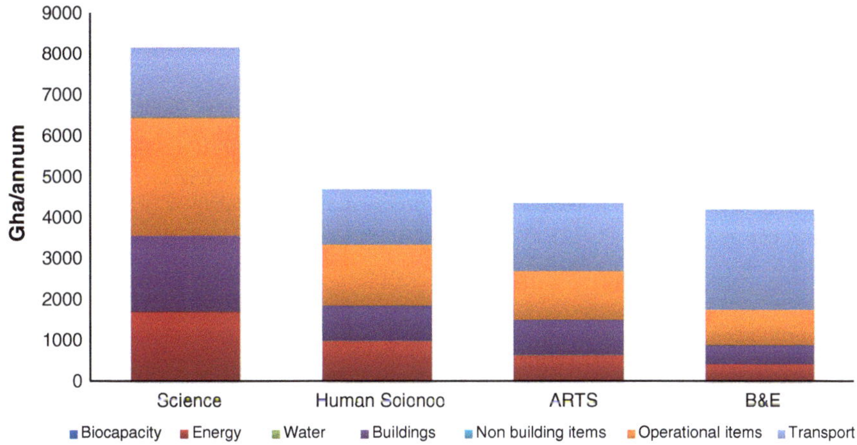

Fig. 7 Absolute global hectares consumed per faculty

Table 6 Summary data collection of faculty EFs in 2011

Faculty	Arts	Business and economics	Human sciences	Science
Useable floor area (m^2)	14,767	5,878	13,929	33,730
Site area (Hectares)	1.9	0.2	0.8	1.6
Academic and support population (FTE)	361	275	445	524
Student population (FTE)	6,112	10,732	4,341	4,323
Energy	1,853,244	1,237,040	2,903,512	3,158,894
Electricity (kWh/year)				
Water (kL/year)	19,672	7,576	13,116	10,175
Buildings	377	179	318	892
Impact (Gm^2/m^2)				
Non building items	NA	NA	NA	NA
Transport	Campus report	Campus report	Campus report	Campus report

consumer of space per person, at more than 2 times the other faculties (Table 7), a result of Science requiring large amounts of space for laboratories and equipment compared to other faculties.

Investigation of the ecological impact in terms of gm^2/FTE revealed B&E to have the smallest ecological impact. However, in absolute terms B&E is similar to Human Sciences and Arts Faculties (Fig. 8). Although there is obvious opportunity for B&E in terms of kWh utilised per m^2 (Fig. 8 and Table 7).

When investigating the faculties in absolute terms, the remaining three units (Human Science, Arts and B&E) have similar footprints (Fig. 7). However, Human Science consumes significantly more energy compared to Arts and B&E (Fig. 7).

Table 7 Faculty preliminary findings

Faculty	Arts	Business and economics	Human sciences	Science
UFA (m^2) per person (FTE)	2.28 m^2/ person	0.53 m^2/ person	2.9 m^2/ person	6.96 m^2/ person
Annual energy (kWh) per person (FTE)	286 kWh/ person/year	112 kWh/ person/year	607 kWh/ person/year	652 kWh/ person/year
Annual energy (kWh) per UFA (m^2)	125 kWh/m^2	210 kWh/m^2	208 kWh/m^2	93 kWh/m^2

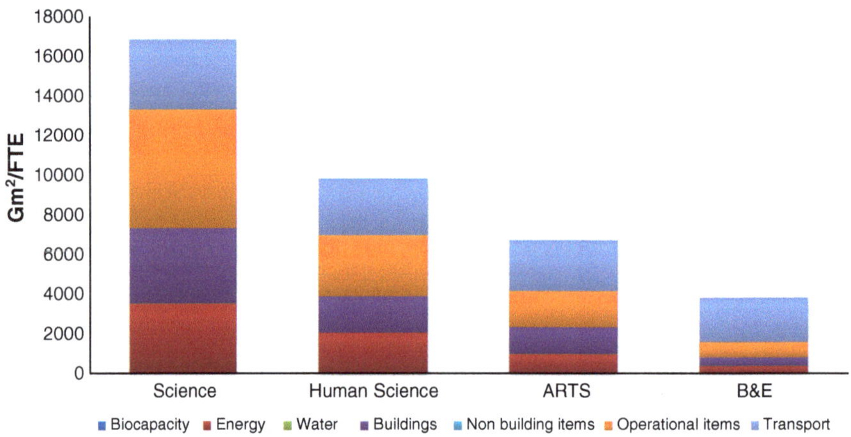

Fig. 8 EF per faculty per person

Human Science is the highest consumer of biocapacity per person compared to Arts and B&E, when in absolute terms all were very similar. Closer investigation of the increased EF on a per person basis revealed that the energy consumption of Human Sciences is significantly larger than Arts in terms of absolute global hectares and $gm^{2/}$FTE (Figs. 7 and 8). Space and energy consumption are institutively related (Fig. 9). Comparison of the UFA utilised per person in the Arts and Human Science Faulty is similar at 2.28 and 2.9 m^2/person. However the kWh/person consumed by Human Sciences is 212 % larger than Arts and 166 % larger in terms of kWh/m^2.

3.2 Campus Ecological Footprint

The EF of the entire Macquarie University campus was determined for the 2011 baseline year to be 1.3 planets. It was identified that Operational items (34 %), buildings (33 %) and energy (18 %) are the leading contributors for the campus's EF (Table 8 and Fig. 10). The annual operational footprint is larger than the capital footprint (Fig. 11).

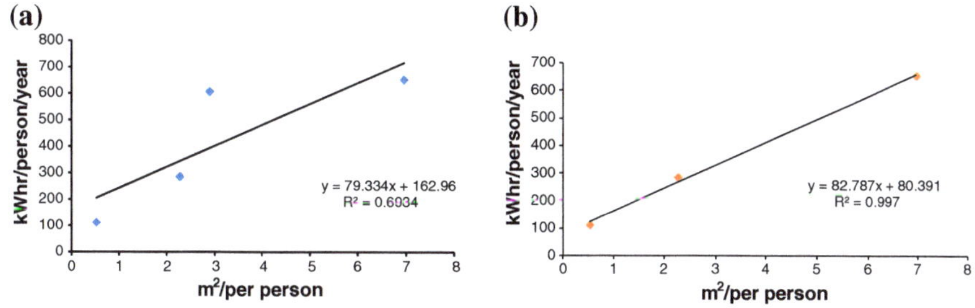

Fig. 9 Correlation between energy consumption and space. Graphical interpretation of Table 6 demonstrating the relationship between energy consumption and space. The outlier in **a** is human sciences and is removed in **b**, demonstrating the expected correlation between space and energy consumption

3.3 2030 Projections

If Macquarie University continues with BAU into 2030 it is projected that on a fast growth trajectory the university will reach 1.8 planets. The Footprint Company has proposed a strategy to get to 1.1 planets by 2030 (Fig. 12). The primary 2030 One Planet recommendations include:

- Reduction of operational expenses (and or shift) focusing on decreased print media and waste expenses e.g. Print media reduction and replace with electronic information provision
- Energy Reduction
- Buildings with less embodied energy and more efficient use and reuse
- Increased space utilisation
- Multi-functional areas

Table 8 Macquarie University campus EF results summary

Footprint aspect	Gha absolute	Gm^2/EFT
Biological capacity	139.9	45.8
Energy	14318.7	4683.9
Water	72.7	23.8
Buildings	26368.9	8625.2
Non building items	3268.8	1069.2
Operational items	27264.3	8918.0
Transport	8043.7	2631.1

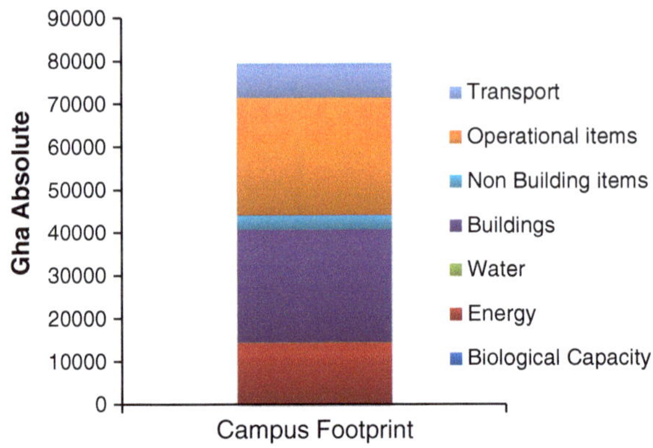

Fig. 10 Macquarie University complete campus EF

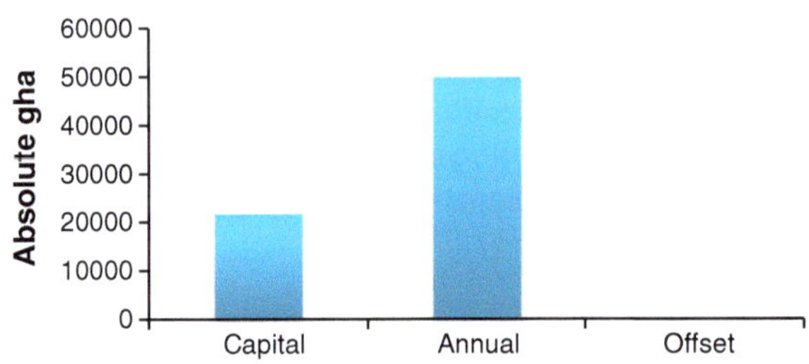

Fig. 11 Macquarie University EF by life cycle assessment. Graphical representation of the relative importance of operational and capital aspects and the benefit achieved through the investment of offsetting activities (if there were to be any). Note that capital is shown as 1/25th of its impact

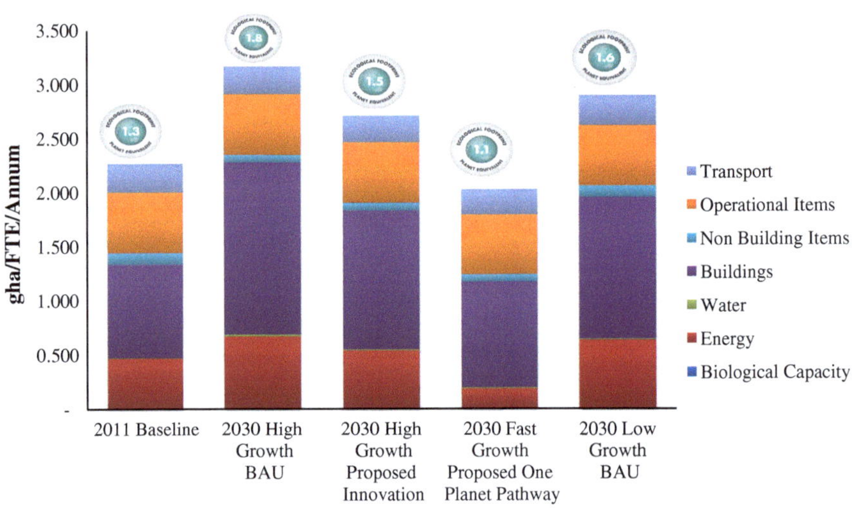

Fig. 12 2030 EF modelled projections

4 Discussion

2014 marks the end of the United Nations Decade of Education for Sustainable Development (Velazquez et al. 2005). Transformative approaches to sustainability across disciplines must appreciate that:

> To transform the world, we must begin with ourselves; and what is important in beginning with ourselves is the intention. The intention must be to understand ourselves and not to leave it to others to transform themselves or to bring about a modified change through revolution, either of the left or of the right. It is important to understand that this is our responsibility, yours and mine...
>
> (Krishnamurti 1954, p. 29)

Setting the intention to decrease individual, building, faulty and campus EFs, with a common goal of a One Planet campus, has facilitated increased understanding of the drivers that make up EFs at a variety of scales and provided increased insight into Macquarie University as a campus.

In order to implement campus wide change the challenge is only partly identifying what needs to be done and primarily capturing the attention of those required for success of the identified changes (Eckel et al. 1999). Universities and individuals are complex creatures, therefore creating a model claiming to investigate all of the intricacies of either, undermines the intricacy of both. EF is unable to account for eutrophication, land erosion, future technological advances and fails to investigate social aspects such as happiness and indoor environmental quality (Wiedmann and Barrett 2010). However, EF allows for goal setting, hypothetical investigation, identification of the key drivers of impact and is a good communication device.

Investigation of EFs at an international level reveals that countries with the largest EF have the highest level of education (Rees 2003). Evidence suggests that a personal contextual change, such as starting university, is a pivotal time to induce changes in behaviors and therefore intention (Caspi and Roberts 2001). Consequently enabling students to be agents of change through targeted education will hopefully lead to long term transformation. Thus the development of teaching models to promote enquiry-driven learning in order to prepare Macquarie students for productive professional and personal lives, which promotes long-term sustainability, is a key imperative of the university (Macquarie University 2013). Segalas et al. (2010) mention that it would be valuable to track how sustainability education affects student's professional lives into alumni. Personal sustained change is still in question at this point and will be investigated in future years. Future studies will take into account that increased income (a likely repercussion of moving from student life to business life) generally leads to increased EFs (Lensen and Murry 2003).

The investigation of the faculties shouldn't primarily be used to compare one faculty to another. Faculties are responsible for different sorts of academic activities and battles about appropriate measures for success and quality are already in debate (Mryglod et al. 2013). Instead of comparing chalk and cheese, the key drivers of each faculty's footprints should be the focus, in the appropriate context of each faculty's goals. Subsequently, targeted sustainability strategies with shared learning for faculties who already demonstrate more sustainable practices should be developed. It is important not to lose sight of the fact that a primary purpose of university faculties is to teach, share and collaborate within the academic community, between disciplines and with broader society. This research challenges current administrative measures, which can stymy and suppress interdisciplinary research and encourages learning across administrative boundaries (Rafols et al. 2012)

Faculty EF assessments provide further granularity of the campus EF and also allow students to visualise how their personal footprint is affected by the studies they undertake. The competitive advantage of sustainability in higher education has been demonstrated (Dobson et al. 2010). However, most EF assessments of other universities are not comparable at this stage, due to different EF methodologies and models being adopted. There has been one preliminary EF assessment comparing 40 universities between Australia and the United States, using only publically available information and TFC™ software, which identified and compared the primary drivers of campus footprints (Bekmann et al. 2013). Previous to this study, assessment of schools and universities has usually found energy to be the primary driver of EF (Conway et al. 2008; Flint 2001; Venetoulis 2001; Li et al. 2008; Wright and Drossman 2002). However, the EF model used in the TFC™ software incorporates the capital or embodied cost of land and its built form. This in turn reveals that the base buildings and fit out are of higher impact than operational energy.

Obtaining an appreciation for the large amount of embodied energy in buildings and fit outs, with respect to other EF aspects, highlights the importance of considering building and material life cycle analysis as part of the construction and refurbishment process. In addition to this, an extended appreciation for the interconnection and dynamic relationship between built form, energy consumption and space utilisation is obtained. EF enables the University to quantify ecological savings between different building designs as well as the savings incurred by reusing materials during refurbishments. The process facilitates gratification for sustainability initiatives, because features and efforts are now quantifiable. EF is generally used to measure how much we are exceeding our fair share, but this method takes a constructive approach as it allows for measurement of how we are tracking towards our fair share and how each effort contributes to the One Planet pathway.

Now that the baseline campus and faulty assessments are complete and high-level projections have been carried out, the University has established what its One Planet targets should be. The next steps of the project will focus on how to achieve these targets in order to commence the University's journey towards One Planet consumption.

5 Conclusions

The future isn't somewhere we are going, but somewhere we are creating. In order to ensure intergenerational equality it is important to measure our impact on the planet and ensure we are doing our part to not consume more than our fair share. Universities are in a privileged position to create a culture of transformative learning that contributes solutions to the world. In order to successfully achieve this a holistic intervention, which focuses on interdisciplinary collaboration, generating engagement across disciplines and departments is vital in ensuring the different dimensions of the university system are working towards the common goal of a One Planet university and society.

Acknowledgments A Macquarie University learning and teaching grant, as well as a learning and teaching grant from the Department of Environment and Geography funded the research presented in this article. The development of the personal EF modules would not have been possible without the free and publicly available personal EF calculator provided by The Global Footprint Network, as well as the time and effort put into the project by The Footprint Company™ to develop a campus calculator. This project would not have been gotten off the ground without the endless help and support of Department of Environment and Geography, Property Sustainability Department and Macquarie's Finance team. A special thanks to Hilary Bekmann, Stuart Browning, Caroline Noller, Tim Ralph and Frank Thomalla in particular, for their time wisdom and guidance and everything else they contributed to the project.

References

Austin JT, Vancouver JB (1996) Goal constructs in psychology: structure, process, and content. Psychol Bull 120(3):338

Bekmann H, Rickards S, Noller C (2013) Ecological footprint benchmarking of 40 tertiary education campuses. In: Proceedings of the 13th international Australasian campuses towards sustainability (ACTS) conference, Sydney, Australia. http://www.acts.asn.au/index.php/2013-conf/conference-proceedings/

Borucke M, Moore D, Cranston G, Gracey K, Iha K, Larson J, Galli A (2013) Accounting for demand and supply of the biosphere's regenerative capacity: the national footprint accounts' underlying methodology and framework. Ecol Ind 24:518–533

Caspi A, Roberts BW (2001) Personality development across the life course: the argument for change and continuity. Psychol Inq 12(2):49–66

Chi G, Stone B Jr (2005) Sustainable transport planning: estimating the ecological footprint of vehicle travel in future years. J Urban Plan Dev 131(3):170–180

Chikoti I (2012) The ecological footprint. Economics for ecology. In: Proceedings from the 18th International scientific conference, April 27–30, 2012. Sumy, Ukraine, pp 23–24. http://iscs.fem.sumdu.edu.ua/data/ISCS_Materials_2012.pdf. Accessed 10 June 2013

Collins L (2012) Sustaining policies: a case study comparison. Proceedings of the 12th annual Australasian campuses towards sustainability (ACTS) conference, 26–28 September 2012. Brisbane, Australia. http://www.acts.asn.au/index.php/2012-acts-conference/proceedings/

Conner M, Norman P (2005) Predicting health behaviour. Open University Press, Berkshire

Conway TM, Dalton C, Loo J, Benakoun L (2008) Developing ecological footprint scenarios on university campuses: a case study of the University of Toronto at Mississauga. Int J Sustain High Educ 9(1):4–20

Cooper T, Scott RH (2009) Sustainability, survival and engagement: implications for curriculum and pedagogy in social professions. eCULTURE 2(1):16. http://ro.ecu.edu.au/eculture/vol2/iss1/16
Cortese AD (2003) The critical role of higher education in creating a sustainable future. Plan High Educ 31(3):15–22
Crawford RH (2008) Validation of a hybrid life-cycle inventory analysis method. J Environ Manage 88(3):496–506. doi:http://dx.doi.org/10.1016/j.jenvman.2007.03.024
Dobson A, Quilley S, Young W (2010) Sustainability as competitive advantage in higher education in the UK. Int J Environ Sustain Dev 9(4):330–348
Eagly AH, Chaiken S (1993) The psychology of attitudes. Harcourt Brace Jovanovich, Fort Worth
Eckel P, Green M, Hill B, Mallon W (1999) Taking charge of change: A primer for colleges and universities. American Council on Education, Washington, DC
Flint K (2001) Institutional ecological footprint analysis-a case study of the University of Newcastle, Australia. Int J Sustain High Educ 2(1):48–62
Global Footprint Network (GFN) (2013) Footprint basics: frequently asked questions. http://www.footprintnetwork.org/en/index.php/GFN/page/frequently_asked_questions/. Accessed 16 June 2013
Gollwitzer PM, Moskowitz GB (1996) Goal effects on action and cognition. In: Higgins ET, Kruglanski AW (eds) Social psychology: handbook of basic principles. Guilford Press, New York, pp 361–399
Graymore ML, Sipe NG, Rickson RE (2008) Regional sustainability: how useful are current tools of sustainability assessment at the regional scale? Ecol Econ 67(3):362–372
Gu XW, Li GJ, Wang Q, Liu JX, Ding Y, Liu JZ (2005) Ecological efficiency of high education—ecological footprint of university campus. J Glaciol Geocryol 3:014
Haberl H, Wackernagel M, Wrbka T (2004) Land use and sustainability indicators. An introduction. Land Use Policy 21(3):193–198
Howitt R, Rickards S (2013) Complete campus engagement via embedding ecological footprinting curriculum. In: Proceedings of the 13th international Australasian campuses towards sustainability (ACTS) conference, Sydney, Australia. http://www.acts.asn.au/index.php/2013-conf/conference-proceedings/
Huyuan L, Yang J (2012) Overcoming organisational resistance to sustainability innovations in Australian univerisites. In: Proceedings of the 12th annual australasian campuses towards sustainability (ACTS) conference, 26–28 September 2012. Brisbane, Australia. http://www.acts.asn.au/index.php/2012-acts-conference/proceedings/
Kim B, Houser LP, Rosenthal A, Neff R (2008) Literature review of methods and tools for quantifying the indirect environmental impacts of food procurement. The Johns Hopkins Center for a Livable Future, Johns Hopkins Bloomberg School of Public Health, Baltimore. http://sustainability.universityofcalifornia.edu/documents/jhufoodprint_rpt.pdf
Kitzes J, Peller A, Goldfinger S, Wackernagel M (2007) Current methods for calculating national ecological footprint accounts. Sci Environ Sustain Soc 4(1):1–9
Klein-Banai C, Theis TL (2011) An urban university's ecological footprint and the effect of climate change. Ecol Ind 11(3):857–860. doi:http://dx.doi.org/10.1016/j.ecolind.2010.11.002
Krishnamurti J (1954) The first and last freedom. HarperCollins, NewYork, p 29
Lensen M, Murray S (2003) The ecological footprint—issue and trends. ISA Research Paper 01–03, The University of Sydney, Australia. http://www.isa.org.usyd.edu.au/publications/documents/Ecological_Footprint_Issues_and_Trends.pdf. Accessed 20 June 2013
Li G, Wang Q, Gu X, Liu J, Ding Y, Liang G (2008) Application of the componential method for ecological footprint calculation of a Chinese university campus. Ecol Ind 8(1):75–78
Lozano R (2003) Sustainable development in higher education. Incorporation, assessment and reporting of sustainable development in higher education institutions. Master of Science thesis in environmental management policy. Lund, Sweden
Lozano R, Huisingh D, Ceulemans K, Lozano F, Lambrechts W, Waas T, … Lukman R (2013) Implementation of sustainability in higher education. From declarations to praxis, or vice

versa? In: Proceedings of the 7th conference of the environmental management for sustainable universities (EMSU)
Macquarie University (2013) Our University: our framing of futures. http://mq.edu.au/our-university/
Maddux JE (1999) Expectancies and the social cognitive perspective: basic principles, processes, and variables. In: Kirsch I (ed) How expectancies shape experience. American Psychological Association, Washington, DC, pp 17–40
Majeau-Bettez G, Strømman AH, Hertwich EG (2011) Evaluation of process-and input–output-based life cycle inventory data with regard to truncation and aggregation issues. Environ Sci Technol 45(23):10170–10177
McNamara KH (2010) Fostering sustainability in higher education: a mixed-methods study of transformative leadership and change strategies. Environ Pract 12(01):48–58
McNichol H, Davis JM, O'Brien KR (2011) An ecological footprint for an early learning centre: identifying opportunities for early childhood sustainability education through interdisciplinary research. Environ Educ Res 17(5):689–704. doi:10.1080/13504622.2011.572161
Mryglod O, Kenna R, Holovatch Y, Berche B (2013) Comparison of a citation-based indicator and peer review for absolute and specific measures of research-group excellence. Scientometrics 97 (3):767–777
National Union of Students (2001) Universities of Australia ecological development charter. Relaunch for Sustainable Universities Campaign. RMIT University website, http://rmit.com/browse;ID=wge5iqekxdr3z. Accessed 17 June 2013
Orr D (1994) What is education for? Earth in mind: on education, environment, and the human prospect. Island Press, Washington, DC, pp 7–15
Pearce JM, Uhl CF (2003) Getting it done: effective sustainable policy implementation at the university level. Plan High Educ 31(3):53–61
Rees WE (1996) Revisiting carrying capacity: area-based indicators of sustainability. Popul Environ 17(3):195–215
Rees WE (2003) Impeding sustainability? Plan High Educ 31(3):88–98
Rafols I, Leydesdorff L, O'Hare A, Nightingale P, Stirling A (2012) How journal rankings can suppress interdisciplinary research: a comparison between innovation studies and business & management. Res Policy 41(7):1262–1282
Segalas J, Ferrer-Balas D, Mulder KF (2010) What do engineering students learn in sustainability courses? The effect of the pedagogical approach. J Clean Prod 18(3):275–284
Velazquez L, Munguia N, Sanchez M (2005) Deterring sustainability in higher education institutions: an appraisal of the factors which influence sustainability in higher education institutions. Int J Sustain High Educ 6(4):383–391
Venetoulis J (2001) Assessing the ecological impact of a university: the ecological footprint for the University of Redlands. Int J Sustain High Educ 2(2):180–197
Wackernagel M, Rees WE (1996) Our ecological footprint: reducing human impact on the earth. New Society Publishers, Gabriola Island
Webb TL, Sheeran P (2006) Does changing behavioral intentions engender behavior change? A meta-analysis of the experimental evidence. Psychol Bull 132(2):249
Wiedmann T, Barrett J (2010) A review of the ecological footprint indicator—perceptions and methods. Sustainability 2(6):1645–1693
World Wildlife Fund (WWF) (2012) Living planet report 2012: biodiversity, biocapacity and better choices. http://wwf.panda.org/about_our_earth/all_publications/living_planet_report/2012_lpr/. Accessed 11 June 2013
Wright EP, Drossman H (2002) The ecological footprint of the Colorado College: an examination of sustainability. Environ Sci 5(1):23
York R, Rosa EA, Dietz T (2004) The ecological footprint intensity of national economies. J Ind Ecol 8(4):139–154. doi:10.1162/1088198043630487

Authors Biography

Sara Rickards is the Sustainability Project Manager (Learning & Teaching) & Eco-Footprint Analyst for Macquarie University. She has investigated the Ecological Footprint of Macquarie University's campus, buildings and faculties. Sara has benchmarked universities both nationally and internationally in terms of resource consumption and efficiency. She is a biomedical scientist and environmental engineer, which gives her the ability to develop transdisciplinary solutions to complex issues. Sara has worked in various capacities including; learning and teaching, consulting, facilities and operations.

Richie Howitt is Professor of Human Geography, Department of Environment and Geography and Director Macquarie-Ryde Futures Partnership at Macquarie University, Sydney. His applied social research addresses social aspects of environmental change, community development and economic change at a variety of geographical scales from the local to the global. He has examined issues of Indigenous rights and social and cultural impacts of resource and infrastructure projects, intercultural capacity of industry, government, community and not-for-profit agencies, and cross-cultural negotiation in planning new projects. His theoretical work focuses on conceptualizing ethical and cultural dimensions of resource systems across geographical scales as a foundations for securing social and environmental justice in sustainable production systems. He is currently researching longer-term narratives of social, economic and environmental change in rural Indigenous communities in New South Wales. He received the Australian Award for University Teaching (Social Science) in 1999 and became Distinguished Fellow of the Institute of Australian Geographers in 2004. He was elected by his academic colleagues to the Council of Macquarie University in 2014.

Sandie Suchet-Pearson is a senior lecturer in Human Geography. Her research and teaching experiences over the last 20 years have been in the area of Indigenous rights and environmental management. She's worked on Cape York Peninsula on community development in the context of a major mining operation, examined the strategies used by Indigenous peoples and local communities to assert their rights in wildlife management in Canada and southern Africa, and her current work focuses on Indigenous self-determination in the context of cultural tourism in North East Arnhem Land, northern Australia.

An Analysis of the Sustainability of Different Methods of Delivering Higher Education

Jonathan Davies

Abstract

Transforming society and the economy to a more sustainable basis remains one of the most significant challenges of the 21st Century. Higher education institutions are considered key in helping to make the transition to a more sustainable society and low carbon economy. This paper examines the sustainability of different methods of delivering higher education with a special emphasis on carbon emission reduction. Traditional higher education teaching and learning involves students living near campus and attending face to face lectures. The paper reviews the progress on reducing campus carbon emissions in the UK higher education sector as well as the carbon implications of alternative methods of delivery such as online distance learning and transnational education, particularly in the form of overseas branch campuses. With more students opting to study via distance learning and at a time when UK higher education institutions are looking to increase international student numbers both at home and abroad, it is important that the environmental, social and economic impacts of these different forms of delivery are understood.

Keywords

Higher education · Sustainability · International students · Carbon emissions

J. Davies (✉)
School of Science and the Environment, Manchester Metropolitan University,
Chester Street, Manchester M1 5GD, UK
e-mail: Jonathan.Davies@mmu.ac.uk

W. Leal Filho et al. (eds.), *Integrative Approaches to Sustainable Development at University Level*, World Sustainability Series, DOI 10.1007/978-3-319-10690-8_5

1 Introduction

The concept of sustainable development was first introduced at the 1972 Earth Summit in Stockholm and was popularised in the late 80s in the Brundtland report which defined sustainable development as "development which meets the needs of the present without compromising the ability of future generations to meet their own needs" (WCED 1987). It has since become one of the most widely used definitions.

Sustainability has grown to be an important agenda internationally and implies that the activities of a higher education institution (HEI) do not negatively impact the environment, are socially just and economically viable, and that they will continue to be so in the future. Universities can be considered 'small cities' due to their size and population (Alshuwaikhat and Abubakar 2008), and over the last few years HEIs have become more conscious of their sustainability performance, particularly their environmental impacts.

The first section of the paper reviews progress on carbon reduction at UK HEIs. One area of higher education that has received less attention is carbon emissions generated from international student air travel. Therefore, the second section analyses the carbon impact of international students using Manchester Metropolitan University (MMU) as a case study and examines the social and economic benefits they bring to the UK. The UK has always attracted international students due to the perceived quality of education, institutional reputations and the English language (Maringe and Carter 2007). This paper comes at a time when HEIs are actively recruiting international students and highlights the conflict between the sector's goal to increase international student numbers and attempts to constrain absolute carbon emissions. If the HE sector is to be at the leading edge of the sustainability agenda, then it must be seen engaging with challenging issues like international student air travel emissions.

Finally, this paper analyses the sustainability of distance learning and transnational education, with an emphasis on carbon impacts. Both have increased in popularity and many HEIs now offer courses online and overseas. Despite these increases, the sustainability implications of different methods of delivering higher education are rarely discussed in the literature.

2 Sustainable Development and Climate Change

Climate change is defined by the United Nations Framework Convention on Climate Change (UNFCCC) as "a change of climate which is attributed directly or indirectly to human activity that alters the composition of the global atmosphere…" (UNFCCC 1992). The Fifth Assessment Report of the Intergovernmental Panel on Climate Change (IPCC 2013) states that it is *extremely likely* anthropogenic activities such as the burning of fossil fuels in transport and electricity generation, are contributing to the unprecedented rise in average global temperatures.

There is a dual relationship between climate change and sustainable development (IPCC 2007). Climate change is influencing both the built and natural environment on which the human population is dependent, thereby affecting social and economic development, while sustainable development policies are influencing anthropogenic carbon emissions that are causing climate change (IPCC 2007).

The need to reduce carbon is a primary concern as its adverse effects are being experienced worldwide (Abolarin et al. 2013). Ignoring the issue of climate change will damage economic growth which in turn will hinder our ability to meet future generations' needs. In order to constrain temperature increase to a maximum of 2 °C then greenhouse gas (GHG) concentrations need to be stabilised at 450 ppm CO_2 equivalent or below (Elzen and Hohne 2008). The UK Government has set its own targets in response to this challenge, that being an 80 % reduction of GHG emissions by 2050 from 1990 levels as stated in the Climate Change Act (2008). The HE sector is considered key in helping to achieve these targets, firstly by reducing carbon emissions from operations (both direct and indirect emissions) and secondly, by leading the transition to a more sustainable society through teaching and research (Disterheft et al. 2012). This paper will be exploring carbon emissions in more detail.

3 UK Higher Education Carbon Emissions

3.1 Emissions scopes

Emission sources are classified into three scopes as defined in the Greenhouse Gas Protocol (WBCSD/WRI 2004). This approach of classification has been adopted by the UK HE sector (Table 1)

Table 1 Emission scopes (HEFCE 2010)

Scope	Description	Example sources
Scope 1: Direct emissions	Direct emissions from sources owned or controlled by the HEI	Direct fuel and energy use e.g. boilers
		Transport fuel
Scope 2: Indirect emissions	Indirect emissions attributed to generation of purchased energy	Purchased electricity
Scope 3: Other indirect emissions	All other emissions that are a consequence of activities of the HEI but occur from sources not owned or controlled by them	Water
		Waste
		Staff commute
		Student commute
		International student air travel

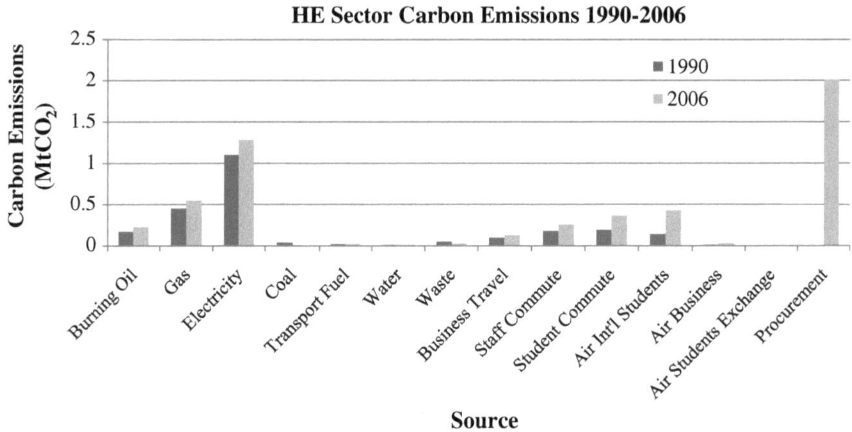

Fig. 1 Graph showing CO_2 emissions by source for the UK HE sector between 1990 and 2006 (*Source* SQW 2009. HEFCE 2012)

3.2 Emission Sources at HEIs

Figure 1 gives a breakdown of UK HE sector carbon emissions by source in 1990 and 2006. It can be seen that in 2006 procurement was the largest source of emissions followed by electricity and gas.[1] However, international student air travel showed the greatest increase, going from 0.140 $MtCO_2$ in 1990 to 0.426 $MtCO_2$ in 2006.

3.3 Carbon Reduction Progress

A review of the UK HE sector was carried out to assess the carbon reduction progress. HEFCE has set sector targets in line with national absolute carbon reduction targets. The sector should achieve a 43 % reduction by 2020 and 83 % by 2050 against a 2005 baseline (HEFCE 2010). Overall Scope 1 and 2 carbon emissions increased by 1.6 % from 2.28 $MtCO_2$ in 2005/06 to 2.31 $MtCO_2$ in 2012/13, peaking at 2.62 $MtCO_2$ in 2009/10 (SQW Energy 2009; HESA 2014). It is clear that more that needs to be done and quickly if the sector wants to achieve its emission reduction targets. One positive point is that the sector has reduced its carbon intensity between 2005 and 2013 from 1.00 T/CO_2 per student to 0.96 T/CO_2 per student. (Table 2)

[1] Procurement data for 1990 was not available.

Table 2 Changes in the UK HE sector carbon emissions and carbon intensity

	2005/06	2009/10	2012/13	% Change
Total students	2,281,235	2,496,645	2,396,050	4.8
Total Carbon ($MtCO_2$)	2.28	2.62	2.31	1.6
Carbon intensity (T/CO_2 per student)	1.00	1.05	0.96	−4

4 Methods of Delivering Higher Education

There are three methods of delivering higher education discussed in this article; traditional, distance learning and transnational education (TNE), as shown in Fig. 2. Traditional higher education teaching and learning in the UK involves students living away from home during term time, on or near campus and attending face to face lectures (Roy et al. 2008). Distance learning can be described as a process whereby tutor and student are separated spatially and by time (Liu 2008). TNE refers to delivery whereby students are based in a country other than the awarding HEI (Alam et al. 2013).

5 Traditional Higher Education

University campuses are not only places to study but places to model sustainable practices such as recycling, energy efficiency initiatives, sustainable building design etc. Many universities are embarking on ambitious campus development programmes with sustainable building design at the heart of projects. Universities have been identified as key in helping to make the transition to a sustainable society and this demonstration of sustainability initiatives influences behaviour and has a long lasting impact on graduates and staff throughout their lives. Universities must also

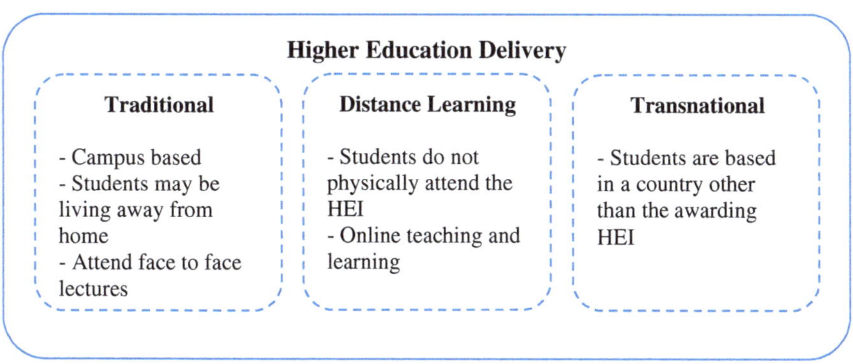

Fig. 2 Methods of delivering higher education (*Source* Roy et al. 2008; Liu 2008; Alam et al. 2013)

look beyond their campus boundaries and build strong relationships with towns and cities in which they are located to help advance sustainability in the community (Bilodeau et al. 2012).

6 Tradition Higher Education: International Students

Traditional higher education sees international students moving to live in the UK to study. The number of international students studying at UK HEIs has more than doubled between 1995 and 2012. In 2011/12 international students accounted for 17.4 % of the total student population (HESA 2014).

International students bring significant economic benefits to the UK and are an important income stream for HEIs, accounting for approximately 10 % of tuition fee income in 2009/10 (BIS 2012). Without international students many postgraduate courses at HEIs would not attract the required number of students to keep them running, as they account for 50 % of all students on taught masters courses and 44 % of full-time doctoral and research masters students (Smith et al. 2010).

International students have a positive impact on the local economy, for example the rental sector benefits from additional students as do businesses like bars and local supermarkets which rely heavily on students (Kelly et al. 2008). Manchester particularly benefits from international students with 17,000 studying at the three universities (MMU, University of Manchester and University of Salford) in 2011/12, accounting for 18 % of the total student population (HESA 2014). In total, including tuition fee income and other spending, international students contributed an estimated £8 billion to the UK economy in 2008/09. This could rise to £17 billion by the year 2025 (Conlon et al. 2011).

The value of international students is not just measured economically. Having 'international' campuses enhances the student experience, increases the diversity of the student population, exposes domestic students to different cultures and helps prepare students for a much more global business world (Altbach and Knight 2011). International students return to their home countries with strong emotional bonds to the UK, developed during their time here, and act as informal ambassadors helping to generate global connections (Mellors-Bourne et al. 2013). The benefits to the UK increase as alumni gain more influential positions as their careers progress bringing potential support to UK economic and social agendas (Mellors-Bourne et al. 2013). Many of the top source countries are also rapidly developing economies and key export markets including China, India and the Middle East (UKCISA 2011).

However, these social and economic benefits also come with environmental costs, namely the carbon emissions generated as a result of students flying to and from the UK.

7 Carbon Impact of International Students

Air travel undertaken by international students has a significant environmental impact in terms of carbon emissions and it is unlikely that there will be a significant step change in aircraft technology in the near future to reduce the carbon impact of flying. The implications for HEIs, who are aiming to reduce their carbon emissions, are very challenging, considering the social and economic benefits international students bring and the drive for growth in this sector.

8 Case Study: Manchester Metropolitan University

Manchester Metropolitan University has around 35,000 students of which 2,700 are international students. International student air travel emissions at MMU increased by 55 % from 3347 T/CO_2e in 2008/09 to 5196 T/CO_2e in 2013/14 (see Fig. 3). They account for approximately 10 % of the institution's total carbon footprint. MMU has a relatively small international student population in comparison with HEIs of a similar size. Consequently it has a target to double the number of international students between 2012 and 2017 in a bid to increase the internationalisation of the University. The demand for UK higher education will continue to increase as developing countries such as Brazil, China, India and the Gulf region in particular change to more knowledge based economic growth and the carbon emissions from the associated air travel will be significant due to the distance of these countries from the UK. MMU, like all UK HEIs, has also committed to carbon reductions. These two commitments are achievable when looked at

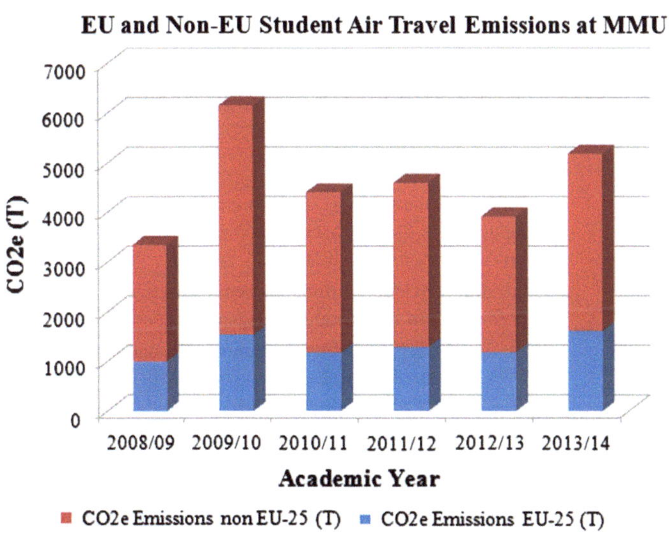

Fig. 3 International student air travel emissions at MMU

separately, but when judged together, they become conflicting policies and raises questions about how to move forward.

One answer would be to ignore the issue of international student air travel emissions and that of Scope 3 emissions in general; however the sustainability agenda has made this no longer a legitimate course of action. Sustainable development demands that organisations such as HEIs look across the breadth of their operations and impacts by calculating a full carbon footprint. HEIs must account for their Scope 3 emissions as they are often the largest source of emissions and offer the greatest potential for carbon reductions.

International students are vital to the higher education sector/MMU and the emissions are unavoidable. Across the sector there is a clear lack of action to address the issue beyond monitoring and reporting. Emissions from international students travelling to study at MMU will increase, meaning carbon reductions must be achieved in other areas, most likely Scope 3. MMU is currently researching alternative ways to compensate for the emissions from international student travel that go beyond typical carbon offsetting schemes like tree planting; rather, schemes which could provide local community support and education, installing onsite renewable energy or improving energy efficiency in the homes of the local community or university staff/students. Actions to compensate for international student air travel emissions will help HEIs engage with the sustainability agenda.

9 Transnational Education (TNE)

UK universities are under pressure to continue growing but there has been a recent flat lining in international student numbers as the Government cuts down on visas for migrants entering the UK. Because of this more HEIs are offering higher education to students in their home countries. Table 3 describes the different types of TNE.

Table 3 Different forms of transnational education

Type of TNE	Description
Branch campus	A HEI from the offering (source) country establishes a fully fledged campus in the host country to deliver its courses to students in that country (Alam et al. 2013)
Collaborative delivery	Education providers in different countries collaborate to offer a single degree program and/or double degree program (Alam et al. 2013)
Validation and franchising	With franchising, the offering country HEI authorises the host partner to deliver its courses and programs. The qualification is awarded by the source country's institution. Education quality and assessments are moderated by the source HEI. Validation is much the same, but the curriculum is determined jointly by the university and its collaborator and may not be exactly the same as the parent institution (Alam et al. 2013).
Distance learning	Distance learning is characterised by the separation of the learner from the tutor. Material is available online from the source institution with guidance available from tutors (Hussain 2007)

10 Carbon Impact of Transnational Education

It has been suggested that educating overseas students via partnerships and foreign branch campuses in the students' home country rather than bringing them to the UK to study may be preferable environmentally (Roy et al. 2008). This would be the case if HEIs did not allow students to come to the UK to study. However, at present, the opening of a foreign branch campus does not provide any carbon reductions to HEIs. There are four main reasons for this, firstly, international branch campuses have been found to be 'international' themselves in their recruitment policies. Most of the branch campuses reviewed recruit students from many different countries, for example, universities based in the UK with campuses in Singapore and Malaysia are now targeting students from countries such as China, South Korea and India (Wilkins and Huisman 2011). Branch campuses in the United Arab Emirates tend to attract very few Emirati nationals because they tend to go to the state universities where they pay no fees, as a result most students are international students (Wilkins et al. 2012).

Secondly, branch campuses increase HEIs brand recognition influencing potential students thinking about coming to study in the UK. HEIs have recognised that branch campuses overseas are a good strategy for expanding their student base and global brand (Wilkins and Huisman 2011).

The third reason is to do with 'study abroad' and exchange programmes at parent institutions. It is likely that opening a branch campus overseas will lead to more UK students taking the opportunity to spend a year or a term studying at the overseas campus therefore increasing carbon emissions associated with international student travel. The UAE campus of New York University is expected to receive an increasing number of US based students participating in a study abroad programme (Wilkins and Huisman 2011). Murdoch University in Australia has launched a programme that allows students at its home campus to spend a year or a term at its Dubai campus and students in Dubai can also spend time studying in Australia (Wilkins and Huisman 2011). Nottingham University offers students the opportunity to spend a year studying at one of its two international branch campuses in China and Malaysia (University of Nottingham 2014). In 2011/12, 174 students participated in the study abroad programme, assuming they made one trip between the UK and Malaysia/China; they contributed 541 T/CO_2e to the university's carbon footprint.[2]

The fourth reason relates to 'flying faculty' provision. Flying faculty sees the parent institution fly out academics to teach in the branch campus or partner institution in short intensive blocks (Alam et al. 2013). Each flight made by academics will contribute to the HEI's carbon footprint.

[2] Calculated using data on the number of students, the distance between Malaysia/China and the UK and the relevant GHG conversion factor.

The establishment of branch campuses and other forms of TNE provision is unlikely to offer any carbon reduction in terms of international student travel emissions at present. If anything due to the 'international' nature of branch campuses and the brand promotion they offer, carbon emissions as a result of student air travel will increase. TNE could offer carbon benefits to HEIs but only by limiting international students to studying at branch campuses or through a partner institution in their home country. However, due to the importance of international students coming to study in the UK both economically and for the student experience, this is an unlikely step any HEI would be willing to take.

11 Distance Learning

Distance learning courses are usually designed to serve people off-campus and are offered to both domestic and foreign students (Hannay and Newvine 2006). They give people who cannot complete a traditional campus based degree due to employment, family commitments, costs, geographical location etc., access to higher education thus helping to address the social equity aspect of sustainability (Hannay and Newvine 2006; Purnell et al. 1996). Many HEIs are transitioning from traditional face to face campus teaching to online distance taught courses in order to maintain competitiveness (Keengwe and Kidd 2010). This reflects the changing demography of students with more mature students entering HE, distance learning suits their lifestyles and there is now demand for HEIs to cater for a larger and more diverse cross section of the population (Garrison and Kanuka 2004).

Distance learners compared to traditional campus based learners have a significantly smaller carbon footprint. Roy et al. (2008) found that distance learning HE courses involve 87 % less energy and 85 % lower CO_2 emissions than full-time campus based courses. This reduction in carbon emissions is possible because distance learning eliminates or reduces infrastructure and activities needed for conventional learning. There is no need for staff and students to regularly commute to and from an institution (Hooi et al. 2011).

Home and overseas students attending university have a significant positive impact on the local economy (Love and McNicoll 1988). Undergraduate student spending alone accounts for up to 10 % of the total economic activity of some cities (Swinney 2011), adding postgraduate, staff and international university business and leisure visitors spending means any drop in student numbers at HEIs due to courses switching to being taught via distance learning is likely to have a negative impact on the economies of the cities in which they are located (Keengwe and Kidd 2010).

Distance learning will have implications in terms of the student experience of learning, particularly in terms of face-to-face academic and social interactions, which is only partially mitigated by social media provision (Bullen 1998). The lack of interaction between students in the classroom often reduces or negatively impacts on the learning experience and if a student is studying from the comfort of their own home they will be missing out on the social side of living on a university campus (Besser and Bonn 1996).

Obviously not all courses can be converted to distance learning but integrating some of the best aspects of distance learning into traditional courses to build a 'blended' learning environment may help to address problems associated with solely distance learning courses. Distance learning has a role to play in a low carbon HE sector but the environmental, social and economic impacts must be taken into account when deciding the mix of campus based and distance learning courses (Roy et al. 2008).

12 Conclusions

This article reviewed the UK HE sector's carbon reduction progress and found that absolute emissions had increased by 1.6 % between 2005 and 2013, highlighting that much more needs to be done and quickly if the sector is to meet HEFCE targets. However, these targets don't include Scope 3 emissions, which have been shown to be highly significant. Even looking at Scope 1 and 2 emissions the sector has failed to achieve reductions despite reducing carbon intensity.

In order to remain competitive many HEIs are increasing the use of distance learning and transnational education. Distance learning offers significant carbon reductions as a substitute to conventional delivery but has a negative impact on local economies and changes the student experience. Transnational education does not offer carbon reductions at present as it is not used as a substitute to the traditional higher education model, if anything due to the 'international' nature of branch campuses and the brand promotion they offer, carbon emissions as a result of student air travel will actually increase.

This article has highlighted the conflict between increasing international student numbers and any attempt to constrain absolute carbon emissions. Air travel is unsustainable; you can't travel by air and not have a significant carbon impact. It is likely emissions from international student travel will account for an increasingly large proportion of total sector emissions in the future and the sustainability agenda makes it no longer possible to ignore the issue. However, the social and economic benefits of international students coming to study in the UK are well known and any reductions in international student numbers will impact upon this. This is a conflict that is not yet being engaged with at a local or national level. Further research exploring innovative ways to compensate for the carbon emissions generated as a result of international student air travel should be undertaken.

References

Abolarin SM et al (2013) A collective approach to reducing carbon dioxide emission: a case study of four University of Lagos halls of residence. Energy Build 61:318–322

Alam F, Alam Q, Chowdhury H, Steiner T (2013) Transnational education: benefits, threats and challenges. Procedia Eng 56:870–874

Alshuwaikhat H, Abubakar I (2008) An integrated approach to achieving campus sustainability: assessment of the current campus environmental management practices. J Clean Prod 16:1777–1785
Altbach P, Knight J (2011) The internationalization of higher education: motivations and realities. J Stud Int Educ 11(4):290–305
Besser H, Bonn M (1996) Impact of distance independent education. J Am Soc Inf Sci 47 (11):880–883
Bilodeau L, Podger J, Abd-El-Aziz A (2012) Advancing campus and community sustainability: strategic alliance in action. Int J Sustain High Educ 15(2):157–168
BIS (2012). Overseas students and net migration. Department for business, innovation and skills. http://www.publications.parliament.uk/pa/cm201213/cmselect/cmbis/425/42502.htm. Accessed 07 March 2014
Bullen M (1998) Participation and critical thinking in online university distance education. J Distance Educ 13(2):1–32
Climate Change Act (2008) Climate change act 2008: Chapter 27. The Stationary Office, London, p 108
Conlon G, Litchfield A, Sadlier G (2011) Estimating the value to the UK of education exports. Department for business, innovation and skills. http://www.bis.gov.uk/assets/biscore/higher-education/docs/e/11-980-estimating-value-of-education-exports.pdf. Accessed 26 Feb 2014
Disterheft A, Caeiro S, Ramos M, Azeiteiro U (2012) Environmental Management Systems (EMS) implementation processes and practices in European higher education institutions - top-down versus participatory approaches. J Clean Prod 31:80–90
Elzen M, Hohne N (2008) Reductions of greenhouse gas emissions in Annex 1 and non-Annex 1 countries for meeting concentration stabilisation targets. Clim Change 91:249–274
Garrison R, Kanuka H (2004) Blended learning: uncovering its transformative potential in higher education. Internet High Educ 7:95–105
Hannay M, Newvine T (2006) Perceptions of distance learning: a comparison of online and traditional learning. J Online Teach Learn 2(1):1–11
HEFCE (2010). Carbon reduction target and strategy for higher education in England. Higher education funding council for England. http://www.hefce.ac.uk/pubs/year/2010/201001/. Accessed 07 Feb 2014
HEFCE (2012). Measuring scope 3 Carbon emissions—supply chain (procurement). http://www.hefce.ac.uk/pubs/rereports/year/2012/scope3carbon/. Accessed 28 April 2014
HESA (2014). Free online statistics. Higher education statistics agency. http://www.hesa.ac.uk/content/view/1897/239/. Accessed 18 March 2014
Hooi K, Hassan P, Jami N (2011) sustainable education: an assessment of carbon footprint at ucsi university and proposed green campus initiative framework. In: Third international conference on information and financial engineering, Singapore
Hussain I (2007) Transnational education: concept and methods. Turkish Online J Distance Educ 8 (1):163–173
IPCC (2007) Intergovernmental panel on climate change fourth assessment report: climate change 2007. http://www.ipcc.ch/publications_and_data/ar4/wg3/en/ch2s2-1-3.html. Accessed 13 Feb 2014
IPCC (2013) Climate change 2013: the physical science basis. Summary for policymakers. https://www.ipcc.ch/report/ar5/wg1/docs/WGIAR5_SPM_brochure_en.pdf. Accessed 13 Feb 2014
Keengwe J, Kidd T (2010) Towards best practices in online learning and teaching in higher education. J Online Learn Teach 6(2):533–541
Kelly U, McLellan D, McNicoll I (2008) The impact of Universities on the UK economy: fourth report. Universities UK
Liu S (2008) Student interaction experiences in distance learning courses. A phenomenological study. Online J Distance Learn Adm 11:1. http://www.westga.edu/~distance/ojdla/spring111/Liu111.html. Accessed 25 March 2014

Love J, McNicoll I (1988) The regional economic impact of overseas students in the UK: a case study of three Scottish Universities. Reg Stud 22(1):11–18
Maringe F, Carter S (2007) International students' motivations for studying in UK HE. Insights into the choice and decision making of African students. Int J Educ Manage 21(6):459–475
Mellors-Bourne R, Humfrey C, Kemp N, Woodfield S (2013) The wider benefits of international higher education in the UK. Department for business, innovation and skills. https://www.gov.uk/government/uploads/system/uploads/attachment_data/file/240407/bis-13-1172-the-wider-benefits-of-international-higher-education-in-the-uk.pdf. Accessed 05 March 2014
Purnell K, Cuskelly E, Danaher P (1996) Improving distance education for university students: issues and experiences of students in cities and rural areas. J Distance Educ 11(2):75–101
Roy R, Potter S, Yarrow K (2008) Designing low carbon higher education systems: environmental impacts of campus and distance learning systems. Int J Sustain High Educ 9(2):116–130
Smith A, Bradshaw T, Burnett K, Docherty D, Purcell W, Worthington S (2010). One step beyond: making the most of postgraduate education. http://www.bis.gov.uk/assets/BISCore/corporate/docs/P/10-704-one-step-beyond-postgraduate-education.pdf. Accessed 18 Feb 2014
SQW Energy (2009) Research into a carbon reduction strategy for higher education in England. SQW Consulting. http://www.hefce.ac.uk/pubs/hefce/2010/10_01/10_01a.pdf. Accessed 22 March 2014
Swinney P (2011) Relationships between Cities and Universities. Centre for Cities. http://www.centreforcities.org/research/2011/05/05/starterforten/. Accessed 25 March 2014
UKCISA (2011) Impact of international students on the UK. UK council for international student affairs. http://www.ukcisa.org.uk/about/impact.php. Accessed 20 March 2014
UNFCCC (1992). Background to the United nations framework convention on climate change. http://unfccc.int/essential_background/convention/background/items/2536.php. Accessed 27 March 2014
University of Nottingham (2014) International students: exchanges, study abroad and summer schools. http://www.nottingham.ac.uk/internationalstudents/exchanges/index.aspx. Accessed 24 March 2014
WBCSD/WRI (2004) The greenhouse gas protocol—a corporate accounting and reporting standard. http://www.ghgprotocol.org/standards/corporate-standard. Accessed 28 April 2014
WCED (1987) Our common future. World Commission on environment and development. Oxford University Press, Oxford, p 400
Wilkins S, Huisman J (2011) International student destination choice: the influence of home campus experience on the decision to consider branch campuses. J Market High Educ 21 (1):61–83
Wilkins S, Balakrishnan M, Huisman J (2012) Student satisfaction and student perceptions of quality at international branch campuses in the United Arab Emirates. J High Educ Pol and Manag 34(5):543–556

Author Biography

Jonathan Davies is a research assistant at Manchester Metropolitan University, undertaking a doctoral study on sustainability in higher education.

Social and Environmental Reporting in the Italian Higher Education System: Evidence from Two Best Practices

Sara Moggi, Chiara Leardini and Bettina Campedelli

Abstract

The increasing need of accountability to key stakeholders, and the demand for new tools that are useful for governance at universities, have led to the development of new forms of reporting, including social and environmental reports (SERs), which are helpful in the implementation of knowledge and building awareness of sustainability issues. SERs are not only a useful tool for reporting but are a way to educate organizations and stakeholders to a greater sensitivity to SD topic, building virtuous circles that involve students, professors, employees, and community. This study presented a picture of the development of SERs in the Italian Higher Education sector, the main frameworks spontaneously developed, and two selected best practices where SER was improved significantly: Università di Macerata and Università di Ferrara. To reach this aim, the main informants involved in the reporting process on the governance level, as the sustainability responsible, or the rectors of the universities were interviewed. The findings outlined the peculiarities of this process in two successful cases, such as governance compliance and the sharing of knowledge.

Keywords

Social and environmental report · Universities · Sustainable development · Italy

S. Moggi (✉) · C. Leardini · B. Campedelli
Department of Business Administration, University of Verona, Via Dell'Artigliere, 19, 37129 Verona, Italy
e-mail: sara.moggi@univr.it

C. Leardini
e-mail: chiara.leardini@univr.it

B. Campedelli
e-mail: bettina.campedelli@univr.it

W. Leal Filho et al. (eds.), *Integrative Approaches to Sustainable Development at University Level*, World Sustainability Series, DOI 10.1007/978-3-319-10690-8_6

1 Introduction: Universities and Sustainable Development

Over the last few decades, several authors have studied sustainable development (SD) in the higher education (HE) system. A widespread range of actions, strategies, and proposals that universities could consider to develop SD in their respective organizations were analyzed. The term "SD" is still misunderstood and is considered as a topic related to the environment. The word "sustainable" has a Latin origin in "sustinere," which is to maintain and to support. Therefore, to be sustainable denotes the capability of upholding a state or situation in several aspects (Leal Filho 2011). Hence the etymology of the "SD" definition given in the Brundtland Commission Report "Our Common Future," which stated: "sustainable development seeks to meet the need and aspirations of the present without compromising the ability to meet those of the future" (WCED 1987, p. 151). The meanings given to this term have considered several aspects of the impact that organizations and people can have on the environment in which they live and work. The environment, as a complex mix of ecosystems, networks, and societies, can be influenced from a widespread range of behaviors, and every activity has an effect on one or more features of the environment. Universities are a part of a national context and have the power to influence the general thinking of the future generation of leaders (Cortese 2003) and represent the culture of the country itself (Sterling 2004). Furthermore, universities are entities that are part of a local community in which students, professors, and employees live together. The universities impact on the environment and society is difficult to manage and measure.

As shown by Velazquez et al. (2006), in the process for understanding the achievement of SD in a university system is pivotal determining the different steps of this process. These phases do not often occur sequentially in the same year because they take time to be implemented in organizations. According to Lozano (2006), SD is a significant change for any society and, similar to all innovations, needs time to improve an organization. This is even more true in an Italian context in which the majority of universities have public ownership (69.79 % at the end of 2012) and the SER remains underdeveloped and with an inconsistent trend.

Following the model derived from Velazquez et al. (2006) study, behind this process of improvement, there is the development of a sustainability mission and vision for organizations, whereby its governance behaves according to SD thinking (Holberg et al. 2008). It follows the creation of a commitment that aims to facilitate SD inclusion in strategies and activities (Lee et al. 2013), such as researching (Wass et al. 2010), teaching (Khan 2013), partnerships, and greening campuses (Koester et al. 2006). Finally, according to Velazquez et al. (2006), a sustainable university model is completed using "appropriate instruments for monitoring, analyzing, and controlling the performance of sustainability initiatives."

The reporting phase of SD actions is a complex process itself that involves several people inside an organization and aims to be accountable and transparent on the performance achieved. In the reporting process, a document is created that might have several titles depending on the content, the origin, and the context in which it was implemented. In this paper referring to social and environmental reports (SERs) we include all the documents that present the performance of three aspects of sustainability using qualitative and quantitative presentations. Following the main idea of the triple bottom line (TBL) presented by Elkington (1997), these reports provide financial, social, and environmental performance.

Several studies on SD have focused attention on the first phase but have paid little attention to the audit reporting process (Karatzoglou 2013). In light of the need for further research on this issue, this study presented two best practices, using two universities that have the longest practice on social and environmental reporting processes and reports in Italy: Università di Ferrara and Università di Macerata. This research contributes to draw the complex worldwide picture of this topic, focusing on process aspects and important evidence derived from interviews, focus groups, and document analysis in regards to these two universities.

In the following section, reporting process phases and the main issue of SD for universities will be presented in this study. This framework has been adopted to understand peculiarities in the cases studied in this paper. The Italian trend on social and environmental reporting will be summarized, giving an overview on the development of reporting practices and main characteristics. Before explaining the results of the two cases analyzed in this study, the methodology applied for analyzing the reporting process and SER of these universities will be described. The paper will conclude with remarks, critiques on the present research, and potential further studies on this issue have been provided.

2 Social and Environmental Reporting Process

Approaches to SERs are varied. One SER approach is "minimal," which is a strictly accounting view. In this approach, the identification of SD issues is entrusted to a widespread range of indicators, giving communication effectiveness, transparency, and comparability for different stakeholders. However, it neglects the importance of the reporting process aimed at sharing mission, values, and knowledge. It is not expected for stakeholders to be involved in the process. The advantages of this approach are the speed of adoption and data collection and comparability over time. However, in this way, SER is quite similar to the financial statement and it derives from a mere internal collection of data, leading to the uselessness of a document.

A second approach is focused on the process and is not limited to the mere adoption of the reporting model, but considers the process itself as the most important point for organizations as a development tool.

In the light of preview researches (Frey et al. 2009; Towns and Cocklin 2006), and the two best practices analyzed in this paper, a hypothesis framework about reporting process phases for SER will be presented (Moggi 2013). The reporting process encountered several barriers similar to those observed by several authors in regards to SD implementation for other university activities (Leal Filho and Wright 2002).

2.1 Expressions of Willingness on the Part of the "Enlightened" Subject

The first real step toward the beginning of a process of social and environmental reporting is linked to the interest of a person who has a good knowledge on SER and understands the importance of reporting any action taken in account of SD practices. Such an individual is usually called "the champion." This person can be a professor who teaches and researches on sustainability or can be an involved administrative staff member who has knowledge in this issue and, "dreaming about or envisioning" on this path (Velazquez et al. 2006), is firmly capable of supporting it. Good SD knowledge is not enough; therefore, it is essential the "enlightened subject" has charisma and the ability to promote these ideas to the governance.

2.2 Commitment Disclosure

In this second phase, the governmental bodies formalize the commitment for the SER, beginning the reporting process. At this stage, it is pivotal to identify the real purpose of the report and its importance in the accountability system of the university. The governance, or a CSR committee, decides if the report will be a mere supplementary document or a management and control tool. In addition, at this phase, the governance of the university is called to re-examine the mission, values, and strategic objectives in order to translate the SD commitment into concrete actions and reporting.

2.3 Definition and Formalization of the Working Group

There are three possible settings: (1) the establishment of a unique committee composed by a few subjects representing several functions; (2) the creation of two levels of committee, the first group is extensive and is called to discuss jointly the various stages of reporting and the second team is smaller and it handles the operatives part of the reporting process, such as data collection; and (3) two groups, the first, at a governance level, decides aims and objectives of the report and the second, more operative than the first, is responsible for data collection and drawing the document. In all these cases, the involvement of governance will be essential to make a formal commitment and, in order to achieve greater impartiality of the instrument, it is important to consider the inclusion of the key stakeholders (internal and external) of the university into the working groups. At this stage of the reporting process, the committee can decide whether to use the counsel of external experts on SD and SER or rely on their own resources.

2.4 Defining the Reporting System

In the fourth phase, the working group decides the best reporting standards to apply to the structure of their report and organizes the data collection process. At this point, a clear mapping of the knowledge of the people involved in the process is performed to help better organize competence and optimize response time during the following phase: information collection. Here, the working group defines the content of the report in terms of indicators and the related responsibilities for evaluating several factors of university activities and implementation of SD strategies, such as the evaluation of research, teaching, intellectual capital and, more in general, the impact of the organization on the stakeholders. To maintain intelligibility of the chosen SER, the group defines quantitative (selection of indicators) and qualitative (e.g., the description of the green projects implemented) information.

2.5 Collection and Processing of Information

This stage focuses on the collection of information and data necessary to report the various areas defined in the previous phase in order to complete the structure determined by the standards chosen. This phase is influenced strongly by the degree of development of information systems, such as the data warehouse's ability to provide (or not) detailed information. Especially during the first year of reporting, this phase appears to be the most difficult step to overcome because the functions involved in the process are not able to provide data; the data either does not exist or the responsibilities of several people are not correctly defined. For these reasons, the commitment phase should be brought to the attention of the whole organization. However, information is already in the information system of the university because data is used to develop other reports, such as documents for research evaluation or records on SD projects.

2.6 Drafting of Document

Once quantitative and qualitative information is collected, the working group selects the data collected with the purpose to obtain an SER that the stakeholders will understand. This selection depends on the university because rarely the reports' content aspired to a unique standard and cherry-picking behavior is common, which is a sign of myopic use of the tool. The balance between content and several parts of the document follows the decision of the working group and, more in general, the soul of the university's vocation on its activities. During the first year of reporting (sometimes named "year zero"), this process often create an experimental document. The working group and governance are only able to evaluate how to better coordinate several people involved with the process, develop stakeholder engagement, and increase the effectiveness of the communication process after the first year has been completed. For these reasons, during the first year of reporting, universities face considerable difficulties and reports are often poor or have gaps. If these difficulties are not overcome, the risk is high that the process will never be recovered.

2.7 Approval of the SER by Governing Bodies

Although some universities do not consider this phase, it is important that the SER is approved by the board of directors after the academic senate has been consulted. In fact, by formalizing and sharing results during the conclusion of a process, stakeholders become more aware of a university's activities and the development and impact of SD inclusion in strategies. A final report could be a helpful tool as a management and control function to begin defining multi-annual programs and new SD strategy goals.

2.8 Communications of the SER

After the operative part of the process, the working group has to present the report to stakeholders. This usually happens by a conference presentation or workshop in which the main aim is to present the university's performance. Recently, several universities have decided to present their SER in a day dedicated to SD issues, taking the advantage to increase both awareness on social and environmental impacts of university activities and improve general SD skills. It is important to remember that SER should answer to the third mission of universities, the diffusion of knowledge among the community, in this case on social, environmental, and financial sustainability and the way to develop projects on SD issues. SERs must be advertised in the major local media and brought to the attention of the scientific community and other universities. Finally, it is pivotal to upload SERs on the respective universities websites, giving proper visibility on the universities home-pages (Nejati et al. 2011).

2.9 Social Audit

A common practice in companies, but with minimal relevance for universities, is the social audit.

The lack of importance given to this practice is linked closely to the use of several standards simultaneously. In light of this, an easy evaluation of reports' content is made difficult.

2.10 Obtaining Feedback

The final phase of the process is evaluation from stakeholders, both on the report accountability value and on the university's activities. This phase could be carried out in complementary or alternative ways. An evaluation survey, which must be completed and returned with suggestions to the university, could be attached to the SER. Another way is to organize focus groups or workshops in order to collect feedback with more interactions, which cannot be obtained using questionnaires.

As underlined by several authors (Lozano 2006; Shriberg 2004), each aforementioned phase describes stakeholder engagement as a pivotal part of the process itself. Starting from stakeholder mapping, university governance has to determine the who, how, and when involved in the development of the document. As demonstrated by preview studies (Rode and Michelsen 2008), the reporting process helps the people involved improve their awareness of SD issues because these subjects, building indicators, and comments on the results help them understand the concrete consequences of the university's activities on the environment and stakeholders, such as their local community.

SERs can become the catalyst to increase knowledge of SD issues because people engaged in the process, even if only in part or in a specific topic, are forced to understand the obtained data. According to Godmann et al. (2014), involving students in the accountability process could be an enriching activity and a "powerful pedagogic device" used to shape SD.

3 The Italian Context on Social and Environmental Reporting

For several years, there has been heated debate on SER at universities in the Italian context. Publication on SER began in 2003 when two HE institutes (Normale di Pisa and S. Anna di Pisa) published the first experimental reports. They presented only a limited summary of social aspects and could not be considered SER. Currently, Italian universities show a growing attention to social and environmental reporting, which has been demonstrated by the presence of several voluntary reports from 2006 that have considered SD issues.

Out of 96 entities declared as institutes of the HE system from the Ministry of Education in 2013, 28.12 % (27 universities) of these organizations have at least at one time declared the intention or actually started the voluntary reporting process.

As shown in Fig. 1, Italian universities have developed different types of voluntary reports. The most common type is the social report (65.38 %) that, despite the name, usually presents the TBL structure. The mission report (7.69 %) is a document that describes a university's performance for the period that coincides with the rector's mandate. The environmental report (3.85 %) limits its focus on accountability to the impact of the university's activities on the natural environment that hosts the organization. The "sustainability report" label is still rarely used

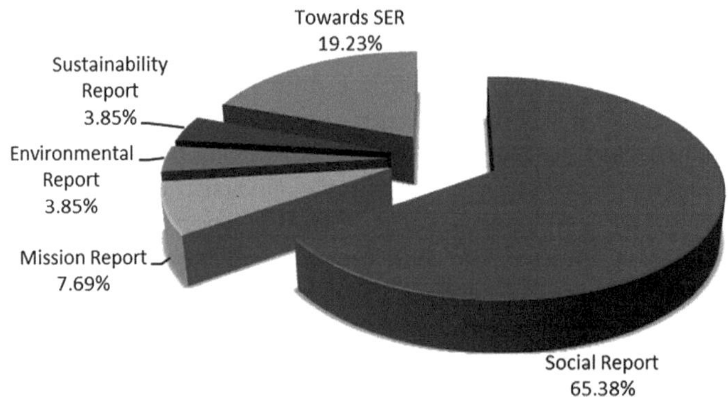

Fig. 1 Types of report founded (updated to December 2012)

Fig. 2 SER and ownership in Italian universities

		SER	
		YES	NO
Ownership	Public	26	41
	Private	1	28

(3.85 %) because Italian standards and practices prefer the word "social report." In the "toward SER" cluster (29.23 %), all the documents that provided a commitment to SER, such as reports expressly declared as beginning points for a future social and environmental reporting process, were collected (Fig. 2).

The comparison of SERs presence in public and private HE institutes revealed a clear preeminence of this practice in public universities, with only one case in private institutes. The trends of published SERs were varied and difficult to measure because several universities decided to report with different periods of reference. Despite this, increasing attention toward this issue was demonstrated with the drafting of SERs and the publication by Study Group for Social Reporting (GBS—Gruppo di studio per il bilancio sociale) of a standard, Guidelines for Universities (GBS 2008), thought expressly these organizations. This interest was renewed in 2013 by this study group: a new team of experts was created with the aim of rewriting these guidelines in light of several updates on aspects undertaken in the first document, such as SD issues, gender analysis, and the role of SER, like a strategic tools for governance in universities to define long-term programs.

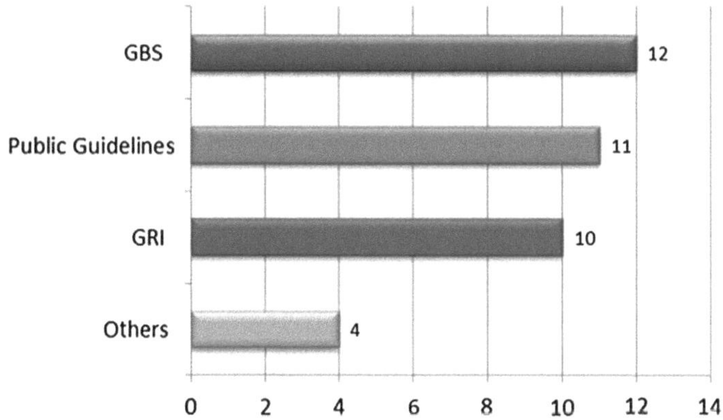

Fig. 3 Number of standards developed (analysis December 2012)

Figure 3 shows the results of the analysis of 18 documents, the most recent SERs published at the end of 2012 by Italian universities. In each report are applied on average 2.7 standards. The GBS standards (general principles, the public sector technical standard, and the universities technical standard) are the most used (12), followed by several general guidelines or manuals for public sector organizations (11) and the GRI (Global Reporting Initiative) standards (10) (general principles and public sector pilot version). In the residual cluster "other," a few universities followed the general approach, such as the Italian project Corporate Social Responsibility—Social Commitment (CSR-SC) published in 2004.

4 Case Studies Analysis

The preliminary analysis of the Italian context was essential to define the best practices described in this article. The two cases selected could be considered as best practices in social and environmental reporting in the Italian HE system because of their constant reporting process throughout the last 10 years. In fact, analyzing the diachronic evolution of the 96 entities declared as institutes of the HE system from the Ministry of Education from 2006 to 2012, the two cases selected are the only two universities that have published SERs for more than 4 years. All the voluntary reports published until now, and the related press reviews, were collected and analyzed for the two SER case studies. To better understand the reporting process, from October 2012 to January 2013, four interviews were performed and one focus group with four employees, were developed with people who were part of the governance body, including CSR managers, rectors or CSR teams, and responsible of students.

4.1 Università degli studi di Ferrara

The Università di Ferrara was established in 1391 in the North of Italy and is one of the most important universities in the region, counting more than 15,000 students (academic year 2012–2013). Despite SD not being declared formally in the university's mission, this university dedicated extensive attention on this issue. The first, and most visible, sign was the website's dedication to sustainability (http://sostenibile.unife.it), where the rector declared that the university was committed to implementing principles of "sustainability" as a pivotal paradigm of research, teaching, and management in order to develop, promote, and increase projects, strategies, and actions consistent concretely with SD. The city of Ferrara is called "university city" because of the strong relation between the university and the local community.

The approach concentrated on long-term projects that increased the sustainability of the university in terms of economic, social, and environmental impacts on the local community and the future impact on financial resources. The attention to SD issues was demonstrated by several actions on teaching, research, and projects and specific initiatives and achievements, such as the establishment of the "deputy rector for sustainable policy" or the annual award for the best thesis on sustainability.

The SER process, still undertaken, began in 2006 with the first "social balance." The improvement of the reporting process and of the SD policies of the university was clearly shown by the last report presented (2012 edition) in April 2013.

4.2 Highlights from the SER

From the beginning, the process was strongly supported by the rector and the university governance. The document was proposed by a champion with the aim of increasing the transparency of the organization, starting a process, now embedded, and managed, by a team of employees with knowledge and experience on the measurement of several aspects undertaken during the reporting process. There were two groups that worked on SER: the control group, composed of board members, who decided the content of the document, and the operative group, that follows the operational part of the process, composed of the responsible of the organization functions and few subjects from other organizations to act as SER experts.

The last report considered two standards for the content of the document: guidelines for the public organizations and GRI. The structure of the SER results as follows:

- Presentation
- History, mission, and identity
- Intellectual capital
- Teaching
- Research, innovation, and relation with entrepreneurs
- Territorial community

- International context
- Sustainability
- Culture
- Health
- Human resources
- Gender report
- Comparison table with GRI

One section of the report was dedicated integrally to SD at the university. This section explained how the organization worked on SD issues through several actions: teaching, good practices in the use of resources (see the development of carbon, water, and ecological footprint tools), energy efficiency, waste control, introduction of green provisions in contracts with external suppliers, maintaining sustainable transportation, improving the quality of the ecosystem, and preserving the native species.

The report presented a good balance between several parts dedicated to the stakeholders and gave an overview of the majority of the activities respectively undertaken. However, engagement and communication was not balanced for some stakeholders. For example, students were involved only in the final phase, which was when the draft of the report was presented to the university board for approval. In that occasion, the representative of the students could express an opinion. Despite this, the students involved in the SD project, and the SER process, declared to be enthusiastic about this role. In general, communication of an SER publication to a stakeholder is provided usually by e-mail to the internal stakeholder and important local authorities. Such information is provided to academia through a presentation aimed to involve external stakeholders, including the local community.

From the beginning, the report has presented an improvement every year. For example, in the last edition, one of the biggest new features in Italian universities was the gender section. However, the weakest part of the reporting process was that it was too long and complex, sometimes exceeding 1 year. For this reason, the intention for future reports is to reduce the number of indicators and the description of several projects, considering only the most important ones. As underscored in interviews, the difficulty will be defining reports' boundaries because SER has become an identity tool for people who make a contribution to reports. This simplification will be combined with a short form and an online version of the SER, with the aim of creating a document that a stakeholder will find easy to read.

4.3 Università degli studi di Macerata

The Università di Macerata was founded in 1290 in Central Italy and is one of the oldest universities in Europe with a marked humanistic vocation in the development of its courses. From here derives the motto of this institute, "innovation through humanism," which describes in few words the declared mission of the university.

For the academic year 2012–2013, this university counted more than 9,800 students, which is considered to be a medium-sized institute.

This university is situated in the hearth of Macerata and is an integral part of the local community. Its importance is demonstrated by the initiatives that the organization fosters for opening the university to the territory and improving culture, in light of the third university's mission.

The SER process began in 2006 with the first report considered as an experiment. Currently, the report is a collection of a few documents and summarizes the activities and the performance of the organization. It could be considered the best way to increase the accessibility of information about the university.

4.4 Highlights from the SER

The SER process for this university started because of the strong intention of the governance. The champion was the rector, with the aim to be more accountable, and the process was embedded thanks to the knowledge and experience of a professor that researches on this field. From the beginning of this process, the management and control office had supervised the reporting. Over the years, the process has involved an increasing number of functions, such research and teaching. For this reason the SER is consider as a coral result of a common work. There are two groups in this process: the coordination group, composed of a few people from the governance, and the working group, which coordinated contributions from numerous people involved in the reporting. The data collection engaged great part of the organization, with the aim to improve the awareness of the SER and knowledge on aspects of accountability.

Several standards were considered in the last report. Each standard was considered for different reasons. GBS and guidelines for the public sector were considered for the general content of the document, AccountAbility 1000 (AA1000) and GRI Guidelines (3.1 version), respectively, for the process of stakeholder engagement and to improve awareness of SD issues. The last report (2012) was divided into three sections: methodology and identity, activities and stakeholder engagement, and dialog with stakeholders. The structure was as follows:

- First section: Methodology and presentation of the university
  - Introduction and methodology
  - Identity
  - Resources
- Second section: Activities in 2012 and developments for 2013
  - Research
  - Teaching
  - Services
- Third section: Dialog with stakeholders
  - Stakeholder engagement

The report presented a structure that focused on economic and social impacts, instead of environmental aspects. This lack of attention to environmental aspects could be explained by the absence in this university of specific knowledge on this issue because humanistic topics prevail in the curriculum. In the methodology section, the report specified the communication plan for the SER. Similar to the report for Università di Ferrara, the report of Università di Macerata was published in a short version, and the university organized a day dedicated to the presentation of the report.

For the long-time experience of reporting, the community considered the university as a pilot for social and environmental reporting. Firms and non-profit organizations asked consultants of the university to begin the process to improve the organization culture on SER.

Also in this report, the weakness was low engagement of the stakeholders during the process. For example the students were involved in a passive way. As such, it became pivotal understand how SER can be used as an instrument to engage stakeholders, such as students or local community, who were not aware of the real meaning of sustainability and the several affects the university has had on social and economic aspects.

This university had demonstrated how the SER could be a tool for rethinking the organization because several measurements were useful for obtaining a better understanding of the organization's performance in a holistic view. In fact, SER provided a moment of reflection where the governance reconsidered the short- and long-term strategies.

5 Conclusion

European universities have recently passed a period of great change that has hampered the development of SER practices. According to Godemann et al. (2014), several factors have had a negative impact on the process, such as reforms and fund cuts. In Italy, as a consequence of a wide reform (D.Lgs 240/2010), the government asked a strong rethinking of each university, influencing deeply the process of reporting, in particular, at the data collection phase, even though the process has been well run. Several structures have been modified, creating the need to identify new meters of measurement and people to involve in the process.

The cutting of funds has fostered the misconception (Leal Filho 2000) that the process is too costly in terms of time and human resources. The paradox is in the failure to understand that the development of environmental and social measurement systems, the reporting and the monitoring of the obtained data, could be a valuable tool for reducing costs and increasing efficiency in universities.

In addition, from the analysis of the SERs published in Italy and the two case studies analyzed, it is clear how difficult it is to understand the tricky word "sustainability" and the measurement of its dimensions (Gray and Milne 2002). In light of this, the GBS has established a group study on SER with the aim of spreading knowledge and experience on this topic and defining a new common standard at universities.

In the two case studies analyzed, the SER was the result of an accountability process on social, environmental, and financial performance and a valid support for the decision-making process. In both cases, the introduction of voluntary reporting partially helped increase awareness of SD issues.

Similar to how SD issues have been improved at other universities, the SER processes for the cases analyzed were embedded because of strong support from governance from the very beginning of the project.

The role of the champion was pivotal; the champion is a person who has a research interest or a personal inclination to an SD topic and its accountability.

To build indicators means to fill in cultural gaps between economic and scientific communities. In fact, when more information is given to the stakeholder, universities improve awareness of SD issues and it is consider a best practice to imitate. As also demonstrated by this study, SER became an important tool to change the organizations and increase collaboration between several courses under a common issue of SD.

A common weakness is the poor stakeholder engagement in the process. Students were not involved during several phases of the reporting; however, these subjects could provide valuable support, such as defining indicators that would be useful for external stakeholders and helping outline a better way to communicate to students who, in the opinion of people interviewed, in the majority of cases, did not know SER existed at their university. One of the ways that was declared to help improve the communication of the report to the stakeholders was to provide a short form and an online version of the report, which could make the SER more accessible.

The main function of SER is to allow universities to provide a greater transparency to stakeholders, bringing more awareness to economic, social, and environmental performance and the related activities and accountabilities of SD projects, which would not have been known otherwise.

The reporting created more awareness and knowledge for the people involved in the process, assuming also the internal function of the identity tool.

This study aimed to contribute to the numerous studies on SD at universities, but has provided a deeper insight into the accountability process undertaken for measuring the sustainability of universities' activities. In light of the results, further studies are needed to understand how people involved in social and environmental reporting really increase their awareness on SD and if this awareness is influenced by the role of the champions and their features. Another critical point underlined by this study is that more attention on SD could be afforded to different dimensions such as social rather that environmental. With the aim to provide more support to universities that decide to start the SER process, in future research, it will be pivotal to understand how the context and vocation of universities can influence SER.

Acknowledgments The authors acknowledge the people interviewed at Università di Ferrara and Università di Macerata for sharing with us their knowledge and experience and their feeling that the path of SER could be developed.

References

Cortese AD (2003) The critical role of higher education in creating a sustainable future. Plann High Educ 31:15–22

Elkington J (1997) Cannibals with forks. The triple bottom line of 21st century business. New Society Publishers, Gabriola Island, BC; Stony Creek, CT, p 416

Frey M, Melis M, Vagnoni E (2009) Recent developments in social and environmental reporting among Italian universities: a critical evaluation of leading edge practices. In: Baldarelli M (ed) Civil economy, democracy, transparency and social and environmental accounting research role: some reflections in theory and in practice deriving from 2nd CSEAR conference-Italy, Mc Graw Hill, p 426

Godemann J, Bebbington J, Herzig C, Moon J (2014) Higher education and sustainable development: exploring possibilities for organisational change. Account Audit Account J 27:218–233

Gray R, Milne M (2002) Sustainability reporting: who's kidding whom? Chart Account J NZ 81:66–70

Gruppo di Studio per il Bilancio Sociale (GBS) (2008) Il Bilancio sociale. Documento di Ricerca N. 7. La Rendicontazione Sociale nelle Università. Giuffrè Editore, p 32

Holmberg J, Svanström M, Peet D-J, Mulder K, Ferrer-Balas D, Segalàs J (2008) Embedding sustainability in higher education through interaction with lecturers: case studies from three European technical universities. Eur J Eng Educ 33:271–282

Karatzoglou B (2013) An in-depth literature review of the evolving roles and contributions of universities to education for sustainable development. J Clean Prod 49:44–53

Khan T (2013) Sustainability accounting courses, talloires declaration and academic research. Int J Sustain High Educ 14:42–55

Koester RJ, Eflin J, Vann J (2006) Greening of the campus: a whole-systems approach. J Clean Prod 14:769–779

Leal Filho W (2000) Dealing with misconceptions on the concept of sustainability. Int J Sustain High Educ 1:9–19

Leal Filho W, Wright TSA (2002) Barriers on the path to sustainability: European and Canadian perspectives in higher education. Int J Sustain Dev World Ecol 9:179–186

Leal Filho W (2011) About the role of universities and their contribution to sustainable development. High Educ Policy 24:427–438

Lee K-H, Barker M, Mouasher A (2013) Is it even espoused? an exploratory study of commitment to sustainability as evidenced in vision, mission, and graduate attribute statements in Australian universities. J Clean Prod 48:20–28

Lozano R (2006) Incorporation and institutionalization of SD into universities: breaking through barriers to change. J Clean Prod 14:787–796

Moggi, S. (2013). La rendicontazione sociale nelle Università italiane: le motivazioni che ne inducono lo sviluppo. PhD Thesis, University of Verona

Nejati M, Shafaei A, Salamzadeh Y, Daraei M (2011) Corporate social responsibility and universities: a study of top 10 world universities' websites. Afr J Bus Manage 5:440–447

Rode H, Michelsen G (2008) Levels of indicator development for education for sustainable development. Environ Educ Res 14:19–33

Shriberg M (2004) Assessing sustainability: criteria, tools, and implications. In: Corcoran Blaze P, Wals AEJ (eds) Higher education and the challenge of sustainability. Problematics, promise, and practice. Kluwer Academic Publishers, Boston, p 355

Sterling S (2004) Higher education, sustainability, and the role of systemic learning2. In: Concoran Blaze P, Wals EJA (eds) Higher education and the challenge of sustainability. Problematics, promise, and practice. Kluwer Academic Publishers, Boston, p 355

Towns B, Cocklin C (2006) Triple Bottom Line (TBL) reporting and strategic change in higher education. In: Leal Filho W, Carpenter D (eds) Sustainability in the Australasian university context. Peter Lang GmbH, Berlin, p 293

Velazquez L, Munguia N, Platt A, Taddei J (2006) Sustainable university: what can be the matter? J Clean Prod 14:810–819
Waas T, Verbruggen A, Wright T (2010) University research for sustainable development: definition and characteristics explored. J Clean Prod 18:629–636
WCED (1987) Our common future. World commission on environment and development. Oxford University Press, Oxford, p 400

Authors Biography

Sara Moggi is a Research Fellow in Business Administration at University of Verona (Italy). Her research interests are in social and environmental reporting and stakeholder theory analysis in the public sector and nonprofit organizations.

Chiara Leardini is an Associate Professor in Business Administration at the University of Verona (Italy). Her research interests are in the area of governance, performance and accountability in nonprofit organization and in public organization.

Bettina Campedelli is a full professor in Business Administration at University of Verona (Italy). Her studies interests are in performance management and reporting. She has a great number of publications on these topics and years of experience in management.

How Do Limbo Dancing, Landlords and Students' Energy Habits Link to EfSD? A Student Green Fund Case Study

Peng Li, Katy Boom and Paul Davis

Abstract

University students who live in privately rented houses account for nearly 42 % of the student population in the UK, and the majority live in relatively old housing stock which is energy inefficient requiring infrastructure improvements to make them more thermally efficient. The student landlord market is a stand-alone sector with some specific challenges including the high turnover of tenants and issues around who pays the fuel bill. This project described and critiqued in this paper is focused on a set of guiding interventions designed to help undergraduate students to save energy at home by changing their behaviour, alongside encouraging property owners to make infrastructure improvements. This paper presents a case study on University of Worcester Students' Union's (WSU) behaviour change project funded through the Higher Education Funding Council for England (HEFCE)'s Student Green Fund (SGF). It aims to develop a cost effective model to assist university students to learn and develop energy saving behaviours. Competition on a bespoke student facing software platform, regular incentives and easy to understand reports are part of a multidimensional approach to this intended behaviour change. The overall goal of the programme design is to identify best or most effective practice and develop opportunities to

P. Li (✉) · K. Boom
Worcester Students' Union, Sustainability Department, Worcester Business School, University of Worcester, Henwick Grove, Worcester WR2 6AJ, England
e-mail: p.li@worc.ac.uk

K. Boom
e-mail: k.boom@worc.ac.uk

P. Davis
Stichting Euregio University College, Aalten, Gelderland, Netherlands
e-mail: paul@davispw.fsnet.co.uk

W. Leal Filho et al. (eds.), *Integrative Approaches to Sustainable Development at University Level*, World Sustainability Series, DOI 10.1007/978-3-319-10690-8_7

engage with wider employability and academic skills in a number of disciplines. This project runs for 2 years from October 2013 in Worcester, UK with Birmingham Guild of Students a partner in year two to test the potential for replication of the same model elsewhere.

Keywords
Energy conservation · Education for sustainable development · Employability · Evaluation

1 Introduction

This paper discusses the policy instruments and intentions of the Energize Worcester (EW) project. EW is currently using numerous different methods to engage with students studying at, and living in shared houses in Worcester, UK. The project aims to encourage and educate those student tenants to learn and adopt appropriate energy strategies and habits, *and evaluate the cost effectiveness of different methods of interventions*. The project commenced in October 2013. Therefore, the evaluation will be based on the learning and experiences of the first 5 months. It is tentative and designed to feed forward in a formative learning manner to guide future actions. It begins with an account of the general context of fuel usage and dwelling choice among students in the UK, and more specifically, in Worcester. This overview provides a rationale for intervening and there follows a brief outline of the development of local policy in this field. That development contained a pleasing element of surprise, the significance of which propelled the design of the intervention under scrutiny here. The terms, underlying logic and purposes of that intervention are then set out.

Beginning, then, with broader context: in the UK, there are approximately 2.5 million students in university in any given academic year. Over 42 % live in privately rented accommodation, and indeed the majority live in shared houses classified as Houses of Multiple Occupation (HMO) (Goodman and Drayson 2014). The quality of that housing provision is an issue. The UK has the oldest housing stock in Europe. According to the English Housing Survey 2012–2013, nearly 86 % of all English dwellings were built before 1991. Older dwellings often consume substantially higher energy to warm them adequately, compared with more energy efficient new built dwellings. The average Standard Assessment Procedure (SAP) score, a housing stock energy efficiency standard using index numbers, for all UK dwellings was 59 in 2012, where the average SAP score for dwellings built before 1991 is 40.2 (BRE 2005a, b; Palmer and Cooper 2013). In addition, the ownership structure of the UK housing stock may present a further issue. Private rental accounts for 18 % of English tenure types, but nearly 80 % of people in full-time education are private renters. Overall, almost 68 % of the entire 16–24 age group are also private renters. A DCLG (2014) report shows the private rented sector has

Table 1 Percentage of energy efficiency measures by tenure types in England

Energy efficiency measures	Owner occupied dwellings (%)	Privately rented dwellings (%)	Social rented dwellings (%)	National average (%)
Central Heating	94	**81**	93	91
Cavity wall insulation	43	**23**	49	40
200 mm or more loft insulation	38	**20**	34	34
Double glaze window	78	**74**	88	79

Source English Housing Survey Headline Report 2013–2013, DCLG (2014)

significantly lower energy efficiency measures compared with other sectors on a number of indicators (see Table 1).

In the UK, domestic energy consumption accounts for almost 30 % of total energy usage. The poor housing stock and increasingly frequent severe weather conditions are likely to increase household energy demand. The Department of Energy and Climate Change (DECC) suggests this is likely to drive more households into fuel poverty (DECC 2014). Reflecting this, the Office for National Statistics finds that the average household energy bill virtually doubled in real terms from 2002 to 2013, despite average energy consumption reducing by nearly 17 % over the same period (DECC 2013).

In general, trends in energy use impact on many aspects of daily life and the statistics indicate that students may endure both relative and multiple deprivations. Students often suffer significantly from fuel debt. This may be explained by two further factors, namely a lack of life experience, and a poor understanding of energy distribution and usage (Bouzarovski et al. 2012; DeWater et al. 2007). In addition, Bouzarovski et al. (2012) suggest that a significant number of university students do not fully understand their rights in relation to their tenancy and property quality, not to mention energy efficiency. A recent survey conducted by the National Union of Students (NUS) has shown 61 % of respondents experience damp in rented houses, 52 % think their homes are uncomfortably cold, and a further 48 % believe their home is poorly insulated and draughty. However, over 76 % of respondents deal with their increasing energy bills by limiting their heating, 66 % claim to wear extra layers in order to deal with the cold at home and 40 % spend time in public facilities, such as libraries, to avoid home heating (Goodman and Drayson 2014).

The indicative evidence of students' generally poor understanding of energy and its effective use suggests a need for learning intervention. Educating students with knowledge about energy efficiency does more than just reduce their environmental impacts in the short term and their expenditure on fuel. The knowledge and skills potentially empowers them to opt for a more sustainable way of living. Given the size of the student population and their importance as future citizens, these choices could potentially influence the future of energy demand across society as a whole (Grønhøj and Thøgersen 2011).

2 Project Background

In recognition of the national evidence on the incidence of fuel poverty among the UK HE student population, local policy makers in Worcester have been pursuing measures designed to improve the lot of students in this important regard. The University of Worcester (UoW) has around 10,000 students and over 30 % of Worcester students live in privately rented HMOs. As with most student housing stocks Worcester's private rented HMOs are often less energy efficient old houses.

In spring 2013, the DECC invited local authorities to bid for funding to instigate the Green Deal (DECC 2010, 2012), an energy efficiency financial package to support domestic housing stock energy upgrades. In Worcestershire, the DECC's funding supported Worcester Energy Pioneers (WEP), a project in which Worcestershire County Council, Worcester City Council and UoW worked together using trained students as energy assessors to survey HMOs in Worcester. Its instrumental goal was to derive enhanced information on aspects of the rented housing stock in the city. The WEP project also indicated, though, that those students who participated in the project tended to develop a stronger sense of energy efficiency, and they were more likely to adopt or even advocate for domestic energy efficiency in the future. These were largely unintended longer-term WEP impacts that prompted the development of more general behavioural modification initiatives, directly stimulating the development of Energize Worcester.

In the UK, university students are generally aware of sustainability issues and want them to be reflected in their institutions, their study and their overall experiences. A series of surveys funded by the Higher Education Academy (HEA), and carried out by the NUS found that 85 % of first year students think their university should actively promote Sustainable Development (SD), and around 60 % of students wish to learn more about it (Drayson et al. 2013).

Among policy bodies, and in recognition of these student attitudes, HEFCE want to support capacity building to develop skills and show through practical steps how SD challenges can be met. HEFCE in their SD in HE public consultation document noted '*With 17,000 universities in the world, higher education is a global enterprise operating collaboratively through the exchange of ideas, students and staff. These connections and the positions of universities in societies mean that higher education has the potential to drive global change*" (2013, p. 7). With this in mind, HEFCE has provided £5 million to support the NUS's Student Green Fund (SGF), translating into 25 student-led sustainability projects (HEFCE 2013). EW is one of the projects. It takes the opportunity presented by SGF to support students living in privately rented houses to learn and adopt appropriate energy habits.

3 Project Aims

EW's publicly broadcast and explicit goal is to work with students in privately rented houses to save quantifiable amounts of energy and money. In pursuing that stated aim, though, an emergent goal (Mintzberg 1978; Mintzberg et al. 1998) was to create a network of skilled students with enhanced employment opportunities, pro-environmental behaviours and energy-saving habits that are likely to endure and whose impact would spread to their friends and family in the future. For the majority of students, this is the first time they will have experienced independent living—a significant lifestyle change, during which they are forming habits likely to stay with them beyond the moment of change itself. If they develop effective energy habits during university life, they are likely to keep those pro-SD habits throughout their lifetime. Behavioural theory provides one explanation of this. The habit discontinuity hypothesis (Verplanken and Wood 2006) suggests individuals are more likely to adopt new habits if they are going through a lifestyle change.

The context in which behavioural modification takes place is also significant, though. The WEP project had indicated that students' energy behaviours in privately rented HMOs are often influenced by their tenancy occupancy agreement. In so-called bill-inclusive HMOs, student tenants often feel indifferent to wasting energy in order to 'get their money's worth'. In contrast, in bill-exclusive HMOs, student tenants are often reluctant to switch on their heating in order to minimise their energy expenditure, despite condensation and mould (Goodman and Drayson 2014).

Such reflections contribute to the forming of relevant goals and a programme theory, in formal evaluative terms (Sidani and Sechrest 1999). The EW project has four key objectives:

1. Import knowledge and skills into the student's problem domain to help develop appropriate energy habits
2. Save energy and carbon in student housing use
3. Provide extra-curricular and curricula opportunities to UoW students thereby building their employability skills for the future
4. Develop strong community partnerships working closely with local authorities, community energy transition groups and landlords in order to engage energy infrastructure improvements to student properties

These objectives align reasonably well against the broad evidence base and implementation theories outlined above—but the fit is not exact.

4 Project Action Plans

One noted problem with many conventional results line readings of how value is produced by public policies is their assumption of a straight-line relationship between what is done by policy-makers and what happens to the evaluation in consequence (Runhaar et al. 2006). There is good reason to believe, though, that a linear aetiology is not wholly appropriate in the present case.

Darby (2010), for example, identifies four types of feedback system based intervention theories as effective to encourage domestic energy efficient habits, namely:

1. Sociological theory. Improve feedback to consumers to make it more visible, encourage users to feel they are in control of their own domestic energy consumption.
2. Economic theory. Encourage appropriate energy habits with incentives, treat energy as a commodity, and offer consumers financial incentives for their commitment.
3. Psychological theory. Psychological stimulus to consumers is likely to improve their understanding of energy, and in turn adjust their behaviour
4. Educational theory. Provide learning opportunities to help consumers develop effective energy use skills that are also likely to encourage behaviour change.

Darby (2010) conducted her work on behalf of the Office for Gas and Electricity Markets (Ofgem) Energy Demand Research Project to study and examine the effectiveness of interventions(AECOM, 2011). Adopting a similar approach, the EW project has developed a tailored action plan for the students as follows:

1. A bespoke online application that provides accurate feedback to participants on their energy consumption over time.
2. Use the existing online application developed during WEP to support retro-spective physical energy efficiency measures in HMOs.
3. Recruit and train students as accredited professional Energy Advocates (EA) to support student tenants with energy advice and provide bespoke property in use energy reports.
4. Recruit and train students as energy assessors to deliver property energy surveys in identified student HMOs.
5. Offer financial incentives to encourage participation and commitment.
6. Engage with students from different academic disciplines, to encourage a sociologically wide range of students to participate in EW.
7. Enhanced use of digital and multimedia communications. Designated website and social media platforms have been created as a primary contact point for student support and instant engagement.
8. Liaise with one of the 'Big Six' energy companies to supply dual fuel smart meters, to be installed in selected houses as part of the real time in-home display strategy.
9. Work with the National Landlords Association to raise EW awareness among the landlord community. This will also involve exploring financing options to help landlords with potential energy upgrades, permitting more energy efficient homes for students.
10. Agree with a partner (the Birmingham Guild of Students) to replicate the programme logic in Birmingham in the second year of its operation to further test the project model. It is also intended to re-run the same project in Worcester for a second year.

At the end of the currently envisaged planning horizon of EW, the different methods will be critically evaluated to accelerate (potentially transferable) learning and establish the most cost effective model. This is with a view to creating a 'product' for NUS to market to universities, colleges and student unions nationwide.

5 Project Progress

This paper is an initial review of the first 5 months of a two-year project.

The Energize online app has been successfully developed and is available for student users to register and input their energy meter readings. The app serves two main functions: firstly to provide students with a better understanding of the houses they live in by showing property profile data; secondly, to inform students of their energy consumption over time.

The app provides instant graphics to inform students of their energy performance (energy consumption in the month against a national average). If students input more information on the type and construction of the property, they will also be given more accurate energy performance feedback against similar types of property in the UK (see Fig. 1). This feedback system can help students to visually track their energy consumption over time (see Fig. 2). As Benders et al. (2005) pointed out, feedback systems, especially software, need to be personalised in order to increase the level of commitment from participants.

There were two reasons for recruiting and training students to function as accredited professional EAs: to enhance their future personal employability, and to use peer-based networks to ease access to the student community. This choice reflects Darby's (2010) insight, that providing learning opportunities helps consumers develop effective energy use skills that are likely to encourage behaviour change. Reliance on peers reflected previous WEP learning indicating that students often preferred to receive help from their peers. Student EAs' main duties are to raise awareness of EW and support students in their homes with energy advice. Therefore, the EAs help students both on- and off-campus; some students will receive house visits and bespoke 'home energy in use' reports.

The EAs have undergone energy awareness training and are invited to develop a strong sense of ownership of the project, plus a willingness to use their new skills and knowledge to help peers. It is anticipated they will also develop a more appropriate lifestyle of their own. Given their centrality, the impact on the EAs and their influence on peers will be closely monitored.

EW has been designed to integrate into the curriculum wherever possible, working closely with students from different academic backgrounds. It offers opportunities to engage with students who are possibly less likely to associate with energy projects. This is exemplified by working with creative arts students to develop an imaginative campaign for the project. In a 'live' brief scenario, art students were given just 24 hours to research and develop a creative marketing campaign. The winning concept campaign and slogan "How Low Can You Go", uses limbo dancing as a symbol

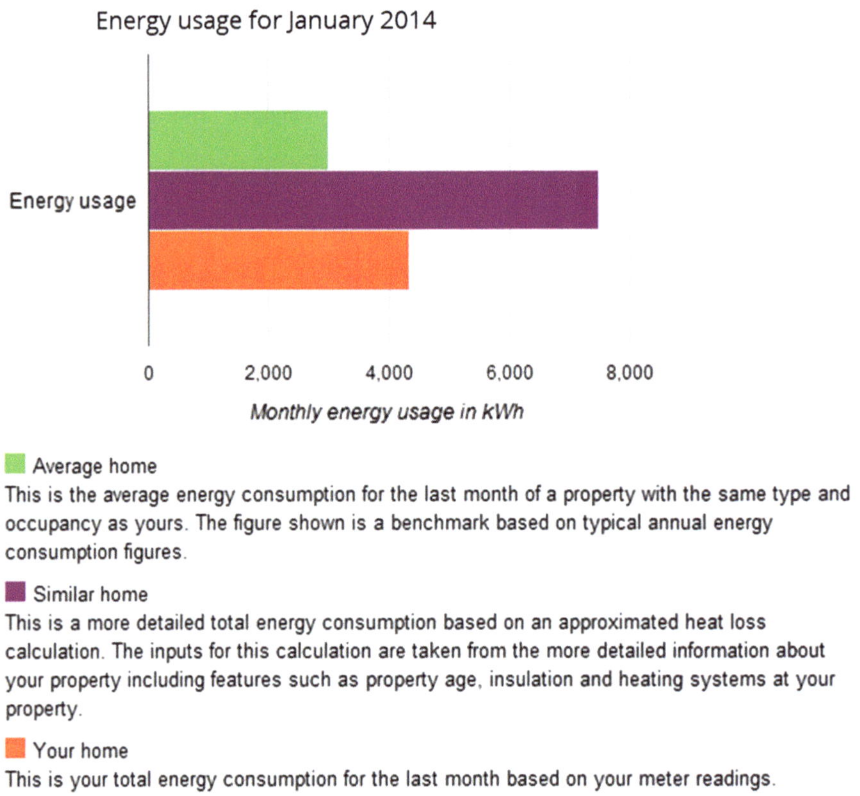

Fig. 1 Example of the energy meter dashboard for student users

for energy/carbon reduction: hence, the title to this Chapter. The orange and black colour theme represents both contemporary trends and warmth. The concept was tested through focus groups and used as EW's communication strategy for engagement and raising awareness.

In terms of Darby's (2010) Economic theory, EW provided incentives, monogramed pencils, rulers, chocolates, and T-shirts as 'ice-breakers' to encourage students to register and commit to the project.

6 Results

A variety of performance indicators (PIs) have been generated to describe aspects to EW. These have not yet been systematically mapped against the project's knowledge, condition, values, attitudes, skills or behavioural (KSAB) assumptions (Alden 2006). The project results for the first 5 months are described through these provisional PIs below.

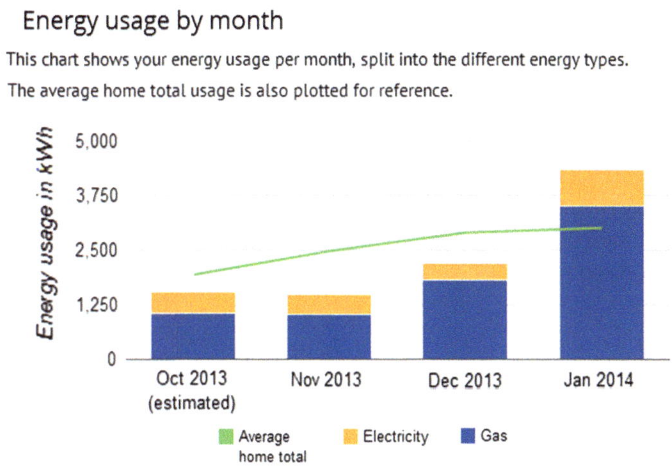

Fig. 2 Example of the energy consumption figure by month

Table 2 Measured performance to date

Key project parameters	Year 1 target	Result to date
No. of students trained as EAs	5	5
No. of households supported by EAs	50	48
No. of households participating in the project	150	48
No. of students living in participating households (ave 4.5/property)	675	216
Unique page views on the website	60,000	3,500
Social media followers	1,000	200

Table 3 Recipients' attitudes analyses (N = 71)

	EW email response rate (%)	Industry average (%)
No. of recipients opened the email	19.6	16.2
No. of recipients clicked the web link on the email	2.2	1.9

Through a promotional stall, a total of 71 students registered, with 48 property addresses recorded. An email was sent through Mailchimp, an advanced mailing system enabling the tracking of recipients' email reactions. Those responses are listed in Tables 2 and 3.

Further engagement was elicited via a targeted email to around 3000 students identified as living off-campus and eligible to access EW services. This used WSU's email system with a message promoting incentives. It did not secure further student registration.

Economic incentives sought to encourage students to input their meter readings and included food vouchers and cash prizes. However, results to date show that this is proving to be a particularly difficult venture. From the 71 registered students, only 9 have spontaneously entered meter readings, 3 of which are EAs. These results are disappointing and the barriers to inputting readings are being investigated further.

7 Project Reflection

The EAs have found their training to be highly positive. They have effected significant improvements in their personal understanding of domestic energy matters. These students have changed their energy habits at home, and have become more confident in talking about domestic energy issues. One EA, who had never previously lived in privately rented accommodation, had very little understanding about energy. When he started searching for a shared house with friends, he was determined to secure a home that was easy and cheap to heat:

> After learning about energy efficiency, I am determined to find an energy efficient house for next year, and I know that is my right.
> –First year EA

EAs will be monitored to assess whether their experience and knowledge has influenced friends' tenure choices. The EAs' peer influence might also impact on the final evaluation of the cost effectiveness of the models, as the training for the accredited energy awareness certificate is relatively expensive.

In their on-campus peer discussions, EAs reported the majority of students expressed concern with energy bills and agreed with the project aims. However, many remained reluctant to participate. Motivating students to engage with energy conservation projects remains a challenge. Focus group discussions were held in early-2014 to explore reasons for this reluctance. Findings include:

- Energy issues are not a priority. Students are less likely to commit to something if it does not have an immediate impact on their lives.
- Low brand recognition. EW is new, students perceived it as an independent organisation not associated with the more trusted and recognisable WSU and UoW brands and did not recognise it.
- Complexity in software use. Some students found the software somewhat more complex than expected. If they found the registration or data input process difficult, they were more likely to give up as a result.

- Low energy literacy rate. Lacking in energy knowledge makes it very hard for students to fully understand energy and this produces a perceived loss of control over home energy.
- Email fatigue. Students receive innumerable emails from UoW, WSU and affiliated clubs and societies, including EW. If they are deemed irrelevant, they are ignored.

How might this situation of expressed concern but inattentive behaviours be explained? The selection bias theory (Clayton and Myers 2009; Frederick et al. 2002), suggests that, if the affect (energy cost reduction) is not going to happen immediately, the threat (high bills) is likely to be discounted. Thus, many students will prioritise course work over housing conditions. Faced with multiple daily demands and with only finitc ability to foreground issues, students appear to be displaying a degree of mindlessness (Seiling and Hinrichs 2005) on this issue of energy use.

Using incentives to keep an issue in the foreground appears of only limited utility. Bell et al. (2013) suggested that financial incentives can only encourage simple commitments. The EW results to date mirror these findings. Drawn to the project by merchandising and allied 'goodie bag' incentives, 71 students initially registered, but these did not translate subsequently into students inputting meter data. It is too early in the evaluation to determine this with certainty, as other technical factors may also contribute: the website user interface may be unsuited for students, for example. Further monitoring is therefore being undertaken.

8 Interim Conclusions

At the time of writing, the project was only 5 months into its two-year lifespan and these are preliminary findings. Some things are becoming clear, though. First among these is the fact that EW has yet failed to reach and change the behaviours of target students on the scale hoped. Students face a knowledge gap in fully understanding domestic energy use and the importance of energy efficiency. Energy literacy could then have a stronger impact on energy habits when citizens feel in control of domestic energy matters (DeWaters et al. 2007; Allcott and Mullainathan 2010). Effecting this cognitive enlightenment is, preliminary EW findings suggest, not proving easy, in the midst of great competition for students' attention. The initial strategy was to deploy personalised messages containing energy information to spark initial interest and then build up to deeper engagement (wherein behavioural change might be effected). Comprehension then precedes behavioural change, but many students may exit before that point is reached. This reflects the fact that EW has finite foregrounding power and that it works in a context of constrained student memory and finite issues management capability (Behr and Iyengar 1985; Thøgersen 2006). Students can be expected to juggle only so much.

The motives of those who have been affected by EW to date are clearly heterogeneous. The EAs are likely to possess pro-social inclinations and their actions need to be understood and evaluated as such (Thøgerson 2011).

National survey data suggest that many of their UoW student peers may share their attitudes, but they do not share EAs' knowledge or behaviours. The recruited EAs have demonstrated greater personal understanding of domestic energy matters. This means greater caution in their energy usage, more confidence in engaging with peers on energy related issues and willingness to use their knowledge to make an impact. This greater readiness and capacity has, though, been expensively attained. The effectiveness of the cascade-training model (situating EAs at its conversational heart) is not yet known, but will be closely monitored.

There are as yet unrealised options for reinforcing EW's underlying logic. This includes the installation of smart meters in selected households. As Darby (2008, 2010) found, in-home display systems present a more dynamic and visual account of energy performance, potentially encouraging more rational usage. Community engagement actions are also in their infancy and may yield partnerships with the local authority, local energy groups and landlords in order to investigate Darby's Social feedback mechanisms. Finally, and deepening EW's behaviourist design, a tighter bundling of its portfolio of initiatives might be used to overcome the thresholds of student attention that may be blocking the project from making further progress (Neuner 2000). Even where EW surmounts these barriers, further behavioural surprises may await. It is plausible to argue, for instance, that students that do learn better energy husbandry approaches may also de-learn those approaches in unexpected ways. Others may feel guilt at violating early promises of participation and deploy complex guilt offsetting behaviours (Steenhaut and Van Kenhove 2006). All represent valid behavioural responses to EW and should be captured in its evaluation.

These various potential challenges are significant. They do not nullify the programme goals or objectives set out earlier, but they significantly adapt them. The ability of EW to enact such change lies in the fact that the project governors have the power largely to set their own goals, within broad parameters and for a finite period of time. EW is therefore a teleonomy (de Laguna 1962) that is willing and able to mould its own emergent strategies. Not only that, but the range of its governing stakeholders is clearly changing over time. Latterly and through new connections, unexpected further opportunities and experiences have been uncovered that are likely to prove relevant to its future development. In short, the governance structure for EW is shifting to reflect emergent organising (Christensen 2004). Finally, the evidence base that was reviewed above is largely derived from national sources or from research done for other purposes, In other words, EW may be characterised as a non-evidence-based programme. These three factors taken together strongly suggest a specific approach to the evaluation of this initiative. Following Shen et al. (2008), EW may be understood through the Fidelity—Adaptation problematic, where a lack of precise and tailored foreknowledge precludes the design of a tightly structured, goals-oriented intervention (see also, Berman 1978). Only an approximate program logic can be adduced for the initiative, its iterative and groping nature is not a policy design failure, and nor is the low take-up to date an implementation failure. The evaluative task is to assess the interaction between the EW initiative itself and its governing stakeholders, where the latter's values and beliefs form an equal part of the evaluative design.

References

Alden J (2006) Measuring the "Unmeasurable". Performance Improv 45(5):7–17
Allcott H, Mullainathan S (2010) Behaviour science and energy policy. Energy policy. http://citeseerx.ist.psu.edu/viewdoc/download?doi=10.1.1.208.5738&rep=rep1&type=pdf. Accessed 14 March 2014
AECOM (2011) Energy demand research project: final analysis. Office for gas, electricity markets. Hertfordshire. Ofgem. https://www.ofgem.gov.uk/ofgem-publications/59105/energy-demand-research-project-final-analysis.pdf. Assessed 10 March 2014
Bell S, McGeevor K, Mocca E, Shaw B (2013) Synthesis report: review of evidence on the use of reward and recognition schemes in enhancing recycling and waste prevention behaviours (EV0528). Policy Studies Institute, Defra, London
Benders R, Kok R, Moll H, Wiersma G, Noorman K (2005) New approaches for household energy conservation-In search of personal household energy budgets and energy reduction options. Energy Policy 34:3612–3622
Behr RL, Iyengar S (1985) Television news, real-world cues, and changes in the public agenda. Publ Opin Q 49:38–57
Berman P (1978) Designing implementation to match policy situation: a contingency analysis of programmed and adaptive implementation. RAND Corporation, Santa Monica, CA
BRE (2005a). Energy use in homes (energy efficiency report): a series of reports on domestic energy use in England. Department for Environment, Food, and Rural Affairs. London, Defra. http://www.bre.co.uk/filelibrary/pdf/rpts/EnergyEfficiency.pdf. Assessed 20 March 2014
BRE (2005b) Energy use in homes (energy summary report): a series of reports on domestic energy use in England. London. Defra. https://www.bre.co.uk/filelibrary/pdf/rpts/EnergySummaryReport.pdf. Assessed 20 March 2014
Bouzarovski S, Petrova S, Kitching M, Baldwick J, Hayman B, Hörschelmann K (2012) Fuel Poverty among young adults in multiple occupancy housing: preliminary research findings and recommendations. Energy, Society and Place Research Unit School of Geography, Earth and Environmental Sciences University of Birmingham. http://espru.files.wordpress.com/2012/06/enyafinalreport.pdf. Accessed 09 April 2014
Christensen CM (2004) Seeing what's next: using the theories of innovation to predict industry change. Harvard Business School Press, Boston
Clayton S, Myers G (2009) Conservation psychology: understanding and promoting human care for nature. Wiley-Blackwell, Oxford, pp 24–26
Darby S (2008) Why, what, when, how, where and who? developing UK policy on metering, billing and energy display devices. Environmental Change Institute, University of Oxford. http://www.eci.ox.ac.uk/research/energy/downloads/darby08-aceee.pdf. Accessed 02 March 2014
Darby S (2010) Literature review for the energy demand research project. Environmental Change Institute, University of Oxford. https://www.ofgem.gov.uk/ofgem-publications/59113/sd-ofgem-literature-review-final-081210.pdf. Accessed 05 March 2014
De Laguna GA (1962) The role of teleonomy in evolution. Philos Sci 29(2):117–131
Drayson R, Bone E, Agombar J, Kemp S (2013) Student attitudes towards and skills for sustainable development. Higher Education Academy (HEA). http://www.heacademy.ac.uk/assets/documents/sustainability/ESD_student_attitudes_2013_v4.pdf. Accessed 14 March 2014
DCLG (2014) English housing survey: headline report 2012–2013. Department for Community and Local Government. ISBN: 978-1-4098-4152-4. London, DCLG
DECC (2010). The green deal: a summary of the government's proposals. Department of Energy and Climate Change, Ref: 10D/996. London, DECC
DECC (2012) DECC local authority funds: DECC local authority competition 2012–2013. Department of Energy and Climate Change, URN 12D/397. London, DECC
DECC (2013) Quarterly energy prices. Department of Energy and Climate Change. URN 14D/276A, London, DECC. https://www.gov.uk/government/uploads/system/uploads/attachment_data/file/296011/QEP_March_2014.pdf. Accessed 20 March 2014

DECC (2014) UK Energy Statistics-2013 provisional data. Department of Energy and Climate Change, Statistical Press Release Ref: 2014/012. London, DECC
DeWaters JE, Powers SE, Graham M (2007) Developing an Energy literacy scale. In: Proceedings of the 114th annual ASEE conference and exposition. Honolulu, HI, June 23–38, 2007, paper number AC 2007-1069
Frederick S, Loewenstein G, O'Donoghue T (2002) Time discounting and time preference: a critical review. J Econ Lit 40:351–401
Goodman J, Drayson R (2014) "Homes fit for study: the states of student housing in the UK" London: National Union of Students. http://www.nus.org.uk/Global/Homes%20Fit%20For%20Study/Housing%20research%20report_web.pdf. Accessed 24 March 2014
Grønhøj A, Thøgersen J (2011) Feedbacks on household electricity consumption: learning and social influence processes. Int J Consum Stud 35:138–145
HEFCE (2013) Sustainable development in higher education: consultation on a framework for HEFCE. London: Higher Education Funding Council for England. http://www.hefce.ac.uk/media/hefce/content/pubs/2013/201331/2013_31.pdf. Accessed 07 March 2014
Mintzberg H (1978) Patterns in strategy formulation. Manage Sconce 24(3):934–948
Mintzberg H, Ahlstand B, Lampel J (1998) Strategy safari: a guided tour through the wilds of strategic management. Free Press, New York
Neuner M (2000) Collective prototyping: a consumer policy strategy to encourage ecological marketing. J Consum Policy 23(2):153–175
Palmer J, Cooper I (2013) United Kingdom housing energy fact file 2013. London: Department of Energy and Climate Change. URN: 13D/276
Runhaar H, Dieperink C, Driessen P (2006) Policy analysis for sustainable development: the toolbox for the environmental social scientist. Int J Sustain High Educ 7(1):34–56
Seiling J, Hinrichs G (2005) Mindfulness and constructive accountability as critical elements of effective sense making. Organ Dev J 23(3):82–88
Shen J, Yang H, Cao H, warfield C (2008) The fidelity-adaptation relationship in non-evidence-based programs and its implication for program evaluation. Evaluation 14(4):467–481
Sidani S, Sechrest L (1999) Putting program theory into operation. Am J Eval 20(2):227–238
Steenhaut S, van Kenhove P (2006) The mediating role of anticipated guilt in consumers' ethical decision-making. J Bus Ethics 69:269–288
Thøgersen J (2006) Media attention and the market for "green" consumer products. Bus Strategy Environ 15(3):145–156
Thøgerson J (2011) Green shopping: for selfish reasons or the common good. Am Behav Sci 55 (8):1052–1076
Verplanken B, Wood W (2006) Interventions to break and create consumer habits. Am Mark Assoc 25(1):90–103

Authors Biography

Peng Li is the project leader of Energize Worcester. He is an Associate member of Institute of Environmental Sciences; he has an MA in Sustainable Development Advocacy, a postgraduate degree with strong professional skills development across a whole range of sustainability aspects. Peng is particularly interested in advocating for sustainability, energy conservation and behaviour change. Working with students in a Students' Union, Peng has experience in sustainability related student engagement in Higher Education.

Katy Boom is Director of Environmental Sustainability at the University of Worcester. She has led the development of wide ranges of sustainability initiatives focusing on the campus, working in the community and embedding sustainability in the curriculum. The University consistently tops the nationally recognised University 'Green League' table and became the first English University to be awarded platinum EcoCampus status. Overseeing the development of students as partners on major sustainability change programmes and building sustainability skills with students and student volunteers Katy's work has won a number of national awards.

Paul Davis is Honorary Professor at Stichting Euregio University, Gelderland, The Netherlands. His research interests include evaluation frameworks and their broad relation to theories of worth. He has published on realist evaluation methods and undertaken numerous commissioned evaluation exercises in local government and for private corporate clients. He is currently investigating the use of deontological systems to shape the conduct of university business schools.

Implementing Sustainability and Social Responsibility Initiatives in the Higher Education System: Evidence from Spain

Manuel Larrán and Francisco Javier Andrades

Abstract

Sustainability and social responsibility issues in higher educational institutions have attracted increasing levels of attention from both the public and policy makers in recent decades. This study analyzes the level of implementation of sustainability and social responsibility issues by Spanish universities in different areas (e.g. curriculum, strategic planning, practices, barriers or drivers and reporting) in order to find out their engagement with the higher education for sustainable development. Key findings show the slow rate of progress of many Spanish universities with regard to implementing sustainability and social responsibility initiatives, a fact which could indicate that there are clearly a number of obstacles to overcome. Barriers that compromise sustainability and social responsibility initiatives in higher education can stem from a lack of institutional policies, research networks or administrative structures to promote and coordinate sustainability and social responsibility at universities.

Keywords

Sustainability · Social responsibility · Spanish universities · Curriculum · Strategic planning · Policies · Barriers · Drivers · Reporting

M. Larrán · F.J. Andrades (✉)
Faculty of Business and Economic Studies, University of Cadiz,
Duque de Najera Avenue, number 8, 11002 Cadiz, Spain
e-mail: javier.andrades@uca.es

M. Larrán
e-mail: manuel.larran@uca.es

W. Leal Filho et al. (eds.), *Integrative Approaches to Sustainable Development at University Level*, World Sustainability Series, DOI 10.1007/978-3-319-10690-8_8

1 Introduction

In recent years, several definitions of a sustainable higher education institution have emerged (Madeira et al. 2011). Alshuwaikhat and Abubakar (2008) argued that a sustainable campus should be environmentally healthy, with a prosperous economy through energy and resource conservation, waste reduction and with efficient environmental management; it should promote equity and social justice and export these values to the community. According to Milutinovic and Nikoli (2014) the vision of higher education for sustainable development is a world where everyone has the opportunity to benefit from a quality education and learn the values, behaviours and lifestyles required for a sustainable future and for positive societal transformation.

In the last 2 decades, an increasing number of higher education institutions have been engaged in incorporating and institutionalising sustainability and social responsibility into their systems (Lozano et al. 2013). Most previous studies have been aimed at assessing students' and deans' perceptions of the factors that contribute towards social responsibility in universities (Lämsä et al. 2007); analyzing the extent to which universities are incorporating social responsibility and sustainability education into business curricula (Christensen et al. 2007); evaluating the factors that could obstruct the implementation of sustainability and social responsibility initiatives in higher education institutions (Velazquez et al. 2005); or examining cases based on specific universities where there are specific sustainability strategies, primarily concerning environmental issues (Lee et al. 2013).

In the specific case of Spain, few studies to date have been undertaken to determine the extent to which universities are implementing sustainability into their functions. Aznar et al. (2011) analyzed the current status for introducing sustainability across the curricula at the University of Valencia. Their findings show the widespread support for introducing sustainability across the university's curricula; however, as might be expected, significant differences occur in how these questions are addressed by staff from various disciplines, including how they relate to departmental perceptions, interpretations and performance of sustainability-related teaching. Garde et al. (2013) analyzed whether social responsibility has become an essential element in activities associated with university accountability, information transparency, and Internet usage. A comparative study of public and private Spanish universities shows that neither is strongly committed to the online disclosure of social responsibility information.

Nevertheless, higher education for sustainable development is still far from being integrated into a holistic and organic manner by university leaders (Lee et al. 2013). In fact, sustainability and social responsibility in universities is at an early stage of the learning process in which much must be learned for sustainable development to become genuinely and fully implemented and for higher education to become a true leader in sustainability (Waas et al. 2010). A number of authors have called for more comprehensive integration of sustainable development into their systems, rather than only as 'add-ons' to existing practices, by engaging in fundamental and radical changes (Fadeeva and Mochizuki 2010). The slow rate of change at universities presents a tremendous challenge to higher education institutions and society to become more sustainable and socially responsible.

On this basis, this paper describes the extent to which Spanish universities are implementing sustainability and social responsibility issues in different areas, such as curriculum, strategic planning, practices, barriers or drivers and reporting. The relevance of this paper relies on two main reasons: first, there is a lack of sufficient studies that investigate sustainable development in higher education in Spain. Second, few studies addressed their attention on providing a holistic and integrated vision of the higher education for sustainable development. Hence, there is a need to give a more comprehensive integration of sustainable development into the university system.

The data collection was based on qualitative and quantitative approaches. First, we analyzed the content of the organizational management undergraduate degrees and the strategic plans published in the web-page at Spanish universities. Second, we sampled the opinions of rectors, senior management members and Social Councils from Spanish higher education institutions with regard to the implementation of sustainability and social responsibility practices and the main barriers and drivers to their implementation.

The paper is structured as follows: first, we introduce the institutional background; second, we analyze the results depending on the different areas; third, we provide a discussion section and finally we highlight the conclusions and limitations of the study.

2 Institutional Background: Special Attention to Spanish Higher Education System

From 1972 onwards a range of international declarations specifically address the importance of higher education for sustainable development (Halifax Declaration 1991; Cre-Copernicus Charter 1994). More recently, the 2012 United Nations Conference on Sustainable Development (UNCSD) held in Rio de Janeiro provided a forum where leaders of the international academic community declared their commitment to the development of sustainable practices for higher education institutions (Rio + 20). In 2013, the Global University Network for Innovation (GUNI) published a new world report titled "Knowledge, Engagement and Higher Education: Contributing to Social Change". One of the main challenges of this report is to propose steps for advancing the contribution of higher education to building a more just, equitable and sustainable society.

In Spain, the higher education system is mainly public. According to the data published by the Conference of Rectors of Spanish Universities (CRUE 2012), Spain has 78 higher education institutions, 50 of which are public universities and 28 are private. In addition, more than 90 % of students are enrolled at public universities. The Spanish higher education system, although regulated by a common framework and coordinated on a national level, are administered within 17 different university systems (one for each Autonomous Community into which Spain is divided). The specific structure of each Autonomous Community is quite diverse, although a common feature is the lack of student mobility and the preference for studying near home. Mobility between the various Autonomous Communities is quite low, around 10 % of first year students (CRUE 2012). With regard to the profile of the Spanish higher education system, there are universities (Granada and Salamanca) with several centuries of history and others (Pablo de Olavide, established in 1997), which were created quite recently. There are some very large universities (Barcelona and Basque Country), medium-sized universities (Malaga and Cadiz), and smaller universities (Pablo de Olavide is the smallest).

Over the last years, there have been several laws which deal with questions related to social responsibility and sustainability from a general point of view, such as the Environmental Protection Law (2006), the Dependency Law (2007), the Equality Law (2007) or the Transparency Law (2007). Moreover, Law 2/2011 on Sustainable Economy notes the need of improving transparency and accountability by any organization. Also, this law states that university education should incorporate skills and innovation-oriented abilities into its curricula, fostering creativity and entrepreneurship.

Focusing on the higher education system, although interest in sustainability and social responsibility issues has been more recent, many Spanish institutions have emphasized the relevance with regard to the higher education for sustainable development. The CRUE has set up a Working Group on Environmental Quality and Sustainable Development, which has issued a set of guidelines to incorporate sustainability into the curriculum. This, in turn, has led to the creation of a special Commission focused on environmental quality, sustainable development and risk prevention (CADEP). This Commission also compiles university experiences and the progress made in the area of environmental management and sustainability, promoting cooperation among universities in these areas. One of the key publications made by this Commission focuses on assessing the implementation of sustainability policies by Spanish universities (Aznar 2011).

The State Board on social responsibility, created by the Spanish Employment Department in 2005, made a working paper on social responsibility and the higher education system in 2011. Its aim was to enhance the implementation of sustainable development and social responsibility in all education levels (Larrán et al. 2014). Also, the Spanish government approved other new legislation with regard to the implementation of sustainability by higher education institutions. For example, the Organic Law 4/2007 on Universities stated the need for incorporating sustainability issues in different higher education areas such as management and accountability (Larrán et al. 2014).

Several years ago, Spanish governing bodies introduced a new way to allocate public universities funding based on performance criteria. By funding universities according to their outputs, rather than inputs, state policymakers in Spain believe they are providing an incentive for universities to improve their quality management and accountability (Vilardel and Álvarez 2010). This performance funding system is known as the contract programme, which is a university policy aimed at promoting institutional responsibility and dialogue between governing bodies and universities (Ribas and Vilalta 2003). One of the main outcomes of this performance funding system has been the greater number of strategic plans or sustainability reports published by Spanish universities in recent years.

Moreover, the Spanish Government developed the 2015 University Strategy based on the need to adapt Spanish universities to the guidelines proposed by the European Higher Education Area. One of the main challenges of this 2015 University Strategy was to promote the implementation of sustainability policies by Spanish higher education institutions. In the context of this initiative, the Spanish Government established a special Commission which was entrusted with the elaboration of a document titled "University Social Responsibility and Sustainable Development" (Ministry of Education 2011).

Although the institutional background has pointed out the relevance of higher education for sustainable development in Spain, the slowing economy has affected the continuity of this process. For example, the special Commission created in 2011 by the Spanish Government in the University Strategy context has stopped working. The economic crisis has led the Spanish government to cut budgets by reducing expenditures. Money has been reallocated to priority goals, and sustainability and social responsibility is not a primary priority for the Spanish government. The lack of financial resources is jeopardizing the future of initiatives related to sustainability.

3 Curriculum

The future success of social responsibility and sustainability will depend on the attitude of future generations, as these generations will influence relations between business and society, whether as citizens, customers, or managers (Muijen 2004). Hence, teaching sustainability, ethics and social responsibility is a critical step toward helping business and management students understand this balance (Kolodinsky et al. 2009). As a consequence, we decided to focus on the curricula of the organizational management undergraduate degrees offered by Spanish universities during the 2010/2011 academic year. The data collection was based on a search of the web-page of Spanish universities in order to analyze the content of the study plans and programmes. The sample of undergraduate degrees analyzed is shown in Table 1.

Table 2 shows the extent to which social responsibility, ethics and environmental sustainability topics are taught in organizational management curriculum depending on stand-alone or embedded subjects. To equip students with alternative views of business, Table 3 reports about the type of training based on compulsory or optional subjects.

An analysis of the findings in Table 2 suggests the lack of presence of social responsibility, ethical and environmental sustainability subjects in the organizational management curriculum in Spain. Comparatively, the results show the greater

Table 1 Organizational management undergraduate degrees offered by Spanish universities

Undergraduate degree	Number	Public	Private
Business and management	83	56	27
Accounting and finance	32	27	5
Economics	42	37	5
Tourism	52	41	11
Public management	16	16	0
Human resources and labor relations	41	41	0
Advertising and public relations	30	16	14
Marketing and research market	25	17	8
Audiovisual communication	45	26	19
Political sciences and administration	21	18	3
Total	387	295	92

Source Adapted from Larrán and Andrades (2014)

Table 2 Extent of social responsibility, environmental sustainability and ethical education

Undergraduate degrees	Stand-alone subjects (%)	Embedded subjects (%)
Social responsibility	24	33
Ethics	32	38
Environmental sustainability	23	33

Source Adapted from Larrán and Andrades (2014)

Table 3 Type of social responsibility, environmental sustainability and ethical stand-alone subjects

Type of subject	Social responsibility (%)	Ethics (%)	Environmental sustainability (%)
Core	11	15	1
Compulsory	21	56	20
Optional	68	29	79
Total	100	100	100

Source Adapted from Larrán and Andrades (2014)

inclusion of ethical education than social responsibility or environmental sustainability training. A more detailed analysis of the results shown in Table 2 suggest that there is a greater presence of social responsibility, ethics and environmental sustainability training by means of embedded subjects in comparison with stand-alone subjects. These results seem to indicate the greater orientation toward the inclusion of sustainability and social responsibility issues by means of horizontal integration (embedded) because of the necessity of providing a interdisciplinary approach towards sustainable development.

With regard to the type of training, while most of ethics stand-alone subjects are compulsory, there is a greater proportion of social responsibility and environmental sustainability stand-alone subjects offered as optional courses (Table 3). These results seem to suggest that social responsibility and environmental sustainability may be considered as a specialized training focused on exploring a concrete topic. Meanwhile, stand-alone subjects on ethics may be related to a basic or elementary training in order to provide a more comprehensive and holistic knowledge.

Also, an analysis of the influence exerted by different factors (public/private, size, social responsibility or sustainability chairs and political orientation) on the extent to which ethics, social responsibility and environmental sustainability topics are taught by Spanish universities was conducted. Findings suggest there are no factors influencing the inclusion of these topics in the organizational management curriculum offered by Spanish universities. According to these findings, many authors found that the lack of specialization in social, ethical or environmental issues by faculty could obstruct the implementation of sustainability in the curriculum (Nicolaides 2006). Hence, there is a need for enhancing the implementation of initiatives directed to providing social responsibility and sustainability training to faculty. To a great extent, there is a still a long way to go before all management students can be said to have had a thorough education in social, ethical and environmental issues.

4 Strategic Planning

Previous studies have highlighted the crucial role of having mission and vision statements to foster the development of higher education for sustainability (Koester et al. 2006). Nowadays, universities worldwide are changing their mission, vision, and educational practices in order to better cope with growing concerns about social and environmental issues and to respond to the proliferating public demand for a sustainable society (Md Shahbudin et al. 2011).

On this basis, we analyzed the strategic plans implemented by Spanish universities with two different aims: first, we examined the extent to which Spanish universities were implementing strategies on sustainability and social responsibility. The second goal was to analyze whether the implementation of strategies on sustainability and social responsibility by Spanish universities were explained by coercive and mimetic pressures. The data collection was based on a search of the strategic plans published in the web-page of Spanish universities. Until 25 July 2012, only 45 universities in Spain had a strategic plan publicly available on their institutional web-page (58 % of the total of Spanish universities) of which 42 belonged to public universities and 3 were private.

The analysis of the strategic plans by Spanish universities was classified in three steps: First, we analyzed the extent to which Spanish universities were implementing strategies on sustainability and social responsibility; second, and based on the global situation of all Spanish universities, we examined whether the implementation of strategies on social responsibility and sustainability were based on uniformity. When this measure would be near to one, the implementation of strategies on social responsibility and sustainability by Spanish universities would be a sign of a behavioural pattern that tends towards standardization; third, we analyzed the viability of the implementation of strategies on social responsibility and sustainability. To analyze the viability we weighted the strategies on sustainability and social responsibility depending on four performance criteria: time frame, budget allocated, indicators for its continuous assessment and those responsible for putting it into practice. These performance criteria were based on assessing the probabilities of successfully putting certain strategies into practice. In the case that all four requirements were fulfilled, the achievement of the objectives of each strategic plan would be more probable.

With regard to the extent to which Spanish universities are implementing strategies on sustainability and social responsibility, the findings show the lack of commitment to sustainability and social responsibility. The results show that the less implemented strategies on social responsibility and sustainability are related to the environment, companies and corporate governance dimensions. Meanwhile, the findings point out that the most implemented strategies on sustainability and social responsibility focuses on continuous improvement, society and students. Focusing on the second step of the analysis, the findings seem to indicate that the implementation of strategies on social responsibility and sustainability are closely linked to standardization and uniformity. In the third step, the analysis of the viability suggest that the success rate with regard to the implementation of strategies on sustainability and social responsibility by Spanish universities is relatively scarce (Table 4).

Also, there are no factors influencing the implementation of strategies on sustainability and social responsibility by Spanish universities. Hence, the research findings seem to indicate that implementing strategies on sustainability by higher education institutions in Spain may not be defined as a relevant strategic and differentiating factor. These findings may be a sign of institutional imitation and

Table 4 Implementation of strategies on social responsibility and sustainability in the strategic plans of Spanish universities

Dimension	Extent (%)	Similarity (%)	Viability (%)
Continuous improvement	51.1	87.9	50
Society	49.2	73.6	44.1
Students	48.9	65.4	45.4
Staff	37.3	74.5	45.8
Corporate governance	28.4	48.6	47.6
Companies	19.3	78.1	47.1
Environment	9.5	14.8	40.6
Weighted global index	31.4	65.7	46.4
Overall global index	34.8	63.3	44.9

Source Adapted from Larrán et al. (2014)

could show a behavioural pattern that tends towards standardization in the strategies of Spanish universities. Therefore, the theory that strategic planning is not really a management tool but rather the consequence of pressure by the government can be inferred.

5 Practices Related to Sustainability and Social Responsibility: Barriers and Drivers

Universities are now expected to engage with communities and benefit the society at large (Lukman and Glavic 2007), as they have the potential to be leaders in all fields of research, teaching and learning, sustainability and community engagement (Stephens and Graham 2010). Velazquez et al. (2006) conceptualize the scope of a sustainable university within the areas of teaching, research, outreach and partnership, and in-campus sustainably practices. According to Lozano (2006), it is broadly reflected into six aspects, i.e. sustainable campus operations, sustainable research, public outreach, cooperation between institutions, curricula and sustainability reporting.

To contribute to fill this gap, this section focuses on analyzing the extent to which Spanish universities have implemented sustainability and social responsibility practices. The main barriers and drivers to implement sustainability and social responsibility practices also were identified. Data collection was based on a survey completed by senior management and Social Councils members from Spanish universities. The number of questionnaires completed was 170 which belonged to 55 Spanish universities. The sample was classified by the following way: 53 questionnaires completed by senior management members and 117 completed by Social Councils members. The survey was measured using a five-point Likert scale (rating scales: 1 = sustainability and social responsibility practice was not implemented, 5 = sustainability and social responsibility practice was strongly implemented).

According to the opinion by respondents (Table 5), the most implemented sustainability and social responsibility practices by Spanish universities are related to students (4.29), followed by the engagement with the implementation of sustainability and social responsibility practices with regard to society (3.92), staff (3.72), continuous improvement (3.47) and corporate governance (3.4). Meanwhile, results show that the less implemented practices on sustainability and social responsibility by Spanish universities are related to environment (2.97) and companies' dimension (2.95). In sum, an analysis of the global results suggests that there is still a long way to go for Spanish universities to incorporate sustainability and social responsibility practices into their mainstream functions. Also, there are no factors influencing the implementation of practices related to sustainability and social responsibility by Spanish universities.

Table 5 Practices related to sustainability and social responsibility depending on the level of implementation

Dimension	Mean value
Students	4.29
Society	3.92
Staff	3.78
Continuous improvement	3.47
Corporate governance	3.41
Environment	2.97
Companies	2.95
Global implementation	3.54

Source Adapted from Larrán (2014)

Table 6 provides a more detailed analysis with regard to the social responsibility and sustainability practices less implemented by Spanish universities. Based on the engagement with the society dimension, there is a lack of implementation of practices related to fostering research on sustainability, followed by practices associated with assessing student satisfaction. Regarding the engagement with staff, practices related to providing sustainability training to faculty are not well implemented. According to the corporate governance, there is a need to enhance the implementation of codes of good governance and codes of conduct. In addition, there is a lack of administrative or political structures focused on sustainability. Finally, practices related to improving sustainability interest in university senior management are not well implemented by Spanish universities.

Based on practices related to environmental sustainability, results show that they are the less implemented ones. Among others features, there is a need to reinforce the implementation of practices related to incorporating measures to reduce noise

Table 6 Practices related to sustainability and social responsibility not well implemented by Spanish universities

Dimension	Practices less implemented
Society	Practices related to fostering research on sustainability
	Practices associated with assessing student satisfaction
Staff	Practices related to providing sustainability training to faculty
Corporate governance	Practices related to implementing codes of good governance
	Practices related to implementing codes of conduct/ethics
	Practices related to implementing administrative/political structures focused on sustainability
Environment	Practices related to incorporating measures to reduce noise for each building
	Practices related to improving the collection of tons of rubbish from university canteens
	Practices related to reducing emission of greenhouse gases
	Practices related to providing information on other significant emissions

Source Adapted from Larrán (2014)

for each building, practices related to improving the collection of tons of rubbish from university canteens, practices related to reducing emission of greenhouse gases or practices related to providing information on other significant emissions. This dimension could be one of the major challenges for Spanish universities regarding their commitment with sustainability.

These findings seem to suggest that there are clearly a number of obstacles to overcome in terms of implementing sustainability and social responsibility practices by universities. Respondents identified several barriers to implementing sustainability and social responsibility practices by universities: lack of sustainability and social responsibility training, lack of financial and human resources and lack of time. Another main barrier to implementing sustainability and social responsibility practices is the lack of pressure from society. Unless society demands major changes in the desired characteristics of graduates and research, a university may find little reason to make transformations and may continue with the status quo. However, respondents also pointed out that there are many drivers to implementing sustainability and social responsibility practices. One of them is related to the need for improving the reputation and legitimacy by universities. On this basis, the publication of sustainability reports may be an important tool to solve the problems of transparency. The second driver identified is related to the fact that the social commitment of universities is intrinsically linked to their inherent nature and the role that universities have played historically.

6 Sustainability Reporting

Although sustainability reporting in universities is still in its early stages, an increasing number of universities are publishing some type of sustainability report following a triple bottom line approach (Lozano 2011). While there are several mechanisms proposed for sustainability reporting, the Global Reporting Inititative (GRI) guidelines are the most extended (Shriberg 2002). The GRI Guidelines offer reporting principles, standard disclosures and an implementation manual for the preparation of sustainability reports by organizations, regardless of their size, sector or location (GRI 2013). In the case of universities, an increasing number of institutions are applying the GRI guidelines to publish their sustainability reports.

Authors such as Benavides et al. (2012) and Gonzalez et al. (2014) found that only 22 of 78 (28 %) Spanish universities have published one or more sustainability reports until 2012. In depth, 13 universities have published more than two sustainability reports, which reflect the greater engagement of these higher education institutions with transparency and accountability. According to the type of university, public higher education institutions in Spain are more engaged in publishing sustainability reports than private ones. Focusing on the guidelines used, 11 of the 22 sustainability reports published by Spanish universities are based on GRI guidelines whilst the other 11 do not have a specific formatting.

7 Conclusions: Implications for Practice

The purpose of this study was to address a gap in research in the emerging field of higher education for sustainability and social responsibility in Spain. The findings achieved may extend to the international audience.

There are some important contributions from this research. First, the findings seem to suggest that despite the wide recognition of this urgent call for responsibility, sustainability and social responsibility are not being integrated by universities into their systems and mainstream university operations. In fact, the findings show a low implementation of sustainability and social responsibility issues by Spanish universities with regard to the curriculum and strategic planning. Also, there is a lack of engagement with the sustainability reporting by higher education institutions in Spain. Finally, practices on social responsibility and sustainability are not well implemented by universities in Spain. These findings seem to indicate the presence of many difficulties to incorporate social responsibility and sustainability in the higher education system.

In addition, there are no factors influencing the implementation of sustainability and social responsibility issues by Spanish universities in their curriculum, strategic planning, practices and reporting. A more in-depth analysis may suggest the key role of the leadership exerted by some universities as a potential driver to implementing sustainability and social responsibility issues. Sustainability champions, often seen as "lone wolves" or "innovators" at their universities, can be important agents for change. This scenario might encourage some universities (leaders) to look for competitive advantages by implementing their sustainability initiatives, a feature which will be gradually adopted by others universities through a process of imitation.

Based on the previous assumptions, the findings seem to indicate that the future implementation of sustainability and social responsibility by Spanish universities would be explained by the influence exerted by institutional and legal factors beyond the individual concern of Spanish universities. Although there is a still a long way to go for Spanish universities to define their main functions adapted to sustainability and social responsibility, there has been a greater implementation of social responsibility and sustainability initiatives by higher education institutions over the last years.

In this sense, there is an increasing interest by university leaders to commit to sustainability by signing higher education declarations. Recently, several international initiatives have sought to promote the implementation of sustainability in higher education institutions. In Spain, the Conference of Rectors of Spanish Universities (CRUE) could play an important role in fostering the implementation of sustainability by Spanish universities. However, it is important to point out that the slowing economy may be fostering the reallocation of financial resources by governments to priority goals. Funding is a major concern for all involved in sustainability initiatives. The lack of financial resources may be an important barrier to enhancing the implementation of institutional initiatives related to sustainability and social responsibility.

Also, greater institutional awareness may be evoked with the implementation of a higher number of administrative or political structures focused on sustainability. The study by Benavides et al. (2012) found that only 16 % of Spanish universities had a specific administrative structure related to sustainability. More specifically, Gonzalez et al. (2014) pointed out that 18 of the 78 Spanish universities contain administrative or political structures related to sustainability. Most universities are bottom-up institutions where individual faculty members make decisions on how best to achieve research and education goals. As such, it is difficult for an administrator to propose changes and achieve consensus among groups of faculty at any level (Ferrer-Balas et al. 2008).

Another possibility to increase the presence of sustainability and social responsibility may be the creation of a greater number of networks for promoting the sustainable development at higher education institutions. In Europe the Copernicus Alliance focuses on promoting the role of sustainable development in European higher education to improve education and research for sustainability in partnership with society. Another European network is the Alliance of Universities for Democracy (AUDEM) created to assist universities from the former communist countries to move to their new socially responsible and proactive roles in democratic societies. At an international level the Regional Centre of Expertise on Education for Sustainable Development seeks to achieve the goals of the Decade of Education for Sustainable Development whilst the Centre for Social and Environmental Accounting Research (CSEAR) is an international, membership-based organization whose purpose is to explore the implications of and the possibilities for social, environmental and sustainability accounting and reporting.

There are some limitations to this study. The study presented here is preliminary and points to further, more in-depth research that needs to occur. Qualitative studies including case studies are required of those Spanish universities that exhibit strategies for sustainability in higher education. Interviews and focus groups with university leadership in individual universities, including curriculum consultants, would help to flesh out the initial findings of this study. Another possible limitation is related to the scope of this study: to a great extent we only analyzed the Spanish higher education context. Therefore, this paper could have been better clarified by comparison to other such studies in other countries, which may be resolved by further research and debate.

References

Alshuwaikhat HM, Abubakar I (2008) An integrated approach to achieving campus sustainability: assessment of the current campus environmental management practices. J Clean Prod 16:1777–1785

Aznar M, Martínez P, Palacios B, Piñero A, Ull M (2011) Introducing sustainability into university curricula: an indicator and baseline survey of the views of university teachers at the University of Valencia. Environ Educ Res 17:145–166

Benavides C, Arranz P, Borrego JM, González MA, Guitart L, Guzmán V, López A, Marchante M, Martín Q, Miravitlles P, Petit E, Qintana C, Santero R, Terrados J, Vallejo E, Aldeanueva I

(2012) Integración de la responsabilidad social en los sistemas de garantía interna de calidad: Hacia una universidad saludable, sostenible y solidaria. Ministry of Education, Spain, pp 1–261
Christensen LJ, Peirce E, Hartman LP, Hoffman WM, Carrier J (2007) Ethics, CSR, and sustainability in the financial times top 50 global business schools: baseline data and future research directions. J Bus Ethics 73:347–368
College Sustainability Report Card (2011) Indicators. http://www.greenreportcard.org/report-card-2010/indicators. Accessed 23 Mar 2014
Conference of Rectors of Spanish Universities (2012) La Universidad española en cifras. http://www.crue.org/Publicaciones/Documents/UEC/LA_UNIVERSIDAD_ESPANOLA_EN_CIFRAS.pdf. Accessed 27 Mar 2014
CRE-Copernicus (1994) CRE-Copernicus declaration. Cre-Copernicus Secretariat, Geneva. http://sustainability.edu.au/content/copernicus-campus. Accessed 26 Mar 2014
Fadeeva Z, Mochizuki Y (2010) Higher education for today and tomorrow: university appraisal for diversity, innovation and change towards sustainable development. Sustain Sci 5:249–256
Ferrer-Balas D, Adachi J, Banas S, Davidson CI, Hoshikoshi A, Mishra A, Motodoa Y, Onga M, Ostwald M (2008) An international comparative analysis of sustainability transformations across seven universities. Int J Sustain High Educ 9:295–316
Garde R, Rodríguez MP, López A (2013) Online disclosure of university social responsibility: a comparative study of public and private US universities. Environ Educ Res 19:709–746
Global Reporting Initiative (2013) G4 Sustainability reporting guidelines: reporting principles and standard disclosures. Global Reporting Initiative, Amsterdam
Global University Network for Innovation (2013) Higher education in the world 5 knowledge engagement and higher education: contributing to social change. http://www.guninetwork.org/guni.report/heiw-5-2013. Accessed 20 Feb 2014
González O, Fontaneda I, Camino MA (2014). Evoluciona la responsabilidad social en las universidades españolas? Working paper held at I International Meeting on University Social Responsibility, Cadiz
Koester BJ, Efli J, Vann J (2006) Greening of the campus: a whole-systems approach. J Clean Prod 14:769–779
Kolodinsky RW, Madden TM, Zisk DS, Henkel ET (2009) Attitudes about corporate social responsibility: business student predictors. J Bus Ethics 91:167–181
Lamsa A, Vehkapera M, Puttonen T, Personen H (2007) Effects of business education on women and men students' attitudes on corporate responsibility in society. J Bus Ethics 82:45–58
Larrán M (2014) Analysis of the implementation of sustainability policies in the Spanish university system. Conference of Spanish Public Social Councils, Madrid
Larrán M, Andrades FJ (2014) Presencia de la responsabilidad social en las titulaciones de grado de las universidades españolas. Forum for Social Councils of Andalusian Universities, Granada
Larrán M, Herrera J, Andrades FJ (2014) La responsabilidad social en la planificación estratégica de las universidades españolas: diferenciación o uniformidad. Forum for Social Councils of Andalusian Universities, Granada
Lee K, Barker M, Mouasher A (2013) Is it even espoused? An exploratory study of commitment to sustainability as evidenced in vision, mission, and graduate attribute statements in Australian universities. J Clean Prod 48:20–28
Lozano R (2006) A tool for a graphical assessment of sustainability in universities (GASU). J Clean Prod 14:963–972
Lozano R (2011) The state of sustainability reporting in universities. Int J Sustain High Educ 12:67–78
Lozano R, Lukman R, Lozano F, Huisingh D, Lambrechts W (2013) Declarations for sustainability in higher education: becoming better leaders, through addressing the university system. J Clean Prod 16:10–19
Lukman R, Glavic P (2007) What are the key elements of a sustainable university? Clean Technol Environ Policy 9:103–114

Madeira AC, Carravilla MA, Oliveira JF, Costa C (2011) A methodology for sustainability evaluation and reporting in higher education institutions. High Educ Policy 24:459–479
Md Shahbudin A, Nejati M, Amran A (2011) Sustainability-based knowledge management performance evaluation system (SKMPES): linking the higher learning institutes with the bottom billions. Afr J Bus Manage 5:9530–9540
Milutinovic′ S, Nikolic′ V (2014) Rethinking higher education for sustainable development in Serbia: an assessment of Copernicus charter principles in current higher education practices. J Clean Prod 62:107–113
Ministry of Education (2011) Social responsibility and sustainable development at Spanish universities. http://www.crue.org/Sostenibilidad/CADEP/Documents/Documentos/24.La_RSU_y_el_desarrollo_sostenible_2011.pdf. Accessed 20 Mar 2014
Muijen H (2004) Corporate social responsibility starts at university. J Bus Ethics 53:235–246
Nicolaides A (2006) The implementation of environmental management towards sustainable universities and education for sustainable development as an ethical imperative. Int J Sustain High Educ 7:414–424
Ribas J, Vilalta JM (2003) La gestió dels contractes-programa entre les universitats publiques catalanes i la Generalitat de Catalunya. Coneixement i Societat 2:86–99
Shriberg MP (2002) Sustainability in US higher education: organizational factors influencing campus environmental performance and leadership. Ph. D. dissertation, University of Michigan
Spanish Dependency Law (2007) Law for the promotion of personal autonomy and care for dependent people. http://www.boe.es/boe/dias/2006/12/15/pdfs/A44142-44156.pdf. Accessed 10 Feb 2012
Spanish Environmental Protection Law (2006) Assessment law of the environmental impact. http://www.boe.es/boe/dias/2006/04/29/pdfs/A16820-16830.pdf. Accessed 10 Feb 2012
Spanish Equality Law (2007) Organic law for the effective equality between men and women. http://www.boe.es/boe/dias/2007/03/23/pdfs/A12611-12645.pdf. Accessed 10 Feb 2012
Spanish Law on Universities (2001) 6/2001 Law on universities modified by 4/2007 Law on universities. http://www.boe.es/boe/dias/2007/04/13/pdfs/A16241-16260.pdf. Accessed 21 Nov 2012
Spanish State Board about social responsibility (2011) Social responsibility and sustainable development and education system. http://www.empleo.gob.es/es/sec_trabajo/autonomos/economia-soc/resposocempresas/docs/110503_GRUPO_DE_TRABAJO_RSE_Y_EDUCACION.pdf. Accessed 3 Nov 2012
Spanish Sustainable Economy Law (2011) Boletín Oficial del Estado. http://www.boe.es/boe/dias/2011/03/05/pdfs/BOE-A-2011-4117.pdf. Accessed 14 Oct 2012
Spanish Transparency Law (2007) http://www.transparencia.espol.edu.ec/documentos/L_acceso.pdf. Accessed 10 Feb 2012
Stephens JC, Graham AC (2010) Toward an empirical research agenda for sustainability in higher education: exploring the transition management framework. J Clean Prod 18:611–618
The Halifax Declaration (1991). www.iau-aiu.net/content/rtf/sd_dhalifax.rtf. Accessed 12 Mar 2013
United Nations (2012) Report of the united nations conference on sustainable development. http://www.uncsd2012.org/content/documents/814UNCSD%20REPORT%20final%20revs.pdf. Accessed 20 Mar 2014
United Nations Educational, Scientific and Cultural Organization (2004) United nations decade of education for sustainable development: draft international implementation scheme (IIS). United Nations Educational, Scientific and Cultural Organization, Paris
United Nations Principles for Responsible Management Education (2010) PRME principles for responsible management education. A global initiative—a global agenda. United Nations Global Compact, New York
Velazquez L, Munguia N, Sanchez M (2005) Deterring sustainability in higher education: an appraisal of the factors which influence sustainability in higher education institutions. Int J Sustain High Educ 6:383–391

Velazquez L, Munguia N, Platt A, Taddei J (2006) Sustainable university: what can be the matter? J Clean Prod 14:810–819
Vilardel I, Álvarez M (2010) Algunas reflexiones sobre la planificación estratégica de la universidad: el caso de la Universitat Autònoma de Barcelona (1998–2010). Investigaciones de Economía de la Educación 5:350–368
Waas T, Verbruggen A, Wright T (2010) University research for sustainable development: definition and characteristics explored. J Clean Prod 18:629–636

Author Biographies

Manuel Larrán is professor at department of finance and accounting in University of Cadiz. He is currently the Dean of the business school. He was the coordinator of the Strategic Plan of the University of Cadiz when he was Vice Chancellor for planning in this institution (2005–2010). He is member from the Commission of the Ministry of Education which has developed the manual on social responsibility at universities. He has published more than 80 papers in national and international prestigious journals and 12 books. Researcher in charge of various projects related to valuation of listed companies, the efficiency of SMEs and social responsibility in SMEs and universities.

Javier Andrades is an assistant professor at department of business and management in University of Cadiz. His PhD focused on analyzing the factors influencing the implementation of social responsibility and sustainability practices (strategic planning and curriculum) by Spanish universities. He is member from the research group related to the value of accounting information in the business management context as well as he is part of different research projects related to higher education for social responsibility and sustainability. He has published 7 research papers in national and international prestigious journals as well as he has published 1 book chapter and 6 books.

Enhancing Knowledge and Learning in Sustainable Developments at Universities

Lim Lan Yuan

Abstract

The green movement in Singapore is said to begin in earnest when the Government introduced the Green Mark Scheme in 2005, and a sustainability legislation in 2008 that mandates buildings to obtain a green mark certification. Professionals in the construction and real estate industry as well as students in design and environment are required to be familiar with the legislation and the requirements for the green certification. This paper discusses the use of an innovative case study and blended teaching approach in delivering topics on sustainable development at the tertiary level. This integrated approach has the advantage of injecting real life situations to students which will enhance their knowledge and learning at a measured pace suited to the students. An exploratory survey with students undertaking sustainability studies was conducted to determine the effectiveness of the approach. The results of the survey are discussed and the implications highlighted. The study confirms the findings of earlier research on the benefits of blended teaching and case studies.

Keywords

Sustainable development · Singapore · Green mark scheme · Case study · Blended teaching

---

L.L. Yuan (✉)
School of Design and Environment, National University of Singapore,
4 Architecture Drive, Singapore 117566, Singapore
e-mail: rstlimly@nus.edu.sg

W. Leal Filho et al. (eds.), *Integrative Approaches to Sustainable Development at University Level*, World Sustainability Series, DOI 10.1007/978-3-319-10690-8_9

1 Background

Since the publication of the Brundtland Report (WCED 1987) and followed by the Earth Summit in Rio de Janeiro in 1992, activities promoting sustainable developments have gathered momentum. The continuing concern with environmental issues arising from global warming and climate change, has led to urgent actions to stem the tide of biodiversity loss, climate destabilisation, resource overuse and other related concerns (Steffen et al. 2011; Orr 2009). Meanwhile in order to get people and students to appreciate the importance of sustainability issues, environmental education research has focused on developing appropriate curricula for schools, tertiary institutions and other contexts (Ardoin et al. 2013; Bartosh et al. 2006). Over the years, education for sustainability has looked at three broad areas of policy (Ryan 2010; Fien 2003), pedagogies (Segalas 2010; Scott and Gough 2003) and research (Ardoin et al. 2013; McKeown 2007; Wright 2007).

Among the construction and real estate practitioners, concerns are expressed on how sustainability and the green movement may impact on construction costs, real estate values and developments. In Singapore, a government initiative was introduced to create greater awareness, and to integrate ecology with real estate to build an environmental friendly urban development under its Green Mark Scheme. Graduates of the School of Design and Environment at the National University of Singapore need to be apprised of the developments in sustainability.

This paper highlights an approach adopted to impart knowledge and understanding of sustainability to tertiary students through a combination of case studies and blended teaching. An exploratory survey was conducted with students to determine the effectiveness of this teaching approach, and the implications and improvements that can be drawn from this study. A significant contribution of this study is the use of innovative teaching methods in delivering knowledge and materials on complicated subjects such as sustainable developments which are difficult to understand and appreciate without life examples.

2 Green Initiatives in Singapore

Singapore has always sought to balance the goals of environmental sustainability and economic growth (NEA 2008). As a small country without natural resources, the approach for Singapore is to focus on optimising the use of the environmental resources such as clean air, land, water and energy. Hence, over the years, actions have been taken to do the following:

- To maintain the good ambient air quality;
- To manage the waste to reduce the need for landfill;
- To conserve water resources; and
- To increase energy efficiency.

These are also part of Singapore's national climate change strategy. One of the potential impacts of climate change is increased energy demand. To mitigate the increase in the amount of energy used by air-conditioning, the Building and Construction Authority (BCA) and National Environment Agency (NEA) implemented measures to further improve energy efficiency in buildings. Launched in January 2005, the BCA Green Mark Scheme is Singapore's initiative to promote environmental sustainability in buildings. The scheme encourages the adoption of green building technologies and innovations to achieve better performance in the following key areas: energy efficiency; water efficiency; site/project development and management, indoor environment quality and environmental protection, and innovation.

Under the Green Mark assessment system, points are awarded for incorporating environmentally-friendly features similar to the United States' LEED and Australia's Green Star. The total number of points indicates the environmental-friendliness of the building design. In order for buildings to be certified green, they must meet minimum standards in each key area. Depending on the overall assessment, a building will be awarded one of the four ratings: Green Mark certified, Gold, Gold plus or Platinum.

In order to encourage the adoption of green building technologies and design practices so as to achieve a sustainable built environment, the Singapore Government set aside S$20 million "Green Mark Incentive Scheme (GMIS)" on 15 December 2006 to offer cash incentives to developers and building owners who made efforts to achieve at least a Green Mark Gold rating or higher.

In addition, the Government introduced legislation on environmental sustainability. The Building Control (Environmental Sustainability) Regulations 2008 which came into operation on 15 April 2008 requires buildings with at least 2,000 m^2 of gross floor area to obtain a minimum green mark score of 50 points. The legislation covers all new developments and existing buildings undergoing major retrofitting works. Major retrofitting works involve major overhauling of air-conditioning system and significant modifications to building façade along with other building services. With effect from 23 May 2008, the incentive scheme was further enhanced by providing cash awards to developers as well as architects and M&E engineers if the buildings were able to achieve at least a Gold rating. The incentive ranged from S$300,000 to $3 million for developers or building owners, and S$50,000 to $100,000 for each architect and M&E engineer. It is envisaged that legislation with green initiatives will help accelerate green building developments in Singapore.

Among the concerns in green buildings and sustainability are how these developments will impact on values and costs, and whether it will motivate developers and property and facility managers to initiate green developments. Studies have shown that green buildings have a tremendous impact on value, image and operational efficiency, and there is a need to assess the enhancement in the value and benefit of green buildings. The challenge for real estate appraisers, for instance, lies in capturing the tangible and intangible benefits of green features and translating them into quantifiable economic value for building owners. It is therefore essential for practitioners as well as tertiary students to be equipped with the necessary knowledge and tools for understanding sustainable developments.

3 Teaching of Sustainability

The development of academic programs for sustainability is inherently challenging (MacVaugh and Norton 2012; Leal Filho 2010). The understanding of sustainability requires a better appreciation of the elements in sustainable developments, and how they will impact on values and costs and the benefits that can accrue from such sustainable initiatives. Several researchers have tried to identify the body of knowledge that is critical to educating students in the environmental, social and economic impacts and outcomes of sustainable developments (Benn and Dunphy 2009; Rands 2009). In addition, it is seen as important to provide students with the skills and capability to act as sustainability change agents such as the development of communication, negotiation, critical analysis and change management skills (Stubbs and Schapper 2011) and the ability to apply these skills in the real world (Brundiers et al. 2010).

Various approaches have been suggested by different researchers. These include the development of an understanding of the sustainability problems, and the holding of an interactive workshop focusing on skills development (Stubbs and Schapper 2011); the delivery of non-technical transferable skills through reflection on the core module taught components (Cruickshank and Fenner 2012); and the application of active learning with substantial use of problem-based learning through the engagement of staff in class and their personal time (MacVaugh and Norton 2012). Different tools have been explored to enable students to better understand and appreciate the concept of sustainability. Role plays and games have been used as activities to impart knowledge on sustainability to students (Cruickshank and Fenner 2012). Dieleman and Huisingh (2006) have described the benefits of playing games as a way of learning about sustainable development. Actual case studies of real life examples have been advocated as another useful way of presenting the sustainable issues.

4 Case Study Approach

Case studies are one of the important modes of teaching focusing on learning by students (Bassey 1999; Merriam 1998). Other ways include problem-based learning (Dobson and Tomkinson 2012; Robinson 1993). For the teaching of subjects such as sustainability or quality of life, case studies provide a useful way of illustrating the subject matter (Barth and Thomas 2012; Corcoran et al. 2004; Lim 2001) although case studies have been criticised for their lack of internal and external validity and a tendency to draw conclusions with insufficient rigour (Dillon and Reid 2004; Kyburz-Graber 2004; Stevenson 2004). A case study may be defined as an empirical inquiry that investigates a contemporary phenomenon within its real-life context (Yin 2003). In case studies, actual data are collected for further exploration and provide opportunities for students to interpret what has been observed in its natural context.

Case studies of sustainable developments and green building projects will provide researchers and students with a pool of materials for understanding the complexity of measuring sustainable developments.

5 Blended Learning

In recent years, various forms of teaching have been introduced to enhance student learning. With the advancement in technology, online learning has been advocated. There are different hybrids of e-learning that has been experimented among which blended learning has shown promise to enhance learning. In simple terms, blended learning refers to the use of a combination of physical face-to-face lectures and e-lectures that are conducted using online technology. Blended learning offers institutions an opportunity to engage in using technology in conjunction with the more traditional delivery. As highlighted by Garrison and Kanuka (2004) blended learning offers transformative potential for students by supporting active and meaningful learning.

While many authors have outlined the benefits of online learning others have questioned whether technological methods are indeed solutions to enhancing learning (Purvis et al. 2011) or appropriate for enhancing the learning outcomes (Barnes 2000). Some suggested that the key to successful use of technology for learning is to implement an effective pedagogy within the learning environment (Williams 2002) and to focus on the specific design of the course and the development of tutor facilitating skills (Hughes and Daykin 2002).

Blended learning has been defined in a number of ways. Singh (2003) describes blended learning as a combination of delivery methods which complements each other and works to support student learning while Sloman (2007) argues that blended learning should not simply be considered in terms of delivery and technology. Despite the different meanings that have been used, there is general consensus that blended learning is a combination of face-to-face learning experiences such as on-campus classroom contact and online learning experiences (Poon 2012). Blended learning is an approach that consists of the integration of asynchronous Internet technology with synchronous face-face learning where the two different methods are optimally integrated in a planned pedagogically manner (Marinagi and Skourlas 2013). Blended e-learning is different from the distance model of online learning in that it combines elements such as web-based delivery and online teaching with various forms of face-to-face contact (Barnum and Paarmann 2002; Young 2002). The advancement in computer technology can now offer students more flexibility in their learning allowing them to learn at a time and place convenient to them.

Although blended teaching has been used only in recent years, research on blended learning has found that students became more excited about the learning process and became more engaged active learners rather than passive learners (Rogers et al. 2010; Wang et al. 2009), hence increasing students' understanding of course materials and improving student grades (Marinagi and Skourlas 2013). These students became the drivers in their own education, taking pleasure in guiding their own learning process (Pollara and Broussard 2011). In addition, many students opted for distance education in order to do their studies at times of their convenience. They wanted to plan their assignment tasks at a flexible time to fit into their normal routine and commitments (Soon 2011), and people can learn wherever they go (Kukulska-Hulme et al. 2007). Blended learning is particularly useful for property education as it provides a useful platform for simulated learning experience which will enhance the student learning experiences by creating opportunities for them to improve their understanding through their own exploration and research of certain issues (Poon 2012).

6 Research Design

In this research, case studies of actual projects were used to illustrate the various sustainable development issues in order to determine whether students find the materials useful in facilitating the acquisition of knowledge and understanding. These case studies were introduced as part of the curriculum on sustainability studies and how green buildings are measured and valued. Examples of such case studies include the description of the building projects, the types of green features that were introduced, the costs and savings arising from the implementation of these features, and the measurement and evaluation of these features, and their impact on their value.

The delivery of the case studies and other materials was carried out by a combination of face-to-face lectures and online technology using a blended approach. Blended learning is a widely used teaching and learning approach in built environment courses (Poon 2013). The scope of this study was confined to 32 students undertaking an undergraduate module related to sustainable development at the National University of Singapore. This experiment is the first that was carried out with the subject group of students. The face-to-face lectures normally span over a period of 12 weeks. For the purpose of this experiment, 4 of the 12 weeks were conducted using online lectures applying the university online technology known as Integrated Virtual Learning Environment (IVLE).

The investigation was carried out through a qualitative and quantitative approach. A qualitative study was first done where the students were asked in an interview to comment on the use of case studies and blended teaching, and the improvements to be made. In the quantitative method, a questionnaire was designed. The questionnaire consists of factors that would affect the blended method of teaching drawn from findings of past research on blended learning. Students were asked to complete the questionnaire. The results of the survey were analysed using basic descriptive statistics.

7 Analysis and Discussion of Findings

This is an exploratory survey carried out with 32 students undertaking sustainability studies. It consists of 15 male and 17 female students within the age group of 21–24 years.

7.1 Interviews with Students

Ten of the 32 students were randomly selected for the interview. The comments given by these students at the in-depth interviews are summarised below under the two main headings of case studies and blended teaching.

(a) Use of case studies

It is easier to understand with case studies.

It introduces realism into the studied materials. We are looking at real life examples.
It allows us to appreciate better some of the complicated issues.
It would be useful if we can have more case studies.
Add in more examples and real life case studies.

(b) Use of blended teaching

(i) Convenient and save time

E-lectures really help to save time and time can be utilised to do assignments/projects. Greatly encouraged.

E-lectures allow us to learn at our own convenience and comfortable pace.
It certainly allows us to read the materials at our convenience and when required.

(ii) Flexibility

Purely e-learning will suit my learning climate better as I can learn at my own pace and desired timing wherever I am most productive.
More e-lectures will better aid my learning as I can take down notes at my own pace—facilitate better learning environment for me.
The online lectures are convenient. We can review as and when we need to.

(iii) Delivery

Difficult to replay video to exact spot for e-learning.

To have more lively e-learning webcasts, make it as engaging as physical lectures.
Need for better audio equipments that help to improve sound and image quality.

(iv) Other comments

It would be best for students to be able to access webcast of all lessons to better enhance their understanding for examination.

It would be good to conduct e-lectures and at the same time, all lectures should be online to enable better learning at our own time and pace.
E-learning is one of the best learning experiences that I had.
I feel that a good mix of e-learning lectures and traditional lectures are good. Instead of a few lectures, it would be good to do a rotation.
It's a good approach given that the trend of higher education is towards that of online learning.
The face to face lecture gives us the opportunities to clarify issues from the online lectures.
If online lecture is the only form it would not be useful. It should be supplemented by face to face lectures or chatrooms that allow us to clarify things.

7.2 Discussion of the Interview Comments

Practically every student was of the view that case studies are useful to enhance learning and understanding. In fact, some students had asked for more case studies and real life examples to be introduced.

On the use of the blended teaching, the positive comments are that the online teaching provides convenience, and saves times particularly for those students who need to travel all the way from home for physical class lectures. E-lectures also allow them to review their materials at their own pace and for purpose of revision. As a result some students had asked for more e-lectures to be held, and a few students suggested that a rotation of the physical and e-lectures be conducted. In the current study students were only given 4 e-lectures out of 12 total lectures.

An interesting comment from students is that with increased technology, online teaching is a trend for higher education, and should be encouraged. One of the students commented that online lecture is one of the best experiences he had encountered.

There are, however, some criticisms of the blended teaching particularly on the delivery of e-lectures. They expressed concern about the availability of good audio equipments which can provide better sound and image quality. They would also like to have control on the e-lectures in terms of stopping at a particular lecture slide and the ability to download the e-lectures. These are technical issues that can be addressed. Another comment is that they felt that the physical lectures tended to be more lively than e-lectures. This occurs because lecturers are able to interact with students in face-to-face lectures much more than online lectures which tended to be a one-way communication. Hence, it is necessary to supplement e-lectures with either face–to-face lectures or chatrooms or tutorials in order to allow students to seek clarification of the subject matter. Studies (Poon 2012) have shown that blended learning cannot totally replace face-to-face contact with students as they require re-assurance and on-going support from lecturers.

7.3 Analysis of Surveys

The first section of the questionnaire survey asks the students about their learning experience with the case studies. They were asked to rate the extent of their agreement on a 7-point Likert scale (with 1 least agreed to 7 most agreed). The results are shown in Table 1.

The results show clearly the effectiveness of case studies in exemplifying concepts and principles. These results are consistent with the interview comments given by the students. They introduce realism and help them to appreciate better the subject on sustainability.

The students were next asked to comment on the use of blended teaching. A list of factors was given for the students' rating. The results are shown in Table 2.

The results show clearly that an important benefit of blended teaching is that it allows the students to learn at their own pace and convenience. These views were also strongly expressed by the students at the interviews. The approach provides them with flexibility and saves time. On the other hand, the survey reveals the possible lack of interaction between the student and instructor and the students themselves as a shortcoming of the approach. These views were also articulated by the students at the interviews. Hence, in order to implement a successful blended teaching approach, provision must be made for tutorials or chatrooms or other forms of physical interaction with the students.

The students were also asked of their learning experience from blended teaching. The results are given in Table 3.

The results show that the students benefited from the learning experience. They understand better and their learning has improved, and are more motivated and likely to learn to acquire knowledge and skill.

8 Implications

An important academic contribution of this exploratory study is the validation of the usefulness of the case study approach and the use of blended teaching as a tool to enhance student learning. It confirms the findings of earlier research on the benefits of blended learning (Poon 2012; Soon 2011; Rogers et al. 2010). This study shows that there are great benefits in using a blended approach with case studies in teaching topics such as sustainability which is easier to appreciate, and in particular for understanding the measurement of it.

Table 1 Use of case studies

Item	Mean	Std Dev
Help understand real life situations	5.88	1.289
Remember concepts better	5.84	1.322
Better illustrate principles	5.75	1.244
Simplify difficult concepts	5.58	1.343

Table 2 Use of blended teaching

Item	Mean	Std Dev
Can learn at own pace	6.41	0.911
Suit own time and convenience	6.41	0.911
Can view lectures any time	6.38	0.907
Take advantage of technological tools	5.78	1.518
Make learning easier	5.38	1.581
Serve to impart information	5.03	1.307
Facilitate student learning	5.00	1.566
Encourage transmission of knowledge	5.00	1.368
Easier to understand materials	4.97	1.581
Introduce latest knowledge	4.94	1.247
Understand better	4.78	1.621
Encourage learning	4.78	1.475
Help motivate students to learn	4.25	1.481
Encourage participation from students	3.81	1.839
Encourage interaction between student and instructor	3.72	1.836

Table 3 Learning outcome

Item	Mean	Std Dev
Understand better	5.13	1.289
Learning has improved	5.00	1.218
Motivated to learn	4.91	1.422
Acquire knowledge and skill	4.88	1.238
More likely to learn	4.88	1.385

The results of this study also show that a good mix of teaching methods are necessary which are able to suit the needs of the different learning styles and preferences of students. Some students have responded in favour of a wholly online teaching environment while others are of the view that face-to-face teaching would provide them with better physical interaction yet a group of students prefer a combination of both.

The study has highlighted the improvements that are necessary in order to ensure a successful implementation of the blended teaching approach. This includes the need to provide for other forms of interaction for the students. The delivery of the online materials using proper equipment to ensure good sound and image quality has to be considered as well as the development of an appropriate online environment.

9 Limitations of Study

One of the limitations in this exploratory study is the small sample size. Future studies of larger sample size or greater number of samples will help to further validate the results of this study.

Another shortcoming of this study is the lack of input from the instructor. Studies have shown that greater resources are required on the part of the instructor, and whether this would impose some impediment to further profileration of the blended approach. Preparing blended learning materials is time consuming and labour intensive (Benson and Anderson 2010). Courses need to be redesigned and instructors need to acquire new teaching and technological skills (Vaughan 2007).

The study was based on students' perception of the blended approach. An important way to validate these perceptions is to compare the actual academic performance of the students with their perceptions to determine whether there is consistency or correlation between the academic performance of the students and their rating of the blended teaching approach. A related issue is to examine the learning needs of students to determine whether these have an influence on the learning outcomes of students on sustainability (Mann et al. 2013).

Despite the shortcomings, this investigation has provided valuable results which will assist in further research in this area.

10 Conclusion

A better understanding of sustainable developments helps in policy decisions and implementation. With urbanisation, climatic change and the deterioration of the environment, the building up of actual cases help to track the trend and direction of the green movement. Case studies provide useful materials for both teaching and research in sustainable developments and for reflection on the issues raised. They offer meaningful insights and rich data. Given the complexity involved in teaching and learning of sustainability development, there is a need for well-documented case studies which will provide not only as useful materials but also addition to research in the field.

The study has revealed potential for the use of the blended teaching approach. However, if education for sustainable development is to be effective, it will need to deliver teaching, learning and research results that make a positive contribution to society's sustainability (Naeem and Peach 2010; Gough and Scott 2001). In addition, greater resources will be required to invest in time and to develop a blended learning environment. Governmental pressures for universities to increase participation and widen access to higher education, and advanced technological development will in future lead to an increased use of online learning in higher education (Jones and Man 2010). The use of technology and the combination with traditional onsite face to face teaching has great potential for future expansion.

References

Ardoin NM, Clark C, Kelsey E (2013) An exploration of future trends in environmental education research. Environ Educ Res 19(4):499–520

Barnes S (2000) What does electronic conferencing afford distance education? Distance Educ 21:236–247

Barnum C, Paarmann W (2002) Bringing induction to the teacher: a blended learning model. T.H.E. J 30:56–64

Barth M, Thomas I (2012) Synthesising case-study research—ready for the next step? Environ Educ Res 18(6):751–764

Bartosh O, Tudor M, Ferguson M (2006) Improving test scores through environmental education: is it possible? Appl Environ Educ Commun 5(3):161–169

Bassey M (1999) Case study research in educational settings. Open University Press, Buckingham

Benn S, Dunphy D (2009) Action research as an approach to integrating sustainability into MBA programs: an exploratory study. J Manag Educ 33(3):276–295

Benson V, Anderson D (2010) Towards a strategic approach to the introduction of blended learning: challenges faced and lessons learned. Brit J Educ Technol 41(6):129–131

Brundiers K, Wiek A, Redman CL (2010) Real-world learning opportunities in sustainability: from classroom into the real world. Int J Sustain High Educ 11(4):308–324

Corcoran PB, Walker KE, Wals AEJ (2004) Case studies, make-your-case studies, and case stories: a critique of case-study methodology in sustainability in higher education. Environ Educ Res 10(1):7–21

Cruickshank H, Fenner R (2012) Exploring key sustainable development themes through learning activities. Int J Sustain High Educ 13(3):249–262

Dieleman H, Huisingh D (2006) Games by which to learn and teach about sustainable development: exploring the relevance of games and experiential learning for sustainability. J Clean Prod 14:837–847

Dillon J, Reid A (2004) Issues in case-study methodology in investigating environmental and sustainability issues in higher education: towards a problem-based approach? Environ Educ Res 10(1):23–37

Dobson HE, Tomkinson CB (2012) Creating sustainable development change agents through problem-based learning. Int J Sustain High Educ 13(3):263–278

Fien I (2003) Towards the UN decade: looking backwards, looking forwards. Dev Educ J 9(3):3–6

Garrison DR, Kanuka H (2004) Blended learning: uncovering its transformative potential in higher education. Internet High Educ 7:95–105

Gough S, Scott W (2001) Curriculum development and sustainable development: practices, institutions and literacies. Educ Philos Theory 33(2):137–152

Hughes M, Daykin N (2002) Towards constructivism: investigating students' perceptions and learning as a result of using an online environment. Innov Educ Teach Int 39:217–224

Jones N, Man ASL (2010) Blended learning: widening participation in higher education. Innov Educ Teach Int 47(4):405–416

Kukulska-Hulme A, Traxler J, Pettit J (2007) Designed and user-generated activity in the mobile age. J Learn Des 2(1):52–65

Kyburz-Graber R (2004) Does case-study methodology lack rigour? The need for quality criteria for sound case-study research, as illustrated by a recent case in secondary and higher education. Environ Educ Res 10(1):53–65

Leal Filho W (2010) Teaching sustainable development at university level: current trends and future trends. J Baltic Sci Educ 9(4):273–284

Lim LY (2001) Quality of life case studies for university teaching in sustainable development. Int J Sustain High Educ 2(2):127–138

MacVaugh J, Norton M (2012) Introducing sustainability into business education contexts using active learning. Int J Sustain High Educ 13(1):72–87

Mann S, Harraway J, Broughton-Ansin F, Deaker L, Shephard K (2013) Seeking richer descriptions of learner's sustainability attributes and learning needs. Int J Sustain High Educ 14 (1):90–100
Marinagi C, Skourlas C (2013) Blended learning in personalised assistive learning environments. Int J Mob Blended Learn 5(2):39–59
McKeown R (2007) Setting the stage for a strategic research agenda for the UNDESD: a joint UNU-UNESCO workshop. J Educ Sustain Dev 1(1):91–96
Merriam SB (1998) Qualitative research and case study applications in education. Jossey-Bass Publishers, San Francisco
Naeem MA, Peach NW (2010) Promotion of sustainability in postgraduate education in the Asia Pacific region. Int J Sustain High Educ 12(3):280–290
National Environment Agency (NEA) (2008) Singapore's National Climate Change Strategy, Feb 2008, NEA
Orr D (2009) Down to the wire: confronting climate change collapse. Oxford University Press, Oxford
Pollara P, Broussard KK (2011) Mobile technology and student learning: What does current research reveal? Int J Mob Blended Learn 3(3):34–42
Poon J (2012) Use of blended learning to enhance the student learning experience and engagement in property education. Property Manag 30(2):129–156
Poon J (2013) An examination of a blended learning approach in the teaching of economics to property and construction students. Property Manag 31(1):39–54
Purvis AJ, Aspden LJ, Bannister PW, Helm PA (2011) Assessment strategies to support higher level learning in blended delivery. Innov Educ Teach Int 48(1):91–100
Rands G (2009) A principle-attribute matrix for environmentally sustainable management education and its application: the case for change-oriented service-learning projects. J Manag Educ 33(3):296–323
Robinson V (1993) Problem-based methodology—research for the improvement of practice. Pergamon Press, Oxford
Rogers Y, Connelly K, Hazlewood W, Tedesco L (2010) Enhancing learning: a study of how mobile devices can facilitate sense making. Pers Ubiquit Comput 14(2):111–124
Ryan A, Tilbury D, Corcoran PB, Abe O, Nomura K (2010) Sustainability in higher education in the Asia-Pacific: developments, challenges, and prospects. Int J Sustain High Educ 11 (2):106–119
Scott W, Gough S (2003) Sustainable development and learning: framing the issues. Routledge-Falmer, London
Segalas J, Ferrer-Balas D, Mulder KF (2010) What do engineering students learn in sustainability courses? The effect of the pedagogical approach. J Clean Prod 18:275–284
Singh H (2003) Building effective blended learning programs. Educ Technol 43(6):51–54
Sloman M (2007) Making sense of blended learning. Ind Commer Train 39(6):315–318
Soon L (2011) E-learning and M-learning: challenges and barriers in Distance Education Group Assignment Collaboration. Int J Mob Blended Learn 3(3):43–58
Steffen W, Persson A, Deutsch L, Zalasiewicz J, Williams M, Richardson K, Crumley C et al (2011) The Anthropocene: from global change to planetary stewardship. AMBIO: A J Human Environ 40(7):739–761
Stevenson RB (2004) Constructing knowledge of educational practices from case studies. Environ Educ Res 10(1):39–51
Stubbs W, Schapper J (2011) Two approaches to curriculum development for educating for sustainability and CSR. Int J Sustain High Educ 12(3):259–268
Vaughan N (2007) Perspectives on blended learning in higher education. Int J E-learning 6 (1):81–94
Wang M, Shen R, Novak D, Pan X (2009) The impact of mobile learning on students' learning behaviours and performance: report from a large blended classroom. Brit J Educ Technol 40 (4):673–695

WCED (1987) Our common future. World Commission on Environment and Development. Oxford University Press, Oxford
Williams C (2002) Learning on-line: a review of recent literature in a rapidly expanding field. J Further High Educ 26:263–264
Wright TS (2007) Developing research priorities with a cohort of higher education for sustainability experts. Int J Sustain High Educ 8(1):34–43
Yin RK (2003) Case study research: design and methods, 3rd edn. Sage, CA
Young JR (2002) Hybrid teaching seeks to end the divide between traditional and online instruction. Chronicle High Educ 48:A33–A34

Author Biography

Dr Lim Lan Yuan has over 40 years of varied experience in the construction and real estate industry engaged in valuation, management, consultancy and training. Currently teaching at the School of Design and Environment, National University of Singapore, he has been involved in sustainability research during the past 10 years. Dr Lim formulated a guide for valuing green buildings in Singapore. He has been a consultant to the World Bank and some UN Agencies on urban development. In addition to his doctorate, he holds a BSc(Estate Management), BSc(Economics), LLB, MBA, MSc(Construction Management) and LLM.

Student Attitudes to Sustainability and Employability Skills: Expectations from the University Curriculum

Julian Priddle, Alison J. Greig and Ella Wiles

Abstract

Students in higher education in the UK may focus on gaining good academic results but neglect skills that underpin successful job applications. This study explored students' understanding of the importance of these skills, and where they felt they were most likely to acquire them. Two data sets are presented. The first comes from an internal survey of all students covering a wide range of satisfaction and experience issues, and in one year included specific questions on employability and sustainability. Data come from a large sample of students (>3,000), but provide little context relating attitudes to experience within and outside the curriculum. To supplement these data, a separate survey was conducted that explored attitudes in more detail. Students (n = 279) at various points in their university career were asked to rate the importance of eleven skills linked to employability and sustainability. Responses indicated that the importance attached to these skills increased consistently as they moved through their course and into employment. Despite an emphasis on delivery of these skills within the university curriculum, students typically felt that they acquired such skills almost exclusively outside the curriculum. For instance, they were almost four times as likely to agree that they developed an 'Understanding of how to act as a responsible citizen' outside the curriculum than within their

J. Priddle
Anglia Learning and Teaching, Anglia Ruskin University, East Road, Cambridge CB1 1PT, UK
e-mail: Julian.Priddle@anglia.ac.uk

A.J. Greig (✉) · E. Wiles
Global Sustainability Institute, Anglia Ruskin University, East Road, Cambridge CB1 1PT, UK
e-mail: alison.greig@anglia.ac.uk

W. Leal Filho et al. (eds.), *Integrative Approaches to Sustainable Development at University Level*, World Sustainability Series, DOI 10.1007/978-3-319-10690-8_10

course. Only skills with a more obvious academic flavour, such as the 'Ability to assess and critique information sources', were seen to derive from their course. These results demonstrate a gap between the perceived skills delivery by the institution and students perceptions of where they acquired them, and imply the need for improvements in skills delivery.

Keywords
Sustainability · Employability · Education · Student attitudes · Skills acquisition

1 Introduction

In UK Higher Education (HE) emphasis is being placed increasingly on students' acquisition of a broader range of skills or attributes beyond their area of study within the taught curriculum (QAA 2013). With the increase in the number of new graduates and the widening of the graduate employment market, students who can demonstrate non-curricular skills may have an advantage in gaining employment. In response to this, universities commonly make efforts to include the provision of appropriate skills within the taught curriculum. There is already good evidence from a series of studies in UK HE that students are aware of the benefits of skills in terms of employability, although they may have different perceptions of the relative values of different skills from those of potential employers (e.g. Archer and Davidson 2008).

Similarly, there is a long-held belief amongst champions of sustainability in education that sustainability should be embedded at all levels of the taught curriculum (Sterling 2001; Jones et al. 2010). It has also been established that students in UK HE value the skills that enable them to be environmentally conscious and 'global citizens' (Drayson et al. 2013), although this may be based on poor understanding of concepts (Kagawa 2007). From this, it might be expected that students would be keen to take advantage of the skills support on offer at their university, especially where this is available as part of their programme of study, is likely to be tailored to their subject area and may be authenticated by their university. Some universities may highlight courses that include sustainable development in order to attract students ('eco-labelling': Boman and Andersson 2013).

This study examined the attitudes of students to a range of skills provision in the areas of sustainability and employability at a single UK university. Students were asked to complete a questionnaire relating to eleven skills, rating how important each skill was at different points in their career and where they thought that they were most likely to acquire the skill. The aim was to discover whether the university was being effective in providing skills training and whether students were taking advantage of this. The detailed survey built on an internal survey of first- and second-year undergraduate students, which had also included questions on employability and sustainability, and this provided a wider context for the more detailed survey.

The outcomes from the two surveys suggest that there is a mismatch between the inclusion of skills for sustainability and employability within the HE curriculum, and students' perception of where they acquire these skills. There is ample evidence from this and national studies that students see these skills as important, but they may be ignoring or failing to take advantage of a deliberate strategy by universities to support their development as employable and socially-aware citizens. The validity of the statement 'Build it and they will come' is dependent on whether the users know what they want, know where to find it, and know what to do with it.

2 Methods

2.1 The Location of the Study

Anglia Ruskin University is located in southern England, with three main campuses in the cities of Cambridge, Chelmsford and Peterborough, and subsidiary campuses and partnership operations elsewhere in the UK and abroad. Anglia Ruskin is a 'post-1992' university, meaning that it achieved higher education status as part of a national programme that allowed a large number of colleges of further education to become universities. It currently has a total student population of c. 31,000, of which c. 21,000 study on the main campuses. The core student population is typical for this type of institution, with a relatively high proportion of older students and part-time students.

2.2 The Student Surveys

University Student Experience Survey (2011–2012): The Student Experience Survey (SES) is undertaken by undergraduate students in the years before graduation. It is an internal survey, but is similar in design to the UK National Student Survey (NSS), and includes some questions that are copied from the NSS. The SES is designed to provide the university with feedback on a wide range of issues affecting student satisfaction, and some questions vary from year to year to capture particular aspects. In 2011–2012, the survey included questions relating to sustainability in addition to standard questions on employability.

Student skills perception survey (2012–2013): Based on the outcomes of the 2011–2012 Anglia Student Experience Survey, students were offered a survey of their perceptions of eleven skills that straddle employability and sustainability. These skills were developed from the competencies of a Change Agent framework (Change Agents UK 2012) and from the Department for Curriculum and Quality Enhancement Graduate and Employability Skills framework (DCQE 2012), but also have considerable overlap with the employability skills set out in the university's Employability Strategy.

Two hundred and seventy-nine students on the Cambridge and Chelmsford campuses were asked how important they thought each skill was at three points in their career:

- Currently
- On graduation
- In employment

The same students were also asked where the thought that they were most likely to acquire these skills. The choices were:

- Throughout my course
- In a particular module on my course
- In the University, outside the taught curriculum (for instance through clubs or societies)
- In life outside the University

Data were collected anonymously, mostly by asking students to complete a paper form in a public area, and were entered by hand to a spreadsheet. All of the four faculties were included, but there was no attempt to obtain samples that were representative of the university as a whole.

2.3 Limitations of the Study

The two surveys were carried out in a single HE institution, and cannot be taken as representative of the sector as a whole. It is not possible to link individual responses in the two surveys and, as will be seen later, responses to the more detailed survey cannot be analysed statistically to explore differences between groups of students.

3 Results

3.1 University Student Experience Survey (2011–2012)

The survey was taken by over 3,000 students. Responses to questions relating to employability skills indicated that a significant proportion of students did not feel that their course was providing them with appropriate skills. Agreement that their course was preparing them for employment varied from 53 to 74 % across the four faculties, with the highest level of agreement coming from the faculty with arguably the greatest level of placement teaching (healthcare and education) (Table 1). Specific skills, including team-working, communication and tackling unfamiliar problems, were also seen as being enhanced by their course by less than three-quarters of students. This is despite a clear commitment to include these in course outcomes and within students' personal development planning (PDP).

Table 1 Responses to questions in the Student Experience Survey 2011–2012, for all faculties and broken down by faculty

Question	All faculties (%)	Faculty 1 (%)	Faculty 2 (%)	Faculty 3 (%)	Faculty 4 (%)
1.7 The course is preparing me for employment	61[a]	54	74	57	53
1.5 The course is developing my team-working skills	64	58	73	59	64
7.1 My course is helping me to present myself with confidence	72	72	77	68	68
7.2 My communication skills are improved	71	70	76	67	70
7.3 As a result of my course I feel confident in tackling unfamiliar problems	67	67	72	64	63
8.6.1 My experience at Anglia Ruskin University is making me more aware of acting as a responsible citizen	51	49	52	53	50
8.6.2 My experience at Anglia Ruskin University is making me aware of using resources efficiently	65	67	65	66	63
8.6.3 My experience at Anglia Ruskin University is making me more aware of global environmental challenges	31	25	26	37	40
8.6.4 My experience at Anglia Ruskin University is making me aware of global and local social challenges	37	36	35	35	48
Number of responses	3,239	695	1,093	899	552

[a]Percentage of students who 'agreed' or 'agreed strongly' with the statement

Four questions were included in the 2011–2012 SES that relate specifically to sustainability and global citizenship. Asked whether their university experience (not just their course) made them more aware of 'acting as a responsible citizen', 'using resources efficiently', 'global environmental challenges' and 'global and local social challenges', as few as 25 % of students agreed (Table 1). Awareness of using resources efficiently prompted the greatest agreement, from about two-thirds of students irrespective of faculty.

These data indicate that a significant proportion of students did not believe that their course prepares them for employment or provides them with employability skills, and that at least half of them did not believe that their university experience increased their awareness of sustainability issues.

3.2 Student Skills Perception Survey (2012–2013)

The detailed survey of students addressed the issues covered in the SES in more detail, looking at students' perception of the importance of various skills as well as where they considered that they acquired these skills. In the first part of the survey, students were asked to indicate the importance of each of eleven skills at three points in their university career—currently, on graduation and in employment. Although the level of importance varied from skill to skill, almost all students viewed skills as being more important in employment than currently or at graduation (Table 2). In one respect, this shows that students know that these skills will make them more employable, although it also demonstrates that they do not perceive the same skills as being immediately valuable in their learning and in their everyday life.

The second part of the survey asked to students to identify where they considered they were most likely to acquire each of the same skills. Again, responses varied from skill to skill, but with one exception, students felt that they acquired skills more outside the curriculum than in it, and that skills delivery within the curriculum was seen as the most important means of acquisition in only around 40 % of cases (Table 3). Outside the curriculum includes university clubs and other social activities as well as life outside the university. The exception was the 'ability

Table 2 Percentage of students who thought the skill either 'very important' or 'extremely important' currently, on graduation and for employment, respectively

Skill	Important now (%)	Important at graduation (%)	Important when in employment (%)
Ability to assess and critique information sources	81	79	77
Effective networker and facilitator	58	75	77
Capable of engaging in self-assessment, self reflection and analysis	81	88	88
Commitment to life-long learning for yourself and others	84	85	88
Strong, engaging and dynamic communicator (written and oral skills)	85	89	92
Critical understanding of environment, social and economic connections	64	76	80
Understanding of how to act as a responsible citizen	81	89	91
System thinkers with the ability to solve problems creatively	77	85	88
Personal organisation and time management	91	90	94
Ability to work in cooperation with others	86	89	94
Ability to be innovative and adaptable	80	89	91

Table 3 Proportions of students who considered they acquired a skill predominantly throughout their course or outside the curriculum[a], respectively

Skill	Acquired throughout course (%)	Acquired outside the curriculum[a] (%)
Ability to assess and critique information sources	64	30
Effective networker and facilitator	34	61
Capable of engaging in self-assessment, self reflection and analysis	48	44
Commitment to life-long learning for yourself and others	30	65
Strong, engaging and dynamic communicator (written and oral skills)	43	52
Critical understanding of environment, social and economic connections	33	59
Understanding of how to act as a responsible citizen	18	79
System thinkers with the ability to solve problems creatively	37	57
Personal organisation and time management	42	56
Ability to work in cooperation with others	27	69
Ability to be innovative and adaptable	27	70

[a]'Outside the curriculum' combines responses from 'university clubs and other social activities' as well as 'life outside the university'

to assess and critique information sources', where students felt that they acquired the skill mainly throughout their course (64 %) or in a specific module (6 %). This skill was also unusual in that students thought that it was slightly more important currently than in employment (Table 2), suggesting that this was identified with their curriculum rather than with the 'outside world'.

In some cases, the researcher was asked to explain what the listed skills meant, indicating that the language commonly used by staff is not necessarily understood by students. The term that was poorly-understood most commonly was 'networking'.

The skills students felt are most important in employment are more likely to be considered as being gained outside the curriculum (Fig. 1, cf. Table 3). It seems apparent that those skills seen to be most important in employment (>90 % agreement) were thought to be acquired mainly outside the curriculum by more than 50 % of respondents. This is further indication that students do not perceive employability skills as part of their current curriculum.

We have examined a more detailed breakdown of the survey data, looking at the results for specific faculties and for year of study. However, the data were not amenable to statistical analysis and we cannot report any significant differences

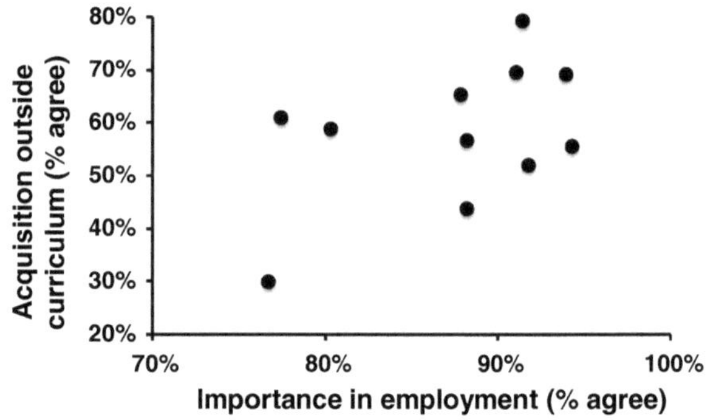

Fig. 1 The proportion of students who considered that they were most likely to acquire a skill outside the taught curriculum plotted against their perceived importance of the same skills in employment

between separate cohorts of students. Some features, however, are of note. For instance, there was no trend for students in the final year of study to ascribe greater importance to skills currently than for first-years, even though both groups would score those skills as being more important at graduation and in employment. Although students in some faculties differed slightly from their peers in other faculties in either perception of importance or where a specific skill was acquired, there was no systematic difference.

4 Discussion and Conclusion

The incorporation of sustainability and employability into the higher education curriculum can take a variety of forms, from embedding within the taught curriculum to compulsory or optional activities. In the case of sustainability, the educational package (ESD or EfS) may include a spectrum of inputs from modification of the curriculum to demonstration of a sustainable approach across the institution (Sterling 2001; Jones et al. 2010; Stewart 2010).

At the time when this study was undertaken, the first of three reports for the UK Higher Education Academy (HEA) on higher education and the sustainable had been published (Bone and Agombar 2011) and this was followed later by two further reports (Drayson et al. 2012, 2013). These were based on studies of student attitudes to a range of sustainability ‘challenges’, and the studies extended gradually through to the data in the third report (Drayson et al. 2013). The consistent message indicated that students were strongly motivated, that they wanted to learn more about sustainable development and that they expected this to be included within their taught curriculum. The report identified a preference for incorporation of sustainable development within the curriculum through reframing the curriculum, rather than through additional content or courses (Drayson et al. 2013).

Similarly, great emphasis has been placed on skills for employability. Collaboration between the Confederation of British Industry, the major UK trade and industry body, and the National Union of Students led to a report that identified clearly the skills that employers expected UK graduates to possess (CBI-NUS 2011). This was in part a response to earlier concerns within industry of the failure of graduates both to offer and understand those skills required by employers (e.g. Edwards 2005; Archer and Davidson 2008). As the graduate job market became more competitive at the start of the century, so universities set about adding 'employability skills' to their core curriculum delivery, commonly through Personal Development Planning (e.g. Yorke and Knight 2006).

This extensive body of research informs current strategic planning at UK universities, including ARU, where ideally:

- Students appreciate the needs to gain skills relating to their future roles in employment and as citizens of a society that places value on sustainability;
- Students expect their taught curriculum to provide those skills, and often to document or authenticate that provision;
- Universities adjust their taught provision to deliver those skills, and
- Universities promote an environment that is itself sustainable and reflective of the world around them.

Against that background, the survey results suggest that this ideal model is not working effectively. The institutional survey indicates that students did not believe that their university course was especially effective in preparing them for employment and was definitely failing to raise awareness of sustainability issues. The more detailed survey on importance of skills and their acquisition suggested that students had a clear appreciation of all of the skills surveyed (covering both sustainability and employability), and that this perception of importance tended to be greater as they anticipated employment. However, they typically expected to acquire these skills outside the curriculum, even where these are clearly incorporated, and emphasised, within the curriculum—for example effective communication. 'Ability to assess and critique information sources' is a core element of the taught curriculum, but even in that case 30 % of students felt that they acquired the skill outside the curriculum.

If the 'ideal model' is not working, where might it be failing? There are four potential 'points of weakness':

- Students are focussed on their subject-learning within the curriculum, so are not amenable to acquiring other skills in that context.
- Students fail to recognize skills delivery within the curriculum, perhaps either through an inability to perceive them or through poor signposting (A similar argument is often advanced for formative feedback on students' work).
- Students have no incentive to acquire skills that they perceive as only being of significant importance after they graduate.
- Delivery of skills within the curriculum is ineffective, perhaps because of lack of staff engagement or understanding of issues.

Students focus on subject learning: Biggs (1987, p. 96) characterised student learning on a continuum from 'surface' to 'deep'. Surface learning occurs where delivery is didactic and students have a passive role, whereas deep learning is student-led. In the former, there is concentration on immediate tasks, especially those that contribute to the students' qualifications. A tendency to surface learning will discourage engagement with tasks that might be considered peripheral, and the skills discussed here might fall into that category.

Students fail to recognize skills delivery: Whilst students have indicated that they prefer skills to be embedded in the existing curriculum rather than being segregated into specialist modules or the co-curriculum (Drayson et al. 2013), this carries the risk that the skills delivery may be obscured. Some universities adopt conspicuous signposting of skills in order to ensure that these are recognized and used by students (e.g. University of Hull LTSU 2010).

Students have no immediate incentive to acquire skills: The detailed survey confirms results of national studies that students are fully aware of the importance of skills, but that this importance is seen to be greatest for employment rather than in their current context. On that basis, students may feel that they can defer acquiring skills, or that their experience once they are in employment may provide a better context for acquiring skills.

Skills delivery is ineffective: Anecdotal evidence indicates that some university staff can be ambivalent about the delivery of skills that do not relate directly to the taught curriculum. This can arise from a range of causes, from a perception that they are poorly-qualified to deliver skills, through a feeling that they should concentrate on the subject curriculum, to a believe that students do not engage.

If the model is failing, does it matter? On the basis of the survey data in this study and reports from national studies (Bone and Agombar 2011; Drayson et al. 2012, 2013), it would be easy to argue that students have a clear understanding of the importance of employability and sustainability skills. So, if they either feel that they are more likely to acquire these skills outside the curriculum, or if they actively choose to do this, then this should be of less concern than if they fail to acquire them at all. After all, students are encouraged to take responsibility for their own learning, and also to achieve their personal development through a combination of curricular-, co-curricular- and extra-curricular learning. However, this raises further potential issues:

- The perception that skills are more likely to be acquired outside the curriculum may simply reflect a lack of recognition of skills provision within the curriculum, rather than an active choice.
- Skills provision within the curriculum can be incorporated within subject learning, giving a specific context that will carry through into work and society. This might be more difficult to assure outside the curriculum, or available only to a subset of students.
- Successful integration of skills delivery within the curriculum can include reflective practice, for instance through personal development planning or student award schemes. Again, this might not be available to some or all students outside the framework offered by the curriculum.

Overall, it is obvious that students are not clear where employability and sustainability skills are present within the curriculum. Currently, there may be some difficulties in delivering such skills as part of the curriculum. These include the perceived practical issues around adding to curricula which may already be full, and making students aware of the support that they receive or that is available to them. Integration of skills for employability and sustainability cannot be undertaken by simply 'bolting it on' but needs to form an integral part of curriculum design. At the same time, the integration of sustainability and employability skills needs to be made both visible and relevant to students.

4.1 Future Prospects

It would be useful to follow up this study with a further survey that aligns with national research (Drayson et al. 2013) and provides a way to explore the expectations and perceptions of students in greater detail. It would also be useful to extend the study beyond a single institution.

Within Anglia Ruskin, a developing emphasis on graduate attributes and students' reflection on these will provide a context for a follow-up study. Further survey data will enable the university to be more effective in building skills provision into the curriculum, by taking the student perspective into account.

References

Archer W, Davidson J (2008) Graduate employability: what do employers think and want? Council for Industry and Higher Education, London, 19 p. http://aces.shu.ac.uk/employability/resources/0802grademployability.pdf. Accessed 7 May 2014

Biggs J (1987) Student approaches to learning and studying. Australian Council for Educational Research, Melbourne 145 p

Boman J, Andersson UP (2013) Eco-labelling of courses and programs at University of Gothenburg. J Clean Prod 48:48–53

Bone E, Agombar J (2011) First-year attitudes towards, and skills in, sustainable development. A report for the Higher Education Academy. Higher Education Academy, York. http://www.heacademy.ac.uk/assets/documents/sustainability/firstyearattitiudes_finalreport.pdf. Accessed 7 May 2014

CBI-NUS (2011) Working towards your future: making the most of your time in higher education. http://www.nus.org.uk/Global/CBI_NUS_Employability%20report_May%202011.pdf. Accessed 7 May 2014

Change Agents UK (2012) Skills of a change agent. http://www.changeagents.org.uk/sites/default/files/downloads/Skills%20of%20a%20Change%20Agent.pdf. Accessed 7 May 2014

DCQE (2012) Curriculum 2012—briefing note 4: graduate and employability skills. Department for Curriculum and Quality Enhancement, London 2 p

Drayson R, Bone E, Agombar J (2012) Student attitudes towards and skills for sustainable development. A report for the Higher Education Academy. Higher Education Academy, York, 47 p. http://www.heacademy.ac.uk/assets/documents/esd/Student_attitudes_towards_and_skills_for_sustainable_development.pdf. Accessed 7 May 2014

Drayson R, Bone E, Agombar J, Kemp S (2013) Student attitudes towards and skills for sustainable development. A report for the Higher Education Academy HEA. Higher Education

Academy, York, 50 p. http://www.heacademy.ac.uk/assets/documents/sustainability/ESD_student_attitudes_2013_v4.pdf. Accessed 7 May 2014
Edwards G (2005) Connecting PDP to employer needs and the world of work. Higher Education Academy, York, 12 p. http://www.heacademy.ac.uk/assets/documents/resources/database/id71_connecting_pdp_to_employer_needs.pdf. Accesses 7 May 2014
Jones P, Selby D, Sterling S (eds) (2010) Sustainability education-perspectives and practice across higher education. Earthscan, London 364 p
Kagawa F (2007) Dissonance in students' perceptions of sustainable development and sustainability: implications for curriculum change. Int J Sustain High Educ 8(3):317–338
QAA (2013) UK quality code for higher education. Part B: Assuring and enhancing academic quality, Chapter B4: enabling student development and achievement. The Quality Assurance Agency for Higher Education, Gloucester, 23 p. http://www.qaa.ac.uk/Publications/InformationAndGuidance/Documents/B4.pdf. Accessed 7 May 2014
Sterling S (2001) Sustainable education—revisioning learning and change (Schumacher Briefings 6). Green Books, Dartington 96 p
Stewart M (2010) Transforming higher education: a practical plan for integrating sustainability education into the student experience. J Sustain Educ 1:13. http://www.jsedimensions.org/wordpress/wp-content/uploads/2010/05/Stewart2010.pdf. Accessed 7 May 2014
University of Hull LTSU (2010) Guidelines for identifying and badging PDP. University of Hull, Hull, 2 p. http://www2.hull.ac.uk/administration/pdf/PDP%20Badging%20Guidelines%20Jan2010.pdf. Accessed 7 May 2014
Yorke M, Knight PT (2006) Embedding employability into the curriculum. Higher Education Academy, York, 28 p. http://www.employability.ed.ac.uk/documents/Staff/HEABriefings/ESECT-3-Embedding_employability_into_curriculum.pdf. Accessed 7 May 2014

Author Biographies

Julian Priddle worked for over 20 years in environmental research, with a focus on oceanic biology and climate. He moved into science education in 2000, and took up his current role supporting learning, teaching and assessment at Anglia Ruskin University in 2012.

Alison Greig is Director of Education for Sustainability at Anglia Ruskin University. She is responsible for embedding sustainability across the curriculum and for the EfS research theme within the Global Sustainability Institute.

Ella Wiles is a researcher for the Global Sustainability Institute at Anglia Ruskin University. She has worked on farmers' perceptions of climate change. She has worked in diverse social projects in the UK and overseas, and is currently involved in urban agriculture.

The authors are grateful to students who contributed to both of the surveys on which this study is based, and to the Director of Anglia Ruskin Student Services for permission to use data from the 2011–2012 Student Experience Survey.

The Application of Ecological Footprint Analysis to Distance Learning University Contexts: A Case Study of the College of Education, University of South Africa

Maicom Sergio Brandão, Patricia C. Silva Leme, Callie Loubser, Johann Dreyer, Soul Shava and Welington Braz Carvalho Delitti

Abstract

The Ecological Footprint (EF) analysis can be considered as an indicator for assessing the environmental impact of universities, since it indicates flow of natural resources consumption and waste generation caused by university activities. This paper presents the estimation of the EF at the College of Education (CEDU) of the University of South Africa (UNISA), a distance learning university. The environmental aspects considered for the EF calculations were electricity, water and paper consumption, transportation and built-up area. The results revealed that electricity consumption itself is the major component of CEDU′s EF. This study contributes to the field of environmental

M.S. Brandão (✉)
Engineering School of Sao Carlos, University of São Paulo,
400 Avenida Trabalhador São-Carlense, São Carlos, SP 13566-590, Brazil
e-mail: maicom.brandao@usp.br

P.C. Silva Leme
Superintendence of Environmental Management, University of São Paulo, São Paulo, Brazil

C. Loubser
Department of Science and Technology Education,
University of South Africa, Pretoria, South Africa

J. Dreyer · S. Shava
University of South Africa, Pretoria, South Africa

W.B.C. Delitti
Dean of the Institute of Biosciences, University of São Paulo,
São Paulo, Brazil

W. Leal Filho et al. (eds.), *Integrative Approaches to Sustainable Development at University Level*, World Sustainability Series, DOI 10.1007/978-3-319-10690-8_11

assessment based on footprint, providing methodological procedures that encourage future research to the almost unexplored field of EF studies applied to distance education universities.

Keywords
Ecological footprint · Open distance learning university · Sustainable development

1 Introduction

The current development model of most countries has a significant effect on the environment because of the way this development prioritizes economic growth and focuses mainly on the consumption of goods and services. Over years, maintaining unsustainable consumer patterns has resulted in serious negative consequences to the environment and human beings due to natural resources having been depleted and a huge level of waste having been generated (Arrow et al. 1995).

However, there is an alternative model to this pattern of development which is focused on sustainability. The concept of sustainability considers not only the economic perspective, but also the environment, the society and its organization in a territorial space, the cultural and political issues and the relationships between all these perspectives (Sachs 1993). One of the important goals of sustainable development is to enable future generations to attain quality of life (WCED 1987).

In a historical perspective, the sustainable development has been promoted since the late 1960s and the early 1970s. During the 1970s, important discussions concerning the future of the planet commenced and many of the discussions still continue today (Turner 1987; Sachs 1993; Du Pisani 2006). As a result of this, many governments have been including sustainability issues into their planning programmes and its effectiveness has been evaluated by indicators of sustainability.

Indicators have been used to collect information from observable and measurable characteristics of issues such as health, climate, the economy, etc. Several public and private organisations use sustainability indicators to monitor and communicate their performance to stakeholders. These indicators provide simplifications of complex phenomena to researchers and managers, playing an important role in decision-making processes concerning sustainability (Hammond et al. 1995; Gallopín 1997; Meadows 1998; Segnestram 2002; Van Bellen 2005).

Literature shows the existence of several indicators of sustainability systems that have been applied in companies, local and national governments and also in universities, e.g. the Global Reporting Initiative (GRI), which provides guidelines to reporting based on indicators of sustainability (GRI 2006), the AISHE: Auditing Instrument for Sustainability in Higher Education (Roorda 2001), etc.

One of these sustainability indicators is the Ecological Footprint, which is based on the question "*how much land do people require to support themselves*?" (Chambers et al. 2001, p. 59). It means that the EF estimates the needed resources of a society or economy and express it as a bio-productive area. The EF considers not only resources consumption, but also waste generation. Therefore, the results achieved from the EF allow inference on whether the consumer patterns are commensurate to the productive capacity of nature (Wackernagel and Rees 1996; Wackernagel et al. 2007).

The EF helps universities to understand their environmental impact, since it simplifies the flow of consumption and waste generation into quantifiable units. As the EF is easily comprehendible, it can be adjusted for didactic purposes in environmental education campaigns. In addition, the EF results can be one strategy to present and inform stakeholders about the environmental performance of the universities (Wackernagel et al. 2007; Amaral et al. 2013).

Consequently, for those and other reasons, several universities around the world have been applying the EF to their contexts, for example, the University of Redlands (Venetoulis 2001) and Ohio State University (Janis 2007) in the USA; the University of Newcastle, Australia (Flint 2001); the University of Toronto (Conway et al. 2008) and Kwantlen University College (Burgess and Lai 2006), Canada; the University of Santiago of Compostela (Álvarez and Heras 2008; Rodríguez et al. 2008) and the University of León—campus Vagaroza (Hernandéz et al. 2009), Spain; the University of Vale, Colombia (González 2010); University Northeastern, China (Li et al. 2008); and the University of Sao Paulo—Sao Carlos campus, Brazil (Amaral et al. 2013). Most of those studies grouped environmental aspects in seven major environmental aspects: energy, water, transportation, waste, buildings, food and consumer goods (e.g. paper).

Furthermore, the universities mentioned above are face-to-face or "traditional" teaching universities. However, few studies have devoted their attention to the application of this methodology in distance learning universities. In this study we attempt to answer the following questions: Would it be possible to calculate the ecological footprint of this type of campus/university? Which environmental aspects should be included or replaced in EF calculations to distant learning university contexts?

Therefore, the present study aims to estimate the EF of the College of Education (CEDU) of the University of South Africa (UNISA), which is an open distance learning university. This paper expects to contribute to the understanding of EF methodology applied to a distance learning campus and whether distance learning education causes less environmental impact compared to traditional face-to-face teaching courses under the limitations of the EF method.

The study represents a relevant contribution to the field of environmental assessment based on footprints, since it provides initial information and methodological procedures that encourage future research in the Ecological Footprint of distance-education universities.

Besides, the study provides information that managers can peruse when dealing with decision making processes. Managers can develop and monitor policies to reduce the environmental impact of the CEDU based on EF data over the years. The CEDU can also use the Ecological Footprint to teach sustainable practices to the university community since the Ecological Footprint is a didactic tool to communicate the environmental impact to stakeholders, i.e. society, the university community, governments, etc. This aspect can be incorporated into environmental education and other related courses offered at undergraduate and post-graduate levels.

2 Methodology Overview

The College of Education (CEDU) at the University of South Africa (UNISA) was chosen as the subject for this case study. UNISA is the first open distance learning university (OPDL) in the world and the largest university dedicated to distance education on the African continent. It has three main campuses: the Muckleneuk campus, in Pretoria, the UNISA Science Campus in Johannesburg as well as the Sunnyside campus in Sunnyside. UNISA enrolls more than 300,000 students, most of them from South Africa and the remainder from Africa and the rest of the world (UNISA 2012). The College of Education is the third largest college at UNISA as far as the number of students is concerned (more than 60,000). The CEDU is located at the Muckleneuk campus and it has two schools and three different centers (CEDU 2012).

The present research estimates the Ecological Footprint to a distance learning university. The EF converts environmental resource consumption into $CO_{2\text{-equiv.}}$ emissions, and then, in global hectares. This is one approach defined by Wackernagel (1994) for energy-land equivalence calculations. The choice of the CEDU as object of this study was motivated by the accessibility of data to calculate the EF and the study was based on the data from 2012.

Different sources of literature were used to establish the conversion factors to estimate the $CO_{2\text{-equiv.}}$ emissions based on natural resources consumption. However, most of these factors were developed for the European contexts (e.g. Chambers et al. 2001; Rodríguez et al. 2008; Defra 2010) and this situation results in a lack of conversion factors appropriate for the South African context. Although the conversion factor to estimate the emissions of $CO_{2\text{-equiv.}}$ resulting from electricity consumption was available to South Africa background, other factors related to transportation, buildings, water and paper consumption were not available (Table 1). For the cases in which specific conversion factors were unavailable, we used factors based on previous EF studies in the literature.

The conversion factors are used to multiply the values of consumption items. For example, suppose that a university consumed 500,000 m^3 of water in one year. The equivalent emissions value is given by multiplying 500,000 m^3 times 0.37 $kgCO_2/m^3$, which is the conversion factor for pumping and treating water. Consequently, the equivalent CO_2 emissions would be 185,000 $kgCO_{2\text{-equiv}}$ or 185 $tCO_{2\text{-equiv}}$. Then, this value is divided by the carbon fixation rate of the average global forest, which is 1.8 tCO_2/ha/year (Wackernagel 1994).

Table 1 $CO_{2\text{-equiv.}}$ emissions conversion factors adopted and references

Category	Conversion factor	References	Observations
Buildings	520 $kgCO_2/m^2$	Rodríguez et al. (2008)	Considers that buildings last 50 years
Electricity	1.054 $kgCO_2$/kWh	Letete et al. (2011)	South-African electricity factor
Paper	1.84 $kgCO_2$/kg paper	Rodríguez et al. (2008)	Virgin fibre
	0.61 $kgCO_2$/kg paper	Rodríguez et al. (2008)	Recycled fibre
Water	0.37 $kgCO_2/m^3$	Chambers et al. (2001)	Pumping and treating
Transportation	0.21 $kgCO_2$/km	Defra (2010)	Taxi
	0.02 $kgCO_2$/km	Rodríguez et al. (2008)	Train
	0.04 $kgCO_2$/km	Rodríguez et al. (2008)	Bus
	0.07 $kgCO_2$/km	Rodríguez et al. (2008)	Motorbike
Transportation (car)	0.20 $kgCO_2$/km	Rodríguez et al. (2008)	Occupation (1 or 2 people)
	0.10 $kgCO_2$/km		Occupation (3 people)
	0.07 $kgCO_2$/km		Occupation (4 people)
	0.05 $kgCO_2$/km		Occupation (5 people)
Transportation of materials	0.297 $kgCO_2$/ton-mile	Environmental Protection Agency (2008)	Truck transportation

This is the approach of EF that deals with carbon emissions and this was the approach that adopted to estimate the EF of CEDU. Other land categories such as agricultural and grazing land were not included in the scope of the present study. There were not available data about food habits of the community and the available conversion factors in literature differ significantly from the African environment and the use of land.

3 Data Collection

The data for EF analysis were collected from direct and indirect sources. The built area of the CEDU, the institutional consumption of electricity, water and paper was derived directly from UNISA documents and interviews with managers. Data related to community paper consumption, the electricity consumption of students and staff members who work from home and the transportation of staff members were collected indirectly by questionnaires. Online questionnaires were sent to Environmental Education course students and staff members of the College of Education, and 115 students returned the questionnaire, which represents low response rate. However, the respondents included students from different regions of South Africa.

The data collected to EF estimations was also used to define a general profile of students and academic staff members of CEDU/UNISA. The information extracted from questionnaires can help managers understand more about the consumption habits of the CEDU community and to promote environmental initiatives to reduce the EF.

4 Methodological Approach

The allocation method was used to define the environmental impact of the CEDU. Allocations were made to estimate the institutional consumption of water, energy and paper of the College of Education. The allocation was made based on the hypothesis that each employee consumes the same portion of each environmental aspect mentioned. Thus, the total annual consumption of water, paper and energy at UNISA was divided equally by all staff members. Afterwards, the individual consumption was multiplied by the total number of the College of Education staff members. Table 2 presents detailed information about the environmental aspects considered to EF and its application to the CEDU (distance-learning context).

Table 2 also outlines the boundaries of this study and provides information that may be used to compare differences between environmental aspects of traditional and distance learning universities.

5 Results

5.1 Ecological Footprint of the College of Education (CEDU—UNISA)

This section presents the results of Ecological Footprint estimations based on data collected from CEDU/UNISA. Figure 1 presents the contribution of EF components in percentage values. The findings of the present case study revel that it is possible to estimate the EF of a distance learning university and, in this case, energy is responsible for the largest part of the EF. Energy consumption in universities is related to infrastructure characteristics. In fact, there are many facilities in buildings that use energy (e.g. lighting, heating, pumping, etc.). As CEDU has a small built area (4560.00 m^2), the institutional electricity did not represent an environmental hotspot. On the other hand, the electricity consumption by CEDU students was responsible for 93 % of the total energy EF component.

Paper consumption comprised more than 4 % of the total EF of the CEDU (Fig. 1). Although the CEDU is part of a distance education system, paper is still used by students and staff members of the CEDU. In fact, most assignments are still in hard copies at UNISA. However, that may change as soon as new online technologies are incorporated in the distance-learning education system.

Of the five analyzed components, transportation was responsible for the third largest part of the CEDU Ecological Footprint (see Fig. 1). Although the impact of the commuting students was not considered in this study, staff-member commuting

Table 2 Environmental aspects to EF of CEDU

Energy
The present study considered that the energy consumption is originated by two main sources that was defined as institutional and CEDU community. The first one is the institutional energy consumption, which refers to the electricity used by the College of Education facilities, such as computers, lighting, etc. at UNISA. The second one is the CEDU′s community consumption that represents the use of electricity outside the UNISA campus by students and staff members who work at home. The electricity consumption is limited to lighting and use of computer at home, considered as an indispensable tool for students in distance-learning universities
Transportation
The environmental aspect related to transportation category is the consumption of fossil fuels. Transportation analysis assumed that only members of the College of Education staff go to UNISA campus (by car or other means of transportation) as all students study from home. The study also considered the transportation of printed materials. The transportation is made by post office trucks, twice a year to different regions of South Africa. Estimations of environmental impact related to transportation of study materials were made considering the geographical distribution of the CEDU students in each region
Buildings
Buildings are part of what Chambers et al. (2001) classify as built land and it includes buildings, roads, etc. In this case study we considered the built area of the College of Education at the AJH Van Der Walt Building. The $CO_{2\text{-equiv.}}$ emissions related to the building construction were included in the study. However, we did not include the roads inside UNISA, because they are a shared place that does not belong exclusively to the College of Education
Paper consumption
Paper consumption is an important environmental aspect also in the context of College of Education (CEDU). There are two main sources of paper consumption: institutional and CEDU′s community. The institutional paper consumption means the internal paper consumption by the staff members of CEDU, which includes paper consumption for operational and academic activities. On the other hand, the CEDU community represents students' paper consumption that occurs outside the UNISA campus (e.g. assignments drafts)
Water
Environmental impact related to water is associated to energy consuming processes such as pumping and treating. Water consumption of the College of Education facilities was included in the study

is environmentally significant (97.6 % of transportation footprint). Transportation of students' materials through South African regions presented the smallest part of the transportation footprint (2.4 %).

The built-up area of the CEDU represents the second smallest EF of the CEDU (0.50 % or 4.36 ha/year). One of the environmental advantages of a distance university is the reduction in the built area. For example, the Ohio State University has more than 700 ha to support more than 70,000 people (Janis 2007) while the CEDU has 0.456 ha and supports almost 70,000 students. In this case, distance-learning universities represent an environmental gain.

Water was the smallest footprint of the CEDU (0.178 or 0.02 %). In universities, water is used for cleaning buildings, watering gardens and green areas, etc. Thus, water consumption is mainly related to university facilities and maintenance.

Fig. 1 CEDU′s ecological footprint components

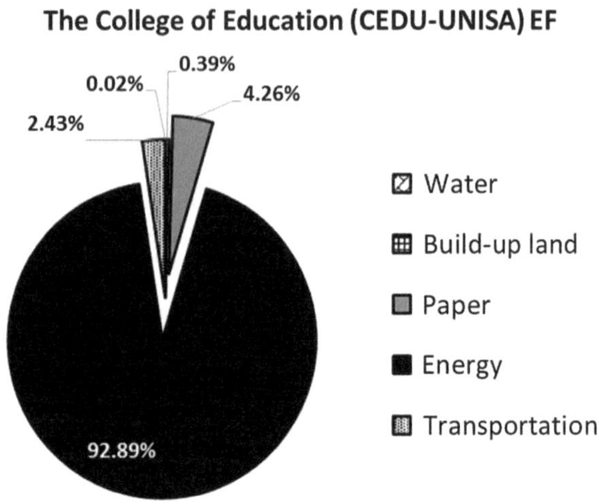

Finally, Table 3 summarizes the findings of this study in terms of CO_2-equiv and hectares per year, and reveal the contribution of each environmental category to the EF.

The total Ecological Footprint (EF) of the College of Education (CEDU) is 7140.27 ha/year. This area corresponds more than 15,600 times the CEDU built area at UNISA. The Ecological Footprint of each CEDU member is 0.103 ha/year or 1,030 m^2/year.

5.2 Students' and Staff Members' Consumption Habits

The present EF study also provided important findings about CEDU community habits. The questionnaire findings helped to draft a profile of a generic CEDU

Table 3 CEDU's ecological footprint components

Category	Item	Conversion to $tCO_{2\text{-equiv}}$	EF (ha)	% EF (%)
Buildings area	AJH Van Der Walt Building (CEDU/UNISA)	47.42	27.62	0.39
Electricity	Institutional CEDU/UNISA	838.23	465.68	6.52
	Community[1]	11,100.80	6167.12	86.37
Paper	Institutional CEDU/UNISA	237.65	132.03	1.85
	Community[a]	310.43	172.46	2.42
Water	Institutional CEDU/UNISA	2.75	1.52	0.02
Transportation	Staff members commuting	305.27	169.60	2.38
	Materials' transportation	7.63	4.24	0.06
TOTAL		12850.18	7140.27	100

[a]CEDU's students and academic staff members who work at home

student. A CEDU student spends between 1 and 15 h a week in front of the computer to do his/her academic assignments. Besides, the average paper consumption of the generic CEDU student is 1–50 sheets a month (58 %). The CEDU students use 50 % of paper from virgin fibres (i.e. primary materials), and 50 % of paper from recycled fibre (i.e. secondary materials).

On the other hand, a generic staff member of CEDU spends eight hours working at the UNISA Muckleneuk campus, and two more hours working at home each day. Most of the staff members use cars fueled by petrol as their means of transportation (93.8 %), do not share their car with anyone (60.5 %), and travel between 20 and 40 km a day (43 %) round trip to UNISA. Finally, a CEDU staff member uses between 1 and 50 sheets of paper a month (56.3 %).

6 Discussion

The present study aimed to estimate the Ecological Footprint of the College of Education- an educational unit of the University of South Africa (UNISA). In this research, methodological pathways were introduced to discuss the differences in environmental impact between face-to-face and distance learning universities in terms of EF analysis.

The findings indicated that electricity consumption was the main component of the CEDU Ecological Footprint (92.89 %). The second major component was paper (4.26 %), followed by transportation (2.43 %), built-up land (0.387 %) and water (0.021 %). The study also provided qualitative information that was used to draft a general profile of students and staff members of the CEDU.

The results demonstrate that, despite the differences between face-to-face and distance education universities, energy consumption is responsible for the largest footprint of the CEDU. Some researches indicated that electricity consumption was the main environmental aspect in the EF of face-to-face universities, e.g. the Colorado College (80 %) (Li et al. 2008), the University of Toronto (69.4 %) (Conway et al. 2008), etc.

However, in the case of UNISA, the biggest part of electricity consumption of the CEDU was originated by the students' computers at their homes. In face-to-face universities, the infrastructure, such as buildings and lighting, is mainly responsible for the electricity consumption (Rodríguez et al. 2008).

Paper consumption was the second major environmental component of the CEDU Ecological Footprint (4.26 %). Although paper is still consumed at CEDU, UNISA has been developing initiatives to reduce the consumption of this material by improving information systems and increasing the percentile of online assignments, as stated by the manager responsible for the assignments section at UNISA during the interviews.

Transportation represents the third largest footprint of the CEDU (2.43 %). In several traditional universities, the transportation category was responsible for the biggest part of the environmental impact (e.g. the University of Newcastle (43 %) (Flint 2001), the University of Redlands (Venetoulis 2001), the University of

Santiago de Compostela (Rodríguez et al. 2008), Kwantlen University College (52.96 %) (Burgess and Lai 2006), Ohio State University (72.24 %) (Janis 2007), etc.

Transportation did not represent a significant environmental impact in comparison regarding the electricity component of EF. In fact, the environmental impact of the university was reduced because students' commuting is not applicable or can be disregarded regarding distance-learning universities. Therefore, we can state that the transportation aspects of distance learning universities represent one of the major environmental gains for the CEDU/UNISA since more than 60,000 students study at home and do not need to use any means of transportation to commute to the university.

Water consumption and built-up land did not present significant results in comparison to other EF components. This fact is also observed in face-to-face universities, where water consumption does not commonly constitute a significant part of universities' footprints (Venetoulis 2001; Conway et al. 2008; Li et al. 2008; Rodríguez et al. 2008).

In this study built-up land had a smaller portion of EF in comparison to face-to-face universities. Therefore, built-up area is also another environmental positive effect of distance-learning on a campus or at a university. Furthermore, the university has also implemented indigenous plant landscaping within the campus premises.

Besides the differences in the values of the EF, there are differences in how each environmental aspect should be addressed in EF calculation on traditional and distance learning universities. Considering electricity consumption, we can state that the use of the computer at home by students should be part of EF calculation on studies based on distance learning universities since students do not commute and depend on the computer to develop their academic activities. Transportation also presents notable difference between traditional and distance learning since only the CEDU staff members commute from their home to the university and UNISA sends study materials to its students by mail. Other environmental aspects, e.g. paper and water consumption, do not differ significantly from traditional universities and the CEDU.

In addition to EF calculations, a generic profile of students and staff-members was created from data collected in this research project. The findings showed that a generic staff member of the CEDU does not differ much from a staff member of a face-to-face university. Most of the staff members of CEDU still work at the UNISA campus and use a vehicle as means of transportation. However, this ecological footprint can be reduced by increasing the number of work-at-home staff or whether staff reduce car use and adopt alternative means of transportation.

On the other hand, a generic CEDU student spends between 1 and 15 h per week in front of the computer observing lectures, and doing university assignments. In this case, the time that the CEDU student would spend in class, he/she spends in front of the computer at home. This exchange can produce positive environmental results, because the per capita Ecological Footprint of the CEDU is smaller than most per capita footprints of face-to-face universities.

However, this study has some limitations. The first limitation is related to uncertainty about data and the assumptions made during the development of the study's scope. The assumptions narrowed the students' environmental aspects to only in electricity, and paper consumption. Waste generation and food consumption are environmental aspects that can be considered in future studies.

The study assumes that all staff members consume electricity and water equally at the UNISA Muckleneuk campus. However, this assumption is reasonable because there are no significant differences between the activities of the colleges of UNISA.

Printer ink and waste production were not taken into consideration because of a lack of available data. However, this component would be important because UNISA prints, on average per year, 600 both-sided pages of study materials for distribution to each student, and also produces waste from management activities. Notably, UNISA uses environmentally friendly ink for its printing.

Another limitation is related to the procedures used to gather data. Online questionnaires are easy to apply to reach people from different geographical areas, but, in this case, they did not provide random sampling. Furthermore, the study does not represent the whole UNISA campus, just one college. However, since there is no significant difference in terms of management between the colleges of UNISA, the results of this study can be used as initial scientific base for future research in other colleges or even the whole UNISA campus.

Thus, we encourage future research to refine methodological issues and include other environmental aspects such as ink consumption, waste generation, energy consumption for heating in students' homes or other aspects that were not mentioned in the previous recommendations. This can also be extended to staff members.

Finally, this exploratory study may contribute to the development of the CEDU of educational and sustainable initiatives that focus on the reduction of its Ecological Footprint. The study also reinforces the commitment of UNISA to sustainability and the society in general.

References

Álvarez NL, Heras DB (2008) Metodología para el Cálculo de la huella ecológica en universidades. In: Congresso Nacional del Medio ambiente, Cumbre del Desarrollo Sostenible, Congresso Nacional del Medio ambiente, Cumbre del Desarrollo Sostenible, Madrid

Amaral R, Heras DB, Leme PS, Malheiros TF (2013) Aplicabilidade da pegada ecológica em contextos universitários. In: Indicadores de sustentabilidade e gestão ambiental, Manole, Brazil, pp 495–522

Arrow K, Bolin B, Costanza R, Dasgupta P, Folke C, Holling CS, Pimentel D (1995) Economic growth, carrying capacity, and the environment. Ecol Econ 15(2):91–95

Burgess B, Lai J (2006) Ecological footprint analysis and review. Kwantlen University College, Canada

CEDU (College of Education). (2012) College of education: learn without limits. UNISA, South Africa

Chambers M, Simmons C, Wackernagel M (2001) Sharing nature's interest: ecological footprint as an indicator of sustainability. Routledge, UK

Conway TM, Dalton C, Loo J, Benakoun L (2008) Developing ecological footprint scenarios on university campuses: a case study of the University of Toronto at Mississauga. Int J Sustain High Educ 9(4)

Defra (2010) 2010 Guidelines to Defra / DECC's GHG conversion factors for company reporting: methodology paper for emission factors. http://archive.defra.gov.uk/environment/business/reporting/pdf/101006-guidelines-ghg-conversion-factors-method-paper.pdf

Du Pisani JA (2006) Sustainable development—historical roots of the concept. Environ Sci 3 (2):83–96. doi:10.1080/15693430600688831

Environmental Protection Agency (2008) Climate leaders: Greenhouse gas inventory protocol core module guidance. Optional emissions from commuting, business travel and product transport. Environmental Protection Agency, US

Flint K (2001) Institutional ecological footprint analysis—a case study of the University of Newcastle, Australia. Int J Sustain High Educ 2(1):48–62. doi:10.1108/1467630110380299

Gallopín GC (1997) Indicators and their use: information for decision-making. part one-introduction. In: Sustainability indicators. A report on the project on indicators of sustainable development, Wiley, Chichester, pp 13–27

González LVA (2010) Aproximación a La Huella Ecológica de La Universidad del Valle, Campus Universitario de Melendez. (Doctoral dissertation). http://opac.univalle.edu.co/cgi-olib?oid=781028

GRI (2006) Sustainability reporting guidelines, Version 3.0. GRI, Amsterdam

Hammond A, Adriaanse A, Rodenburg E, Bryant D, Woodward R (1995) Environmental indicators: a systematic approach to measuring and reporting on environmental policy performance in the context of sustainable development. World Resources Institute

Hernandéz PA, Álvarez JM, Fernández JF, Sanz CM, González GA, Calabuig EL (2009) Huella ecológica del campus de Vegazana (Universidad de León). Seguridad Y Medio Ambiente 113:38–51

Janis JA (2007) Quantifying the ecological footprint of the Ohio State University. (Thesis). https://kb.osu.edu/dspace/bitstream/handle/1811/28365/Janis?sequence=1

Letete TM, Mungwe NW, Guma M, Marquard A (2011) Carbon footprint of the university of cape town. J Energy S Afr 22(2):1–12

Li GJ, Wang Q, Gu XW, Liu JX, Ding Y, Liang GY (2008) Application of the componential method for ecological footprint calculation of a Chinese niversity campus. Ecol Ind 8:75–78. doi:10.1016/j.ecolind.2007.01.007

Meadows D (1998) Indicators and information systems for sustainable development. The Sustainability Institute

Rodríguez RL, Iglesias JT, Álvarez NL (2008) Impacto ambiental em centros da USC. Universidade de Santiago de Compostela, Spain

Roorda N (2001) AISHE: auditing instrument for sustainability in higher education. DHO

Sachs I (1993) Estratégias de transição para o século XXI. In: Para pensar o desenvolvimento sustentável, Editora Brasiliense, Brazil, pp 29–56

Segnestram L (2002) Indicators of environment and sustainable development theories and practical experience. The International Bank For Reconstruction And Development, Washington, DC

Turner RK (1987) Sustainable global futures: common interest, interdependency, complexity and global possibilities. Futures 19(5):574–582

UNISA (University of South Africa) (2012) Annual Report 2011. UNISA, South Africa

Van Bellen HM (2005) Indicadores de sustentabilidade: Uma análise comparativa. Editora FGV, Brazil

Venetoulis J (2001) Assessing the ecological impact of a university: the ecological footprint for the university of redlands. Int J Sustain High Educ 2(2):180–197. doi:10.1108/14676370110388381

Wackernagel M (1994) Ecological footprint and appropriated carrying capacity: a tool for planning toward sustainability. (Thesis). http://elk.library.ubc.ca/handle/2429/7132
Wackernagel M, Rees W (1996) Our ecological footprint: reducing human impact in the earth. New Society Publishers, Canada
Wackernagel M, Wermer P, Goldfinger S (2007) Introduction to the ecological footprint: Underlying research question and current calculation strategy. Global Footprint Network, US
WCED (World Commission on Environment and Development) (1987) Our common future. Oxford, Oxford

Authors Biography

Maicom Sergio Brandao is an undergraduate student from the University of Sao Paulo with a part-time scholarship at North Carolina State University (USA). His major is Industrial Engineering. His main research interests are environmental management in higher education institutions focused on indicators of sustainability and food waste management.

Patricia C. Silva Leme is an environmental educator at University of São Paulo (Brazil, www.sga.usp.br). She has degree in Biology at Federal University of São Carlos, Brazil (UFSCar), Master Degree and PhD in Education at UFSCar (2008) with part-time scholarship at Universidad Autónoma de Madrid, Spain (UAM). In the last years she has been researching and publishing in the areas of training of environmental educators and address environmental education and sustainability in universities.

Callie Loubser is a professor in Science and Environmental Education at the University of South Africa. His main research interests are in the evaluation of environmental education programmes and the development of programmes for the conservation of biodiversity.

Professor Johann Dreyer joined the University of South Africa after a teaching career of 10 years, and has been involved in the training of teachers for the past 21 years. Johann read several papers at international and national conferences and has published several articles and book chapters in the field of Environmental Education.

Dr Soul Shava is an environmental education and sustainability specialist. He holds a PhD in Environmental Education from Rhodes University, South Africa. Dr Shava has research interests in indigenous knowledge representation in knowledge generation process at modern institutions and local community interfaces and its application in community development contexts and environmental education processes in southern Africa. His other current research interests include community based natural resources management (CBNRM), traditional crops and indigenous food plants and their use by local communities to achieve food security and resilience to climate change impacts. Dr Shava had published widely in his areas of research interest.

Professor Welington Braz Carvalho Delitti has degree in Biology at University of São Paulo (1977), Master in Biology at University of São Paulo (1982) and Ph.D. in Biology, University of São Paulo (1984). He is professor at Department of Ecology (Institute of Biosciences /USP) and has experience in Ecology, with emphasis on functional ecology, acting on the following topics: nutrient cycling, ecology, rain forest, air pollution and savannas. Currently, he is the Dean of the Institute of Biosciences at University of São Paulo.

Indicator-Based Analysis of the Process Towards a University in Sustainable Development: A Case Study of the University of Tübingen (Germany)

Simon Meisch, Nikolas Hagemann, Johannes Geibel, Elisabeth Gebhard and Moritz A. Drupp

Abstract

In order to contribute to sustainable development (SD), complex and heterogeneous institutions such as universities need instruments to define SD goals as well as to assess and to communicate their SD performance. We discuss the potential of sustainability indicators to provide structure and guidance for SD processes of universities, analyse the suitability of the sustainability indicator set Nachhaltigkeitscheck 2.0 (*Sustainability Check 2.0*), developed for German

S. Meisch (✉)
Junior Research Group Ethics of Science in the Research for Sustainable Development, International Centre for Ethics in the Sciences and Humanities, University of Tübingen, Wilhelmstr. 19, 72074 Tübingen, Germany
e-mail: simon.meisch@uni-tuebingen.de

N. Hagemann
Geomicrobiology, Center for Applied Geoscience, University of Tübingen, Hölderlinstraße 12, 72072 Tübingen, Germany
e-mail: nikolas.hagemann@uni-tuebingen.de

J. Geibel
Department of Economics, Free University Berlin, Garystraße 21, 14195 Berlin, Germany
e-mail: johannes.geibel@fu-berlin.de

E. Gebhard
Department of Geography, Kiel School of Sustainability, University of Kiel, Hermann-Rodewald-Straße 9, 24098 Kiel, Germany
e-mail: stu116089@mail.uni-kiel.de

M.A. Drupp
Research Group on Environmental, Resource and Ecological Economics, Department of Economics, University of Kiel, Wilhelm-Seelig-Platz 1, 24118 Kiel, Germany
e-mail: drupp@economics.uni-kiel.de

W. Leal Filho et al. (eds.), *Integrative Approaches to Sustainable Development at University Level*, World Sustainability Series, DOI 10.1007/978-3-319-10690-8_12

universities, and suggest a revised version: the *Sustainability Check 3.0*. We apply this new set to the bottom-up driven SD activities at the University of Tübingen. We understand this as a case study that is relevant and may be exemplary for possible processes of transformation in other German and possibly international full-scale universities. Finally, recommendations will be offered of how SD processes at universities can be measured and steered more effectively based on inclusive stakeholder participation.

Keywords
Sustainable universities · Sustainability indicators · Institutional change

1 Introduction

Being social laboratories and major providers of higher education and research, universities can contribute to societal transformations towards sustainable development (SD). Many universities have already recognised their responsibility for SD and taken initial steps, be it on the operational level with an environmental management system or on the level of research and teaching. In order to follow this path systematically, it is necessary to transfer SD conceptions into concrete target systems and indicators. Indicators also promote comparability and mutual learning of universities (Kopfmüller et al. 2001; Müller-Christ 2013b; Renn et al. 2007).

Although sustainability indicators (SI) play an important role in SD transformations, there has been no systematic discourse on this subject in Germany so far (Müller-Christ 2013a, b). The international discourse on assessing SD at universities has recently intensified (Ramos and Pires Moreno 2013; Lozano 2010; Mader 2012). Although many policy statements and declarations promoting SD in university contexts have been made since the 1990s (Shriberg 2002; Disterheft et al. 2013; Jenssen 2012), the development of cross-institutional assessment tools is a relatively new field (Shriberg 2002). Existing approaches can be roughly clustered in four categories according to their functions (Jenssen 2012; Ramos and Pires Moreno 2013): First, certified environmental management systems (e.g. Eco-Management and Audit Scheme (EMAS); EcoCampus) provide an environmentally-focused assessment framework and can be seen as the origin of more holistic approaches. Second, self-assessment tools and questionnaires (e.g. Alternative University Appraisal (AUA); University Leaders for Sustainability Questionnaire; Greening Campus Manual) have the capability to provide institutions with a quick overview of their sustainability performance (Abdul Razak et al. 2013; Shriberg 2002). The depth of these assessments varies from a purely environmental focus to more holistic approaches. Third, whole-system benchmarking tools cover a broad range of sustainability issues and provide a rating system (e.g. Sustainability Tracking, Assessment and Rating System (STARS); Sustainability Reporting Card (SRC)). Fourth, only a few certified integrative sustainability assessment tools, such as e.g. Auditing Instrument for Sustainability in Higher Education (AISHE), Assessing Responsibility in Sustainable

Education (ARISE), Learning in Future Environments (LiFE) currently exist (Yarime and Tanaka 2012; Boer 2013; Abdul Razak et al. 2013; Nguyen 2011).

This paper aims to contribute to this international discourse by discussing an indicator set proposed for assessing sustainable university development in Germany: the Nachhaltigkeitscheck 2.0 (*Sustainability Check 2.0*) (Müller-Christ 2011, 2013c). This set will be analysed with regard to its strengths and weaknesses and modifications will be suggested. Subsequently, the *Sustainability Check 3.0* will be applied to the University of Tübingen. We understand this as a case study that is relevant and may be exemplary for transformation processes in other German and possibly international full-scale universities. The case study follows a threefold approach: First, existing activities and facilities will be made transparent; second, goals of the university will be presented and evaluated; and third, developmental deficits and opportunities will be identified. In a concluding discussion, we show that in contrast to other universities often highlighted in the German discourse on sustainability at universities, e.g. Lüneburg, Eberswalde and Hamburg (Schneidewind 2012), the development in Tübingen is driven by *bottom-up* processes in various networks and forums (Drupp et al. 2012; Roosen-Runge et al. 2012), which also calls for a *bottom-up*, stakeholder-based development of sustainability indicators.

2 Analysing the *Sustainability Check 2.0*

2.1 The *Sustainability Check 2.0*

Müller-Christ (2011, 2013c) developed a SI set (*Sustainability Check 2.0*, cf. Fig. 1) that is meant to help universities in their attempt to make their SD commitment transparent and to evaluate their SD activities.

This SI set consists of three objectives each of which is subdivided in different fields of action. Each field of action is composed of several manifestations. The first objective asks what universities can contribute to SD with regard to research, teaching and transfer ('Universities for Sustainable Development'). The second one deals with the resources universities depend on by considering their environmental footprint (air, water, soil, etc.), energy consumption, participation structures, health management and family-friendly structures ('The Sustainable University'). In addition to these two objectives, there is a meta-perspective that asks for a university's SD management. It specifies whether and how a university (a) commits itself to sustainable development (guiding principle), (b) provides resources (time, money, attention) to this end (resources), (c) practices its sustainability management in a systematic way (management systems) and (d) communicates its commitment to sustainable development (communication). As represented in Fig. 1, each of the different manifestations are assigned to one university-internal field of action.

For an internal self-evaluation of universities, a five-point grading scale to evaluate SD performance was assigned to all fields of action (cf. Fig. 2). The lowest mark (1) indicates that no activity is present in this field of action, the highest mark (5) indicates that the university took considerable action in this field.

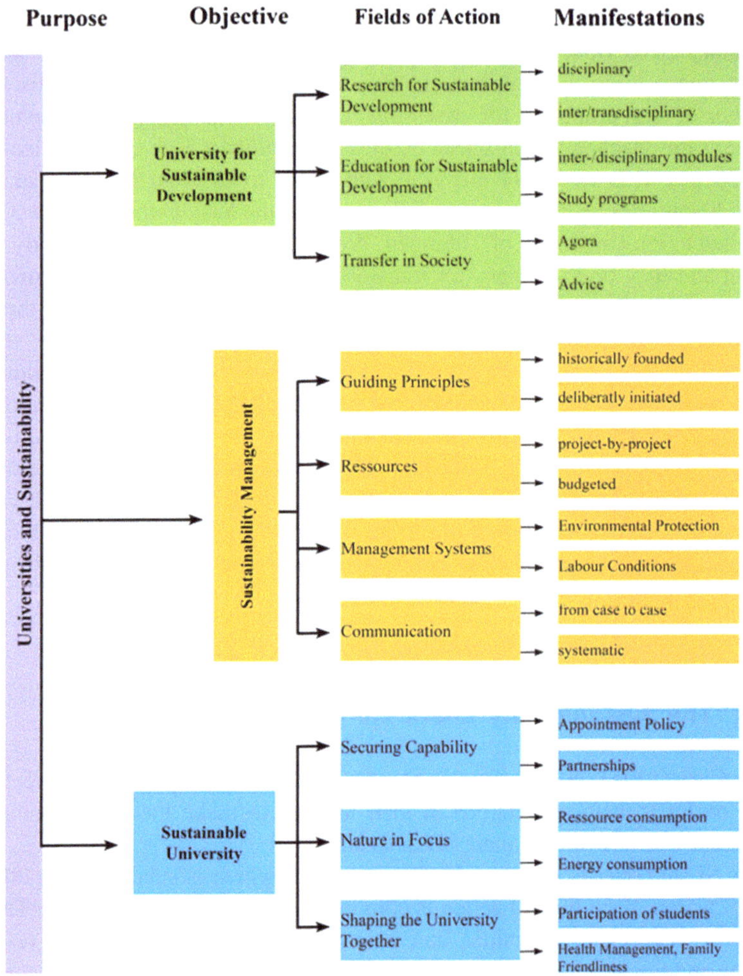

Fig. 1 Graphical representation of the sustainability Check 2.0 (own translation, source: Müller-Christ 2013c, p 65)

2.2 Evaluating the *Sustainabity Check 2.0*

We identified several weaknesses of the *Sustainability Check 2.0*. First, there are inconsistencies regarding linguistics and content between the graphical and the tabular representation of the indicator set which cannot be discussed here (cf. Hagemann and Meisch 2014). Second, we question the suitability of the five-point grading scale and therefore suggest a new scale for assessment. As the *Sustainability Check 2.0* summarises too many manifestations in some fields of action, it is not clear how an evaluation is supposed to take place at all. One would need additional preference judgements in order to evaluate whether the existence of one manifestation or the

University Internal Fields of Activities	1	2	3	4	5
Research: Disciplinary research in the field of sustainabiliy is internally supported. Inter- and transdisciplinary research in the field of sustainability is internally supported.	No support in place.	Single projects receive support.	Research projects and research institutions with internal funding are clearly visible.	Main research activities in the field of sustainability are clearly represented and consolidated.	Sustainability is the main research focus of the university.
Teaching: Courses embracing the topic of sustainability with a disciplinary or interdisciplinary approach are offered in study programs and in compulsory as well as elective modules.	Not existing.	Few courses exist.	A visible number of courses exists.	A consolidated offer of different courses with medium scope exists.	A comprehensive and permanent offer of courses exists.
Transfer: Sustainability issues are represented in life-long learning programs as well as in consulting and discussion meetings with the public.	No such offers exist.	Single offers exist.	A visible number of offers exists.	A consolidated offer with medium scope exists.	Sustainability is the leading topic of the transfer.
Guiding Principles: The guiding principle of sustainability is formulated and is accounted for in the decisions of the university.	No guiding principles exist.	A sustainability based guiding principle exists.	Research projects and research institutions with internal funding are clearly visible.	University committees direct systematically the implementation of the sustainability guiding principle.	The sustainability guiding principle is taken into account at all decision-making levels.
Ressources: Ressources are provided on a temporary or permanent basis for the implementation of sustainability.	No ressources are provided.	Ressources are occasionally released for single projects.	Ressources are regularly released for single projects.	Ressources are provided systematically for all departments and institutes.	Ressources for sustainability are permanently budgeted.
Management Systems: The university disposes a certified environmental management system (e.g. EMAS) and its social performance is certified.	No management systems exist.	Processes for the introduction of environmental and social certification systems are running.	Management systems exist.	The university sets ambigous targets within its environmental and social management systems.	After several runs successful and effective environmental protection measures and social improvements have been implemented.
Communication: A systematic sustainability reporting is in place.	There is neither systematic communication, nor reporting taking place.	Single subjects related to sustainability appear in the university´s communication.	A first seperate sustainability report is published.	Sustainability reporting is taking place on a regular basis.	Sustainability reporting is institutionalized and is conducted compliant to internal regulations.
Ensuring Performance: The university is aware of the material and immaterial ressources its overall perfomance depends on and invests time, money and attention for their regeneration.	The university only administers the governmental subsidies.	Investements in the preservation of ressources take place from case to case.	The permanent provision of single ressources is considered systematically (e.g. in appointment policy).	The university places itself as cooperative partner within its social environment.	The university operates a systematic ressource management.
Nature in Focus: Energy saving measures and measures to reduce the overall ressource consumption are implemented consistently in the operations of the university.	No measures exist.	Single projects in the field of facility management and office supplies exist.	Single projects aiming at a change of consumer behaviour of users exist (e.g. mobility management).	Energy and resource measures are clearly visible and permanantly followed up with effective outcomes.	Energy and ressource consumption are accounted for in all decision-making processes at all levels (including research).
Shaping The University Together: The campus is understood as a human living space and developed towards sustainability in a participatory way.	No participation possibilities.	Single student activities are supported. First health care and equal opportunities measures exist.	Student initiatives in the field of sustainable development are supported. Assigned contact persons for socially acceptable working conditions exist.	Student involvment is clearly visible across campus. There are permanent positions for questions concerning child care, healt care and equal opportunities.	Students and employees are an integral part of the process towards sustainability. The aim to create socially acceptale working conditions is accounted for in all campus wide processes.

Fig. 2 Overview of university-internal fields of action and their possible manifestations (own translation, source: Müller-Christ 2011, 2013c, pp 66–67)

absence of another allows giving a specific grade. For instance, how would the indicator 'Shaping the University Together' look like when we find no form of participation (grade 2) but very good child-care opportunities (grade 4)? Building an average seems counterintuitive, as the two are incommensurable. Third, some indicator fields encompass too many different indicators (e.g. 'Shaping the University Together') while others need to be complemented (e.g. 'Transfer').

With regard to a revised *Sustainability Check 3.0*, we suggest two improvements. Both changes are inspired formally and substantially by the German National Sustainability Strategy (Bundesregierung 2012).[1] First, another form of representation seems more suitable. Specific SD goals shall be defined by the respective university for each indicator field and regularly assessed. This also implies critically taking stock of the status quo at the beginning of the assessment. Second, we make a case for differentiating individual indicator fields more rigorously. It will be shown that some fields need to be complemented while others need to be split into two different fields. Within this paper, a comprehensive discussion of every indicator field is not possible. However, two examples ('Transfer' and 'Shaping the University Together') will be given in order to indicate how a revision of the indicator set would look like.

The indicator field 'Transfer' consists of the indicators 'advice', 'societal discourses' and 'lifelong learning'. It might be argued that these indicators support the idea of science and society as being separated from each other. This notion has been criticised for many reasons and a different type of university was demanded. Therefore, one could add an indicator called 'University in the Midst of Society', that is supported by social groups and whose scientific activities are oriented by concrete socio-political goals (de Haan and Meisch 2012, p. 58). One might also argue to rename the indicator field into 'Interaction with society'.

With regard to other indicator fields, such as 'Nature in the Focus' and 'Shaping the University Together', a more rigorous differentiation is suggested. Using the example of 'Shaping the University Together', it will be demonstrated how an indicator field could be extended in analysis of the indicator set of the German National Sustainability Strategy. So far, the field is described (Müller-Christ 2013c, p. 67) as follows: "The campus is understood as a humane living space and developed towards sustainability in a participatory way." Participation and humane living space can be seen as related concepts. In the current version, 'participation of students and employees', 'health management', 'family-friendliness' and 'social acceptable working conditions' are used as indicators. However, it makes sense to treat 'Participation' and a 'Humane Living Space' as separate indicator fields, because they have their own specific characteristics which cannot be merged. Broadly speaking, participation is about politics; humane living space relates to policies.

[1] We also follow the German National Sustainability Strategy linguistically by using 'indicator fields' instead of 'fields of action' and 'indicators' for 'manifestations'.

The new indicator field ‘Participation’ needs indicators for measuring the participation of students, scientific and non-scientific employees and other stakeholders, respectively. However, for a substantial participation, it is important that individual groups are involved in decision making, but also that they can initiate new policies. The new indicator field ‘Humane Living Space’ could be based on the existing field as well as extended in analysis by the indicator set of the German National Sustainability Strategy. That would result in new indicators such as ‘health and nutrition’ (quota of adult smokers, proportion of people with obesity), ‘perspectives for families’ (all-day care for children aged 0–2 and 3–5), ‘gender equality’ (salary differentials between men and women), ‘integration’ (immigrants with a degree) and ‘employment’ (increasing the labour force participation rate). Individual indicators would have to be adapted to the situation at universities. A reduction of the quota of smokers can be a measure to generally improve the health of university members. The same would be true for people with obesity. Most likely, both data sets would have to be collected by asking university members in a survey. In case this would be impossible or too difficult, participation in university sports classes by members of the university in question could be an alternative. Generally, developing and accessing this indicator can greatly benefit from experiences in university health management. Gender equality and all-day care also contribute to a humane university. Indicator development can benefit from experiences in the previous work for gender equality and equal opportunities. For universities to become humane living spaces, integration plays a role as well. The number of new students from non-academic families might be chosen as a possible indicator. This would comprise different target groups of young people (e.g. so-called working-class children). With regard to employment, the number of scientists with permanent contracts might be an indicator for a humane living space university (especially in the German context).

The present paper suggests comparing existing SI sets in order to inspire a SI set specific to universities. It does so by using the example of the German National Sustainability Strategies. However, it acknowledges that there are other possible sets too (Hagemann and Meisch 2014; Yarime and Tanaka 2012). All suggestions for modification were incorporated in a new overview (Fig. 3). By using the outlook of the German National Sustainability Strategies, the *Sustainability Check 3.0* refrains from using the five-point grading scale. Instead, we suggest that concrete goals have to be defined for each indicator by the respective university. The application of the *Sustainability Check 3.0* therefore always evaluates the existence, adequacy and pursuit of such goals.

3 Representation and Evaluation of Sustainability Activities at the University of Tübingen

The modified *Sustainability Check 3.0* is used to assess the developments at the University of Tübingen. It is not attempted to name goals for the university, because that would be the task of the university understood as polis. This paper examines

No.	Field of Action/ Indicator Field	Specification / Indicator	Aim
I. University for Sustainable Development			
1	Research for Sustainable Development		
1a		Disciplinary	
1b		Interdisciplinary	
1c		Transdisciplinary	
2	Education for Sustainable Development		
2a		Disciplinary Modules	
2b		Interdisciplinary Modules	
2c		Study Programs	
3	Transfer in Society		
3a		Agora	
3b		Advice	
3c		Lifelong Learning	
3d		Citizen University	
II. Sustainability Management			
4	Guiding Principles		
4a		Historically Founded	
4b		Deliberatly Initiated	
5	Ressources		
5a		Project-by-Project	
5b		Budgeted	
6	Management Systems		
6a		Environmental Protection	
6b		Labour Conditions	
7	Communication		
7a		From Case to Case	
7b		Systematic	
III. Sustainable University			
8	Securing Capability		
8a		Appointment Policy	
8b		Partnerships	
9	Nature in Focus		
9a		Proportion of renewable energies in end energy consumption	
9b		Proportion of electricity from renewable energy sources in electricity consumption	
9c		Increase of Settlement and Traffic Areas	
9d		Biodiversity and Landscape Quality	
9e		Intensity of Passenger Transportation	
9f		Organic Farming	
10	Participation		
10a		Participation of Students	
10b		Participation of Academics	
10c		Participation of non-academic employees	
10d		Participation of external stakeholders of the university	
10e		Influence on policy by university members	
11	University as human living space		
11a		All-Day Care for Children	
11b		Salary Differentials between men and women	
11c		Share of new students from non-academic families	
11d		Share of organic and fair-trade foods and products in canteens and cafeterias	
11e		Frequented sport offers by university members	
11f		Share of people with disabilities	

Fig. 3 Sustainability Check 3.0. (own representation, source: Müller-Christ (2011, 2013c, p 66f)

whether the university has already set adequate goals in the indicator field identified above or at least made efforts in the respective directions.

The University of Tübingen is a classical full-scale university with a broad diversity of subjects. It encompasses about 28,000 students in 280 degree courses with 450 professors. In 2008, the university started to integrate SD systematically into research, teaching and management. An initial and substantial impulse was given by the students' initiative Greening the University (SIGU) with the symposium *Greening the University—Perspectives for a Sustainable University* (SIGU 2009). The symposium can be regarded as the starting point for introducing the environmental management system EMAS as well as the course programme *Studium Oecologicum* (Drupp et al. 2012). Meanwhile, more and more members of the university's status groups support the university's transformation to SD.

Initiated by SIGU, the *Studium Oecologicum* is an interdisciplinary course programme aimed at integrating Education for Sustainable Development (ESD) into the curriculum. Organisationally, it is part of the programme on key competencies and open for students of all faculties. In 2009, the *EMAS* process was initiated at the University of Tübingen with financial support from Baden-Württemberg's Ministry of the Environment. It became validated in 2011. Measures within the scope of EMAS are developed by a group of auditors including all status groups. Environmental guidelines constitute the normative core of the university's environmental policy. At the University of Tübingen, these include a comprehensive SD conception that transcends conceptions of environmental conservation (saving energy, reducing emissions etc.) that very often inform environmental policies. It also explicitly includes research and teaching. In 2010, the university's *Advisory Board for Sustainable Development* was founded as another platform for discussion, in addition to the group of EMAS auditors. Professors, representatives of the president's office, the administration, central teaching establishments and the non-professorial teaching staff as well as students are board members. It deals with areas of the university's daily routine as well as education and research for SD. It links actors and existing individual projects and launches its own projects. In the moment, the board is about to build up facilities for a strategic SD consulting of the university's presidency.

4 Classification and Evaluation of the Activities at the University of Tübingen

4.1 Objective 'University for Sustainable Development'

The indicator field 'Research for Sustainable Development' addresses the question whether and how universities offer incentives (funding, institutional cooperations etc.) for its members to deal with SD questions and to perform inter- and transdisciplinary research. At the University of Tübingen, there are two interdisciplinary research bodies that deal with SD issues: the International Centre of Ethics in the Sciences and Humanities (IZEW) and the Geo- and Environmental Centre (GUZ).

Furthermore, there are numerous individual projects such as the Junior Research Group 'Ethics of Science in the Research for Sustainable Development'. The activities of these smaller projects are to be made visible and linked by means of the Advisory Board for Sustainable Development. The Sustainability Award rewards the best student theses that deal with SD issues. The prize singles out student research on SD but is also meant to promote research on SD via student demand. So far, there is no database that comprises SD research projects. The newly founded *Tübingen School for Sustainability* will create a structure that connects SD research projects and creates added value. In this context, an overview of relevant research projects will be compiled.

The indicator field 'Education for Sustainable Development' encompasses different formats of teaching and learning. Concerning the range of *subject-specific SD* modules, no comprehensive evaluation of about 300 different degree courses exists so far. The Studium Oecologicum offers classes on SD and ESD and can be credited in most degree courses. Within the project 'Erfolgreich Studieren in Tübingen' ('Studying Successfully in Tübingen') funded by the German Federal Ministry of Education and Research, research is done on the integration of ESD into subject-specific teaching. However, a comprehensive ESD programme or goal of the university cannot be identified.

The indicator field 'Transfer' encompasses the interaction with society and thus overcomes the university-society dualism. For the indicator 'Agora' that shows whether a regular exchange at eye level exists between science and society, different institutions and formats of events can be named (Gibbons 1999). In the university's Studium Generale, lecture series on SD topics are frequently offered, which are explicitly addressed to the public outside the university and which seek discussion with the public. Furthermore, there is the annual Sustainability Lecture given by a well-known public figure. This event is connected to the presentation of the Sustainability Award. However, there are no structures that systematically promote science-society-interactions.

For the indicator 'Advice', no established structure for the university can be recognised. Individual projects aim to advice society and politics. A comprehensive overview of the advisory activities by individual scientists concerning SD issues does not exist. The same is true for the two indicators 'Citizens' University' or 'Lifelong Learning'.

4.2 Objective 'Sustainability Management'

The University of Tübingen has a number of constitutive documents committing it to SD. In addition to the guiding principle of the university ('Leitbild für die Universität Tübingen'), there are the constitution ('Grundordnung') and the environmental guidelines ('Umweltleitlinien'). Immediate guidance for actions is offered by the environmental guidelines, which in a broad interpretation encompass research and teaching as well. Their implementation takes place in the EMAS process.

The indicator field 'Resources' mentions time and attention next to financial resources. The involvement in the Advisory Board for Sustainable Development as well as in the group of EMAS auditors is time consuming for all participants and claims a lot of attention. The university's presidency is represented in both committees linking the commitment of many employees to university politics. For the EMAS process, the university spends resources, e.g. to fund an environmental coordinator. The advisory board does not get budgetary means and has to raise money for projects. For a number of student projects, resources in the form of land or manpower were provided (the student projects 'Klimagarten' or 'Bunte Wiese'). Recently, the Tübingen School for Sustainability grants seed money for small and innovative SD projects.

In the indicator field 'Management Systems', the university successfully implemented the EMAS environmental management system, which unfortunately does not extend to the university hospitals. The latter are certified with the *audit berufundfamilie* acknowledging family friendly employers.

Through EMAS, the university is committed to systematic 'Communication': The environmental report has to be published and revised annually. This does not yet include the activities of the Advisory Council for Sustainable Development or other activities of exemplary research and teaching. In its environmental programme, the university sets the goal of reporting all activities concerning SD on a website.

4.3 Objective 'Sustainable University'

The indicator field 'Securing Capability' deals with the economic foundations of a university with regard to its financial resources as well as its appointment policy. Even if it is the university's self-interest to ensure its continued existence and to get additional funding, the mere orientation at the quota of third-party funding is questionable. There needs to be a discussion on what research is conducted with those funds, as public debates about animal experiments, the civil clause (*Zivilklausel*) and other issues indicate (Meisch 2012).

The EMAS environmental programme of the university names binding goals for most of the indicators of 'Nature in the Focus' ('energy productivity', 'climate protection', 'species diversity', 'mobility'), which, however, do not follow a comprehensive plan. With reference to 'renewable energies' and 'agricultural land use', goals are being discussed. In spite of projected investments in construction, there is no goal for 'land consumption'.

This paper further aims at differentiating the indicator field 'The University as a Humane Living Space'. It assesses the following indicators: 'all-day care for children', 'salary differentials between men and women', 'proportion of new students from non-academic families', 'proportion of organic and fair-trade food in canteens', 'participation in sports offers for university members' and 'proportion of people with disabilities'. Like every public sector entity, the University of Tübingen has representatives for equality of opportunity and for disabled persons. A systematic discourse about the indicators of gender equality, equality of opportunity and severe disability within the sustainability strategy of the University of Tübingen has not taken place so far. The student service of the university offers for a growing number of childcare facilities. Concerning the proportion of new students from non-academic families, no figures could be found, and the university does not seem to have set itself any further goals in this regard. The proportion of organic and fair-trade food in the canteens is limited and undocumented. All in all, in this indicator field, there are individual measures, but no goals. In individual areas, which are determined by law, such as gender equality, there are activities, but they are not discussed as part of a sustainability strategy.

The EMAS process created a culture of 'Participation' at the University of Tübingen. As it includes research and teaching next to management, students as well as employees also take part in the group of auditors. So far, this committee has taken up suggestions for environmental measures from within the university. This openness facilitated participation of individuals and initiatives in the EMAS process. The Advisory Board for Sustainable Development is equally open for suggestions and initiatives. Sustainability activities at the University of Tübingen are characterised by a particular commitment of students. With many diverse projects, they seize the opportunity of actively shaping their environment at university—be it through a self-governed garden on the campus of the university or the setup of their own educational concepts. However, it has to be noted that the mentioned student projects involve less than one percent of the student body directly.

5 Discussion and Conclusion

The paper aimed to promote the discourse about universities in SD. We presented an existing SI set for universities—*Sustainability Check 2.0*—and suggested improvements regarding its consistencies and content, thus leading to a modified version: the *Sustainability Check 3.0*. We applied this set to the SD activities of the University of Tübingen, a case study that is relevant and may be exemplary for possible processes of transformation in other German and possibly international full-scale universities.

In the synopsis of the developments at the University of Tübingen, it became obvious that these activities cannot be described as a coherent *top-down* process. Rather, it is a diffuse procedure, partly structured by the EMAS process, which is mostly driven in a *bottom-up* fashion, i.e. by individual actors and their affiliations (Drupp et al. 2012). This process—or more precisely, these processes—is an obvious contrast to universities which were presented as exemplary e.g. in Schneidewind (2012). Due to its transformation into a foundation under public law in 2003, Leuphana University Lüneburg has extended possibilities for independent actions, which allowed for a reset-like new orientation towards sustainable development. The same applies to the Eberswalde University for Sustainable Development which also has the status of a flagship university with regard to SD in Germany. It looks back at an almost 200-year-long tradition in forestry. In its present structure as Eberswalde University of Applied Science, it was only founded in 1992. Furthermore, both universities only cover a limited curriculum. Their manageable size and their emergent, comparatively flexible structures are good prerequisites for a top-down processes of change such as the two universities have experienced in the past few years in a very successful and coherent way. In individual cases, e.g. for smaller universities, this model may be conferrable. Furthermore, it is beyond question that Lüneburg and Eberswalde have achieved considerable innovations in the fields of research and teaching for sustainable development.

However, classical full-scale universities such as Heidelberg, Cologne, Munich or Berlin with considerably more than 25,000 students can hardly be successfully reorganised according to this model. In their case, the 'ship will have to be reconstructed on open sea' (Gropper and Meisch 2012). Therefore, strategies for the promotion of sustainable universities have to strengthen the bottom-up factor and rely on many small corrections instead of radical reorganisation, and furthermore involve students and employees as early and as broadly as possible. In light of this, any external top-down sustainability indicator set—such as the presented and modified *Sustainability Check 3.0*—may not serve its full purpose, as the described bottom-up processes should also lead to discussions on the guiding principles for SI development. This leaves important and fruitful room for further research.

References

Abdul Razak D, Sanusi ZA, Jegatesen G, Khelghat-Doost H (2013) Alternative University Appraisal (AUA): reconstructing universities' ranking and rating toward a sustainable future. In: Caeiro S, Leal Filho W, Jabbour C, Azeiteiro UM (eds) Sustainability assessment tools in higher education institutions, Springer International Publishing, p 432

Boer P (2013) Assessing sustainability and social responsibility in higher education. Assessment frameworks explained. In: Caeiro S, Leal Filho W, Jabbour C, Azeiteiro UM (eds) Sustainability assessment tools in higher education institutions, Springer International Publishing, p 432

Bundesregierung (2012) Nationale Nachhaltigkeitsstrategie. Fortschrittsbericht 2012

De Haan G, Meisch S (2012) Bildung für nachhaltige Entwicklung. Bildung für eine zukunftsfähige Volluniversität. In: Greening the University e.V. Tübingen (eds) Wissenschaft für nachhaltige Entwicklung!, Metropolis, p 447

Disterheft A, Caeiro S, Azeiteiro UM, Leal Filho W (2013) Sustainability science and education for sustainable development in universities: a way for transition. In: Caeiro S, Leal Filho W, Jabbour C, Azeiteiro UM (eds) Sustainability assessment tools in higher education institutions, Springer International Publishing, p 432

Drupp MA, Esguerra A, Keul L, Loew Beer D, Meisch S, Roosen-Runge F (2012) Student initiatives for universities in sustainable development. In: Leal Filho W (ed) Sustainable development at universities, New Horizons, Peter Lang Scientific Publishers, p 994

Gibbons M (1999) Science's new social contract with society. Nature 402(6761 Suppl):C81-4

Gropper S, Meisch S (2012) Das Schiff auf offener See umbauen. Nachhaltigkeit Lernen an Hochschulen. In: Tübingen V (eds) Greening the University e Wissenschaft für nachhaltige Entwicklung!, Metropolis, p 447

Hagemann N, Meisch S (2014) Kriterien für nachhaltige Hochschulen—am Beispiel der Universität Tübingen. In: Tremmel J (ed) Generationengerechte und nachhaltige Bildungspolitik, Springer International Publishing, p 325

Jenssen S (2012) Sustainability at universities. An explorative research on assessment methods and tools for sustainability implementation at universities. Master Theses at Maastricht University, International Centre for Integrated Assessment and Sustainable Development (ICIS), p 118

Kopfmüller J, Brandl V, Jörissen J, Paetau M, Banse G, Coenen R, Grunwald A (2001) Nachhaltige Entwicklung integrativ betrachtet—Konstitutive Elemente, Regeln, Indikatoren edn. Sigma, Berlin, p 432

Lozano R (2010) The state of sustainability reporting in universities. Int J Sustain High Educ 12(1):67–78

Mader C (2012) How to assess transformative performance towards sustainable development in higher education institutions. J Educ Sustain Dev 6(1):79–89

Meisch S (2012) Verantwortung für den Frieden: Welche Fragen stellen sich bei der Umsetzung von Zivilklauseln? In: Nielebock T, Meisch S, Harms V (eds) Zivilklauseln in Forschung. Lehre und Studium, Hochschulen zum Frieden verpflichtet, Nomos Verlag Baden-Baden, p 344

Müller-Christ G (2011) Nachhaltigkeit in der Hochschule: Ein Konzept für die interne Selbstüberprüfung". In: Deutsche UNESCO-Kommission e.V. (ed) Hochschulen für eine Nachhaltige Entwicklung. Nachhaltigkeit in Forschung, Lehre und Betrieb, Deutsche UNESCO-Kommission, p 73. http://www.bne-portal.de/index.php?id=1806&no_cache=1. Accessed 04 June 2014

Müller-Christ G (2013a) Einleitung: Die Latte hoch liegen lassen oder das Recht auf Ungeduld erhalten? In: Deutsche UNESCO-Kommission e.V. (ed) Hochschulen für eine Nachhaltige Entwicklung. Ideen für eine Institutionalisierung und Implementierung, Deutsche UNESCO-Kommission, p 67. http://www.unesco.de/7949.html. Accessed 04 June 2014

Müller-Christ G (2013b) Landkarte einer nachhaltigen Entwicklung von Hochschulen. In: Baumert S, Schlütter K, Stoppe S, Zlotowski M (eds) Nachhaltigkeit. Auf den Spuren eines Begriffs und seiner Bedeutung im universitären Kontext, Lit Verlag, p 132

Müller-Christ G (2013c) Nachhaltigkeitscheck 2.0. In: Deutsche UNESCO-Kommission e.V. (ed) Hochschulen für eine Nachhaltige Entwicklung. Ideen für eine Institutionalisierung und Implementierung, Deutsche UNESCO-Kommission, p 67. http://www.unesco.de/7949.html. Accessed 04 June 2014

Nguyen T (2011) Nachhaltigkeitsbewertung von Hochschulen. In: Deutsche UNESCO-Kommission e.V. (ed) Hochschulen für eine Nachhaltige Entwicklung. Nachhaltigkeit in Forschung, Lehre und Betrieb, Deutsche UNESCO-Kommission, p 73. http://www.bne-portal.de/index.php?id=1806&no_cache=1. Accessed 04 June 2014

Ramos T, Pires Moreno S (2013) Sustainability assessment: the role of indicators. In: Caeiro S, Leal Filho W, Jabbour C, Azeiteiro UM (eds) Sustainability assessment tools in higher education institutions, Springer International Publishing, p 432

Renn O, Deuschle J, Jäger A, Weimer-Jehle W (2007) Leitbild Nachhaltigkeit. Eine normativ-funktionale Konzeption und ihre Umsetzung, VS Verlag für Sozialwissenschaften, Wiesbaden, p 222

Roosen-Runge F, Drupp MA, Keul L, Esguerra A (2012) Bottom-up Engagement für Nachhaltige Entwicklung: Erfahrungen, Chancen und Grenzen. In: Greening the University e.V. Tübingen (eds) Wissenschaft für nachhaltige Entwicklung!, Metropolis, p 447

Schneidewind U (2012) Nachhaltige Wissenschaft auf Länderebene: Perspektiven eines unterschätzten Reformprogramms. In: Greening the University e.V. Tübingen (eds) Wissenschaft für nachhaltige Entwicklung!, Metropolis, p 447

Shriberg M (2002) Institutional assessment tools for sustainability in higher education. Int J Sustain High Educ 3(3):254–270

[SIGU] StudierendenInitiative Greening the University Tübingen e.V. (2009) Greening the University: Perspektiven für eine nachhaltige Hochschule. Oekom Verlag, München, p 160

Yarime M, Tanaka Y (2012) The issues and methodologies in sustainability assessment tools for higher education institutions. a review of recent trends and future challenges. J Educ Sustain Dev 6(1):63–77

Authors Biography

Dr. Simon Meisch is heading the Junior Research Group "Ethics of Science in the Research for Sustainable Development" at the International Centre for Ethics in the Sciences and Humanities (IZEW, University of Tübingen). After studying Political Science and Modern German Literature in Tübingen and Edinburgh (Scotland, UK), he did his Ph.D. in Political Science. From 2009-2012, he was researcher in different ethical projects at the IZEW ("Value Isobars—The Landscape and Isobars of European Values in Relation to Science and New Technology" (EU-Projekt in FWP7) and "Taking responsibility"). Currently, he is member of the Advisory Board for Sustainable Development of the University of Tübingen. In 2009, he was chairman of the student initiative Greening the University e.V.. His fields of interest are Ethics in the Sciences and Humanities (with a special focus on water, water research and water governance), concepts of sustainable development and education for sustainable development.

Nikolas Hagemann studied geo-ecology and ecosystem management at the Universities of Tübingen, Hohenheim and Stellenbosch. He is a PhD Candidate at Department of Geosciences at University of Tübingen working on biochar and a fellow of the Rosa Luxemburg Foundation. He has co-led the student initiative Greening the University e.V and is a member of the expert group "Science for Sustainability" of State Ministry for Science, Research and Arts Baden-Württemberg, Germany.

Johannes Geibel studied International Economics (B.Sc.) at the Universities of Tübingen and Cape Town, South Africa. He is currently doing his master's in Public Economics at the Free University Berlin. From its beginning, he has been part of the student's initiative Greening the University and become actively involved as a chairman. In 2012, he joined netzwerk n e.V., a network of student initiatives aiming at promoting and supporting a student-driven transformation of universities towards sustainable development. Since 2013, he acts as its chairman representing the network on conferences and political talks.

Elisabeth Gebhard graduated in international economics and political science at the Universities of Tübingen and Lyon III. She is currently enrolled at the Kiel School for Sustainability for her Master´s in Sustainability Economics and Environmental Management. She has co-led the student initiative Greening the University e.V. and worked in several research projects related to participation processes in institutions and urban development.

Moritz A. Drupp studied economics and political science at the Universities of Tübingen, Hagen and Tufts and obtained an MSc in Environmental Economics from the London School of Economics. He is a PhD candidate at the University of Kiel and conducts research on various issues related to sustainability economics. Amongst others, he has published books and papers relating to sustainability science and sustainability indicators. Previously, he was a member of the advisory board for sustainable development of the University of Tübingen, advisor to the Student Council for environmental issues, and has co-led the student initiative Greening the University e.V.

Carbon Stored in a Sustainable University Building: Bringing Education to Practice

Bruno V. Kobiski, Eloy F. Casagrande and Gabriel Pendleton

Abstract

Nowadays, humanity's presence on Earth brings alarming statistics regarding the environmental impacts associated with its needs. This research aimed to conduct a survey of CO_2 emissions of materials that were used in the construction of the Green Office of UTFPR. This was accomplished by analyzing the partial energy life cycle as well as the amount of carbon stored in these materials in order to subtract CO_2 emissions from these values. Obtained values were compared to two forest types. One forest type was an Araucaria forest and other was a Pinus taeda plantation. Six materials were analyzed: Plastic-wood used on the deck and also on the pergola; a wood framing structure formed by panels of Oriented Strand Board, Pinus taeda studs and "I" beams also in Pinus taeda, windows of Eucalyptus urograndis; and a staircase made of Erisma ucinatum wood. The data was adjusted to allow the equivalence of the real carbon stored in vegetation types area. The results showed that the Green Office can equate to 555.85 m^2 of Araucaria forest and 706.16 m^2 of a plantation of Pinus taeda. Difficulties including the lack of data and lack of processes inputs and outputs controlling

B.V. Kobiski (✉) · E.F. Casagrande
Federal Technological University of Paraná, Silva Jardim Avenue, 807, Curitiba, Brazil
e-mail: Kobiski@gmail.com

E.F. Casagrande
e-mail: eloy.casagrande@gmail.com

G. Pendleton
Rochester Institute of Technology, 1 Lomb Memorial Drive, Rochester, NY 14623, USA
e-mail: bkpk_6@hotmail.com

W. Leal Filho et al. (eds.), *Integrative Approaches to Sustainable Development at University Level*, World Sustainability Series, DOI 10.1007/978-3-319-10690-8_13

that can give support to such data were present. This research concluded that the real carbon stock can be admitted as sustainability criteria for decision making in civil construction as environmental consideration increases.

Keywords
Sustainability · Real carbon stock · Sustainability in civil construction · Carbon balance

1 Introduction

Since the dawn of civilization humans have sought shelter in built environments. Technological advances in construction did not cause a detectable environmental impact at a global level because of low population density and a small number of people on the planet required a small amount of raw materials from nature.

From the nineteenth century on, with the advancement of medicine and technology in general, populations started to grow very fast and are still growing today. Thus, the huge demand for materials has made construction inputs to be extracted intensely causing latent environmental impacts, bringing out concerns about scar city of these inputs and also creating several consequences for community health (Kibert and Guy 1997).

Sustainability is a current paradigm and has great importance in construction where there is still a conservative building style. However, there has been much thought about alternatives to help minimize the amount or increase the variety of materials used in the buildings, causing economic advantages in a highly competitive sector and reducing major impacts on raw materials extraction (Fabricio 2002).

A lot of energy is used in the manufacturing of construction materials. Great importance should be given to emissions of greenhouse gases (GHGs) that are associated with these materials and their activities in the country (Tavares 2006; Stachera and Casagrande 2007). Much less energy is required to bring down, cut, and transport wood related to the most used materials in construction such as cement, mortar, plaster, red ceramic, and steel (Stein 1989). Therefore, in a context in which sustainability parameters are considered, alternatives that enhance the use of wood in the construction industry has a great incentive.

Forests are large pools of carbon since photosynthesis in plants allows for retention of carbon dioxide and releasing of oxygen (Renner 2004). Forests store carbon and can store it for many decades. This process is contrary to the current dilemma with GHGs and with this in mind a study about a wood frame construction as a carbon sink was made. The aim of this article was to understand how and if a wood frame construction can act as a forest in terms of carbon store.

2 The Green Office

The UTFPR's Green Office is located at 807 Silva Jardim Avenue in the city of Curitiba, Paraná, and has the approval of the Campus Curitiba direction to develop policies and practices aimed at sustainability. This initiative was conceived by TEMA Group—Technology and Environment, which is a collaboration of students, teachers and researchers. The mission of this building is to implement a series of programs to reduce the environmental impacts caused by academic activity and also propose solutions to environmental problems through scientific research (Escritorio Verde Online 2011).

Several programs are currently in development to increase the environmental efficiency of Campus Curitiba as a result of the effort to achieve the objectives prevised in "Pact 21". This pact consists of an official document rectified by UTFPR and other higher education institutions in Paraná in order to develop actions according with the principles of Agenda 211. Actions in development are:

- **CAZA**—Carbon Zero in the Academy—Principles of greenbuilding
- **REZTO**—Zero Waste: Technological and Organic
- **TRECO**—Treating Electronic and Computer Waste
- **BUYING GREEN**—making purchasing decisions that help to create a sustainable university
- **GREEN LABEL**—green certification for technology, processes, and materials
- **GREEN CONSULTANCY**—service offer by the UTFPR professionals and the Students Company to companies that want to become sustainable
- **PRIZE "GREEN CITIZENSHIP UTFPR"**—*a prize given every year by the UTFPR to a person or group that helps to achieve sustainability*

3 Students Company

Since the beginning of its construction, the Students Company encouraged the direct participation of graduate and undergraduate students which has since resulted in the formation of the first interdisciplinary Students Company of the university called "Econsultoria". Around twelve students have learned about all the building processes and now, under supervision of a teacher, they are able to do consult people who may have an interest in green building.

4 The Green Office Headquarter

The Green Office project of UTFPR aims not only to implement an environmental management program for the campus, but also to promote the integration of environmental education, practical activities on campus, and visibility of sustainable projects that can be replicated. One example would be its headquarters built with the principles of sustainable construction:

- Bioclimatic guidelines applied in the architectural design
- Wood-frame System—double-walled structure in OSB (Oriented Strand Board) panels and treated pine
- Thermal insulation acoustic blankets of recycled PET and recycled tires
- Windows in wooden frames and double glazing
- Green Roof—use of local vegetation in two modules of the building
- Natural lighting through skylight intelligent systems and LED lights
- Use of a thermodynamic solar heating system for water
- Solar energy generation using photovoltaic panels to supply up to 80 % of energy consumed (the building can also charge an electric car)
- *Rainwater collection system* for toilets and cleaning
- Use of raised floor made of recycled polypropylene covered by modular *carpet* of approximately 45 % *recycled* content
- Mezzanine floor of certified wood (FSC—Forest Stewardship Council)
- Outside deck made of wood plastic (made of sawdust and recycled plastic)
- Bamboo furniture for the coffee break area

The study of the building as a carbon sink was made to complement all the other environmental qualities of the UTFPR's Green Office. This building is illustrated by Fig. 1.

Fig. 1 UTFPR's Green Office. *Source* Escritório Verde Online 2011

Table 1 Embodied energy for different wood materials

Material type	Embodied energy ($MJ.m^{-3}$)
Wood—dried in the oven	2.100
Wood—dried outdoors	300
Wood—laminated Timber	4.875
Wood—medium density fiberboard	5.850

Source Tavares (2006)

5 Methodology

The study of the Green Office as a carbon sink was created based on a comparison with a phytophysiognomy (*Pinus taeda* settlement) and a vegetal formation (*Araucaria* Forest). The following four steps were followed. The first step consisted of the GHG emissions inventory of the wood materials used in the Green Office building based on the methodology proposed by Tavares (2006). Then the amount of carbon stored in these materials was quantified and a relationship between them was made in order to subtract emissions values from the amount of carbon stored. The third step consisted of evaluating the carbon stock on the two types of vegetation. The final step was to make a comparison between them. It was then possible to quantify how much area of forest the Green Office would be equal to in terms of carbon stock.

6 CO_2 Emissions Inventory

The Tavares (2006) methodology for calculating the amount of CO_2 emissions was followed in this work. This methodology is based on the embodied energy and the type of energy used to process the analyzed material. When it comes to wood, Tavares (2006) lists the embodied energy in the Table 1

The analyzed materials were: plastic wood used on the deck and also the pergola which has a composition of 35 % low density polyethylene, 50 and 15 % binders; a wood frame structure formed by panels of Oriented Strand Board—OSB, amounts of *Pinus taeda* and beams also in *Pinus taeda*; *Eucalyptus urograndis* windows, and a wooden staircase made of *Erisma ucinatum* (cedar).

Each material was classified following the information from Table 1 along with the volume of the material utilized and the emission factor (derived from the same methodology). As a result of combined emission factors from different uses of energy, it was possible to reach the CO_2 emissions in the materials. This information is shown in Table 2.

Table 2 Variables considered for CO_2 emissions calculation

Material	Quantity (m^3)	Embodied energy (MJ)	Emission factor (tCO_2.MJ)
Oriented strand board	10.5	51187.5	0.000069
I Beams	5.5	1,650	
Studs	10.1	3,030	
Plastic Wood	1.2	5,850	
Windows	6.9	14.,490	
Staircase	0.75	3656.25	

Source *Author*

7 Carbon Stored in the Green Office

Stored carbon is defined as the quantity of carbon present in a volume of material mass. The collected data had values in cubic meters. In this case it was necessary to idealize an equation that represents the amount of carbon. As the volume was given in m^3, density values in $kg.m^{-3}$ and values of carbon content in the wood in $tC.t^{-1}$, the carbon stored was given by the following equation:

$$tC = \sum Ri\,.\,Vi\,.\,di$$

where:

tC Tons of carbon stored
Ri Carbon content in the wood for each species
Vi Volume used in construction for each material
di Density of the material
i Type of material

In order to establish a relation with CO_2 emissions, the amount of carbon stored was transformed into "CO_2 removed from the atmosphere", by using the stoichiometric reaction of photosynthesis:

$$\mathrm{C} + \mathrm{O_2} \rightarrow \mathrm{CO_2}$$

Thus, the values of CO_2 emissions could be subtracted from the values of CO_2 removed from the atmosphere and for this value the name given was "real carbon stored".

8 Carbon Stored in the Phytophysiognomy and the Vegetal Formation

The definition of forest carbon stock per hectare of vegetation type constitutes a major part of this work because through this it was possible to compare vegetation with the Green Office.

The work of Watzlawick (2003) was chosen as a reference point for carbon stock estimation in *Pinus taeda* formation in the municipality of General Carneiro, in Paraná state, discriminating different parts of trees (leaves, live branches, dead branches, stem bark, stem wood and roots). In his research, the roots were collected from a depth of 0.5 m and tree density in that area was 300 trees per hectare. An average carbon stock value in different ages of *Pinus taeda*, with 14, 19, 21, 22, 23 and 32 years old was taken.

The same methodology was implemented in the work of Watzlawick (2003) in order to provide information about carbon stock in an Araucaria forest, with density of 590 trees per hectare.

9 Data Comparing

At this stage of the study, a comparison was made between the amount of real carbon stored in the materials analyzed and selected vegetation types which correspond to the *Pinus taeda* formation and the Araucaria forest, both situated in General Carneiro, PR.

The mass ratio between carbon dioxide and carbon is approximately 3.6. The amounts of carbon stocks were multiplied by this value so that the corresponding CO_2 removed from the atmosphere by woody material could be obtained.

After conversion, the amount of CO_2 removed from the atmosphere was compared with CO_2 emissions that were calculated through the methodology of analysis of Tavares (2006) which considers the energy life cycle of materials. At this stage the CO_2 balance was defined by the following equation:

$$CO_2\ Balance = \sum\nolimits_i (A_i - B_i)$$

where:

A Amount of CO_2 absorbed from the atmosphere by the material type i (tCO_2)
B Amount of CO_2 emitted in the manufacturing of the material type i (tCO_2) process
i type of material

For the comparison with the vegetation types it was necessary to convert the balance of CO_2 in tonnes of carbon stored by dividing the value obtained for 3.6. After this the real carbon stored value was obtained.

Finally, the amount of real carbon stored was related to the vegetation types. It was then possible to define a corresponding value in hectares of vegetation type, as shown in the following equation.

$$Va_i = \frac{C_r}{A_i}$$

where:

Va_i equivalent area for i vegetation type (ha);
Ai Amount of carbon stored in i vegetation type ($kgC.ha^{-1}$);
Cr Actual quantity of carbon stored in UTFPR's Green Office;
i typology of the forest system;

10 Carbon Stored in UTFPR's Green Office

Carbon stock was calculated based on carbon content in each vegetal species that was verified as present. *Eucalyptus urograndis, Erisma ucinatum, Pinus taeda* types of wood were found in the Green Office materials, the carbon content was verified in scientific papers and were chosen as 506.6 $gC.kg^{-1}$ (Balbinot et al. 2007), 475 $gC.kg^{-1}$ (Neves 2000) and 457 $gC.kg^{-1}$(Sanquetta 2012), respectively.

Along with the density of the materials and the volume utilized, the carbon stored in the UTFPRs Green Office and the amount of CO_2 removed from the atmosphere was calculated. Results are shown in Table 3.

11 CO_2 Emissions Inventory

The same materials were analyzed in the CO_2 emissions inventory in order to calculate the CO_2 balance in the UTFPR's Green Office. Those emissions and also the carbon dioxide balance are shown in the Table 4.

Table 3 Carbon stock and CO_2 removed from atmosphere in the analyzed materials

Material	Carbon stored (tC)	CO_2 removed (tCO_2)
Oriented strand board	2.887	10.39
I Beams	1.502	5.41
Studs	2.73	9.83
Plastic wood	0.39	1.41
Windows	1.82	6.55
Staircase	0.203	0.73
Total	9.52	32.34

Source Author

Table 4 Carbon dioxide balance

Material	CO_2 removed (tCO_2)	Emissions (tCO_2)	CO_2 balance (tCO_2)
Oriented strand board	10.39	3.53	6.86
I Beams	5.41	0.11	5.30
Studs	9.83	0.21	9.62
Plastic wood	1.41	0.40	1.01
Windows	6.55	1.00	5.55
Staircase	0.73	0.25	0.48
Total	32.34	5.51	26.83

Source Author

Table 5 Real carbon store

Material	CO_2 balance (tCO_2)	Real carbon stored (tC)
Oriented strand board	6.86	1.91
I Beams	5.30	1.47
Studs	9.62	2.67
Plastic wood	1.01	0.28
Windows	5.55	1.54
Staircase	0.48	0.13
Total	26.83	7.45

Source Author

The real carbon stored was calculated by using the stoichiometric relation between carbon and carbon dioxide, as shown in Table 5.

12 Carbon Stock in Vegetation Types and Data Comparison

According to Watzlawick (2003), the values for carbon stock in the *Pinus taeda* formation and in the Araucaria Forest are 134.03 $tC.ha^{-1}$ and 105.5 $tC.ha^{-1}$. These values consider the carbon present above ground, below ground, and also the leaf litter.

The comparison of the vegetation types with the real carbon stored in UTFPR's Green Office is shown in the Fig. 2.

This figure considers the amount of carbon stored in one hectare of each vegetation type but just 150 m^2 of the Green Office, which corresponds to its built area. This reveals that the Green Office real carbon stock (7.45 tC) can equate to 0.050 tC/m^2, while the vegetation types equate to 0.013 tC and 0.011 tC/m^2 for Araucaria Forest and *Pinus taeda* formation, respectively. Thus, the UTFPR's Green Office can store 5 times more carbon than the vegetation types analyzed.

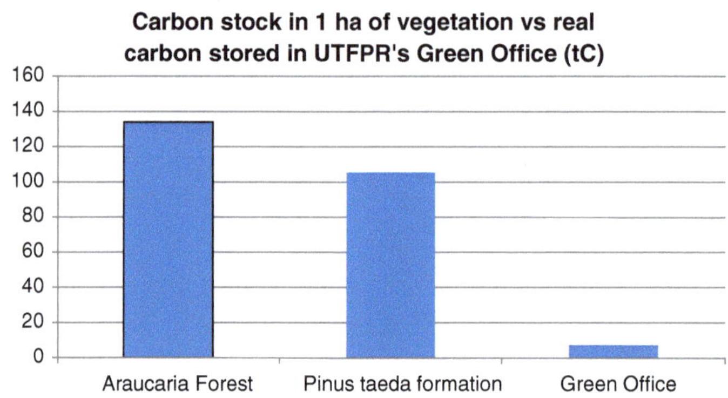

Fig. 2 Carbon stock comparison. *Source* Author

Table 6 Relation between green office building and the vegetation types

Vegetation type	Carbon stock ($tC.ha^{-1}$)	Equivalent vegetation área of the Green Office's real carbono stock (m^2)
Araucaria forest	134.3	555.85
Pinus taeda formation	105.5	706.16

Source Author

On the other hand, when an area comparison was made it was discovered that the UTFPR's Green Office can have great importance on carbon stock, as shown in the Table 6.

13 Conclusions

The results show that a wood frame building can be very significant when it comes to carbon store. With 150 m^2 of built area, the Green Office can store the same amount of carbon of 555.85 m^2 of Araucaria Forest, and 706.16 m^2 of *Pinus taeda* formation.

Although it is a significant value, it is important to notice that the carbon dioxide emissions can significantly modify the carbon stock of a building. Thus, when this type of calculation is made, it is also important to measure the emissions associated with the materials.

This article emphasize that the comparison with the forest types occurs only in terms of carbon sink, being despised the infinite variables and benefits brought by vegetal areas like preservation of biodiversity, mitigation of climate factors and others.

Lack of data is still a problem in GHG emission estimation, and it can significantly impact on the results. Hence, to provide good quality data, it is important to add more registry control to the processes. Also the embodied energy research shall be amplified to give more precise data and contribute positively for the carbon balance issues.

Finally, in the future, this research can work as a base to clean development mechanism projects in civil construction, once that it deals with carbon stock. The entrepreneurs, who want to construct buildings and houses with sustainable techniques, can also use this data in order to make decisions on what type of construction method shall be used.

References

Balbinot R, Valerio AF, Sanquetta CR, Caldeira MVW, Silvestre R (2007) Estoque de Carbono em Plantações de Pinus spp. em Diferentes Idades no Sul do Estado do Paraná. Revista Floresta 38(2):317–324

Escritorio Verde. www.escritorioverdeonline.com.br. Accessed 26 Mar 2014

Fabricio MM (2002) Engenharia simultânea no projeto de edifícios. 317 f. Tese (Doutorado em Engenharia Civil e Urbanismo)—Universidade de São Paulo, SãoPaulo

Kibert CJ, Guy B (1997) Developing sustainable communities and buildings: planning, design, and construction. Unpublished Course Material for BCN 6585 Principles of Sustainable Development and Construction, Center for Construction and Environment, University of Florida

Neves JCL (2000) Produção e partição de biomassa, aspectos nutricionais e hídricos em plantios clonais de Eucalipto na região litorânea do Espírito Santo. Tese (Doutorado) Rio de Janeiro

Renner RM (2004) Sequestro de Carbono e a Viabilização de novos reflorestamentos no Brasil. 132 f. Dissertação (Mestrado em Ciências Florestais)—Setor de Ciências Agrárias, Universidade Federal do Paraná, Curitiba

Stachera T, Casagrande E (2007) Avaliação das emissões de CO_2 na construção civil: um estudo de caso da habitação de interesse social no Paraná. IX ENGEMA—Encontro Nacional sobre Gestão e Meio Ambiente, Curitiba

Tavares SF (2006) Metodologia de análise do ciclo de vida energético de edificações residenciais brasileiras. Tese (Doutorado)—Universidade Federal de Santa Catarina, Programa de Pós-Graduação em Engenharia Civil: Florianópolis, p 225

Watzlawick LC (2003) Estimativa de Biomassa e Carbono em Floresta Ombrófila Mista e Plantações Florestais a partir de Dados de Imagens do Satélite IKONOS II. Universidade Federal do Paraná. Curitiba, Tese (Doutorado)

Author Biographies

Bruno Victor Kobiski is a master degree student at the Federal Technological University of Paraná and volunteer of the Green Office project for 3 years in which he participated in several projects dealing with sustainable construction and Green House Gases emission. He is also graduated in Environmental Chemistry by the Federal Technological University of Paraná.

Eloy F. Casagrande Jr has obtained his PhD in Mineral Resources Engineering and Environment, from the Universitty of Nottignham, UK, 1996. He did his Post Doctoral in Technological

Innovation and Sustainability, at the Technical University of Lisbon, 2007. He holds a Bacharelate in Industrial Design from the Catholic University of Paraná, Brazil, 1983. He is a teacher of the Post-graduation Program in Technology in the Technological University of Paraná—UTFPR (master and doctor degrees), in the sustainable developent area, in Curitiba, Brazil, since 1997 till present

Gabriel Pendleton is a graduate student of Environmental Science at the Rochester Institute of Technology—RIT. Rochester, NY, United States of America.

Developing and Managing Integrated [Interdisciplinary/Transdisciplinary] Graduate Programs in Environmental Science and Management in a Collaborative Context

Michal Bardecki

Abstract

This paper draws on the lessons from 15 years of experience at Ryerson University with graduate programs (master's and doctoral) in Environmental Applied Science and Management (EnSciMan). The program incorporates 80 faculty members from 21 departments and schools in each of the university's six Faculties. The paper has three principal objectives. First, to outline the conceptual framework for the environment programs which, in contrast to traditional, highly specialized "I-shaped" curriculum developed within the traditional academic structure, follows a "T-shaped" curriculum: providing problem-solving and research depth in one area while incorporating overall breadth in perspective and skills. Second, since the programs operate independently outside, but concurrent with, existing academic departments and schools, to discuss the challenges in developing cooperation and collaboration for integrated (interdisciplinary/transdisciplinary) graduate programs within a collaborative context. Finally, in this context, to describe the key metrics used to evaluate the degree of program success.

Keywords

Graduate education · Environmental applied science · Environmental management · T-shaped curriculum · Canada

M. Bardecki (✉)
Environmental Applied Science and Management, Ryerson University,
350 Victoria Street, Toronto, ON M5B 2K3, Canada
e-mail: bardecki@ryerson.ca

W. Leal Filho et al. (eds.), *Integrative Approaches to Sustainable Development at University Level*, World Sustainability Series, DOI 10.1007/978-3-319-10690-8_14

1 Introduction

A remarkable assortment of neologisms has been proposed to embody the nature of the relationships developed when practitioners of different disciplines join toward common purpose: multidisciplinary, metadisciplinarity, postdisciplinary, supradisciplinarity, extradisciplinarity, transdisciplinarity, de-disciplinary, intradisciplinarity, paradisciplinary, juxtidisciplinarity, antidiscplinary, pluridisciplinarity, among others. Even with those terms regularly used (multidisciplinary, interdisciplinary and transdisciplinary), there remains considerable inconsistency in use, including conflicting meanings regularly used in Europe and North America—for example, the usage of "interdisciplinarity" in North America largely subsumes the ideas associated with "transdisciplinarity" in Europe (Fischer et al. 2011; Huutoniemi et al. 2010). To avoid the rhetoric associated with these terms, in this paper alternate terms are employed. Research or education undertakings that involve bringing together practitioners from a range of disciplines with each operating within the domain of his or her subject with little in the way of integration are termed "collaborative." "Integration" is used to describe a systematic or holistic assimilation of the contributions of disciplines, which may involve the re-organization and construction of knowledge beyond that of the traditions of academic disciplinary structure.

Given that the total environment consists of and is impacted by various interconnected frameworks (physical, biological, social, legal, political, economic, and cultural), environmental education is widely seen as requiring such approaches (Foster 1999; McNeill et al. 2001; Hiwasaki and Arico 2007; Fortuin and Bush 2010). The UNESCO-sponsored Tbilisi Intergovernmental Conference on Environmental Education in 1977 recognized this: "Environmental education is the result of a reorientation and rearticulation of the various disciplines and of various educational experiments (natural sciences, social sciences, arts and letters, etc.) providing an integrated perception of the environment and fostering more rational environmental action replying appropriately to social needs" (Tbilisi Intergovernmental Conference on Environmental Education 1978).

The curricula of the M.A.Sc. and Ph.D. programs in Environmental Applied Science and Management at Ryerson University embody one approach to connecting disciplinary frameworks in environmental education: the T-shaped model. The details of the curriculum structure and content are not addressed in this paper. Rather the focus is on the challenges in developing cooperation and collaboration for the program and to describe the key metrics used to evaluate the degree of program success.

2 The T-Shaped Curriculum

In response to disciplinary traditions in academia, demands for accreditation, and the developing need for specialist training for the workplace, universities generally have offered curricula organized within traditional academic structures. Students acquire and graduate with highly specialized and deeply developed "I-shaped" expertise. Such skills have been generally incentivized and their development sought by individuals and rewarded by organizations. However, in the 1990s an alternative model emerged, first within the field of computer technology, recognizing the value of those with "T-shaped" skills in the workplace (Guest 1991; Heinemann 2009). Figuratively, the horizontal crossbar represents an ability to apply knowledge across disciplines and an understanding of fields outside one's principal area of expertise, as well as complementary skills of communication, institutional knowledge, and the ability to solve problems collaboratively (Fig. 1).

Organizations as diverse as the World Bank and IBM have adopted the idea (Donofrio et al. 2009; Hansen 2010; World Bank 2012), and recent papers have argued the case that the goal of shaping a T-shaped professional should guide program development and delivery in a wide variety of discipline areas at universities (e.g., Bitner and Brown 2008; McIntosh and Taylor 2013). From the point of view of the university, a T-shaped curriculum would provide problem-solving and research depth in one area while incorporating overall breadth in the understanding of a range of other fields and developing complementary skills seen as valuable to students in the development of their careers. There are manifest benefits to students as employers often seek those who can perform as "environmental integrators" (i.e., managing and coordinating projects, working in multidisciplinary teams and networking effectively) (Thomas 1992; Giacomelli et al. 2003; Vincent and Focht 2009). In addition, students possessing skills as both "specialists" and "generalists" may be better able to adapt to the inevitable fluctuations in the job market (Krozer 2005). In many ways, the concept of a T-shaped curriculum is echoed by the call by Klein (2008) for a "new quadrangulation of disciplinary depth, multidisciplinary breadth, interdisciplinary integration, and transdisciplinary competencies" (p. 406).

Many undergraduate interdisciplinary programs are focused on general education and/or allow for a high degree of individualization. Others provide their own curricular base, offering general courses without prior exposure to disciplinary-based knowledge—with questions often raised about their rigour (Newell 1992; Soule and Press 1998; Holley 2009). There are risks that students may be exposed

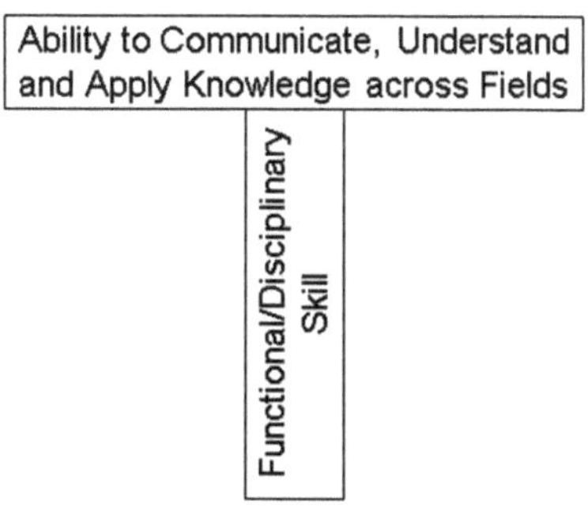

Fig. 1 T-Shaped skills

to little more than "drive-thru learning" in several disciplines. That is one reason why many of the most successful interdisciplinary programs are found at the graduate level. It may require considerable time to develop a level of understanding in a discipline; it is much more effective to build a T-shaped curriculum by building on a strong foundation of specialized knowledge. For students seeking professional development, approaches such as these are often particularly attractive since they offer the promise of bridging to workforce relevance.

3 Environmental Applied Science and Management (EnSciMan)

The smallest of Toronto's three universities (all public), Ryerson University brands itself as an "urban university with a mission to serve societal need and a long-standing commitment to engaging its community." It identifies itself as a leader in career-focused education and in innovation and entrepreneurship. Ryerson was founded in 1948 as Ryerson Institute of Technology and evolved over time to become the sole Polytechnic in the province. Coincident with its transformation into Ryerson University, the first graduate programs were launched in 2000. Currently, the university offers more than 100 undergraduate and graduate programs with an enrollment of 27,300 undergraduate and 2,300 graduate students. In addition, the Raymond Chang School of Continuing Education, with circa 70,000 enrollments annually, is Canada's largest and most successful continuing education program. Entrance to undergraduate programs is highly competitive: almost ten applications are received for each of the 6,500 first-year undergraduate positions.

In 2000 students enrolled in the university's first independent graduate program, the 2 years Masters of Applied Science (MASc) program in Environmental Applied Science and Management (EnSciMan). The program was established through the combined efforts of faculty members from schools and departments across the university who taught in the field and were involved in on-going environmental research. The success of the MASc program, in publishing research and successful student graduation, resulted in the approval in 2008 of a Ph.D. program. The initial cohort of doctoral students enrolled in September 2009 with the first convocating in October 2013.

EnSciMan operates as a collaborative effort among its faulty in a horizontally-organized university-wide form outside the traditional academic departments and schools. With the exception of one cross-appointed faculty member, all faculty were hired by disciplinary departments and schools without direct input from EnSciMan; they continue to fulfill teaching, research and administrative responsibilities within their home departments and schools. Although when the program was established there were few options, today virtually all faculty members are involved in Master's and PhD programs within their disciplines.

Administratively, the Director of EnSciMan reports directly to the Dean of Graduate Studies. The principal administrative connection to the university's departments and schools is in the process of negotiating teaching release of individuals to teach courses—EnSciMan has access to a budget to compensate departments for faculty teaching in the program. There continues to be strong institutional and functional support for the program from the offices of the Dean of Graduate Studies and Provost.

Each faculty member associated with EnSciMan has opted to be involved—there is little formality involved—and each determines his/her form and level of involvement. For a minority, EnSciMan is their primary graduate focus and they form the base of the core group active in governance through the Graduate Program Council. Democratic cooperation has been found to be more appropriate than hierarchical forms of governance in collaborative programs (Hollaender 2003).

EnSciMan exhibits many of the characteristics indicated in the literature as key conditions for the success of interdisciplinary initiatives at universities (Shepard et al. 1985; Hollaender 2003; Newswander and Borrego 2009):

- Inception as faculty initiative, rather than a top-down administrative approach,
- A fluid organization and flexible involvement,
- A supportive administration,
- Consistent yet flexible funding,
- A short reporting structure and ease of access to administration, and
- A core of "gluons": individuals acting as the core of the interdisciplinary initiative.

Nonetheless, key challenges have been faced in promoting collaboration and integration.

The problem of melding individual objectives into a collaborative effort in collaborative initiatives has been noted (Jeffrey 2003). This is particularly important given the variety of epistemologies, norms and cultures, and methodological approaches among different disciplines (National Academy of Sciences 2005; Osborne 2013). Even in research, collaborative work may not the norm. In practice, the program is creating "its own, enclosed collaborative paradigm" (Fischer et al. 2011, p. 354) outside that of the university's departments and schools.

A particular challenge is in providing a curriculum structure responsive to students from diverse academic backgrounds. Typically students take most of their courses in Year One, including a required course in each of environmental science and environmental law (or policy). Although choices of course electives are flexible, including options from other graduate programs, students are required to balance their selection from those based in science and engineering and those in environmental management. The Seminar requirement, managed by a faculty team, brings all students together in the Fall semester of Year Two. EnSciMan courses generally are taught by individual faculty members (and by external practitioners) from a disciplinary perspective and students expected to integrate knowledge from each disciplinary base and are challenged to develop the capacity to apply it and to understanding fields outside their principal area of expertise. Independence in curriculum control has been important in encouraging this.

The M.A.Sc. is a research-based degree with the expectation that students will enhance their existing knowledge and proficiencies and foster specialized and deeply developed research expertise, often, but not always, within the framework of their disciplinary background. In the M.A.Sc. program student research requires the production of a Proposal after about 8 months in the program. Much of Year Two is focused on the research undertaking. The Ph.D. program has a conventional set of milestones.

All EnSciMan students receive scholarships and fellowships from funds directed by the program and they may be supported by stipends from faculty research accounts. Students also compete in university-wide for graduate assistantships (i.e., teaching assistant and other support). Their success varies widely by department: in some cases all students supervised by their faculty members are assured positions. Additional program funds are available for research assistantships, and to encourage departments and schools to hire students as graduate assistants. The overall result is that EnSciMan students are funded at levels comparable to and competitive with students in disciplinary programs at Ryerson and elsewhere.

Interdisciplinary programs, if they evolve to be dominated by one field, may be subsumed into existing departmental structures. Alternatively, over time they may become more identified as academic disciplines in their own right and be transformed into the departmental structure in universities (Metzger 1987; Brint et al. 2009). In such cases, stand-alone environmental programs may be as limited in their external collaboration as those associated with other disciplines (Fischer et al. 2011). In response to the risk, within EnSciMan there has been an explicit effort to subsume that disciplinary and extra-disciplinary space at the university related to graduate education concerning the environment and site the program as the locus of environmental collaboration and integration at the university. Wherever possible, collaborative links are lengthened, strengthened, and added: active outreach assures that several newly-hired faculty join the program each year; co-supervisions involving faculty from different departments is encouraged; and new research initiatives have developed among faculty first linked through their EnSciMan involvement. With their focus on application, university research centres, particularly the Centre for Urban Energy, Centre for the Study of Commercial Activities, and Centre for Studies in Food Security, act as points of contact between university and outside world; students benefit from various links with and through them.

The lack of a common physical location and limited dedicated space for EnSciMan remains a challenge. However, many students have access to computer, office, laboratory, and other space within their supervisors' departments and schools.

4 Metrics of Integration and Collaboration

Generally, with the difficulty in assessing interdisciplinary programs, there is a problem in identifying the body of knowledge and skills to be embodied; it may be best to focus on specific unique objectives (Field and Lee 1992). Based on information in the periodic review of the program (Bardecki and Pushchak 2014), six such objectives are reported here:

- Encouraging a diverse disciplinary background of faculty;
- Having a student body with a diversity of educational and life experiences;
- Encouraging cross disciplinary student-supervisor links;
- Encouraging co-supervisions and other form of faculty interaction, particular across disciplines;
- The production of research publications;
- Success in obtaining employment related to their studies.

Given the relatively short history of the Ph.D. degree, this discussion focuses on the Master's program.

As of March 2014 there are 80 faculty associated with EnSciMan. They come from each of the university's Faculties and from 21 diverse academic departments and schools (Table 1). In the program's inaugural year, 2000, the faculty were drawn from three engineering departments, the departments of chemistry and biology, geography, and economics, and the schools of public health, and urban and regional planning. Newly-hired faculty continue to join the program each year, as do current faculty with changing research interests.

With a goal of imbedding an integrated student body in a collaborative cloud of faculty, part of the selection process for each cohort of students is to assure a broadly-based student body (Table 2). The benefits of their interactions in and out of the classroom are seen as central to the program's success. In addition, students differ in their experience: in terms of the time between last degree and entry into EnSciMan, just over a third of students enter directly from their undergraduate program; one third have been out of school for at least 3 years; and more than one in six has been in the workplace for at least 5 years.

One outcome of the diversity of faculty and students is the development of cross disciplinary student-supervisor links. Table 3 illustrates the range of such links. One third of students are supervised by faculty in the same discipline area as their undergraduate degree and another 11 % by supervisors from closely aligned disciplines. Even considering those students who earned broadly-based degrees, 40 % of the links can be characterized as "radical" (i.e., connections involve faculty and students from disparate disciplines). From the point of view of faculty, the attraction often is the opportunity to be involved in supervising good students, frequently with different backgrounds and skills than available within their own disciplines (e.g., students with skills in geographical information systems to work in research with

Table 1 Faculty Participating in EnSciMan (2013–2014)

Faculty of Arts (15)
Department of Economics (1)
Department of Geography (8)
Department of History (1)
Department of Philosophy (1)
Department of Politics and Public Administration (2)
Department of Sociology (2)
Faculty of Communication and Design (1)
School of Fashion (1)
Faculty of Community Services (10)
School of Nutrition (1)
School of Occupational and Public Health (5)
School of Urban and Regional Planning (4)
Faculty of Engineering and Architectural Science (25)
Department of Aerospace Engineering (1)
Department of Architectural Science (6)
Department of Chemical Engineering (4)
Department of Civil Engineering (8)
Department of Mechanical and Industrial Engineering (6)
Faculty of Science (19)
Department of Chemistry and Biology (18)
Department of Physics (1)
Ted Rogers School of Management (10)
Department of Entrepreneurship and Strategy (1)
School of Business Management (2)
Ted Rogers School of Hospitality and Tourism Management (3)
Ted Rogers School of Information Technology Management (4)

The numbers indicated the number of faculty members from each academic unit

Table 2 Academic backgrounds of EnSciMan MASc students

Sciences	39.9 %
Environmental specializations	27.0 %
Social sciences	14.9 %
Engineering	12.8 %
Resource studies (forestry, water mgt.)	1.8 %
Others (e.g., commerce, education, nursing, technology mgt.)	5.3 %

civil engineers, policy analysts involved in studies of energy use in architecture). Such interdisciplinary collaboration has increased over time among both faculty and students.

Table 3 Student-supervisor pairings by discipline in EnSciMan M.A.Sc. Program (2013–2014)

Legend: Same discipline; Closely aligned disciplines; Distant disciplines: "radical"; Integrated [interdisciplinary] connection

Faculty		Students: Geography	International Development	Environment and Resource Studies	Biology	Chemistry	Earth Science	Agriculture	Architecture	Environmental Engineering	Chemical Engineering	Mechanical Engineering	Information Technology Mgt.	Management	Journalism	Occupational and Public Health	SUM
Arts	Economics			1													1
	Geography	4		5	1		1			1			1		1		14
	Political Science			1	1								1				3
Science	Biology and Chemistry				7	1				1							9
Engineering and Architectural Science	Architecture			1													1
	Chemical Engineering				2			1			3						6
	Civil Engineering	1		1					1								3
	Mechanical Engineering			1	1							1					3
Ted Rogers School of Management	Business Management		1														1
	Entrepreneurship and Strategy	1			1	1	1										4
	Hospitality and Tourism Mgt.			1													1
	Information Technology Mgt.										1		1	1			3
Community Services	Occupational and Public Health			3	2		1			1						1	8
	Urban and Regional Planning			3													3
	SUM	6	1	17	15	2	3	1	1	3	4	1	3	1	1	1	60

Co-supervisions enhance faculty-to-faculty collaboration. One-fifth of students' research has been co-supervised with a noteworthy range of faculty pairings of supervisions, e.g., Mechanical Engineering–Politics, Urban Planning–Chemistry and Biology, Economics–Civil Engineering, Civil Engineering–Business Management, Chemistry and Biology–Geography, Chemistry and Biology–Mechanical Engineering, Geography–Business Management, Fashion–Mechanical Engineering, Hospitality and Tourism Management–Geography. In addition, among the 15 individuals from government, private industry, NGOs, and other universities who are Associate members of EnSciMan there have been a number of co-supervisions. Such co-supervisions have been found to enhance the breadth of students' skills (Blackwell et al. 2009).

Although about 40 % of M.A.Sc. research has resulted in academic publications, to date no assessment has been undertaken of interdisciplinarity in students' theses or research publications such as those proposed by Nash et al. (2003). Nash et al. (2003) also recommend following the career path of individuals to assess the success of interdisciplinary programs. A 2013 survey was able to trace over 90 % of graduates. Of those, 99 % (i.e., all but two individuals) held positions related to environmental science and management. Just less than one-third of them worked in each of government and private companies. Another 20 % were employed in consulting. The remainder worked for NGOs or quasi-governmental agencies, in education or were self-employed entrepreneurs. In addition, 7.5 % of Master's graduates successfully entered Ph.D. programs. There are several options available for students to seek third-party accreditation, both in Canada and the United States, but these are not widely recognized, so few graduates opt to apply.

5 Conclusions

Vincent and Focht (2009) suggest that there are three dimensions differentiating environmental programs: "orientation to curriculum design (liberal arts vs. professional training)"; "curriculum breadth versus depth"; and "fixed versus flexible core competencies." The EnSciMan curriculum has developed to offer professional training with a flexible approach to specific core competencies. However, the answer to the long-standing debate between curricular breadth and depth (Clayton 1976; Newell 1992) has been to embrace both. The T-shaped curricular model recognizes the value in knowledge and technical expertise developed in a disciplinary context. However, "disciplinary specialization is still necessary but not necessarily sufficient" (Weiler 2007, p. 151). A key is the ability to interact with others from different disciplines and understand relevant fields outside one's principal area of expertise (Giacomelli et al. 2003).

The principle objectives of integrated graduate programs are to enhance students' appreciation of interdisciplinary approaches to addressing problems, to broaden their understanding of and expertise in fields outside their principal area of expertise while strengthening that expertise. However, as Casey (2010) suggests, there are a number of particular challenges in managing these programs Although integrated programs may attract more academically innovative students (and faculty) (Brint 2005), more effort is required of students due to the challenges associated with breadth of expertise required and the reliance on student's integration of concepts. With faculty bound to disciplinary schools and departments, curriculum tends to follow disciplinary norms rather than leading innovation. Similarly, evaluation and assessment of interdisciplinary research undertakings is a key challenge with different perspectives among the discipline areas. Nonetheless, it is possible for a program to play a networking role, bringing together disciplinary-based faculty and students from a variety of backgrounds to offer integrated, interdisciplinary programs that cross traditional disciplinary boundaries.

References

Bardecki MJ, Pushchak R (2014) Environmental applied science and management at Ryerson: a fifteen-year retrospective, ENSCIMAN Occasional Paper 14-01. Environmental Applied Science and Management, Ryerson University, Toronto, p 44

Bitner M, Brown S (2008) The service imperative. Bus Horiz 51(1):39–46

Blackwell AF, Wilson L, Street A, Boulton C, Knell J (2009) Radical innovation: crossing the knowledge boundaries with interdisciplinary teams, Technical Report Number 760. University of Cambridge Computer Laboratory, Cambridge, p 124

Brint S (2005) Creating the future: 'new directions' in American research universities. Minerva 43:23–50

Brint S, Turk-Bicakci L, Proctor K, Murphy S (2009) Expanding the social frame of knowledge: interdisciplinary, degree-granting fields in American four-year colleges and universities, 1975–2000. Rev High Educ 32(2):155–183

Casey BA (2010) Administering interdisciplinary programs. In: Frodeman R, Klein JT, Mitcham C (eds) Oxford handbook of interdisciplinarity, Oxford University Press, Oxford, pp 344–359

Clayton K (1976) Environmental sciences/studies: a decade of attempts to discover a curriculum. Area 8(2):98–101

Donofrio N, Spohrer J, Zadeh HS (2009) Research-driven medical education and practice: a case for T-shaped professionals. Almaden Research Center, IBM, San Jose. http://www.ceri.msu.edu/wp-content/uploads/2010/06/A-Case-for-T-Shaped-Professionals-20090907-Hossein.pdf. Accessed 26 Sept 2011

Field M, Lee R (1992) Assessment of interdisciplinary programmes. Eur J Educ 27(3):277–283

Fischer ARH, Tobi H, Ronteltap A (2011) When natural met social: a review of collaboration between the natural and social sciences. Interdisc Sci Rev 36(4):341–358

Fortuin IKPJ, Bush SR (2010) Educating students to cross boundaries between disciplines and cultures and between theory and practice. Int J Sustain High Educ 11(1):19–35

Foster J (1999) What price interdisciplinarity? Crossing the curriculum in environmental higher education. J Geogr High Educ 23(3):358–366

Giacomelli P, Travisi C, Nava M (2003) Are graduates in environmental sciences potential managers of the environment? Some problems and examples in the north of Italy. Int J Sustain High Educ 4(11):9–16

Guest D (1991) The hunt is on for the renaissance man of computing". The Independent, London, Sept. 17

Hansen MT (2010) "IDEO CEO Tim Brown: T-shaped stars: the backbone of IDEO's collaborative culture" chiefexecutive.net January 21. http://chiefexecutive.net/ideo-ceo-tim-brown-t-shaped-stars-the-backbone-of-ideoae%E2%84%A2s-collaborative-culture. Accessed 20 Oct 2013

Heinemann E (2009) Educating T-shaped professionals. In: Proceedings of the fifteenth americas conference on information systems, San Francisco, California August 6–9, 2009. http://aisel.aisnet.org/amcis2009/693. Accessed 5 Jan 2012

Hiwasaki L, Arico S (2007) Integrating the social sciences into ecohydrology: facilitating an interdisciplinary approach to solve issues surrounding water, environment and people. Ecohydrol Hydrobiol 7(1):3–9

Hollaender K (2003) Success factors in inter- and transdisciplinary research: selected results from the program Urban Ecology. In: Tress B, Tress G, van der Valk A, Fry G (eds) Interdisciplinary and transdisciplinary landscape studies: potential and limitations, DELTA Series 2, Wageningen, Netherlands, pp 91–99

Holley KAE (2009) Understanding interdisciplinary challenges and opportunities in higher education, ASHE Higher Education Report 35(2). Jossey Bass, San Francisco, p 131

Huutoniemi K, Klein JT, Bruun H, Hukkinen J (2010) Analyzing interdisciplinarity: typology and indicators. Res Policy 39(1):79–88

Jeffrey P (2003) Smoothing the waters: observations on the process of cross-disciplinary research collaboration. Soc Stud Sci 33:539–562

Klein JT (2008) Education. In: Hirsch Hadorn G, Hoffmann-Riem H, Biber-Klemm S, Grossenbacher-Mansuy W, Joye D, Pohl C, Wiesmann U, Zemp E (eds) Handbook of transdisciplinary research, Springer, Dordrecht, Germany, pp 399–410

Krozer Y (2005) The life-cycle of environmental professionalism. Greener Manage Int 49:43–55

McIntosh BS, Taylor A (2013) Developing T-shaped water professionals: reflections on a framework for building capacity for innovation through collaboration, learning and leadership. Water Policy 15:42–60

McNeill D, García-Godos J, Gjerdåker A (eds) (2001) Interdisciplinary research on development and the environment, Centre for Development and the Environment, University of Oslo, Oslo, p 56

Metzger WP (1987) The academic profession in the United States. In: Clark BR (ed) The academic profession: national disciplinary and institutional settings. University of California Press, Berkeley, pp 123–208

Nash JM, Collins BN, Loughlin SE, Solbrig M, Harvey R, Krishnan-Sarin S, Unger J, Miner C, Rukstalis M, Shenassa E, Dubé C, Spirito A (2003) Training the transdisciplinary scientist: a general framework applied to tobacco use behavior. Nicotine Tobacco Res 5(S-1):S41–S53

National Academy of Sciences (2005) Facilitating Interdisciplinary Research. National Academies Press, Washington, p 306

Newell WH (1992) Academic disciplines and undergraduate interdisciplinary education: lessons from the School of Interdisciplinary Studies at Miami University, Ohio. Eur J Educ 27 (3):211–221

Newswander LK, Borrego M (2009) Engagement in two interdisciplinary graduate programs. High Educ 58:551–562

Osborne T. (2013). "Inter that discipline!". In: Barry A, Born G (eds) Interdisciplinarity: reconfigurations of the social and natural sciences. Routledge, London, pp 82–98

Shepard K, Yeo G, McGann L (1985) Successful components of interdisciplinary education. J Allied Health 14:297–303

Soule ME, Press DL (1998) What is environmental studies? Bioscience 48(5):397–405

Tbilisi Intergovernmental Conference on Environmental Education (1978). Toward an action plan: a report on the tbilisi conference on environmental education. U.S. Government Printing Office, Washington, p 33

Thomas IG (1992) Integrators: an outcome of environmental education. Environmentalist 12 (4):261–266

Vincent S, Focht W (2009) US higher education environmental program managers' perspectives on curriculum design and core competencies: implications for sustainability as a guiding framework. Int J Sustain High Educ 10(2):164–183

Weiler CS (2007) Meeting Ph.D. graduates' needs in a changing global environment. EOS 88:149–151

World Bank (2012) What is the purpose of the Young Professionals Program? http://web.worldbank.org/WBSITE/EXTERNAL/EXTJOBSNEW/0,,contentMDK:23131984~menuPK:8479833~pagePK:8453902~piPK:8453359~theSitePK:8453353,00.html. Accessed 12 Dec 2013

Author Biography

Professor Michal Bardecki received his doctorate in geography from York University (Canada) in 1981 after completing B.A. and M.Sc. degrees at the University of Guelph. He holds the position of Professor in the Department of Geography in Ryerson University and has been Director of the graduate programs in Environmental Applied Science and Management for the past 5 years. Dr. Bardecki teaches two large undergraduate classes: Global Environmental Issues, and Viva Las Vegas! Each course requires students to integrate materials from a wide disciplinary base. His principal areas of current research focus on environmentally-based tourism in Nepal, environmental interpretation, media and environment, and wetland evaluation

Reinforcing Sustainable Development in Schools in Malta: A Potential Approach by the University of Malta to Meet the Country's Sustainable Development Challenges

Cynthia Caruana and Mark Mifsud

Abstract

Various important historical documents have emerged which have left an impact on the world. One of these is Agenda 21, which is an important tool for policy-makers and environmental managers for bringing about Sustainable Development. For a long time it has been widely established that education plays a highly significant role in Sustainable Development, and inevitably this noteworthy discussion leads to school contexts. In this scenario, the validity of using Agenda 21 is highlighted and presented as a tool that can be used by environmental managers within the school whereby students from a very young age can start acting and promulgating Sustainable Development. With this in mind, Sustainable Development is expressed as four-pillared and intercultural management is viewed as an important field for social cohesion and justice, particularly within the school environment. The study also provides a strategic intercultural and environmental management plan tailor-made for a particular school in Malta. Both quantitative and qualitative research techniques were carried out to establish the context and to increase the success rate for this, long-term plan inculcating intercultural and environmental management practices. To reinforce Sustainable Development in schools, this study recommends among others, to enhance links between schools and the community, to form a database as reference and inclusion of interculture in the National Strategy for Sustainable Development in Malta. Considering the focus on innovative pedagogies and

C. Caruana (✉) · M. Mifsud
Education Department, CEER, University of Malta, Msida MSD 06, Malta
e-mail: cynthia.caruana.2@ilearn.edu.mt

M. Mifsud
e-mail: mark.c.mifsud@um.edu.mt

W. Leal Filho et al. (eds.), *Integrative Approaches to Sustainable Development at University Level*, World Sustainability Series, DOI 10.1007/978-3-319-10690-8_15

applied sustainable development, this paper will be useful for lecturers and teachers that intend to develop strategic sustainable development plans in their communities or institutions.

Keywords
Local agenda 21 · Intercultural and environmental management plan · Mixed methods · Education for sustainable development

1 Introduction—The Importance of Education for Sustainable Development

The initial document inspiring the work behind this paper is Agenda 21 which emerged from the United Nations Earth Summit of 1992. This non-binding document consists of a detailed action plan for sustainable development containing 115 programme areas. This document reiterates that efforts on multiple fronts are needed to create a more sustainable world (UNCED 1992). Furthermore, Chap. 36 'Promoting education, Public Awareness and Training' describes education as one of the fronts with high potential for advancing Sustainable Development efforts. Education for Sustainable Development was reinforced in 2002, at the World Summit on Sustainable Development (WSSD), when the decade 2005–2014 was declared to be the International Decade of Education for Sustainable Development (IDESD). A reaffirmation of the will of governments to work in favour of sustainable development was one of the key concepts emerging from the Johannesburg Declaration on Sustainable Development (UN 2002). The civil society and the business world were also encouraged to carry out collaboration initiatives facing specific problems and lead to concrete results to improve the life conditions of the world's peoples.

Since the coining of the term 'Sustainable Development', it has been steadily integrated into policies, projects and phrases, at times attaching misconceptions to its real meaning. According to Lemonick (2009), the term sustainable is a 'simple' term with a high 'conceptual heft'. Nonetheless, environmental issues are being considered as part of development strategies and there is wide consensus that there can be no sustainable development for our planet without the eradication of poverty. Following this line of thought, "culture is not given enough space within development issues", and there is no "systematic attention to cultural and intercultural diversity at all intervention levels of development policies" (Elame' 2004). In this regard, the three pillars of Sustainable Development (social solidarity, effectual economy and environmental responsibility) without consideration of culture and interculture deprive humanity from peace, progress and social justice, without which no development can take place.

Interculture may be defined as "an eco-socio-cultural dynamic process aiming at promoting inter-ethnic harmony among people, becoming aware of our cultural and biological differences, deeply rooted in our social, economic and ecological practices" (Elame' 2004). Concern about emerging environmental issues is highly evident. Once cannot however dispute that intercultural issues are also on the rise and their effects are devastating. These issues have the potential to undermine global peace and security. This line of thought places interculturality solidly within discourse of Sustainable Development. Furthermore it gains strategic importance within the school context, a place where children from a very young age can understand and weave ties among diversity. The importance of Education for Intercultural Understanding and Dialogue gained worldwide impetus with the Copenhagen Conference held in October 2008, uniting more than 140 experts from all over the world (UN 2009). In his forward, Director General Koichiro Matsuura seeks to highlight the timely importance of focusing on the role of education in building intercultural dialogue and understanding in the current global environment.

It is vital therefore to find ways how to integrate Education for Sustainable Development and interculturality in current educational systems. This paper introduces how through its involvement with Ca' Foscari, the University of Malta has started to address this important challenge. This paper outlines the study undertaken in order to provide a tangible, tailor-made environmental and intercultural plan for a secondary school. This is highly relevant as it provides a practical way how students at universities can pursue, apply and promulgate sustainability. Considering the focus on innovative pedagogies and applied sustainable development, this paper will be useful for lecturers and teachers that intend to develop strategic sustainable development plans in their communities or institutions. The paper concludes with an evaluation of the study's limitations together with recommendations as to how this innovative study can be further enhanced in order to achieve the goals of IDESD.

2 Malta as a Background for Implementing Education for Sustainable Development

Malta is considered as a Small Island Developing State due to its limitation of natural resources and size which contribute to economic, environment and social problems. For instance a high population density and the tourist industry create a burden on Malta's waste management efforts. The large influx of illegal migrants is also presenting great challenges to Malta. The quality of the limited water aquifers is undermined by bad agricultural practices whilst the 'Not in my back-yard' (NIMBY) Syndrome is rampant within Maltese society. In spite of Malta's climate, the island is still almost dependent on non-renewable energy resources. The economic growth may also be low compared to other European countries due to lack of diversification. In spite of this scenario, Malta's culture and heritage can, in no way, be described as limited or limiting. Malta's name is cognizant with the country's rich cultural background spanning 7,000 years.

It is in the school's best interests to adopt a strategy of intercultural eco-management. If we want our schools to help the young become conscious citizens, they cannot be just a learning space where students only acquire theoretical knowledge.

The schools' organisation and standards of reference both on a teaching and administrative level should make of them a benchmark for environmental and intercultural governance. This absolutely highlights the importance of management plans that address the intercultural and environmental issues within the school. These could be integrated within the school's development programmes. A Senior Management Team member from each school needs to be professionally trained to become a competent intercultural and eco-manager in order to develop effective management plans which could eventually lead to our schools following the process of School Agenda 21.

This plan, if in place and managed well, provides the supportive infrastructure to two newly introduced cross-curricular themes of Education for Sustainable Development and Education for Diversity within the National Curriculum Framework in Malta, which was launched on 14th February of 2013 (Times of Malta 2013). Albeit towards the end of the IDESD, this significant step has been met favourably by institutions such as the Centre for Environmental Education and Research (CEER). "CEER welcomes the proposal of Education for Sustainable Development as a cross curricular theme and hence one of the priorities of the curriculum" (CEER 2012, p. 6)…"A very positive step towards the formation of a citizenry which is aware and responsible towards the environment and is a fitting national contribution towards the UN Decade of Education for Sustainable Development" (CEER 2012, p. 7).

3 The Role of Higher Education

The University of Malta was involved in the setting up of an applied Master degree in Intercultural and Environmental Management of Schools, named the 3EMI. It was led by the University Cà Foscari of Venice and co-financed by the European Commission in the framework of the Lifelong Learning Programme (LLP). The project partners included the Youth Centre of Haskovo (BG), the University of Malta (MT), KCDKO—Administration and Management Centre of Krakow (PL), Centro de Formação dr. Rui Grácio (PT), Filocalia Foundation (RO), Çankırı National Education Directorate (TK). All these institutions are members of the consortium set up for the European project COMENIUS 3EMI.

The driving philosophy behind the project was the emerging importance in political, sociological, and educational discourse of natural environments, globalization, migration mobility, sustainable development and intercultural dialogue. The main principle was that school management has to be developed in order to be able to cope with the evolved and interconnected environmental and intercultural requirements. Additionally, students required exposure to this new dimension in order to effectively answer to the various challenges of interculturality and sustainable development throughout their lives. The main aim of the project was to improve environmental and intercultural management skills in principals and the school management team. The Master training programme (including e-learning and face-to-face stages) was designed and tested with a sample of about 150 school managers from Italy, Bulgaria, Malta, Portugal, Romania, Turkey.

The course was aimed at developing new breed of professionals, called intercultural eco-managers of schools, able to project and develop interventions and policies on sustainable development in an intercultural perspective. In order to reach these objectives, the project proposed:

1. a training needs analysis of 700 school principals and executives from the different partner countries,
2. to develop a 1st Level Master course in Intercultural eco-management of schools,
3. to ensure an e-learning blended training course for at least 150 European school managers,
4. to help every participant in designing an intercultural eco-management action plan for their own school.

The 3EMI project was held inside a virtual campus to give attendants (school managers and teachers with work-related duties) the chance to overcome disadvantages due to time limits, distance or disability. The virtual campus specially created for the 3EMI project used the Moodle software for the online display of the learning activities. In addition to the main online learning environment provided for the masters there were a number of face to face meetings with respective national and international tutors in order to ensure better understanding of concepts. This satisfied the more informal side of the course and resulted in a blended type of learning which is innovative and very effective. Students were actively tutored to create research which could be applied to the context of their school systems in order to produce tangible improvements. The research and plans produced in this paper are, in fact a direct result of this Master programme.

4 Research Methodology

In order to provide an intercultural and environmental plan for the school, research methodology techniques were used to collect data and establish the current intercultural and environmental situation within the school, since this is highly significant for the success rate of the plan. Participant Observation was primarily used from the months of February till June 2010. The information obtained with this technique related to what is currently happening in the school and main events in the school calendar falling within the period of study including Prize Day and Parents' Day were analyzed. Speeches and in-house circulars also had their content analyzed, together with informal interviews with school stakeholders such as members of SMT, teachers, students and ancillary (cleaning) staff. The use of this method was very appropriate since it gave an excellent insight into the school's running and established the context for the strategic plan (Table 1).

Table 1 Research methodology plan

Time of study	Research technique	Activities carried out	Objectives
Sequential time of study. App. 2 months per technique	Surname study	Consultation with school registers from previous years (1982 till 2009)	To determine if there are any changes in culture based on foreign students' surnames and if there are any emerging trends along the years
	Participant observation	Analysis of events including Prize Day, Parents' Day, SDP Day; Analysis of discourse during meetings; Analysis of circulars and internal memos	To determine the current situation as regards environment and interculture. What importance, if any, are these aspects given?
	Informal interviews	Discussions with: SMT member Ancillary (cleaning) staff Teachers in staffrooms and special rooms; home economics teachers, science teachers	To find out policies that outline management techniques and if these are sustainable
			What type of products are used for cleaning and the practices used during their work as regards waste management and use of resources
			To get information about levels of awareness and participation in school practices
	Questionnaires to students	Pilot study with one class, then real study with 254 students responding to the survey	To obtain insight into students' lifestyles and attitudes and to determine how they react to situations related to intercultural eco-management
	Questionnaires to teachers	Pilot study with five teachers, then real study with 62 teachers, for which 48 questionnaires were returned to trainee	To obtain insight into teachers' lifestyles and attitudes and to determine how they react to situations related to intercultural eco-management

In order to reinforce the research method, two sets of questionnaires were also formulated, primarily used in a pilot study, edited and then given to students and teachers. This quantitative method gave further insight into the way these stakeholders think, which are the most important issues that need to be addressed and furthermore how would they react to future actions or proposals. With all this data

Table 2 Questionnaire respondents

Stakeholder type	Number of questionnaire respondents
Students	254
Teachers	48

in hand, the trainee could execute the project and provide the strategic plan which is tailor-made for the school under study.

The school lacks organized documentation as regards different cultures. There is space for representation from different cultures and minority groups but there are no cross-cultural strategies observed. In conjunction with participant observation, a student survey was conducted and analysed using SPSS version 14.0. Out of 581 available students, a total of 254 students participated, giving a 95 % confidence level and 4.5 confidence interval.

To review the teachers' opinions as regards the school's environment and interculture, a survey was administered to teachers. 4 teachers outrightly refused to fill in the questionnaire when they were approached by the trainee and 10 questionnaires were not returned. The survey finally yielded 48 replies, 31 females and 16 males (Table 2).

5 Data Collection and Analysis

The school under study, made up of 5 distinct blocks, is centrally located in one of Malta's heavily urbanised areas, which often means heavy congestion at all times of the day. Parking space is often a problem in the school, so much so that cars are often parked too close to the main waste separation site or may obstacle paths to the different blocks. Its strategic location makes it important as an examination centre, premises for setting of national evening courses and for foreign students to learn English. Although, this is important as the building is being utilised throughout the whole day and year, there could be an element of neglect by external users, particularly since they are not taken on board on issues of sustainability within the school premises. The distinct blocks present a problem of lack of communication and exchange between staff members who may only meet each other during infrequent meetings.

6 Results

Participation was low, so much so that only 30.7 % think that participating in school activities is important. Furthermore, only 36.6 % of the respondents know whether the school takes part in environmental initiatives, and 39.8 % agree that the students are involved in thinking up ways how to help the school's environment. 68.9 % do not know the meaning of the term sustainable development and 83.1 %

Table 3 Teachers' reply to the question: have you ever dealt with the subject of differences in culture in any of your lessons at school?

		Frequency	Percent	Valid percent	Cumulative percent
Valid	Yes	26	54.2	54.2	54.2
	No	21	43.8	43.8	97.9
	Not sure	1	2.1	2.1	100.0
	Total	48	100.0	100.0	

have never heard of School Agenda 21. 43.7 % strongly agree that there is more to be done as regards the school's environment. 43.3 % are unsure whether the current system of waste separation is effective, implying that there is a serious lack of communication and information being channeled to the students as stakeholders. The subject of differences in culture has come up during school subjects for 68.5 % of the survey respondents. This was mainly due to languages, History, PSD, Geography, Social Studies and Religion. Analysis of data collection has shown that links between the school and parents as well as the community need to be properly established. Students are not pro-active and there is not enough awareness and/or information about current projects is not being properly disseminated. On a positive note, there is agreement that the school attempts to involve all students in the school, whatever their culture is and that there is a good introductory procedure when there are new entrants. Exactly, 54.2 % of teacher respondents have dealt with the subject of differences in culture in their lessons. Out of these, 15.4 % of teachers involved more than 15 lessons on this theme. Fifty percentage of respondents strongly agree that the school attempts to involve all students in the school, whatever their culture is. Only 16.7 % strongly agree that there is a good introductory procedure where there are new students in the school. In the light of all this, it can be stated that with the school's high ethnic capital, it has very valid potential to cross-culturalisation, which however needs to be specifically addressed by cross-cultural management (Table 3).

54.2 % of teachers have discussed sustainable development with their students but only 8.3 % have heard of School Agenda 21. Twenty-nine point two percentage strongly agree that there is more to be done as regards the school's environment, whereas 62.5 % agree. Sixty-six point seven percentage are aware of intercultural/environmental activities occurring in the school and 56.3 % are willing to participate in a system of car pooling/sharing and a car free day. Twenty-five percentage are willing to voluntarily participate in teams to discuss how interculturality and the environment at school can be improved (Tables 4, 5, 6 and 7).

55 % of students strongly agree that their parents are involved in thinking up ways how to help the school's environment, whereas 53.5 % of parents attend Parents' day, school meetings, school activities etc. Ten point four percentage of teacher survey respondents strongly agree that there is a good link between the parents and the school, implying that this is a serious limitation. Sixteen point seven

Table 4 Teachers' reply to the question: have you ever discussed sustainable development with your students?

		Frequency	Percent	Valid percent	Cumulative percent
Valid	Yes	26	54.2	54.2	54.2
	No	18	37.5	37.5	91.7
	Not sure	4	8.3	8.3	100.0
	Total	48	100.0	100.0	

Table 5 Teachers' reply to the question: have you ever heard of School Agenda 21?

		Frequency	Percent	Valid percent	Cumulative percent
Valid	Yes	4	8.3	8.3	8.3
	No	37	77.1	77.1	85.4
	Not sure	7	14.6	14.6	100.0
	Total	48	100.0	100.0	

Table 6 Teachers' reply to the question: are you willing to participate in a system of car pooling/ sharing and a car free day?

		Frequency	Percent	Valid percent	Cumulative percent
Valid	Yes	27	56.3	56.3	56.3
	No	13	27.1	27.1	83.3
	Not sure	8	16.7	16.7	100.0
	Total	48	100.0	100.0	

Table 7 Teachers' reply to the question: are you willing to participate in a team to discuss ways hot interculturality and the environment at school can be improved? (voluntary activity, weekly during mid-day break)

		Frequency	Percent	Valid percent	Cumulative percent
Valid	Yes	12	25.0	25.0	25.0
	No	13	27.1	27.1	52.1
	Not sure	23	47.9	47.9	100.0
	Total	48	100.0	100.0	

percentage of teacher respondents and 10.2 % of student respondents think that there is a good link between the school and the local community, thereby showing another serious limitation (Tables 8, 9 and 10).

Table 8 Students' reply to the question: do your parents attend Parents' day, school meetings, school activities?

		Frequency	Percent	Valid percent	Cumulative percent
Valid	Yes	136	53.5	53.5	53.5
	No	2	0.8	0.8	54.3
	Not always	116	45.7	45.7	100.0
	Total	254	100.0	100.0	

Table 9 Students' reply to the question: there is a good link between the school and the local community, e.g. local council, other companies

		Frequency	Percent	Valid percent	Cumulative percent
Valid	Strongly agree	26	10.2	10.2	10.2
	Agree	71	28.0	28.0	38.2
	Unsure	107	42.1	42.1	80.3
	Disagree	39	15.4	15.4	95.7
	Strongly disagree	11	4.3	4.3	100.0
	Total	254	100.0	100.0	

Table 10 Teachers' reply to the same question: there is a good link between the school and the local community, e.g. local council, other companies

		Frequency	Percent	Valid percent	Cumulative percent
Valid	Strongly agree	8	16.7	16.7	16.7
	Agree	15	31.3	31.3	47.9
	Unsure	20	41.7	41.7	89.6
	Disagree	4	8.3	8.3	97.9
	Strongly disagree	1	2.1	2.1	100.0
	Total	48	100.0	100.0	

The school aims to foster social responsibility by organizing certain events aptly shown in the annual school magazine *Il-Malju* (2010, pp. 24–50), for example: career visits, sporting events, a visit to the House of Representatives and l-'Arka' Trade Fair shop. However these events are sporadic and not fixed dates on the school calendar. Social aspects such as smoking and alcohol were the focus of workshops and talks on a form level. The intangible cultural heritage is also given

Table 11 Teachers' reply to the question: do you know whether the school takes part in intercultural/environmental activities?

		Frequency	Percent	Valid percent	Cumulative percent
Valid	Yes	32	66.7	66.7	66.7
	No	4	8.3	8.3	75.0
	Not sure	12	25.0	25.0	100.0
	Total	48	100.0	100.0	

importance in the annual Prize Day musical. A number of competitions and/or exhibitions are also presented to celebrate art in its variety of formats. This positive aspect is undermined by the lack of communication within and beyond the school. A number of teachers are sometimes unaware of what goes on and the school website is not regularly updated (Table 11).

An analysis of the School Development Plan (SDP) document shows several aims high on the sustainability ranking list such as linking with other schools within the college and stimulating awareness of a safe school environment. Unfortunately, during the study period, these aims were not met, possibly due to no specific actions pertaining to them within the plan. Although it is a starting point, the school's SDP is limited and does not address all the challenges which need to be faced by a modern school housing global citizens.

There are no internal communications of sustainability reporting or actions to be taken as regards sustainable mobility. Similarly, there is no intercultural communication plan, mediation plan or intercultural management plan. Green Procurement is often exercised but these activities are not communicated to the rest of the school and as a result they cannot serve as a positive example to the rest. Although teachers often forward a list of items which they require, there is no effective participatory budgeting.

7 The Intercultural and Environmental Management Plan for a School in Malta

The contextualization carried out has shown that there are no environmental and intercultural management instruments being used. There are some attempts at environmental measures but these are sporadic and not consistent. To address this lacuna, a strategic plan which introduces intercultural and environmental management instruments was formulated. As final products these management tools should raise awareness of all stakeholders involved and offer them participatory opportunities; constantly and effectively communicate with all actors; stimulate sustainable practices by taking environmentally sound cost-effective and socially responsible decisions that also celebrate culture in its tangible and intangible form of all the cultures present in the school; regularly monitor the actions undertaken and link to the external community and also informally and formally guide students into a sustainable lifestyle which will be second nature to them.

Following a feasibility study, it was concluded that due to a number of constraints, the plan is feasible over a number of years. For this reason, the plan will last 5 years. Some projects, depending on the priority can start at the beginning of the first year, second year etc. These priorities could be determined by using SWOT analysis (Poggipollini 2010, p. 10). Funding might be problematic, hence the school might wish to organise some fund-raising activities. The project is also feasible only if there is a good means of communication between stakeholders, so this should be the primary action within the plan. The contextualised plans produced in this study include: Green Procurement Plan, Waste Management Plan Sustainable Mobility Plan, Sustainability Reporting and a Cross-Cultural Communication Plan, a Cross-Cultural Mediation Plan and an Intercultural Management Plan.

8 Conclusion

Environmental management and Intercultural management plans are non-existent in the school under study. Though one cannot generalise, through one of the author's experience as a peripatetic teacher around schools in Malta, this was highly characteristic up to some years ago in the majority of schools. Through courses such as the 3EMI discussed in this paper, this lacuna can be addressed through the provision of holistic plans and other similar projects that can emerge. In view of the Maltese context within a global dimension, it is wise to include aspects of interculturality within the educational platform of the school. The initial auditing of the school context and interviewing the stakeholders prove essential for gaining insight into the school's needs thereby allowing for the appropriate adjustment of the management plans according to their needs. In this case, the relative sample of stakeholders involved was quite small. More effective stakeholder involvement could have been achieved through a longer timescale.

The study performed can be supplemented with other proposals to face the challenge of Sustainable Development in schools. Different forms of support would need to follow, namely, financial aid, expert assistance, guidelines on effecting management plans and staff development training. This would boost the quality of education within our schools to a higher sustainable level. There should be linking with international entities for teachers and students. There is also a missing link between schools and the local community. This needs to be addressed as it is believed that local governance strategic plans from the local councils should be given more importance and be seen as management tools that can be applied to most localities and with some changes, to schools.

A database with experiences and active discussions can be set up, whose aim is to build a bank of good practices to which other schools can refer. This is useful in avoiding Fragmentation. Bad practices need not be repeated by other schools as this could lead to wasting time, a precious resource that sustainability can ill-afford. Positive practices can be emulated and improved thereby ensuring more successes and eventually more encouragement and support for sustainable development projects. This database may also contain statistics, trends, and winners of awards from any programmes currently taking place at our schools, thereby raising awareness and increasing ownership.

Future research in the area should focus on the development of a set of sustainability indicators for Malta's schools. These indicators should eventually be used in other pilot studies to gauge their applicability and effectiveness within our schools. The effectiveness of such tools to promote ESD and for student empowerment should also be studied in order to render ESD in schools more effective.

References

CEER (2012) Feedback to the consultation process held with regard to the draft national curriculum framework: University of Malta and Higher Education Institutions. Appendix V, 6 August, 2012 Valletta, Malta. https://curriculum.gov.mt/en/Resources/The-NCF/Documents/Consult_Docs/Analysis of Feedback_University.pdf. Accessed July 2014

Elame E (2004) Interculturaliser le developpment durable. In Actes du colloque "Developpment durable: lecons et perspectives" Volume 1, Agence Universitaire de la Francophonie (AUF), pp 71–80, France

Lemonick M (2009) Top 10 Myths about Sustainability. In Scientific American Earth 3.0, March 2009. Government Publication, USA

Poggipollini P (2010) Sustainable development and new participation practices. 3EMI Course, Module 2.1. University of Ca' Foscari, Venice, Italy

Saint Ignatius College Girls' Junior Lyceum (2010) Blata l-Bajda, *Il-Malju 2010*: *Magazin Annwali tal-Liceo tal-Bniet Blata l-Bajda.* Blata l-Bajda, Malta

UNCED (1992) Agenda 21: programme of action for sustainable development. Rio declaration on environment and development. United Nations, NY

United Nations (2002) Johannesburg declaration on sustainable development. Johannesburg, South Africa, 26 August–4 September 2002. United Nations, New York

United Nations (2009) Investing in cultural diversity and intercultural dialogue among civilizations: the Copenhagen Agenda, the Copenhagen conference of education for intercultural understanding and dialogue 21–22 October 2008. Paris, France

Authors Biography

Cynthia Caruana is an Education for Sustainable Development Teacher in Maltese state, church and independent schools. She has studied Biology Methodology, Islands and Small States Studies and Intercultural and Environmental Management in Schools at the University of Malta and at the University of Ca Foscari. She currently promotes sustainability in schools and their communities.

Mark Mifsud is a Lecturer at the Centre for Environmental Education Research at the University of Malta. He is a Chartered Biologist, a Chartered Environmental Scientist, a Chartered Science Teacher, a Certified Wildlife Photographer and a Prince II Certified Practitioner with more than 20 year experience in Environmental Sciences, Environmental Education and Sustainable Development. Mark is presently the coordinator of the Masters in Education for Sustainable Development and his main areas of research include youth and the modelling of environmental knowledge, attitudes and behaviour.

Commitments of University Leaders to the Talloires Declaration: Are They Evidenced in Industrial Design Teaching and Learning?

Mariano Ramirez Jr

Abstract

By the act of signing the Talloires Declaration, university presidents and chancellors worldwide committed themselves to educate young people towards being more sustainability-focused, with a lofty ambition to change the world for the better. A good number of these higher education institutions offer degrees in industrial design, a career which has often been implicated for its sizeable contributions to the mountains of short-lived products and packages in landfills around the world. This paper investigates whether the promises made in this declaration are permeating in the education of industrial designers in signatory universities. It looks into the curricula and student portfolios of industrial design degrees and uses content analysis to understand how sustainability thinking is integrated into product design pedagogy.

Keywords

Industrial design · Design education · Education for sustainability · Talloires declaration

In 1990, presidents, rectors and vice-chancellors from 22 universities convened in Talloires, France during an international conference on the role of universities in environmental management and sustainable development. The conference participants acknowledged the urgent need for sustainability leadership in higher education, and agreed to commit themselves to a bold ten-point agenda for incorporating

M. Ramirez Jr (✉)
Australian School of Architecture and Design, University of New South Wales,
Anzac Parade, Kensington, Sydney, NSW 2052, Australia
e-mail: m.ramirez@unsw.edu.au

W. Leal Filho et al. (eds.), *Integrative Approaches to Sustainable Development at University Level*, World Sustainability Series, DOI 10.1007/978-3-319-10690-8_16

environmental literacy and sustainability into academic practice and other critical activities of tertiary education. This historic document became known as the Talloires Declaration and was the first official statement targeted to 'greening' higher education and initiated by university leaders. From the original group of 22 institutions in 1990, the number of signatories has grown to 470 universities in 54 countries in 2014.

The Association of University Leaders for a Sustainable Future (ULSF) functions as the secretariat for the signatories of the Talloires Declaration. It defines a truly sustainable college or university as one that emphasizes in its curriculum and research the concepts of ecological soundness, social justice and economic viability for current and future generations; one that prepares its graduates to contribute to an environmentally healthy and equitable society; and one that functions as a sustainable community, embodies responsible consumption of energy, water, and food, and supports sustainable development in its local community and region (ULSF 2009).

The endorsement of the Talloires Declaration not only symbolizes the recognition from higher education of the adverse impacts of human activities on the environment and on the well-being of global society, but also emphasizes the huge potential and responsibility of universities in shaping future citizens armed with the values to contribute to an ecologically responsible and socially just global community.

1 Background

Some of the Talloires Declaration signatory institutions offer degrees in industrial design or product design. Amongst the design disciplines, industrial design is often been singled out as being directly implicated in the global environmental crisis (Ramirez 2007), mainly due to its practitioners' tendency to promote or encourage mass-consumerism, technical or fashion obsolescence, short-lived convenience items, and inefficient use of resources throughout the product lifecycle.

One of the early practitioners and professors of industrial design, Victor Papanek, criticized his own profession as being one of the most harmful of all, arguing that industrial designers 'concoct the tawdry idiocies hawked by advertisers', 'create a whole new species of permanent garbage to clutter up the landscape', and 'choose materials and processes that pollute the air we breathe'. He also lamented that 'much recent design has satisfied only evanescent wants and desires, while the genuine needs of man have often been neglected by the designer' (Papanek 1971).

In 1954 the US industrial designer Brooks Stevens announced that the mission of industrial design was 'instilling in the buyer the desire to own something a little newer, a little better, a little sooner than is necessary'; he referred to this contentious business model as 'planned obsolescence' (Adamson 2003; Dannoritzer 2012; Leonard 2010). He argued that intentionally shortening the lifecycles of products—by making them break easily or quickly go out of style—so as to encourage mass-consumerism was beneficial for America's postwar economy, as it stimulates the manufacturing industry. This pronouncement degraded the value of industrial design as a shallow marketing tactic to trick consumers into purchasing more stuff. Planned obsolescence has been blamed for the short-lived products around us, the rise of the throwaway culture, and consequently the swelling of landfills. One social critic of that time, Vance Packard, accused industrial designers, in connivance with irresponsible businessmen, as being 'waste makers' by negligently promoting a culture of premature product obsolescence and systematically making consumers 'wasteful, debt-ridden, and permanently discontented' (Packard 1960).

This paper now seeks to understand if industrial design education institutions, particularly those who have endorsed the Talloires Declaration, demonstrate that they are practicing what they have signed up to do. Public sustainability declarations are often critiqued for rewarding signatories with accolades before they have actually carried out their promises (Bekessy et al. 2007), and not holding them accountable when they don't deliver. This study intends to uncover whether these promises are permeating in the education of industrial designers in signatory universities, and whether students are now getting trained to become more socially responsible and ecologically conscientious, compared to the design practitioners that Papanek was censuring in the 1970s.

2 Methodology

The study started with the gathering of data from the official websites of the Talloires Declaration signatories. The collected data were entered into a spreadsheet. The internet-based 'Google Translate' machine-translation service was utilized to interpret textual information from non-English language websites. Following this a shortlist of signatory institutions offering degrees in industrial design or product design was made.

Digital versions of the university catalogues, bulletins, prospectuses, project exhibition brochures, and handbooks of undergraduate and postgraduate courses, were downloaded. Web pages of the university's industrial design programs were captured as PDF documents. These documents were consulted for the curricula, syllabi, mission-vision statements, program and course descriptions, and galleries of student work; altogether they helped paint a clearer picture of the pedagogical content of the design degree offerings.

A thematic content analysis was carried out, examining the textual and visual contents of the official websites of the tertiary institutions. Websites have been shown to be a reliable vehicle for communicating an organization's profile; they also provide the most far-reaching medium for promoting an organization's activities, service offerings and capabilities to the world. Website content analysis as a research approach has been used successfully to investigate the online communication of corporate social responsibility issues (Capriotti and Moreno 2007).

A limitation of this method is that websites represent only one approach for advertising the offerings, activities and services of higher education institutions. A website might not exhaustively cover the full range of activities or accomplishments of the organization. Understandably, some institutions are not diligent with regularly updating materials on their websites. Nevertheless, organizations have been shown to extensively communicate corporate responsibility issues such as environmental action on their websites (Capriotti and Moreno 2007). It is acknowledged that interviewing industrial design program leaders and academic staff may provide deeper insights that are not published on their website, but due to time constraints this method was not pursued in this paper.

This research deliberately considers only information that is available online, taking the viewpoint of a prospective student who would be investigating a degree program for potential study, and later on basing study decisions partially from data published on the university websites. Understandably, some institutions post comprehensive information on their websites while others do not. This situation is no different from hardcopy publications: some printed prospectuses are packed with detailed information, including curriculum listings and course descriptions, while others publish relatively shallow information on what their degree is about.

The study is only interested in coursework degrees, wherein students undertake a formally structured program of study, with regular classes that should be attended and courses or subjects that must be successfully passed to obtain the degree. It excludes research degrees, which in comparison are more self-directed, typically do not have pre-scheduled group classes, tend to be focused deeply on one's specific interest, and have a substantial scholarly thesis as its main outcome. This paper believes that a research degree is flexible enough to be tailored to studying sustainable design if a student proposes to, while in a coursework degree learning sustainable design would only occur if the program stipulates this topic in its curriculum as a core unit, as a module, as a major project, or as an elective.

A coding pattern was devised to analyze the data, whereby it was noted whether or not the program overview, objectives, or mission-vision statement mention aspects of sustainability; whether the curriculum offers courses or modules for learning sustainability or sustainable design; and whether or not the student design projects in the online galleries show evidence of consideration of social responsibility or ecological sustainability. Eight possible combinations were found, summarized in Table 1. The last column summarizes the strength of the evidence of sustainability coverage; those which have no evidence of sustainability in their program overviews, course descriptions and student projects are rated zero, while those which have evidence in all three indicators are rated 3. Degrees wherein one 'yes' was present were rated 1, while those with two 'yes' responses were rated 2.

Table 1 Coding scheme for assessing sustainability evidences in industrial design degrees

Code	Sustainability statement in program overview	Sustainability courses in curriculum	Sustainable design projects shown	Strength of sustainability coverage	Undergraduate courses described by this code	Postgraduate courses described by this code	Overall courses described by this code
YYY	Yes	Yes	Yes	3	21	7	28
YYN	Yes	Yes	No	2	26	21	47
YNY	Yes	No	Yes	2	5	2	7
NYY	No	Yes	Yes	2	3	0	3
NYN	No	Yes	No	1	20	13	33
YNN	Yes	No	No	1	14	13	27
NNY	No	No	Yes	1	5	4	9
NNN	No	No	No	0	23	16	39

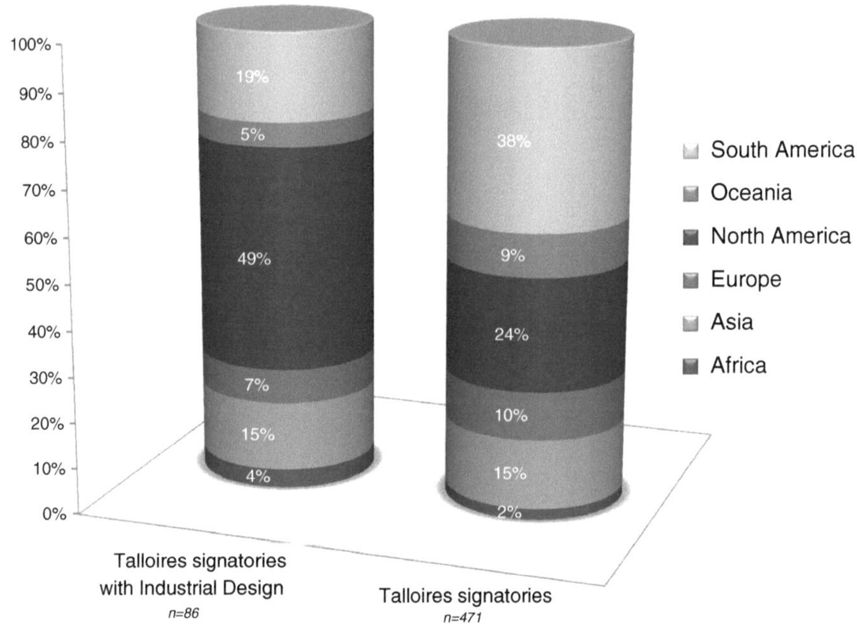

Fig. 1 Comparison of geographic distributions by continent of Talloires Declaration signatories and Talloires Declaration signatories with offerings in industrial design

3 Results

The websites of all the 471 Talloires Declaration signatory institutions were considered for this study. Of these, only 86 (18 %) were found to have relevant offerings in industrial design or product design (Fig. 1). The universities and colleges shortlisted for further analysis in this study come from 22 countries from around the world. The subset of countries and continents in the study somewhat differs from the total set of Talloires Declaration signatories: the proportion of North American institutions in the study is double that of the actual proportion of North American signatories, while the proportion of South American institutions studied was only half of the actual signatories from that continent. Brazil, the United States, and Colombia had the most number of institutions represented, but the majority of academic degrees came from the United Kingdom, the United States and Australia (Fig. 2).

3.1 Degrees

Overall, a total of 193 academic degrees related to industrial design were found available amongst the 86 eligible Talloires signatories. Of these 60 % were undergraduate and 40 % were postgraduate. The undergraduate programs were

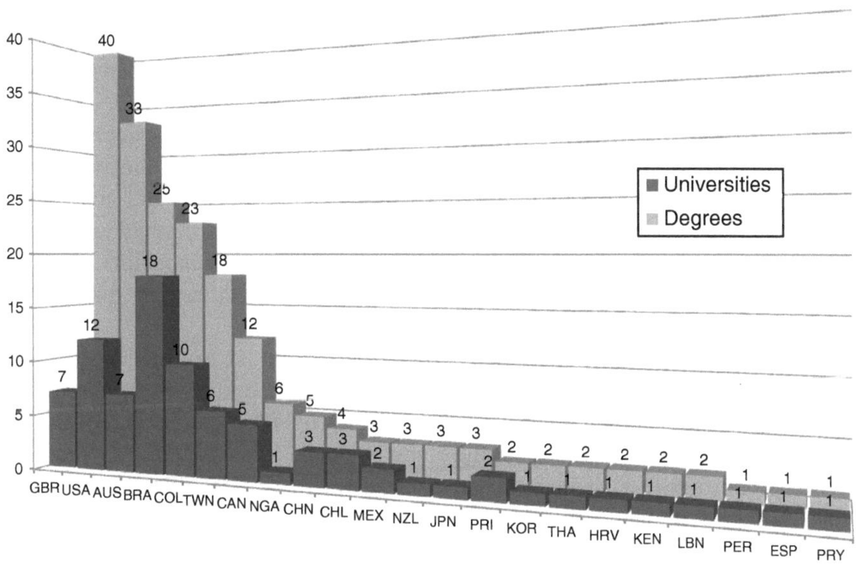

Fig. 2 Country origins of the universities and degrees included in the study

mostly bachelor's degrees (or licentiates or 'professional titles' in Spanish-heritage countries). While the bulk of undergraduate degrees were majors in 'industrial design' and 'product design', other related majors included were 'three dimensional design', 'design and technology', 'design for industry', and 'furniture design'. There were also a few majors in 'product design engineering', 'product design innovation', and some dual degrees in industrial design and engineering. In cases where strategic management and environmental science degrees showed significant coverage of industrial design training, these were also included.

In the postgraduate area, the majority studied were master's degrees in design, industrial design, or product design. Other postgraduate concentrations included design management, design studies, design innovation, product development, strategic design, service design and product design engineering. There were also some sustainability-specific master's degrees: 'Sustainable Futures' from University of Plymouth; 'Design Innovation and Citizenship' from University of Glasgow; 'Sustainable Product Development' from University of Strathclyde; and 'Design and Urban Ecologies' from Parsons The New School for Design.

For the purpose of this study, if a degree was available as a 1½-year master's degree, 1-year postgraduate diploma and ½-year postgraduate certificate, these academic awards are counted as three separate degrees. Similarly an undergraduate qualification which was available as a 3-year bachelor's degree and extendable as a 5-year integrated master's degree was regarded as two separate programs. Some universities had programs which were available both as a bachelor of science and as a bachelor of art, with some differences on final year requirements; these were treated as being two different degrees.

3.2 Overall Sustainability Inclusion

On average only 14 % of the programs studied showed strong evidence of sustainability coverage (Yes = 3) in their program overviews, their course descriptions, and in their gallery of student works (Fig. 3), based on the coding scheme detailed in Table 1. Thirty percent show evidence of two of the three sustainability indicators (Yes = 2). Those which show evidence in just one indicator total 36 % (Yes = 1). Therefore 80 % of the 193 industrial design degree offerings were found to have demonstrated at least one proof of incorporation of sustainability. 20 % do not show any indicators of sustainability aspects, at least not in their online information (Yes = 0).

A higher proportion of undergraduate programs show strong evidence of sustainability (Yes = 3) than the overall mean, and this proportion is double that of their postgraduate counterpart. There is an equal proportion of undergraduate and postgraduate degrees which do not have any evidence of sustainability (Yes = 0).

The proportion of sustainability coverage as evidenced in all three indicators looks strongest in the industrial design degrees in the United States, Mexico, Australia, Canada and the United Kingdom (Fig. 4). However, it must be noted that the actual number of degrees studied in Canada and Mexico are small (Fig. 5).

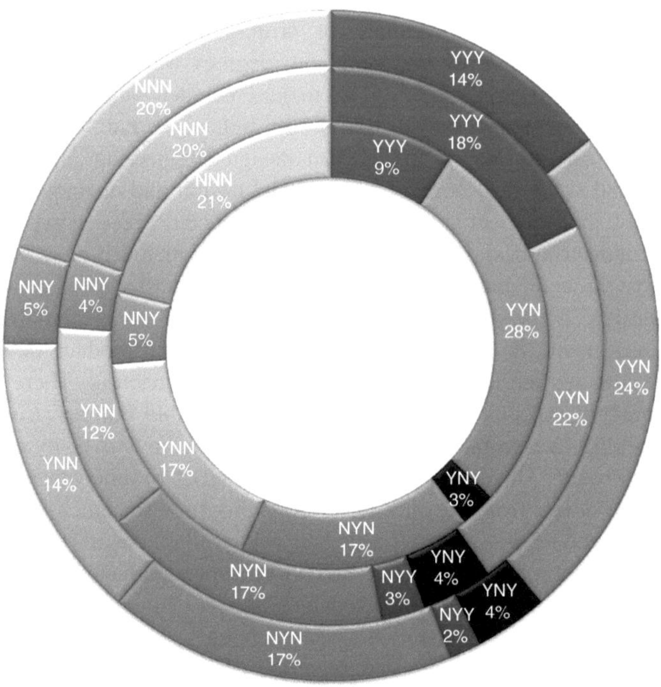

Fig. 3 Summary of sustainability coverage in industrial design in Talloires Declaration signatories. (*Outer-ring* represents overall data; *middle ring* is for undergraduate data; *inner ring* is postgraduate)

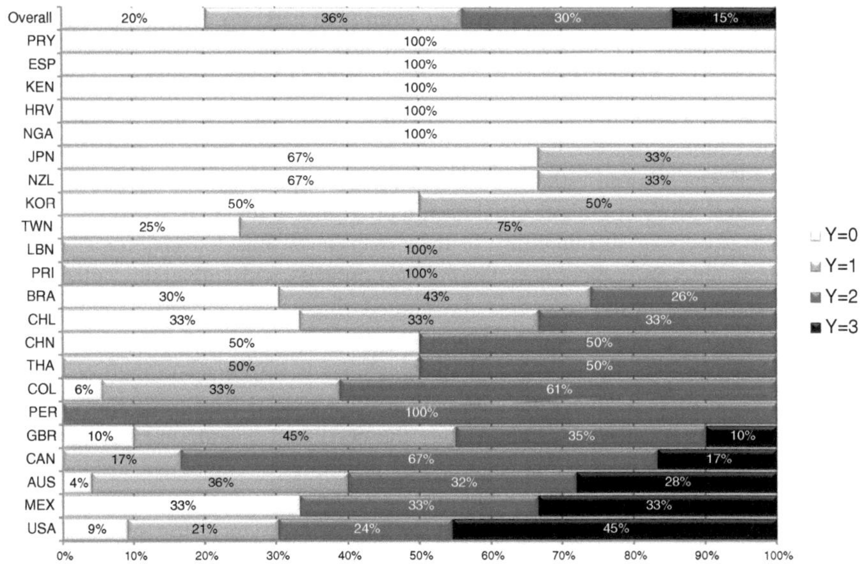

Fig. 4 Sustainability coverage in industrial design degrees by country, showing proportions

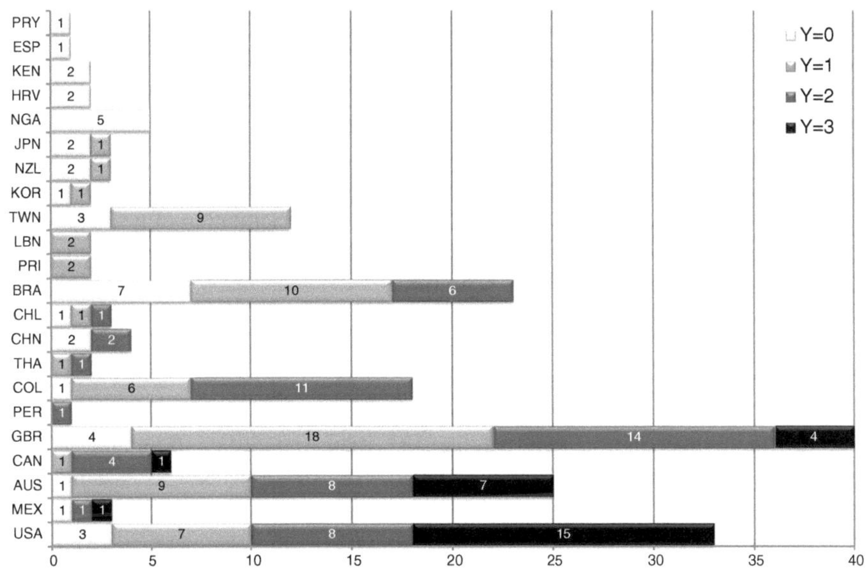

Fig. 5 Sustainability coverage in industrial design degrees by country, showing actual values

The lone universities representing Spain, Paraguay, Croatia, Kenya and Nigeria didn't show any evidence of sustainability in their program overviews, curriculum, and student projects; the few representatives from Puerto Rico and Lebanon had evidence in only one indicator and the one from Peru had two indicators. The South American countries of Colombia, Brazil and Chile, which had many representative institutions, had a reasonably good showing of mid-level sustainability coverage (Yes = 2 + 1). Three-quarters of the Taiwanese programs had one indicator, and half of the Chinese degrees had two indicators. The programs from New Zealand, Japan and Korea showed rather weak coverage with only one indicator.

On average, 56 % of the industrial design programs in the 86 universities studied mentioned aspects of sustainability in their overviews, mission-vision statements, and program aims and objectives. Among the countries with universities offering at least 15 industrial design degrees, the United States, the United Kingdom and Australia have the largest numbers and largest proportions of design programs which profess in their overviews that sustainability is important to their academic activity. 73 % of U.S. undergraduate and postgraduate design programs claimed that their academic offerings acknowledge the need for contemporary designers to be educated within a system that is guided by principles of ecological sustainability and social responsibility. The adoption of technological innovations and the embedding of pleasurable user experiences into new products for the market are central to industrial design education and practice, and while all programs talk about covering these fundamental learning objectives, more than half the industrial degrees worldwide are now aiming to equip their graduates with a strategic mindset that engages with the wicked problems challenging humanity and the natural environment. There is an expressed vision to prepare future practitioners who are human-centered and are global citizens, keen to genuinely improve people's lives while respecting the ecosystem that sustains us. Another mission commonly mentioned is the inculcation of an ethical conscience that would enable students to think twice about the extended influences of their design decisions. There is also a growing move to consider the broader application of the possibilities for design thinking beyond its commercial role and more into the facilitation of positive social change. For some universities, sustainability has become a core value defining their teaching, research and operational activities.

3.3 Courses, Subjects, Modules

Slightly more than half (57 %) of the 193 industrial design programs studied actually had a course, subject or module focused on or related to sustainability aspects. Amongst the countries which have at least 15 industrial design degrees studied, the United States (82 %), Colombia (78 %) and Australia (76 %) showed the highest proportions of programs which in fact offered sustainability as a course of study in their officially published curricula.

Ecological sustainability in the design curriculum is invariably offered under such course titles as 'ecodesign', 'ecological design', 'sustainable product design', 'design for environment', 'green design', 'ecofriendly packaging' and 'design for sustainable futures'. In many cases the topic is embedded as a core outcome or

Table 2 Students' sustainability projects

Topic	Student projects
Children and education	Baby monitor to prevent sudden infant death syndrome; lightweight respiratory unit for kids; hospital-and-home infant incubator; multisensory indoor play for autistic spectrum kids; toy to encourage social interaction and physical activity.
Crime and personal safety	Automotive crash avoidance system; seat that prevents bag stealing.
Developing regions	Outreach vaccine backpack for developing countries; smartphone attachment for affordable eye examinations; prenatal care kit for rural and remote Australia; modular water filtration kits; low-cost and intuitive retina camera for eye disease; water transport and purification device for Third World; lighting system for remote African communities; waterless toilet for urban slums.
Disaster and lifesaving	Emergency communications support unit for search-and-rescue operations; Tracker for locating people trapped in buildings; on-site decompression chamber for scuba divers; portable emergency ultrasound; fire victims air mask; self-assembling emergency housing.
Elderly and disabled	Travel luggage for independent wheelchair users; mnemonic aid for traumatic brain injury patients; smartphone-connected diabetes management device; portable non-invasive multi-parameter tester for diabetics; night-time warning system for stroke victims; shared communications booth for retirement village residents; aqua aerobics exercise toy for elderly; self-operated arthritic pain relief system; renal failure lifestyle garment; improving wheelchair access on trains; lawnmower for ageing population; hands-free nebulizer for severe asthma; adaptive tableware for dementia sufferers; multipurpose garden tool for elderly; design and end-of-life planning.
Homeless and disadvantaged	Outdoor jacket with integrated sleeping bag to keep rough sleepers warm and safe; Red Cross emergency alert kit; furniture range for Salvation Army crisis housing services.
Public and occupational health	Electrical shock prevention for construction workers; occupational therapy food tray; emergency medical bag; ergonomic hip basket for coffee picking; accessible workstation for disabled in the workplace; firefighter walkie-talkie.
Environmental protection, nature conservation	Electric car for 2020; agricultural aerial reconnaissance; remote camera for monitoring marine environments; atmospheric water generation for subsurface crop irrigation; alternative hydroponic gardening; upcycling Chinese factory leftovers; rentable compact garden shredder; autonomous vehicle sharing system; rainwater energy generator; domestic food waste minimizer; park-as-pharmacy; human powered vehicles; air-powered motorcycle.

(continued)

Table 2 (continued)

Topic	Student projects
Enabling sustainable behaviors	Recovery of compostable waste from in-sink waste disposal units; personal carriage for plastic bag-free grocery shopping; secure storage compartment for bicycles; shopping trolley for parents with prams; recycling receptacle for small living environments; do-it-yourself retrofit water efficient shower; illuminated jackets and bike lights for night cyclists; vertical mount for bikes on the commuter train network; urban agriculture in a carton tube; soap from waste oil initiative; rethinking urban transportation in a world without oil; smartphone app that provides easy-to-follow guidelines that make help people make small, permanent life changes towards reducing personal environmental impact.

learning objective of mainstream 'product design studio' courses. Another learning opportunity for students is to cross-enroll in such electives as 'alternative/renewable energy systems', 'appropriate technology', 'biomaterials', 'cleaner production', 'design for X', 'environmental impact assessment', 'environmental resource management', 'industrial ecology', 'life cycle engineering', 'manufacturing and the environment', 'solar design', 'sustainable development', and 'sustainable engineering'; these options are typically offered outside the school of design, for instance in the school of engineering or school of science.

Social and cultural sustainability are covered via such courses as 'democratic design', 'design across boundaries', 'design and cultural identity', 'design and society', 'design for cultural groups', 'design for social responsibility', 'environmental ethics', 'inclusive design', 'local community design', 'social design', and 'sustainability and social impact'. There are also courses on 'contemporary issues in design' in which the evolving role of design and of the designer in modern-day society are discussed; and also 'design and business' which not only acknowledge the significant role of design in furthering an organization's moneymaking agenda but also emphasize the need for design to be economically viable and to be mindful of the sustainability pressures which affect the current business environment. There is a perceivable growth in the offerings on 'sustainable products and services', 'systems thinking' and 'service design' as strategies for expanding industrial designers' innovation skills into creating solutions that are less materially intensive and that consider the complex interrelationships of the elements in the system in which design solutions operate.

There are a few emergent courses on 'sustainable design thinking', which, as a human-centered process, can be used to empathize with people about the effects on their lives of the unsustainability of contemporary production and consumption, and provide impetus to ideate and co-create meaningful solutions with them.

The remaining 43 % of programs without identified courses in sustainability are of two types: those whose published curricula do not appear to include any courses related to design for sustainability, and those whose websites do not exhibit any list of courses.

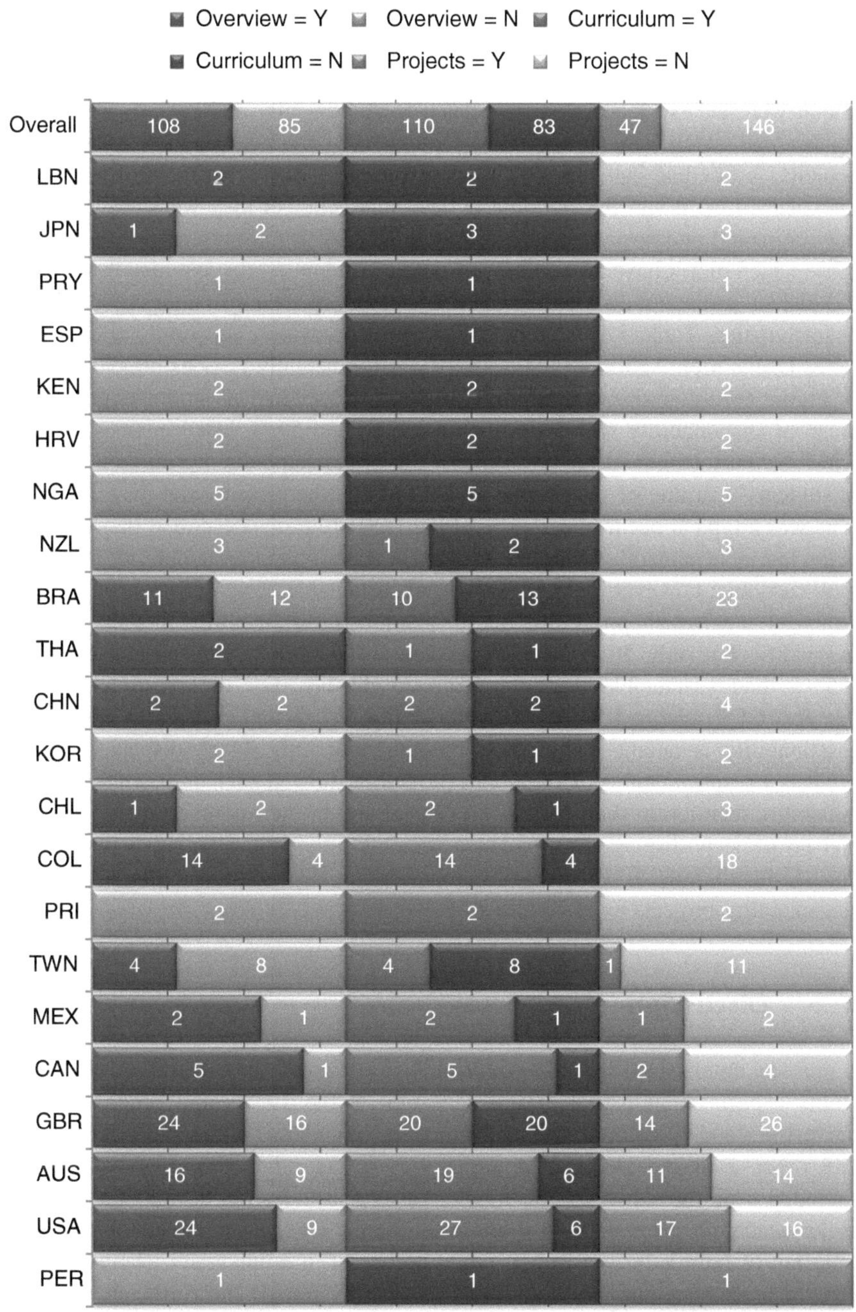

Fig. 6 Presence of sustainability attributes in program overviews, curricula, and student projects

3.4 Student Work

Only a quarter of the 193 industrial design programs displayed student projects which were clearly focused on or related to ecologically and socially sustainable futures, and which were appropriately captioned to communicate their sustainability benefits. A few had galleries and images of student works but most of these were unexplained, so it was not possible to determine whether they were designed with environmental and social impacts in mind. The 76 % majority comprise not only those which didn't have any online galleries of student works, but also those which did display project images but had no explanation of their sustainability aspects or characteristics. Table 2 lists some student projects which apparently address Talloires Declaration Action #3 (educate for environmentally responsible citizenship).

Figure 6 further shows that while 56 % of the studied programs claimed coverage of sustainability within design education, and 57 % listed courses of study which actually taught sustainability, only 24 % showed evidence on their websites of students' learning of sustainable design through the projects that they displayed. This was a typical occurrence. 73 % of the U.S. programs described themselves as believing in sustainability, and 82 % of them actually had sustainability courses in their curricula; however, only 52 % showed sustainable design projects of students. Amongst the British programs, 60 % mentioned sustainability in their program overviews, but only 50 % actually offered sustainability courses, and even less (35 %) had sustainable design projects to display. For the Australian design degrees, 64 % claimed belief in sustainability, 76 % had courses to ensure that sustainability is taught within their program, but only 44 % displayed student projects on sustainability. Colombian programs showed high proportions of sustainability mention in their overviews (78 %) and in their course titles (78 %) but had no projects displayed to evidence students' authentic learning in sustainable design. Institutions in 15 countries, representing 73 programs, didn't have any sustainability projects exhibited on their websites.

4 Discussion

It must be emphasized that this paper only investigated the integration of sustainability aspects in industrial design education in the 193 design programs in 86 universities and 22 countries which formed the boundaries of the study. There were 385 other Talloires Declaration signatories in 32 other countries which were excluded from the study, for the simple reason that industrial design is not offered as an academic program into those institutions. This paper aimed to capture a snapshot of how industrial design might be taught in a university where its president or leader has publicly committed to actively pursue the furtherance of sustainability in its teaching and research activities. Its findings should not be misconstrued to apply to all Talloires Declaration signatories, nor do they relate to the non-design programs being offered in those institutions.

While the data for the design programs studied do not seem to paint a strong picture of sustainability integration for the institutions, it must be stressed that the study only looked at the academic side of the university's overall activity. They do not reflect other actions occurring in the university in terms of campus greening, such as their attempts to save energy and water, or minimize waste.

To be fair, a good number of the universities in the study have been found to be quite involved in sustainability as they have promised in the Talloires Declaration. The United Kingdom's largest student campaigning network, People and Planet, awarded the University of Plymouth and the Liverpool John Moore University – both Talloires Declaration signatories – 'first class' in its 2013 Green League ranking of 143 British universities (People and Planet 2013). One aspect investigated by the Green League was the commitment to integrate sustainability into the curriculum in the corporate or strategic plans, in the environmental policies, and in the teaching and learning strategies of these institutions; it also checks if support or training was made available to all staff to help them in this curricular sustainability integration, and if there were review and reporting processes to monitor. All the British universities in this paper received first class, second class, or third class honors on the People and Planet Green League tables, which also looks into 12 other criteria such as environmental policy, carbon management, waste recycling, student engagement, energy and water efficiency, transport emissions and sustainable food.

In the United States, various metrics have been used to evaluate America's greenest universities. Three institutions in the study—namely Appalachian State University in North Carolina, Grand Valley State University in Michigan, and University of Colorado Boulder—earned Gold ratings in the Sustainability Tracking, Assessment and Rating System™ (STARS) scheme of the Association for the Advancement of Sustainability in Higher Education, which asks participating universities to gauge their progress towards sustainability in their academic, engagement, operations, planning and administration activities (www.aashe.org). These three institutions were also placed in the top 41 amongst 162 universities in Sierra Magazine's 'Cool Schools' 2013 ranking scheme, which evaluates 80 aspects of sustainability teaching, research and operations. In the curriculum greening area, the Sierra criteria included: sustainability course curriculum and identification; sustainability-focused or related courses; learning outcomes, immersive experiences, literacy assessment on sustainability; undergraduate program in sustainability; and engagement of, and incentives for, staff and departments in sustainability research (Andrews 2013). Eight of the 12 U.S. institutions studied were also included in the 'Guide to 332 Green Colleges' (The Princeton Review 2014b), and nine have also signed up to the American College and University Presidents' Climate Commitment (www.presidentsclimatecommitment.org).

Of the 86 Talloires Declaration signatories studied, 11 are among the over 10,000 participants in the United Nations Global Compact, the world's largest corporate citizenship initiative, which encourages organizations to embrace and support sustainable and socially responsible policies, via its ten principles in the areas of human rights, labor, the environment and anti-corruption (www.unglobalcompact.org).

Four have joined the list of 5489 worldwide endorsers of the Earth Charter, which is a declaration of 16 fundamental ethical principles for 'building a just, sustainable and peaceful global society in the 21st century' and for 'inspiring in all people a new sense of global interdependence and shared responsibility for the well-being of the whole human family, the greater community of life, and future generations' (www.earthcharterinaction.org). Eight are among the 198 members of the Cumulus International Association of Universities and Colleges of Art, Design and Media, whose Kyoto Design Declaration in 2008 pledges to create sustainable designs and human-centered societies, through collaboration of institutions, businesses and individuals (www.cumulusassociation.org).

The industrial design institutions which are Talloires Declaration signatories represent only a minor subset of the universities around the world with academic programs in design. Ramirez (2007) found 836 higher education institutions offering degrees in industrial design and product design worldwide, so the 86 universities in this present study represent only a tenth of the total. However it is encouraging to know that this present study, which found 80 % of the 193 industrial design degrees showing at least one indicator of sustainability, appears to show a marked improvement to the findings of a similarly designed study 2 years earlier, where 278 (69 %) of all the 402 industrial design programs in the English speaking world (Australia, Canada, UK, Ireland, USA, and South Africa), demonstrated evidence of at least one sustainability indicator (Ramirez 2012).

The Talloires Declaration's main ambition is to ensure that all university graduates are environmentally literate; it is worth noting that nowhere in this historic document does the word 'social' appear at all. When the declaration was drafted 24 years ago, the prevailing thought might have been that environmental awareness would lead to knowledge and understanding, which in turn would lead to concern, and which would ultimately motivate pro-environmental actions (Fien 2003). This assumption has been argued to be very questionable, and it has been shown that a good understanding of the subjective dispositions of people and of societies is more important to achieve humanity's goals of a more positive environmental future; one suggestion put forward the re-emphasis of the social dimensions of change in the Talloires Declaration (Adlong 2013). It's not enough to gain education *about* the environment; effective and transformative education *for* the environment must enable people to take actions and to change personally to result in better environmental outcomes (DEH 2005). For these reasons, this paper considered sustainability in the design curriculum as not only those which addressed environmental protection or nature conservation, but included those which enabled behavioral change through design (Table 2). Moreover topics of social responsibility such as addressing the needs of developing regions, the elderly and disabled, the elderly and disabled, children, and those affected by crimes and disasters, were considered as encompassing elements of the human-centric dimensions of sustainability. In the tabulation of curricular subjects, it was promising to increasingly find the knowledge of sustainability being covered, applied and experienced as projects within design studio courses; again, this is a marked departure from past studies where sustainability literacy in the curriculum was achieved by delegating its teaching to the environmental science or social science departments of the university (Ramirez 2006). This goes back to the distinction between education *about* the environment and education *for* the environment, and how application of knowledge in the latter yields more lasting educational outcomes.

The act of publicly declaring adherence to sustainability by any large organization is often viewed with suspicion as 'greenwash'; this is particularly the case when commonly known polluting businesses such as the automotive, chemical and petroleum industries highlight their environmentally positive but trivial actions to cover up their significantly negative activities (Beder 2002). Unfortunately greenwashing does occur in academia as well (Greer and Bruno 1997), whether it's 'taking minor steps to adopt the appearance of sustainability but avoiding the really difficult changes' (Carlson 2006) or not making any real attempt or progress to implement a sustainability declaration within their institution (Wright 2003). It is too easy to follow the trend of committing to a public document that professes belief in sustainability pursuits and then forget about it because no one is going to punish you for not implementing it; like broad statements of intent marked in stone in a university's sustainability policy, they lead to nowhere if there is no budgeted action plan or specific directive to guide the institution's day-to-day activities (Wright 2003). Over 1,400 universities worldwide have signed one or more of the 31 existing declarations on sustainable higher education (Grindsted 2011), but many of these haven't actually acted on what they have committed, due to either the lack of incentive structures or the absence of regulation, policing or monitoring by the secretariats of these declarations. While it is true that many signatory universities are actually leading in campus environmental performance, there is also evidence on the inaction of the majority of signatories regarding what they pledged to do. The act of endorsing dominates the headlines in the university's webpage for some time, but for many it is somewhat rare to find follow-up news to prove that their obligations are actually being met and true gains are being achieved. They have received the reward of positive publicity even before any real action has eventuated, and the absence of accountability, tracking or public scrutiny (Bekessy et al. 2007) makes it easier to not do anything that's genuinely going to make a difference.

On a less cynical level, there is evidence that many of the Talloires Declaration signatories have established sustainability research centers, groups or institutes that would contribute to the generation of new knowledge for advancing a more positive future. Of note are the DESIS (Design for Social Innovation and Sustainability) Labs at Northumbria University and Parsons The New School for Design, which are two of 43 such school-based 'labs' scattered around the world and united by the DESIS network; these are 'groups of professors, researchers and students whose design and research activities are oriented towards social innovation' (www.desis-network.org).

5 Conclusion

One of the principal purposes of this paper was to verify if industrial design institutions which have pledged themselves to the Talloires Declaration actually have proof of advancing sustainability literacy beyond the greenwash. The analyses do show that 80 % of the industrial design degrees in the study show that they cover, or intend to cover, aspects of sustainability within their undergraduate or postgraduate academic offering. However, this is poorly evidenced in the amount of sustainable design projects actually displayed on the websites. Design projects are creative demonstrations of the application of learnt sustainability principles, and comprise one of the most concrete confirmations of sustainability coverage within design education. They also serve as inspirations for many about the innovation that happens within the walls of a university, so it is best to capture more of these in official websites.

Websites are an effective method for displaying sustainability commitments and capabilities of universities. With the increasing internationalization of higher education worldwide, prospective students from afar could be attracted to enroll in programs which match their interest, passion and ideology. A recent survey of 14,150 college applicants and parents in the United States revealed that 61 % would 'strongly', 'very much', or 'somewhat' factor a college's commitment to environmental issues in their decision to apply or attend a particular school (The Princeton Review 2014a). This substantiates the perception that sustainability made explicit in a university's academic, research and operational functions is important to students.

Being a signatory to the Talloires Declaration is one way to signal to the rest of the world that the institution values sustainability, and that it pledges to all their students that they will be at least environmentally literate when they graduate, if not environmentally passionate. These universities have already made their first leap of faith by endorsing the declaration; the next big thing to do now is to ensure that an implementation plan is there to guide them on their journey and ensure that their sustainability ambitions do get off the ground.

But certainly for sustainability to have a lasting imprint amongst graduates, universities have to go beyond education *about* sustainability, and aim to inculcate lifelong learning *for* sustainability. As the Chinese proverb goes, "Tell me and I'll forget, show me and I'll remember, but involve me and I'll understand". By its very nature, design education already involves students by challenging them with innovation projects in response to problem briefs; therefore 'involved' learning should effortlessly occur in the teaching and learning of design. Embedding sustainable design thinking in industrial design is trickier, as it requires passion and enthusiasm on the part of lecturers and tutors, and might even require the conversion of some design lecturers who might be climate-change deniers. As employees of a Talloires Declaration signatory, all academic staff would need to align with the sustainability mission-vision of the institution so as to effectively impart a common message to the students we are trying to educate. As more batches of sustainability-literate young people study in Talloires Declaration universities and practice responsibly after they graduate, then the vision of a more promising future could be within reach little by little, but a clear institution-wide action plan and progress monitoring system would be prerequisites to this.

Acknowledgment The author thanks Maria Charissa Maliglig Ramirez for her invaluable assistance in gathering and tabulating sustainability data from the websites of Talloires Declaration signatory institutions.

References

Adamson G (2003) Industrial strength design: how Brooks Stevens shaped your world. MIT Press, Cambridge 300p

Adlong W (2013) Rethinking the talloires declaration. Int J Sustain High Educ 14(1):56–70

Andrews A (2013) Cool schools 2013: the valedictorians. Sierra Mag 98(5):31–32

Beder S (2002) Greenwashing. In: Barry J, Frankland EG (eds) International encyclopedia of environmental politics. Routledge, Abingdon 544 pp

Bekessy SA, Samson K, Clarkson R (2007) The failure of non-binding declarations to achieve university sustainability: a need for accountability. Int J Sustain High Educ 8(3):301–316

Capriotti P, Moreno Á (2007) Corporate citizenship and public relations: the importance and interactivity of social responsibility issues on corporate websites. Public Relat Rev 33(1):84–91

Carlson S (2006) In search of the sustainable campus: with eyes on the future, universities try to clean up their acts. Chronicle High Educ 53(9):A10

Dannoritzer, C. (2012). The light bulb conspiracy: the untold story of planned obsolescence [videorecording]. Televisió de Catalunya, Barcelona, 75 min

DEH (2005) Educating for a sustainable future: a national environmental education statement for Australian schools. Australian Government Department of Environment and Heritage, Canberra 32p

Fien J (2003) Education for a sustainable future: achievements and lessons from a decade of innovation, from Rio to Johannesburg. Int Rev Environ Strat 4(1):5–20

Greer J, Bruno K (1997) Greenwash: the reality behind corporate environmentalism. Third World Network, Penang 258p

Grindsted TS (2011) Sustainable universities: from declarations on sustainability in higher education to national law. Environ Econ 2(2):29–36

Leonard A (2010) The story of stuff: how our obsession with stuff is trashing the planet, our communities, and our health: and a vision for change. Free Press, New York 352p

Packard V (1960) The waste makers. McKay, New York 306p

Papanek VJ (1971) Design for the real world: human ecology and social change. Thames & Hudson, London 378p

People & Planet (2013) Green league 2013: driving UK universities' transition to a fair and sustainable future. People & Planet, Oxford

Ramirez MJ (2006) Sustainability in the education of industrial designers: the case for Australia. Int J Sustain High Educ 7(2):189–202

Ramirez MJ (2007) Sustainability integration in industrial design education: a worldwide survey. In: ConnectED international conference on design education, Sydney

Ramirez MJ (2012) Inclusion of environmental and social aspects of sustainability in industrial design education. In: Leal Filho W (ed) Sustainable development at universities: new horizons. Peter Lang, Frankfurt, 994p

The Princeton Review (2014a) College hopes and worries: survey report. http://www.princetonreview.com/college-hopes-worries.aspx. (Last accessed 23 April 2014)

The Princeton Review (2014b) Guide to 332 green colleges. Center for Green Schools, Washington 217p
ULSF (2009) Sustainability assessment questionnaire (SAQ) for colleges and universities. University Leaders for a Sustainable Future, Wayland
Wright TSA (2003) A tenth year anniversary retrospect: the effect of the Halifax Declaration on Canadian signatory universities. Can J Environ Educ 8:233–248, Spring

Author Biography

Mariano Ramirez Jr is a Senior Lecturer in the Industrial Design program at the University of New South Wales (UNSW) in Sydney, Australia. He studied Sustainable Technology and Product Development at the Delft University of Technology in the Netherlands. Prior to joining UNSW in 2001, Dr Ramirez was Academic Chair for Industrial Design at the De La Salle University College of Saint Benilde and at the Mapúa Institute of Technology in the Philippines; and Senior Industrial Design Specialist at the Product Development and Design Center of the Philippines. He has worked with a number of Philippine manufacturers on innovative product, package and graphic development approaches, and is a regular contributor to international conferences and academic journals on design and sustainability. He is joint convenor of the Oceania chapter of the Learning Network on Sustainability (LeNS). His research projects include studying the integration of environmental and social sustainability aspects in design education and practice, optimization of product lifetimes, and enabling sustainable consumption behaviors through design.

From Consumer to Citizen: Engaging Students with Participative Methods in Design

Antony Johnston

Abstract

This paper takes as its staring point a debate series organised by the University of the Arts London in 2013 to explore the relationship between enterprise and sustainability within the creative arts. Traditionally designers have attempted to generate and respond to the desire of consumers through the creation of products. Increasingly this is perceived as unethical in that it contributes to an unsustainable world. In response designers are drawing upon design methodologies that refigure the consumer in different ways that include: user, public, citizen and co-designer. Such approaches draw upon participative methodologies requiring greater use of relational skills such as empathy and facilitation. This creates new challenges for design educators in terms of how to foster such skills which are tacit rather than formal. This paper will consider some concrete examples from design curricula to explore how educators are adapting teaching and learning approaches.

Keywords

Design · Sustainability · Participation

A. Johnston (✉)
Centre for Learning and Teaching in Art and Design, University of the Arts London, 272 High Holborn, London WC1V 7EY, UK
e-mail: a.johnston@arts.ac.uk

W. Leal Filho et al. (eds.), *Integrative Approaches to Sustainable Development at University Level*, World Sustainability Series, DOI 10.1007/978-3-319-10690-8_17

1 Introduction

Design represents a fundamental and far-reaching human activity (Wahl and Baxter 2008); which according to Graedel et al. "80 % of a product's environmental and economic costs [are] committed by the final design stage before production begins" (1995, p. 17). Broadly defined, design is "the expression of intentionality through interactions and relationships" (Wahl and Baxter 2008, p. 73) and can be considered in terms of two interrelated processes, the first material and the second immaterial. The material dimension of design is embodied in "cultural artifacts, institutions, patterns of production, and consumption" (ibid.); whilst the immaterial describes "the 'metadesign' of our conscious awareness, value systems, world-views, and aspirations" (ibid.). The immaterial dimension is significant as it defines the intentionality behind materialized design (ibid.). For Wahl and Baxter design taken holistically is to think and act materially whilst being conscious of the immaterial dimension that shapes what is defined as intentional (2008).

The traditional view of design has been one in which a material problem is perceived and following a process of investigation, usually incorporating testing, trial and error, a material is solution posed. Such a perspective emphasises the materiality of design, but eschews the immaterial dimension. Such a distinction can be helpful for describing how incongruences between the intention of a designer and their material consequences emerge:

> industrial design has put murder on a mass-production basis. By designing criminally unsafe automobiles that kill or maim nearly one million people around the world each year, by creating whole new species of permanent garbage to clutter up the landscape, and by choosing materials and processes that pollute the air we breathe, designers have become a dangerous breed (Papanek 1972, p. ix).

Whilst we may challenge Papanek's polemic many recognise the incongruence between values and behaviours which often appear to lead to contradictory intentions and practices. For many, design "was conceived as an art of giving form to products for mass production… [and] has been firmly embedded in consumer culture" (Margolin 1998, p. 83) ever since. Design approaches are often caught within the same unsustainable production-led model that has essentially remained the same since the 19th Century (Thackara 2005).

Within this context design educators are adapting pedagogic approaches and curricula in order to meet the challenge of an environmentally and socially responsible design. This paper will explore a number of ways in which educators are attempting to address the sustainability of design with their students; which has implications for both what is taught and how. This represents an important problem as similarly to other disciplines didactic approaches rarely lead to 'behavioural change'. In response educators are increasingly exploring approaches that draw upon participation and collaboration; which require the development of relational skills that can design people back into the design process (Thackara 2005). Some of the dynamics that create tensions and barriers for design-educators teaching sustainable design will be explored and following a brief discussion of design for sustainability, two curriculum case studies will be drawn on to illustrate.

2 Barriers to Sustainable Design Education

Despite the interrelated issues around sustainability and design, tutors often feel ambivalent in how they can respond. From my experience as an academic developer it is not unusual for tutors to express views similar to:

> I do not want to be seen to be pushing an agenda
> There is a danger of evangelising here.

Such perspectives arise from beliefs that sustainability is external to tutors' discipline and that their 'interest' is a personal one and therefore should be viewed suspiciously. Furthermore this represents a significant barrier to effecting change as tutors hold a particular view that shapes their understanding of the world and yet they perceive that they can distance themselves from this in terms of what they communicate to their students. Such a perspective takes an atomistic view of a person and their activity as a teacher. Drawing upon the definition of design above, it could be argued that tutors attempt to 'bracket' the immaterial dimension, by acknowledging that it is there and trying to minimise its effects. It is argued here that sustainable design and teaching requires an approach that integrates this rather than removes it. In both design and teaching the dynamics that challenge this are manifold, two are considered here, the first looks at design's shift from a modern to postmodern activity and the second considers significant trends in the UK HE system.

Modernist design was dominant during early-mid 20th century, is ahistorical and rooted in the rational pursuit of progress through science and technology. It was imagined that design could conceive of "infinite possibilities, [that] would enable the emancipation of humanity to take place: emancipation from ignorance, poverty, insecurity and violence" (Leonard 1997, p. 6). This 'grand narrative' however began to unravel when increasingly the notion of objective ahistorical truths leading inexorably to human progress and fulfilment terminated in its opposite. Adorno and Horkheimer offered an early and powerful critique of how belief in the enlightenment myth of rationality and progress culminated in the 'Final Solution' (1997). Such critiques paved the way for more postmodern sensibilities to emerge that have highlighted the manifold ways that power circulates and can subjugate people.

The importance of postmodern critiques of modernism are that they challenge the notion of an integrated knowing subject; and posit "the impossibility of transparent communication of meaning and the historicity of truth and knowledge that is local, partial and subject to continual reinvention…" (Brookfield 2000, p. 34). The position of a 'de-centered' subject can lead to feelings of disempowerment and the rejection/denial of certain types of knowledge and authority. As a result, sustainability becomes one further signifier in an endless circulating chain and loses any connection with its empirical and substantive dimensions. As Jucker notes:

> perhaps, there has never been a broader based and more carefully corroborated scientific consensus on any issue, involving thousands of the world's top climate scientists, backed up by the most elaborate computer modelling (Jucker 2014, p. 27).

Yet despite this consensus sustainability continues to be perceived as a personal expression or understanding of the world and in the mind of a reflective teacher can if propagated represent an abuse of power or 'agenda pushing'. Such a perspective is reinforced via recent trends in HE.

The second process concerns the development of HE which in recent years is characterised by discourses of marketization and managerialism; leading to the metaphor of HE as mass-producer of 'products' generated through learning that is goal-orientated and instrumental rather than developmental and transformational (Parker 2003, p. 529). Critics of the marketization of HE argue that individualistic processes of coming to know are prioritised as the value of knowledge is considered redundant for a market economy requiring courses that are:

> primarily vocational, [where] new 'delivery' technologies are welcomed…experiential learning is valued and 'core skills' are particularly prized. In 'fast capitalism' the content of HE courses is relatively unimportant because knowledge so quickly dates (Trowler 2001 pp. 11–12).

This process sets up what seems to be an abstracted individualistic process that is well reflected in terms of constructivist approaches to pedagogies. For many critics, constructivism has become a dominant discourse in HE because it fits an atomistic, individualistic, self-directed, outcome oriented approach and with its emphasis upon the construction of learning eschews the historical and symbolic dimensions of knowledge (Jucker 2014). Trowler indicates how disciplinary practices can both mitigate and enhance such discourses:

> Progressivist academics in art and design subjects…tend to subscribe to a story about their discipline and the pedagogy that 'must' accompany it. This stresses the need for intensive and extended project work supported by tutors, and ideally leads to a 'gestalt' experience. The studio system of working in which the tutor's attention is individualised, the Atelier system, encapsulates this ideology (2001, p. 13).

Whilst this 'story' is told such academics typically experience an erosion of disciplinary practices, for instance in relation to studio learning. Increasingly the accompanying pedagogies of art and design must contend with increasing numbers of students and fragmented and atomised practices which make providing a gestalt experience increasingly difficult.

3 Design for Sustainability

The issues above, beg the question in what way can design be sustainable? Environmental and sustainable concerns are not new in design disciplines; however in the last 40 years they have gathered momentum. Increasingly there is a questioning of the social purpose of design as opposed to, or in addition to its economic functioning. In response it is argued that sustainable design needs to problematize production-led models that lack progressive visions for society (Thackara 2008). Design therefore needs to "disengage…from consumer culture…and participate in projects for the welfare of humankind both inside and outside the market economy" (Margolin 2002, p. 98).

This raises the further question of how this can be realised, especially against claims that designers have "…never strongly threatened the underlying premise of design practice that the role of the designer is to work within the system of consumer culture and to provide service to his or her clients." (Margolin 1998, p. 85). Furthermore, as indicated above in terms of the unintended consequences of design, the process by which design intentionality comes about within consumer culture only appears to fail the vision of design to "reinvent design cultures so worthwhile projects are more clearly identified and likely to be realized" (ibid. 86). In order to realise visions for a sustainable design, designers need mechanisms: to reflect on how their intentionality is shaped, to refigure design as a process where the outcome is not an artefact but the encapsulation of a social vision, to engage in cultural practices that promote sustainable visions for design and to engage ethically; which following from the above, is a dialectical process between self and others within a community.

3.1 Realising Design for Sustainability

There are multiple ways in which designers seek to be more sustainable and can be viewed on a continuum between more modern or postmodern conceptions of design: hi-tech industries, can be seen as a further development of an expert designer as master over the world, such a model is in danger of being caught in the same patterns of thinking that produced the initial problem (Einstein cited in Jucker 2014, p. 10). 'Additive design', maintains the traditional processes and methods of design whilst adding sustainable elements and subtracting those that are un- or less sustainable; which might be through the transposition of materials that are ethically sourced and less resource intensive, rare and polluting. The danger of this latter approach is that it does not appear to accept the ultimate issue of environmental limits.

One caveat for both these approaches is that they are often accompanied by the discourse of cradle-to-cradle production which aims to develop products through "wholly beneficial industrial systems driven by the synergistic pursuit of positive economic, environmental and social goals" (Braungart et al. 2006, p. 7). Through such a framework materials are never consumed and energy for production comes from renewables (ibid.). The realisation of cradle-to-cradle however has been criticised as "the industrial creation of every object…requires more natural material than is contained in its final form" (Lettenmeier et al. 2009, p. 8).

Here I focus on co-design which significantly reorients the designer's role, where he/she engages in collaborative design processes with potential users of a service or product. This provides the potential to overcome the problems with modernist design by reducing the distance between designer and user and through being in a dialectical and relational engagement with a community to enable a positive social vision to emerge (Cross 1972). The reduction of 'distance' between designer and 'user', reconstructs design as a series of collective acts leading to co-design and co-creation. Co-creation describes any form of collective act whilst co-design describes "collective creativity as it is applied across the whole span of a design process" (Sanders and Stapper 2008, p. 2). Building on the definition of design earlier, co-design can be considered as a collective expression of intentionality, one that would need to negotiate multiple intentions at the material dimension and potentially multiple differing value-systems and world-views that shape these intentions at the immaterial level.

For Sanders and Stapper the co-design process adds a 'fuzzy font end' to the design process. Whereas traditional design could be considered as starting out from a problem defined by the designer, the fuzzy front end problematizes this and instead of problem-solving, the process commences with problem-finding. The purpose of this 'fuzziness' is to "determine what is to be designed and sometimes what should not be designed and manufactured" (ibid., p. 3). The change in orientation, from outcome to process, requires a reworking of the roles actors play in the design process. The 'user' is no longer perceived as a passive recipient of an outcome to address their needs, but an active participant in defining the design brief as well as participating throughout the later development and realisation stages. The design process here becomes refigured as a collective, relational and reflective research process and inquiry:

> In co-design…roles get mixed up: the person who will eventually be served through the design process is given the position of 'expert of his/her experience', and plays a large role in knowledge development, idea generation and concept development. In generating insights, the researcher supports the "expert of his/her experience" by providing tools for ideation and expression. The designer and researcher collaborate on the tools for ideation because design tools are very important in the development of the tools. The designer and researcher may, in fact, be the same person (ibid., p. 8).

4 Relational Skills

The implications of the above sections indicate that design-teachers and students must develop approaches in order to negotiate different demands if they are to realise design for sustainability. Whilst the implications are many, here relational skills are considered by drawing upon early person-centred theory of learning; which derives from a person-centred view of the psychotherapeutic relationship. According to Rogers, empathy is inter-relational and developmental, if one can be sufficiently empathetic and successfully communicate this to another person then it will in turn promote empathy in another (Rogers 1957). In terms of teaching, it is argued that empathy supports student learning through helping to develop an embodied and congruent sense of knowing (Rogers 1967). In terms of co-design it may further help move through difficult and conflictual periods in a participative process as the perspective of the other can both be understood and clarified; which in turn promotes recognition and understanding.

Empathy is recognised as increasingly important in design and education for sustainability as it is essential if we are to treat others with compassion, build solidarity in face of the scale of climate change and inequality, and to resolve conflicts between competing global objectives (Eriksson 2006). Whilst empathy's importance is understood it is seldom defined; Roger's defines empathy as an understanding of an other's:

> awareness, of his [sic] own experience, To sense the client's private world as if it were your own, but without ever losing the 'as if' quality'…When the client's world is this clear…and he [sic] moves about freely, then he can both communicate his understanding of what is clearly known to the client and can also voice meanings in the client's experience of which the client is scarcely aware (1957, p. 226).

In this sense empathy within the design process allows designers to place themselves in the context of others and to see their world as if they are that person, this is explicit and tacit, conscious and unconscious and opens up the world of material design, as well as intimating at the world of the immaterial. The dual function in maintaining the *as if* quality ensures the designer is both aware of their own world and what they perceive as that of others. The designer's sense of the other's world can then be 'checked' through communication and ideation methods. If participation is to be genuine and non-coercive (see Arnstein 1969) then it seems that empathy is an essential attribute for designers.

5 Case Studies in Teaching Design for Sustainability

The following two case studies reflect some of the work undertaken at the University of the Arts London in order to build capacity for and embed sustainability within the curriculum. Both cases illustrate the ways in which design-educators are attempting to negotiate disciplinary and institutional barriers and opportunities to create learning experiences that promote ethical and responsible ways of being. Each case highlights how a design process can engage students in participative systems that encourage them to think beyond their own context as individual consumer and instead engage them as citizens actively pursuing positive social visions through design.

5.1 Case Study 1: Conscientious Communicators

This case study outlines how an assessment strategy can be used to realign teaching and learning for design for sustainability. This intervention took place on second year BA (hons.) 'Collaborative Project' unit taught across graphic and media design disciplines (240 students). Conscientious Communicators represents a series of practical assessments that:

> place environmental and socially responsible thinking at the heart of…[the] curriculum… and foster a set of values and challenges for all academic levels and disciplines to consider (Temple and Hanrahan 2013, p. 1).

However, given the critiques of constructivism it might be argued that through focusing upon assessment only that this perpetuates an unsustainable curriculum through reinforcing atomisation and compartmentalisation. As indicated above 'resisting' dominant discourses in HE can be difficult. In this case such discourses are challenged by engaging effectively with them and drawing upon their productive power; if sustainability challenges and resists dominant discourses within HE then such resistance is partly determined by such discourses (Apple 1982). An effective resistance needs to engage with dominant discourses as it is through this circuit that power for change can be harnessed. The focus upon outcomes in HE often results on an emphasis upon assessment as that which makes an outcome observable, in this way by defining an assessment in terms of the values and approaches of design for sustainability then significant portions of the curriculum in terms of knowledge and teaching, learning and assessment strategies necessarily become realigned.

Conscientious Communicators engages predominantly with the process of design and categorises these within five general, but interrelated areas:

1. Thinking—Responsible innovation committed to what and how we create
2. Materials—Substrate intelligence which seeks to eliminate waste and develop new materials
3. Process—Exploring alternative production methods and energy use, leading to reduced impact on global resources
4. Message—Communication engaging, persuading and provoking with inspirational messaging
5. People—Changing attitudes and behaviours in regard to community and responsibility (Temple and Hanrahan 2013, p. 3).

Whilst each in isolation may be considered further atomisation, each can be seen to lead onto the other creating a complex, manifold and holistic perspective on design for sustainability that provides a plurality of entry points through which teachers and students can situate themselves. These assessment briefs raise problems that are able to provide a platform to think through and question design holistically, that is through both its material and immaterial dimensions. The focus upon process immediately raises questions of metadesign, for instance:

> The world faces some big challenges—from poverty, to human rights abuse, to environmental exploitation. In many cases, the processes by which we go about things (farming, distribution, manufacture, transportation, energy, education, social cohesion) are flawed and an investigation and reinvention of how we operate is required (ibid, p. 19).

Students are not posed the question whether there are big challenges, they are not asked to construct an individual understanding that inequalities exist, but rather these are taken as accepted in response to "the almost unanimous warning from scientists that our current lifestyles are causing untold harm to the ecosystem that support our existence…." (ibid., p. 12). Students and teachers are asked to respond to a practical problem that is material and caused by human practices in terms of action and inaction. In doing so a person/student needs to take up a position in relation to an object of knowledge through which their perspective can form a dialectical part. Critiques of constructivism seem to point to the view that there is no Other to provide this dialectic and as a result there is a self-referential ego constructing individual knowledge; which in turn implies a liberal ethical position. This requirement, to take up a position in relation to something, is important as it intimates an intersubjective domain where knowledge is created in response to an experience of the world and others. This constructs material practices as also relational and perspectival; which can highlight the immaterial dimension of design, but only if the other aspects of the curriculum are aligned with the assessment.

Such a procedural view needs to be supported through teaching and learning strategies that draw upon and encourage co-design approaches through: group learning and working and experiential approaches involving 'live' projects that challenge students to generate positive social value. Design here becomes orientated by a social vision and unlike modernist design remains subject to certain dynamics that are more in-keeping with the postmodern world without becoming relativist. One of the key aspects of ethical design is to take up a position, not of master or dominance over a problem, but as being subject to (see Loewenthal 2011). Designers, teachers, educators are all subject to many things for instance, knowledge, language, others, the world/environment, etc. Conscientious Communicators helps situate students as subject to knowledge and in doing so implies students as being subject to others and the world. This has the capacity to challenge reductive neoliberal discourses and to foster a more relational ethics that puts the other first as a being (human or more-than-human) whom one must respond 'responsibly' to (see Tangyin 2008). Art and design pedagogies are often expressed in dyadic terms as:

> 'A kind of exchange'…[that] expresses the predominantly dialogic nature of teaching in these subjects in which students' experience is central to teachers' concerns and learning is seen as a partnership (Shreeve et al. 2010).

Conscientious Communicators moves beyond this conception as this is not simply a dual relationship between teacher and student, but plural and not about an experience of, but experience in relation to. If designers are to move beyond modernist design then they need to also give up the phantasy of dominance and mastery over objects. But how can this be achieved? The immaterial dimension of design indicates the tacit dimension of our experience and as with environmental education simply pointing out how this leads to unsustainable behaviours does not necessarily result in a change. Design as a material process provides opportunities to engage in experiential and active learning that is rooted in practical activity, this aspect will be drawn out briefly in the following case study.

5.2 Case Study 2: Inside Outside

This case study considers an intervention outside of the formal curriculum through an optional and participative course. Blake and Salvadori developed the 'Inside Outside' project to explore sustainable printmaking (2013). The project drew upon experiential learning to expand upon what is considered as the studio; this was achieved by exploring printmaking in external locations. Each locale was intended to help participants think about the relationship between themselves, the environment and others, the locations were an urban park, a derelict housing estate and a rural coastal headland. Along with drawing upon experiential learning the project created a learning community through flattening out the hierarchies between teacher and student, as well as between students, as participants were drawn from a variety of courses and levels.

The project challenged assumptions of the designer as master over 'his' world in terms of control over the creative process and materials. As Trowler indicates above the 'story' of the studio as the container in which a designer works and engages in professional activity is incredibly important (2001). However whilst such a strong disciplinary boundary can be productive it can obstruct the development of different types of relationships. For instance the relationship between designer and the world can be one in which the world is perceived as more than a material resource, but as one which also has a part to play, as one participant comments she was able "to accept spontaneity and unplanned developments and accidents as a creative benefit" (ibid.). Which is in contradistinction to how one might work within a studio as one would immediately discard 'a mistake' and start again in order to meet an industrial ideal (ibid.). The project modelled the ethos of participative and co-design approaches, through focusing upon:

> cooperation where staff would not be staff, but would be part of this learning group. Our ethos was that a shared learning environment would foster collaboration as learning by doing, rather one in which teacher leads and student follows (ibid.).

Participative approaches build empathy in multiple ways: in terms of understanding the tacit dimensions of a creative piece of work, self and environment: "being on the spit of land by the sea, watching the change in weather and tide brought me closer to the marks I made and colours used" (ibid.). And in terms of empathy for others, "it made me travel to another epoch and imagine the lives of the people who used to live there. I could really feel those lives around me while printing" (ibid.).

6 Conclusion

This paper aimed to briefly contextualise and explore how design education in a particular setting has attempted to become more sustainable and ethically responsible; in doing so it highlighted some of the barriers and opportunities that exist within design as a discipline, the context of HE and how educators conceive of their role. Design for sustainability has attempted to address some of the problems with design; however tensions remain. It was argued above that more sustainable approaches involve shifting the role of the designer from a classic agent who is master of the world to one who is answerable to a community; and which fosters a different kind of ethical responsibility.

The case studies illustrate how teachers draw upon participative design approaches in order to structure learning experiences that subvert the typical role of student as passive recipient of knowledge; which in turn creates space for a more relational ethics to emerge. The second case study in particular shows how through engaging with environmental and social issues within experiential and group settings that empathy is fostered and how it can bring someone into relation with the material world and other beings. Whilst the effectiveness of such approaches may lie in the extent to which they are embedded within the socio-historical context of the discipline such approaches are still useful for other educators to consider. Both case studies reveal the tensions of embedding sustainability in the curriculum. Conscientious communicators represents a wide ranging intervention impacting a large number of students however it is able to do this through working within the frame of dominant discourses. The potential problem with this is that whilst sustainability may be considered holistically within the context of the unit of study there is the danger that sustainability becomes another compartmentalised part of the curriculum within a course of study. In contrast Inside Outside represents a standalone course that is able to holistically embody the principles of sustainability and participative approaches to learning; with their focus upon emergent curriculum, exploration and negotiation, but in doing so must lie outside a student's formal curriculum. These are contrasting approaches and whilst it may be possible to mediate between the two to derive a 'compromise' position some aspect of each will necessarily be lost. Here we can begin to see how power and discourses shape the curriculum; whilst both possibilities exist this may not be an issue; however in constituting part of the formal curriculum Conscientious Communicators is embedded and so has an enduring power not enjoyed by Inside Outside.

This paper highlights a number of areas for further investigation. Firstly, to consider how art and design pedagogies need or are changing in relation to the challenge of sustainability. Secondly, the extent to which such interventions are effective in terms of students activity beyond the interventions. Lastly, how participative methods create spaces for a relational ethics to emerge and how this might challenge neoliberal conceptions of practice and education. The latter would be of particular interest in the context of the two case studies which ultimately engage with dominant HE discourses either by working within or 'outside' them rather than by directly attempting to transform them.

References

Adorno T, Horkheimer M (1997) Dialectic of enlightenment: Verso

Apple M (1982) Education and power. Routledge, NY

Arnstein S (1969) A ladder of citizen participation. JAIP 35(4):216–224

Blake J, Salvadori B (2013) Inside outside: sustainable printmaking. http://insideoutsideprintmaking.blogspot.co.uk/, (last accessed 30/3/2014)

Braungart M, McDonough W, Bollinger A (2006) Cradle-to-cradle design: creating healthy emissions strategy for eco-effective product and system design. J Cleaner Prod 1–12

Brookfield S (2000) The concept of critically reflective practice. In: Wilson A, Hayes E (eds.) Handbook of adult and continuing education (new ed). Wiley, San Francisco, pp 33–49

Cross N (ed) (1972) Design participation: proceedings of the design research society's conference 1971. Academy Editions, London

Eriksson E (2006) Empathetic understanding of the existential situation of fellow human beings as a field of knowledge. In: Holmberg J, Samuelsson B (eds) Drivers and barriers for implementing sustainable development in HE. UNESCO, Paris

Graedel T, Allenby B, AT&T (1995) Industrial ecology. Prentice Hall/AT&T, New Jersey

Jucker R (2014) Do we know what we are doing? Reflections on learning, knowledge, economics, community and sustainability. http://rolfjucker.net/20140116_Do%20we%20know_incl%20Strachan_webversion.pdf. Accessed 5 Jan 2014

Leonard P (1997) Postmodern welfare: reconstructing an emanicipatory project. Sage, London

Lettenmeier M, Rohn H, Leidtke C, Schmidt-Bleek F (2009) Resource productivity in 7 steps. Wuppertal Institute for Climate, Environment and Energy. http://epub.wupperinst.org/files/3384/WS41.pdf. Accessed 30 March 2014

Loewenthal D (2011) Post-existentialism and the psychological therapies: towards a therapy without foundations. Karnac, London

Margolin V (1998) Design for a sustainable world. Des Issues 14(2):83–92

Margolin V (2002) The politics of the artificial. University of Chicago Press, Chicago

Parker J (2003) Reconceptualising the curriculum: from commodification to transformation. Teach High Educ 8(4):529–543

Papaneka V (1972) Design for the real world: human ecology and social change, 2nd edn. Rev. Academy Editions, Chicago

Rogers C (1957) The necessary and sufficient conditions of therapeutic personality change. In: Kirschenbaum H, Henderson V (1989) The Carl Rogers reader. Houghton Mifflin Company, New York

Rogers C (1967) The interpersonal relationship in the facilitation of learning. In: Kirschenbaum H, Henderson V (eds) The Carl Rogers reader. Houghton Mifflin Company, New York

Sanders E, Stappers J (2008) Co-creation and the new landscapes of design. CoDesign. http://www.maketools.com/articles-papers/CoCreation_Sanders_Stappers_08_preprint.pdf. Accessed 30 March 2014

Shreeve A, Sims E, Trowler P (2010) 'A kind of exchange': learning from art and design teaching. High Educ Res Dev 29(2):125–138

Tangyin K (2008) Reading Levinas on Ethical Responsibiltiy. In: Tze-wan Kwan (ed) Responsibility and commitment: eighteen essays in honor of Gerhold K. Becker Edition Gorz

Temple S, Hanrahan T (2013) Conscientious communicators. UAL, London. http://workflow.arts.ac.uk/artefact/file/download.php?file=88690&view=12948. Accessed 30 March 2014

Thackara J (2005) In the bubble: designing in a complex world. MIT, Cambridge

Trowler P (2001) Captured by the discourse? The socially constitutive power of new higher education discourse in the UK. http://eprints.lancs.ac.uk/13852/1/Trowler_Captured_by_the_Discourse.doc. Accessed 30 March 2014

Wahl D, Baxter S (2008) The designer's role in facilitating sustainable solutions. Des Issues 24 (2):72–83

Author Biography

Antony Johnston is a Curriculum Development Coordinator at the University of the Arts London; where he develops strategies for embedding sustainability in the curriculum. He has been teaching within design and media disciplines for 13 years and within HE teacher development for 7. He leads an Education for Sustainability unit on a Higher Education Academy accredited MA Academic Practice.

Integral Theory: An Expanded and Holistic Framework for Sustainability

Robert M. Fleming

Abstract

The meme of sustainability continues to gain traction in almost every corner of the planet, and yet, there is still wide spread resistance against the basic premise. While the Triple Bottom Line has proved useful and has aided in the movement forward, deeper and more holistic frameworks are required to engage wider and more diverse audiences. For those working in fields dominated by subjective thinking and qualitative analysis, such as design, sustainability presents a cold, neutral and unbending proposition. This paper will examine the promise of Integral Theory as a framework for sustainability that unites the subjective and objective aspects of human existence. Furthermore, Mark DeKay's Integral Sustainable Design and Barrett's Whole System Change Methodologies, which both use Integral Theory as their foundation, will also be discussed. Finally, the proposed Quadruple Bottom Line will be introduced as a new and superior framework to what is used in practice today. In this way, the likelihood of the adoption of sustainability as THE core meta-value for humanity will be increased, thus spurring deeper levels of environmental regeneration, social equity, economic viability, and cultural significance. Examples of how this leap forward has transformed curriculum development and campus operations will be shared.

Keywords

Education · Sustainability · Sustainable design · Integral theory · Curriculum

R.M. Fleming (✉)
College of Architecture and the Built Environment, Philadelphia University, School House Lane and Henry Avenue, Philadelphia, PA 19144, USA
e-mail: flemingr@philau.edu

W. Leal Filho et al. (eds.), *Integrative Approaches to Sustainable Development at University Level*, World Sustainability Series, DOI 10.1007/978-3-319-10690-8_18

1 Introduction

The purpose of this paper is quite simple: to make the argument that the Triple Bottom Line (TBL) framework is limited in its ability to capture and drive the full potential of sustainability as the foundation of a fundamentally new societal world view. The new Age of Integration possesses far deeper and far more profound possibilities than our current imaginations can account for—partially due to ineffective or incomplete frameworks. Accounting methodologies such as the Triple Bottom Line offer corporations the ability to pursue greening initiatives but they do not serve to inspire the imagination or instill hope in the million if not billions of souls seeking something better. Nor can it serve as the catalyst for a movement from casual and lightweight greening to a deeper and more authentic sustainability. Authentic sustainability includes the triad of the TBL with: Cognitive empathy replacing *people*; Economic viability replacing *profit;* and Ecological regeneration replacing *planet*. Furthermore an entire dimension of human motivation is absent from the TBL requiring the addition of a fourth variable of *experience* to the equation. Experience includes the dimensions of human existence which cannot be quantified and yet we know they exist in the interior of our beings, invisible to the eye but ever-present in our actions and behaviors. These include spiritual, phenomenological, aesthetic, sensory and biophilic motivations. As a response, a more holistic and inclusive framework is needed. The term Quadruple Bottom Line (QBL), developed in collaboration with Anne Sherman, forms a more comprehensive and compelling framework for sustainability. In order to explore the opportunities associated with the QBL, this paper will examine in some detail the efficacy of Integral Theory, Integral Sustainable Design and Barrett's Whole Change Methodology as underlying and powerful tools in shaping the Quadruple Bottom Line. Examples of how the frameworks are used to benefit university education will be shared.

2 Origins and Context

Before answering these questions we begin with the requisite historical narrative which explores cultural evolution not as a slow and gradual rise to new levels of complexity but rather through a series of hyper-accelerated jumps in human consciousness. The jump from dispersed hunter gatherer cultures to centralized agrarian societies and then to industrialized nations correlates well to the convergence of new energy sources and the invention of new communication technologies. Jeremy Rifkin argues in his book The Empathic Civilization: The Race to Global Consciousness in a World in Crisis that "The convergence of energy and communications revolutions not only reconfigures society and social roles and relationships but also human consciousness itself." (Rifkin 2009) The early twenty-first century, as characterized by unprecedented sharing of information via wireless networks and by the emergence of renewable energy technologies, demarcates a threshold from an industrial world view to a new integral world view. This includes a jump from and mechanistic conception of nature as immutable and infinite to a Gaia inspired view of nature as alive, intelligent and, most of all, fragile in the hands of humankind.

Additional context includes climate change. Not since the dawn of the age of Agriculture has the climate changed so greatly and so quickly. This reality has never been more clear today with dramatic shifts in weather patterns, unprecedented global rise in temperatures (for the last 12,000) years), extreme weather events and sea level rise to name a few. Some have argued that we have entered a completely new and unpredictable climatic period in history called the Anthropocene (Zalasiewicz 2010). In addition, the specter of peak oil weaves in and out of the headlines as Canada pursues Tar sands and oil companies continue to drill in deep waters risking calamities such as the BP oil spill. While fossil fuels remain available, the difficulty in extraction, transportation and processing suggests that prices will continue to rise. But even if climate change and peak oil are acknowledged as real, human action and response is still limited. Competing motivations for profit, comfort and convenience dominate daily activities for some, while others, stuck in poverty spend their time living day to day. Hence, the inclusion of social equity in the Brundtland Commission (WCED 1987) and the clarion call for the fight against poverty in Lester Brown's Plan B 4.0 (Brown 2009). These and many other forces have helped to shape contemporary understandings of the Triple Bottom Line (Elkington 1997). If in fact we have entered a new world view based on deeper levels of integration coupled with dramatic environment changes, a new context for sustainability has emerged. The diagram below left in Fig. 1 contains a well-worn, but sadly, a poor representation of said integration. The diagram speaks to a fragmented, divergent and incomplete set of societal imperatives that intersect in rare occurrences to reach sustainability. The diagram below right in Fig. 1 provided by Elkington's version of the TBL is far superior in the way it expresses a sense of integration or at least balance between the "lines" and therefor is better suited for today's drive towards deeper levels of integration. However, it also presents a potentially incomplete view of the totality of sustainability, leaving the variable of *experience* unrepresented.

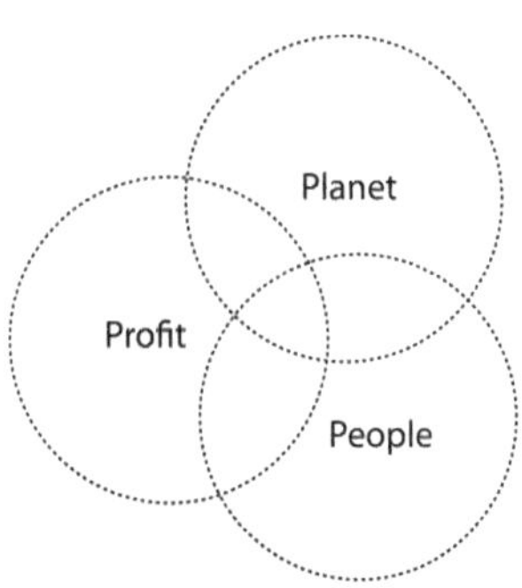

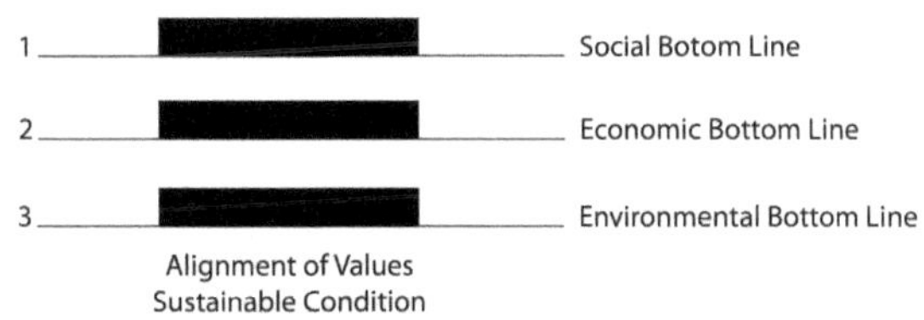

Fig. 1 The Triple Bottom Line *Source* (*Left*) Re-drawn and adapted by author from a standard diagram (*Right*) Elkington (1997)

3 Integral Theory

The framework of Integral Theory, invented by Wilber (2000) serves as way to bring the dimension of *experience* into sustainability in a seamless and integrated way. It is different from the highly valuable and better understood systems thinking approach to sustainability which tends to focus on the set of relationships in the context of an objective, measurable, interrelationships of systems and their impacts. Integral Theory, perhaps, falls under the larger umbrella of "Soft Systems" as developed in part by Peter Checkland (Checkland and Poulter 2006) among many others and could have been influenced by the work of Enid Mumford (Stahl 2007), who advocated for the participatory design process and the role of ethics in systems thinking.

It should be noted that the use of Integral Theory in this discussion is limited on the first of first of the five elements of the AQAL approach—standing for all quadrants, all levels, all lines, all states and all types (Esbjörn-Hargens 2009). Integral Theory is only beginning to be respected in wider academic circles and Ken Wilber himself has received his fair share of criticism from other integral thinkers, (Visser 2003) especially in Jeff Meyerhoff's book, *Bald Ambition: A Critique of Ken Wilber's Theory of Everything*. But if Wilber's frameworks are taken strictly at face value and stripped of their origins and influences, the potential for this holistic method for communicating sustainable design to students as an audience, for example, is *the* clearest and most compelling model developed to date.

Integral Theory, as shown in Fig. 2a offers such a framework.

The right column focuses on the objective, measurable, quantifiable aspects of reality—the world and individuals as viewed from the exterior. The left column focuses on the subjective, the immeasurable and the undefinable aspects of reality as understood from an interior perspective. The difference between the two sides

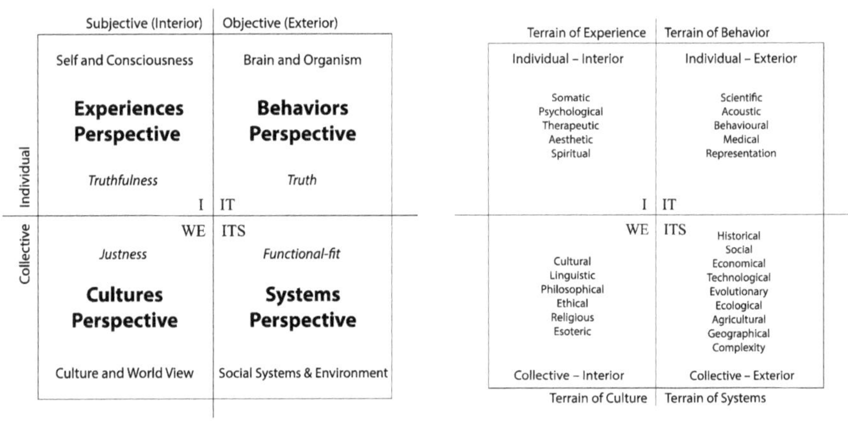

Fig. 2 **a** Ken Wilber's integral theory, *Source* Wilber (2000). **b** Ecological perspective via integral theory. *Source* Esbjörn-Hargens (2009)

can be understood as the *map* (right side—objective view) and the *map maker* (left side—subjective view). The rows reflect either an "individual" point of view or a "collective" point of view. The difference between the rows is critical to the success of Integral Theory because common experience has shown that a person acts and behaves very differently than groups of people.

As shown in Fig. 2a, the lower left quadrant **Cultures Perspective** contains the collective world views of a particular culture, organization or larger society—in short the "WE Perspective". This is the zone that expresses the larger societal consciousness. Cognitive empathy, social equity, and cultural heritage are all located within this quadrant. Figure 2 above right was created by Sean Esbjörn-Hargens, a leading scholar in Integral Theory. His placement of the *ethical* in the lower left quadrant reinforces the inclusion of social equity in the Cultures Perspective.

In contrast the upper left quadrant: **Experiences Perspectives** deals with personal consciousness, an interior experience in non-quantifiable terms—in short the "I Perspective". Beauty, which can never be universally agreed upon, is an example of something that is experienced differently by each human being. Universal consensus is impossible to attain and remains purely subjective in nature. Lance Hosey argues that, "If it's not beautiful, it's not sustainable (Hosey 2012)." His point is well taken as the laws of attraction, so necessary for the regenerative processes of nature, can also be harnessed by humans to imbue sustainable design projects with high quality aesthetics in addition to the typical quantitative aspects of a green project. The logic is simple: those buildings and structures that have a sense of intrinsic beauty often become associated with deep cultural significance thereby leading to a longer life of said building. Historic preservation then serves as a form of sustainability (WBDG) and is an excellent example of how the fourth bottom line of experience for the proposed QBL can drive sustainability initiatives.

However, the LEED® Green Building Rating System, Passivhaus and other systems do not account for beauty and/or cultural significance in their criteria. And for good reason: aesthetics cannot be measured quantitatively and therefor do not have a place in those systems. However, in recent years the Living Building Challenge and others have begun to accommodate the ambiguity of beauty as integral to the evaluation process.

In the upper right quadrant the **Behaviors Perspective** contains the objective view of the individual or of an isolated object or a specific point in time. All of these are addressed as something which has material qualities that have measurable observable behaviors (location and size). This is better understood as a set of individual *facts* that can be accurately measured and recorded. Examples include: the amount of rainfall on a given day or the rate of cars that pass an intersection, or the weight of bird. Sean Esbjörn-Hargens, places the Scientific in this quadrant (see Fig. 2 on the right above) and rightly so, as experiments are rigorously tested to insure accuracy. It should be noted that the term "Behaviors" has proven extremely difficult for students to grasp and recently, I have used the simple term of "facts" to describe this quadrant's content.

The lower right quadrant **Systems Perspective** (ITS) describes the observable social and environmental systems that lead to the "facts." For example, the amount of rainfall on a given day is the result of complex weather systems understood as a series of dynamic relationships that ultimately lead to average rainfall amounts—a specific fact which is contained with the upper right quadrant. Or, the interactions between work hours, available public transportation and settlement patterns might explain the "rate of cars passing a give intersection". Lastly, the weight of a bird can be explained by complex ecological forces that, over time, explain how a bird comes to have a certain weight. To say that this quadrant reflects a "systems" perspective would be an understatement.

The subjective lower left and upper left quadrants clearly impact any system. In the example above regarding traffic patterns, settlement patterns do not occur solely as a logical response to opportunities for better access to food, water, infrastructure, or affordable housing. The US suburban phenomenon of the 1950s and 1960s had as much to do with racial tensions as it did with the advent of cheap cars and cheap gas (Schaefer 2008). While this seems like an obvious connection, I also speculate that the "experiential draw" of early suburban communities with fresh air, open space, and views to the sky may have been as much a driver of new settlement patterns as any of the other factors.

Brown (Brown 2005) made the first link between Integral Theory and sustainable development in a much cited article in 2005: *Theory and Practice of Integral Sustainable Development*. In the article he explores the importance of aligning human values towards a common purpose—sustainability. This approach, he argues is key to enabling the more practical and quantifiable aspects of sustainable development which deals with systems and performance. In 2014, these connections are more than obvious, but the Integral Theory framework has proven remarkably useful in teaching and learning because it allows the subjective and objective realms to be equated *and* unified in a comprehensive framework that is potentially superior to the Triple Bottom Line.

The four quadrants of Integral Theory map well onto the four components of the previously discussed Quadruple Bottom Line (social, experiential, economic and environmental). For example, the environmental bottom line maps well to the lower right systems perspective. The social bottom line maps well to the Cultures Perspective, especially when the word "ethical" is used as one of the defining attributes. The experiential bottom line directly matches the experiences perspective. And lastly, but a bit more awkwardly, the economic bottom line connects to the behaviors perspective.

4 Integral Sustainable Design

More recently Professor Mark DeKay developed the framework of Integral Sustainable Design (ISD) (DeKay 2011). Clearly based on Integral Theory, ISD offers an opportunity to finally organize the typically disparate entry points to sustainable design into an interdependent, cohesive and inclusive set of principles that can

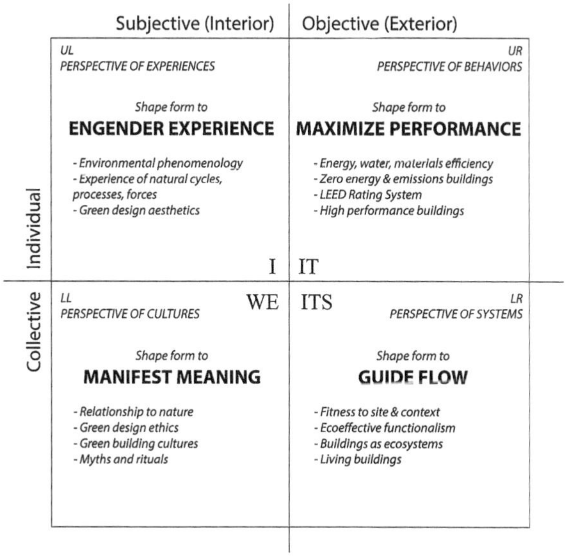

Fig. 3 The four perspectives of integral sustainable design. *Source* DeKay (2011)

easily be translated into strategies which can all be fused together into one integrated system—a sustainable built project (DeKay 2011).

In Fig. 3 below, Integral Theory has been re-purposed by Mark DeKay as a set of "imperatives" that help to guide the actions of a designer. ISD resolves that age old question in architecture: Does form follows function? Here the four divergent perspectives are adopted simultaneously and used to move towards a deep version of sustainable design that transcends the more predictable energy efficient design approach to generate a much richer, more nested and potentially superior definition of sustainable design (DeKay 2011). As an example, a university campus architect can now make a logical and perhaps compelling argument for beauty in design as intrinsic to the larger goal of sustainability.

5 Barrett's Whole Change Methodology

This last foundational system looks at how we can move from the ideal of sustainability as a static framework to an active process of identifying the core values and behaviors necessary to achieve higher levels of integration and ecological performance. Stuart Walker, author of Sustainable by Design writes, "It [sustainability] represents much more than simply an analytical approach to environmental auditing or business accountability. It also represents a way of acknowledging our values and beliefs and ascribing meaning to our activities (Walker 2006)." Figure 4 below left illustrates the relationship between values and behaviors in the process of enacting deep change in organizations such as universities and colleges. The values and beliefs of a single individual can lead to set of sustainable actions and behaviors, which in turn can begin to influence the collective values of groups,

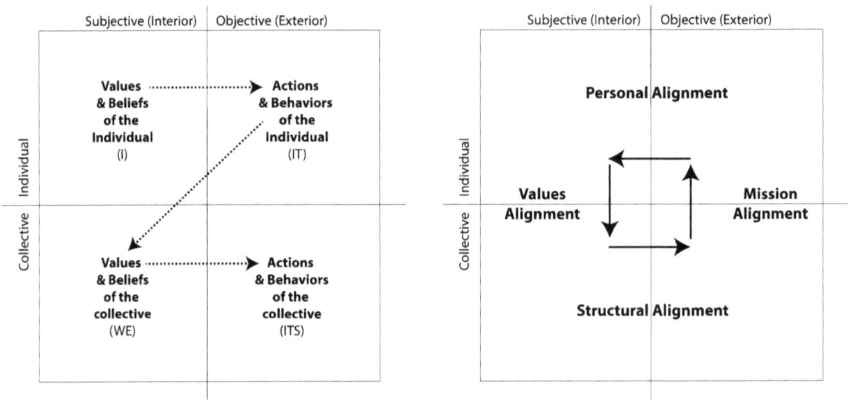

Fig. 4 *Left* Barrett's four quadrants of human systems (*Right*) Barrett's four conditions for whole system change, *Source* Barrett (*accessed 2012*)

which in turn can be translated into definable and measurable policies and procedures. The process can work in reverse. For example, the American College and University President's Climate Commission offers a set of pre-determined collective behaviors such as constructing LEED® Silver Buildings and committing to carbon neutrality by a certain year (ACUPCC). In that way, the stated behaviors begin to shift the values of university staff. At that point sustainability initiatives become less onerous and more likely to occur.

Figure 4 above right offers an important framework for program administrators and faculty seeking to build a more cohesive and more consistent approach to sustainable design education that moves from the greening of the curriculum to a more aligned and integrated approach. Assuming that an organization can move towards sustainability (and most have on some level), then the expectation among the stakeholders (those that are touched by the activities of the organization) will be to see alignment between the stated values of the organization and its actual behaviors. For example, if a university proclaims that sustainability is a core value or a top priority, then students and faculty would expect to see a variety of activities including, but not limited to: the construction of green buildings, the use of sustainable landscape practices, installation of energy management programs, and of course the evolution of the curriculum to more directly address sustainability.

6 Examples of Integral Theory in Practice at a University

Integral Theory, Integral Sustainable Design and Barrett's Four Conditions for the Whole System Change serve as the basis for a new Masters Degree in Sustainable Design (MSSD) at Philadelphia University. To say that the MSSD program is "new" is a misnomer as the program is now in its 8th year of operation. However, relative to other design programs around the world which are content to tweak

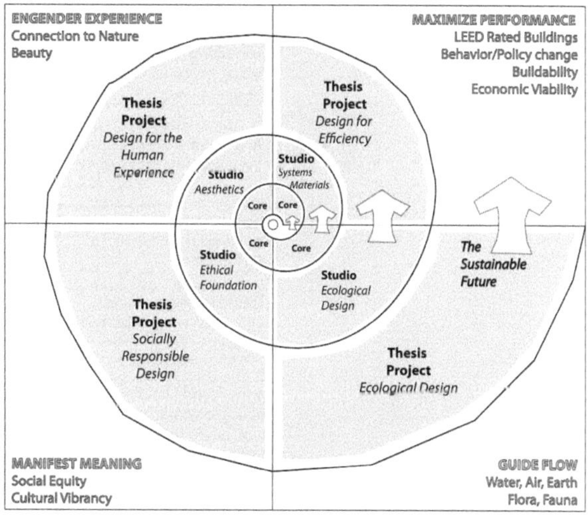

Fig. 5 Curriculum map for the MS in sustainable design program. *Source* Drawn by Author

existing curricula in a superficial "greening" process, the founders of the MSSD Program started with a complete re-definition of higher education in support of authentic sustainability. The use of Integral Theory and Integral Sustainable Design provided a framework within which a holistic and interconnected curriculum was developed and implemented. In Fig. 5 below the four quadrants are expressed via Integral Sustainable Design, which was explained earlier. At the center of the diagram stands the start point for the student. In this case there is no real beginning and the end simply gets wider as the student learns increasing amounts of knowledge suggesting multiple avenues of career exploration. The Core Courses stand in the center and each has an emphasis on one of the quadrants but still includes all perspectives in the course. The Core Courses provide a foundation upon which a sustainability studio can build upon. The studio shown as four segments in the diagram below has learning outcomes that express each of the four perspectives. The 2-dimensional aspect of the diagram does not reflect the simultaneous aspect of how the perspectives are actually delivered. Lastly, through the Thesis Project, the students are asked to again express their project from the four perspectives at all times with assessment metrics in place to insure the holistic nature of their project.

7 The Sustainable Design Studio

The design studio itself involves the usual challenges associated with an inclusive and cooperative pedagogy. For example, the vast array of different disciplines and background of the students required the professors to develop a clear studio pedagogy that could accommodate such diversity. Figure 6 below illustrates how the design studio is organized. At the bottom of the period is Consciousness. This forms the platform upon which the rest of the course pedagogy rests. For some this proves difficult so the use of core values is stressed. In other words, as educator's

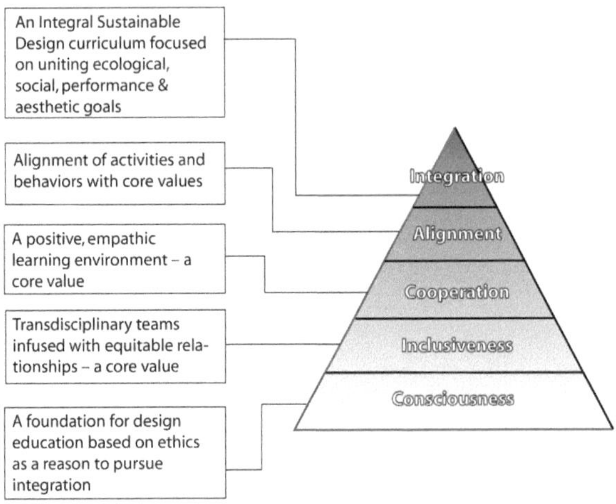

Fig. 6 Studio framework for the MS in sustainable design program. *Source* Fleming (2013)

we cannot always change a student's belief system or consciousness, but we can ask them to temporarily align their values to the project at hand, thereby creating a platform by which students can cooperate on a project.

Therefore an inordinate but necessary amount of time is spent on establishing the core values. The Quadruple Bottom Line Serves that process well as it stands as shorthand for a more sophisticated and powerful Integral Theory which, despite great efforts remains more difficult for the students. The four elements then of sustainability: Economic Viability, Ecological Generation, Deep Empathy and Cultural Significance serve not only as reminders of the goals of the project but also as a grading rubric. At the top of the pyramid lies integration—the elusive and difficult to reach level of design integration.

8 New Methodologies for Evaluation of Design Work

One of the biggest changes in the MSSD program was the elimination of Design Juries. For those not familiar with design school pedagogy, the design jury has been around for centuries. In the 20th century the jury moved from behind closed doors without the presence of the students themselves to public venues (Salama and El-Attar 2010). However, the jury itself, while accepted as a ritual at almost every design school in the world is a very poor educational methodology due to its emotional brutality (Salama and El-Attar 2010). In Fig. 7 below left, the individual motivations of the students are typically based on fear or the desire to dominate the group (Upper left quadrant). This leads to unsustainable behaviors including long nights, bad health habits and low scores in non-studio courses (Fleming 2013). These behaviors are reinforced by the institution which accepts the jury as a collective behavior (methodology) but also from a cultural perspective as a badge of honor or as a rite of passage that must be endured to graduate from design school (Fleming 2013).

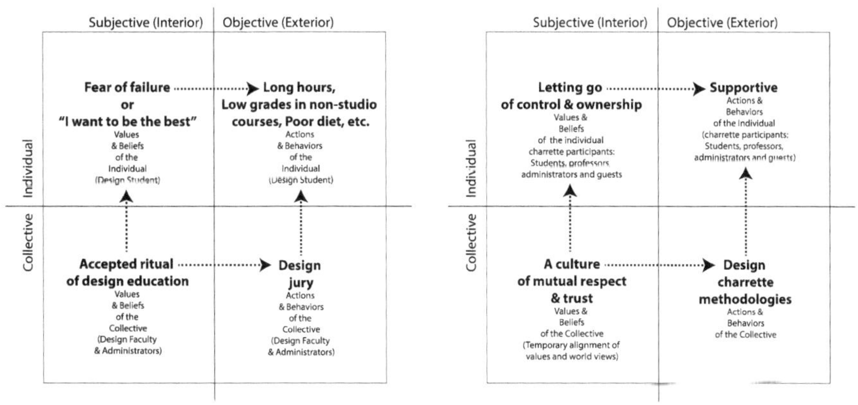

Fig. 7 *Left* Collective behaviors of design programs inscribed with Barrett's Whole Change Methodology. *Right* Charrette methodology used to better pursue sustainability. *Source* Fleming (2013)

Sustainability, which requires empathy, inclusion and cooperation, makes the traditional design jury, in short, an educational disaster. Faculty members in standard design programs argue that the jury "simulates the toughness of the real world (Anthony1991)." The actual assignments in studio lack a construction budget and often ignore gravity thereby having little resemblance to the real world of design (Fleming 2013). In Fig. 7 above right, the jury is replaced with the *collective behavior* of the design charrette (lower right quadrant). Briefly, the design charrette is a process used in the design professions to engage all stakeholders early in the process in an equitable environment. The studio courses in the MSSD program use this methodology as the primary education strategy for both design processes and design review. The process stands in stark contrast the jury. In the upper left quadrant, the motivations, values and beliefs of the student are expressed by "letting go of control and ownership", a very difficult task for anyone, especially a designer. But the quantitative driving forces behind sustainability such as efficiency, requires that the ego driven idiosyncratic design ideas to fade away in favor of collaboratively developed solutions. Students learn supportive behaviors which are reinforced by grading mechanisms that reward cooperation and penalize selfishness.

In the end, the MSSD program remains a work in process. The diagram below in Fig. 8 illustrates Barrett's Whole Change Methodology inscribed within Integral Theory which helps to communicate internally and externally the goals of the program thereby helping to unite disparate groups of students and faculty in an inclusive model and sets the stage for cooperative behaviors.

The lower left quadrant centers on the question "why the program exists." The lower right centers on the functional operation of the program. The upper right focuses on the objective assessment of the program and finally, the upper left emphasizes the subjective but no less important aspects of the "look and feel" of the program. While the diagram remains in the aspirational, the framework is essential

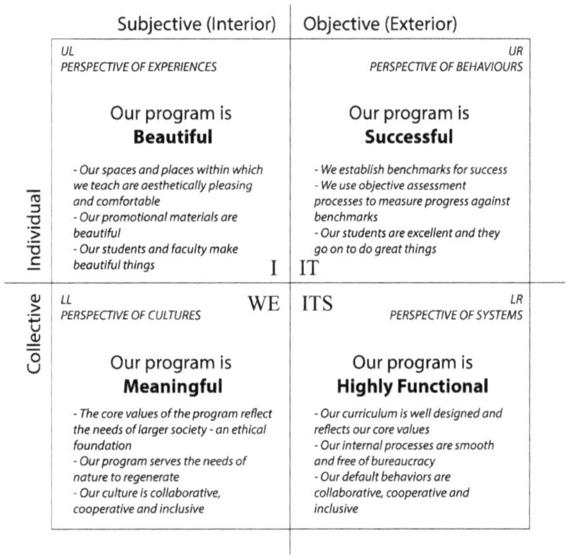

Fig. 8 Aspirational diagram for the MSSD Program *Source*. Fleming (2013)

in a sustainable design program where confusion continues to drive uncertainty. A clear and compelling mental map for both students and faculty helps to reduce confusion and create deeper levels of alignment of values. Considering that the vast majority of assignments require group work, a framework that is both inclusive and comprehensive is critical.

When interacting with outside visitors and charrette participants the students use the Quadruple Bottom Line as a more straightforward expression of authentic sustainability. The diagram below helps to communicate intent for the designers and faculty with the flexibility expressed in the bottom of the diagram for designers to make sure that the students do not get locked into a specific style but rather to pursue beauty as a critical and equally important component for sustainability. In the Fig. 9 below, sustainability is portrayed as a giant "tent" (in plan view) which suggests a structure that is open (inclusive) while the "roof" helps to unite the visitors under a common refrain.

9 Early Results

It has only been a few years since Integral Theory has been used directly in the curriculum of the MSSD program. In the theory course that all students take at the beginning of the program, the use of Integral Theory has provided entry points to sustainability for a diverse set of students. In the first year, every student performs a case study on a famous building using the four quadrants as content areas for the information they gather. The assignment purposeful mixes in historically famous structures from around the world along with the more predictable sustainable

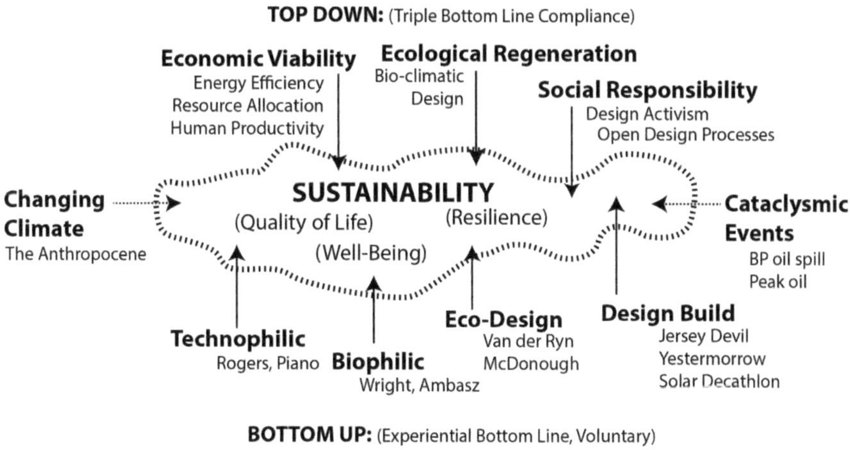

Fig. 9 Quadruple bottom line diagram. *Source* Fleming (2013)

structures. Then a peer to peer analysis of each project is conducted where students rate the relative success of each project by providing a score for each quadrant and an overall composite score. This is completed in Google Docs (Fig. 10) so that all of the projects can be studied in comparison. In the 2014 case studies which are almost complete, the difference in score between the renovations to Hagia Sophia and the Heathrow Airport renovation for example, are clear. The actual scores are less important than the process used to obtain the scores as students learn that sustainability in design is a much broader endeavor than say, using the a green building rating systems as the definer of a project's greenness, hence Hagia Sophia scores quite well.

In the sustainable design studio which is offered in the following semester to the course described above, the engineering students gravitate towards the systems quadrant but learn about and begin to appreciate the experience quadrant (aesthetics). The designers, who are typically dominated by subjective and often irrational decisions, now get to see the quantitative impacts of their choices because they are forced to consider the performance based quadrant in their work via energy simulation. Students are asked to present their work via the quadrants and then conduct peer to peer evaluation of each other's projects using the same exact system described earlier. Finally, all MSSD students conduct an independent thesis project prior to graduation. Interestingly and happily, we are beginning to see the students

	Spring 2014 Group		FACTS Behaviors/Performance			Systems			Culture			Experience		
score	Name of Project	Name of Designer	Enter Score Below	Total	Avg.	Enter Score Below	Total	Avg.	Enter Score Below	Total	Avg.	Enter Score Below	Total	Avg.
	Exeter Library	Kahn	8 8 7	23		9 9 8	26		8 8 7	23		8 8 9	25	
9.0625	Hagia Sophia	Isidore of Miletus & Anthemius of Tralles	9 8 8 8	33	8.25	10 9 9 9	37	9.25	10 9 9 10	38	9.5	9 9 9 10	37	9.25
6.9375	Heathrow Airport	Richard Fairey	5 7 7 6	25	6.25	8 8 7 8	31	7.75	7 8 7 7	29	7.25	5 8 7 6	26	6.5
	Comcast Center	Robert A.M. Stern	10 7 8 8	33		10 8 9 7	34		9 7 8 8	32		9 8 8 9	34	
	Barnes Foundation	Tsien williams	8 8 8 9	33		10 9 8 9	36		10 8 9 8	35		9 8 8 9	34	
	Jatiyo Sangsad Bhaban	Louis I. Khan	8 8 9 6	31		10 9 8 7	34		8 8 9	25		7 7 5 7	26	
	Dilworth plaza philadelphia	Olin	9 8 8	25		10 7 9	26		8 7 8	23		8 8 9	25	

Fig. 10 Integral theory case study project peer to peer analysis results. *Source* Google Docs file

use Integral Theory without prompting to organize their projects and communicate their outcomes. For example, one thesis student, Saglinda Roberts, designed four houses, each from the perspective of a single quadrant. She harvested knowledge from each design and completed her thesis by designing a "composite" house with the goal of maximizing the total success of the project. She then solicited feedback on the final design via the same peer to peer evaluation process and shared those results as part of her conclusions.

10 Conclusion

For universities seeking to pursue sustainability more deeply and to engage a broader range of students and faculty more deeply, Integral Theory helps to build a broader and more holistic vision of sustainability. Such visions align well with an emerging Age of Integration that will demand the multi-lens perspective offered by Integral Theory. DeKay's Integral Sustainable Design and Barrett's Whole Change Methodology are but two of many initiatives built upon Integral Theory. Others are sure to follow as society moves from a "green" approach based on the Triple Bottom Line to an "Integral" approach expressed more simply by the Quadruple Bottom Line which is more comprehendible and accessible than Integral Theory. This paper was admittedly horizontal in nature and was meant to spur conversation and incite interest in the topic of Integral Theory as the basis for a more evolved and more effective foundation for sustainability.

References

Anthony K (1991) Design Juries on trial: the renaissance of the design studio. Van Nostrand Reinhold, New York p 3

Brown B (2005) Theory and practice of integral sustainable development: Part 2–Values, developmental levels, and natural design. AQAL 1(2)

Brown L (2009) Plan B 4.0, Mobilizing to save civilization, 1st edn. W.W

Checkland P, Poulter J (2006) Learning for action: a short definitive account of soft systems methodology, and its use practitioners, teachers and students. Wiley, New York

DeKay M (2011) Integral sustainable design: transformative perspectives. Earthscan, London p 11

Elkington J (1997) Cannibals with forks: the triple bottom line of 21st century business. Capstone, Oxford

Esbjörn-Hargens S (2009) An overview of integral theory: an all-inclusive framework for the 21st Century. Integral Research Centre. http://www.integralresearchcenter.org/sites/default/files/integraltheory_3-2-2009.pdf. Accessed 8 May 2012

Fleming R (2013) Design education for a sustainable future. Routledge/Earthscan, London

Hosey L (2012) The shape of green: aesthetics, ecology and design. Island Press, Washington p7

Rifkin J (2009) The empathic civilization: the race to global consciousness in a world in crisis. Penguin, New York, pp 34

Salama A, El-Attar ME (2010) Student perceptions of the architectural design jury. Special volume: Design education: explorations and prospects for a better built environment. Archnet-Int J Architect Res 4(2–3):175

Schaefer R (ed) (2008) The encyclopedia of race, ethnicity, and society. SAGE publications, New York
Stahl B (2007) Ethics, Morality and Critique: An Essay on Enid Mumford's Socio-Technical Approach. J Assoc Inf Syst 8 (9) Article 3
Visser F (2003) Ken Wilber: thought as passion. State University of New York, Press, Albany
Walker S (2006) Sustainable by design: explorations in theory and practice. Earthscan, London p 27
Wilber K (2000) A theory of everything: an integral vision for business, politics, science and spirituality. Shambhala, Boston
WCED (1987) Our common future. World commission on environment and development. Oxford University Press, Oxford, New York 400 p
Zalasiewicz J (2010) The new world of the anthropocene. Environ Sci Technol 44(7):2228

Author Biography

Rob Fleming is a LEED® Accredited Professional and a registered architect with over 18 years' experience of teaching, research and practice in sustainable design. He is the co-creator and Director of the award winning Master of Science in Sustainable Design Program at Philadelphia University. Rob recently authored the book Design Education for a Sustainable Future published by Routledge/Earthscan. Rob is a Sustainability Fellow at Re:Vision Architecture in Philadelphia where he consults and facilitates integrated design charrettes.

Interdisciplinary Environmental and Sustainability Education and Research: Institutes and Centers at U.S. Research Universities

Shirley Vincent, Antje Danielson and B.S. Rica Santos

Abstract

Interdisciplinary environmental and sustainability focused institutes and centers (IESICs) serve a crucial role in bridging the knowledge needs of society and the knowledge production capabilities of universities. They facilitate research, administer academic programs, support campus sustainability initiatives, and engage in collaborative problem-solving with internal and external partners including students, faculty, staff, public and private sector organizations, citizen scientists, other colleges and universities, and governmental institutions from local to global. Few studies have examined the roles and structures of research institutes and centers and none have investigated IESICs specifically. This chapter describes the results of the first empirical study of IESICs in the United States. The data were obtained from a census of IESICs at research universities and a survey completed by a representative sample of 340 directors. The results reveal that IESICs comprise approximately 8 % of all U.S. research institutes and centers and fall into seven distinct categories, each exhibiting distinctive characteristics. Findings discussed include the types of IESICs, their primary

S. Vincent (✉) · B.S.R. Santos
National Council for Science and the Environment, 1101 17th Street NW, Suite 250, Washington, DC 20036-4722, USA
e-mail: svincent@ncseonline.org

B.S.R. Santos
e-mail: rsantos@ncseonline.org

A. Danielson
Tufts Institute for the Environment, Tufts University, Miller Hall, 210 Packard Avenue, Medford, MA 02155, USA
e-mail: antje.danielson@tufts.edu

W. Leal Filho et al. (eds.), *Integrative Approaches to Sustainable Development at University Level*, World Sustainability Series, DOI 10.1007/978-3-319-10690-8_19

goals, their funding sources, and how these attributes are related to their operational and administrative structures.

Keywords
Research institutes and centers · Interdisciplinary research · Transdisciplinary research · Environmental education · Sustainability education

1 Introduction

Research institutes and centers (ICs) play an increasingly important role in higher education in the U.S. ICs have traditionally served to support research focused on the needs of external sponsors or research that did not fit into departmental structures because of its interdisciplinary/transdisciplinary nature, the magnitude of the research task, the cost, and/or the need for continuity that did not fit well with traditional academic cycles (Stahler and Tash 1994). Today, ICs occupy a pivotal and expanding role as boundary spanning organizations that provide the organizational context for interdisciplinary, transdisciplinary and applied university research that directly addresses pressing societal needs (Geiger 1990).

In the U.S., the growth in the number of ICs over the last three decades has been extraordinary. This rapid expansion has made ICs a common feature in higher education; they often outnumber departments at research universities (Jacobs and Frickel 2009). The most recent *Research Center Directory* lists over 15,000 research centers in the U.S., most based at research universities (Gale Research 2011). This is in contrast to the approximately 1,500 identified in a census conducted in 1980 (Friedman and Friedman 1986). Increasing recognition of the importance of sustainability-oriented problem solving centered on understanding and managing complex linked environmental, social and economic challenges, is leading to a steady stream of new or restructured ICs involved in interdisciplinary environmental and sustainability research and education (IESICs).

Only a few studies have examined the roles and functions of institutes and centers. No studies have focused specifically on IESICs and how their operational and administrative structure influences integrative sustainability research and education. To investigate this question, the Council of Environmental Deans and Directors and the National Council for Science and the Environment designed an initial study to establish a baseline dataset and understanding that will be used to facilitate further research.

2 Methodology

Our study was designed to provide foundational information about the number, structure, activities and resources of IESICs in the United States. We first conducted a census to identify all institutes and centers focused on the environment and/or sustainability at research universities in the U.S., followed by an online survey of directors. Analysis of the data from the census and the survey allowed us to characterize the population and explore relationships between various attributes.

Census. The Carnegie Classification of Institutions of Higher Education is widely used in higher education research in the U.S. The Carnegie Classification system categorizes institutions primarily based on the highest degree conferred, the number of degrees conferred, and the level of research activity. The census included all academic institutions classified as doctorate-granting universities—institutions that award at least 20 doctorate degrees per year. A search of these 297 universities' websites and catalogs was conducted during the spring of 2013 to identify IESICs located at each university. Although we strove to identify all IESICs, we may have missed some if their focus wasn't sufficiently clear based on the materials we examined. We limited our current study to research universities because the vast majority of IESICs are located at these institutions.

Survey Sample. A survey of U.S. IESIC directors was conducted May–July 2013. All 1,122 IESIC directors identified during the census were invited to participate. Completed survey responses were received from the directors of 340 IESICs, a response rate of 28 %. The sample size was sufficient to measure correlations between attributes with a power of 0.90 to detect a 0.20 effect (small-moderate) size at $\alpha = 0.05$; statistical frequencies have a margin of error of ±5 %. The survey included questions addressing three sets of characteristics: operational structure, activities and resources (see Appendix A for the questionnaire).

The representativeness of the sample was assessed by comparing four defining program attributes between the sample and target population at $\alpha = 0.05$: institution basic Carnegie class, institution control (public or private-not-for-profit), institution U.S. census division (region of the United States), and IESIC type (seven categories —see the typology discussion in the next section). The sample was representative on all four parameters.

Relationships. Two nonparametric statistical tests ($\alpha = 0.05$) were used to explore relationships among the types of IESICs and their attributes (answers to the survey questions): the Wilcoxon Mann Whitney t test and Kruskal-Wallis one-way analysis of variance by ranks (KWANOVA). The Mann Whitney test is used to test for differences between two independent groups. Kruskal-Wallis is a non-parametric test of the difference in the shape or location (central tendency) of populations underlying two or more groups.

3 Results and Discussion

3.1 IESIC Typology

During the census we discovered that IESICs at research universities can be classified into seven main types and twenty-six subtypes based on their primary focus (Table 1). Broad environmental and sustainability IESICs are those with a comprehensive focus on coupled human-natural environmental systems and/or

Table 1 IESIC categories

Institute/center category	Number	Proportion of total (%)
Broad environmental and sustainability focus		
Environment	93	13
Sustainability	30	
Place/region/biome	37	
Category total	160	
Energy and climate change focus		
Energy technology	236	24
Climate/climate change	57	
Category total	293	
Natural systems focus		
Freshwater aquatic systems/watersheds	64	22
Marine/coastal systems	72	
Forests	15	
Earth systems/geosciences	48	
Ecology/conservation	48	
Natural resources/lands management	24	
Category total	271	
Human wellbeing focus		
Human heath, risk assessment and management	85	13
Security	13	
Population studies	4	
Agriculture and food	38	
Education and outreach	19	
Category total	159	
Societal systems focus		
Policy and economics	32	10
Law	27	
Society and behavior	25	
Business and finance	35	
Category total	119	
Technology and informatics focus		
Engineering and technology	75	10
Modeling and informatics	20	
Geospatial technology and remote sensing	21	
Category total	116	

(continued)

Table 1 (continued)

Institute/center category	Number	Proportion of total (%)
Built environment focus		
Built environment	56	8
Sustainable cities/communities	38	
Land use/landscape design	9	
Category total	103	
Total	1,221	100

sustainability, including those targeting a particular place, region, or biome. The other six categories are IESICs with a more narrow focus: (1) energy and climate change, (2) natural systems such as aquatic systems or forests, (3) human wellbeing such as risk assessment or sustainable agriculture, (4) societal systems such as environmental policy or law, (5) environmental and sustainability research technology and informatics, and (6) sustainable built environments/communities. IESICs focused on energy and climate change are the largest group, followed by those focused on natural systems.

Our classification system is based on our census observations and is imperfect—many individual IESICs could be classified into more than one category. As a result, caution is required in interpreting the survey findings regarding the seven categories. We hope to gather additional data in the future that will allow us to classify IESICs using an improved empirically-derived schema.

Statistical tests reveal that each category has its own unique set of characteristics, each exhibiting from 13 to 36 statistically significant differences when compared pairwise (using the Mann Whitney t test or the KWANOVA test) with the other categories (results not included here). Distinguishing characteristics were found in all three broad sets of parameters: structure, activities and resources (Vincent et al. 2014).

3.2 Differences Between Institutes, Centers and Similar Units with Other Names

One of the key differences among the seven categories of IESICs is the proportion of institutes and centers in each category. The majority of U.S. IESICs are named ‘center,’ about a third are named ‘institute,’ and a small proportion do not include either institute or center in their name but instead use titles such as academy, agency, collaborative, consortium, initiative or network. These entities are similar in structure and function to centers and institutes and therefore are included in the collective term IESICs used in this article.

Our survey sample included 218 centers, 99 institutes and 23 units with other names; a ratio representative of the total population. Most institutes are found in three categories: broad environmental and sustainability, energy and climate

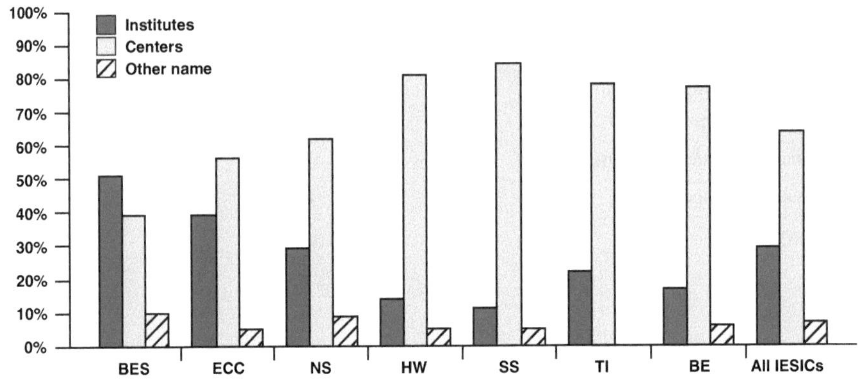

Fig. 1 IESIC categories and name types *BES* Broad Environmental and Sustainability, *ECC* Energy and Climate Change, *NS* Natural Systems, *HW* Human Wellbeing, *SS* Societal Systems, *TI* Technology and Informatics, *BE* Built Environment

change, and natural systems. These three categories have a broader research focus, indicating that institutes tend to have a more expansive research mandate than centers (Fig. 1).

The broader research focus of institutes is also reflected in significantly higher numbers of affiliated faculty. The average number of formally affiliated faculty for institutes is 59, compared with 22 for centers and 24 for entities with other names.

Institutes are also significantly more likely than centers to be administratively located at the primary university level with a director that reports to upper administration (president/chancellor, chief academic officer or chief research officer). Centers are more often located within a secondary or tertiary level—within a college or shared by colleges with directors that report to one or more college deans, or within a department with directors that report to the department chair/head. Institutes are also more likely to be housed within their own building or suite of offices than centers.

Institutes and centers are similar in their allocation of resources; they both focus about half of their activities on research and about a quarter on education of students. The IESIC units with names other than institute or center are distinctive—they focus their activities more on education and less on research compared with either institutes or centers.

3.3 IESIC Roles in Sustainability Education and Research

IESICs facilitate integrated research that address coupled human-nature systems research, support collaborative sustainability problem-solving efforts with a wide range of partners, and advance campus and community sustainability initiatives (Krizek et al. 2011). We found that the majority of IESICs devote most of their resources and activities to three goals—research, education and community

Table 2 IESIC category and mission/goals

IESIC category	Mission/goals									
	Research		Education		Outreach/ continuing education		Campus sustainability		Other[a]	
	% = proportion with goal | μ = mean proportion of resources/activities devoted to goal									
	%	μ (%)	%	μ (%)	%	μ (%)	%	μ (%)	%	μ (%)
Broad environmental and sustainability $n = 57$	98	52	90	29	88	19	53	10	16	12
Energy and climate change $n = 72$	97	59	82	25	88	15	24	12	19	33
Natural systems $n = 73$	97	58	89	23	82	25	12	8	8	28
Human wellbeing $n = 37$	97	48	84	30	92	24	27	9	16	23
Societal systems $n = 44$	96	42	80	33	82	30	18	10	18	44
Technology and informatics $n = 27$	100	62	85	23	78	16	19	12	11	32
Built environment $n = 30$	100	53	67	25	77	28	33	8	23	27
All IESICs $n = 340$	98	54	84	26	84	22	26	10	16	28

[a]Economic development; provision of services and goods; partnership coordination; policy analysis and advising; publishing

outreach/continuing education—but missions vary for different categories of IESICs (Table 2). For example, societal systems IESICs place less emphasis on research and more on education, outreach and policy advising.

IESICs are at the forefront of interdisciplinary education. Degree programs administered by IESICs are a growing trend; the number offering IES degree programs increased 6 % from 2008 to 2012 (Vincent et al. 2012). While only a third of IESICs operate formal academic programs, student education is a primary goal for the majority.

A third of IESICs administer some type of academic program; the proportion ranges from a high of 44 % of broad environmental and sustainability IESICs to a low of 26 % of energy and climate change IESICs. Broad environment and sustainability and technology and informatics IESICs are the most likely to offer undergraduate programs and master's degree programs, and they are the only two groups that offer all types of programs—undergraduate and graduate degrees, undergraduate and graduate certificates and minors, and professional/continuing education certificates (Fig. 2).

Community outreach and post-graduate continuing education is also a widely held goal for IESICs; between 77 and 92 % are engaged in these activities. For most IESICs outreach and continuing education involves a smaller proportion of their activities, ranging from an average of 15 % for energy and climate change focused IESICs to 30 % for societal systems focused IESICs.

Campus sustainability is not widely held as a primary goal for most IESICs; however, it is a goal for over half of the broad environmental and sustainability IESICs and about a third of the IESICs focused on the built environment and human wellbeing. The average level of resources/activities devoted to campus sustainability is low, averaging 12 % or less.

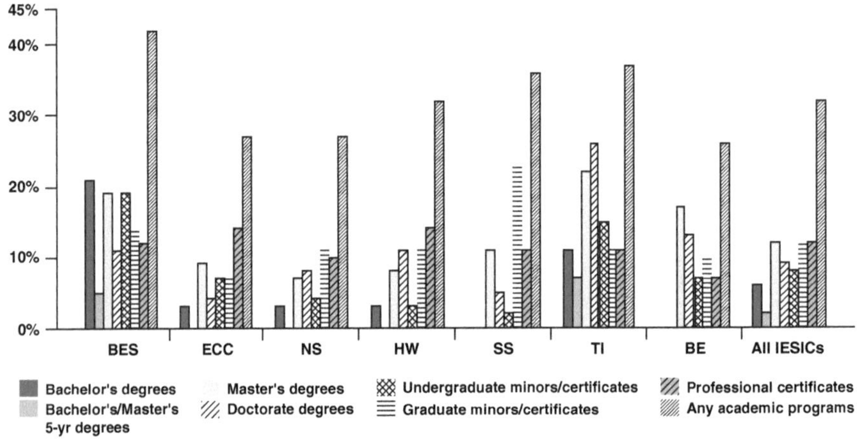

Fig. 2 IESIC category and academic programs *BES* Broad Environmental and Sustainability, *ECC* Energy and Climate Change, *NS* Natural Systems, *HW* Human Wellbeing, *SS* Societal Systems, *TI* Technology and Informatics, *BE* Built Environment

3.4 IESIC Funding

Bozeman and Boardman (2013) classified ICs into four types based on their primary source of fiscal support: university ICs, National Science Foundation (NSF) and other federal agency ICs, state ICs, and other ICs. They found that university ICs are by far the most common, comprising about three-fifths of all ICs. These ICs are created by universities and sustained by a combination of university resources and individual investigator grants, foundations, and industry funds. NSF and other federal agency ICs make up about a quarter of all ICs and are created and funded by the NSF and other federal agencies under their various programs. State ICs include about a tenth of all ICs and are created by special state programs and supported by state appropriations. Most state-funded ICs are focused on technology-based regional economic development. The remaining few ICs are those that do not fall into one of the other three groups, such as non-profit organizations formally affiliated with universities. Although we did not categorize IESICs according to this typology, only 2 % of the survey respondents reported that half or more of their budget was from state or federal funding (not counting short-term grants and contracts), indicating a smaller proportion of IESICs are government-sponsored compared with all ICs.

Funding for ICs and IESICs has transitioned over time. The earliest ICs, established in the 1900s, were primarily funded by philanthropic foundations and donors, although some of the engineering centers developed flourishing contract research relationships with industry. A relationship between ICs and the U.S. federal government developed during World War II when the government contracted universities to perform war-related research. By the end of the war, this partnership was viewed as essential for national security and economic competiveness.

Following the Soviet launch of the Sputnik satellite in 1957, the federal government initiated massive investments in basic scientific research and education. This abundance of funding was channeled primarily through NSF and the National Institutes of Health and was targeted toward more basic departmental-based research. Both ICs and departmental research thrived due to the greatly increased funding, but the balance of university research shifted away from sponsored IC research. The federal research funding boom reached its peak in the 1960s and has been declining as a proportion of GDP. As federal funding has waned, public and private sector-sponsored support has picked up. Support for scientific research in the United States has kept pace with the size of the U.S. economy, comprising from 2.2 % to 2.8 % of GDP, but the proportions have reversed from two-thirds of total support from the federal government and one-third from the public and private sectors, to two-thirds from public and private sectors and one-third from the federal government (Press 2013).

For IESICs there appears to be a growing trend of foundations and other philanthropic donors providing substantial support—recent examples include a $25 million challenge grant from the James F. and Marion L. Miller Foundation to support the Institute for Sustainable Solutions at Portland State University, a $20 million gift from an anonymous donor to establish the Environmental Initiative at Georgetown University, and a $27.5 million grant from the Rob and Melani Walton Fund of the Walton Family Foundation to support the Global Institute of Sustainability at Arizona State University.

Funding sources are directly tied to the missions and goals of IESICs. On average, IESICs spend 54 % of their resources and activities on research (Table 2). The other half of their resources and activities are devoted to other goals—education, community outreach/continuing education, campus sustainability, and provision of services.

About a third of the IESICs in our survey reported they received most or all of their funding from short-term grant or contract funding. These IESICs focus more of their resources and efforts on research and were likely created as the result of a single grant or contract. A small proportion, 12 %, reported they currently had no funding. These were also likely created from a single award and have been unable to find additional support after their grant or contract expired.

The remaining IESICs rely on diverse funding sources, including institutional appropriations, endowments and other long-term directed sources, gifts from donors, and fees for products and services (Table 3). Research grants and contracts are still the most commonly reported sources of funding; this funding also makes up the largest average proportion of all IESICs' budgets.

Different categories of IESICs have different funding source patterns. The broad environmental and sustainability IESICs and the natural systems focused IESICs are more likely than the other types to receive funding from institutional appropriations; this funding supplies over half of overall funding—53 % on average—for the broad environmental and sustainability group. This group is also more likely to receive funding from endowments and donor gifts.

We also noted a relationship between funding sources and the location of the IESIC (Table 4). IESICs at the primary university level with directors that report to the president/chancellor, chief academic officer, or chief research officer are more likely to receive support from direct appropriations, potentially reflecting a relationship between institutional funding and the importance that high-level administrators assign to interdisciplinary environmental and sustainability research and education at their institutions.

IESICs with higher average proportions of funding from short-term grants and contracts most often report either directly to faculty (department chairs) or to the chief research officer, indicating a relationship between these IESICs and the importance of individual investigator grant funding and/or enhanced institutional grant writing support.

Table 3 IESIC category and budget sources

IESIC category	Budget source									
	Institutional appropriations		Endowments and other long-term funding		Short-term contracts and grants		Donor gifts		Other[a]	
	% = proportion having as a budget source | μ = mean proportion of budget from source									
	%	μ (%)	%	μ (%)	%	μ (%)	%	μ (%)	%	μ (%)
Broad environmental and sustainability $n = 57$	74	53	39	29	74	43	42	13	12	59
Energy and climate change $n = 72$	54	41	22	15	76	72	28	16	7	51
Natural systems $n = 73$	70	34	27	28	80	65	30	13	12	40
Human wellbeing $n = 37$	41	41	22	23	87	59	35	26	22	46
Societal systems $n = 44$	61	42	34	51	68	47	39	16	11	64
Technology and informatics $n = 27$	48	44	15	50	82	75	15	19	4	5
Built environment $n = 30$	57	44	13	31	77	62	27	13	13	54
All IESICs $n = 340$	60	42	26	31	77	61	32	16	12	50

Fees for goods or services; federal or state funding other than grants and contracts; other unspecified sources

Note 12 % of all IESICs currently do not have a designated budget

Table 4 IESIC budget sources by reporting office

Director reports to	Budget source									
	Institutional appropriations		Endowments and other long-term funding		Short-term contracts and grants		Donor gifts		Other[a]	
	% = proportion having as a budget source | μ = mean proportion of budget from source									
	%	μ (%)	%	μ (%)	%	μ (%)	%	μ (%)	%	μ (%)
President/Chancellor $n = 6$	67	36	33	50	75	59	42	9	17	18
Chief Academic Officer $n = 20$	75	50	33	27	75	51	26	16	11	56
Chief Research Officer $n = 19$	83	39	24	10	81	65	36	10	7	38
Dean of a College $n = 138$	62	40	30	32	82	59	38	19	12	36
Deans of 2 + Colleges $n = 21$	48	45	24	53	62	56	29	24	24	63
Department Chair/Head $n = 41$	20	45	10	60	66	82	17	10	2	100
2 + Dept. Chairs/Heads $n = 2$	50	10	50	1	100	88	50	14	0	–
Steering Committee $n = 5$	40	65	0	–	60	57	0	–	0	–
Other[b] $n = 22$	46	40	18	14	64	58	27	7	23	85
All IESICs $n = 340$	60	42	26	31	77	61	32	16	12	50

[a]Fees for goods or services; federal or state funding other than grants and contracts; other unspecified sources
[b]Board of directors, advisory board, multiple administrators, other administrator

4 Conclusion

Our study reveals a set of findings that provide a foundation for further research into IESICs. As a group, IESICs share similarities, but on an individual level they may differ markedly from each other. They can be categorized into seven main types based on their scope and focus; some are broadly focused on sustainability while others have a more targeted focus related to specific aspects of sustainability. Although not discussed here, each type has its own set of defining characteristics (Vincent et al. 2014).

IESICs typically fulfill several roles in their universities, facilitating interdisciplinary research, providing interdisciplinary education for students and career professionals, engaging with a variety of partners on applied sustainability projects and transdisciplinary research, and supporting campus sustainability initiatives. A growing number of IESICs serve as the administrative home of interedisciplinary environmental and sustainability degree programs. Undergraduate students, graduate students, faculty, university staff, external partners and career professionals all benefit from the educational and research opportunities provided by IESICs.

There are differences in goals and funding sources for different types of IESICs and for IESICs located at different levels within the university. IESICs located at the primary level with directors that report to top administration have more institutional, endowment and donor support, which may make them more stable incubators for research and teaching on complex coupled human-natural systems. IESICs located within colleges or departments rely more on short-term funding from grants and contracts. Since federal grant funding is declining in the U.S., these IESICs may be more vulnerable to funding issues.

Broad environmental and sustainability IESICs are especially likely to offer academic programs and have a broader reach across the campus with more affiliated faculty from more diverse fields, which make these IESICs especially well-suited to preparing students for careers at the science-policy, science-management, and policy-management interfaces.

IESICs provide advantages for both students and faculty. Many university research-faculty split their time between ICs and departments. IC-affiliated researchers are motivated to participate by the prospect of doing research that is more intellectually interesting and important and believe that their participation positively influences their own research (Rhoten 2005, Bozeman and Boardman 2013). IC-affiliated faculty have also been shown to provide more fiscal support for undergraduate and graduate students, and they have greater involvement with teaching undergraduates than their non-affiliated peers, thus strengthening both the research and teaching missions of universities (Bozeman and Boardman 2013).

However, there are downsides to researchers when participating in IESICs —"role strain" faced by affiliated researchers caused by the tensions and limitations inherent in current universities' structures and reward systems (Boardman and Bozeman 2007). These constraints are of special concern for IES researchers and can negatively impact the effectiveness of interdisciplinary research (Pfirman 2011).

A recent study of IC effectiveness concluded the main impediment to effective integrative research is the lack of systematic implementation of university structures that explicitly support interdisciplinary work (Rhoten 2005). The study found both extrinsic attention (funding agencies, leadership) and intrinsic motivation (faculty, students) are sufficient to support research that is more heterogeneous, interdisciplinary, fluid, and horizontal. The key constraints are academic research communities that do not adequately accommodate interdisciplinary work in their institutional structures and systems of professional standing for faculty researchers. The consequence is that many ICs have a "tendency to become a nexus of loosely connected individuals searching for intersections, as opposed to cohesive groups tackling well-defined problems" (Rhoten 2005:9).

We identified two common models for IESICs: (1) broadly interdisciplinary institutes located at the primary administrative level, with strong institutional financial support and a broad mandate to tackle complex coupled human-natural systems problems; (2) more narrowly defined centers managed at the college or department level that rely more on short-term grants and contracts. More research is needed to understand how the operational and administrative structures of each of these predominant models impacts IESIC effectiveness in achieving various goals, developing sustainable fiscal support, supporting faculty participation and collaborative integrative research, and educating students and post-graduates.

5 Appendix A—Survey Questionnaire

Center/institute name:
University:
Director name:
Director email address:

1. Where is the center/institute located in the administrative hierarchy of your institution?

- Primary level—the center/institute is a primary level administrative unit.
- Secondary level—the center/institute is located administratively within one college (school/division) or shared by two or more colleges (schools/divisions).
- Tertiary level—the center/institute in located administratively within a department or shared by two or more departments.

Other (explain):

2. Which disciplines, professional fields and external groups are typically involved in the center's/institute's projects?

- Environmental science and studies
- Life sciences
- Physical sciences
- Applied sciences/engineering
- Natural resources management/agriculture
- Social sciences)
- Humanities
- Professional fields (e.g. law, business, public administration)
- Governmental agencies or organizations
- External public or private organizations
- Other higher education institutions

3. To whom does the center/institute director report?

- President or chancellor (administrator in the office of the president or chancellor)
- Chief academic officer (administrator in the office of the provost or vice-chancellor)
- Vice president for research or similar position
- Dean of one college/division/school
- Deans of two or more colleges/divisions/schools
- Chair/head of one department
- Chairs/heads of two or more departments
- Steering committee

Other (explain):

4. What are the primary goals of the center/institute? Please indicate the proportion of the center's/institute's activities/resources devoted to each area. Proportions should add to 100 %.
Research ____%
Education ____%
Outreach/continuing education____%
Campus sustainability ___%
Other (explain):

5. Does the center/institute administer any academic programs? Check all that apply.

- Baccalaureate degree(s)
- Accelerated 5-year baccalaureate/master's degree(s)
- Master's degree(s)—MS/MA
- Master's degree(s)—Other/professional (e.g. Master of Environmental Management)
- Master's degree(s)—Professional Science Master's™

- Master's degrees specifically designed for working professionals (e.g. executive master)
- Doctoral degrees(s)
- Undergraduate minor(s)
- Graduate minor(s)
- Undergraduate certificate(s)
- Graduate certificate(s)
- Professional/continuing education certificate(s)

6. Does the center/institute have its own physical space?

- The center/institute is located in its own building
- The center/institute is located in its own distinct space (suite with a separate entrance)
- The center/institute is located within another space (e.g. college dean's office)
- The center/institute does not have its own distinct/dedicated space

The center/institute space includes the following (check all that apply).

- Reception area
- Offices for administrators/staff
- Offices for faculty
- Workspace/offices for students
- Informal meeting place/lounge
- Conference room(s)
- Laboratories or other technical facilities
- Computer labs

Other (explain):

7. Please indicate if the center/institute supports the following administrators/faculty and staff. Check or complete all that apply.

- Director/executive director (Full-time FTE in the center/institute)
- Director/executive director (Part-time FTE in the center/institute)
- Associate/assistant director (Full-time FTE in the center/institute)
- Associate/assistant director (Part-time FTE in the center/institute)
- Other administrator(s) (Full-time FTE in the center/institute)
- Other administrator(s) (Part-time FTE in the center/institute)

____Number of full-time staff
____Number of part-time staff
____Number of core faculty (Full-time FTE in the center/institute)

____Number of joint faculty (Part-time FTE in the center/institute or temporary release from unit)
____Number of participating faculty (formally affiliated with the center/institute)

8. Identify the proportion of the center's/institute's budget that comes from the following sources (average over last 3 years). Proportions should add to 100 %.

- The center/institute does not have its own budget

Non-directed funds (institutional appropriations) _____%
Long-term directed funds (e.g. endowments) ______%
Short-term directed funds (e.g. grants, contracts) _____%
Donor gifts ____%
Other (explain):

References

Bozeman B, Boardman C (2013) Academic faculty in research centers: neither capitalism's slaves nor teaching fugitives. J High Educ 84(1):88–120

Friedman RS, Friedman RC (1986) Sponsorship, organization, and program change at 100 universities. Institute for Policy Research and Evaluation, University Park

Research Gale (2011) Research Centers Directory, 40th edn. Gale Publishing, Farmington Hills

Geiger RL (1990) Organized research units—their role in the development of university research. J High Educ 61(1):1–19

Jacobs JA, Frickel S (2009) Interdisciplinarity: a critical assessment. Ann Rev Soc 35:43–65

Krizek KJ et al (2011) Higher education's sustainability imperative: how to practically respond? Int J Sustain High Educ 13(1):19–33

Pfirman S (2011) Interdisciplinary hiring and career development: guidance for individuals and institutions. National Council for Science and the Environment, Washington, 74 p

Press (2013) What's so special about science (and how much we spend on it?). Science 342:817–822

Rhoten D (2005) Interdisciplinary research: trend or transition. Items Issues 5:6–11

Boardman C, Bozeman B (2007) Role strain in university research centers. J High Educ 78 (4):430–463

Stahler GJ, Tash WR (1994) Centers and institutes in the research university. J High Educ 65 (5):540–554

Vincent S, Bunn S, Sloane L (2012) Interdisciplinary environmental and sustainability education: results from the 2012 Census of U.S. four-year colleges and universities. National Council for Science and the Environment, Washington, 34 p

Vincent S, Santos R, Cabral L (2014) Interdisciplinary environmental and sustainability education and research: institutes and centers at research universities. National Council for Science and the Environment, Washington, 140p

Authors Biography

Dr. Shirley Vincent is the Director of the Center for Environmental Education Research at the National Council for Science and the Environment. She received her doctorate in environmental science from Oklahoma State University where she researched stakeholder participation in environmental decision making and core competencies for tertiary environmental education. Her work includes landmark longitudinal studies of interdisciplinary environmental and sustainability degree academic programs (see the list of reports on NCSEonline.org). She has held numerous leadership positions, is a frequent speaker on environmental and sustainability education topics, and consults nationally and internationally on environmental and sustainability education and related topics.

Dr. Antje Danielson has been a researcher, consultant, entrepreneur, lecturer, and student mentor in environmental geosciences and sustainability since 1993. She is an Assistant Professor and currently leads the Tufts Institute of the Environment. She is also the President of the Council for Environmental Deans and Directors at the National Council for Science and the Environment. She received her PhD in geochemistry in 1989 from the Freie Universität in Berlin, Germany. Between 1991 and 2005 she worked at Harvard University in the Department of Earth and Planetary Sciences, the Center on the Environment, and the Green Campus Initiative on teaching and research related to greenhouse gas emissions reductions from energy consumption in the U.S. From 2005 until 2008 she was the Deputy Director for Sustainability at the Centre for Research into Earth Energy Systems at Durham University in the UK, working on carbon capture and sequestration and creating a campus sustainability program, which was awarded the Green Gown Award in 2008. She joined Tufts University in 2008. She is currently interested in the mechanisms that underpin interdisciplinary research and collaboration. In 1999/2000 she also co-founded the car-sharing company Zipcar.

Rica Santos serves as a Research Assistant in the Center for Environmental Education Research at the National Council for Science and the Environment. She is a recent graduate of Santa Clara University in California, where she earned a BS in political science and environmental studies. She has worked as a research assistant for projects in education research and sustainable development agriculture and policy.

Transition Towards Sustainable Development: The Role of Universities

Angelo Riccaboni and Francesca Trovarelli

Abstract

Sustainability has become worldwide recognized theme. Social and environmental problems are known for some time and the challenge now is to accelerate the social process of change that leads to sustainable development. Starting from the analysis of the transition from present social systems towards systems more sustainability oriented, this paper examines the role that Universities play as facilitators in driving and accelerating this transition process. In particular, this study suggests that planning for sustainability within Universities promotes a synergic action with internal (students and academic community) and external (civil society, policy makers and industry) stakeholders. Moreover exchange of knowledge and experience (cooperation of research centers, and creation of local and global networks) highlights new dynamics, roles and responsibilities. Several evidences are emerged from the study of the case of University of Siena. University of Siena's current effort is to be leader in promoting sustainability. Its activities (teaching, research, training and advisory services) evolved by implementing a sustainable strategic plan and by participating in important international projects, such as UN Sustainable Development Solutions Network. Inter alia this paper will be useful to foster the exchange of experiences acquired by successful initiatives.

A. Riccaboni · F. Trovarelli (✉)
Department of Business and Law, University of Siena, 53100 Siena, Italy
e-mail: trovarelli@unisi.it

A. Riccaboni
e-mail: riccaboni@unisi.it

W. Leal Filho et al. (eds.), *Integrative Approaches to Sustainable Development at University Level*, World Sustainability Series, DOI 10.1007/978-3-319-10690-8_20

Keywords
Transition management • Universities • Sustainable development • Management system

1 Introduction

The search for sustainable development is a trend that pervades and involves the contemporary world. More than ever the human society is facing challenges connected to three categories of change: environmental, social and economic change. Within last decades a drastic acceleration of these changes combined with the phenomenon of globalization and technological evolution has resulted in increasing complexity and dynamicity in society. As a consequence the path towards the sustainable development is becoming a relevant topic; issues of sustainability are not only of concern to scholars and professionals, but they are also troubles of each person in terms of daily worries. Since the behavior and the style of living of everyone have to be modified in order to cope the challenges for a sustainable world, it is the cultural change that gains importance and that plays a significant role.

It could be pointed out that this very important role has long been underestimated. Just to make an example giving validation of this, it suffices to recall that the first World Climate Conference was held in 1979. That was exactly 35 years ago and, even though so much time has passed, climate change is today the most important environmental problem that still remains to be faced. This means that, even if importance and consequences of climate change are already known for a long time now, the evidences coming from scientific area and the offering of environmentally friendly technologies have been not enough to push politics and societies in order to give a significant change to their behavior. Not even the good will of individuals seemed to have had effectiveness. Until now, the long expected transformation towards sustainable development did not receive proper cultural response from industrialized countries, and from emerging and poor regions of the world. Only in recent times, they gave the impression to remove any cultural resistance and to be aimed at a more sustainability-oriented conduct.

Moving to this new positive direction, the ways to behave and to react from the different societies, and therefore from their social and environmental politics, are affected not solely by economic factors, but also by social and cultural traditions. Even if issues related to sustainable development must be able to attract the attention of public opinion, they do not need to be addressed in an alarmist tone nor they may be set aside, given the prominence to other issues such as the financial crisis. It would be wrong to uniquely promote financial and economic policy forgetting environmental and social issues.

The cultural change is usually a global phenomenon of rather vague nature and indefinite duration. Nevertheless, the call for cultural transformation at societal level is becoming central nowadays and the transition towards a sustainable development seems to be accelerated (Balas et al. 2010). Solutions are urgently requested to overcome the reasons of any delay, solving the conflicts in responding to new trends in favor of sustainable development and giving way to search for developing change in cultures and for cooperating at local, national, European and global levels.

Based on of the above considerations, new and interesting topics are discussed among those who are willing to share their cultural heritage, to compare different opinions, and to exchange their knowledge. And, when the fundamentals of culture are in doubt, it could be necessary to work building an international scientific network. On the one hand, several international networks and local initiatives are beginning to take account of these issues through dissemination of information and promotion of activities and projects in the field of sustainable development. Within the global context, there has been a proliferation of codes of conduct, guidelines, declarations, through which multilateral and non-governmental organizations, trade unions, and business require companies, institutions, and also individuals to participate in programs that secure social and environmental standards, to carry out activities of monitoring, to certify and encourage good practices. On the other hand, universities appear as ideal and traditional places where to develop cultural change and to promote the necessary cooperation among various actors of a change towards ideals of sustainability. In literature the role of universities in advancing practices and values for the evolution of cultures has been widely observed in many and different disciplinary domains.

This paper contributes to explore if and how universities can help to raise awareness and to facilitate cultural change towards more conscious engagement of humane society on the increasing urgency to face sustainability challenges. In doing so, it is here of interest to analyze the relation between transition management and universities and to investigate whether and, if so, how they intervene in the path of transition towards the sustainable development. In particular, we want to focus on the implementation of management accounting systems inside the university in order to plan the change and to facilitate the transition. To this aim, the paper is based on the case of the University of Siena. Like other Italian universities, University of Siena is asked to face the emergency of an economic crisis combined with a growing demand of solutions that allow to give answer to issues related to sustainability. University of Siena has adopted management tools that plan strategic goals leveraging initiatives that are connected to sustainability issues and that involve internal and external stakeholders.

The paper is structured as follows. In following section the concept of transition is illustrated by drawing on the relevant literature. The theory of transition management is then used to conceptualize the role of universities in favor of social and cultural change. Afterward, the case study is introduced and first evidences are argued. Specifically, it is presented how activities and strategic planning within university can support internal and external transformations towards sustainable development. In the final section main findings are discussed.

2 Transition Management

Introduced by Rotmans et al. (2000), the transition's concept indentifies a gradual and continuous process of societal transformation whereby society, or a subsystem of society, undergoes a fundamental change. Transitions are system transformations from an equilibrium state to another. In the period between the two equilibrium states, the system passes throughout the sequence of the four phases of pre-development, take-off, acceleration, and stabilization[1]. Following an S-curve, in transitions the system transformation moves from initial equilibrium dynamics through a period of instability and rapid developments reverting to relative stability again, at a higher level of systemic change.

An acceleration in the path of transformation could be caused by one-time events, such as huge ecological disaster or a crisis, even if it is not originated by such events only. Instead, slow changes can determine a fundamental modification; undercurrent events may accelerate the transformation process. Transitions are results of both endogenous and exogenous developments (Loorbach 2007).

Transition is also a multilevel concept that describes the interplay between: the macro- (or landscape) level, the meso- (or regime) level and the micro- (or niche) level[2]. A transition only occurs when the three levels start to coincide and interact, and a modulation between the different change-processes is achieved (Loorbach 2001).

In literature, the concept of transition has also been introduced into the field of sustainability and governance (Rotmans et al. 2000; Loorbach and Rotmans 2006). According to Loorbach (2007), transition management is a new mode of governance for sustainable development. 'As our world is constantly changing, the challenge for governance and policy is to deal intelligently with all sorts of processes of change and to redirect and accelerate these processes in a desired

[1] (Loorbach 2007; Kemp and Loorbach 2003). The nature and speed of change differ in each of the transition stages: in the predevelopment phase, there is very little visible change on the societal level but there is a lot of experimentation; in the takeoff phase, the process of change gets under way and the state of the system begins to shift; in the acceleration phase, structural changes take place in a visible way through an accumulation of sociocultural, economic, ecological, and institutional changes that react to each other; during this phase, there are collective learning processes, diffusion and embedding processes; in the stabilization phase, the speed of societal change decreases and a new dynamic equilibrium is reached (Van der Brugge 2006).

[2] The transition management must work within the three different levels. The model gives insight in the process of regimechange: at the macrolevel, economy, demography, worldviews, and culture are inert, but important variables; at the mesolevel, organizations and infrastructure, but as well values, belief systems, norms, and unwritten rules and practices form a regime that is most likely to be aimed at protecting its own existence and status quo, rather by optimization than by innovation; at the microlevel, initiatives and innovations from individuals or individual organizations challenge the existing regime (Geels 2002).

direction. This challenge becomes especially important in light of the ambitions for sustainable development.' (Loorbach 2007, p. 11) A shared sense of urgency for fundamental changes needs to be felt from human society (societal transitions), as well as strong and inspiring long-term sustainability visions and societal innovation strategies. Transition management does not want to control the uncertainties of change but steer and redirect facilitating the choices of different actors towards more sensitiveness to sustainability.

In this sense, societal transitions are processes of change that alter configuration and practices of a societal system, and that finally modify its culture. These processes take a very long time to materialize at system's level, although partial processes can occur almost suddenly (i.e., fundamental changes in thinking or radical innovation). A societal transition results from interacting changes in all societal domains: economy, ecology, institutions, technology and welfare (Loorbach 2007).

The type of governance activities that are identified as relevant to societal transitions are: strategic, tactical, and operational activities (Kemp et al. 2007). Strategic activities include processes of vision development, strategic discussions, long term goal formulation. They focus on changes in the culture of the societal system, at the landscape level and in a long term scale. Tactical activities include processes of agenda-building, negotiating, networking, and coalition building. They focus on the structure of institutions, at the regime level and in a mid term scale. Operational activities include processes of experimenting, project building, and implementation. They focus on practices, concrete actions and projects, at the niche level and in a short term scale.

All these activities are addressed to different output which co-evolve throughout the processes[3], and are carried out by respective actors (multi-level governance). In transition management new roles and new practices are required: processes of societal transformations can be directed combining the individual interests of the different actors with collective interest of sustainable development (network-steering and self-steering). Actors of each level can be generally categorized as follows: government (central and local), businesses, non governmental organizations, knowledge institutes and citizens (Loorbach, 2001). In transition management, the control is distributed rather than being top-down control, since every actor is managing or influencing at least some part of the system (Kemp et al. 2005; 2007; Loorbach 2010). Moreover a wide number of participants within governance of the transition process allows to exploit more opportunities identifying a vision

[3] Kemp et al. (2007, p. 7) highlight that 'transition management is about organizing a sophisticated process whereby the different elements of the cyclical transition management process may co evolve: the joint problem perception, vision, agenda, instruments, experiments and monitoring through a process of social learning about radical innovations and new systems.'

[4] Developing a set of low carbon UK energy system transition pathways and exploring different governance patterns, Foxon et al. (2010, p. 3) suggest how the different kinds of actors 'create a broadly defined "action space" in which the current energy regime sits. Different kinds of relationships between actors exist and different forms of transition may develop, depending on the evolving balances of "power" between these actors'.

that will be representative for all. In this sense, a larger involvement of actors creates a natural consensus and reduces resistance to cultural change[4].

In order to have a societal transformation, and particularly transformation towards sustainable practices and values, it is crucial the participation of the society itself. In the next section it will be pointed out the gap in literature because lack of proper consideration to the role of universities within the process of transition towards sustainable development. In particular, it will be investigated how changes inside the single universities can give support to societal transformation.

3 Social and Cultural Change: What Is the Role of Universities in Transition Towards Sustainable Development?

Although universities can be considered actors with significant influence on societal transformations towards sustainability, in the transition theory opportunities and challenges for universities have not completely been explored. To the authors' knowledge, little has been investigated about the role of universities in transition management and, specifically, in transition towards sustainable development.

Stephens et al. (2008) have been the first that within the literature on transition management start to consider the role played by universities. They see universities as important agents in advancing more sustainable practices in different cultures and context and suggest to reflect on potentials that universities have as change agent in any particular region or place. 'The potential for higher education to be a change agent accelerating a transition toward sustainability is dependent on a variety of factors including the current position, structure, and arrangement of higher education within its society as well as the location-specific sustainability challenges and opportunities facing a given community or region' (Stephens et al. 2008, p. 322). The authors identifies five critical issues to be considered in assessing these potentials. The five critical issues include: the dominant sustainability challenges of the region, the financing structure and independence, the institutional organization, the extent of democratic processes, and the communication and interaction with society.

More recently, Schneidewind and Augenstein (2012) present a multi-level perspective on transition applied to a national science system. Based on insights about the German science system, they have shown that 'the existing niche strategies are not sufficient to put substantial change pressure on the existing regime. Only coalitions of niche actors and activities will be able to increase the pressure on the regime' (Schneidewind and Augenstein 2012, p. 13). In particular, they highlight the importance of niche coalitions of frontrunner universities and research institute.

Given the paucity of literature on this topic, we will base part of our review on more general ambits in order to finalize and develop our following reflections on how universities can be of help in transformations and changes, and to find correlation with the specific theoretical domain of our interest.

Traditionally conservative and resistant to change, universities begin to recognize themselves as important driver of change in transforming societies, both through incorporating sustainability into internal system and by educating decision-makers, leaders, entrepreneurs and academic (within research, education, operational management and outreach dimensions). Understanding the nature and the dynamics of such transformative process relates to think about collaborating with other universities; creating alliances and networking; developing internal participation, fostering transdisciplinarity; making sustainable development an integral part of the institutional framework; creating on-campus life experiences.

During the last decades several international conferences and declarations testify the increasing interest on the role of universities as multiplayer towards sustainability. However, despite the many years efforts from many individuals and groups within the formal educational system and from single universities, education for a just and sustainable world is starting only now to be considered a high priority. The change in mind-set necessary to achieve this vision is a sustained, long-term effort to transform education at all levels (Cortese 2003).

In addition, on the one hand, participatory design is presented as the most effective means to change values within university community, on the other hand, an open approach that help university members to develop and adopt new social norms can be transformed into shared rules for sustainable life-styles (Balas et al. 2010).

For the reason that universities should start to be sustainability leaders and drivers of change, present and future generations need to be better understood and trained, so that 'professionals who are well versed in sustainable development can effectively educate students of all ages to help make the transition to sustainable societal patterns. In order to do so, university leaders and staff must be empowered to catalyze and implement new paradigms that ensure sustainability values throughout the entire university system' (Lozano et al. 2013). In addition, participation in university context refers to involvement of students, faculty and staff giving the institutional community to put into practice principles of sustainability in their academic, personal and professional life, and to be engaged in institutional change processes too (Disterheft et al. 2012).

Besides, several authors highlight that universities have to be considered significant contributors to the pursuit of regional sustainability initiatives (Stephens et al. 2008). Their ways to explain how universities can contribute are very similar and synthetically include: implementing internal environmental management systems and practices; supporting students and researchers to cope with the future multidisciplinary complex challenges of sustainability; being source of technical expertise in promoting the engagement of local authorities and other societal stakeholders to implement regional sustainability plans; and creating new research and teaching agenda as centers of development of the sustainability science as an innovative scientific field (Kuratzoglou 2013).

An emerging and possibly new academic function is related to the 'co-creation for sustainability' (Trencher et al. 2013). In this sense, the powerful trend of universities reaching across campus boundaries to form partnerships with government, industry and civic organizations to drive an urban sustainability transformation.

The approaches adopted by universities in the perspective of sustainability are different and are made up of single projects and initiatives, as well as all of elements of management (mission, values system, strategic objectives, governance system and stakeholder engagement) and organizational mechanisms. In particular, the performance measurement system and its related key performance indicators, the evaluation system, the reporting system and the different tools used to communicate the commitment towards sustainability (sustainability reports) (Mio 2013).

As previously anticipated, the aim of this paper is to evaluate if internal institutional changes can enhance the potentials that universities have externally, as change agent in society. Particularly, the attention here is paid on if and how institutional management system can support the path towards the societal transformation. In this regard, a representative case will be presented in the next section.

4 The University of Siena

The University of Siena, founded in 1240, is a medium-sized Italian university, which has an excellent reputation nationally and internationally recognized thanks to the high quality of teaching and research. Rich in history and with a great tradition of learning, in eight centuries the University of Siena has produced knowledge in every field of study.

For some years now this University commits itself to implement sustainable policies with the intention to be a leading institution in promoting sustainability by providing teaching, research, training and consulting to businesses and institutions. Supported actions are aimed at the development of knowledge and environmental awareness at the local scale, thanks to the relations with the Region, and the international community that have been established within research programs.

A number of activities has been undertaken and several projects promoted in order to spread sustainability policies. Since 2009 administrative staff members, professors and researchers and students constituted a spontaneous working group, named Unidea, whose aim is to promote initiatives towards a Green campus. The group intends to develop educational projects to promote sustainable development, mainly with concrete action in university facilities. Some of the proposals of the group are related to the introduction of policies to support some best practices connected to the water supply, the waste minimization and paper use minimization, the recycling and reuse, and the promotion of sustainable mobility options.

Since 2011, the University of Siena established a Network called Ne.S.So. (Network Siena Sustainability), whose purpose is to implement its commitments to teach sustainable developments, to encourage research on sustainable issues and to improve the actions towards a green campus.

Among other actions promoted in the agenda towards sustainability, in 2013 a special course on sustainability was activated in favor of all University students regardless of course of study to which they are enrolled and also opened to external people. Absolutely innovative in the Italian university system, this teaching course is organized according to an interdisciplinary approach that crosses environmental, economic, legal, energy, urban planning, sociology domains in order to provide a solid base on the theme of sustainable development.

As for the research, the University of Siena is working on an observatory on the sustainability in order to introduce a model based on statistical indicators that could be used firstly at local level, but to be extended in the future to other territories and regions.

Most of these activities undertaken by the University see the participation of both internal community and external actors such as the local governments (Provincia di Siena, Comune di Siena), the regional government (Regione Toscana), the national government (Ministero dell'ambiente, MATTM), and the economic actors of the territory, such as the associations of enterprises (Camera di commerico, Confindustria).

Consistent with the growing commitment of the University of sustainability issues is the important role played within the project Sustainable Development Solutions Network. The UN SDSN, promoted by United Nations, was born with the aim to accelerate the identification of solutions for sustainability in the development at local, national and global level. As part of this project, the University of Siena has been commissioned to create and coordinate MED Solutions, the first Regional Network for the Mediterranean. The University is acting as a hub for the Mediterranean Region and coordinating the activities of a network of universities, research centers and companies interested in the identification and promotion of proposed solutions to the Mediterranean Basin.

One of the core goals of the MED Solutions Network is to strengthen local and national problem solving for sustainable development by empowering universities and research centers to play more active and effective roles. To this end the network aims to ensure that every Network's member university and research center in any part of the world can access the knowledge needed to stay up to date and to be an effective support to government, civil society, and the business sector in local and national problem solving.

The commitment described is totally transparent and provides for the fondamental engagement of the whole community including students. In this sense, students of the University of Siena constitute the Greening USiena network, that is part of a worldwide student network and that deals with ecology, conservation and sustainability. Greening USiena aims to reduce the environmental impact of university facilities, to create a healthy cultural climate of aware student associations within the community of Siena, to participate in the initiatives of the World Student Environmental Network, to promote and organize seminars for students, working groups and orientation activities, to collaborate with the academic institution for the creation and management of events and, national and international, projects through which students actively engage in concrete activities that have an attention towards sustainable issues.

Finally, it has been recently approved by the governing bodies of the University of Siena the planning document that defines the strategic objectives of policy action and development for the next few years. The Strategic Planning Document

2013–2015 was elaborated after an open debate intended to stimulate the contributions of teachers, administrative staff and students, and aimed to identify reasonable projects.

Validating a willingness to act as a support in local activities and to promote a strategic vision which aims to cooperate at regional level, the university leaders said:

> Beyond the administrative compliance, the University of Siena wanted to take this opportunity to reinforce its commitment to the university planning in the perspective of a strategic vision. That can be configured as a new cultural approach in management. Furthermore, the specificity and uniqueness of the territory certainly supports the intention of the University to play a central role in the improvement of the city and the territory itself, at the same way as it has already happened with a positive participation in numerous projects shared with local governments.

In the Strategic Planning Document 2013–2015, sustainability is one of the strategic objectives of the University together with employability, internationalization and enhancement of student services. These objectives will be achieved with interventions on the three main institutional activities, which are research, teaching and all the initiatives to support the economic and social development, represented primarily by the transfer of technology and expertise.

And again with a view to future it has been confirmed that:

> The University will continue in the organization of cross-training initiatives in the field of sustainable development and also promote the direct link between teaching activities and the results of the research projects of major importance.

The program is the result of a long and detailed effort that involved all members of the university community, government bodies and, educational and scientific units. Thus, for each strategic objective, including the sustainability, departments and administrative structures of the University are committed to specific projects, which will be measured at each stage of implementation on the bases of specific indicators identified in the 3 years from 2013 to 2015.

5 Conclusion

In the last years an increasing number of universities is moving away from any resistance and is beginning to make more systemic changes towards sustainability by re-orienting their education, research, campus facilities management and community outreach activities, all simultaneously or a subset thereof. In some cases, they are doing educational reforms towards efficiency, accountability, privatization, management and control that are favorable to such a re-orientation (Wals 2014).

Despite sustainability has not fully permeates all disciplines, curricula, scholars and university leaders, the number of rules and declarations are increased, and the scientific literature presents frameworks in the implementation of sustainability principles and investigates on examples of good practices in disseminating principles and values of sustainability.

Also the University of Siena has shown a serious commitment on activities and practices addressed to sustainability topics. Based on the analysis of this case study, we have seen that social actors can stimulate a transition towards sustainability. We have gained insight in how transitions dynamics are working and how institutional approach of transformation towards sustainable development can be used to describe and analyze transitions of society (local, national and international). In particular, it emerges how management tools can be adopted by university not only to involve internal actors in building and sharing values of sustainability, but also to raise awareness externally and to facilitate the cultural change at social level.

In our analysis, the development of planning tools defines strategic lines and allows comparative analysis on the effects of sustainability. It also offers the necessary support to institutional administrators to make informed long-term decisions in regard to economic, environmental and social issues. Indicators for sustainability have been identified in order to monitor and improve the local and possibly affecting national situation. The implementation of such planning has carried out in consideration of appropriate investment. Each University's internal unit has identified projects and initiatives for the achievement of specific objectives that are included in the whole plan of the University. The budget has to comprise the recognition of the sustainability goals identified through the activities planned by the single units. And, in this way, the University is committed to invest in supporting actions for sustainability in order to facilitate the dissemination of its principles.

Moving from the evidences emerged from the case analyzed, our research indicates that the transition toward sustainable development seems currently in the take-off stage, and near the acceleration stage where structural changes take place in a more visible way through an accumulation of socio-cultural and institutional changes. The process of transformation gets under way and the state of the system begins to move. Although there is still not enough alignment between the strategic vision, actor configurations and practical implementation, there are not important barriers from universities to change and to be agent of societal change. It is also emerged that a vision of the future appears crucial in realizing the transition. Finally, the establishment of networks of and alliances with other actors has become an indispensable condition for success because hardly any effective effort to make a transformation towards sustainability can be undertaken by a university in isolation.

References

Balas DF, Lozano R, Huisingh D, Buckland H, Ysern P, Zilahy G (2010) Going beyond the rhetoric: system-wide changes in universities for sustainable societies. J Clean Prod 18: 607–610

Cortese AD (2003) The critical role of higher education in creating a sustainable future. Plan High Educ 31(3):15–22

Disterheft A, Ferreira da Silva Caeiro SS, de Miranda Ramos MR, Azeiteiro UM (2012) Environmental Management Systems (EMS) implementation processes and practices in European higher education institutions e Top-down versus participatory approaches. J Clean Prod 31:80–90

Foxon TJ, Burgess J, Hammond GP, Hargreaves T, Jones CI, Pearson PJ (2010) Transition pathways to a low carbon economy: linking governance patterns and assessment methodologies. In: Paper, IAIA10 Conference Proceedings, The Role of Impact Assessment in Transitioning to the Green Economy, 30th Annual Meeting of the International Association for Impact Assessment, 6–11 April, International Conference Centre, Geneva, Switzerland
Geels FW (2002) Understanding the dynamics of technological transitions: a coevolutionary and socio-technical analysis. Centre for Studies of Science, Technology and Society, Universiteit Twente, Enschede
Karatzoglou B (2013) An in-depth literature review of the evolving roles and contributions of universities to education for sustainable development. J Clean Prod 49:44–53
Kemp R, Rotmans J (2005) The management of co-evolution of technical, environmental and social systems. In: Weber M, Hemmelskamp J. (eds) Towards environmental innovation systems. Springer-Verlag, Heidelberg, pp 33–57
Kemp R, Loorbach D (2003) Governance for sustainability through transition management. In: Paper for open meeting of the human dimensions of global environmental change research community, Montreal, Canada, 16–19 Oct
Kemp R, Loorbach D, Rotmans J (2007) Transition management as a model for managing processes of co-evolution towards sustainable development. Int J Sustain Dev World Ecol 14 (1):78–91
Loorbach D (2002) Transition management: governance for sustainability. In: Paper for the conference governance and sustainability 'new challenges for the state, business and civil society', Berlin, 30/09-01/10
Loorbach D (2007) Transition management: new mode of governance for sustainable development. International Books, Utrecht
Loorbach D (2010) Transition management for sustainable development: a prescriptive, complexity-based governance framework. Governance 23(1):161–183
Loorbach D, Rotmans J (2006) Managing transitions for sustainable development. In: Olshoorn X, Wieczorek AJ (eds) Understanding industrial transformation. Views from different disciplines. Dordrecht, Springer
Lozano R, Lukman R, Lozano FJ, Huisingh D, Lambrechts, W (2013) Declarations for sustainability in higher education: becoming better leaders, through addressing the university system. J Clean Prod 48: 10–19, Special volume
Mio C (2013) Towards a sustainable university: the Ca' Foscari experience, Palgrave Pivot
Rotmans J, Kemp R, Van Asselt M (2000) Transitions & transition management: the case of an emission free energy supply. ICIS (International Center for Integrative Studies), Maastricht, The Netherlands
Rotmans J, Kemp R, Van Asselt M (2001) More evolution than revolution: transition management in public policy. Foresight 3(1):15–31
Schneidewind U, Augenstein K (2012) Analyzing a transition to a sustainability-oriented science system in Germany. Environ Innov Soc Trans 3(3):16–28
Stephens JC, Hernandez ME, Román M, Graham AC, Scholz RW (2008) Higher education as a change agent for sustainability in different cultures and contexts. Int J Sustain High Educ 9 (3):317–338
Trencher GP, Yarime M, Kharrazi A (2013) Co-creating sustainability: cross-sector university collaborations for driving sustainable urban transformations. J Clean Prod 50:40–55
Brugge R, van der Rotmans J, Loorbach DA (2005) The transitions in Dutch water management. Reg Environ Change 5:164–176
Wals AEJ (2014) Sustainability in higher education in the context of the UN DESD: a review of learning and institutionalization processes. J Clean Prod 62:8–15

Author Biographies

Angelo Riccaboni, Ph.D. is Rector of the University of Siena and Professor of Management Control Systems. He was Visiting Scholar at USC (Los Angeles), INSEAD, LSE, University of Wales at Bangor (UK), Columbia Business School and DePaul University Chicago. Angelo is member of the Leadership Council of the UN Sustainable Development Solutions Network and Chair of Med Solutions, the UN SDSN Regional Center for the Mediterranean. His research activities, with a strong international profile, are mainly focused on the issues of corporate governance and management accounting systems, and specifically oriented to the field of change in universities. You can reach Angelo at riccaboni@unisi.it.

Francesca Trovarelli is a research fellow at the University of Siena, Italy. Her research interest is devoted to systems of governance and management control, with particular reference to organizational change and management in universities. You can reach Francesca at trovarelli@unisi.it.

Integrating Sustainability into the Curriculum of Adult Education Studies: A Journey Across Disciplines

Violeta Orlovic Lovren

Abstract

Global orientation towards implementing sustainability in education programmes and processes at all the levels is being reflected in the most relevant documents in the field of adult education. While there is awareness of that need in declarations and documents at the national level, in the context of an early stage of implementation of sustainable principles, there is yet a lot to be done to incorporate those principles at the higher education level in Serbia. The paper presents some steps taken towards the integration of sustainability into the curriculum of adult education studies at the Faculty of Philosophy, University of Belgrade. Approaching sustainability from a lifelong learning perspective, it describes the context and goals of developing syllabus for the subjects of Environmental Adult Education and Ecological Andragogy for undergraduate and Ph.D. students at the Department for Andragogy of the Faculty of Philosophy, University of Belgrade. The aim of this contribution is to inspire discussion on needs and opportunities for increasing awareness and understanding sustainability in future adult educators in the course of higher education, for the sake of strengthening capacities for sustainable development across the borders of countries, sectors and disciplines.

Keywords

Adult education · Andragogy · Sustainability · University of Belgrade

V.O. Lovren (✉)
Department for Andragogy, Faculty of Philosophy, University of Belgrade,
Cika Ljubina 18-20, 11070 Belgrade, Serbia
e-mail: violeta.orlovic@f.bg.ac.rs

W. Leal Filho et al. (eds.), *Integrative Approaches to Sustainable Development at University Level*, World Sustainability Series, DOI 10.1007/978-3-319-10690-8_21

1 Adult Education, Lifelong Learning and Sustainability: Global Perspectives

Internationally relevant documents for integrating sustainability into adult education have been developing noticeably since the 1990s, influenced by the global acceptance of this concept. Lifelong learning, interdisciplinarity, partnership, multiculturality and empowering are promoted as key topics of education for sustainability (UNESCO 2004). Hence it comes as no surprise that these trends influence conclusions from the most relevant conferences of experts and decision-makers in the field of adult education at the global level.

Thus, for example, according to the well-known Hamburg Declaration (5th International Conference on Adult Learning), "...education for environmental sustainability should be a lifelong learning process which recognises that ecological problems exist within a socio-economic, political and cultural context.... Adult environmental education can play an important role in sensitising and mobilising communities and decision-makers towards sustained environmental action' (UNESCO 1997, p. 4).

The signatories of the Declaration commit themselves to work on "...promoting the competence and involvement of civil society in dealing with environmental and development problems... by integrating environmental and development issues into all sectors of adult learning and developing an ecological approach to lifelong learning" (UNESCO 1997, p. 12).

Adoption of UN resolution no. 57/254 regarding the Decade of education for sustainable development (2005–2014) and its close connection to the documents and processes significant for adult education, such as the Dakar Framework for Action and the UN Literacy Decade (http://unesdoc.unesco.org), helped exercise further influence on global policies and relevant documents in this area of education.

This influence is obvious not only in the choice of topics, but also in the overall approach to organising most relevant meetings of experts in the field of adult education. The last, 6th International Conference on Adult Education, was organized on the basis of the "culture of sustainability" principle, which means that organisers invited the participants to observe their environmental impact, and activities such as planting trees were also undertaken in order to compensate for one's own contribution to pollution through the meeting preparation and realisation. Documents created there also emphasize the importance of adult education "...for building equitable, tolerant, sustainable and knowledge-based societies" (UNESCO 2009, p. 2). The participants concluded that despite the diversity of programmes for adult education and learning, most countries predominantly focus on programmes of professional and vocational education and training, recognising the obvious lack "...of more integrated approaches to adult learning and education to address development in all its aspects (economic, sustainable, community and personal)" (Ibid. p. 10).

The dedication of adult education proponents and experts to the cause of integrating knowledge into all aspects of life is neither new nor coincidental. Original principles of the concept of lifelong learning, organically embedded into the development of the theory and practice of adult education, refer to learning as a lifelong, but also a life-wide process. Concepts of lifelong learning and sustainability possess a whole range of common dimensions, such as: the interdisciplinary holistic approach and its integration into education at all levels; the development of the capacity for facing changes and managing them; the dimension of equality; respecting the diversity of people and the environment, and the principle of intersectoral collaboration—in both cases seen as a prerequisite for the quality of education, i.e. integrating sustainability into all segments of life (Orlović-Lovren 2012).

Taking into consideration the similarity of these concepts and their connectedness to globally agreed developmental goals, it is clear that adult education, as a segment of lifelong learning, also represents an integral part of sustainable development. "Learning for Life", as stated by Thomas, is one of the principles of Education for Sustainability, which means that "...sustainability is linked to lifelong learning and not confined to formal schooling" (Thomas 2009, p. 252).

Why is then the integration of sustainability into andragogy and adult education slow and insufficient? It was only in the final years of twentieth century that there was an increase in the number of authors aiming to examine and demonstrate multiple links and possibilities for adult education and learning in this field. Part of the reason certainly lies in nowadays widely accepted assessment of the concept of sustainable development as too broad to be operationalized and implemented. Naturally, that in turn adds to confusion in defining sustainability, in the same way as fragmentary tackling of only some of its aspects does. Quite a few authors, however, also blame the dominance of economic interests in all political spheres—including the educational one, which contributes to a greater representation of the professional education in comparison to others. In the opinion of authors of socio-ecological orientation, whose approach strongly relies on the social theory postulated by the Brazilian educator Paulo Freire, the greatest challenges for the development of sustainability lie in economic aspect and are manifested by means of the so-called corporative globalisation, i.e. prioritising programmes whose aim is "learning for earning", where programmes focusing on social and ecological aspects of life, or sustainable development, would be neglected (Summer, according to: Hill and Clover 2003).

One of the important messages conveyed by these authors expresses the necessity for experts and practitioners in adult education to raise awareness about the significance of integrating ecological, social and economic aspects together with transformative approach into adult education programmes. We deem this to be of utmost importance as well for all programmes at university level, particularly those relevant to the teaching of adult education experts, who will in future work as advisers supporting the learning process, or teachers in one of the fields of adult education.

2 Sustainable Development and Education in Strategic Documents in Serbia

In the last decade, as part of social reforms and influenced by European trends, Serbia has seen an improvement in establishing a normative basis for integrating the concept of sustainability into education.

The Strategy for Sustainable Development of Serbia (Official Gazette, 57/08), was elaborated for the period between 2009 and 2017. The improvement of the quality of education in the way described by this strategy in all components of sustainable development is anticipated to take place by means of the following mechanisms: in the domain of the "economy of knowledge", there is a planned quality advancement that will "…ensure better efficiency of education at all levels"; in the sphere of socio-economic conditions, it is predicted that securing a complete inclusion and easy access to education for all vulnerable groups of people—be it children, particularly from vulnerable groups, or adults without qualifications, in the domain of inclusion into primary education.

Regardless of the tendency to integrate the principle of sustainability into all segments of society, the emphasis here also lies on economic efficiency on the whole, even when it comes to the field of education. Despite promoting quality and equality in all segments of education, the focus is significantly more placed on the system of formal education. Action plan for implementing this strategy much more explicitly envisages activities in the field of adult education as part of measures for improving sustainability.

In implementing the Strategy so far, the results are achieved within the reform of pre-school and primary education, principally by means of integrating elements of ecology and environment protection into the curriculum. At the secondary education level, where the reform has only commenced, for certain vocational profiles new subjects such as Education for Sustainable Development (for the profile of Recycling Technician, Department of Metallurgy) have been introduced (Radojević et al. 2011).

After signing the Ministerial Declaration of UNECE region at the Fifth Conference "Environment for Europe" in Kiev, in May 2003, Serbia accepted and translated UNECE Strategy for education for sustainable environment (UNECE 2005). Starting from the stance that "…the development of sustainable society should be seen as a continual process of learning, exploring problems and dilemmas, where appropriate answers and solutions could be changed in accordance with increasing our experience", the emphasis is on the need for reorientation of both formal and non-formal education in accordance with the principles of sustainability (UNECE 2005). Participants' agreement with these statements were also confirmed by signing of the Declaration "Building Bridges to the Future" at the ministerial meeting of UNECE region held in Belgrade in 2007 (http://www.unece.org/env/esd/belgrade), whereby Serbia committed itself to including sustainability in educational programmes at all levels.

Concordantly with European strategic documents, a document called "The Strategy for the Development of Education in Serbia until 2020" was created according to which "…the duty of the educational system of the Republic of Serbia is to timely provide good quality and efficiency of Serbian population in accordance with the expressed or recognised developmental orientations in the direction of sustainability and to respond to educational needs of each Serbian citizen throughout their entire life" (Official Gazette 2012, p. 5). Although this document includes social, economic, and to a certain extent ecological aspects of development —particularly at the level of Ph.D. studies, emphasizing of paramount need for interdisciplinary research and development of scientific research staff—sustainability as a concept and sustainable development as a developmental objective are not explicitly promoted in Serbian educational projections for the future.

Concisely stated strategic trends also have a certain impact on acknowledging the importance of integrating sustainability into different areas of higher education in Serbia. More than 20 faculties in Serbia founded departments or groups for studying environmental issues both at undergraduate and postgraduate levels (UNECE 2005).

3 Integrating Sustainability into Higher Education Programmes at the University of Belgrade

The University of Belgrade is a state university consisting of 31 faculties, divided into groups for social sciences and humanities, medicine, science and technology (http://www.bg.ac.rs/).

Within the group for scientific and technological orientation, education of future environmental experts mostly takes place at the Faculty of Biology, Faculty of Chemistry and Faculty of Technology and Metallurgy. Courses at these faculties do not include a single subject dealing explicitly with the education for sustainable development. Subjects which contain the term "sustainable development" in their name can be found in the curriculum for Environmental Engineering within undergraduate and master studies at the Faculty of Technology and Metallurgy (http://www.tmf.bg.ac.rs), whilst at the Faculty of Architecture there are following subjects at master studies: Architecture and the Environment (with elements of sustainable development), and Sustainable Urban Development at master studies in urban planning (http://www.arh.bg.ac.rs). Both Faculty of Agriculture (http://www.agrif.bg.ac.rs) and of Organizational Science (http://www.fon.bg.ac.rs) offer subjects covering environmental and socio-economic aspects of sustainability, while the study programme of the Faculty of Forestry (http://www.sfb.bg.ac.rs) comprises subjects dealing with sustainable development (Environmental engineering and land resources department).

Study programmes at the Faculty of Geography, and in particular its Department for Geospatial and Environmental Science, comprise a whole range of such courses at undergraduate level: Sustainable Development, Human Ecology, Economic

Environmental Evaluation. At master level, students can choose a course called Education for sustainable development, in the context of methodical knowledge and orientation towards educating children and young adults (http://www.gef.bg.ac.rs/).

Based on the analysis of the programmes of ten faculties which can be classified as social sciences and humanities at the University of Belgrade,[1] Lončar concludes that not a single of these faculties in their undergraduate programmes contains any subject with "sustainable development" in its name, while elements of education for sustainability can be anticipated in few interdisciplinary subjects which treat ecological aspects from a legal point of view (The Faculty of Law), economic and ecological aspects of security (The Faculty of Security) or social aspects of sustainability (The Department of Sociology at the Faculty of Philosophy) (Lončar 2011).Within the 2009/2010 programme, the master studies in sociology at the Faculty of Philosophy includes a course in Contemporary Issues of Sustainable Development, which contains the topics of social policy measures and the principles of sustainable development (www.f.bg.ac.rs).

At the Faculty of Political Science, there is the interdisciplinary optional course in Social and Political Ecology. Elements of education about sustainable development are included in the subject called Ecological Policies of the European Union (Nadić 2011).

This Faculty also houses the Centre for Ecological Policies and Sustainable Development (http://cepor.fpn.bg.ac.rs/), which is a scientific unit devoted to research, publication and organising actions and campaigns with the aim of developing ecological awareness in students. In 2011, this Centre coordinated a symposium dedicated to the University and sustainable development. Participants came from different fields of study, unanimously promoting the need for implementing the concept of sustainable development in higher education programmes at the University of Belgrade and in the rest of Serbia.

Judging by the previous brief overview, there are examples of implementing the concept of sustainability not only at faculties educating experts in the environmental field, but also in other academic institutions. These efforts mostly boil down to individual lecturers' initiative, mainly representing minor changes within the curriculum. There is a noticeable lack of organised initiative at the University and its poor coordination in cooperation with the authorities and the community regarding the introduction of sustainability into the University in all its aspects (Table 1).

---

[1] The Faculty of Law, The Faculty of Economics, The Faculty of Orthodox Theology, Teachers' Training Faculty, The Faculty for Special Education and Rehabilitation, The Faculty of Security, The Faculty of Sports and Physical Education, The Faculty of Political Science, The Faculty of Philosophy, and The Faculty of Philology.

Table 1 Preview of sustainability aspects in study programme of faculties within the University of Belgrade

Scientific area groups/faculty	Environmental aspects of sustainability	Socio-economic aspects of sustainability	Environmental education based on sustainability concept	Sustainable development as a separate subject	Education for sustainable development as a separate subject
Social sciences and humanities		✓		✓	
Faculty of philosophy/sociology					
Faculty of philosophy/andragogy		✓			
Faculty of law	✓				
Faculty of security	✓	✓			
Faculty of political Science	✓	✓			
Science	✓				
Faculty of biology					
Faculty of chemistry	✓				
Faculty of geography/department for geospatial and environmental science	✓	✓		✓	✓
Technology	✓	✓		✓	
Faculty of technology and metallurgy					
Faculty of architecture	✓	✓		✓	
Faculty of agriculture	✓	✓			
Faculty of forestry	✓	✓		✓	
Faculty of organizational sciences	✓	✓			

4 Integrating Sustainability into Adult Education Studies at the Faculty of Philosophy

The Department for Andragogy at the Faculty of Philosophy in Belgrade was established 35 years ago and its studies at undergraduate, master and Ph.D. levels include general and specific issues related to adult learning and education. The group enrols 35 undergraduate, 20 master and 2 Ph.D. students annually.

The objective of the study programme is "… the development of creative skills which enable the implementation of particular concepts, organisation, methods and procedures regulating adult education, acquiring specific skills of managing and organising processes of adult education whilst including all aspects and spheres of an adult individual and the needs of their social milieu" (http://www.f.bg.ac.rs/andragogija/odeljenje).

Current programme of studies includes compulsory and optional subjects whose content covers a really broad range of "aspects and spheres of adult life", i.e. their roles in family, societal, professional, managerial and civic sector, in creating free time and cultural life, as well as ability and opportunities of adults to learn.

The programme of studies is orientated towards the development of the following general competences in students: understanding, critical analysis and assessment of andragogical phenomena, knowledge and ideas; familiarisation with basic methods, techniques and procedures for collecting, organising and analysing data in adult education and learning; abiding by ethical and professional norms in practice and an adequate attitude towards the development of one's own profession (http://www.f.bg.ac.rs/andragogija/odeljenje).

The principles of lifelong learning as well as modern concepts which represent the foundations of the processes of learning and teaching (transformative and experiential learning, learning orientated towards the learner, towards problem solving and the development of critical thinking) are deeply interwoven in planning and realisation of the study programmes.

The content of existing subjects includes the issues of sustainability in social and economic aspect (regardless of the fact that sustainable development is not explicitly mentioned), so we can conclude, as other authors did as well (Lončar 2011) that this study programme possesses potential for integrating the concept of sustainability into the curriculum. As a first step towards that goal we see the integration of environmental aspects (as the third necessary dimension of the concept of sustainable development) into the programme of andragogy studies.

Starting from there, two new subjects are incorporated for the first time in the new study programme, planned to commence in the course of the 2014/2015 academic year: at undergraduate level, an optional course in Environmental Adult Education, and Ecological Andragogy at Ph.D. level.

The objective of the course in Environmental Adult Education is focusing on the significance and characteristics of environmental education as a field of interdisciplinary andragogical action, a segment of sustainable development and lifelong learning. Upon completion of this course, students should be able to identify characteristics of environmental adult education, to improve their critical stance towards ecological issues and solutions and to recognise mechanisms and possibilities of educators' actions for raising environmental awareness and management. It is also expected that they can create programmes for education and learning in the environmental field, based on the principles of sustainability, then the measures for social support of individuals and groups in acquiring knowledge and competences in the environmental and sustainability fields and create systemic solution to educational actions in this area of study.

The course in Ecological Andragogy is designed for Ph.D. students with the aim of integral study of the environment as the surroundings adults have whilst learning, including all domains of such a milieu. The approach for creating this course is based on the principles of sustainability, in terms of understanding the surrounding as a prospect for learning, through the interaction of an individual with the environment in the broadest sense of the word. The course content includes, amongst other things, the issues of perception and potential for learning in the natural, working, local and wider social environment, as well as the relations between sustainable styles of living and adult learning. In their research and practical activities, the students will be able to apply methods and techniques of ecological analysis when addressing issues of adult learning and education (Table 2).

Incorporating this content into the curriculum of andragogy studies, we do not intend to "simply add a list of sustainability concepts that might be added on to an existing curriculum" (Thomas 2009, p. 251), but to contribute to the interdisciplinarity of the entire study programme and to enable students to comprehend contemporary global challenges included in other fields of study, from another critical perspective. Including such an integrative subject into the curriculum has been facilitated by the profile of the entire study programme which unites knowledge from the domain of lifelong education, adult learning and development, knowledge management, social and economic aspects of education and environmental education.

Table 2 Aspects of sustainability in study programme of andragogy at the Faculty of Philosophy

Faculty of philosophy/andragogy	Environmental aspects of sustainability	Socio-economic aspects of sustainability	Environmental education based on sustainability concept	Sustainable development as a separate subject	Education for sustainable development as a separate subject
Current study program (since 2009/10)		✓			
Study program to start from 2014/2015 school year	✓	✓	✓		

Analysing the concept of education orientated towards sustainable development, authors today rightfully differentiate between education ABOUT sustainable development (ESD) and education FOR sustainable development (EfS), stating ".. that ESD focuses on the learning process required to support sustainable development, while EfS emphasises building our capacity to live more sustainably" (Thomas 2009, p. 250). It is also highlighted that education for sustainability or sustainable development,[2] i.e. the educational process in this field "...must focus on elements relating to the processes of learning, rather than the accumulation of knowledge—to develop graduates with capabilities to improvise, adapt, innovate, and be creative." (Ibid. p. 245).

Sharing such beliefs, we wish to point out that the key competences of andragogy students this particular study programme is aimed at, such as critical analysis, team work, creative use of methods and techniques of teaching and research—significantly overlap with those we aim at within the domain of education for sustainability. Therefore, we believe that a separate subject—Education for Sustainable Development is not necessary in the structure of andragogy studies. The evolution of an "ecological point of view" and awareness of the relations between education and all aspects of sustainable development serves the purpose of enhancing competences and personal responsibility of andragogy students but indirectly also to their creation within a wider community, via the future professional activities and working with various groups of adults. "Environmental adult education adds another critical lens, an ecological lens through which we can address environmental problems and give voice to the needs of those who are most affected.... The appropriation by adult educators of a concept normally used solely by biologists and environmentalists can help to create synergy between formal, non-formal and informal education and enhance our understanding of how learning environments contain hidden curricula that influence participation and cognitive development" (Hill and Clover 2003, p. 14, 79). Hence it is logical to require that the initial education of expert adult educators at university level include social, economic, cultural, discursive, political and ecological spheres (Clover et al. 2010).

We are by all means at the beginning of the process and about only one of the aspects of sustainability of higher education—introducing new academic subjects. This should lead towards influencing the entire andragogy curriculum with an even more mature critical stance and values propagated by sustainability. Regardless of a certain lack of clarity when it comes to what sustainable society and sustainable future are, a common viewpoint of the concept of sustainability and the approach maintained at the group for andragogy is surely that sustainable development cannot exist without continuous learning in all stages and spheres of life. Such learning does not represent a mere transmission of knowledge, but contributes to the development of the capacity of students "as agents of change" (Tilbury, according to: Thomas 2009, p. 251), so that they would in turn be able to develop such capacities in others.

[2] In this paper we use the terms *sustainable development* and *sustainability* as synonyms.

Convinced that the process would be even more effective if, prior to the programme implementation, we examine the previous knowledge and activities of students in the spheres of ecology and sustainable development, we conducted a survey of the 3d- and 4th-year students of andragogy in February 2014. It included 49 out of the total number of 60 students enrolled.

The first question (out of three) asked them to define sustainable development. Students were able to choose between four possible answers provided: (1) the philosophy of the protection and development of our environment; (2) a concept which propagates maintaining a balance between poverty and economic development; (3) a developmental concept which equally considers economic, environmental and social aspects; and (4) a doctrine which involves equalisation of the development of different world regions.

According to their answers, a great majority of students demonstrate a general understanding of the essence of sustainable development, opting in 73 % of the cases for the definition number 3. They are followed by other 20 % of students who chose answer number 2, whilst a significantly smaller percentage of students selected other answers. The issues of poverty and economic development are present in the content of andragogy studies within quite a few academic subjects and it does not come as surprise that 20 % of students, without prior knowledge about ecology, opted for such an answer.

Aiming to find out how students perceive global developmental challenges and their (1) connectedness with adult education, (2) coverage in the study programme, and (3) personal interest in addressing them, we provided a list with 7 challenges, taken from the Centre for Strategic and International Studies, whose experts define them as "Seven revolutions": Governance (Corporate Citizenship, Civil Society, Strategic leadership), Conflict (Terrorism, Mass destruction, Change of skills needed), Economic (Globalisation, Rapid Economic Development, Inequality), Information (Information economy, Technology and Knowledge Exchange), Technology (Technology development and backlash), Resource (Nature resources, Overconsumption, Food, Water, Energy) and Population (Growth, Migration, Generation gap) (www.csis.org). The results shown in the following graph represent ranges of each challenge in relation to the three dimensions observed, which students marked on the scale from 1 (lowest) to 5 (highest) (Graph 1).

Information, Population and Governance are in the first place when it comes to their relevance to adult education, and, according to our students, are followed by Economic, Technology, Resource and Conflict. Resource, Conflict and Technology issues are in their opinion least represented in andragogy studies, whilst Information, Governance and Population have the highest coverage. Students estimate that they are most interested in Information, then Population, followed by Technology issues, and least of all for Resource!

These results imply the following possible factors of such a choice made by students: the topics which the study programme covers most (Information, Population and Governance) are at the same time the ones students consider most relevant for adult education, and two of them (Information and Population) are those they are personally most interested in! Similarly, the lowest interest of

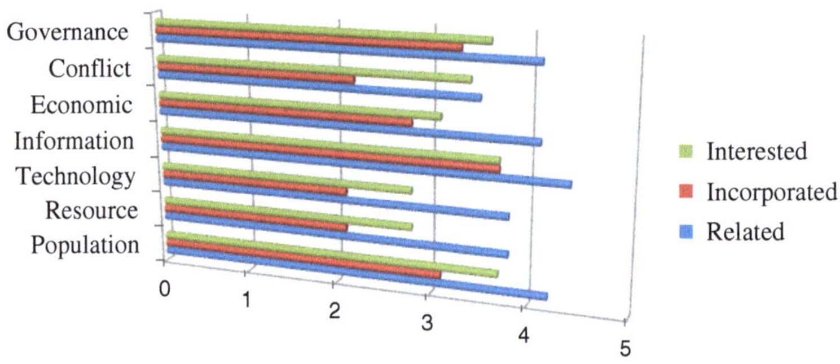

Graph 1 Students' estimations of seven challenges

students in issues to do with Resource coincides with the lowest estimated coverage of these topics compared to all the others within the study programme. Hence it appears that students assess the relevance of topics by relying on what the programme authors deemed relevant to include in the andragogy studies, whilst Resource, due to it being insufficiently represented in the content of this study programme, seems to be an unfamiliar field in which educational needs and interests of students could not develop.

When it comes to personal activities in the fields of the environment and sustainable development, more than 50 % of students admit to be absolutely inactive, a little more than a third of students claim to read a lot about it, whilst only few (8 %) participates in projects, i.e. actions within civil society (4 %) in that field.

5 Concluding Remarks

By accepting international declarations in the area of sustainable development and education for sustainability, as well as through creating its own strategic documents, Serbia demonstrated its interest in integrating sustainability into all levels of education. That involves a commitment to systematic integration of sustainability into the programmes of higher education. However, it occurs sporadically, without coordination by the Rectorate or the competent ministry. The examples of introducing new academic subjects in study programmes at faculties within all faculty profiles at the University of Belgrade exhibit a positive trend, although they are rarely accompanied by other aspects of University sustainability. Bearing that in mind, we can conclude that the University of Belgrade, according to the categorisation introduced by Leal Filho, is still at level one, i.e. that apart from few individual changes to the curriculum at different faculties and some research, there are no other relevant activities, such as Institutional Sustainability Policies, Mobilising staff and students, Training of Staff, Inclusion of Sustainability in Continuing Education and Extension (Leal Filho 2009, p. 315).

Having in mind that the real immersion of the sustainability principle into the University is a complex and multidimensional process, introducing changes to faculties' curricula is considered to be a modest but significant step.

Including the content which would cover the missing dimensions of sustainability, e.g. introducing Environmental Adult Education and Ecological Andragogy into the study programme for andragogy in Belgrade, should not represent another burden but an opportunity to accept certain values, and bring changes to one's own actions and those performed by the community via the future professional work on adult education.

In order for this process to be successful, it is necessary for lecturers to cooperate, to "open doors" of their "own" fields of study to other disciplines' standpoints and to the integration of sustainability at all levels, but a broader societal support is required as well. Exchange of experiences between faculties both at the national and international levels, related to integrating sustainability, initially into academic subjects and then elsewhere, provides collegial support which is invaluable, particularly in the absence of a broader societal and organised initiative.

References

Clover D et.al (2010) The nature of transformation: environmental adult education. University of Victoria, Victoria

Hill L, Clover D (eds) (2003) Environmental adult education: ecological learning, theory, and practice for socio-environmental change. Jossey-Bass, San Francisco

Leal FilhoW (ed) (2009) Sustainability at universities—opportunities, challenges and trends. Frankfurt am Main, Peter Lang

Loncar J (2011) Obrazovanje za odrzivi razvoj na fakultetima drustveno—humanisickih nauka. In: Pavlovic V (ed) Univerzitet I odrzivi razvoj. Fakultet politickih nauka, Beograd, p 241–263

Nadic D (2011) Mesto I uloga obrazovanja za odrzivi razvoj u obrazovanju politikologa. In: Pavlovic V (ed) Univerzitet I odrzivi razvoj. Fakultet politickih nauka, Beograd, p 113–125

Orlovic-Lovren V (2012) Sustainable development and lifelong learning concepts: two frames for one view on adult education. Andrag Stud 1:9–23

Official Gazette (2008) Strategija održivog razvoja Srbije. RS 57/08

Official Gazette (2012) Strategija razvoja obrazovanja u Srbiji do 2020. 107/2012

Radojevic D et.al (2011) Nacionalna strategija odrzivog razvoja—Obrazovanje za odrzivi razvoj. In: Pavlovic V (ed) Univerzitet I odrzivi razvoj. Fakultet politickih nauka, Beograd, p 99–113

Resolution on Education for Sustainable Development (2014) http://unesdoc.unesco.org. Accessed 20 Feb 2014

Thomas I (2009) Critical thinking, transformative learning, sustainable education, and problem based learning in universities. J Transform Edu 7:245–264. http://jtd.sagepub.com/content/7/3/245.refs.html. Accessed 15 Feb 2014

UNESCO (1997) The hamburg declaration: the agenda for the future. In: Proceedings of the 5th international conference on adult education. http://www.unesco.org/education/uie/confintea/. Accessed 23 March 2014

UNESCO (2009) Global Report on Adult Learning and Education (GRALE). Institute for Lifelong Learning, Hamburg. http://www.unesco.org/en/confinteavi/grale/. Accessed 3 March 2014

UNESCO (2004) United Nations decade for sustainable development 2005–2014: draft international implementation scheme

UNECE (2005) Strategija razvoja održivog obrazovanja, Vilnius. http://www.unece.org/fileadmin/DAM/env/esd/strategytext/StrategyinSebian.pdf. Accessed 15 March 2014
Web page Department for Andragogy (2014) Faculty of philosophy. http://www.f.bg.ac.rs/andragogija/odeljenje. Accessed 21 March 2014
Web site Faculty of Agriculture (2014) http://www.agrif.bg.ac.rs. Accessed 21 March 2014
Web site Faculty of Architecture (2014) http://www.arh.bg.ac.rs. Accessed 20 March 2014
Web site Centar za ekolosku politiku I odrzivi razvoj. http://cepor.fpn.bg.ac.rs/. Accessed 21 March 2014
Web site Faculty of Forestry (2014) http://www.sfb.bg.ac.rs. Accessed 20 March 2014
Web site Faculty of Geography (2014) http://www.gef.bg.ac.rs/). Accessed 21 March 2014
Web site Faculty of Organizational Science (2014) http://www.fon.bg.ac.rs. Accessed 21 March 2014
Web site Faculty of Philosophy (2014) www.f.bg.ac.rs. Accessed 21 March 2014
Web site Faculty of Technology (2014) http://www.tmf.bg.ac.rs. Accessed 20 March 2014
Web site of University of Belgrade (2014) http://www.bg.ac.rs/. Accessed 20 March 2014
Web site Center for Strategic and International Studies (2014) www.csis.org. Accessed 10 March 2014

Author Biography

Violeta Orlovic Lovren received her doctorate in Andragogy from the Faculty of Philosophy, University of Belgrade, in 2011. Her interdisciplinary orientation builds on more than 15 years of work, study and research in the field of nature conservation, environment and sustainability from education perspective. She is active member of IUCN Commission on Education and Communication and World Commission on Protected Areas, as well as LEAD International and Hubert Humphrey Fellow. Lives in Belgrade, where she is teaching at the Department for Andragogy, Faculty of Philosophy, University of Belgrade.

The Impact of Withdrawing a Structured Initiative Aimed at Engaging Departments in Sustainable Activities, at a UK University

Sally R. Lampkin

Abstract

This paper investigates the impact on departmental engagement in sustainable activities following the withdrawal of a structured campus-wide initiative at a UK university, 20 months after it had been introduced. The research questions included: What changes did the initiative bring? What, if anything, replaced the influential factors once the initiative was withdrawn? What has been the impact on engagement since the withdrawal? How did people perceive changes to the engagement level? What or who has been critical to any successes? What factors are most likely to influence engagement positively in the future? The study compared and contrasted academic departments and support services. Some predictable results were found, along with quite surprising ones such as the impact of losing recognition and support for undertaking activities and of copying the behaviour of others. Communication, co-ordination and the provision of authority were found to play key roles. There are case-studies in the literature examining sustainability initiatives in the workplace; few, however, report the impact of withdrawing a scheme less than 2 years after its introduction.

Keywords

Engagement · Sustainability · University · Framework · Communication · Coordination

---

S.R. Lampkin (✉)
Bristol Business School, The University of the West of England,
Coldharbour Lane, Bristol BS16 1QY, UK
e-mail: sally.lampkin@btinternet.com

W. Leal Filho et al. (eds.), *Integrative Approaches to Sustainable Development at University Level*, World Sustainability Series, DOI 10.1007/978-3-319-10690-8_22

1 Introduction

Universities can provide a favourable environment and opportunity for engaging people in sustainability. They have experts in the latest research and understanding concerning the issues, and a dynamic and transient population who are open to change and eager to embrace new ideas. Engaging people in a workplace has a different dynamic to how and why people choose to behave in their domestic lives, as it can be perceived that the employer is responsible for such issues (Campbell 2007) and, as there is no direct benefit or return to the employees, less importance may be placed on such activities. In this case study, all the departments at a UK university had been involved in an initiative that focused on changing sustainability practices, which was then withdrawn because the university's senior management considered that the level of engagement was sufficient for the departments to continue without the formal process. The departments were encouraged to maintain the new practices and extend them where possible. So what has happened to that engagement level since the initiative was withdrawn? There are a number of case-studies published that examine sustainability initiatives in the workplace. Few, however, report the impact of withdrawing such a scheme less than 2 years after its introduction; a situation other institutions and organisations around the world may well find themselves in due to limited resources, accountability and/or changing priorities.

2 Background

Engaging people in sustainable activities in the workplace is multi-faceted and requires an understanding of the interplay between the employer and the employees. Social exchange theory (Homans 1961; Blau 1964) posits that decisions between 2 parties are governed by the balance of rewards and costs for actions taken. When individuals receive economic and socio-emotional resources from their organisation, they feel obliged to respond in kind and repay the organisation, reflecting a two-way relationship between the employer and employee (Saks 2008). For employees to engage in new behaviours they need to feel that there is some benefit in the efforts they are being asked to make, along with adequate provision of facilities, resources and support, which outweigh the costs that will occur as a result. When the organisation fails in such provision, the employees are more likely to withdraw and disengage themselves. Sustainability can be perceived as a peripheral consideration by universities as very little funding is directly linked to such activities. There are league tables, such as the People and Planet Green League, which publicise an award level for individual institutions although participation is currently not mandatory. Hence, the stance taken by an employer, who is responsible for leading by example and providing guidance, support and inspiration, can greatly influence employee compliance to new practices and is itself influenced by resource availability and changing priorities.

To facilitate engagement across a university campus, it is important to understand the group dynamics between and within the various departments, which essentially act as autonomous units and in their own self-interests. There is a growing appreciation that although people behave selfishly, they are inherently social creatures and do care about the welfare of the groups to which they belong (Haidt et al. 2008). Academic departments, for example, have a common focus regarding research and teaching roles, and sustainability in the workplace may be considered relevant only to departments that specifically focus on subjects such as environmental management, corporate social responsibility or alternative energy sources. Patterns of individual behaviour do not simply combine to determine the behaviour of a group as a whole. Individuals within a group may engage in a number of behaviours that are contingent upon the behaviours of their peers, such as copying (Kennedy 2009) or learning by observing the behaviour of others (refer to social learning theory; Bandura 1977) and emergent patterns depend on how the agents within the group interact and the dynamics between, for example, competiton and cooperation (Goldstone and Gureckis 2009). Ideally, the engagement increases until the moment, or tipping point, when a sufficient number of adopters of the new behaviour (the critical mass) is reached, so that the rate of adoption becomes self-sustaining and creates further growth (see, for example, Kiron et al. 2012).

The attitude and behaviour of individuals also affects the engagement process. There are a nubmer of theories that attempt to categorise individual behaviour and capacity for change. According to Defra's (2007) Segmentation Model, for example, people range from those who engage willingly in sustainability activities at work and home to those who are totally uninterested in such issues and inactive. The Stages of Change model (Prochaska and DiClemente 1983) describes a series of stages that individuals go through when changing a behaviour, encompassing ignorance or indifference, premeditation regarding ways to engage, adjustments once the change has been made and the new behaviour ultimately becoming the norm; relapses can occur at any stage and periods of consolidation are often required. In their self-determination theory, Deci and Ryan (1985, 2000) identified different motivational states that affect whether or not an individual's engagement can be expected to be maintained once external triggers are removed. An internalised motivation, for example, is driven by internal characteristics such as the satisfaction, or importance, of undertaking activities and does not require external incentives, such as financial rewards or recognition.

This paper presents the findings from a UK university where a sustainability initiative was introduced and then withdrawn 20 months later, primarily due to its success in engaging departments and increasing the number of participants. The departments were encouraged to maintain the new practices and extend them where possible. The initiative consisted of departmental champions and green teams, a structured and progressive set of activities, central co-ordination and an annual achievement award.

The key research questions were:

(1) What were the key factors of the initiative's success?
(2) What has been the impact on engagement since the initiative was withdrawn?
(3) What factors are most likely to influence engagement positively in the future?

3 Methods

Each department was invited to participate in the research via the staff member who had been the main contact when the initiative was running and those departments that accepted completed an on-line survey. The aims of the survey were to ascertain the factors that were perceived to be influential in engaging staff whilst the initiative was running, the factors that had influenced any changes since the initiative was withdrawn and factors most likely to increase engagement in the future. The departments that completed the survey were then invited to participate in an interview to discuss, in greater depth, the processes that had been employed since the initiative was withdrawn and the influencing factors of any changes. The interviews were a semi-structured format, lasted up to 1 hour and focused on the impact of withdrawing the initiative on engagement levels and the factors that were likely to improve that engagement in the future, based on the survey responses for each department. Observations were also made via a walkabout in various departments, to gain an overall impression of the visibility of sustainability related activities. Finally, all the departments that chose not to participate in the interviews, or in the research at all, were asked for the main reason why they did not. As the university wanted to remain anonymous, the results are presented in a way that accommodates that.

4 Results and Discussion

There was a 64 % response rate for the survey, split evenly between academic and support services departments. There was a 52 % response rate to participate in the interviews (33 % of the initial sample), two thirds of which were support services departments and a third academic ones. The data collection took place between October 2013 and February 2014.

The results showed that there were six factors that had been equally influential in the success of the initiative; each respondent selected either three or four of them and no single factor stood out as more crucial than any of the others. These factors, of which the respondents had selected either three or four, were regular communications with a like-minded network of staff, departmental co-ordination, university-wide co-ordination, a structured and progressive framework, support from senior staff and an achievement award. All of these factors were expected to play a role; the fact that their importance was perceived as similar regarding the success of the initiative reflected, perhaps, that engaging groups of individuals within autonomous units required a range of inputs and approaches. Providing information and the size of a green team were not considered particularly important relative to the other factors. It is generally accepted that knowledge on its own does not necessarily result in people taking action (Moser 2006; Eden et al. 2008).

There was a common perception by the departments that a dip in engagement levels occurred immediately after the initiative was withdrawn, as a result of the loss of drive, focus, energy and authority that the initiative had provided. For the people involved, the withdrawal sent out a message that what the staff had been doing was no longer valued and that bigger things, such as the campus heating system, mattered more than the efforts and awareness raising by individuals and departments. Such a dip, or relapse, can occur when changes in behaviour are not full developed and people are unable to maintain the new practices without external incentives and facilitators (refer to Prochaska and DiClemente 1983) or when employees feel that the costs for the efforts they arc being asked to take outweigh the benefits (refer to Homans 1961; Blau 1964). The perception that individual efforts no longer mattered was very different to the message that the university senior management had wanted to communicate, which was to show their gratitude for the efforts so far and that, because of the success of the initiative, they believed that departments could maintain, if not extend, the changes already made without the need for an ongoing formal process (personal communication, October 24, 2013). It appears that, in this instance, the university hierarchy underestimated the value, purpose and power of the initiative in engaging others and the ongoing need for a formal structure to embed the changed practices.

A lack of time was the most common response provided to explain why two-thirds of the departments still had a lower engagement level compared to when the initiative was withdrawn, even though some recovery from the initial dip had occurred. A lack of time was also the only reason given, out of a possible eight, for why departments had chosen not to participate in the research at all or only completed the survey. The perception of having no time was associated with having no direction or coordination to progress the relevant activities within departments: "I struggled enough to find the time to follow a list of actions, never mind having to spend time thinking about what actions should be being taken." Substantiating this, those departments that did currently have a staff member whose role included sustainability concerns found that their engagement level was increasing and it was much easier to action the changes that the department wanted to make.

The results revealed a common perception that whether or not people chose to engage in sustainability activities was down to the individual and the importance of such matters were to them; this was the second reason given to explain the decrease in the engagement levels. This perception, and the fact that group mechanisms such as critical mass and learning by observing others were largely discounted, reflected the infancy of the engagement process: "Others doing it only adds a bit to the overall engagement." It was felt that even though it was possible to continue the awareness raising, there was little point because people were invariably aware of the issues and it was down to each person to decide whether or not to engage; this was the case even in smaller departments where changes might have been more straightforward to implement. Furthermore, the results of a recent staff survey in which 90 % of respondents believed that all staff were responsible for addressing sustainability at the university, as opposed to only senior management or the Estates department, suggested that feeling responsible was not enough on its own to motivate people to take action. In the few departments where the engagement level was increasing, specific individuals were cited as making ongoing efforts because such matters were important to them as individuals (refer to Defra (2007) segmentation model), so they continued to remind people, 'name and shame', and update posters and other information sources. Thus, it was perhaps the type of individual and their ability to motivate others that influenced whether or not engagement levels were maintained.

One surprising result was the ongoing reliance by some of the departments that had been substantially engaged in the process, on ongoing senior management support and achievement recognition. For example, one department with a strong green team, effective channels of communication and a Gold award had become less engaged than departments that achieved Bronze. There was a belief in that department that as the university had withdrawn its support there was no point maintaining the efforts that were being made:"...gone off the radar, with nothing driving it, no incentive to keep things going, so it's all stopped bar a bit of recycling but very little." A second department that also achieved Gold considered that recognition of their efforts by the university hierarchy would markedly improve the level of engagement and embedding of sustainability practices in the long term. In contrast, two other departments, one that achieved a Silver award and the other Gold and perceived increasing levels of engagement, emphasised the importance of an inward facing team approach to their efforts centred around communication and awareness raising, and discounted the need for recognition from outside the department. They also discounted any benefit from introducing a competitive element, such as the provision of information on the progress of other departments. The variation in the ability to remain engaged indicated that it is essential to understand the focus of, and rationale driving, actions within individual departments regarding this particular process and how the group dynamics manifest.

Even though the critical mass was evidently not reached in any department before the initiative was withdrawn, a level of engagement had been successfully achieved and the role of departmental champions was seen as crucial to this success. Daily practices of turning off lights, computers and other appliances, ensuring windows were shut and radiators set appropriately were evident across the university. The literature generally portrays a positive outcome from introducing champions; see, for example, Taylor et al. (2012) for a discussion on the different types and contexts of champions and Hargreaves (2011), who described the subtle shifts in behaviour that can occur following an intervention. Identifying individuals as champions can also have a negative impact. The results in this study showed that although champions were perceived as crucial in achieving and maintaining the initial engagement, there was a feeling by some staff that there was no need to get involved as specific people had been allocated to do so, reiterating the idea that addressing sustainability was optional. The results showed the importance in getting the right type of champion, someone who "...empowered others and facilitated them to act" and the need to remove such a role at a certain point to enable further engagement and a wider acceptance of new behaviours.

The analysis of the descriptive statistics revealed consistent relationships in the comparison between academic and support services departments: a higher mean average value for the latter regarding the number of people in a green team (4.89), the award achieved (3.0, on a 4 point scale: nearly bronze, bronze, silver, gold) and the perceived engagement status (by green team, 2.13, and staff, 1.88, on a 3 point scale:

decreased, stayed the same, increased), compared to academic departments (3.78, 2.13, 1.50 and 1.38 respectively). These results do not necessarily imply that people in academic departments were less interested in sustainability issues, although this may have been the case. It does suggest, perhaps, that for support services such issues were considered equally with issues such as ethics and equal opportunities and that time to undertake such activities was more acceptable than in academic departments where only research and teaching was perceived as core business. It is perhaps significant, in the light of these findings, that none of the academic departments involved in the study perceived that their engagement level had increased since the withdrawal of the initiative and these departments were much harder to recruit.

For many departments the key factor for engaging people was communication. Examples of such activities university-wide included communications about initiative developments and energy usage, a network of like-minded people and regular sharing of good practice between the departments. These exchanges were disseminated within the departments via newsletters, emails, posters, notice boards and departmental meetings, with updates highlighting successes and areas of improvement and information on what to do. Rotating the emphasis on different sustainability issues was found to be very beneficial to focus efforts and highlight particular aspects e.g. 1st month lights and computers, 2nd month healthy eating, 3rd month recycling and so on. For a couple of the support services departments in particular, communications was considered the underlying factor of their ability to maintain, if not increase, engagement levels: "It's about understanding what needs to be communicated and how the rest of the staff will best get that message"; "We found displaying our energy use as it was happening particularly useful." The following comment succinctly highlighted one of the communication challenges in engaging academic staff: "Academics don't talk to each other. If they were outgoing people and people-focused people, they probably wouldn't be academics!"

The initiative provided a channel of authority via its formal process and framework of activities, which facilitated the engagement as it reduced the readiness of others to ignore what people were asking them to do. Since the withdrawal of the initiative, some departments had inadvertently introduced mechanisms that served to replace that sense of authority and formality. For example, one of the a senior managers specifically focusing on sustainability issues and energy usage reduction, a volunteer student in the SU with a particular interest in sustainability. Auditing also served to maintain people's involvement as being monitored enhanced the motivation to continue the changed practices. Some departments perceived auditing as crucial to long term engagement, suggesting a mix of regular small internal audits and, then on occasions, more formal and extensive external ones. It was suggested that the internal ones could be undertaken by students as a way of utilising a readily available resource and engaging students in the specific sustainability related issues facing the university. The support services departments felt that there was potential to incorporate a student voice in sustainability issues whereas the academic departments generally did not as they felt students had other priorities in their lives such as completing their studies, socialising and getting a job. This difference in opinion may simply be reflecting the perceived purpose of the respective departments in the student experience at university. Another mechanism that the results indicated would provide that sense of authority for engaging others was demonstrable support from the university senior management, as discussed earlier.

The final factor that was deemed critical to all stages of the engagement process, and ongoing maintenance, was having a central co-ordination; this was the case even for those departments that had achieved Gold and were increasing in engagement currently. This was because that co-ordination provided a specific allocation of time within the university to develop and progress the relevant issues and maintained a grasp on the bigger picture university-wide. It meant that the whole process was gelled together and people could get on with their specific departmental issues whilst someone else was co-ordinating the departmental activities in a united university wide manner. Departments had a sense of what others were doing, an awareness of the top down support and recognition and shared good practice. There is an immense width and depth to sustainability practices and central co-ordination was considered essential in order to change the behaviour of individuals as well as the dynamics of the various groups and the university as a whole—without such co-ordination, the overwhelming perception was that the tipping point could not be reached. The extent to which the level of engagement had reduced over the past 18 months was surprising, both within departments and the total number of departments, and it is likely that on-going central co-ordination would be beneficial in pulling together the somewhat disconnected pockets of engaged people in a way that could give those involved a sense of unification to build on.

5 Recommendations

(1) Set up a communication channel campus-wide to maintain the current level of activities and staff involvement. Examples include monthly email updates, an on-line forum or wiki and a regular meeting time/place with an open invitation.
(2) Use simple, practical, reinforcing messages and actions aimed at You with an underlying sense of value and belonging, making a difference and having the capacity to act, and specific to people's priorities and aspirations so they appeal to the emotional, as well as rational, levels.
(3) Create a team of staff from across the university to identify the most resourceful and least time consuming way forward, focusing on communication, co-ordination and the provision of authority.
(4) Use specific sustainability related themes campus-wide to frame students and staff activities and interventions e.g. one particular issue per month or semester.
(5) Re-establish actions such as sustainability considerations being a regular agenda item on all departmental meetings and time being allocated for sustainability activities.
(6) Ensure that interventions and/or support demonstrated by the university senior management is communicated effectively, so the intended message is received.

6 Conclusion

The evidence showed that even though there was excellent engagement in sustainability related activities across the campus whilst the formal initiative was running, many people had become disengaged as a result of its withdrawal. Further data collection from stakeholders, such as staff who were not engaged, the university senior management and students, would extend the understanding regarding the mechanisms behind this change. The perception that whether or not people engage was down to the individual is indicative of the infancy of the process to date, and it is important for institutions to consider the impact of high level decision making regarding how to address the sustainability issue long term.

This paper highlighted a number of lessons. First, that a range of inputs and approaches, rather than one or two, increases the likelihood that such initiatives are successful. Second, that decision makers understand the critical factors to any engagement process, to ensure appropriate replacements are introduced when necessary; examples include the authority provided by the initiative, the ongoing need for central co-ordination and specific time allocations, especially in academic departments. Third, the use of champions needs to be planned carefully; they were deemed crucial for the initial engagement and a hindrance to the more widespread engagement and embedding of practices. Fourth, attention must be placed on ensuring all communications are effective and that the intended message is the received message.

The author believes that what has occurred in this UK case study could happen at any university world-wide and the lessons need to be heeded to ensure that any loss of momentum experienced does not supersede the ongoing efforts, and the realisation of the huge potential that introducing such initiatives, and the embedding of sustainable practices has, both within university communities and their spheres of influence further afield.

References

Bandura A (1977) Self-efficacy—toward a unifying theory of behavioural change. Psychol Rev 84 (2):191–215

Blau PM (1964) Exchange and power in social life. Transaction

Campbell JL (2007) Why would corporations behave in socially responsible ways? an institutional theory of corporate social responsibility. Acad Manage Rev 32(3):946–967

Deci EL, Ryan RM (1985) Intrinsic motivation and self-determination in human behaviour. Plenum Press, New York

Deci EL, Ryan RM (2000) The "what" and the "why" of goal pursuits: human needs and the self-determination of behavior. Psychol Inq 11:227–268

Defra (2007) A framework for pro-environmental behaviours: a report by the behaviours unit. Defra, London

Eden S, Bear C, Walker G (2008) Understanding and (dis)trusting food assurance schemes: consumer confidence and the 'knowledge fix'. J Rural Stud 24:1–14

Goldstone RL, Gureckis TM (2009) Collective behavior. Top Cogn Sci 1:412–438

Haidt J, Seder P, Kesebir S (2008) Hive psychology, happiness, and public policy. J Leg Stud 37:S133–S156
Hargreaves T (2011) Practice-ing behaviour change: applying social practice theory to pro-environmental behaviour change. J Consum Cult 11(SI):79–99
Homans George (1961) Social behavior: its elementary forms. Harcourt Brace Jovanovich, New York
Kennedy J (2009) Social optimization in the presence of cognitive local optima: effects of social network topology. Top Cogn Sci 1(3):498–522
Kiron D, Kruschwitz N, Haanaes K, von Steng Velken I (2012) Sustainability nears a tipping point. MIT Sloan Manage Rev 53(2):69–74
Moser SC (2006) Talk of the city: engaging urbanites on climate change. Environ Res Lett 1:1–10
Prochaska JO, DiClemente CC (1983) Stages and processes of self-change of moking: toward an integrative model of change. J Consult Clin Psychol 51(3):390–395
Saks AM (2008) The meaning and bleeding of employee engagement: How muddy is the water? Indus Organ Psychol Perspect Sci Pract 1(1):40–43
Taylor A, Cocklin C, Brown R (2012) Fostering environmental champions: a process to build their capacity to drive change. J Environ Manage 98(1):84–97

Author Biography

Sally R. Lampkin has spent over 20 years in the health and sustainability sectors in the UK and overseas, as a clinician, manager and, most recently, an academic. These activities have involved individuals, communities and strategic planning, based on many of the principles that underpin psychology, behaviour change and social influence. She received her doctorate in environmental psychology from the University of East Anglia in 2010 where she researched the individual motivation to mitigate climate change. Her current interests focus on sustainability in the HE curriculum, engaging people in the sustainability agenda in the workplace and the impact of digital technology at home on sustainable lifestyles. She wishes to thank all the people who contributed to making this paper possible.

Students' Vision on Integrating Sustainable Development at Tallinn University of Technology

Marija Klõga and Alvina Reihan

Abstract

Tallinn University of Technology (TUT) is one of the largest universities in Estonia with total enrolment of approximately 13,000 students. The university provides three levels of higher education, practically in all engineering fields needed in Estonia. Over the past years, the integration of sustainability aspects into the university curriculum has become increasingly important worldwide, as well as at TUT. Thus, since 2009, the course "Environmental Protection and Sustainable Development" has become compulsory for many students of technical faculties. In this paper, the authors aimed to analyse students' attitudes and expectations for this course, as well as the level of integration of sustainability aspects into selected educational programs at TUT. Interdisciplinarity is a key factor to obtain viable results in the implementation of sustainable development at universities. Therefore, special attention was given to defining problems and weaknesses of the current situation regarding the integration of sustainability aspects into university educational programs and, based on students' vision, coming out with some practical ideas for embedding overall sustainable development processes at TUT.

Keywords

Sustainable development · Higher education · Assessment · Curriculum

M. Klõga (✉) · A. Reihan
Tallinn University of Technology, Ehitajate Tee 5, 19086 Tallinn, Estonia
e-mail: marija.kloga@ttu.ee

W. Leal Filho et al. (eds.), *Integrative Approaches to Sustainable Development at University Level*, World Sustainability Series, DOI 10.1007/978-3-319-10690-8_23

1 Introduction

Sustainable development is considered as the most crucial and important challenge faced by the human society in the 21st century (Mintz and Tal 2013). Many definitions and interpretations of sustainable development exist, but all of them reflect the idea that social, environmental, and economic progresses must be attainable within the limits of our earth's natural resources (SDI 2014).

Education for sustainability is defined as education aiming to link knowledge, perspectives and actions to help people build a healthy and fair future for their communities and the planet (ISC 2014). In this regard, universities have a great responsibility in increasing awareness, knowledge, skills and values needed to create a just and sustainable future, since they prepare most professionals, decision makers, and consumers of tomorrow (Cortese 2003; Karatzoglou 2013; Orr 1995). The Baltic 21 has also highlighted the role of education as a means to achieve broader objectives on sustainable development (Baltic 21 2004). Universities have unique academic freedom, critical mass of students and diversity of skills to create new ways of thinking and shift people's values and attitudes towards sustainability (Cortese 2003; Sedlacek 2013). However, embedding of sustainability issues into the university curriculum and fostering overall sustainable development at universities is, indeed, a challenging task. Even in countries with strong education systems, this process needs a lot of effort from different university levels (BUP 2009). Some of the barriers in this process might be: overcrowded university curriculum, limited staff awareness and expertise, limited institutional drive and commitment, etc. (SDHI 2005).

The concept of sustainability and education for sustainability is still new in many universities worldwide, including Tallinn University of Technology (TUT). TUT is the second largest university in Estonia in terms of student enrolment (the total number of students is approximately 13,000), and the only technical university in the country. There is one core course at TUT for teaching sustainable development —"Environmental Protection and Sustainable Development" (EPSD), which, in 2009, was restructured and expanded from another TUT course—"Environmental Protection." The course "Environmental Protection" was carried out at TUT since the 1990s and initially dealt with mainly environmental degradation and pollution problems. It had a very specific focus on the major natural processes, conservation of biodiversity and provision of technical solutions for air, water and land pollution prevention and cleanup. In order to enrich the course content with a more sustainable approach and to embed overall sustainable development within the university curriculum, TUT students' organisation "Sustainable Development Club" initiated change of its general concept in 2009. Hence, the course "Environmental Protection" changed its name to "Environmental Protection and Sustainable Development" and focused more on multiple global problems such as proper and fair use of natural resources, maintenance of ecosystems, well-functioning human society and world economy issues. The overall task of the course became more ambitious and complex. Its long-term learning outcomes now included developing in students a deeper understanding of functioning not only natural but also social and economic environments, and the ability to connect diverse scientific, technical and social factors in human development.

Despite the fact that many papers have been published about the approaches for teaching and learning sustainable development, not many studies have been conducted to assess the current state of sustainability integration across universities (Lozano and Young 2013; Quist et al. 2006; SDHI 2005). Similarly, not much research has been done on this topic at TUT. For this reason, it was interesting to investigate, for the first time, how sustainable development is seen and perceived by students at TUT and come out with some good practice to help academic staff strengthen the role of sustainable development at the university.

Therefore, the present research included the following tasks:

- Identify the definitions of sustainable development used by students from particular faculties;
- Identify the key drivers in students' perception for sustainable development;
- Identify students' expectations for the core course of sustainable development (EPSD) learning outcomes;
- Identify present content and connections between sustainable development and specific educational programs;
- Identify the barriers and some possible solutions in embedding sustainable development at TUT.

2 Method

2.1 Research Setting

The present study was conducted at TUT at the beginning of spring semester 2014 and was based on survey data.

It is often claimed that there is no single concept of the sustainability issue (Mintz and Tal 2013; Kagawa 2007). This was an important barrier while developing appropriate survey questions for assessing the current state of sustainable development at TUT. For this reason, the survey questions were designed with slightly different focus points. Thus, the first two questions were more general and asked the students to define the term "sustainable development" and grade the key drivers of this process. The next question was about students' expectations for the current EPSD course learning outcomes. Finally, the last three questions asked the students about present integration of sustainability aspects into their educational programs and possible ways to improve this process.

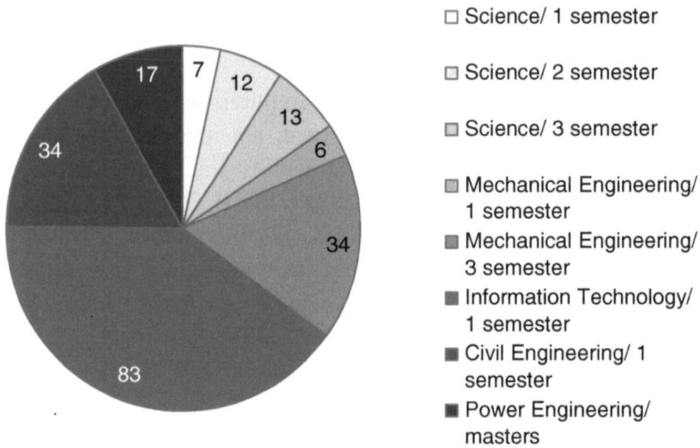

Fig. 1 Distribution of respondents according to the faculty and semester of study

2.2 Data Analysis and Respondents' Background

The data for this study was gained from student responses to both open- and closed-ended survey questions. All respondents were informed that the survey was anonymous and voluntary, and the only additional information required was students' gender and their faculty abbreviation. The questionnaires were distributed personally at the end of EPSD practical sessions.

The collected data was analysed using descriptive statistics.

Altogether, 231 students from five university faculties and four different years of study were studied. The number of respondents to the survey was 206, equivalent to 89 % of all students who attended the EPSD practical sessions. The gender dispersion showed that there was a greater proportion of male respondents (71 %) than female (29 %).

The EPSD course was obligatory for first, second and third semester students from the faculties of Science, Mechanical Engineering, Information Technology (IT) and Civil Engineering. Therefore, these students were the initial target audience to investigate. However, some Master's degree students from the faculty of Power Engineering were attending this course as elective and their 17 answers were also included in the survey data collection (Fig. 1).

3 Results

There are many different ways of explaining the term "sustainable development". For this reason, the first aim was to identify students' perceptions and understanding of this definition. The most frequent answer among students to the question: "What is sustainable development?" was (with slight modifications) *rational use of present natural resources, keeping in mind the needs of future*

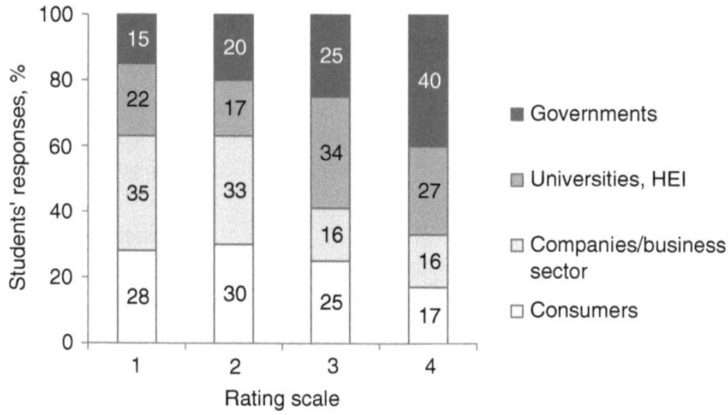

Fig. 2 Importance of driving forces of sustainable development (grading system: *1* the most important, *4* the less important)

generations. This definition was in accordance with the most widely accepted one from the World Commission on Environment and Development, presented in 1987. It defined sustainable development as "Development that meets the needs of the present without compromising the ability of future generations to meet their own needs" (WCED 1987). However, the definition given by most students had more technical approach, as was considering an appropriate use of natural resources as a key factor for ensuring a sustainable future. Other interesting interpretations included: *maintenance of ecosystems; use of better, environmentally-friendly technologies; reduction of energy usage; live inexpensively; development accomplished with decreasing ecological footprint of consumption,* etc.

Secondly, the students were asked to grade the key driving forces of sustainable development in modern society (Fig. 2).

Figure 2 shows that most respondents identified the companies/business sector as the most significant driver for implementing sustainable development in practice. The consumers and their daily choices were ranked as the next key drivers for fostering sustainable development. The role of universities was given the third priority, whilst the role of governments was considered as the most negligible.

Only students from the Science faculty ranked first place the role of universities in the process of fostering sustainable development (Fig. 3).

This ranking can lead to an interesting conclusion. Perhaps, all students understand their own responsibility and involvement in the process of sustainable development. The students from strong technical faculties put at the first place the business or industrial sector, keeping in mind that they are also future professionals and leaders in this field. Similarly, students from the Science faculty put at the first place the role of universities and science in this process, as they were the people more likely to work, and be the driving force, in this field in the future. Other driving forces students mentioned were different active environmental organisations such as Greenpeace or World Wildlife Fund, citizen campaigns for the environment, awareness raising from media resources, etc.

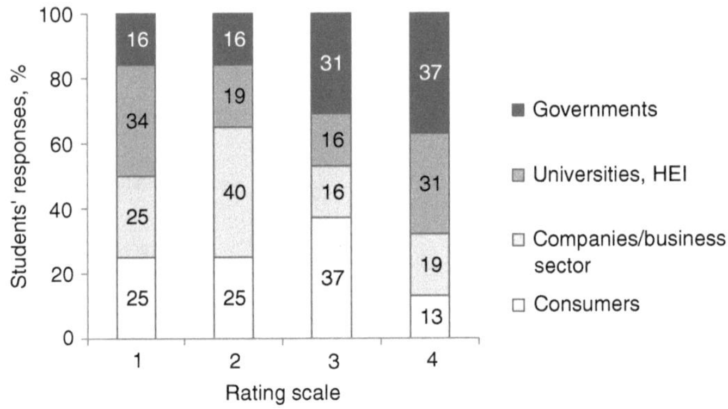

Fig. 3 Importance of driving forces of sustainable development by students from the Science faculty (grading system: *1* the most important, *4* the less important)

The next question asked the students to rank the main learning outcomes of the EPSD course according to their expectations (Fig. 4).

It must be noticed that all of these learning outcomes are expected of university graduates in general and, in particular, EPSD course; all of them must be considered to promote the sustainable ways of life (Mintz and Tal 2013). However, students come to university from different backgrounds and with different expectations, so it was interesting to identify their opinion about the importance of these outcomes for a specific EPSD course. As seen in Fig. 4, students gave most importance to the development of basic knowledge, but very close to it was also ranked motivation raise. Interestingly, students of the IT faculty gave the highest importance to motivation raise. Perhaps it shows that sustainability aspects appear rather seldom in IT study programs and, therefore, its students need some extra motivation and inspiration to empower them and strengthen their willingness to act for sustainable development after the EPSD course is finished.

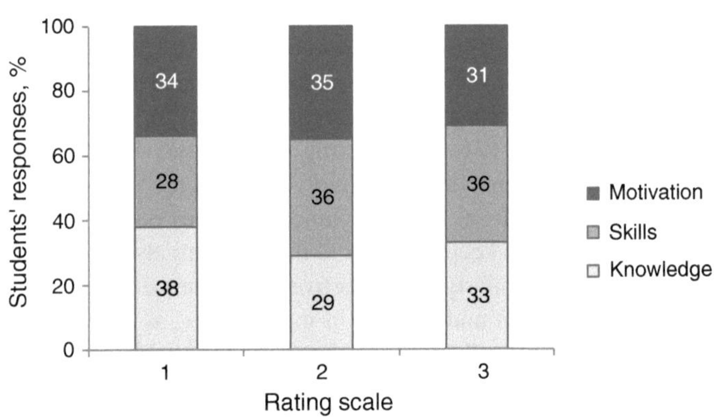

Fig. 4 Students expectations for the main learning outcomes of the EPSD course

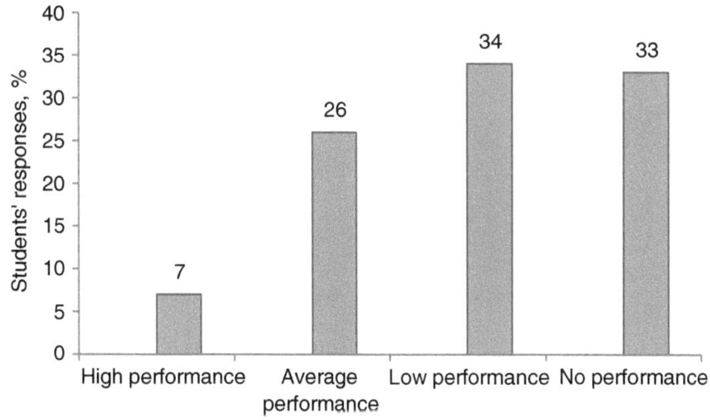

Fig. 5 Performance of sustainability aspects in other teaching courses (besides EPSD)

A single course might not be enough to achieve in-depth knowledge and understanding in many studying disciplines, even more so for the multidimensional and complex topic of sustainable development. In order to better understand the current level of integration of sustainability aspects into the university curriculum, students were asked whether sustainability aspects appear in other teaching courses (besides EPSD) of their faculties. Generally, respondents reported that performance of sustainability issues in other teaching courses was low or there was no performance (Fig. 5).

The students from Science and Mechanical Engineering faculties gave higher scores to the performance of sustainability aspects in their other teaching courses (Fig. 6a, b). The students from the Power Engineering faculty also gave higher scores to its performance. However, it might be hard to draw any conclusions based on these results due to the small number of respondents (Fig. 6c). The lowest score given to the performance of the sustainability aspect in educational programs was from the IT students (Fig. 6d), and their rating played a key role in the formation of the overall picture (Fig. 5), as the absolute majority of the respondents were studying at the faculty of IT.

However, when the students were asked to define the exact subjects (if applicable) in which the sustainability issues appear, they gave a variety of responses. The longest list was provided by the students from Mechanical Engineering and Power Engineering faculties with subjects dealing mainly with today's realities of industrial processes. Thus, the courses mentioned by the students from the Mechanical Engineering faculty included: Integrated Product Development, Vehicle and Mechanics Dynamics, Welding and Welded Structures, Strength of Materials, Heat Management, Heat and Power Devices, Design of Power Plant Equipment, etc. Students also commented that the mentioned sustainability aspects were mostly related to efficient usage of materials (resources) and waste minimisation. In some courses, equipment for alternative production or "green" energy sources were also discussed.

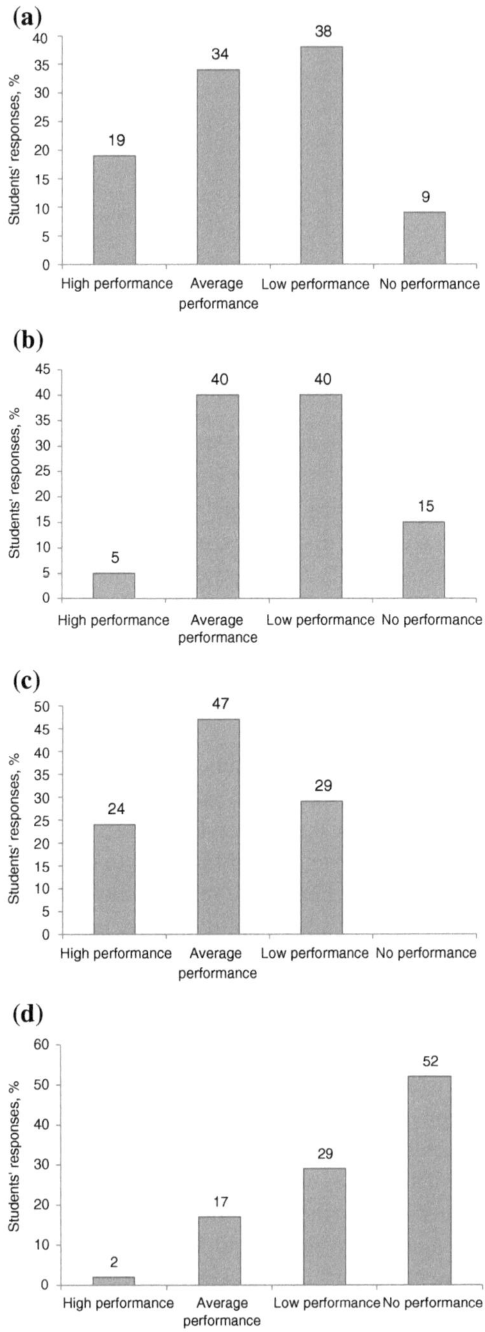

Fig. 6 Performance of sustainability aspects in other teaching courses (besides EPSD) by students within four faculties. **a** Students of the Science faculty. **b** Students of the Mechanical Engineering faculty. **c** Students of the Power Engineering faculty. **d** Students of the IT faculty

Students from the Power Engineering faculty mentioned courses such as Hydrology and Hydrotechnical Constructions, Environmental Protection of Mining, Electricity Supply, Vision Electricity Supply, Wind Power, Implementation of Wind Energy, etc. In general, students from the Power Engineering faculty emphasised that sustainability aspects, related mainly to energy use (such as CO_2 emissions, global energy market, energy politics), are part of most of their specialisation courses.

Students from the Science faculty mentioned courses mainly containing environmental issues of sustainability, such as Baltic Sea, General Oceanography, General Meteorology, Geology of Estonia, Ecology, etc. Students also clarified some more specific topics mentioned during these courses, such as eutrophication of the Baltic Sea, climate change, ecosystems and extinction.

Much more humble responses were provided by the students from the IT and Civil Engineering faculties. Students from the IT faculty could mention only some topics related to sustainability issues being discussed during very basic courses. For example, during practical sessions in Physic, Discrete Mathematics, Ergonomics, Micro- and Macroeconomics, Computers students were asked to complete some exercises on environmental topics (for example, calculation of air emissions, comparison of energy loss in normal vs energy efficient house, compilation of energy use in electric and gasoline cars). In regards to sustainable aspects of working within the IT sector, students mentioned that almost all of their studies are conducted online and, therefore, the use of natural resources in their case is minimal (for example, almost no paper consumption).

Students from the Civil Engineering faculty mentioned courses such as Physics and Chemistry of Materials, Micro- and Macroeconomics, Geodetic Surveying. In their comments, they clarified that taught topics concerned hazardous elements in nature, use of environmentally friendly materials for construction and building planning and environment. According to the university curriculum from the Civil Engineering faculty, it is very likely that more specialised courses (given during second and third academic years) would cover sustainability issues more precisely.

The last question was about students' suggestions or ideas on better integration of sustainability issues into the university curriculum in order to enhance the overall education for sustainable development at TUT. This question actually reflected a big challenge, faced by many universities today, on how to efficiently place sustainability into university programs and inspire students to implement in practice principles of sustainability.

Students' responses to this question were created in three different ways. Thus, some respondents were quite satisfied with the current level of sustainability content at university programs, and, in their opinion, there was no need for further integration of these aspects into the university curriculum. Other students gave some suggestions for the improvement of current EPSD course. Finally, some other students provided advice for the better integration of sustainable issues into the university curriculum, as well as other campus activities (Table 1).

Table 1 Suggestions for the improvement of current EPSD course and better integration of sustainability issues into university curriculum and campus activities

Students from the *Science faculty*	
EPSD: To tell and discuss more about real technological solutions that could be implemented now or in the future; more practical approaches on how to preserve the environment; speak more about functioning of megacities; more stress on "green chemistry" principles	**Integration/campus activities:** Increase the use of e-learning to use fewer resources like printed textbooks or notebooks; organisation of excursions and open-air lectures
Other interesting thoughts: Principles of sustainable development should be taught from childhood so that university going students would have sufficient background knowledge about the sustainable development concept	
Students from the *Mechanical Engineering faculty*	
EPSD: To connect the handled topics more precisely to a mechanics curriculum; to talk more about green energy sources in energy production; how to practically preserve natural resources during industrial production; explain how the attitudes towards environment have changed since past decades and what is expected to happen in the near future; to show more practical examples of implementation of sustainability aspects	**Integration/campus activities:** Different courses could mention how to be more energy efficient and preserve nature in their specific area; pollution and waste generation should be mentioned more, and then describing the use of different technologies; to talk and analyse different possibilities of sustainable production at lectures related to technological processes; in each subject make a single lecture related exactly to the sustainability aspects of this specific course; to increase the capacity of teaching staff, so that they would know and be able to talk about sustainability issues in their lectures; to use less paper
Other interesting thoughts: Every educated person instinctively understands our life connection and dependence on the environment, and it is expected that the higher education provided by TUT ensures sufficient knowledge and intelligence to understand this connection and contribute to future development with this regard; the EPSD course must be obligatory for the first year students to create the right attitude and consider other course topics in later years in the light of sustainability	
Students from the *IT* faculty	
EPSD: To show how to implement sustainable principles in day-to-day life; to connect this course more with business logic; to show the practical use of sustainability principles; to teach more about the ecological footprint of the IT sector	**Integration/campus activities:** To write programs for environmental smart applications; to use less paper and electricity at the university campus; to switch-off computers in the computer class after work is completed; to provide practical work at the environmental enterprises; to organise info events or fairs at TUT on sustainability topics
Students from the *Civil Engineering* faculty	
EPSD: To teach which materials are more sustainable to be used in building processes; to use more practical examples, and then introduce principles of sustainable development	**Integration/campus activities:** Emphasise the interaction between built and natural environments; organisation of field-trips to see the application of theoretical knowledge into practice

(continued)

Students from the *Power Engineering* faculty	
EPSD: To show the practical implementations of sustainable principles, which the students can apply in their daily life; stress the amount of emissions from fossil fuel use, and stress the need for "green" energy resources; introduce possible improvements for industrial production steps	**Integration/campus activities:** At the end of every course, tell the students about the future vision of the discussed topics in the light of sustainability; to make a separate short course (1ETC) about sustainability of the main vital aspects of human existence, such as water, energy, agricultural production, etc.
Other interesting thoughts: Hard to do integration due to already overloaded curricula	

4 Discussion

The concept of sustainable development can be interpreted in many different ways. The survey findings showed that students of all faculties gave a variety of definitions for what sustainable development is, but, in most cases, with a strong technical approach. These results lead authors to an idea that interpretation of the sustainability concept greatly depends on respondents' work/study activity and reflects the sphere of life they have responsibility for. The findings of the next question, about identification of key drivers for fostering sustainable development, also proved this working conclusion. Similarly, students gave the highest rating to the driver, which had the biggest connection to their professional areas and of which they probably had the biggest knowledge base.

The students' expectations for running EPSD course learning outcomes were also identified. According to Baltic University Programme report, "the main objective of education for sustainable development is that everybody should acquire the relevant knowledge and be motivated to work for and practice sustainable development (BUP 2009)". Thus, the three key components: knowledge, skills and motivation, should be equally considered in the learning outcomes of any course, and particularly in EPSD. However, the survey results identified knowledge development as the biggest need, perhaps due to strong technical background of all respondents. Motivation was also considered as a very important feature, especially by students from the IT faculty. One option to increase students' motivation could be identifying real-life, practical problems of sustainable development and finding concrete solutions after provision of theoretical basis. The need for showing the practical use of sustainability principles in real-life was also mentioned by many other students from different faculties. Some interactive teaching methods that could be used in this regard might be organisation of open debates, discussions, small-group work, forum play or role play, etc. during the EPSD course (BUP 2009).

The research survey also identified big differences in curricula connections in response to the sustainability agenda depending on respondents' educational programs and year of study. During the first year of study, some educational programs have already introduced to students the relevant sustainability themes such as green energy, CO_2 emissions, global energy market, climate change, etc. (for example, the Science faculty). In other educational programs, the deeper introduction is very likely to happen in the later years of study, for example, in the Civil Engineering faculty. For some faculties, the introduction of sustainability issues was not taught during the first year of study, and it is not very likely to happen in the future (for example, the IT faculty). The general picture of introducing principles of sustainable development at TUT curriculum is rather patchy with big gaps, especially in the areas of economic and social sustainability. Overall, three main categories of response to sustainable development were identified in different educational programs:

- Educational programs that have made some progress in the integration of mainly environmental sustainability issues, such as the Science faculty.
- Educational programs that have made some progress in the integration of mainly technical sustainability issues, such as faculties of Mechanical Engineering, Civil Engineering and Power Engineering.
- Educational programs where the integration of sustainability issues is rather limited, such as the IT faculty. In this faculty, the interest towards sustainability issues has been identified, but it is very likely that the integration process of these aspects into the IT curriculum is much more difficult.

Concerning the progress of the current EPSD course, students from different faculties were asking to include some more specific topics into its curricula related to the background of their specific educational programs. Achieving this very good suggestion is indeed a challenging task, as during the EPSD lectures, students from different faculties are combined in one classroom (up to 200 students) and only during the practical sessions they are separated into smaller groups (15–25 students) according to their faculties. For this reason, the implementation of some interactive teaching methods during the EPSD lectures mentioned above could be also problematic. Therefore, it would be recommended that during the practical sessions the academic staff use the chance and implement a more interactive approach and talk about more specific themes and then do practical exercises with the students. Also, survey respondents mentioned some concrete topics, which they would like to know more about during the EPSD course, such as the analysis of emissions from burning of different types of fossil fuel, "green" energy potential, sustainable materials used in building construction, etc. These topics are definitely essential and, in fact, they are all covered in the current EPSD course given at TUT; but, as the survey was made at the very beginning of the semester, the students have not yet started handling them.

For further progress of the integration of sustainability aspects into the university curriculum, students suggested the organisation of some free time activities related to the promotion of sustainable development at the university campus, such as organisation of different students events or exhibitions related to sustainability topics; organisation of field trips and excursions to familiarise them with good examples of the implementation of sustainability aspects in practice; developing e-learning for practical use to reduce the ecological footprint of students and fostering capacity-building among the academic staff.

5 Conclusion

Students' interpretations of sustainable development had a clear technical approach. As the most important drivers in the sustainable development process, students identified the companies/business and consumer sectors, which are probably reflecting their biggest professional areas of responsibility and knowledge.

Students' expectations towards specific EPSD course learning outcomes were the highest in the fields of knowledge creation and motivation rise. In open-ended survey questions, students also stressed the need for more practical insights and examples of sustainable development practices that would increase their understanding and motivation to implement the sustainability principles in real-life. The authors believe that this suggestion could be implemented, not only during the practical sessions of EPSD course, but also in any other educational course of TUT, for embedding education for sustainable development within the whole university.

Students also reported that sustainability issues were handled in their educational programs (with the lowest presence in the IT faculty) mainly in the area of nature protection and engineering. Therefore, designing future strategies for the integration of sustainability aspects into the university curriculum would be recommended to emphasise the social and economic aspects of sustainable development in order to keep its overall concept balanced. Need for university staff capacity-building in the area of sustainability was also identified in respondents' comments. In addition, outcomes from the previous informal discussions with the students revealed that innovative aspects of sustainable development need to be studied more thoroughly during EPSD course, and a tutorial on sustainability in the national language is very necessary.

In open-ended remarks some respondents came out with interesting ideas that the concept of sustainable development should be the logical outcome at the end of any university study programme, not just at the end of one specific EPSD course. The authors think that this is one of the most important findings of the conducted survey. This is a clear recognition of the value of an interdisciplinary approach for sustainable development, which should extend restrictions of any narrow specialization. Education for sustainable development should be integrated within the overall university curriculum and other campus activities in order to achieve valuable and long-term results in this field. More interactive and practical teaching methods during EPSD course, increased staff awareness and improved dialogue between colleagues across faculties and departments in order to create a framework and work together to make the most out of everyone's input would enhance education for sustainable development at TUT and perhaps in many other universities.

It must be mentioned that this paper was rather a quick assessment of the current status of sustainable development introduction at one university in Estonia, and not a complex comprehensive review. This research is only the first attempt to analyse this complex issue, and further research is needed to elucidate our preliminary results in more detail.

References

Baltic 21 (2004) An agenda 21 for education in the Baltic sea region—Baltic 21E. Baltic 21 series no 1/2002. www.baltic21.org/?publications. Accessed 27 Apr 2014

BUP (2009) Education for change: a handbook for teaching and learning sustainable development. Baltic University Programme, Uppsala University, Uppsala 74 p

Cortese AD (2003) The critical role of higher education in creating a sustainable future. Plan High Educ 31(3):15–22

ISC (2014) Educating for sustainability. Institute for sustainable communities. http://www.iscvt.org/where_we_work/ukraine/article/sustainability.php. Accessed 23 Mar 2014

Kagawa F (2007) Dissonance in students' perceptions of sustainable development and sustainability, implications for curriculum change. Int J Sustain High Educ 8(3):317–338

Karatzoglou B (2013) An in-depth literature review of the evolving roles and contributions of universities to education for sustainable development. J Clean Prod 49:44–53

Lozano R, Young W (2013) Assessing sustainability in university curricula: exploring the influence of student numbers and course credits. J Clean Prod 49:134–141

Mintz K, Tal T (2013) Sustainability in higher education courses: multiple learning. Studies in educational evaluation (in press, corrected proof, available online 15 Dec 2013)

Orr D (1995) Education for the environment: higher education's challenge of the next century. Change 27(3):43–46

Quist J, Rammelt C, Overschie M, de Werk G (2006) Backcasting for sustainability in engineering education: the case of Delft University of Technology. J Clean Prod 14(9–11):868–876

SDHI (2005) Sustainable development in higher education: current practice and future developments. A report for The Higher Education Academy, Heslington, York, 85 p

SDI (2014) What is sustainable development? Sustainable development information. http://www.sustainabledevelopmentinfo.com/. Accessed 21 Mar 2014

Sedlacek S (2013) The role of universities in fostering sustainable development at the regional level. J Clean Prod 48:74–84

WCED (1987) Our common future. World commission on environment and development. Oxford University Press, Oxford 400 p

Author Biographies

Marija Klõga is a Ph.D. student and teaching assistant in Department of Environmental Engineering at Tallinn University of Technology (TUT). Her research focuses on the state of the environment and environmental protection (water quality and factors determining it, selfpurification processes in rivers, methodology of water monitoring). Since 2009 Marija Klõga gives practical sessions of Environmental Protection and Sustainable Development course at TUT to students of full-time attendance and gives lectures on the same subject to correspondence students.

Dr. Alvina Reihan is Associated Professor in Department of Environmental Engineering at Tallinn University of Technology (TUT). She educated in Estonia and Russia, and improved her skills on numerous International Postgraduate Courses. She earned her Ph.D. in technical sciences from TUT in 2008. Alvina Reihan research interests include changes in surface water hydrology, climate change impact on water resources, environmental protection and sustainability aspects in environmental issues and higher education. Since 2004 she delivers lectures on Environmental Protection and Sustainable Development at TUT and Tallinn College of TUT. In addition Alvina Reihan teaches General Hydrology, Engineering Hydrology and Applied Hydrology at TUT.

Sustainable Development as a Framework for Ethics and Skills in Higher Education Computing Courses

Neil Gordon

Abstract

The impact of sustainable development on the curriculum remains variable, and in some disciplines the inclusion is considered by some to be inappropriate or not relevant. This paper considers the ways in which sustainable development can be embedded within the curriculum, with the dual aims of showing how it can be made both relevant to students within the context of their discipline, and how sustainable development can provide a framework for developing an appreciation of the legal, social, ethical and professional (LSEP) aspects of the discipline and to develop awareness of sustainability values in students, i.e. the meaning and aspects of sustainable development. The paper focusses on a case study in embedding sustainable development within Computer Science degree programmes, where the LSEP requirements are recognised by accrediting bodies and by many employers as essential characteristics and skills in graduates. The paper will describe how sustainable development provides an overarching framework within which to explore these issues. Moreover, the paper will include some examples of how this is successful in engaging students who may otherwise struggle to appreciate the LSEP topics. The success will be demonstrated through some objective data showing the impact of this approach to students understanding and acknowledgment of sustainability and how this may be applied to other disciplines and national contexts.

Keywords

Sustainable IT · Green computing · Framework for professional values

N. Gordon (✉)
Department of Computer Science, University of Hull, Cottingham Road,
Hull HU6 7RX, UK
e-mail: n.a.gordon@hull.ac.uk

W. Leal Filho et al. (eds.), *Integrative Approaches to Sustainable Development at University Level*, World Sustainability Series, DOI 10.1007/978-3-319-10690-8_24

1 Introduction

We consider the general context of Sustainable Development within Higher Education, and in particular the different aspects from a UK perspective. We will then look at how to map these on to the professional and ethical skills relevant to many disciplines, and in particular to the case of computing.

The concept of Sustainable Development (SD) has become established (Hopkins and McKeown 2002), and education is identified as a way to promote and support SD from an international strategic viewpoint. SD has been adopted and adapted by a variety of national government and government-related organisations. Considering the UK perspective, recent UK governments have developed the following key areas that underpin the international SD agenda, from a UK perspective, the UK SD guiding principles (DEFRA 2011) are summarised as:

1. Living within Environmental Limits
2. Ensuring a Strong, Healthy and Just Society
3. Achieving a Sustainable Economy
4. Using Sound Science Responsibly
5. Promoting Good Governance.

Within the educational context, SD is considered throughout the different stages, with Higher Education (tertiary education) being the final formal educational environment for SD skills and knowledge to be potentially promoted, as graduates move on to play their roles within society. This paper considers some of the issues around delivering SD within the curriculum, and in particular within the context of Computer Science within the UK setting. However, the ideas and issues are transferrable to other educational levels, to other national settings and to different disciplines. The English funding council for HE (HEFCE 2005) summarised the UN (2010) as

> development which meets the needs of the present without compromising the ability of future generations to meet their own needs (HEFCE 2005, p. 9).

Many countries are increasingly focussing discussions and policy about the benefits to individuals and societies from Higher Education, and as part of this, there is an increasing recognition of the graduate attributes or skills that students develop, and can then apply within their later careers and other societal impact. Such graduate attributes (O'Connor et al. 2011) can show how such attributes are important to communities and link to wider civic, social and moral issues. This is relevant to most disciplines, and in particular to those with a strong professional ethos, such as computing.

2 Sustainable Development and Higher Education

In Higher Education in many countries, the autonomy of institutions is seen as paramount and provides the confidence in the type of skills and approaches developed in graduates. Within numerous countries, this means that curriculum and the outcomes for students are controlled by institutions, with loose oversight—possibly through quality review processes—by government organisations. In terms of SD, this can mean that the choice to engage with any part of the SD agenda is limited, or lacks cohesion. For example, national policies around carbon reduction, taxation and energy dependence can encourage or require that academic institutions —as large users of power and significant contributors to carbon pollution—adopt rigorous approaches to energy monitoring, management and reduction as a priority and thus engage with that facet of the SD agenda. However, the impetus for curriculum engagement with SD is typically much looser and lenient, which can mean that institutions lack the stimulus to deal with this part of the Education for SD program. In the UK context, elements of Higher Education policy is set by the individual states own funding councils, which adopt different approaches to expectations for SD (Gordon 2009b).

A further dimension to potential engagement with SD in Higher Education stems from the differing nature of disciplines. In some, the expectation and requirement to include topics and issues that fall into the SD remit is clear; examples here include geography, biology environment and earth sciences. For other disciplines, the links can be made, though maybe less frequent—such as engineering and chemistry. Within the sciences, physics, mathematics and computer science can all be linked to SD, though the nature of the links and the motivation varies. Moving away from the Science, Technology, Engineering and Mathematical (STEM) disciplines, many others have clear potential motivating links to SD—including the social and governance issues (politics, social sciences), the economic drivers and consequences related to SD (business and logistics), as well as the wider civic and social concerns (education, politics and health). International aspects can motivate links to disciplines that focus on national concerns, such as nation based studies. Areas such as history, archaeology and drama can all be linked to SD, with considerations about change and the portrayal of change in societies, and how we can learn from past decisions and events. The UK Higher Education Academy provides support for education in universities, and includes resources for many subjects that demonstrate and provide case studies of how SD can be related to specific disciplines.

The discussions so far have shown how there are numerous drivers for SD within Higher Education (Gordon 2010b). However, as noted this may well be purely framed as guidance and suggestions to practitioners. Echoing the autonomy of HE institutions, disciplines within HE are typified by their own communities of practice (Becher and Trowler 2001), who themselves have ownership of the nature of the content and practice of teaching within the Higher Education framework. With regards to SD, the examples in the previous paragraph reflect some of the views of how different disciplines relate to SD, and thus as to how practitioners may expect to demonstrate their own approaches. The decision of if and how to do such may be supported by disciplines—perhaps through subject expectations and requirements —and then these cascade down to the approach of distinct departments and the individual practitioners themselves. The viewpoint and approach of the individual can affect how much they do—or do not—engage with the SD subject and agenda,

Table 1 Students' views on SD skills (HEA 2013)

% of students who agreed (very/somewhat)	Survey statements: students asked how much they agreed with
24	The importance of the environmental approach of the institution in selecting it originally
41	The university's approach to global development
19	Thought their course should improve their understanding of people's relationship to nature
51	That they should have skills to consider medium/long term planning
46	They should understand approaches to using resources efficiently
32	Should have developed skills for whole system thinking

though in is still not uncommon for individual practitioners to have little exposure to or awareness of SD and how they could use it within their own teaching.

Considering the student perspective, recent surveys of student attitudes (for example, Drayson et al. 2013) shows that students expect and want the skills related to Sustainable Development. Table 1 summarises some results of students' expectations at the author's institution (based on data from the HEA 2013 review of SD attitudes, across campus, but with a majority of computing students).

3 Professional Practice and Sustainable Development

Following on from the earlier consideration of graduate attributes, these can be further considered in terms of the professional development of students, as they become graduates and move into various roles in society. Supporting the focus on professional practice and employment, many disciplines have professional expectations and requirements, in particular where Higher Education (degree) courses are accredited and are expected to lead to direct professional practice. Such accreditation may require that degree programmes include content—and corresponding learning outcomes and assessment methods—that would ensure students have met and appreciate the professional aspects of their chosen discipline. Such professional aspects include acting in an ethical manner, and appreciating the ethical framework within which they function. This content typically complements the subject content itself, and includes the wider set of material about how the future graduate should act as a professional. They may include the legal framework within which they will be expected to operate. These topics may provide motivation and context in which to consider the ethical and moral issues and situations that the graduate will be expected to handle, and the wider social impact and social environment in which the student will later be living and working. The legislative and behavioural framework is sometimes referred to as the Legal, Social, Ethical and Professional (LSEP) features.

Beyond accreditation, the demand for students with skills has been identified by a number of reports (e.g. Cade 2008), where the need for graduates with skills related to environment and social responsibility was a key point. The idea of responsible employers needing equally responsible employees can fit well into the SD remit, with SD potentially providing a basis within which to develop these particular attributes (Gordon 2009a).

Having considered the wider context of SD and Higher Education, we now focus on the case study of SD within the Computer Science curriculum.

4 Computer Science and Sustainable Development

As a discipline, Computer Science has a strong ethos of professional development, with a focus on developing students as future practitioners. In common with many other professional disciplines—from health and nursing, through subjects such as law and engineering—the accrediting bodies for degree programmes require evidence that students are familiar with, and should abide by, the professional values for the subject. This is common across the range of sciences, where requirements for professional scientists are typically formalised through codes of practice or requirements for practitioners to follow. In computing, the various national professional and discipline bodies—including the British Computer Society (BCS) and Association for Computing Machinery (ACM) and IEEE Computer Society specify issues that a graduate should demonstrate. Whilst they come under a variety of acronyms and labels, one summary is the Legal, Social, Ethical and Professional (LSEP) values. These are values and concepts that students should be familiar with and able to demonstrate an understanding and appreciation of, especially the requirements of society and of the impact of their discipline and activities on society. These societal concerns are recognised by groups such as the Computer Professionals for Social Responsibility (CPSR 2009), an organisation focussing on the impact of computing on society. The potential to build

In terms of curriculum issues, the ACM (2013) report on Computer Science curricula 2013 includes the social and professional practice expected of a computing student, with sustainability a core feature, developing from earlier expectations that students understand "cultural, social, legal and ethical issues inherent in the discipline of computing" (ibid). The BCS course accreditation guidelines also identified the LSEP values as being core requirements, and now includes environmental and sustainability aspects (BCS 2012) as examples, within the wider LSEP topics. These requirements for course accreditation and approval also reflect the codes of practice for the respective organisations in terms of behaviours of graduates who work as computing professionals.

The 2008 HEFCE report (2008) on Sustainable Development in Higher Education in England noted how contrasting activity around Sustainable Development seemed, and empirical evidence from more recent reviews and projects reinforces this as something that remains an issue, with variable engagement with the SD agenda. Again, as noted in the HEFCE report, for some institutions and for individual practitioners, Sustainable Development is not considered important and lacks links to the curriculum.

Empirical evidence of staff perceptions identifies some of the typical barriers to including SD within local curricula, which can be summarised as

- Lack of relevance to the main subject—sometimes in spite of the accrediting bodies expectations;
- Overfull curricula—finding space for new material;
- Fear of indoctrination—recognising the balance between providing advice on what the issues are, distinct from forcing a view on what is the right answer;
- Attitudes—as noted earlier, SD engagement is potentially dependent on the individual practitioners own personal perspective and attitude.

The Quality Assurance Agency, responsible for standards in UK universities, provides content expectations through discipline specific benchmark statements. The QAA Computing Benchmark for undergraduate courses (QAA 2007) and the more recent Master's degree benchmark (QAA 2011) provide explicit links to these areas. The 2007 guidance declares the expectation that students demonstrate cognitive skills: "Professional considerations: recognise the professional, economic, social, environmental, moral and ethical issues involved in the sustainable exploitation of computer technology and be guided by the adoption of appropriate professional, ethical and legal practices" (QAA 2007, p. 5); whilst the 2011 report specifying under subject content that students should have an "understanding of professional, legal, social, cultural and ethical issues related to computing and an awareness of societal and environmental impact". (QAA 2011, p. 9).

In the context of preparing students for work, the requirements by employers and by accrediting bodies is also a potential incentive: within computer science, a number professional certificates and practice encourage or require evidencing aspects around environmental awareness, or of cultural and societal impact, with

- concerns around data centres (European Commissions 2009);
- utilising IT to support low carbon economies (Climate Group 2008);
- dealing with the waste from IT (WEEE 2006).

5 A Framework for LSEP

Developing from the discussion on the expectations amongst students, accrediting bodies and quality agencies of the inclusion of issues around legal, social, ethical and professional practice, we now consider how sustainability can provide an effective framework for this.

As already demonstrated, the practicing computer scientist will be expected to demonstrate an understanding of the impact of their work on society and the environment. Furthermore, depending on their specialism—be it data centres or commissioning new Information Systems, there is an expectation of both developing appropriate solutions and of potentially gaining evidence of continuing professional development related to this (e.g. green IT certification (BCS, 2009). Power usage—the carbon footprint of IT—and the potential of computer science to address this, through efficient design or through improving the efficiency of other human activity—may provide case studies and examples around which to develop the impact of computing on the environment (Gordon 2010b). This utilisation of computer science in addressing these types of problem can be considered as responsible use of science. Such professional aspects can be built into the curriculum, utilising sustainability related concepts in exploring the nature of being a professional. In a similar vein, the issues of waste and inefficiency can provide concrete examples through which to explore ethical and moral dimensions (Gordon 2010a). The societal impact of technology—with concerns around the digital divide and the opportunities for computing to introduce new approaches to democracy and governance link directly to the SD topic of social issues, and can consider how social computing may—or may not—contribute to stronger societies. Legal topics around waste, data protection and information freedom can be related to these aspects too—with the hardware and information systems related to the first of those, and social aspects considered alongside the last two. Considering these overall topics, the motivation for LSEP can be clearly linked to the social, economic and environmental aspects of SD. Moreover, with the additional concerns of responsible science and good governance, we can develop the following framework (Fig. 1) for SD within the LSEP expectations for computer science.

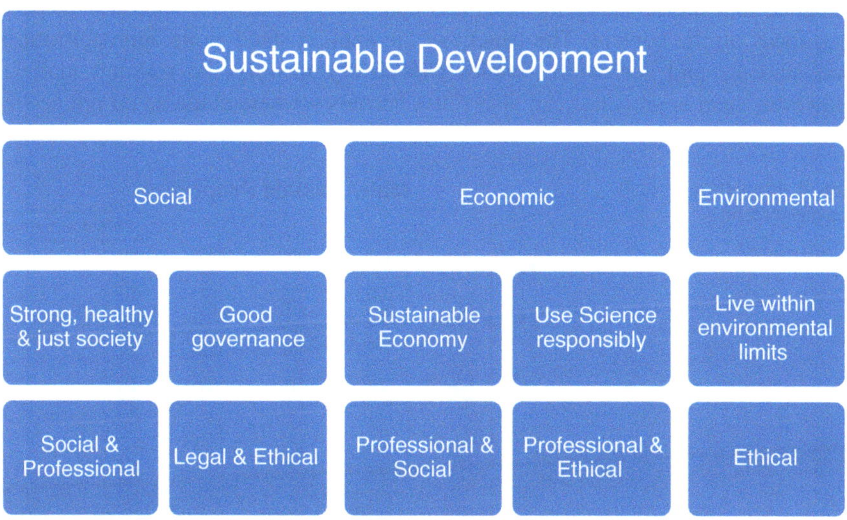

Fig. 1 A framework for LSEP and SD in Computing

6 Case Study of the Impact of Embedding Green Issues Within the Computing Curriculum

In order to demonstrate the potential gains from an educational perspective of embedding sustainable and green computing issues within computer science, we now consider some data arising from a first year undergraduate computer science module (circa 180 students per cohort, over a 5 year) period.

7 Issues Within Computing Education

The stereotypes of a typical computing student does have some resonance with experience, especially in the UK where the gender imbalance (Scragg and Smith 1988) in computing courses is recognised as a weakness and risk to the discipline. Figure 2 shows intake proportions for the author's own department, they are reflective of the typical (approximately 90 % domination of male students in computing disciplines). Moreover, beyond the gender bias within the discipline, the technical focus for many students can seem at odds to the topics that fall within the LSEP and SD remit. Social aspects may be considered by many students as peripheral, when their main interest is to work on their latest assignment or programming project by themselves. Ethics—being a more philosophical concept—can be considered by some students as of only limited interest and relevance. The notion of professionalism itself—bringing together the other topics, may well be considered by some as extraneous. The study summarised below provides some data on the positive impact of motivating LSEP material through SD topics.

When including new material within teaching, there is an issue about whether to integrate it within existing modules and courses, or to include specialist modules that focus on the content. The benefit of the explicit stand-alone approach can be that students and accrediting bodies can clearly identify the relevant material. However, such approaches can also cause barriers—where students do not see the

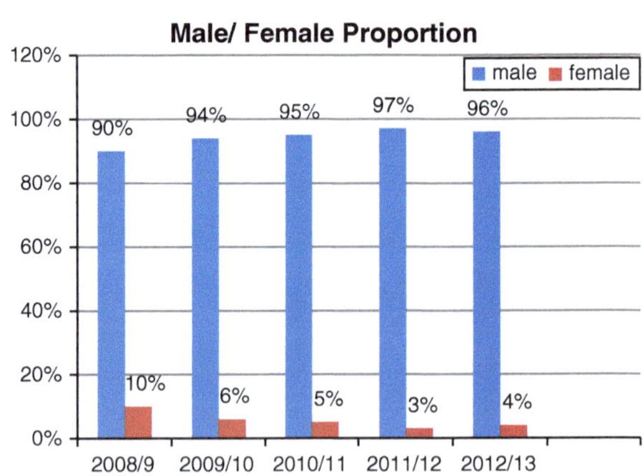

Fig. 2 Proportion of male/ female students

relevance of the material, or colleagues are reluctant to take on the teaching of the content. The benefits of integrating ethics and social responsibility into the core curriculum (Martin and Weltz 1999) are that teaching staff and students will meet it, and it offers the opportunity to place the material in context.

8 Evidence of Impact

This case study considers an embedding approach, with LSEP and SD material embedded within existing core computing content, related to professional and IT skills. Considering the 5 years of the study, student engagement with the module has generally improved. There is some indication from assessment that students have greater recognition of the relevance of the LSEP material—motivated by examples and links to SD. It also appears that this material has in general improved the engagement—at least as measured through attainment—of the students based on their end of section assessments. The results for the female students is more varied—whilst it was postulated that they may respond to the social aspects more strongly than their male counterparts, the overall results are not markedly better as illustrated in the figures below.

Figures 3, 4, 5, 6 and 7 show the results of assessments at the end of a semester's teaching Students performance within the module was split between the IT content, explicit LSEP content, and combined material taught under the auspices of SD, that is motivating the IT and LSEP themes through the context of SD related examples.

As can be seen from Figs. 3, 4, 5, 6 and 7, the achievement in IT material was generally higher than the LSEP content, for all categories of students—apart from the most recent cohort where the figures were similar. There was no pattern between attainment in IT versus LSEP when considered by gender. However, the marks for

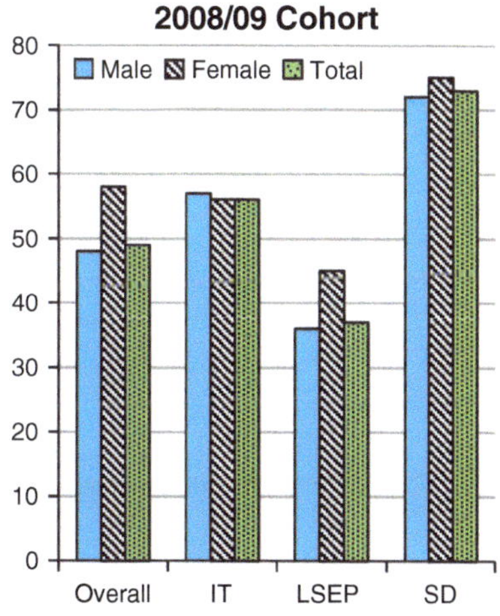

Fig. 3 Assessment results 2008/2009

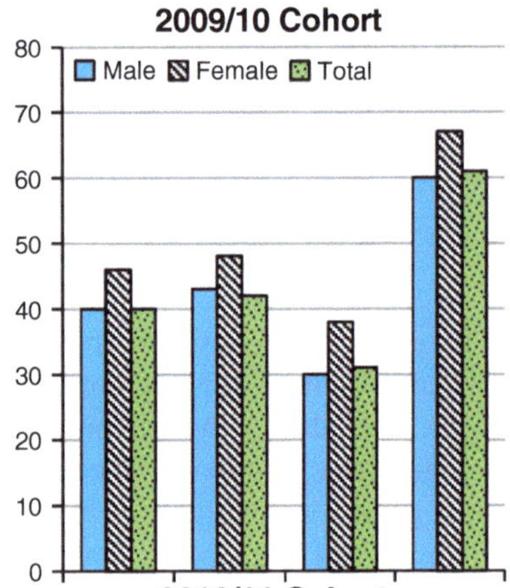

Fig. 4 Assessment results 2009/2010

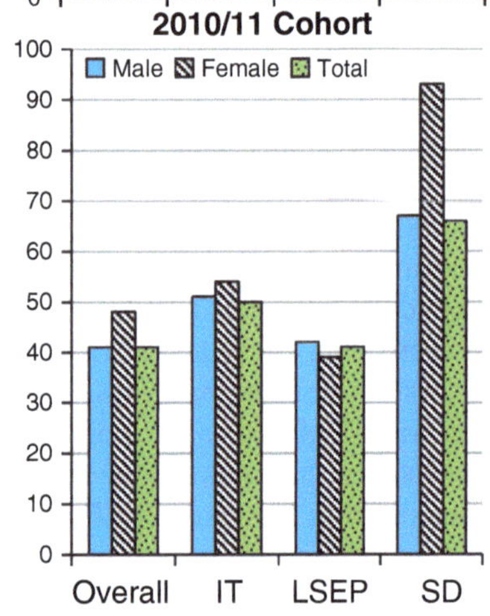

Fig. 5 Assessment results 2010/2011

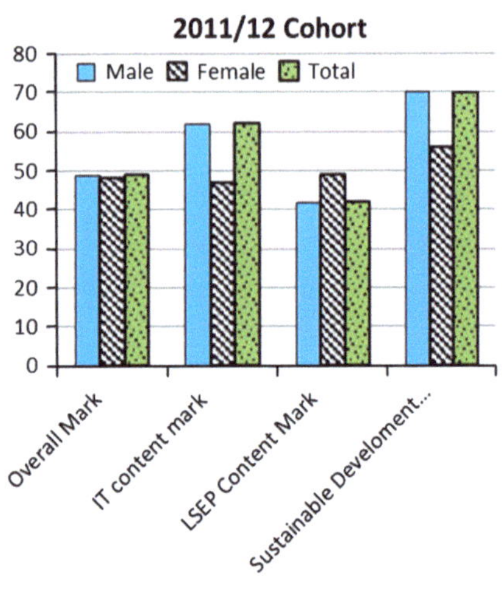

Fig. 6 Assessment results 2011/2012

Fig. 7 Assessment results 2012/2013

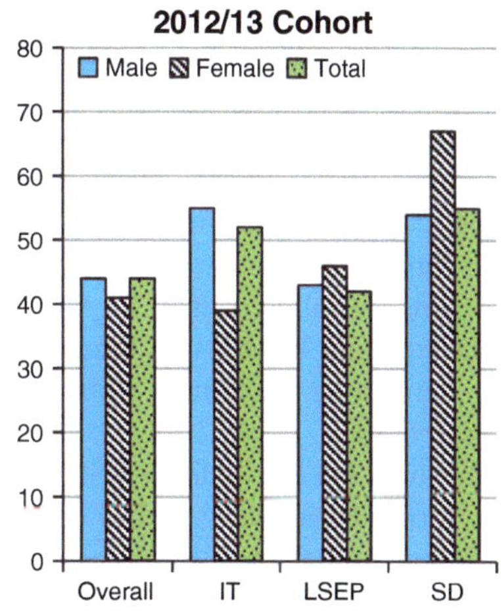

the SD motivated material are consistently and substantially higher than that for the separate IT and LSEP streams.

Overall, this demonstrates that the students appear to connect with the material most effectively when the material was combined, exceeding the performance in the separate material by a significant margin.

The overall profiles for the material taught to 5 different cohorts demonstrate an consistent pattern, with generally a higher mark in IT material than LSEP, and with a significantly higher mark in Sustainable Development than the other two. From the data considered in the figures, there is some evidence that providing a strong and coherent context for the delivery of LSEP material linked to IT, situated within the framework of Sustainable Development can improve students' engagement with material, and their performance as measured through assessment. The data provided indicates that this can have a greater effect for female students in computing, see Figs. 3, 4, 5 and 7. However, Fig. 6 shows that the impact of delivering LSEP material through the context of SD was not as great, though there was still a positive effect.

9 Conclusions

As we have considered in this paper, SD is both expected and required to be included within the computing curriculum at universities, at least to some extent. As noted in the wider context, the impact of SD on the curriculum is variable. The discussion of how SD can be used to motivate LSEP and related topics within computer science discussed ways that could be applied to other disciplines. A key element of this approach to embedding SD within a course—whether as a stand-alone module or placed within other modules (Pattinson and Gordon 2011)—is that it demonstrates how SD can be made both relevant to students within the context of their discipline, and how sustainable development can provide a framework for developing an appreciation of the legal, social, ethical and professional aspects of their discipline. The framework described and illustrated in the chapter is transferrable to other disciplines—where the underpinning ideas of professional behaviours are key elements of the graduate attributes expected from students successfully completing recognised courses.

The success of this approach has been illustrated through review of repeated application for several cohorts—involving approximately 1000 students over 5 years—where the positive impact of motivating LSEP concepts, by linking them to core discipline content (in this case IT within computing) within the context of SD examples has led to improved attainment. Whilst the focus of the case study is on computing, such motivation through examples linked to SD is relevant to other disciplines.

Whilst the embedding of SD was shown to be successful in terms of improving performance in LSEP and IT learning, the material was not explicitly labelled as SD, and end of module surveys of students indicated that many students were not aware they had met Sustainable Development in spite of demonstrating the skills and attributes related to it.

References

ACM (2013) Computer Science Curricula 2013. Association for computing machinery. http://www.acm.org/education/CS2013-final-report.pdf. Accessed 29 Mar 2014

BCS (2009) The green I.T. foundation" british computer society. http://www.bcs.org/server.php?show=nav.10479. Accessed 08 Apr 2014

BCS (2012) Guidelines on course accreditation. British Computer Society. http://www.bcs.org/upload/pdf/hea-guidelinesfull-2012_1.pdf. Accessed 29 Mar 2014

Becher T, Trowler P (2001) Academic tribes and territories: Intellectual enquiry and the culture of disciplines. McGraw-Hill International

Cade A (2008) "Employable graduates for responsible employers. Report to the Higher Education Academy, York, Higher Education Academy

Climate Group (2008) SMART 2020: enabling the low carbon economy in the information age. http://www.smart2020.org/_assets/files/01_Smart2020ReportSummary.pdf. Accessed 08 Apr 2014

CPSR (2009) Computer professionals for social responsibility. http://cpsr.org/. Accessed 08 Apr 2014

DEFRA (2011) Guiding principles for sustainable development. Department for Environment, Food & Rural Affairs, U.K. Government, http://sd.defra.gov.uk/what/principles/. Accessed 29 Mar 2014

Drayson R, Bone E, Agombar A, Kemp S (2013) Student attitudes towards skills for sustainable development. http://www.heacademy.ac.uk/assets/documents/sustainability/Executive_summary_2013-4.pdf. Accessed 24 Mar 2014

European Commission (2009) EU code of conduct for data centres. http://iet.jrc.ec.europa.eu/energyefficiency/ict-codes-conduct/data-centres-energy-efficiency. Accessed 08 Apr 2014

Gordon N (2009a) "Improving student awareness of sustainable development and related employability issues through embedded course content". In: Proceedings of the 2009 higher education academy annual conference. http://www.heacademy.ac.uk/resources/detail/events/annualconference/2009/papers/Neil_Gordon. Accessed 03 Feb 2014

Gordon N (2009b) Sustainable development and social responsibility—making it professional". In: Proceedings of the 2009 higher education academy subject centre for information and computer sciences, University of Kent, August 2009, pp 21–24
Gordon N (2010a) Education for sustainable development in Computer Science. ITALICS. http://journals.heacademy.ac.uk/doi/abs/10.11120/ital.2010.09020003. Accessed 03 Feb 2014
Gordon N (2010b) Sustainable information technology awareness. ITALICS. http://journals.heacademy.ac.uk/doi/abs/10.11120/ital.2010.09020004, 9(2):17–20
HEFCE (2005) Sustainable development in higher education. Higher Education Funding Council for England. http://www.hefce.ac.uk/pubs/hefce/2005/05_01/. Accessed 08 Apr 2014
HEFCE (2008) HEFCE strategic review of sustainable development in higher education in England. Higher Education Funding Council for England. https://www.hefce.ac.uk/pubs/rereports/year/2008/sdhefcestrategicreview/. Accessed 25 Mar 2014
HEA (2013) 2013: Student attitudes towards and skills for sustainable development. Higher Education Academy. http://www.heacademy.ac.uk/resources/detail/sustainability/2013_student_skills_final_report. Accessed 20 Mar 2014
Hopkins C, McKeown R (2002) Education for sustainable development: an international perspective. Education and sustainability: responding to the global challenge, IUCN—The World Conservation Union, pp 13–24
Martin C, Weltz E (1999) From awareness to action: integrating ethics and social responsibility into the computer science curriculum. ACM SIGCAS Comput Soc Archive 29(2):6–14
O'Connor K, Lynch K, Owen D (2011) Student-community engagement and the development of graduate attributes. Educ Train 53(2):100–115
Pattinson C, Gordon NA (2011) Green IS in teaching: specialist or generalist?. http://sprouts.aisnet.org/11-7. Accessed 03 Feb 2014
QAA (2007) Benchmark statement on computing. Quality Assurance Agency. http://www.qaa.ac.uk/Publications/InformationAndGuidance/Documents/computing07.pdf. Accessed 08 Apr 2014
QAA (2011) Subject benchmark statement: Master's degrees in computing. Quality Assurance Agency, http://www.qaa.ac.uk/Publications/InformationAndGuidance/Documents/QAA386_Computing.pdf. Accessed 29 Mar 2014
Scragg G, Smith J (1988) A study of barriers to women in undergraduate computer science. SIGCSE Bull. 30(1):82–86
WEEE (2006) Waste electrical and electronic equipment regulations. The Stationary Office, Statutory Instruments

Author Biography

Dr. Neil Gordon is a lecturer in the Department of Computer Science at the University of Hull, where he is Director of Taught Postgraduate Studies. His doctorate was in applied mathematics, considering the use of computer algebra, and provided the basis of later work at the interface of mathematics to computing. As a lecturer in computer science, his teaching duties range from first year core modules through to specialist masters' level computing courses. Earlier work as an educational technology advisor developed into his research interests in Technology Enhanced Learning, including support for Peer Assessment. Recently he has been involved in work on Education for Sustainable Development, working with the UK Higher Education Academy on a number of projects, especially relating the relevance of Sustainability to professional development.

Action Research from a Critical Perspective: Analysis of an Environmental Education Program for Employees at a Brazilian Public University

Daniela Cassia Sudan and Vânia Gomes Zuin

Abstract

This article presents some results from an ongoing research on an Environmental Education (EE) program for employees across seven campuses of a public Higher Education Institution (HEI) in São Paulo State, Brazil. The program is being developed with 17,000 administrative staff through a capillary process system. The research involves the university and its role regarding the global environmental crisis and how it can establish a context that favors the participants' educational activities to build a sustainable university. The objective of this work was to investigate this collective construction process based on an emancipatory EE program proposal. To reach this aim, a qualitative action research approach, including focal group techniques, participant observation and the analysis of complementary documents were conducted. Based on action research method in a critical perspective, this work discusses the ways in which such frameworks can contribute to understand this capillary EE process highly complex, involving several individuals of the HEI under study.

D.C. Sudan (✉) · V.G. Zuin
Federal University of São Carlos, Rodovia Washington Luiz, Km 235, São Carlos, SP, Brazil
e-mail: danisudan@usp.br

V.G. Zuin
e-mail: vaniaz@ufscar.br

D.C. Sudan
São Paulo University, Av. Trabalhador Saocarlense 400, São Carlos, SP 13566-590, Brazil

W. Leal Filho et al. (eds.), *Integrative Approaches to Sustainable Development at University Level*, World Sustainability Series, DOI 10.1007/978-3-319-10690-8_25

Keywords
Action research • Critical environmental education • Higher education institution

1 Introduction

This work is a qualitative research performed in a Brazilian public Higher Education Institution (HEI) in the field of non-formal Environmental Education (EE). The study involved a collaborative action research, being the researcher also an employee at the studied institution since 2001, as an educator in the Superintendence of Environmental Management (SEM), a department directly connected to the Dean's office. The mission of SEM is to handle the many existing initiatives of outreach, research, teaching and management considering socio-environmental topics and to seek the sustainability of this institution (USP 2014a, 2014b). There is a common understanding that SEM plays a fundamental role in building sustainability at the HEI, conducting socio-environmental education programs for employees (technicians, lecturers and professors) and students. Based on this mission, SEM´s staff proposed an EE program for the university community that would broaden its theoretical-practical repertoire and, simultaneously, encourage the university community to contribute to building a sustainable university.

The critical and emancipatory perspective considered in this EE program is presented in the current Brazilian literature, as well as in the HEI project's political-pedagogical project:

> [...] is essentially characterized by a complex understanding of the environmental issue, a critical attitude towards the challenges of civilization crisis; understands that children are not the priority audience in EE; [and] considers democracy, dialogue and participation as essential to building sustainability and to change situations that go against collective wellbeing (USP 2014a, p. 3).

This HEI has a community of approximately 120,000 people (16,839 technical employees with primary, secondary and Higher Education levels), working across seven campuses in São Paulo, Brazil (USP 2013). The aims of the EE program are: to contribute wich EE of the university community aiming at building a sustainable university in a continuous, connected and emancipatory way; to internalize sustainability in the university management sectors; to change the organizational culture based on agreed socio-environmental values, which should be included in the university's official documents; to support USP's employees in broadening their view/perception/analysis and possibilities of socio-environmental activities in their workplace and everyday life (USP 2013).

In this sense, a program for all administrative staff was proposed during a three-year period (2013–2015). The proposal was based on other experiences developed at the university, and on an Education Program for Environmental Educators by the Federal Government that was performed in 2005 across Brazil (Ferraro Junior and

Sorrentino 2005). The EE program was initiated in 2013 under the coordination of SEM. The first action was to establish the EE Working Group with 25 specialists (lecturers, professors and technicians). The methodology involves two fundamental and connected aspects: an education program in a capillary system and the "participant-action-research" or "people who learn participating" (PAP, as per the acronym in Portuguese, maintained in its original form in this article because of its resonant feature) in a capillary network working in several levels—Working group PAP 1, 2, 3 and 4. The role of SEM technicians coordinating the EE Program (Working Group included in the capillary network as a collective educator—PAP1, with approximately 25 individuals) is to review and improve the initial political-pedagogical project, promoting its implementation in the campuses, as well as to encourage and guide a group of employees that hold the 2nd leadership position (PAP2, 200 individuals). PAP1 and PAP2 are expected to compose, together, a "learning content menu" with information and activities on EE and sustainability that should be developed with PAP3 and PAP4 on the seven campuses (USP 2013). The development of socio-environmental education practices by PAP1, PAP2 and PAP3 is an essential factor in their workplace. To reach all employees within the three-year period, each PAP2 must reach, through an educational intervention, about 30 PAP3 from their campuses. These, in turn, will be committed to developing an activity involving another 30 university employees (in average) (PAP4).

It should be highlighted that what differentiates one PAP from another in the capillary network is not a hierarchy of knowledge, chief positions or level of education, but the moment at which each person arrives in the educational program and the number of hours dedicated on the program and to monitored practices.

Within this context, the objective of this paper is to discuss some of the initial results from the evaluation of this capillary EE program at the HEI of interest, considering a critical perspective, involving the educational actions of the participants aiming at building a sustainable university. To achieve this objective, a qualitative action research approach has been employed, taking into account the active participation of the author in the conception of the university's project and her role as a facilitator in the EE Working Group created by the SEM. The study also included focal group techniques (Zuin 2011; Kind 2004; Gondim 2003), participant observation, and the analysis of official complementary documents. The relevance of this work was the analysis of an ample EE program that comprehends more than 16,000 employees, considered an innovative approach firstly conducted in a public HEI in Brazil. In this way, it is necessary to understand how the action-research frameworks can contribute to the praxis of the desired emancipation in this capillary program. Finally, by looking at the demands presented by the program's coordination group after 1 year of its creation, it is important to mention the possible contributions that this study can make to the program as a whole.

2 Methodology

According to Thiollent (2003), action research is a participative methodology guided by "objectives of transformation (that are) relevant in a specific context" with "groups, institutions, and small or large communities":

> It is a type of social research with an empirical basis, conceptualized and performed with close association to an action or solution of a collective problem and in which the researchers and participants representing the situation are involved in a cooperative or participative way (p. 14).

The method can be considered adequate for this research considering that the coordination group (PAP1) as well as the university employees in PAP2, PAP3 and PAP4 are involved in conducting and reconstructing the EE program at the institution. Although the researcher and technical team of SEM worked as moderators of the project development and showed greater dedication to its execution, the concepts, discussions and general deliberations were defined taking the views of each participant into consideration, which were presented in periodical meetings. Thiollent (2003) also highlights that action research emphasizes the socio-political aspects of interpersonal relationships without disregarding the psychological and existential reality, and underlines that "the structural aspects of the social reality cannot remain unknown, because action only occurs within a collection of structurally determined social relationships" (p. 9) and in "the process of investigation and action, argumentation (or deliberation) plays an essential role, opposite the traditional concept of research, in which logical-formal and statistical values are valued" (p. 13). In action research, the interpretations of the observed reality and the transforming actions are also objects of deliberation.

Some risks exist in studies such as these, as pointed out by Thiollent (2003), which should be consciously avoided. These include abandoning the scientific criteria, the political manipulation, and activism with which the actions developed, among others. Therefore, the researcher's intensive involvement with those processes, despite not impeding the performance of quality research, requires greater attention, supervision by the adviser, and a profound methodological foundation in the development of the study in order to assure scientific rigor in the data collection and analysis. It should be considered that this research requires permanent transparency and attention to its objectives, in a collaborative approach, being performed with the participants' consent.

Regarding the contributions from the action research method to the EE Program for employees of this Brazilian HEI, it is important to mention: its condition and requirements as a participative project that also collaborate with the actions, decisions, and intentions of the activities performed by the PAP groups; its direction towards enlightenment and the solution of concrete problems, aiming at questioning the existing gaps and proposing/performing some possibilities of solutions in the field of socio-environmental sustainability; it can contribute to identifying the necessary knowledge that should be produced based on the problems found in the action or between the actors involved; and encourages the plan of action and collective praxis, increasing the awareness of people about what is done in collaboration as an antithesis of simply conducting normative "mechanical actions" in EE, which are common in many education and environmental management programs. Thus, considering the qualitative action research approach, some of the data discussed in this work were obtained by focal group technique (PAP 1 and 2), participant observation and also by the analysis of official and complementary documents of the process.

3 Results

3.1 Critical Perspective and Contributions of the Action Research Project

Considering the emancipatory perspective aimed, how could the Critical Theory be employed to understand this proposed EE Program? What is the potential of education in social transformations? The Critical Theory emerged among a group of intellectuals from the institute for social research in Frankfurt (Germany) interested in studying the characteristics of German workers within a context of intense social transformations in Europe and the USSR in the 1930s. The Critical Theory, therefore, assumed the role of invigorating the dialectic concept, understanding Marxism as an unfinished body of truths, thus, subject to being reviewed, complemented and reconstructed.

Since its beginning, the Critical Theory has been intensively employed to analyze social processes that also encompasses exploitation of workers, slavery, and extreme poverty in capitalist society, explicating the reproduction of these conditions, interested instead in overcoming domination, given its "constant interest to suppress class domination" (Pucci 1994, p. 31). It also established a different use of the theory-practice relationship, redefining the social class categories using broader concepts, with contributions from philosophy, sociology, and psychoanalysis fields.

The movement also known as Frankfurt school was additionally influenced by empiricism, which Adorno (1986) mentions when pointing out the positive facts that these lines of thought may have brought at that time, despite their not being largely accepted in those times. The main highlight, however, is on the influence of Kant and Hegel on Critical Theory in rescuing the emancipatory concept in the discussion about the active dimension of reason and critical self-reflection as the foundation of social processes as is the case of educational contexts, presented in the following topics.

3.2 How Critical Theory Understands Education from an Emancipatory Perspective?

The comprehension of the emancipation concept in this case, from the Critical Theory perspective, refers to the role that the EE Program can add to the institutional project, aiming at a critical praxis within its purposes. The complex idea of emancipation surpasses the discussion of thoughts of Enlightenment and the understanding of emancipatory reasoning. Under the influence of Kant, Critical Theory rescues the illuminist emancipatory reasoning, establishing "Aufklärung" or "enlightenment" as the central key. Enlightenment is understood as a slow historical and personal process involving the "overcome minority," which represents the "inability to understand" and "courage to make use of one's understanding without influence from another". Then, the individual is accountable is also responsible for his own liberation from minority (Pucci 1994, p. 19).

The illuminist reasoning developed by the bourgeoisie, therefore, contained the emancipatory and instrumental dimensions according to Adorno and Horkheimer, (Pucci 1994) but as the bourgeoisie imposed its domination over other social classes, it increased the privilege of the instrumental dimension, losing its liberating potential. With the appearance of monopolist capitalism, with the colonialist intensification, the scientific revolutions and the predominance of instrumental reasoning became universal at the service of progress and workers' exploitation with the objective to increase capital reproduction. From this perspective, to be considered a scientific work, it must be presented in the form of measurable and quantifiable data, the basic premises of positivism. The one-dimensional society based on the technoscience became an instrument of reproduction and domination. There is a *"thingification"* of reason, its reification in the relationship with the environment; so, the human beings must learn to dominate it and, through it, dominate other human beings. Positivism (and its instrumental reasoning) now assumes the place of enlightened reasoning, emphasizing the utility and standardization of everything in a totally administrated world, denying even art. Within this context:

> [...] an eagerness to seek truth beyond the fact, of denouncing totalitarianisms, wherever they may be, of enlightening the darkness of ignorance, barbarism, fetishism, ideological manipulation, of questioning everything that casts shadows on the power of awareness, the space of freedom, the statement of man's individuality and autonomy (Pucci 1994, p. 30).

The so called culture industry is an exemplary presentation of instrumental reasoning, and, for Critical Theory, it became "the most sensitive instrument of social control and in the sale of cultural goods, in the decadence of culture and progress of barbarism", understanding that "the different technologies, such as the radio, television, cinema and music, were connected to the domination and could easily be used as instruments of power" (Pucci 1994, p. 32, 44).

3.3 Culture History as the Appearance of Instrumental Reasoning

In the text "The Culture Industry: the enlightenment as the deception of the masses", Adorno (1986) explains how the cinema, radio and magazines form a system that promotes excessive information, causing the universal presence of an alienated spirit, resulting in people's dependency as opposed to liberation, enlightenment and emancipation. The elements of the culture industry led to an intense standardization, improvement and mass production of commodities and consumption across the world, starting in liberal industrial countries, at the service of the most powerful sectors of steel, oil, electricity and chemistry, thus strengthening the capitalist system. In this ideology of business, "the budgets in the culture industry do not bear the slightest relation to factual values, to the meaning of products" (p. 116).

According to Adorno, the culture industry—by means of cinema, radio, magazines, music, and today, by the apps, network and virtual communities etc.—promoted a violent stimulus for consumption and established a "cultural semi-formation" and occupying the complete prediction of facts and happenings, leaving no room for spectators' imagination and thoughts, and atrophying their creativity "in a way that any intellectual activity by the spectator is made impossible if they want to avoid missing the facts that quickly parade in front of them" (p. 119). "Everything that is made public was first adjusted" to the approval of the majority already on the first contact, where the "advantages and disadvantages that acknowledged individuals discusses serve only to perpetuate the illusion of the competitor and the possibility of choice." For consumers, nothing is left besides classifying that which had not been anticipated in the production schema (Adorno 1986, p. 116). Therefore, the culture industry presents an accelerated succession of things that are apparently new but that essentially are always the same thing; "imitation is somewhat absolute", composing a system of "esthetical barbarism" and "a system of non-culture" (p. 121), that can be extrapolate to some learning contexts.

But why does the culture industry win so much territory without any resistance? Adorno argues that the control that the culture industry has over people is mediated by "entertainment" (p. 128), rewards instead of demanding efforts, produces merchandise for entertainment—aiming to achieve the leisure and happiness of all—and also promises pleasure. That pleasure, however, is never truly achieved; it is always delayed in some form of "organized cruelty" (p. 129). In this apparent entertainment, though humorous, laughter comes to ridicule the dominated everyday life itself and break any possibility of criticism towards the system and idea of resistance. Thus, each person is established merely as an eternal consumer, an object of the culture industry (p. 132).

This crisis in the development of culture, strengthened by the culture industry, results in a problem that can be defined as "semi-formation" (a word used by Adorno—"Halbbildung") that will only be understood if confronted with the perspective of emancipation. For this reason, Adorno develops the theory of "semi-formation", which is hegemonic in relation to cultural education, i.e., a deformed formation in which the problem lies in separating culture from spirit and praxis. The culture of the spirit, alone, loses the idea of reality and, the opposite focuses exclusively on the everyday life, losing the criticism about reality. Hence the importance of maintaining the tension between these two dimensions—culture of spirit and praxis.

Considering the idea of cultural education, one must postulate the situation of humanity being free from exploitation. The more reality shows such conditions, the more the system will impede cultural education. Adorno states "the only remaining possibility for culture to survive is the critical self-reflection about semi-formation, to which it necessarily was converted" (Adorno 2010, p. 39). Education plays a fundamental role in developing critical self-reflection. The frankfurtian authors highlight the role of education as a contributor to cultural education that permits the constitution of a critical awareness regarding cultural, economic, social and historical reality, and, most of all, the understanding of culture industry mechanisms.

Pucci (1994) proposes that Critical Theory contributes to the construction of a pedagogic theory in the following aspects: the educational/formative function of reflecting, to go beyond what is repetitive, beyond the commodity, in a search for autonomy; Kant's self-determination, overcoming the limits of semi-culture; there is an invaluable pedagogical potential in enlightenment; critical reflection results in a transforming action in the search for a humanity free from status and exploitation; the rescue of the cultural education as a pedagogical postulate of emancipation; problematized culture in the monopolist capitalist society; education as the subjective historical appropriation of culture being constituted; the importance of education and the school's responsibility in the de-barbarization process against the 'blind' identification with the collective; barbarism de-educates, de-forms and is deeply present in the dominant social relationships; development through education in three complementary ways (awareness regarding the subjective mechanisms; becoming aware of the coldness within and identifying the reasons that caused it; an education that no longer rewards pain); the transition from ignorance to knowledge by critical enlightenment; critical enlightenment about semi-culture; assimilate the past as enlightenment (interpret the contradictory human history). The role of the collective intellectuals in the process of de-barbarization occurs not only by the formal school but also by other means, including the media, thus "the Critical Theory is a sign of individual and collective resistance, resistance through reason, culture, education/formation, [and] art" (Adorno 2010, p. 33).

4 Conclusions

In one year of collective EE program proposed by PAP1, based on the frequent participation of PAP2 in on-site meetings and in the virtual communications, the following two questions emerged: the need or lack of information regarding the program inside and outside the university, also emphasizing the need for this type of communication by email, creating blogs, websites, fan pages on Facebook, and producing videos and other media; and difficulty in texts and deeper theoretical information according to PAP2 and also identified by PAP1, who often stated that the solution would be to facilitate the access for all by providing simple, short texts that could be accessed on the internet. Sometimes, this issue is related to complaints against the philosophy or principles behind the EE program, or the "thinking" involved, arguing that the proposal should focus exclusively on the practical education of technical employees and not on the program's philosophy or delving into theoretical aspects.

Considering this discussion about theory and practice dimensions, thinking is not an innocent activity but, rather, a dangerous topic that influences the practice of individuals and leads them to doubt the traditional forms of culture (Pucci 1994). The relationship between theory and practice, avoiding the deterministic unity of theory and practice and providing a real articulation of both, seeking an emancipatory policy, according to the frankfurtians, since:

> […] emphasizes the autonomy of the theory in relation to practice but reaffirms that, just as practice does not occur without theory, the latter is not independent from practice. However, if the practice were the criterion of the theory, it would not achieve what is intended; if practice followed the theory guidelines, it would become doctrinally hard and would fabricate the theory (Pucci 1994, p. 44).

In view of the questions pointed out in the introduction of this article, observed in the process of reconstructing the EE program by PAP1 and PAP2, it was possible to observe that Critical Theory bears uncountable contributions to the discussion on socio-environment, action-research, the theory and practice issue, communication, and environmental education as it:

- analyzes the education program based on the concept of semi-formation and how it relates to the idea of technical rationality and hegemonic pragmatism also in the university domain, with non-faculty employees as those who carry out tasks for which you do "not need to read or produce knowledge", leaving this function exclusively to teachers.
- permits a critical reading of the concept of the culture industry and how it relates to the escalating demand within the university and in general society by the disclosure of everything that one does (and, perhaps, what one does not do), seeking, in record time, to meet the premise of "appear to be" or "if you are not seen, you do not exist".
- contributes to overcoming a naive analysis of sustainability in green products disseminated by the culture industry, visible in many approaches that reduce people to mere consumers, and the belief that the maximum that they can reach on this path is that of green consumers, provided their socio-economic condition allows it, because of the highest price of products with the required characteristics.
- offers support for a critical reading of the concept of sustainable development and of how this concept rewards the praxis of environmental management at the

institution. Permits critical understanding of the production system based on business and on the exploitation of human beings and environment and on increasing the production of commodities and consumption, that cold integrate the analysis of the unsustainable form of extraction and use of natural resources; the encouragement towards consumption that is connected to planned and induced obsolesce of the products and impacts of the accumulated trash due to the high disposal of containers and other superfluous objects of the culture industry.
- critical reading of the role of EE in the HEIs and its contradictions in the search for the emancipation and critical education for the university community.
- critical reading of the history of the constitution of universities in Brazil and the impacts of capitalism on Brazilian higher education, among other contributions that deserve a deeper approach in the development of a doctoral study.

The urgent need to develop and analyze large-scale socio-environmental educational programs for HEI employees is presented in view of the current civilizatory crisis. In this sense, it should be proposed models that do not render to market expectations but, rather, permit political, critical and emancipatory education. In this aspect, this study offers contributions as it proposes to evaluate the capillary model that may be extrapolated to other HEIs.

The research method and its principle of encouraging the active participation of those involved, allied with the objective of action research of contributing to a better balance of the situation at the university as long as it is connected to the development of a critical sense, tends to provide a politicized results instead of apolitical, ahistorical or decontextualized theoretical-practical actions at the studied university. Similarly, considering the objective of the action research (how to obtain information with difficult access and share it in order to increase knowledge in some situations) reverberates in the perspective of the Critical Theory of semi-formation, valuing the cultural rescuing among those involved as opposed to the apparent rule that (non-faculty) employees only follow orders and arçe unable to think and produce knowledge. The research also has contributed to produce innovative knowledge that allows for advancing the search to make the HEI a sustainable and emancipatory place, thus broadening the understanding of the specificities of this space, which is the public university.

References

Adorno TW (1986) "A indústria cultural: o esclarecimento como mistificação das massas". In: Adorno TW, Horkheimer M (orgs.) Dialética do esclarecimento: fragmentos filosóficos. 2 Ed., Tradução de Guido Antonio de Almeida. Editora Jorge Zahar, Rio de Janeiro, pp 113–156

Adorno TW (2010) "Teoria da semiformação". In: Pucci B, Zuin AAS, Lastoria LAA (orgs.) Teoria critica e inconformismo: novas perspectivas de pesquisa, pp 8–39

Ferraro Junior LA, Sorrentino M (2005) "Coletivos Educadores". In: Ferraro Junior LA (org.). Encontros e caminhos: formação de educadoras(es) ambientais e coletivos educadores. Ministério do Meio Ambiente, Brasília, pp 57–69

Gondim SMG (2003) Grupos focais como técnica de investigação qualitativa: desafios metodológicos. Revista Paidéia, Ribeirão Preto 12(24):149–161
Kind L (2004) Notas para o trabalho com a técnica de grupos focais. Revista Psicologia em Revista, Belo Horizonte 10(5):124–136
Pucci B (1994) Teoria critica e educação: a questão da formação cultural na escola de Frankfurt. Petrópolis, Editora Vozes, p 197p
Thiollent M (2003) Metodologia da pesquisa ação, 12th edn. Editora Cortez, São Paulo 136p
USP (University of São Paulo) (2014a) "Projeto Político Pedagógico de Formação Socioambiental de Servidores Técnico-administrativos da Superintendência de Gestão Ambiental da Universidade de São Paulo". www.sga.usp.br. Accessed Jan 2014
USP (University of São Paulo) (2014b) "Superintendence of Environment Management Report", 2012–2013. http://www.sga.usp.br/. Accessed 15 Jan 2014
USP (University of São Paulo) "Statistical Yearbook" 2013. https://uspdigital.usp.br/anuario/AnuarioControle. Accessed 23 Jan 2014
Zuin, V.G. (2011). "Delineamento da Pesquisa: o estudo de caso". In: A inserção da dimensão ambiental na formação de professores de química. Editor Átomo, Campinas, pp 83–91

Author Biographies

Daniela Cássia Sudan Holds an undergraduate degree in Biological Sciences, Master's of Education and is currently a doctoral student in Education. She has been an Environmental Educator at USP since 2001. She has experience in the field of Education, with emphasis on Environmental Education, and works primarily with the following themes: social-environmental sustainability, shared management and minimization of solid residue, collective educators and the educating of new environmental educators.

Vânia G. Zuin Dr. Zuin is a professor at the Federal University of São Carlos (UFSCar, Brazil). Her background is in Science (Analytical Chemistry—São Paulo University, Brazil, and Università degli Studi di Torino, Italy, 2001) and Education (Environmental Education—São Paulo University, Brazil, 2010), with major interests in developing green methodologies to determine bioactive high-value organic substances extracted from Brazilian agroindustrial wastes. Her research also focuses on the development and dissemination of Green Chemistry and Environmental Education material for universities.

The Sustainable Campus Observatory: A Comprehensive Framework for Benchmarking University Performance Toward Sustainability

Mathias Bouckaert

Abstract

This paper aims to address the lack of global assessment approaches for benchmarking the sustainable performance of higher education institutions. Such approaches are critical for universities wishing to analyse and improve their performance through the identification of best practices among their peers. Over the past several years, comparative assessment has become a central concern in the academe, especially with the introduction of global rankings. Beside rankings, two other broad categories of assessment can be differentiated: specific research projects and sustainability assessment frameworks. In a first step, a comparison of the 3 identified categories is undertaken on the basis of four criteria for success: relevance, measurability, holism and digestibility. We observe that while the three types of methods differ in their strengths, none of them manages to meet all the criteria. In a second step, we introduce a new assessment tool: the Sustainable Campus Observatory. Built on three pillars, i.e. institution's missions, operations and external outreach, it proposes a benchmarking solution to university stakeholders wishing to identify areas for improvement. Lying at the intersection of existing evaluation practices, the Observatory integrates the multiple dimensions of sustainability in HEIs through a comprehensive framework that ensures intelligibility and comparability of results.

Keywords

Assessment · Benchmarking · Sustainable development · University · Performance

M. Bouckaert (✉)
Sustainable Campus Chair, REEDS Laboratory, University of Versailles Saint-Quentin-en-Yvelines, 5-7 Bd. D'Alembert, 78280 Guyancourt, France
e-mail: mathias.bouckaert@uvsq.fr

W. Leal Filho et al. (eds.), *Integrative Approaches to Sustainable Development at University Level*, World Sustainability Series, DOI 10.1007/978-3-319-10690-8_26

1 Introduction

The imperative to establish sustainability as a guiding principle in the management of our societies is now widely acknowledged (IPCC 2013). In this regard, higher education institutions (HEIs) play a leading role due to their collective nature along with their position at the forefront of knowledge. Their missions of education and research are driving forces of tomorrow's society and their operations must meet the highest requirements of exemplarity.

In this context, evaluation is key for any institution seeking to understand its past performances and to correct possible deficiencies. Along with a clear communication of results, it also enables them to compare their performances with those of their peers and identify ways for improvement.

The assessment of university performances towards sustainable development gave rise to a wide range of research. Yet, literature remains weak because of a lack of baselines as of a solid common methodological framework (Wright 2002; Corcoran et al. 2004; Swearingen White 2009; Vaughter et al. 2013).

Developing a universal and concrete action plan for HEIs to reach sustainability is a complex task. The majority of proposals emphasize the necessity to be multidimensional, trans-disciplinary, based on long-term perspectives and gather all stakeholders to co-construct adapted solutions (Yarime and Tanaka 2012). In addition, due to its integrated nature, sustainability must be in line with HEIs core missions of education and research as with their own imperatives of excellence and financial health (Hua 2013).

Existing tools developed to measure university performance in sustainability often fail to fully address these issues for different reasons. Firstly, they lack to capture the multidimensional nature of a sustainable HEI by focusing too narrowly on its environmental externalities (Vaughter et al. 2013; Yarime and Tanaka 2012). Secondly, they hardly provide mechanisms for cross-institutional comparison because they rely on a high number of indicators, often qualitative (Shriberg 2002; Karatzoglou 2012). Finally, current assessment tools do not address sufficiently the systemic essence of sustainability (Shriberg 2002), nor do they take into account the own interests and core missions of universities that are yet at the heart of any sustainable institution.

There is therefore a need for the improvement of assessment methods. This paper seeks to contribute to this process. It is structured around two main sections. A first part will analyse the different types of tools developed to assess university performance. They will ultimately be reviewed with regards to four criteria: relevance, measurability, holism and digestibility. The second part will introduce a new framework, the Sustainable Campus Observatory: a benchmarking tool designed to provide comparative assessments of sustainable institutions on the basis of a comprehensive and multidimensional perspective.

2 Towards a Typology of HEIs External Assessment Methods

Evaluation is needed for every organization implementing actions in order to reach a goal. This reflexivity allows for analysis and corrections of past operations to reach better future performance. When combined with effective communication, external assessment also provides valuable information on the performances of similar institutions.

This section aims at reviewing the various methods that have been developed to measure university performances. Due to the large variety of approaches we will focus on external assessments, thereby excluding internal management and monitoring systems as well as political evaluations. The methods considered are grouped into three main categories according to their respective purposes and methodologies: global rankings, specific research assessments and sustainability assessment tools. Each of these categories is briefly described below.

2.1 General Rankings

General academic rankings appeared in the beginning of the years 2000. The most famous ones are the Shanghai Academic Ranking of World Universities (*ARWU*), the Times Higher Education World University Ranking (*THE*) and the QS World University Ranking (*QS*). Every year, ranking organizations develop a classification of "world's best universities" based on several indicators that never exceed 13 as for the *THE* ranking. Each set is then compiled into a single composite indicator forming the final score assigned to each institution.

The *ARWU* ranking is often regarded as the most consistent ranking among those three (Altbach 2010). Methodologically, it measures mainly research performance through quantitative indicators. The other two rankings (THE and QS) are regarded as more subjective because of their reliance on reputation indicators. Nevertheless, they investigate beyond research performance by incorporating other criteria such as internationalization or teaching quality. Table 1 shows the relative weights assigned to the different criteria in each of the three rankings.

The introduction of international university rankings gave rise to a lot of controversies with criticisms questioning their validity (Billaut et al. 2010; Enserink 2007; Florian 2007).

Regarding assessment methodology, rankings are accused of favouring research over other university missions. As shown in Table 1, it represents on average nearly half of an institution's score while teaching never accounts for more than 20 % of final scores in the three tools considered.

Because of the indicators used to measure research performance, global rankings are also charged with favouring English language publications as well as natural sciences at the expense of humanities and social sciences (Rauhvargers 2013).

Table 1 Relative weights of specific dimensions in global university rankings (in % of final score)

	ARWU[a] (%)	The ranking (%)	QS ranking (%)	Average (%)
Reputation[b]	0	33	50	27
Teaching	15	15	20	17
Research	85	42	20	49
Internationalisation	0	7.5	10	6
Knowledge transfer	0	2.5	0	1

Remarks

[a]The *ARWU* ranking includes an indicator of institutions effectiveness, the "Performance per Capita", accounting for 10 % of the final score. It amounts to the overall score divided by the number of full-time-equivalent academic staff. Since this indicator covers both research and teaching its weight as been equally shared between the two dimensions

[b]Indicators of reputation focus on research and education. In this table they are considered separately to emphasize their specific nature. Reputation is measured by surveys among academics (for *THE* and *QS* rankings) and/or private sector employers (*QS* Ranking)

Indicators selected to assess teaching quality are also limited (Altbach 2010). The *ARWU* ranking measures teaching through the number of Nobel Price or Fields Medal winners among alumni. Considering *THE*, more than half of its teaching score is based on doctoral education. One might wonder about the relevance of using such indicators to measure teaching performances since they seem to be further related to research activities.

Alongside methodological weaknesses, a second point of controversy concerning global rankings regards their impact. The introduction of global rankings was followed by extensive media coverage and increased consideration from academics and government officials. In universities, these have led to changes in the actions and objectives of various stakeholders in order to reach better positions in rankings. New phenomena appeared such as cross-citations between researchers or multiple publications of a single scientific contribution with only few changes at the margin. Such practices harm the evolution of science by removing the value of the citation index as an indicator of quality and by flooding the literature with worthless repetitions—a problem we could refer to as "infobesity".

Several recent initiatives have emerged in response to these problems. Among them we can highlight the U-Multirank project (Van Vught and Ziegele 2012). Its originality is to be much more comprehensive than existing rankings by taking into account a larger amount of assessment criteria. It does not provide any aggregated final score but results in multidimensional profiles developed for each institution.

The main contribution of the U-Multirank project relies in its exhaustiveness and appropriate response to various stakeholders, especially students. However, the absence of a single final score may be a threat for its media outreach compared to existing rankings that have the advantage of being highly readable.

2.2 Specific Research Assessments

A second category of initiatives aiming at assessing HEIs performances can be identified in the field of highly specialized research projects. These are usually focused on specific topics and fall within many scientific disciplines ranging from experimental psychology to public management and socio-economic analysis.

This section is not indented to provide a comprehensive state of scientific knowledge on HEIs assessments but rather seeks to give an overview of the wide diversity of research conducted on this subject. Although projects may differ both in their object and discipline, they all share two characteristics in common: a high degree of specialization and scientific quality.

A first field of analysis worth noting is the study of teaching and learning performances. Scientific disciplines covering this topic are often grouped under the name of educational sciences and gather approaches such as psychology, sociology or neuroscience.

Educational sciences have led to the development of assessment methods to understand how well universities are teaching their students and, therefore, how could they improve their activities. The *PISA* and *AHELO* studies undertaken by the OECD provide a well-known illustration of such initiatives.

The Programme for International Student Assessment, *PISA*, aims at assessing the quality and efficiency of school systems worldwide (OECD 2010a). Although it is not targeted towards higher education, the study provides a good example of teaching outcomes measurement.

The equivalent of *PISA* for higher education is rather to be sought in the *AHELO*—Assessment of Higher Education Learning Outcomes—Programme (OECD 2010b). The initiative is still new and has not yet resulted in actual assessment. The OECD just completed its first feasibility studies. Unlike *PISA*, *AHELO* does not lead to any classification or ranking. Its aim is to measure what student know and can do once they graduate by testing their generic—and sometimes specific—skills in a comprehensive and comparable manner.

The tool will be of direct benefit to universities that take part in the program. Once the assessment is completed, institutions will be provided with anonymous data that will enable them to compare their performance against that of their peers and find new ways for improvement. One of *AHELO's* main originality lies in its design around the concept of educational processes added value. Thereby, a university welcoming B+ student and turning them into A+ graduate will be deemed differently than another attracting A+ students to finally produce A+ graduates (OECD 2010b).

In addition to the measurement of teaching outcomes, universities are also evaluated relative to other activities. This is of course the case of research performance assessed through bibliometric studies with indicators of publication and citation (Moed et al. 1985). Economic analysis has also been utilized to measure HEIs efficiency in their teaching and research activities with regard to their level of funding or staff (Kuah and Wong 2011; Kempkes and Pohl 2006).

Besides those core missions, an increasing attention is given to the issue of HEIs external outreach. Research and assessments are conducted on topics such as the regional economic impact of universities (Uyarra 2008; Kelly et al. 2004) or their social engagement with local communities (Hart et al. 2007).

Therewith, the economic and managerial literature has been particularly interested in the study of university involvement in innovation and technological development. Arguing that knowledge transfer has become the third mission of HEIs beside teaching and research (Etzkowitz et al. 2000; Wissema 2009), new kinds of assessments have emerged. They focus on issues such as business incubator management (Mian 1997) or university-industry collaboration (Monjon and Waelbroeck 2003).

A final category of HEIs performance assessments lies in research projects focussing on sustainable development. Those initiatives are discussed in the following section.

2.3 Sustainability Assessment Tools

Sustainability assessment and management frameworks for HEIs appeared in the early 1990s but really took off in the years 2000. The main purpose of these tools is to support sustainable development initiatives in universities and campuses. To this end they provide assessment methods based on indicators that allow institutions to detect new lines of action by analysing their performance over time and with respect to those of their peers.

In a recent review of 16 HEIs sustainability assessment tools, Yarime and Tanaka (2012) classified the whole set of indicators into five broad categories: governance, education, research, campus operation and outreach. They found that two categories—governance and campus operations—gather more than 80 % of all indicators (Yarime and Tanaka 2012). Shriberg (2002) reached the same conclusion putting forward that assessment methods mainly focus on operational eco-efficiency while we need to consider sustainability across all aspects of the university.

Several key dimensions relative to universities therefore appear to be underrepresented in sustainability assessments. This is particularly the case of education and research. While there is a clear consensus on the need to take them into account, the heart of the debate lies in the question of what is to be measured.

Considering education, the main difficulty lies in defining the expected outcomes of education for sustainability. While existing tools measure the number of courses oriented towards sustainability, many authors call for a greater focus on the actual "skills for sustainability" acquired through education (Vaughter et al. 2013; Davis et al. 2003; Wiek et al. 2011; Rieckmann 2012).

Similar challenges are encountered when dealing with the assessment of research for sustainability. Existing assessments mostly consider input indicators, i.e. faculty participation rate or the number of grants awarded for research projects addressing sustainable development issues. However, even if output indicators were applied, the lack of a common definition for sustainability research remains a significant obstacle. We cannot be confined to assessing "sustainability labelled" research projects knowing that this amounts to reject other initiatives addressing future challenges that are still unknown today.

Besides the question of what is to be measured, the design of assessments tools and processes is a second important issue. Existing tools often include more than 150 indicators and rely on institutions to collect their data through self-reporting. This implies a range of difficulties in terms of complexity for university staff in charge of the assessment, incompleteness of reports, inability to verify statements accuracy, incomparability of results between institutions and consequently a potential loss of credibility for assessment final outputs. Some authors also note that the complexity and time-consuming nature of assessments may cause reluctance for institutions to engage in such efforts (Beringer 2006; Glover et al. 2011).

As it was noted above, incomparability of findings reduces their ability to provide a real addition to knowledge (Karatzoglou 2012). With regard to assessment tools, comparability is not just a means to rank institutions. It is also a necessary condition to understand their true performances and find areas for improvement by facilitating the identification and communication of best practices.

However, comparability is of value only if the results compared are themselves relevant. Rauch and Newman (2009) point out that while we need specific metrics and a clear idea of what is to be measured, assessment frameworks are to be kept flexible to allow for institutional differences. On the other hand, too much flexibility can result in complex assessments that lack comparability and have limited informative value (Fonseca et al. 2011). These findings suggest an opposition between comparability and relevance when the latter requires a certain level of flexibility.

2.4 Comparison of Assessment Frameworks in the Light of Four Criteria of Success

Three assessment methods were presented above: global rankings, scientific studies and sustainability assessment frameworks. Each has its own strengths and vulnerabilities. In a study on sustainability assessment frameworks, Shriberg (2002) emphasizes five conditions the ideal evaluation tool should meet. It must: (1) identify important issues while being as specific as possible, (2) be calculable and comparable, (3) move beyond eco-efficiency to consider true sustainability, (4) measure processes and motivations and (5) stress comprehensibility. In this paper, we build on Shriberg's conditions to highlight four criteria to evaluate the appropriateness of an assessment tool: relevance, measurability, holism and digestibility.

Relevance implies that a tool focuses on important and actual issues to capture the different dimensions of a sustainable institution in the most appropriate manner. Withal, this should be done wisely so as to meet the second condition of *measurability* that requires a clear and viable methodology ensuring the feasibility of assessments and the comparability of results.

The third criterion, *holism*, reflects the tool's ability to catch the comprehensive and systemic nature of a sustainable university. The holism criteria suggest that sustainability is more than correcting environmental externalities. It takes into account every dimension of HEIs, including finances, efficiency or teaching performances, as well as the links that bind them together.

The last criterion of success is the *digestibility* of an assessment and its output. It aims at ensuring the tool's visibility and impact. Even with the most complex modelling framework, the final result must remain simple and comprehensive to attract people's interest.

Table 2 gives a reading of the three types of assessment previously identified with regard to the four success criteria. The analysis focuses on categories as a whole; it does not provide information on the individual characteristics of specific tools, which obviously differ within groups.

As it can be seen in Table 2, none of the tool categories succeeds in meeting the four criteria for success at once. Considering global rankings, their strengths lie in their clarity, concreteness and high comprehensibility. Nevertheless, due to poor methodologies, their weak performances in relevance and holism make them inadequate to properly assess what are the "best universities".

For their part, academic research projects are most of the time relevant in their specific fields of analysis. This degree of specialization, however, makes the inclusion of the multi-dimensional nature of HEIs difficult, hindering the fulfilment

Table 2 Compliance of HEIs assessment tools with the criteria of relevance, measurement, holism and digestibility

	Relevance	Measurability	Holism	Digestibility
Global rankings (*ARWU, THE & QS*)	No	Yes	No	Yes
	Significant biases, weak methodology	Definite and objective indicators except for reputation surveys (*THE & QS*)	Research oriented, favour English publications & natural sciences, absence of basic dimensions	Highly intelligible, single final result
Specific research assessments	Yes	Yes	No	No
	Constant efforts to reduce biases, high consideration for rigor and relevance	High complexity but compliance with scientific protocols and concern for reproducibility	Often one-dimensional, highly specialized	High complexity, no incentives for dissemination outside the academe
Sustainability assessment frameworks	No	No	Yes	No
	Biased indicators, input oriented, incompleteness of reports	Self-reporting, incomparability, high complexity for non-specialists	Multi-dimensional albeit undue weight on governance and eco-efficiency	High complexity, limited transparency

of the holism criteria. Furthermore, even with a high level of transparency, specific assessments lack visibility due to their great complexity and the absence of incentives to spread knowledge beyond the academic sphere.

Lastly, sustainable assessments represent the most advanced category in terms of holism but do not fully address this criterion due to an excessive focus on governance and eco-efficiency. They often lack the minimum levels of relevance and comprehensiveness required to measure the true sustainability of institutions. Deficiencies in measurability and visibility are due to the large number of indicators included and their reliance on universities self-reporting. Yet, such initiatives are still young and we can expect them to evolve over time.

3 The Sustainable Campus Observatory

University assessment initiatives, whether branded as "sustainable" or not, show several deficiencies. There is thus a need for the construction of new tools or the improvement of existing ones. The Sustainable Campus Observatory seeks to bring a contribution to this process.

3.1 Context and Conceptual Background

The development of the Observatory took place at the Sustainable Campus Chair of the University of Versailles. The ultimate purpose of the project was to design a benchmarking tool enabling a relevant assessment and effective communication of university performances toward sustainability.

Originally developed for the improvement of industrial management processes, benchmarking is understood as "the search for those best practices that will lead to superior performance for a company" (Camp 1989). When applied to HEIs, we can define benchmarking as a method of comparison that enables institutions to evaluate their performances with respect to those of their peers in order to identify and learn from best practices.

The methodology adopted for the development of the observatory was based on two complementary approaches: an analysis of existing assessment tools and the definition of the sustainable university. The resulting modelling is presented hereafter.

3.2 Methodology

Three main areas were identified for the conceptualization of institutions: university missions, operations and external outreach. Furthermore, the concept of sustainability is divided into two general principles: the *responsibility* of institutions towards society and the imperative for *development*. Each of the broad areas is divided into 2 sub-themes from which the key objectives are derived in relation to

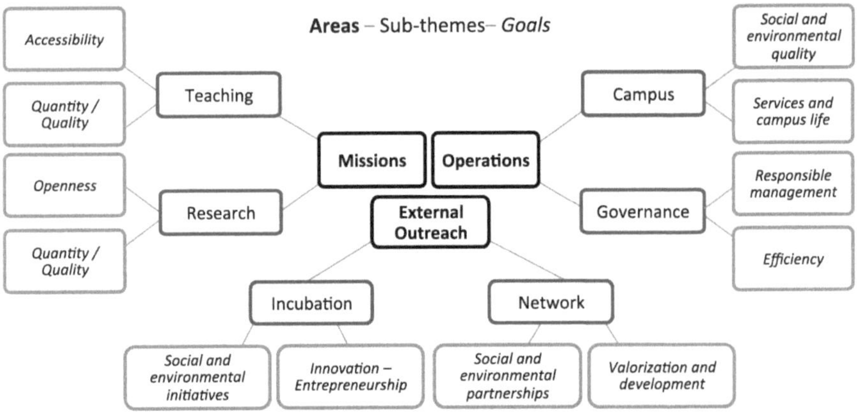

Fig. 1 General structure of the Sustainable Campus Observatory

the assumed principles of sustainability. The global structure of the observatory is outlined in Fig. 1. It aims at providing an exhaustive and coherent representation of the various dimensions of the sustainable university.

The first main area "Missions" covers the core activities of the university, namely teaching and research. Delivering quality education and conducting cutting-edge research is the primary purpose of any institution. In addition to being a key requirement for HEIs' own long-term survival, those activities are central for the sustainability of our societies. Research is at the heart of future innovations while education is probably the most powerful tool to ensure intergenerational equity and long-term development.

The second focus of the observatory is on "Operations" and covers all the functions supporting the core missions of institutions. Allegorically, it can be seen as the fertile ground on which research and teaching can be carried out optimally. The operations dimension therefore brings together issues of governance, campus management, student life and work conditions.

The third area considered in the observatory is "External Outreach". It includes all the activities of the university explicitly intended to contribute to the creation of social, environmental and economic value for what lies beyond its walls.

The sub-themes identified in the 3 main areas are broken down into key objectives. In addition to being derived from the various dimensions of universities, each objective is bounded to one of the two principles of sustainability, namely *responsibility* or *development*. If we take the example of "Teaching", the goal of delivering quality education to a large number of students (the "Quantity/Quality" objective) is in line with the development component of sustainability. On the other hand, the second objective focuses on accessibility and aims at ensuring the right for everybody to receive a quality education regardless of his or her social and financial conditions. Hence, accessibility is the application of the responsibility principle on universities' teaching mission.

Each of the key objectives are then broken down into a set of indicators designed to reflect the actual performances of institutions keeping in mind the requirements of relevance and measurability. Thereby, performances on the teaching quality/quantity objective are determined by a range of information collected on issues such as teaching evaluation programs, graduate employment rates, the use of digital technology, or the success rate of students adjusted for selection practices.

3.3 Perspective and Expected Impact

The goal of the Sustainable Campus Observatory is to be useful for all university stakeholders, be they students, faculties, external partners or public administrations. It also seeks to meet the four success criteria identified above. Relevance and holism are addressed through a global and systemic design while indicators are selected on the basis of their realism and contextualized according to local conditions. Therewith, a particular focus has been put on feasibility by limiting the number of indicators and ensuring the greater ease for data collection. Finally, a clear and simple presentation of results from a multidimensional perspective is intended to meet the condition of digestibility.

The question of whether it can be beneficial to provide a ranking of universities based on a single composite indicator is still pending. A priori, despite the significant benefit they bring in terms of comprehensibility, these practices remain highly subjective. Nevertheless, rankings' visibility confer them the biggest impact and could be very valuable in promoting sustainable development at universities (Fadeeva and Mochizuki 2010). The main obstacle to overcome would then be to ensure the legitimacy of the method. It could be achieved with continuous improvement based on external feedbacks or by a validation of the tool through a collaborative process gathering representatives from the entire university community.

4 Conclusion

The need to support our societies and universities towards a more sustainable development is recognized by all. Sustainability is not limited to environmental responsibility but incorporates social and economic dimensions on an equal footing. In order to address the challenges of tomorrow, we need new methods of management and assessment based on multidimensional and transdisciplinary perspectives.

With respect to higher education, external assessment tools were divided into three broad categories: global rankings, specific research projects and sustainability assessment frameworks.

An evaluation of the three classes of tools has been conducted on the basis of four success criteria: relevance, measurability, holism and digestibility. It followed that while each category of tools has its own advantages, none is able to meet the four conditions.

In response to these findings, we introduced a new comprehensive benchmarking tool for universities: the Sustainable Campus Observatory. It is designed to facilitate the identification of best practices for university stakeholders through the assessment and comparison of HEIs performances toward sustainability.

Its methodology relies on the principles of holism and complementarity. This has two main implications: first sustainability can only be implemented and assessed through the integration of all dimensions of the institution; and second a university can only be sustainable for society if it is already sustainable for itself.

The main contribution of this approach is its ability to capture the interconnectedness of all the components of the sustainable university. Consequently, it enables the identification of synergies reconciling the interests of society and those of the institution.

Ultimately, this new scheme was designed to meet the four success criteria. It therefore lies at the intersection of the three categories of assessment tools identified, bringing together the relevance of academic projects, the comprehensiveness of sustainability assessment frameworks and the digestibility of global rankings.

References

Altbach PG (2010) The state of the rankings. Inside Higher Ed, 11 Nov 2010

Beringer A (2006) Campus sustainability audit research in Atlantic Canada: pioneering the campus sustainability assessment framework. Int J Sustain High Educ 7(4):437–455

Billaut J-C, Bouyssou D, Vincke P (2010) Should you believe in the Shanghai Ranking? Scientometrics 84(1):237–263

Camp RC (1989) Benchmarking: the Search for industry best practices that lead to superior performance. American Society for Quality Control (ASQC), Milwaukee, 299 p

Corcoran PB, Walker KE, Wals AEJ (2004) Case studies, make-your-case studies, and case stories: a critique of case-study methodology in sustainability in higher education. Environ Educ Res 10:7–21

Davis SA, Edmister JH, Sullivan K, West CK (2003) Educating sustainable societies for the twenty-first century. Int J Sustain High Educ 4(2):169–179

Enserink M (2007) Education. Who ranks the university rankers? Science 317(5841):1026–1028

Etzkowitz H, Webster A, Gebhardt C, Terra BRC (2000) The future of the university and the university of the future: evolution of ivory tower to entrepreneurial paradigm. Res Policy 29 (2):313–330

Fadeeva Z, Mochizuki Y (2010) Higher education for today and tomorrow: university appraisal for diversity, innovation and change towards sustainable development. Sustainable Science 5:249–256

Florian RV (2007) Irreproducibility of the results of the Shanghai academic ranking of world universities. Scientometrics 72(1):25–32

Fonseca A, Macdonald A, Dandy E, Valenti P (2011) The state of sustainability reporting at Canadian universities. Int J Sustain High Educ 12(1):22–40

Glover A, Peters C, Haslett SK (2011) Education for sustainable development and global citizenship: an evaluation of the validity of the STAUNCH auditing tool. Int J Sustain High Educ 12(2):125–144

Hart A, Maddison E, Wolff D (2007) Community-university partnerships in practice. National Institute of Adult Continuing Education (NIACE), Leicester 224 p

Hua Y (2013) Sustainable campus as a living laboratory for climate change mitigation and adaptation: the role of design thinking processes. In: König A (ed) Regenerative sustainable development of universities and cities. Edward Elgar Publishing Limited, Cheltenham

IPCC (2013) "Climate change, The physical science basis". WG I contribution to the IPCC fifth assessment report—summary for policymakers, Twelfth Session of Working Group I, Stockholm, 27 September 2013, 36 p

Karatzoglou B (2012) An in-depth literature review of the evolving roles and contributions of universities to education for sustainable development. J Clean Prod 49:44–53

Kelly U, McLellan D, McNicoll I (2004) The economic impact of UK higher education institutions. Universities UK, University of Strathclyde, UK

Kempkes G, Pohl C (2006) The efficiency of German universities—some evidence from non-parametric and parametric methods. Appl Econ 42(16):2063–2079

Kuah CT, Wong KY (2011) Efficiency assessment of universities through data envelopment analysis. Procedia Computer Science 3:499–506

Mian SA (1997) Assessing and managing the university technology business incubator: an integrative framework. J Bus Ventur 12(4):251–285

Moed HF, Burger WJM, Frankfort JG, Van Raan AF (1985) The use of bibliometric data for the measurement of university research performance. Res Policy 14(3):131–149

Monjon S, Waelbroeck P (2003) Assessing spillovers from universities to firms: evidence from French firm-level data. Int J Ind Organ 21(9):1255–1270

OECD (2010a) PISA 2009 at a glance. OECD Publishing, Paris 99 p

OECD (2010b) AHELO assessment design. Group of national experts on the AHELO feasibility study, Directorate for Education, Institutional Management in Higher Education Governing Board, Paris

Rauch JN, Newman J (2009) Defining sustainability metric targets in an institutional setting. Int J Sustain High Educ 10:107–117

Rauhvargers A (2013) Global University rankings and their impact—report II. European University Association, Brussels 86 p

Rieckmann M (2012) Future-oriented higher education: which key competencies should be fostered through university teaching and learning? Futures 44(2):127–135

Shriberg M (2002) Institutional assessment tools for sustainability in higher education: strengths, weaknesses and implications for practice and theory. Int J Sustain High Educ 3(3):254–270

Swearingen White S (2009) Early participation in the American college and university presidents' climate commitment. Int J Sustain High Educ 10:215–227

Uyarra E (2008) The impact of universities on regional innovation: a critique and policy implications. Manchester Business School Working Paper, p 564

Van Vught F, Ziegele F (2012) Multidimensional ranking. The design and development of U-Multirank, vol 37. Springer, Dordrecht, 194 p

Vaughter P, Wright T, McKenzie M, Lidstone L (2013) Greening the Ivory Trower: a review of educational research on sustainability in post-secondary education. Sustainability 5:2252–2271

Wiek A, Withycombe L, Redman CL (2011) Key competencies in sustainability: a reference framework for academic program development. Sustain Sci 6(2):203–218

Wissema JG (2009) Towards the third generation university: managing the university in transition. Edward Elgar Publishing, Cheltenham 272 p

Wright TS (2002) Definitions and frameworks for environmental sustainability in higher education. High Educ Policy 15:105–120
Yarime M, Tanaka Y (2012) The issues and methodologies in sustainability assessment tools for higher education institutions: a review of recent trends and future challenges. J Educ Sustain Dev 6(1):63–77

Author Biography

Mathias Bouckaert is a Ph.D. candidate at the Sustainable Campus Chair of the University of Versailles since October 2011. His research covers the methods for applying the precepts of sustainable development on university campuses. More precisely, the thesis aims to develop specification for the assessment of sustainable campuses in France: a typology of campus models that adapt to the changing uses of the future, serving academic and scientific excellence.

Science and Policy for Sustainable Development: Bad Scientific Practice and Plagiarism

Joost Platje and Eckhard Burkatzki

Abstract

Objective and honest science supports sustainable development by identifying problems and solutions as well as the provision of knowledge and information, enabling the redefinition of developmental goals. Bad scientific practice and plagiarism do not only lead to the production of useless publications, but may also have wrong decisions regarding policy for sustainable development as a consequence. The aim of this article is to create a theoretical framework for analysing the causes and consequences of bad science and plagiarism. Furthermore, the author will discuss a checklist for usefulness of policy research (Kampen and Tamas 2014), enabling policy makers as well as scientific reviewers to reduce the transaction costs for identifying bad scientific practice and plagiarism.

Keywords

Sustainable development · Scientific methodology · Plagiarism · Checklist for policy research

J. Platje (✉)
Wrocław School of Banking, Ul. Fabryczna 29-31, 53-609 Wrocław, Poland
e-mail: jplatje@uni.opole.pl

E. Burkatzki
IHI Zittau/TU Dresden, Markt 23, 02763 Zittau, Germany
e-mail: burkatzki@ihi-zittau.de

W. Leal Filho et al. (eds.), *Integrative Approaches to Sustainable Development at University Level*, World Sustainability Series, DOI 10.1007/978-3-319-10690-8_27

1 Introduction

Sustainable development aims at inter- and intra-generational equity, which, according to the definition provided in the so-called Brundtland Report (WCED 1987), should put particular emphasis on the needs of the still huge amount of very poor people on this planet. However, in more developed countries there seems to be a tendency to focus on environmental issues when this does not threaten their own development (Platje 2011). This may lead to conflicts and trade-offs as environmental policy may negatively influence socio-economic development and the other way round. Furthermore, there always exists the inherent problem related to scarcity of resources which makes it impossible to achieve all goals at the same time (Lomborg 2004, p. 1). Even though some problems may be so evident that, as economics Nobel Prize winner Amartya Sen (1999) argues, there would be a general agreement that they need to be solved, reality is often different. Furthermore, looking at the exploitation of more and more natural resources e.g., shale gas, sand oil (see Yergin 2011), there still exists the tendency of sacrificing natural resources and the environment for short-term economic interests (see Keijzers 2003). It may be argued that economic sustainability is and will remain for a long time the fundament for sustainability in other areas (Platje 2011). This may lead to overemphasis on research in the field of economic performance as demand for such research is likely to be large.

Furthermore, this short digression shows there are immense problems with creating and implementing a policy for sustainable development, as, although synergy and win-win effects may be achieved, the trade-offs between goals have as a consequence that to be sustainable should not be interpreted as to be harmonious. Science is of incredible importance in this context. When being objective and carried out by honest scholars and efficiently functioning research institutes that are not influenced by particularistic interests science supports sustainable development "as it facilitates the identification of problems and solutions as well as the provision of reliable knowledge and information, facilitates institutional change and helps to redefine developmental goals (Platje 2011, p. 98)."

However, in reality science is less objective than is often thought. This will be discussed with help of a model of factors influencing scientific research in Sect. 1. Furthermore, due to high transaction costs and the factors discussed in the model, there exist incentives for opportunistic behaviour. The analysis in this article will focus on human and social sciences involved in creating knowledge as well as policy advice for sustainable development. Due to different reasons discussed, practice in these sciences are not always in accordance with what can be called "good science". Bad science ranges from plagiarism to misinterpretation or manipulation of data, the use of wrong research methods, the lack of a clear aim, etc.

The consequences of such practices may be serious, and may range from the production of useless articles and books to wrong advice for policy measures, which may have serious negative social, economic and environmental impact. While the latter may be more serious than the first, even when a useless paper is produced not harming society, the resources used to support such science are wasted and could have been used for more useful purposes. The model presented explains some of the factors leading to such a situation, and will be discussed in the context of the importance of the governance of scientific research. An underlying aim is to stimulate awareness of threats to good scientific practice and the sources of these threats, which may help us, scientists, to deal with the issue and to be conscious of our own practices. The ideas presented in this article need further development and analysis in the future.

After the discussion of determinants of opportunistic behaviour in science, a checklist for the usefulness of policy research will be discussed based on the work of Kampen and Tamas (2014). Such a checklist may provide information on the usefulness of science, and may be an instrument to reduce the cost of searching knowledge on a scientific issue, while providing signals for reviewers regarding the quality of an article.

2 Determinants of Opportunistic Behaviour in Science

A reason for scientists to engage in plagiarism or more sophisticated forms of bad science is obtaining benefits at a low cost. There can be a wide range of benefits of producing what will be called for simplicity "scientific snake oil"[1] (see Kampen and Tamas 2014). In fact, the fundamental question may be whether scientists are really interested in gaining and creating knowledge, or whether other factors are of such importance that they resign from the quest for gaining knowledge or pretend to create knowledge by way of producing such scientific snake oil (see Schopenhauer 2004). Motivations may be the aim of obtaining an academic degree, save time and be able to produce more articles or books (in particular important in the context of a "publish or perish" highly competitive environment (see Landreth and Colander 1994, p. 8)), obtaining better scores in university or research institute evaluations, etc. These motivations are related to the aim of obtaining academic reputation, which is a form of symbolic capital (see Bourdieu 1992).

Some determinants influencing the quality of science, providing incentives for increased normativity or bad scientific practices, are presented in Table 1. First of all, the demand for scientific research may influence the outcome, as the organization paying for the research may not be willing to employ the scientist in the future when outcomes are not in accordance with the interests of the customer. Thus, here the question is whether the patron is really interested in objective research. When, as argued in the introduction, economic interests prevail, there are incentives to show that continuous economic growth can support sustainable development. Showing that economic activity in developed countries should be reduced in order to support sustainable development and an increase in the quality

[1] Of course this notion is a simplification, as there is a wide range of scientific misconduct ranging from criminal activities and cheating (plagiarism, making up your own data) to mistakes made due to a lack of skills and/or education.

Table 1 A simple model of causes and consequences of bad science

Factors influencing "bad science"	Effect on quality of research
Demand/research budget	Choice of the topic influenced by the priorities of the funder. This may lead to a neglect of environmental and social issues compared to economic issues in the interest of large funders, and lead to research of topics in the interest of highly developed countries (see Woźniak 2000)
	Fear of loss of contracts may lead to use of research methods and interpretation or manipulation of facts which leads to outcomes satisfying the funder. Such behaviour may lead to excessive claims, which may lead to wrong developmental policies
Mental model and worldview of the researcher	Ideology and worldviews may lead to the refusal of use of models for analysis and research as well as the manipulation of data. Worldviews, beliefs, etc. may lead to different interpretation of the same facts. This factor becomes more important in case of weak scientific governance and lack of knowledge on and/or use of proper research methods
Fruitfulness of the method	May cause a kind of path-dependency, where incentives for searching for better methods are weak. This may lead to satisfactory, though not necessarily best science. Simple methods may lead to strong results, which provides disincentives for using system approaches, which are the basis of analyzing processes in the field of sustainable development
Social factors, culture	Depending on the organizational culture in a scientific organization, pressure for good scientific practice may differ. There seems to be a tendency for "loosening" scientific standards, leading to lower quality science (see discussion of organizational models below). Due to the perceived importance of different issues of sustainable development, incentives exist to forget about the difference between scientific research and policy advice
Others use the method	A successful method is not necessarily the best one. High costs of acquiring knowledge on new methods may hamper the progress in scientific knowledge, while the received view in science or a certain research institution may lead to taboos of lack of funding for unconventional approaches
Taboos, codes of conduct, ethical issues	It may be "not done" to research different topics, due to ethical constraints, taboos, etc. Ethical standards may deteriorate in scientific organizations
Political factors, legal constraints	Political factors and legal constraints rather determine the research agenda, while eventually limiting the use of certain research methods. Political pressure may lead to false interpretation of data, excessive certainty of claims for political reasons, etc

Source authors' own elaboration, extended from Platje (2011, p. 108)

of life in less developed countries is unlikely to obtain massive support. Would a medicine producer be interested in finding out its cure for a disease is in fact useless or has negative side effects, losing in this way revenues? Is a government interested in research showing their policy on sustainable development is improper, which may lead to loss of contracts in the future for the researchers? Is a company interested in research giving evidence for that its production methods and policy are anti-social and anti-environmental? While such incentives do not mean that all scientists engage in opportunistic behavior and make exaggerated claims in order to keep the funder satisfied, it is important to find out whether such a tendency exists in order to deal with the problem. Funders in fact determine to a large extent the research agenda, and as such do not influence the scientific practices. However, when output is required to obtain funds, there is an incentive to speed up the research, at the expense of good scientific practice.

All determinants discussed in Table 1 are in one or the other way related to the motivation of the researcher. When for different reasons a student does not think seriously about his/her education and a scientist is not really interested in obtaining knowledge (e.g., the diploma, publication, reputation effects are more important than the knowledge as such), methodology of science is likely to become weaker. In general it may be argued that when less strict methods are used and /or when scientific governance structures are weak (like in the anarchistic model discussed below), the worldviews, political agendas, etc. become more important for interpretation, adding to the inherent normativity of sciences involved in policy for sustainable development. The results of research are likely to become less and less useful in case of increased problems with "bad science". Furthermore, when less knowledge is produced, it becomes more difficult to solve problems. When producing false knowledge, policy for sustainable development may have harmful effects and lead to wrong investments as well as development of wrong and harmful policy.

Scientific malpractice is also likely to have a negative impact on innovation and the development of knowledge. It may be that in countries or regions with a weaker institutional setting (e.g., large corruption), which as such already hampers economic development as well as policy for sustainable development (North 1990; Platje 2011), the probability of plagiarism and bad science increases. As a consequence, the difference in levels of development may increase, as the knowledge generation in the weaker institutional setting remains behind the knowledge generation in stronger institutional settings. However, increased economic performance may lead to increased support for scientific research on this issue. While sustainability research increases in importance, much research may be financed by large economic interest groups neglecting environmental and social issues, or support research showing that such negative effects do not exist. This tendency may be counteracted by an increased level of transparency and opportunities for stakeholders to monitor and check research results. In other words, increasing the probability for discovery of plagiarism and bad science may be a good strategy for counteracting the development of bad science. Furthermore, the intrinsic motivation of the scientist is elementary, as this prevents many of the discussed problems.

It may be useful to discuss the issue from the point of view of organizational theories. Science takes place within organizations with their own dynamics and structures, providing specific incentives for good scientific practice. Scientists are trained within these organizations, and learn there the working tools for their further career (see Landreth and Colander 1994). Scientific and organizational goals may contradict, and time pressure may appear to obtain quick results. A corporate culture or ethics may support or hamper the development of good science. In Table 2, four organizational models are presented in the context of incentives for bad science. Only a general outline is provided here, which may be useful for deeper analysis of corporate governance as a determinant of bad science.

Corporate governance of research institutes may be an important factor in the incentives for good scientific practices and creating policy for eliminating bad scientific practice. This is related to a question which requires deeper research. Are universities themselves really interested in improving the quality of science? The

Table 2 Four organisational models and incentives for bad science

	Organisational goals and preferences		
		Clear and shared (consensus)	Unclear and/or disagreed upon (conflict)
Organisational processes/ technology	Transparent/ clear	*The rational model:* the problem of bad science is likely to be small, as transparency reduces incentives for such activity, while truth finding and knowledge creation is a common goal, implying that incentives for bad science are very weak or even do not exist.	*The political model:* strong interest groups may have aims conflicting with the goal of good science, in particular when their financial situation and reputation depends on the acquisition of grants. Their mental models may also influence what is allowed to be researched, and the way in which it is researched.
	Ambiguous/ unclear	*The social system model:* while there is reduced or lack of transparency, incentives for bad science strengthen. However, incentives for learning are strong. In this model, the incentives for bad science are determined by the organizational goals, where either bad or good science is the result. This theory may explain differences between universities as well as between faculties/research units, etc.	*The anarchistic model:* the goal ambiguity and lack of transparency weaken incentives for good science. Conflicts do not necessarily reduce incentives for good science in case of healthy competition. However, it can lead to a lack of proper scientific discussion, and a disinterest in dealing with bad scientific practice, in particular when networks with high process-based trust develop.

Source Authors' own elaboration based on theory drawn from Ellström (1983)

general idea is similar to incentives for business to change the organizational structure and production processes when results are satisfactory. While competition between universities may provide incentives for good science, the existence of unobserved behaviour related to the issue of asymmetric information has the opposite effect. An instrument like blind peer review is useful for improving the quality of science. However, a question is to what extent different networks are featured by process-based trust (trust based on repeated interaction between individuals (see Raiser 1999)) and are closed to new, innovative approaches by new scientists. While the movement of scientists between different universities or research institutes during their academic life reduces this problem, it may result in a lack of organizational commitment. Furthermore, like with a company selling a product with a high perceived quality, while the consumer faces high transaction costs of finding out the quality (e.g., high costs of obtaining information on the way a product is produced and what are the ingredients), there are incentives to reduce costs at the expense of the real quality (assuming the perceived quality remains the same). Similarly, when the costs of obtaining information on scientific work are high, which is in particular the case with advanced research methods, similar processes may appear.

An interesting approach towards university organization is what Cohen et al. (1972) call the "organized anarchy". Such an organization is characterized by unclear and ambiguous goals, while the governance structure is not really understood by many members and many of them only work short-time or part-time. Part-time engagement prevents the creation of closed clubs within a scientific organization, which may cause disinterest in new and innovative ideas (compare Raiser 1999; Raiser et al. 2001). This does not prevent the development of inter-university networks, which may have similar features. As the organizational goals are supposed to be ambiguous, while the governance structure is not really understood by many of its members, it is unlikely that a cooperative approach will be used to deal with issues of bad science. Only when real problems appear, action may be undertaken. However, there may be a tendency of improvisation and waiting until the last moment with making decisions.

According to Cohen et al. (1972), decision-making on the organized anarchy can be compared to a "garbage can". They argue that decision-making is often accidentally, and based on drawing problems and solutions, which came up through time, from a metaphorical garbage can, while little discussion takes place and many existing problems are not really considered. It is unlikely that good scientific practice will quickly appear on the organizational agenda, as the peer review of colleagues may be assumed to fulfil its task. Furthermore, accusing colleagues of bad scientific practice is unlikely to have positive effects on the reputation of the person in question in the organization, as the authority of colleagues is questioned as well as their fundamental professional attitudes. Scientists are often specific individuals with their own interests, preferences and ideologies, which tends to lead to goal conflicts. Furthermore, preferences, ideologies and worldviews may also influence the view of science as well as on ethical issues. In such an organizational structure, it may be expected that issues of scientific malpractice are only dealt with when the problem is becoming very pressing.

The scientific organization may also be not always really interested in dealing with low scientific standards or scientific malpractice. Incentives for what can be called scientific quality management may weaken when the existence of a university depends on the number of students and Ph.D.'s promoted. Furthermore, a problem in policy-making for improving the quality of science is that often small interest groups prepare important decisions. Hence, the question is whether the most powerful groups are willing to change the existing status quo, when their own reputation and income depends on the reputation of the scientific organization, which may be damaged when malpractice is made public. Furthermore, scientific issues are complicated and not always fully understood by all parties, while the members of faculty boards, scientific boards, senates as well as administrators such as rectors and deans tend to have a lack of time for deep analysis due to the wide range of other tasks they have to fulfil.

3 A Short Checklist Regarding the Usefulness of Research

In the overload of information available for researchers and policy makers in any area of science, it may be necessary to apply instruments in order to make an initial selection of articles which may be potentially useful for sustainability sciences as well as policy for sustainable development. However, while instruments on the one hand reduce the cost of search of existing knowledge, it may also lead to exclusion of potentially useful research. For example, the reputation of an author, journal or research institute may provide information on the quality of the research as well as the potential relevance, when the reputation is based on experience. This may lead to exclusion of original research by less known authors in journals which are not so highly ranked. While the number of times quoted provides some information on the popularity of an article, it does not guarantee high quality. The short checklist on the usefulness of an article presented below includes the most basic ones from a list developed by Kampen and Tamas (2014): is the work plagiarized, is there an aim or research question, is there a conclusion, is the logic and structure of arguments clearly explained, are references regarding ideas and data provided, is there enough information on the research method and on techniques of data processing and interpretation. Also with these indicators the reader should be careful, as, for example, lack of information on the research method does not directly disqualify an article as such, as this may be according to the policy of the journal the article is published.

First of all, the question is whether the work is plagiarized or not? When yes, the research can be rejected, and the authors should be blacklisted as their research does not contribute anything to the development of knowledge. Of course, different levels of plagiarism can be determined while the consequences also differ. It may be that the research is original, but that part of the work is auto-plagiarism. When the methodology section is copied word for word from another paper of the same author, this may not necessarily reduce the value of the research. For this reason, a limit of, say, 10 % of the text being allowed to be derived from earlier publications may be set by journals. However, when, for example, the introduction is a compilation of different quotes without quotation marks, which one of the authors has observed reviewing many articles, this may be an indicator that the research is not original or that the body of the text also does not contain proper research, as the researcher has not acquired the skills to write independently.

A second important indicator for the usefulness of science regards the question whether there is an aim or research question formulated in the article or not. The moment no clear aim is formulated, it is highly likely the article lacks focus, while not clearly dealing with a problem regarding sustainable development. Such articles are unlikely to contain original research. When the aim or research question is improperly formulated, this does not necessarily mean the content is improper or irrelevant. For example, a research question "how to support vegetarianism" is improper, as the authors probably have a political agenda which they are likely to be willing to prove. While a research question in a form of "what are the determinants of vegetarianism, and which determinants have the strongest impact on the choice of eating meat or not" does not exclude a political agenda, it may be less likely that factors are excluded that could lead to results which contradict the personal or political aim or worldview of the researcher. Furthermore, when neglecting important determinants which could support the increase in demand of meat, there always exists the possibility that the model used for analysis is incomplete and does not show all interactions and feedback loops (compare Sterman 2000). Another issue is that the researcher is likely to try to find data supporting his/her ideas, while not looking for possibilities for falsification. At this moment, due to the elimination of falsifiers, it may be that the researcher finds evidence for a stimulant of vegetarianism, which in reality is not effective, blocking research in another direction. Furthermore, it should not be forgotten that, generally speaking, the scientist presents facts, probabilities, possibilities, etc., while it is the task for policy makers or other stakeholders having a political agenda to base policy on the research results.

Is there a conclusion? A proper introduction and conclusion provide the reader with quick, relevant information. In fact, the abstract of the paper should already provide enough information for the reader in order to decide whether the research is potentially useful of not. A lack of or a poor conclusion may indicate that the author has not critically interpreted the results of the research. The conclusion should make a clear relation with the aim(s) and research question(s) of the article. Furthermore, the reader should be careful with excessive strong claims, over-interpretation of data, etc. The moment a scientist does not show the limitations of his or her research, this may indicate neglect of careful interpretation of, for example, empirical data and the inherent uncertainty in science (see Kahneman 2011).

Is the logic and structure of arguments clearly explained? The moment a proper introduction and conclusion lack, this may be a sign of poor logic of the arguments in the article. A reason is that when an author is not able to clearly express the arguments in these parts of the article, this may also be the case in the rest of the article. In this case, a reader should be aware of more potential shortcomings of the article.

Are references (sources of information) regarding ideas and data provided? Is there enough information on why a research method has been chosen, and how data were processed and interpreted? While lack of references may be a signal of plagiarism, together with information on the research method, data processing and interpretation it enables verification and controllability of data, arguments, etc.

4 Concluding Remarks

While an interdisciplinary science dealing with policy for sustainable development is inherently normative, the objectivity and use of proper scientific method may be negatively influenced by a wide range of factors. The identification of such factors is elementary, as it provides a framework for improving scientific practice and scientific discussion. A scientist should be completely aware of the incentives for opportunistic behaviour and the resulting impact on science itself as well as on policy for sustainable development. While it may be argued that useless science is less harmful than outrights cheating, resources are wasted which could be used for more useful purposes. Furthermore, it should not be forgotten that the result of science in the form of valid knowledge is a kind of public good. The moment knowledge is used for supporting sustainable development and improving the quality of life, no-one can be excluded and many positive effects (e.g., environmental improvement) are non-rival as the advantage for one person or group does not reduce the potential benefits for other persons or groups. However, following such simple logic it can be argued that negative effects of policy based on improper scientific knowledge also is characterized by non-excludability and non-rivalry. Furthermore, as science is featured by fundamental uncertainty (see Kahneman 2011), not only bad scientific practice but also useless science may undermine the necessary trust. Without such trust in scientific, verifiable, knowledge, discussion may focus too much on the reliability of the research instead of on discussion on the basis for policy for sustainable development. The shortlist discussed in this paper is the basis for deeper research, and may be an instrument for reviewers and policy makers to make an initial selection of useful research. However, the indicators are not meant for immediate rejection of research not fulfilling the conditions, but should signal potential problems with research. Besides the fact that, for example, papers without a proper research question and well developed conclusion may have an interesting content, among others, one element which needs elaboration is the way in which ideas and knowledge are communicated and how the communication is perceived by the reviewers and policy makers. Even when a checklist for the usefulness of science seems to be objective, the reviewers and policy makers (in fact decision-makers on the usefulness of the scientific work) are influenced themselves by their culture, mental models, tastes and preferences, etc. And maybe that should be the starting point for every scientist. Not only being critical towards what others produce, but in particular being critical towards what we do ourselves.

References

Bourdieu P (1992) Homo Academicus. Suhrkamp, Frankfurt a.M (orig. 1984)
Cohen MD, March JG, Olsen JP (1972) A garbage can model of organizational choice. Adm Sci Q 17(1):1–25
Ellström P-E (1983) Four faces of educational organizations. High Educ 12(2):231–241
Kahneman D. (2011) Thinking fast and slow. Farrar, Straus & Giroux, New York
Kampen J, Tamas P (2014) Should I take this seriously? A simple checklist for calling bullshit on policy supporting research. Qual Quant 48(3):1213–1223
Keijzers G (2003) Creating sustainable directions: collaborative stakeholder approach of governments and business. Erasmus Universiteit, Rotterdam
Landreth H, Colander DC (1994) History of Economic Thought, 3rd edn. Houghton Mifflin Company, Boston
Lomborg B (2004) Introduction. In: Lomborg B (ed) Global crises, global solutions. Cambridge University Press, Cambridge
North DC (1990) Institutions, institutional change, and economic performance. Cambridge University Press, Cambridge
Platje J (2011) Institutional capital: creating capacity and capabilities for sustainable development. Wydawnictwo Uniwersytetu Opolskiego, Opole
Raiser M (1999) Trust in transition. EBRD Working paper 39. EBRD, London
Raiser M, Haerpfer C, Nowotny T, Wallace C (2001) Social capital in transition: a first look at the evidence. EBRD Working paper 61. EBRD, London
Schopenhauer A (2004) De Wereld als Wil en Voorstelling (The world as will and representation), Part 1. Wereldbibliotheek, Amsterdam
Sen AK (1999) Development as freedom. Anchor Books, New York
Sterman JD (2000) Business dynamics: system thinking and modelling for a complex world. Irwin / McGraw Hill, Boston
WCED (1987) Our common future. Oxford University Press, Oxford
Woźniak T (2000) Propaganda Scjentystyczna – funkcje społeczne przekazów popularnonaukowych (Scientific Propaganda - social functions of popular science). Wydawnictwo IfiS PAN, Warszawa
Yergin D (2011) The quest: energy, security, and the remaking of the modern world. The Penguin Group, New York

Author Biographies

Prof. Dr. hab. Joost Platje is profesor at the Wrocław School of Banking (Poland). He obtained a Ph.D. from the University of Groningen (The Netherlands) and his habilitation thesis (University of Economics, Wroclaw, Poland) dealt with institutional capital as a factor determining capacity and capabilities for sustainable development. His research focuses on institutional foundations of science and sustainable development.

Dr. rer.pol.habil. Eckhard Burkatzki (1965) is assistant professor at the Chair of Social Sciences at TU Dresden—International Institute Zittau (Germany). His fields of interest in research and teaching are methods and methodology of empirical social research, determinants of ethical decision making and sociology of economic crime. Recent publications dwell on perceived utilities and costs of plagiarism in high- and low-trust student samples.

Achieving Sustainability in Applied Marine Science Education: Using European Experiences in Russia

Michael B. Shilin and Tatjana R. Eremina

Abstract

The significance of reorganization of the educational processes within Applied Marine Sciences (AMS) is related to the increasing relevance of environmental issues in coastal EU countries and Russia, and to changes within the labour market. Negative anthropogenic changes in marine and coastal ecosystems, with the background of climate change, could result in serious consequences for future generations. Only competently trained professionals who understand the relationship between theoretical and practical multidisciplinary problems of AMS can find the sustainable solution to these problems. Training of such specialists should be based on the latest international achievements. A newly developed curriculum should be oriented upon the changed labour market within the EU and Russia.

Keywords

Education · Applied marine science · Curriculum · Labour market · Dissemination

M.B. Shilin (✉) · T.R. Eremina
Department of Oceanography, Russian State Hydrometeorological University, 195196 Malookhtinsky Pr. 98, St. Petersburg, Russia
e-mail: shilin@rshu.ru

T.R. Eremina
e-mail: tanya.er@gmail.com

W. Leal Filho et al. (eds.), *Integrative Approaches to Sustainable Development at University Level*, World Sustainability Series, DOI 10.1007/978-3-319-10690-8_28

1 Introduction: Project "Education in Applied Marine Sciences"

The new socio-economic conditions prevailing in Russia in the early twenty first century aided its integration into the global economy whilst the appearance of the new labour market resulted in prerequisites for a higher education reform. The reform is aimed at overcoming the following problems:

- lack of competency-based close cooperation between experts, researchers and teachers;
- training on obsolete (non-renewable) programs;
- weak differentiation of three levels within the curricula;
- lack of well-formed and unified transparent standards for assessment of educational process at different universities;
- little use of modern information and communication technologies;
- insufficient focus on the development of the system related to environmental safety of the marine and coastal environment.

During 2011–2014 the TEMPUS *eMaris* project "Education in Applied Marine Sciences" had been implemented by six partner universities (PU) from the EU and Russia: Russian State Hydrometeorological University (RSHU, St. Petersburg, Russia), Immanuil Kant Baltic Federal University (IK BFU, Kaliningrad, Russia), State Marine Technical University (SMTU, St. Petersburg, Russia), University of Klaipeda (KU, Lithuania), University of Cadiz (UCA, Spain), University of Bremen (UNI HB, Germany). The other partners were Academic Association of Universities for Hydrometeorology (AAU, St. Petersburg, Russia), and St. Petersburg Branch of P.P. Shirshov Institute of Oceanology of Russian Academy of Sciences RAS (SPBIO, St. Petersburg, Russia) (Eremina et al. 2014). The main goal was to achieve sustainability in AMS education within Russia under new socio-economic conditions using the EU experience. The overall wider objective was to continue the reform of the Russian higher education system in the area of AMS to comply with the Bologna Declaration and the demands of Strategic Framework for European Cooperation in Education and Training (ET 2020). The significance of the reorganization of the educational process in the field of AMS is related to the increasing relevance of environmental issues. Global environmental change and economic development of society, including marine and coastal sectors, are closely related to each other (Shilin and Khaimina 2014). Negative changes in marine and coastal ecosystems, caused by human activity and climate change, can lead to serious consequences for future generations. In the future, only competently trained professionals can find a correct and adequate solution to these problems (Karlin et al. 2012; Eremina et al. 2014). Training of such specialists and development of new curricula should be based upon the latest international achievements (Ryzhov 2013; Wishnevsky et al. 2013; Eremina et al. 2014; Gordeeva et al. 2014).

The main objectives of the *eMaris* project were:

- to develop internationally recognized, effective working and sustainable competency-based three-level curricula (Bachelor/Master/Doctor) and syllabi in AMS according to the best Bologna practices;
- to create and implement a collaboration platform for development of sustainable academic environment and to enhance networking among higher education and research institutions across the PU;
- to enhance the syllabi for AMS with innovative learning methods (mobile lectures, e- & m-learning modules, "training through the research" approach, etc.);
- to implement the developed competency-based three-level curricula in AMS at Russian universities (RSHU, IK BFU, SMTU);
- to use the results and outputs of the project in all European PU.

2 Research of Labour Market for Experts in Applied Marine Sciences: Case of Russia

Since 1990 there has been a 2.3 fold decrease in the number of employees within the scientific sector managed by the Russian Academy of Sciences (RAS). The average age of a researcher is now about 60 years. Annual inflow of young scientists to RAS institutes accounts for less than 4 % of the total number of research associates. In the last 6–7 years certain measures have been implemented to improve the age structure of scientific personnel, to attract and fix talented youth at the academic institutes and to ensure scientific and administrative growth of young scientists. Analyzing the dynamics of age structure within Academy staff over the last 5 years indicates a growth within the proportion of younger researchers (30 years old and younger, and 30–40 years old). Although the average age of "doctors of science" is nearly 65 years, the number of doctors below 40 years old steadily grows. Average age of "candidates of science" is over 50 years, but the number of candidates within this age category below 40 years old is growing. In total, annually about 900 people who finished their postgraduate study (about 2 % out of number of RAS research associates) stay for paid employment at RAS institutes.

The macroeconomic parameters corroborate a satisfactory present state of the labour market in Russia. Employment level is remaining high (over 65 %), whilst unemployment level is low (below 7 %). At the same time, dynamics within employment structure reflects the absence of movement towards modernization. Modernizing and restructuring of the science-based economy must generate new job opportunities of high quality. Such new employment positions should be more technologically advanced, requiring a higher level of education and qualification.

According to the official statistics, young people under the age of 30 make about 35 % of the able-bodied population of Russia. Over 50 % of young people employed in science-based companies have a second job in various alternative forms of employment. Young people are actively participating in entrepreneurship. About 70–80 % of registered companies of the alternative economy sector are organized by people under 30 years old. At the same time, the absence of mechanisms regulating the job placement of university graduates leads to serious problems. Qualified young professionals often change their specialization which in future may result in a misbalance within the professional structure of the labour force. Priority is not given to intellectual work, but rather low intensity work, aimed at gaining significant and immediate financial profits. If the existing trends within reproduction of qualified staff does not change, a growth of unemployment can be expected. Mostly it will be amongst young university graduates who do not progress into further education.

Educational institutions are still weakly oriented upon market requirements and they do not study placement of their graduates within the market. Most of the university graduates (90 %) are aimed at working activity, however, only 60 % of graduates plan to work in their specialization field. The remaining 30 % want to work in a different field.

During 2000–2011, place of work was changed by 19.4 % of scientists, including 15.9 % of researchers, 18.5 % of university professors and 34 % of industrial enterprise staff. "Candidates of science" have been changing their place of work more often than "doctors of science". The main reasons are: low salary (31.5 %) and lack of prospects of career growth (23.5 %). Nearly 40 % of experts with a scientific degree have a second place of employment. The change of area within scientific activity is typical for "candidates of science" of 50–59 years and 30–39 years of age.

Thus, based on the obtained results, the following complex measures directed towards the improvement of age structure of scientific personnel can be advised:

- motivation and attraction of youth;
- initiative of formation of Master courses;
- competition for medals of the Academy with awards for young scientists;
- growth of salary rate for youth;
- development of a departmental system which includes additional provisions for pensions.

Currently, total number of researchers working in the field of marine sciences in RAS institutions is composed of about 1,000 people. If the RAS funding within the next few years will not be cut down significantly, and the economy of Russia will develop according to the basic (inertial) scenario, the Academy's need for young AMS scientists will make at least 20 people annually. This represents 2 % out of the total number of RAS scientists.

3 Research of Labour Markets in the Baltic Countries Region

The EU Baltic Sea Region (BSR) is made up of nine economically developed countries. It has almost 85 million inhabitants representing 17 % of the EU population. The BSR has actively embraced many opportunities reinforcing EU integration and increasing regional coherence.

The BSR has a relatively large offer on higher education possibilities. Almost every Baltic university offers some sort of marine-related course. Such courses include oceanography, marine ecology, marine physics, shipping, navigation, fishery, as well as maritime policy and spatial planning. Besides the traditional maritime industry sectors such as shipbuilding, navigation and fishery, the so-called New Maritime Sector is emerging globally as a powerful future labour market. It includes a variety of disciplines such as marine energetics, offshore construction and exploration, aquaculture, bio-energy, maritime safety and law, production of hydrocarbons, desalinisation, blue biotechnology, marine spatial planning, modern harbour management, and environmental engineering and protection.

The labour market of BSR strongly depends on economic situation. Economy growth increases the number of jobs and the need for specialists. In times of economic recession the demand for specialists decreases. Therefore EU education and training policies have gained impetus since the adoption of the Lisbon Strategy in 2000, with the EU's overarching programme focusing on growth and jobs. As each BSR state is responsible for its own education and training systems, policies are designed to support national actions. In the future, universities of the BSR can benefit from sharing experiences.

Training the next generation of marine scientists presents new challenges in the context of important policy changes. The EU Innovation Strategy recognizes that education is essential for the creation of an innovation-oriented society. The unified professional qualification system, expressed in credits (ECTS), can help research and indicate the level of labour markets in the BSR. ECTS is a standard for comparing the study attainment and performance of students of higher education across the EU and other European countries.

In general, the BSR maritime educational offer is large, but does not always reflect the job market conditions. It would be beneficial if all Baltic higher education institutions would cooperate with each other and with labour market agencies. This would facilitate their graduates to pursue careers in the Maritime Sector.

Well educated specialists in AMS are needed within the labour market. However, there is a lack of common methodology for the evaluation of specialist knowledge. There is no unified system that will help to determine how many graduates are employed in the Maritime Sector. The development and implementation of a common assessment methodology is therefore needed.

Due to the increasing migration and mobility of specialists within the BSR, it is required to develop the pan-European equivalent of student competencies. The unified system would equalize the competitiveness opportunities and enhance the mobility of professionals in AMS.

4 Survey of Employers and Professional Associations in Russia

In order to carry out an analysis of the labour market in AMS in Russia, a questionnaire was prepared for employers and for higher education institutions that train relevant staff. Questionnaires were sent to 106 state and private companies in order to find the need for experts in AMS, application of modern technologies, and to discover the directions of improvement within expert training. More than 50 answers were received from organizations of St. Petersburg, Kaliningrad, Murmansk, Rostov-on-Don and Krasnodar region. Number of staff within these organizations varied from 6 to 525 people (average 122 people).

According to the survey, only in 10 % of studied organizations the modernization and introduction of new technologies was not planned. In over 50 % of organizations, new technologies are planned to be introduced and new experts are going to be invited. Moreover, the further development of the organization and recruitment of new staff is expected in 10 % of organizations.

Answers to the question "How would you assess the salary level of experts in AMS" reflected the fact that salary is one of the key aspects of choosing the work place by an employee. Only in 15 % of organizations the salary level was higher than the average in the region. Usually, the salary of experts in AMS is predominantly lower than the average in the region.

When answering the question about the sufficiency of the university training to graduates in the field of AMS, employers and heads of departments noted the level of training of young experts as "moderate," but with a low degree of practical experience. No respondent stated the maximum degree of qualification of young experts. Thus, the general preparedness of experts in the field of AMS straight after graduating from a university can be considered as "partially sufficient."

All respondents noted the high level of theoretical training of graduates. Ten percent of respondents noted the low level of practical preparedness. The survey showed a good degree of provision of young experts by necessary information and literature sources in universities.

It is notable that the qualitative level of the teaching staff (TS) is estimated as being rather high by 50 % of respondents.

According to the level of outside study and training within EU universities, the majority of respondents indicated the "middling" level of this parameter. This might be the result of the absence of traditions.

The assessment of graduates' experience of independent research activity upon graduating from universities is estimated as "moderate." According to the requirements of the Russian State Educational Standard (SES), the number of "hours" /ECTS allocated in syllabi to research projects is quite low, and this skill can be undeveloped in principle.

A number of questions were directed towards the detection of the need to improve graduate training. Such need was noted by about 50 % of respondents. The laboratory base is desirable to be improved according to 40 % of respondents. All respondents indicated that the provision of modern equipment is necessary in all universities.

Improvement of information and software support of education is more "obligatory" than "desirable." Only 5 % of respondents indicated the improvement of information and software support of educational process to be unessential. Rising of TS skill level is on the opposite "desirable" but not "obligatory". Only 15 % of respondents declared the need of TS professional development. This means that experts working in the sphere of university education have high TS level.

When asked about the improvement of the organization of practice during training, more than 50 % of respondents indicated that improvements should be "obligatory". The majority (75 %) of respondents pointed to a need of training directly at specialized companies. Moreover, over 60 % of respondents indicated a need of students' involvement in research.

Personal qualities, competences and abilities of experts in AMS were studied as well. It was shown that a true professional should not only have a diploma on graduation from a specialized Higher Education Institution. He/she also has to possess a number of personal competences formed, on the one hand, during the training, and on another—being an integral part of personal development (Table 1).

For the assessment of the degree of importance of practical skills, a scale ranging from 1 to 5 was chosen (Table 2). A personal quality, with which the maximum number of interrelations of skills and other qualities is observed, is the ability to "solve non-standard problems." The ability to "think critically" is in second place by number of significant links, and "research skills"—in the third.

It was necessary to find out if modernization and application of new technologies in AMS is foreseen in institutions, so the students would be qualified for that. All the employers indicated that new technologies would be installed and new specialists would be required, except for one respondent with low financial support. In the future the demand for specialists in AMS is likely to increase, although an accurate prediction is not possible at the moment. At the present time, there is a demand for students with Bachelor, Master and Ph.D. qualifications, but Master and Doctoral levels are more appreciated. Also, the importance of experts with high qualifications in AMS was pointed out.

Summarizing the responses, in general, employers are quite satisfied with the preparation of students in universities. Employers were asked to evaluate in scores (where 1 is the lowest, 5 is the highest) the quality of training of graduates in AMS, based on their personal experience. The results obtained were quite homogeneous. All of the answers for theoretical training were 3 or 4, which demonstrated high

Table 1 Assessment of importance of personal qualities, competences and abilities of experts in applied marine sciences

Indicator	Score
Initiative and entrepreneurial spirit	3.60
Ability to work independently	4.05
Ability to work in a group	4.57
Orientation on career growth and professional development	4.16
Communication with non-experts	4.05
Ability to effectively represent oneself and results of one's work	4.51
Ability to solve non-standard problems	4.78
Ability to adapt to changing conditions	4.00
Readiness and capability of further training	4.72
General intelligence and culture	4.10
Ability to work in an international context	4.16
Ability of understanding other cultures	3.42
Ability to extract and analyze information from various sources	4.73
Ability to think critically	4.47

level of students' theoretical knowledge. However, the results related to practical training were lower. The employers are in agreement about the need of additional practical training of graduates. It was recommended that universities should pay attention to practical programs in all links of the three-level curricula.

The employers have proposed to improve the following aspects of the education of young experts in AMS, in the following ways:

- knowledge of the English language;
- knowledge of computer technologies, methods of data and information processing;
- ability to analyze and interpret maps, charts and graphs, and to conduct monitoring of natural conditions in real time usage of radar and satellite data;
- awareness of the latest scientific and technological achievements in the area of AMS;
- knowledge and ability to apply principles, methods and forecast schemes, understanding of the principle of numerical model action.

As a conclusion it should be marked that all of the employers were in agreement about the high level of the graduates' theoretical knowledge. At the same time all employers highly recommended the improvements of students' practical training through their involvement in research. Experts with Master and Ph.D. degrees are

Table 2 Assessment of importance of practical skills in various fields of activity for experts in applied marine sciences

Indicator	Score
Ability to interpret and carry out a survey	4.79
Skills to present information to an audience	4.26
Experience of research work	4.52
Awareness of the latest scientific and technical achievements in AMS	4.42
Ability to estimate risks connected with consumer use of provided information and products	3.63
Ability to apply computing skills and knowledge of methods of data and information processing	4.73
Abilities to analyze and interpret maps, charts and plots, to integrate all available data to make a summary diagnosis, to carry out monitoring of environmental conditions in real time with the use of radar and satellite observations	4.95
Knowledge and abilities to apply principles, methods and forecasting schemes, understanding of numerical models' principle of action	4.31
Understanding of assessment, control and quality management principles	3.63
Skills of information collection from primary and secondary sources, including Internet search	4.63
Knowledge of computer technologies	4.79
Knowledge of English language	4.47
Ability of development and management of projects	3.68
Leadership skills	3.31
Knowledge of the main marketing methods and procedures	2.95
Abilities to apply innovative management principles	2.78
Skills of human resource management	3.26

mostly needed within the labour market, so it is necessary to develop high quality programs on Master and Ph.D. levels.

The results of the questionnaire were homogeneous among the employers from different institutions.

5 Ways to Make Education in AMS Sustainable

As a result of the collaborative *eMaris* project, a new paradigm for AMS education in Russia was created. It dwells on integration, harmonization and aggregation of various types of quality-controlled e-learning components. The related contextual content, such as mobile lectures and test-centers, was developed. The EU universities (UCA, KU) have successfully transferred their experience in the development of advanced training programs and promoted innovative learning technologies (UNI HB). Russian universities have applied the experience of EU partners and developed and implemented a new three-cycle educational model (RSHU, IK BFU, SMTU). It adapted achievements of scientific research institutions for the development of doctoral programs (SPBIO) and performing a wide implementation of project results at the national level (AAU).

The educational programs of the EU and Russian PU were compared. Master curricula of the PU differ significantly from each other. These differences are "natural," because teaching programs are focused on practical needs of the marine industry in such geographically diverse countries as Russia, Spain, Lithuania and Germany. A comparative analysis of a set of competences confirmed the possibility of development of competence-based integrated Master programs.

The comparative analysis of the differences in Doctoral training at PU served as a basis for formulation of recommendations for training programs compilation. These recommendations included specialized courses in marine sciences, the study of modern research methods, the role of modern communications, etc. The curricula and syllabi developed during the project are directly related to Russian national priorities and, in particular, to marine and coastal environmental problems (Shilin and Khaimina 2014).

The three-level enhanced curricula and syllabi were developed by the retrained staff of PU for Bachelor (240 ECTS, 4 years), Master (120 ECTS, 2 years) and Doctoral level (180 ECTS, 3 years) in AMS. It took into account the results of the surveys of labour market and competencies. The following modules were created: (1) Applied marine ecology; (2) Fisheries management and assessment; (3) Numerical modeling of marine environment and hydro-ecosystems; (4) Operational forecasting modeling of natural and anthropogenic stress-factors. New textbooks on these topics for M.Sc.-level were published using the best international practices. The subjects of new syllabi for all three levels were provided by innovative technologies and e-modules based on the collaborative platform SAKAI (Wishnevsky et al. 2013; Gordeeva et al. 2014).

Also, training programs and methodological guidelines for the third (Doctoral) cycle were developed. The structure of individual plans for post-graduate training was proposed including targets, training time and acquisition of professional skills.

The joint experience of PU was used in order to:

- shift from individual to structured training programs;
- move to multilateral (including international) educational guidance (supervision);
- obtain "practices" at different universities;
- support mobility at the doctoral level;
- organize international summer schools and scientific conferences, etc.

As a result, a collaborative Network was developed (*eMaris* Network). The collaboration platform aims at different target groups (students, professors, administrative staff, project members etc.).

Russian universities will use the hybrid learning model. In such model, the traditional full-time (face-to-face) education is supported by the Web-oriented educational technologies (e-& m-learning). Students get access to the collaborative platform which incorporates constantly updated learning content, system of online tests, and access to the various online libraries and other services. The hybrid form of training allows them to make the educational process more flexible. For example, it is possible to modify the educational processes by selecting the desired learning module /study area. This model will give more chances for women including students, teachers and research staff. It is based on an omnipresent approach, continuous access via the Internet, mobility and better transparency for the job competence in career planning.

One of the results of the project is the expansion of the application of developed programs and possibility to use them in different practical cases (Eremina et al. 2014; Shilin and Khaimina 2014).

The implementation of distant information exchange will expand the number of students both with a full degree assignment and with separate modules of educational programs of certain levels.

As such, the main outputs of the project, and at the same time the tools for making the educational process sustainable, are:

- new three-level curricula and syllabi;
- retrained teaching staff;
- new textbooks which summarized the best World experience;
- e&m-learning modules;
- mobile lectures;
- structured doctoral program approach;
- methodological guidelines;
- *eMaris* Network and collaboration platform for the AMS community.

To ensure that the project outcomes are being used after its finalization, it is important to organize the dissemination of the knowledge on three levels. They include: the institutional level within each of the PU, the inter-institutional level to facilitate continuous communication between PU, and the external level.

The first level of communication was achieved by arranging information seminars and an electronic newsletter system in PU. Project materials were distributed and peer-reviewed within the AAU dissemination system. This involved practically all institutions in Russia where marine sciences are taught. After assessment and positive feedback, the materials were passed onto the Intergovernmental Oceanographic Commission of UNESCO and to the World Meteorological Organization to ensure their international use. Wider dissemination and feedback from interested target groups will help to achieve the highest quality of developed syllabi and new teaching materials. This was ensured through approval first by AAU and then by the Ministry of Education and Science of the Russian Federation.

When developed and in place, the online collaboration platform will be kept operative by academic and research communities and will not require much further investments. The maintenance costs will be covered from research and education projects. Better interaction between students and teachers from different countries will greatly improve the quality of M.Sc. and Ph.D. education and research.

6 Conclusion

The framework of the *eMaris* project was the first attempt to analyze, evaluate and compare the educational programs in the field of AMS in different EU and Russian universities. As a synthesis of this work, the effective working and sustainable competency-based three-level curricula (Bachelor/Master/Doctor) and syllabi in AMS were developed according to the best Bologna practices.

The developed curricula and syllabi will be used in Russian universities for a long time. The textbooks, learning technologies, new collaboration platform, etc. are expected to be a significant component of educational processes in all the PU.

The final stage of the project involves the student exchange between PU to enable the students' learning on the basis of the newly developed curricula. This activity will contribute to the enforcement of inter-university links making the project sustainable. The collaborative work between PU will also continue after the project has ended.

References

Eremina T, Khaimina O, Shilin M (2014) International educational collaboration of Russian Universities in applied marine sciences. In: Proceeding of XX international scientific and methodical conference "Modern education: content, technologies, quality," St. Petersburg, State Elektrotechnical University Publishing House, vol 1, pp 89–90

Gordeeva S, Novopashenny I, Ogurol Y, Ryzhov V, Zhao J (2014) Sakai CLE for blended learning model. In: Proceeding of international academic conference on education, teaching and e-learning, prague, Feb–March, pp 157–158

Karlin L, Shilin M, Eremina T, Ershova A, Suzyumov A (2012) Studying sustainability through research with the floating university program. Sustainable development at universities: new horizons. Environmental education, communication and sustainability, vol 34. Peter Lang Internationaler Verlag der Wissenschaften, pp 723–731

Ryzhov VA (2013) Concept and perspectives of use of online testing and distance learning system. In: Proceeding of Sino-Russian workshop for naval architecture and ocean engineering, Zhoushan, China. Zhejiang University Publishing House, pp 12–13, 27–29 Oct 2013

Shilin M, Khaimina O (2014) Applied marine science. Textbook. St. Petersburg, Russian State Hydrometeorological University Publishing House, p 85

Wishnevsky M, Novopashenny I, Zhao J, Ogurol Y, del Valls Casillas A (2013) Experience with Sakai CLE in the framework of international educational TEMPUS project *eMaris* in the field of applied marine sciences. In: CALMet X conference and EUmetical workshop, Touluse, France, pp 64–65

Author Biographies

Professor Michael Shilin is working at Russian State Hydrometeorological University (RSHU, St. Petersburg) since 1989. He is a co-author of the Concept of Environmental Education at technical universities in Russia, and participant of a number of international projects. In 1996–2006 he participated in the UNESCO "Baltic Floating University" project, developing the

"Training through the Research" principle. He is the author of a number of teaching courses and original textbooks: "Natural and Technical Systems in the Coastal Zone" (2008), "Environmental safety of the Technosphere" (2010), "Applied Marine Ecology" (2011). He has over 140 articles in Russian and international journals.

Dr. Tatjana Eremina is Assistant Professor at RSHU, St. Petersburg. She graduated this University in 1979 and received a Ph.D. in Physics in 1997. Since 1998 she was the Team Leader of the UNESCO "Baltic Floating University" project. For 7 years she was a leading teacher of the Sustainable Water Management course of International Baltic University Programme at the Uppsala University (Sweden). Recently, Dr. Eremina has been involved in a number of scientific projects devoted to environmental study of the Baltic Sea. Since 2011, she has been the Leader of an on-going educational project "*eMaris*" of TEMPUS program. She has 85 articles in Russian and international scientific journals.

Analysis of Sustainability Content into Undergraduate Engineering Curriculum at the Engineering School of Sao Carlos, University of São Paulo, Brazil

Bárbara Pacheco Lopes, Laís Sanchez Assumpção, Yovana M.B. Saavedra, Fabio Neves Puglieri, Affonso Azanha, Patrícia Silva Leme and Aldo Roberto Ometto

Abstract

Nowadays it is important that universities contribute to improving society through sustainability, preparing future professionals with the knowledge, capabilities, values, attitudes and abilities to act on Sustainable Development

B.P. Lopes (✉) · L.S. Assumpção
Biology Sciences, Federal University of São Carlos, Rodovia Washington Luís, km 235, Jardim Guanabara, Saint Charles, SP, SP 310, Brazil
e-mail: ba.lopes08@gmail.com

L.S. Assumpção
e-mail: assumpcao_lais92@hotmail.com

Y.M.B. Saavedra
Environmental Engineering Sciences, Engineering School of São Carlos, University of São Paulo, Trabalhador São-Carlense Blvd, Arnold Schimidt 400, Saint Charles, SP, Brazil
e-mail: ybarrera11@yahoo.es

F.N. Puglieri · A.R. Ometto
Department of Production Engineering, Engineering School of São Carlos, University of São Paulo, Trabalhador São-Carlense Blvd, Arnold Schimidt 400, Saint Charles, SP, Brazil
e-mail: puglieri@usp.br

A.R. Ometto
e-mail: aometto@sc.usp.br

A. Azanha
Mechanical Production Engineering, Engineering School of São Carlos, University of São Paulo, Trabalhador São-Carlense Blvd, Arnold Schimidt 400, Saint Charles, SP, Brazil
e-mail: affonsoazanha@gmail.com

P.S. Leme
USP Recicla, Engineering School of São Carlos, University of São Paulo, Trabalhador São-Carlense Blvd, Arnold Schimidt 400, Saint Charles, SP, Brazil
e-mail: pazu@sc.usp.br

W. Leal Filho et al. (eds.), *Integrative Approaches to Sustainable Development at University Level*, World Sustainability Series, DOI 10.1007/978-3-319-10690-8_29

(SD). In 2013, the Engineering School of São Carlos (EESC), University of São Paulo (USP), Brazil, and the Environmental Management Superintendent (EMS), initiated an education for sustainable development project, with the aim of analysing the content of nine undergraduate engineering curricula with a view proposing more environmental-oriented disciplines. In this context, the educational policy project was reviewed and a curriculum analysis done, using an analytical method consisting of the identification of sustainability radicals, such as soc*, sustent*, ecolog* and natur* was performed. The results show that Environmental Engineering courses present the largest sustainability integration with 57 % of environmental-oriented disciplines, followed by Production Engineering courses with 13 %. On average, not considering Environmental Engineering course, only 5.7 % of the EESC engineering courses include some sustainable content, which indicates the necessity for change to the curricula.

Keywords
Higher education · Engineering · Curriculum change · Education for sustainable development

1 Introduction

The discussion around the need for undergraduate students to be educated under ethical and responsible principles regarding issues in sustainable development is increasing, and university curricula play an important role in achieving this goal.

In order to foment educational acts to encourage behavioural changes in the present and future generations, United Nations Educational, Scientific and Cultural Organization (UNESCO) promoted the Decade of Education for Sustainable Development for the years 2005–2014. The UNESCO goals include a quality standard for sustainable development in education, considering environmental integrity, economic feasibility, and social justice as guidelines to be reached (UNESCO 2002).

Higher Education Institutions (HEIs) have great relevance in the current scenario of socio-environmental crisis upon the necessity of new relations construction between society and nature. According to Evangelinos et al. (2009), considering these new challenges, HEIs contribute to the goal of sustainability, providing intellectual leadership and showing opportunities to create a sustainable society. Wright and Wilton (2012) say that sustainability leadership at HEIs must include the understanding and participation of all institutional actors: employees, academics and students.

In addition to modifications to the policies as well as managerial and organizational practices at the HEIs, changes are also necessary in students' curricula. It can also help to educate professionals, by providing them with resources to assist societies demands that have developed under unsustainable models, in a holistic and preventive way regarding the environmental issues and decision-making.

In an international context, several HEIs include sustainable content in their curricula. Some universities, such as Michigan University (United States), Gävle University (Sweden), Cambridge University (United Kingdom), Delft Technical University (Netherlands) and New South Wales University (Australia) have developed and applied their knowledge to include sustainability, presenting advances and the importance of assessing its feasibility, and also the need of activities to include sustainability in the curricula and institutions (Uliana 2007).

In a study conducted by Chinese researchers, a sample of HEIs in China was surveyed in order to analyse the standard of "green curriculum". They used a list of 14 keywords to identify the institutions courses, and their findings suggest that "green education should be integrated into any academic discipline" (Xiong et al. 2013).

At *Tecnologico de Monterrey*, a Mexican university, a new bachelor's degree in engineering was developed, which focused on the demands and engineering skills for sustainable development. In this case, a tool called STAUNCH® (Sustainability Tool for Assessing Universities' Curricula Holistically) was used to systematically assess the curriculum elaboration (Lozano and Lozano 2013).

At the Georgia Institute of Technology, in the United States, the curriculum of Civil and Environmental Engineering was evaluated using not only STAUNCH tool, but also their students' perception. In this case, using both approaches provided a more holistic overview of the contribution to sustainable development, as well as detecting gaps between sustainability content in curricula and sustainability pedagogy. The results showed that the curriculum mostly covered the environmental dimension to sustainability and the crosscutting themes.

In the Brazilian context, there is the National Policy for Environmental Education (Política Nacional de EducaçãoAmbiental—PNEA), which states that education must be characterized by a process in which individual and society build social values, knowledge, attitudes and competences, related to environment conservation and common good, essential to quality of life and sustainability. Furthermore, PNEA considers the environment as an essential and permanent component of national education, which should be presented in all levels and modalities of education processes (PNEA 1999).

In this case, the environmental-oriented curriculum is one of the alternatives to move towards education for sustainability. In Brazil, there are some other examples from Universidade Federal de São Carlos (UFSCar), Universidade Estadual de Campinas (Unicamp), Universidade Estadual Paulista (UNESP—Rio Claro), among other South-American, for instance, Rede de Ambientalizacíon Curricular de los Estudios Superiores (ACES). European universities have also been developing curriculum models which integrate sustainability. The research into greening the curriculum exposes the challenges, the opportunities and the different stakeholders in the process (ACES 2003).

The concept "environmental-oriented curriculum" or just "greening curriculum" means to set up, in the education system, complex changes series, which can include innovations, technical knowledge and behaviour. ACES network aims to promote experiences and define models to interpret social and environmental issues that may contribute to professionals' education, which will be able to promote benefits for society and the environment (Freitas and Souza 2011).

Knowledge of the environment and its conditions and structures, locally and globally, is not enough. The processes of greening curriculum cannot be attempted without paying attention to human history, politics, and culture. In this sense, it is necessary for structural and organisational changes to enable an interdisciplinary focus within the undergraduate curricula, facilitating a global planning composed of goals and contents to understand the complexity of environmental issues. Moreover, it should be facilitated by greater flexibility of the curriculum, and decentralisation, while bearing in mind the problems that may occur during the process (Muñoz 1996).

At the Engineering School of Sao Carlos (EESC), research shows low presence of sustainable content in engineering curricula and difficulties for this to be surpassed, indicating the need for changes to the curricula (Jabbour and Santos 2006; Uliana 2007; Geromini 2010; Pavesi 2011; Ometto et al. 2012).

Therefore, the present research shows its relevance because allowed EESC's curricula mapping, regarding sustainability content inclusion. Additionally other contributions of this study are in the adequacy and application of ACES methodology for Brazilian case. Finally, the authors expect the results from this study to bring important information to be used and disseminated in engineering courses from other universities, as a model to be adopted.

In this sense, this paper aims to analyse the environmental content level in all disciplines from nine engineering courses at EESC. The results will be a starting point to develop strategies towards further education for sustainability in engineering courses.

2 Methodology

The methodology adopted to perform this study was initially based on an exploratory bibliographic review to outline the theories in the field, which then supported the data collection, data analysis and discussion. This allowed the authors to identify several initiatives and methodologies adopted in Brazil and in other countries regarding the inclusion of sustainability in the engineering curriculum. After this first step, an extensive document analysis was performed. The document analysis included the Pedagogics and Policies Projects (PPPs), the objectives, content, evaluation methods and bibliography of mandatory, optional and elective disciplines from nine engineering courses offered at Engineering School of Sao Carlos (EESC): Aeronautical Engineering; Civil Engineering; Computer Engineering; Manufacturing and Materials Engineering; Production Engineering; Electrical Engineering; Mechanical Engineering; Mechatronic Engineering; and Environmental Engineering.

Adapted from ACES methodology, five radicals, in Portuguese, were used in order to assist the characterization of the level of sustainability present in engineering curriculums at EESC. The radicals were "ambient" (for environment), "sustent" (sustainability), "natur" (nature), "ecolog" (ecology) and "soc" (society). Each time these radicals were found in the engineering curriculums documents, a thorough reading of the document was performed (Junyent and Geli 2003). From this first analysis, it was possible to categorise the disciplines in three categories: (1) *Environmental-oriented disciplines*, i.e., those disciplines presenting at least one sustainability radical, or those disciplines showing clear evidences of sustainable content. (2) *Potential disciplines to be environmental-oriented*, category that comprises disciplines where the indicators are present, but there are no guarantees related to the inclusion of sustainable content or these elements are not included. Also, that the discipline could add socio-environmental topics. (3) *Non-environmental-oriented disciplines*, characterised by a lack of indicators of education for sustainability.

The following step consisted of a second classification to all sustainable disciplines in five different categories: (1) *Correction*, which includes those disciplines that include solutions for environmental problems and specific knowledge about corrective actions as definitive solution. (2) *Remediation*, which includes those disciplines that encourage their students to solve socio-environmental problems with remediation actions. (3) *Prevention*, which includes those disciplines that consider a set of specific knowledge to prevent environmental problems generation. (4) *Society-environment relation*, which includes those disciplines comprising content involving society-environment relation. (5) *Ecological relation*, which includes those disciplines that bring a knowledge of living beings and their relation to the surrounding environment.

The choice for a quantitative–qualitative analysis method allowed the authors to explore the official guidelines and documents related to the graduate courses in a complete and reflexive way, resulting in a preliminary diagnosis for base strategies and to go on towards the creation of a more sustainable curriculum in engineering at EESC. The next topic presents a diagnosis of the inclusion of education for sustainability within undergraduate curriculum at EESC.

3 Results and Discussion: Engineering Diagnosis at EESC

3.1 Engineering Curriculum at EESC

The process to include environmental sustainability in the undergraduate engineering disciplines at EESC—USP is part of "Sustainable EESC" Program, aiming for an institutional policy development that understands environmental education and sustainable practices at EESC both within administration and management and its teaching and extension activities, in a broad and integrated way (EESC SUSTENTAVEL 2014).

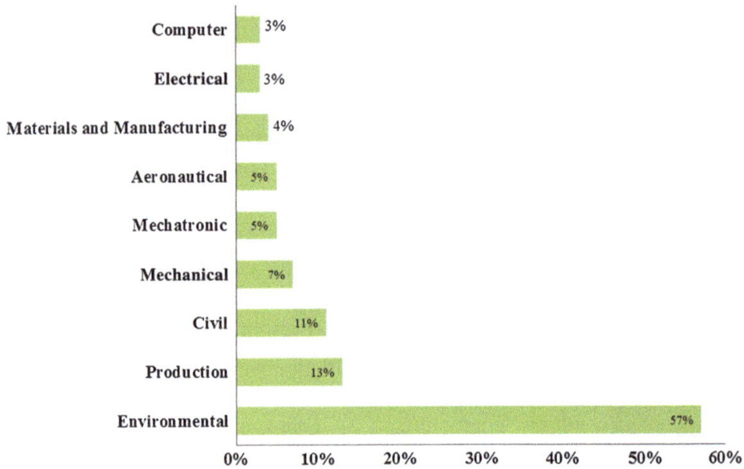

Fig. 1 Environmental sustainability inclusion average in disciplines from nine engineering courses at EESC

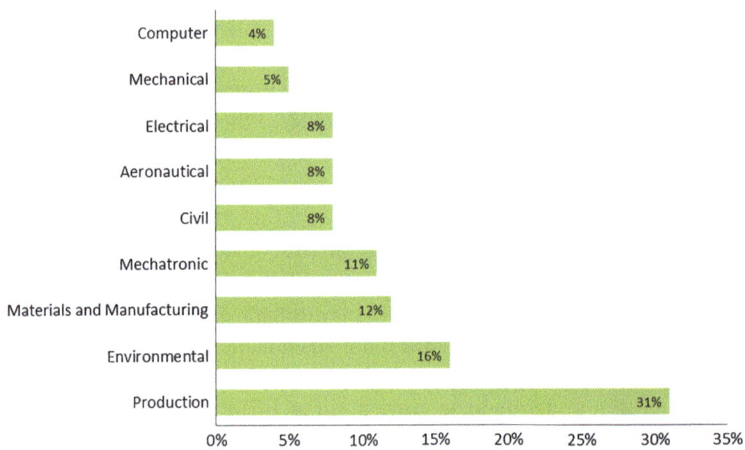

Fig. 2 Engineering courses with disciplines presenting higher potential to include environmental sustainability

The diagnosis presented in this paper points out that although environmental content is present in the nine engineering curricula reviewed, the feasibility and the insertion levels vary. It is possible to notice from Fig. 1 the different levels in which sustainability is included in engineering courses at EESC. At same time, Fig. 2 shows those engineering courses with higher potential to include environmental sustainability.

Although the results presenting some variation among the engineering courses were expected, it is possible to confirm that this is related to each career. There is a perception that those engineering courses with lower levels of environmental sustainability content present unclear opportunities to embed sustainability, while the possibility for the opposite to occur is also plausible.

As expected, Environmental Engineering is the course with the highest level of environmental content inclusion, showing an average of 57 % of its disciplines with some topics related to environmental issues and sustainability. These disciplines are presented in nine out of ten semesters. The lowest percentages of disciplines with environmental orientation was about 37 and 44 %, and the highest levels got percentages between 60 and 78 %, considering mandatory and optional disciplines at same time.

Analysing only the optional disciplines, an increase of environmental content level is noticed, and the same trend is verified in the other engineering courses. The Environmental Engineering course had 16 % of its curriculum rated as "potential" for greening, and this ratio also increases when analysing optional disciplines in Environmental Engineering and other engineering courses at EESC.

The Production Engineering course presented 13 % of its disciplines with environmental orientation, and 31 % of its disciplines as "potential" to include environment content. For Civil Engineering, the rates were 11 and 8 % respectively. It is argued that in Production Engineering, the diagnosis considered changes to the curricula made in this course recently, which are not present in the other engineering courses reviewed. In the case of the Production Engineering curriculum restructure, proposals to include environmental content were partially implemented.

The efforts to embed environmental issues in Production Engineering began with Jabbour and Santos (2006) who conducted a case study to analyse the opportunities to include environmental sustainability in this course. They also indicated that the concepts related to environmental education were not being considered within the Production Engineering curriculum at that time.

More recently, according to Ometto et al. (2012), the experiences with the pilot study in the Production Engineering Department are associated to the identification of topics and contents present in the undergraduate curricula in order to enable inclusion of environmental content in the disciplines in an integrated way. Sixteen disciplines were reviewed and then specific references were proposed. Furthermore, an optional discipline entitled "Life Cycle Engineering", attended by an average of ten students, became mandatory. Currently, as mandatory, Life Cycle Engineering is attended by 40 students.

In Civil Engineering, the environmental oriented disciplines are available only from the sixth semester, while the potential disciplines can be found during the entire course. It shows how important the necessity of changing the engineering curricula to include environmental topics is, providing more integration with specific knowledge disciplines. In this case, other departments such as Hydraulic Engineering, a department in charge for Environmental Engineering, offer modules with environmental content. Some examples include "Sanitation 1 and 2," and "Residual water treatment," justifying an intermediate level of environmental sustainability in the Civil Engineering undergraduate course.

Some homogeneity is also noticed in environmental content levels in disciplines from Aeronautical Engineering, Computer Engineering, Mechanical Engineering, Mechatronic Engineering, Electrical Engineering and Materials Engineering. The indexes for these engineering courses vary from 3 to 7 % considering mandatory and optional disciplines. The disciplines defined as environmental-oriented are similar in several occasions. For example, a discipline entitled "Humanities and Social Sciences" is mandatory for five from the six engineering careers mentioned.

Besides "Humanities and Social Sciences", and excepting a few disciplines with environment content specific to each engineering course, the other environmental-oriented disciplines are related to environmental management or to topics about pollution and solid waste treatment. Offering environmental-oriented disciplines, the departments of Architecture, Hydraulic Engineering and Production Engineering are the most representative.

In this preliminary diagnosis, the disciplines aggregating both social and environmental topics were put in categories but not in a homogeneous way, i.e., there are disciplines that present more sustainability indexes to be considered to be more in-depth sustainability-wise than other disciplines. This will be discussed in topic 3.2.

The undergraduate engineering courses with intermediate level of feasibility and low levels of environmental content need to be modified in order to change from "potential" status to become environmental-oriented. In those engineering courses where environment content is less easy to include, it is important to consider the disciplines' distribution along the course and which sustainable aspects they will embrace, aiming for the maximization of learning.

Accordingly, some organisational structure difficulties need to be overcome. As presented by Pavesi (2011) the departmental structure still has great influence at the university, minimising the flexibility of the curricula. The same author also points the need for structure reformulation, knowledge and the professors' values relating to sustainability. As Ometto et al. (2012) say the integration of environmental considerations into the institution is a gradual process.

Pavesi and Freitas (2013) assert that the process to develop a green curriculum is necessary, and the efforts range from people intermediation, coordinating teams actions and enhancing collective structures that guarantee a coherent pedagogical project. In this way, it would be interesting to see a management of the stresses created by the need of curriculum reformation and teaching methods, and also the external pressures from social, cultural, political and economic dimensions.

3.2 Sustainable Disciplines Aspects and Pedagogic and Policies Projects

Considering the analysis that resulted from the classification process discussed so far, and understanding the path that was followed to green the curriculum at EESC, a new group of sustainable disciplines has been created. Following the same idea defined initially of five radicals of sustainability and documents that were read, the aspects that had been contemplated in EESC disciplines was inferred.

Emphasis was determined by disciplines, firstly for Environmental Engineering curriculum and then the assessment was extended to the remaining courses.

As the Pedagogic and Policies Projects (PPP) aim for the education of a "professional with capacity to solve engineering problems related to the environmental impacts by prevention, correction or remediate actions", and assuming that is possible to adopt reactive, preventive and proactive postures, the aim was changed and then five different categories were defined: "*prevention*", "*correction*", "*remediation*", "*society-environment relation*" and "*ecological relation*".

Indeed, the topics that cover the environmental issues in the Environmental Engineering course have a strong emphasis on pollution, environmental management and problem solving. However, the presence of these themes is not proportional. The current analysis shows that the category least included was "ecological relation" (7 %), while the category "remediation" is included in 29 % of the disciplines. Considering the reactive characteristic of "*remediation*" and "*correction*", and summing up the percentage of both aspects, we find that 52 % of Environmental Engineering disciplines have reactive emphasis. "*Prevention*" is present in 26 % of disciplines, and the category "*society-environment relation*" represents 15 %. On the other hand, the radical "ecolog" appears only once, "soc" appears twice, while "ambient" is found 53 times. The radicals "sustent" and "natur" appeared 8 times.

In the category "*society-environment relation*", the disciplines are mainly humanities, but also from other education areas. The fundamental disciplines from biology are included in the "*ecological relation*" category. Analyzing the environmental-oriented disciplines from the other undergraduate engineering courses, the following relation presented in Fig. 3 was obtained.

It is possible to observe that the ratios found in the Fig. 3 are similar to the results obtained for the Environmental Engineering, i.e., it shows a more reactive position. About 50 % of the categories covered by Fig. 3 demonstrate a reactive position, while 29 % have a more preventive focus. The category "ecological relation" is missing.

According to the sustainability indicators "soc" and "ecolog", the disciplines follow a pattern in the curricula, in their PPPs, and in their purposes. The indicator "soc" is present in a small number of disciplines, as the same way as "sustent", "natur" and "ecolog". In the PPPs, the indicator "ambient" was often present, and as discussed by Pavesi (2011), in spite of the presence of environmental concerns, just a few PPPs turns that in directives and curricula practice.

In this sense, this paper strengthens the note made by Geromini (2010), that there is a low presence of environmental content offered at EESC, and the relations between environment and society are low in the curriculum. We agree that the approach to include environmental topics into engineering disciplines reveals a strong scientific and technological bias, without considering yet, the cause-effect relation between human activities and the environment, encompassing social, political and ethical aspects as well.

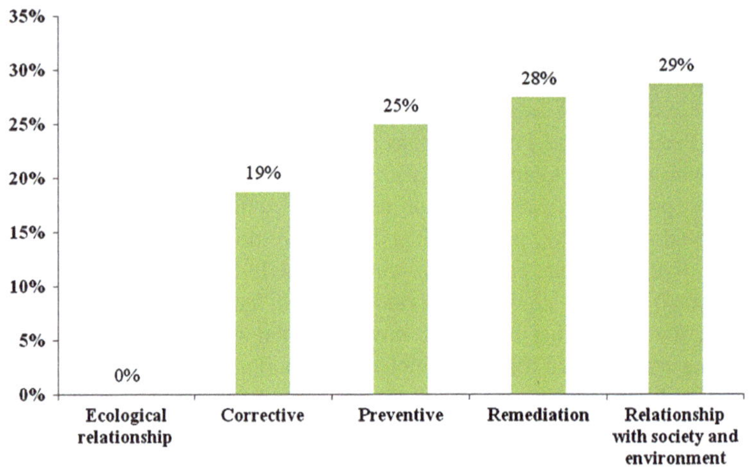

Fig. 3 Categories from the disciplines classified as "sustainable" at EESC engineering courses, excluding Environmental Engineering

4 Conclusion

In the conclusions session, some analysis and contributions for research continuity and improvement are described.

From the discussions presented in this paper related to nine undergraduate engineering courses at EESC, we can assert that except for Environmental Engineering, only Production and Civil Engineering demonstrate better level of environment-oriented disciplines when compared to the other engineering courses.

It is important to notice that disciplines classified as "potential" refer both to those fundamental and specific disciplines that show more opportunities to include changes to environmental issues. Only by this mapping process is it possible to formulate proposals for environment-orientation, improving the engineering curricula and education.

It is necessary to identify connections between social, environmental and specific technical content in engineering curricula, and furthermore, to relate them with traditional disciplines contents. In this sense, sustainability concepts could be integrated into EESC's egress engineer education.

Moreover, it was often observed in the curricula that engineering students are encouraged to adopt reactive actions. This means that the knowledge developed by engineering students demonstrates professional actions related to existing problems and impacts. Thus, it is necessary that Engineers' education should be planned considering more proactive solutions as well, aiming for problems and impacts prevention.

Not only USP, but also many other universities in Brazil and in the world are facing the challenges to include sustainability in their undergraduate courses, developing diagnosis, experiences, and adopting strategies. Among the challenges found, there is a need for professors' skills and capacity. For many times, professors are not prepared to correlate disciplines specific content with environmental issues.

Promoting actions such as short-time courses, lectures and projects related to environmental issues could be more encouraged at EESC. These actions should be focused on professors' learning and how to include environmental sustainability content into engineering curricula. In this sense, through a "continuous learning" process, professors would be more able to direct their knowledge to preventive and proactive solutions in the undergraduate classes, and prioritising, for instance, pollution prevention, waste management planning, etc.

In spite of limitations imposed by a documental analysis performed in this paper, it was possible to consider important inferences and indications for a deeper study. To continue the present study, questionnaires application with courses coordinators and lectures will be conducted. It aims to understand their sustainability perception and know the objectives and methodologies used in each discipline, since this is a fundamental process towards more a sustainable curriculum.

This preliminary diagnosis can collaborate with the current actions in progress at EESC for sustainability content inclusion in engineering curricula. This process may bring meaningful results for engineering education, allowing the development of skills needed to overcome the challenges to build sustainable societies. Moreover, it is necessary to develop interdisciplinary strategies and curricula changes to promote sustainability integration at EESC curricula.

References

Escola de Engenharia de São Carlos (EESC) (2014) Programa EESC Sustentável. http://www.eesc.usp.br/portaleesc/index.php?option=com_content&view=article&id=188&Itemid=349. Accessed 28 Jan 2014 [In Portuguese]

Evangelinos KI, Jones N, Panoriou EM (2009) Challenges and opportunities for sustainability in regional universities: a case study in Mytilene, Greece. J Clean Prod 17:1154–1161

Freitas DE, Souza ML (2011) O Ensino Superior No Brasil: Desafios Para Ambientalização Curricular. La educación superior en Brasil: retos para la ambientalización curricula. Available at: Visões e Experiências Ibero-Americanas De Sustentabilidade Nas Universidades, pp. 130–135 [In portuguese]

Geromini MMH (2010) Um panorama sobre a temática ambiental nos currículos dos cursos de graduação oferecidos pela Escola de Engenharia de São Carlos da Universidade de São Paulo. Curso de Especialização em Educação Ambiental e Recursos Hídricos—Centro de Recursos Hídricos e Ecologia Aplicada (CRHEA) [In portuguese]

Jabbour CJC, Santos FCA (2006) Entre Desafios e Oportunidades: Reflexões Acerca da Inserção da Dimensão Ambiental no Currículo de Engenharia de Produção. Revista Gestão Industrial 2 (4):75–86 [In portuguese]

Junyent M, Geli AM (2003) "Características de La Ambientalización Curricular: Modelo ACES". Volume 2. Proceso de Caracterización de La Ambientalización Curricular de los Estudios Superiores, Universitat de Girona—Red ACES [In portuguese]

Lozano F, Lozano R (2013) Developing the curriculum for a new bachelor's degree in engineering for sustainable development. J Clean Prod 64:136–146
Muñoz GCM (1996) Principales tendencias y modelos de La Educación ambiental en el sistema escolar. Revista Iberoamericana de Educación—Número 11—Monográfico: Educación Ambiental: Teoría y Práctica". http://www.rieoei.org/oeivirt/rie11a01.pdf/. Accessed 2 Feb 2014 [In portuguese]
Ometto AR, Saavedra YMB, Puglieri F, Uliana RB, Musetti M (2012) Ambientalização do curso de Engenharia de Produção: Caso da EESC-USP. XXXII Encontro Nacional de Engenharia de Produção, ENEGEP, Bento Gonçalves, RS, Brazil [In portuguese]
Pavesi A (2011) A Escola de Engenharia de São Carlos (EESC-USP) Diante do Desafio da Ambientalização Curricular Dos Cursos De Graduação. Revista de Ensino de Engenharia 30(1):24–34 [In portuguese]
Pavesi A, Freitas D (2013) Desafios Para a Ambientalização Curricular No Ensino Superior Brasileiro. IX Congreso Internacional Sobre Investigación En Didáctica De Las Ciencias. Girona [In portuguese]
Política Nacional de Educação Ambiental (PNEA) Lei n° 9.795, 27 de Abril de 1999. http://www.planalto.gov.br/ccivil_03/leis/l9795.htm. Accessed 23 June 2013 [In portuguese]
Uliana R (2007) Um Panorama da Evolução da Abordagem e da Adoção da Sustentabilidade na Formação do Engenheiro. Relatório parcial apresentado ao Programa Ensinar com Pesquisa pelo Departamento de Engenharia de Produção da Escola de Engenharia de São Carlos da Universidade de São Paulo, São Carlos, Brazil [In portuguese]
United Nations Educational Scientific and Cultural Organization, UNESCO, Education for Sustainability—from Rio to Johannesburg: Lessons Learnt from a Decade of Commitment (2002) http://www.unesco.org/new/en/jakarta/education/education-for-peace-and-sustainable-development-psd/education-esd/. Accessed 20 Feb 2014
Wright TSA, Wilton H (2012) Facilities Management Director's conceptualizations of Sustainability in Higher Education. J Clean Prod 31:118–125
Xiong H, Fu D, Duan Ch, Liu Ch, Yang X, Wang R (2013) Current status of green curriculum in higher education of Mainland China. J Clean Prod 61:100–105

Authors Biography

Bárbara Pacheco Lopes is an undergraduate student in Biology Sciences by Federal University of Sao Carlos (UFSCar), developing a research entitled "Science teaching, professors' education and environmental-oriented curriculum" under Professor Denise de Freitas' supervision, and working for the Environmental Management Superintendent (EMS) at EESC, USP, with Professor Aldo Ometto.

Laís Sanchez Assumpção an undergraduate student in Biology Sciences by Federal University of Sao Carlos (UFSCar). She is working for the Environmental Management Superintendent (EMS) at EESC, USP, with Professor Aldo Ometto.

Yovana M. B. Saavedra has degrees in Environmental Engineering Universidad Manuela Beltrán—UMB, in Colombia. She has Master's degree in Production Engineering. Currently, she is a Ph.D. Candidate in Environmental Engineering Sciences at EESC, USP. She has experience in Environmental Management, Strategies for End of Life Products (Recycling, Repair, and Remanufacturing), Life Cycle Assessment (LCA), Material Flow Analysis (MFA), Industrial Ecology and Management Life Cycle Product.

Fabio Neves Puglieri is a production engineer from Federal University of São Carlos (UFSCar). He got his Master's degree in Production Engineering from EESC, USP, and currently he is a Ph.D. Candidate in Production Engineering at EESC, USP. Puglieri's research comprises Sustainable Business Strategies, Strategic Planning, Life Cycle Management (LCM), Life Cycle Assessment (LCA) and Design for Environment (DfE).

Affonso Azanha is an undergraduate student in Mechanical Production Engineering at EESC, USP. He has experience in Lean Manufacturing tools such as Value Stream Mapping (VSM), Total Productive Maintenance (TPM), Poka Yoke development, among other shop floor techniques to improve efficiency in production flow.

Patrícia Cristina Silva Leme is biologist from Federal University of Sao Carlos (UFSCar). She has Master's degree and a Ph.D. in education by UFSCar. She is an invited researcher at College of Education, da University of South Africa (2013–2015), and educator at USP working with the Environmental Management Superintendent (EMS).

Aldo Roberto Ometto is a chemical production engineer from Federal University of São Carlos (UFSCar). He has Master's degree in Environmental Engineering Sciences and a Ph.D. in Hydraulic Engineering, both from EESC, USP. Currently, he is a professor at EESC, USP, working with researches related to Engineering and Life Cycle Management of Products.

Working Together, Driven Apart: Reflecting on a Joint Endeavour to Address Sustainable Development Within a University

Chris Shiel and Amanda Williams

Abstract

A holistic and transformational approach to Sustainable Development within a university requires systemic change and embraces new ways of working. Champions must challenge silo mentalities, develop new processes to encourage synergies across university functions, and strive to re-align systems and goals towards the common endeavour of sustainability. But how easy is this to achieve? It is well documented that working across disciplines presents challenges but forging a synergistic relationship between the environmental management function of Estates and an academic champion for ESD is not only logical but might be an easier place to explore how two roles can work together to achieve change. This paper provides a reflective account of such an alliance, outlining a joint endeavour to address sustainable development. An analysis is provided of those factors which impede such working and the different role tensions that make working together challenging. It will also consider the benefits of collaboration, as the perspectives from the operational and academic domains provide a broader context for understandings, access to different forums, an ability to tackle conflicting agendas together and an opportunity to genuinely effect change, providing mutual support through shared perseverance. The paper will conclude by questioning the extent to which progress made will

C. Shiel (✉)
Faculty of Science and Technology, Bournemouth University, Fern Barrow, Poole, Dorset BH12 5BB, UK
e-mail: cshiel@bournemouth.ac.uk

A. Williams
Environment and Energy Manager Estates, Bournemouth University, Studland House (S501), Christchurch Road, Bournemouth BH1 3NA, UK
e-mail: awilliams@bournemouth.ac.uk

W. Leal Filho et al. (eds.), *Integrative Approaches to Sustainable Development at University Level*, World Sustainability Series, DOI 10.1007/978-3-319-10690-8_30

endure, if the benefits of this synergy are not acknowledged by university leadership.

Keywords
Sustainable development • Reflective learning • Integrative approaches

1 Introduction

A holistic and transformational approach to sustainable development within a university requires systemic change and new ways of working (Sterling et al. 2013; Wals and Corcoran 2006). Champions need to challenge silo mentalities, develop new processes to encourage synergies across university functions, and strive to re-align systems and goals towards the common endeavour of sustainability. The aim would be to move beyond one dimensional approaches, such as campus-greening (which is important but not enough on its own), and initiatives where 'integrating sustainability' result in the development of a single module (an 'add-on', or 'package of knowledge' response (Haigh 2005) and again insufficient), to a response which requires whole-institutional change, systemic transformation, and a radical re-thinking of the purpose of education.

But how easy is this to achieve? In a sector that is "notoriously resistant to change" (Wals and Blewitt 2010, p. 57) the vision that universities should play a key role in contributing towards sustainable development, remains largely unrealised. The evidence suggests (at least within the UK) that while a few institutions exemplify attempts at a holistic approach and are exploring institutional change (evidencing the emergence of what Walls and Blewitt describe as "third-wave sustainability"), there are few examples of what might truly be called 'the sustainable university' (Sterling et al. 2013); many universities find it easier to focus on campus greening/environmental management (Leal Filho 2010), rather than engage with a more ambitious and integrative endeavour. Addressing sustainability across campus, curriculum and community (Jones et al. 2010) not only continues to be a big challenge but is such, that champions will confront what can seem like insurmountable hurdles, when they seek to transcend organisational boundaries.

'Transcending boundaries' and sharing learning with multiple stakeholders has to be at the heart of sustainable development (the solutions to many of the problems that the world faces will not be solved by a single discipline, or one group of people alone); finding ways to incorporate academic and practitioner knowledge is important for sustainability research (White 2013) and must be part of an integrative approach to sustainability within a university (given that universities comprise both academics and practitioners). However, it is not always easy to get academics to work collaboratively with their own academic colleagues, let alone to align education and research, with the interests of professional services/administrative staff (Sharp 2002).

The challenges of inter-disciplinary working (in both research and education) are well documented (Holley 2009; Wade and Stone 2010; Richter and Paretti 2009; Whitfield 2008). Barriers to collaboration are often cited as: a lack of resources to support interdisciplinary working; lack of supportive academic reward systems; contrasting academic cultures in different disciplines; different departmental policies and procedures and; decentralised budget strategies. While creating the right conditions for academic inter-disciplinary working has received attention, there is very little written about the challenges of collaboration between academic and professional services staff, where it might be expected that the barriers (cultures, policies, budgets) may not be very different to those experienced in academic inter-disciplinary working but may be even more challenging, to the extent that they involve practitioner and academic perspectives and quite different professional identities, role demands and operational responsibilities.

As those seeking to uphold an integrative approach to sustainability can hardly avoid such boundary crossing issues and will certainly experience the tensions of seeking to bridge organisational divides, this paper focuses on that topic, providing specific reflection on the relationship between an academic champion for sustainability and the champion responsible for environmental management, in light of their experience of collaboration to secure an integrative endeavour. It is well known that environmental management in the UK, (led by environmental managers, often in the estates function of a university) has made far greater progress than curriculum change. Sterling and Scott (2008) suggest that this is in part due to legislation and financial incentives which have spurred environmental management, but not withstanding this, perhaps a further explanation might be that the work has been led by a particular professional group, operating separately and under different organisational constraints to academic colleagues. Perhaps developing and maintaining an authentic relationship between environmental champions in Estates and academic champions for ESD (although the most logical place to begin an integrative endeavour) is more challenging than appreciated. An integrative approach to sustainability will surely falter unless it capitalises on the synergy between these areas of activity?

There is very little written in the sustainability literature about the tensions involved in making such relationships work, even by those universities who claim to have developed ‘integrative’ or ‘holistic’ approaches. Perhaps integration is sometimes more reflected in how case studies are written up; a post hoc presentation maybe, where separate activities (estates, curriculum, research, community) are audited, summarised and marketed, as more of an integrative whole than is actually the case? There are certainly very few case studies which suggest that systemic transformation and radically new ways of working have been achieved. The reputational benefits of being seen to be a sustainable university may not only be drivers for campus greening (Savelyeva and Park 2013) but to some extent, may also influence the way case study authors present their achievements in a more positive light; to say ‘things are not quite as they seem’ and to include a more negative commentary presents some risks. However, critically evaluating what is not working and sharing with others the challenges is an important endeavour; if we are not critical then we can only blame ourselves when visions are not realised. As Selby and Kagawa warn, those involved in education for sustainability should be wary of the dangers of striking a 'Faustian bargain'; accommodation within the system, may be "tantamount to trimming on our worldview for short-term influence" (Selby and Kagawa 2011).

It was a concern that the authors might be trimming on their worldviews, and a desire to engage in critical reflection to understand how a better and more congruent way forward might be achieved, that inspired this paper. The reflection reported, is set within the context of a university that has been one of the leaders in the UK in attempting to implement an integrative approach. The relationship between two champions has driven success, however despite external recognition, both have become increasingly aware of the difficulties; a number of tensions impede collaborative working, and will detract from the real vision for a sustainable university, in the sense outlined in Sterling et al. (2013).

This paper is innovative in that it deploys a reflective learning process to explore (through the lens of an academic/practitioner relationship) the difficulties of taking forward an integrative approach to sustainability, something which is rarely revealed (Velasquez et al. 2005). As the assumption behind a holistic approach to sustainable development is that collaboration is required, not just across discipline boundaries but across the academic/professional services divide, the reflection on experience will be of relevance to those seeking to develop integrative approaches and such cross-boundary relationships.

A brief account of the context is provided initially, before consideration is given to those factors which have contributed to success. Reflection on the conditions that impede collaboration and the different role tensions that have made the journey challenging are then considered.

2 The Context

Bournemouth University (BU) is a medium-sized UK university, inaugurated in 1992, with around 17,000 students, including 1,800 from non-EU countries, 650 academic staff and 800 professional and support staff. The vision for the university includes the aim of "inspiring our students, graduates and staff to enrich the world" and the bold statement: "we will ensure our environmental credentials are held in high esteem" (BU 2018).

The 2012–2018 Strategic Plan refers explicitly to "a holistic approach to SD" (p. 30), the need to "ensure that graduates develop a global perspective and understand the need for sustainable development by seeking to embed sustainable development across the curriculum" (p. 19) and the need to "ensure BU operates an affordable, sustainable and secure estate" (p. 53). The goal of implementing a holistic approach and the journey towards becoming a sustainable university (in the sense used by Sterling et al. 2013) has been driven by champions at BU with varying degrees of success, since the late nineties. The approach arose from an ambition to develop

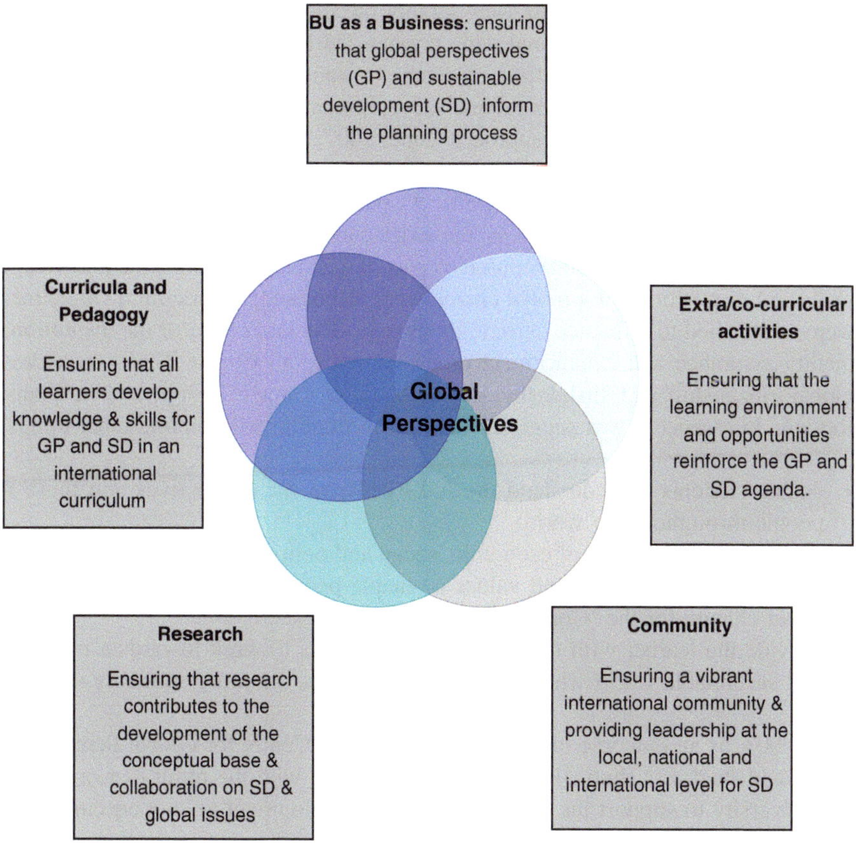

Fig. 1 Global perspectives and sustainable development in a global university (Shiel and Mann 2005)

graduates as 'global citizens who understand the need for sustainable development' (Shiel and Bunney 2002; Shiel 2007); it was instigated by a group of champions who sought to begin a discussion about how the university might make a better contribution to a world, where globalisation and unsustainable development requires futures-thinking and a better educative response. In parallel, another group (with overlapping membership) sought to explore energy saving (developing effective measures to benchmark progress) and environmental activity with campaigns such as 'turn-it-off'. Early initiatives sought to engage the support of senior leaders in a change process, and to inspire students and staff to engage with an agenda that would impact upon curriculum, campus and community—an approach which is not dissimilar to the '4C' model at Plymouth University (Jones et al. 2010, p. 7) and has been taken forward by other UK universities.

Up until 2005 developments were largely piecemeal and opportunistic; beyond 2005 funding for an institutional wide project, led by an academic champion and with the support of two part-time research assistants (one with an environmental background, one with an international development background), helped drive a more strategic and integrative approach (Shiel 2007, 2011) based on the concepts of developing global citizenship and addressing sustainable development (Fig. 1 is an early example of the model suggested). As part of the strategy which emerged, a range of projects were initiated in the extra-curricular sphere (an easier starting place); a holistic model was proposed as part of a Strategic Report; the knowledge, skills and dispositions of a global citizen were elaborated; 'Curriculum Guidelines' were established to influence curriculum change. The latter (part of the institutional quality assurance and enhancement process) require all Course Teams to address global citizenship and sustainable development at Course Design and in Course Review. In essence it was suggested that the curriculum will (among other things):

- enable students to understand the links between their own lives and those of people throughout the world;
- increase understanding of economic, social and political forces which shape life;
- develop skills, attitudes and values to enable people working together to bring about change for the 'Common good';
- provide the learner with the knowledge and skills to work towards a more just and sustainable world where power and resources are more equitably shared.

In 2008, as an outcome of the 2005 strategy, the Centre for Global Perspectives was established as a 'hub' (Petford and Shiel 2008) with the remit to work across the university to support the agenda, lead staff development to support curriculum change and implement projects to enhance activity across the institution.

A number of these projects (securing Fairtrade status for example and leadership development for sustainability) have involved collaboration with staff in Estates; outcomes from research (students' attitudes to sustainable development, for example) have also fed back to Estates just as environmental initiatives were fed in the other direction. Reciprocity and sharing have been at the heart of collaboration.

In relation to the environmental management of the Estates, 'campus-greening' activities gained momentum with the appointment of a dedicated Environmental Officer in 2005 and an Energy Officer in 2006. Activities initially focused on three target areas: energy efficiency, travel planning, and waste management and recycling. The environmental programme has since developed considerably to include a wider range of impact areas including carbon management, water reduction, biodiversity management, sustainable construction and sustainable procurement. Significant investment has been made in carbon management projects such as a biomass heating project, voltage optimisation and building management systems. It was fortuitous that one of the first Research Assistants on the 2005 strategic project (previously referred to), went on to become the institutions Environment and Energy Manager; her success in this new role enabled greater co-ordination in taking forward sustainable development across the academic (curriculum and research) and professional service domains (estates) than might otherwise have been possible. This meant that progress in developing environmental sustainability across the Estates proceeded in parallel (and sometimes faster) with developments in research, the curriculum, and the extra-curricular sphere.

The Environment and Energy Team now consists of 4.6 FTE (full time equivalent) staff, whereas the Centre for Global Perspectives (previously four staff), ceased to function in 2012, as the agenda became embedded across Schools—a cross-university academic function was no longer considered necessary in the light of the new Strategic Plan (BU 2018). This has made taking forward the academic aspects of global citizenship and education for sustainability more challenging, as the role of the original champion (and subsequent Director of the Centre) became re-located to a Faculty. The removal of a 'formal' function has since meant that the Environment and Energy Team has reduced access to the academic agenda; coordination of education and research for sustainability has stalled.

The institutional 'Environment Strategy Group' continues to have oversight of the environmental agenda; the Environment and Energy Manager ensured that the academic champion was included in the membership, albeit that the group primarily focuses on an Estates agenda, however over time, the education agenda has been acknowledged but not the research agenda. An academic champion on a group whose remit is 'estates' has been important in reminding the committee, that sustainable development encompasses a broader remit, something which is too easily overlooked.

Over time, an integrative approach to sustainable development (although never as fully integrated as originally conceived in 2005), has meant that BU has been perceived as one of the greener universities in the UK (with a 'first-class' award, 4 years in a row in the UK Green League), and as one of the early adopters of a holistic approach, where environmental concern is just one part of a broader agenda. Initiatives at BU have been rewarded by external recognition both locally and nationally, and include:

- AIBEAT Earth Charter Award—Engagement in Sustainability 2013
- EcoCampus Gold Award 2011
- Gold Sound Impact Students' Union Award
- Green Gown Awards: Transport 2005; Energy Efficiency 2004
- Finalists (nationally) in the following: Green Gown Awards—Sustainable procurement 2011; Green Gown Awards—Promoting Positive Behaviour 2011; Times Higher Education Awards—Outstanding Contribution to Sustainable development in 2007 and 2011 (both in relation to recognising an innovative and holistic approach)
- Dorset Business Awards—Environmental Excellence—2007

On the face of it, this is something to be proud of; the attempts to develop a holistic approach to sustainable development look like they have been quite successful. So if that is the case why should two of the original champions of the approach feel the journey has been a battle and the gap between vision and reality remains? What are the tensions that have contributed to a feeling that things are not quite right; what is stopping the full vision being realised?

3 Method

A critical inquiry methodology, participative action research and cooperative inquiry have supported the development of a holistic approach at BU (Shiel 2013), which with the aim to mobilise change and action, falls under the umbrella of 'new paradigm research' (Reason and Rowan 1981) in that:

> It seeks to bring together action and reflection, theory and practice, in participation with others, in the pursuit of practical solutions to issues of pressing concern to people, and more generally the flourishing of individual persons and their communities (Reason and Bradbury 2001, p. 1).

However the purpose of this paper is not to reflect on the overall change methodology, nor to present a case study of BU which glowingly describes success, but to contribute to an emerging dialogue (see Sharp 2002 and also Sterling et al. 2013) about what has been working, what is making further change difficult, what else needs to happen that might accelerate progress. The overall aim is to address the questions posed in the previous section and to understand what lessons might be drawn which others might find useful.

In a journey that has sometimes been more about action with little time for reflection, what seemed important was to create space for the change agents to individually and collaboratively engage in reflection, share realities, concerns and feelings, and find ways to integrate academic and practitioner knowledge to contribute to future capacity building (White 2013).

Boud et al. (1985, p. 3) refer to reflection as "a generic term for those intellectual and affective activities in which individuals engage to explore their experiences in order to lead to new understandings and appreciation." The reflective process is normally triggered by an experience and the need to resolve and clarify an issue (Boyd and Fayles 1983; Steinaker and Bell 1979). Experiential learning theorists generally agree that reflection then moves through a number of stages (see Moon 2000, p. 28 for a summary table) towards processing new ideas, resolution, transformation and possible action.

In order to develop a sense of the current situation, and out of a desire to formulate possible actions to progress the agenda (if not achieve emancipatory outcomes), the change agents decided to adopt a more formalised approach to 'reflection-on-action' (Schön 1983, 1987). Two formal periods of reflection were scheduled where both participants (the academic champion and the environment champion) were able to focus on the questions set, raise concerns (personal and organisational), and explore emotions. The explicit purpose of reflection at the first meeting was to consider at a local level, what factors have supported progress, what constitute hurdles, and how might greater synergy be achieved in the future? Systematic reflection (on personal experience and that of others) begins with clarification of the issues and moves towards developing new insights through interpretation (Moon 2000). An outcome of the first meeting was the need for a template that each individual could take away and populate later with further reflection to enable sense-making. At the second meeting the purpose was to discuss and review the template, share perspectives of the barriers and the different role tensions that make working together challenging, and to begin to formulate possible courses of action; a third meeting formalised conclusions.

In between meetings, reflection was on-going and communication continued via e-mail and telephone conversation.

The next section will present the outcomes from the reflective process and offer analysis and discussion of those factors which have contributed to a successful collaboration between the environmental management and academic endeavour, and those which serve to block progress.

4 Outcomes from the Reflective Process and Discussion

Dewey (1933) suggests that reflection can be uncomfortable, is related to emotions and involves exploring doubt and uncertainty; all those things were evident at the first meeting where anger, frustrations, and general weariness made it difficult to begin to formulate more practical outcomes. It would have been easy at that point to simply give up, rather than find a way beyond a flood of negativity. Structuring discussion around 'critical incidents', for example, 'I felt like this when…' and reviewing those things that had caused frustrations ('when you did this…') but trying to use neutral language helped to create a more positive discussion. It also helped to focus on those qualities that the other admired and areas, where collaboration had worked very well ('I have appreciated the benefits of you acting as a coach…'; 'admired your tenacity in leading change…') before re-visiting issues from the past that had been more difficult to make sense of (when one was recognised for work but not the other). Open and honest conversation, on current personal circumstances and organisational constraints, contributed to move the discussion beyond emotions to a more analytical perspective. Agreement of a framework to take away and populate with reflection was also a useful outcome to move reflection further.

At the end of the first meeting it was much easier to see where working together had been highly effective; that bringing together an academic perspective with a practitioner had yielded huge benefits but that organisational constraints did not support those who wish to transcend boundaries.

At the second meeting, the focus was largely on considering the organisational constraints but also, given that the change agents were both dealing with personal frustration some attention was focused on developing personal coping strategies in a context of organisational change. The populated framework was developed further see Fig. 2; it was agreed that the categories made sufficient sense to form the structure for further reflection and discussion. Finally, discussion focused on the

	Personal	Organisational	
Factors which have contributed to change and success	• Background of working together in Centre • Common interests at start; shared frustrations • Shared Values – ecological concern, social justice • Shared Vision – Holistic Approach to SD • Passion for shared projects: Leadership for SD, Fairtrade, raising awareness of students & staff; campaigning • Support when things are challenging • Complementary knowledge and skills • Proud of others achievements / academic mentor • Ability to resolve differences through communication • Friendship has endured • Access to different knowledge • Open and honest communication	• Senior Management Support – but this is transient • Board level buy-in/engagement – particularly around carbon • The right words in Vision, values and strategy • Different opportunities to influence: Member of Estates SMT; Member of Faculty and ESEC • Oversight of Estates development proposals - sign off of plans EET membership on design teams • Access to Chief Operating Officer - quarterly meeting • Annual presentation to Board (Carbon) • Access to Professoriate, Education Committees, Research Committees	• Rio +20 • UN Decade of ESD • Role of HEFCE & HEA • Funding Opportunities e.g. LGMF • External Recognition including awards • People & Planet Green League • Government policy/lead
Factors which have detracted; potential hurdles	• Ownership/Boundaries • Different pressures/organisational demands • Communication breakdowns • Different ways of dealing with stress • Gender?	• Changes in leadership: Loss of senior champions (Director of EIS, former DVC, changing priorities of Director of Estates) • Lack of sustainability capability at senior level • Lack of **highly visible authentic** leadership for SD • New line managers (different management styles) • Gender issue: UET all men; Estates – male dominated & adversarial BUT skills required for sustainability tend to be more female • Failure of academic staff to acknowledge professional services staff as knowledgeable practitioners. • Lack of willingness of academics in some areas to engage when their expertise would be useful to Estates. • Inability to influence the curriculum from within Estates, especially since the loss of the CGP. • Competing priorities: Student experience/Financial Sustainability/Estates development • Conflicting values • Focus on Estates Development above all other Estates Functions • Lack of stable long term strategy of estates development - goal posts keep moving • Committee burden • Immediacy of operational role requirements	• Neo-liberalism • Lack of interest in Rio+20; ineffectiveness of DESD • Lack of strong lead from HEFCE/departure of champion/loss of steering group. • Uncertainty over capital funding • Changes to CRC removed reputational element • Removal of cap on student numbers • Less emphasis on SD in sector • Change of government & weakening of priority

Fig. 2 Framework for reflection

outcomes of the reflective process (see Moon 2000), which might simply be resolution (accepting the status quo), empowerment and transformation (which might be personal and/or include a way forward to challenge the organisational context) and/or practical actions.

The discussion that follows elaborates further on the categories deployed within the framework. As this is the outcome of reflection the first person (singular and plural) may be used where appropriate.

5 Personal

It is undoubted that much of the success of an integrative approach at Bournemouth has been down to the personal nature of the relationship between the change agents, and the personal qualities which have enabled not only the relationship to sustain but progress to be made, sometimes in adverse contexts. This concurs with Bartlett and Chase (2004) who highlight the importance of personal relationships and perseverance in taking forward sustainability and Acevedo et al. (2012, p. 390) who suggest that, in their institution "personal attitude toward collaboration was a key factor." But it is also worth reinforcing a point made by Moore et al. (2005) which stands as a reminder to all champions of this agenda that although "energy and commitment are high but so is the danger of burn-out" (Moore et al. 2005).

It seems critically important that champions find ways to retain their vision, and sustain energy and commitment while at the same time sustaining self. Developing relationships between champions fulfils a personal support function and serves to sustain tenacity; scheduling time to reflect on the relationship and the challenges is important. Further, as Sharp (2002) reminds "the practice of reflection, humility and reciprocity are essential in maintaining positive relationships" when seeking to institutionalise sustainability.

The most critical factor to the success of our collaboration has been the ability to maintain a friendship, in contexts where at different times it has seemed as if we have been pitched against each other; where one or the other has been lauded by the organisation; where one might have had access to more interesting opportunities and one might have been over-burdened with organisational demands and; both, at different times might have felt that the other was accruing more benefits. Both at different times have also questioned the rationality of the organisation, so a useful reminder is that: rationality within universities is a myth which inhibits systemic transformation and that "the reality of organizational irrationality" (Sharp 2002, p. 136) is a prevalent form of stress for staff.

A stressful journey as change agents and champions has been supported by working quite often outside of the formal system, building trust and networks (many externally) and support structures. At the base of collaboration have been shared values and beliefs: a common vision of a sustainable university; a belief that education needs to play a leading role in contributing to a better world and; a passionate concern for the environment and social justice. Essentially, 'starting on the same page' has been a big advantage to collaboration; reminding ourselves that we are still on the same page (and not delusional), while seeking to take that vision into different parts of the university has sustained collaboration. The same message but coming from different voices with access to different forums and participation in different networks has built momentum for change.

The benefits of different knowledge bases and different understandings of the academic and professional domains of a university have also been important factors. At times academic knowledge (particularly theoretical arguments but also familiarity with academic drivers) has been useful to engage academics, to persuade leaders, and to demonstrate the broader links between sustainable development and other academic institutional drivers, for example, internationalisation and employability (Shiel et al. 2005). At other times practitioner knowledge has been useful in anchoring what at times might be an idealist or theoretical perspective, in the realities of organisational life. Practitioner knowledge has also brought to the fore hard evidence in relation to environmental performance indicators, the legislative drivers, and the experience gained from working in an estates function that is largely male-dominated.

Complementary styles of working (but different personality types) and different strengths in leading change have also played an important role and allowed one, or other to come to the fore in different situations, with different audiences. Both champions might be described as 'activists' (a strength for leading a change agenda and a danger within a university context) and as a consequence both are used to campaigning and speaking out; both have been able to utilise these skills in different contexts, where 'being professional' requires drawing on particular styles and deploying different professional repertoires.

The only barriers that were considered under 'personal' arose from an honest consideration of feelings in relation to one or two critical incidents, for example, where a senior leader attributed all success down to the environment team and chose not to recognise the academic contribution. Although anyone championing sustainable development will realise that no one person (or single part of the organisation) can own the agenda, and what is actually a mark of success is when everyone owns it, we are all human. It is only natural to feel some anxiety when collaborative contribution is ignored. University environments too frequently create situations which pitch individuals and different functional areas in competition with each other, which poses challenges for collaborative relationships; finding ways to collectively celebrate achievements and avoiding the danger of competing, are important to progress a change agenda which involves an integrative approach. Always speaking positively of the other's work and achievements, particularly in wider audiences is also helpful where it contributes towards a 'pygmalion effect' (Rosenthal and Jacobson 1968).

Finally at a personal level, there is something to be considered relating to gender. Both champions felt that being female had been an advantage and a constraint; there was a natural affinity for inclusive ways of working, collaboration and participatory approaches but this was at odds (particularly in relation to estates management) with a culture of autocracy, and at times, macho-management. On occasion, a male dominated culture, characterised by an adversarial approach, had been exhausting to challenge; sometimes being female meant that 'voice' seemed less significant. Although sex differences in leadership styles is an area of research that is contested, there is some evidence to suggest that women tend to favour a more democratic and participative style and use less an autocratic or directive style (Eagly and Johnson 1990) than men. In a later study (Eagly and Johannesen-Schmidt 2001, p. 791) results showed that women "exceeded men on three transformational scales: the attributes version of idealized influence, inspirational motivation, and individualized consideration." They are also more likely to positively reward performance.

While our reflection on gender was not conclusive it seemed an important issue to highlight, as it merits further research. As much of the literature on leadership for sustainable development suggests that what is required is challenging "patriarchal thinking" (Doppelt 2010) and developing more participatory ways of working (Marshall et al. 2011), then perhaps the role of gender in relation to leading sustainability initiatives merits exploration.

6 Organisational

Unfortunately most of the negatives that arose during the process of reflection fell under the organisational heading. It was also reflection on this category and in particular the history of organisational constraints, which more strongly tapped the emotive aspects of reflective learning (for example, anger and frustration).

The organisational constraints which have detracted both from the holistic approach, and the joint endeavour that underpinned it, have been constant and varied. Most critical have been changes in leadership, particularly the loss of senior champions, changing coalitions and transient leadership support, and lack of sustainability capacity (knowledge and awareness) at a senior level. Sharp (2002, p. 129) refers to the "complexity of the environmental imperative"; a shared frustration for the change agents, has been the failure of university leadership to fully appreciate the immediacy or complexity of this imperative. Despite the collaborative working of the two champions and several projects, including one to enhance leadership understanding of sustainable development and to enable them to consider role-modelling leadership for sustainable development (Shiel 2013), leaders are still guilty of several of Doppelt's "sustainability blunders" (Doppelt 2010, p. 49–54), particularly "patriarchal thinking that leads to a false sense of security" and a "siloed approach to environmental and socio-economic issues", the latter being a mechanistic approach where organisational structures inhibit collaboration. Developing and sustaining the 'awareness' of sustainability which Ballard (2005) suggests must include not just the nature and urgency of the issues but also an awareness of how worldviews (on both sustainability and leadership) impede change, has been an impossible task. A context where senior leaders have so many other pressing concerns has not helped but staff turnover has also meant that key staff, who were supportive of the original approach, are no longer within the organisation. Changes in personnel at the most senior level have contributed to a "failure to institutionalise sustainability" (Doppelt 2010, p. 54).

Perpetual change is a feature of higher education but one which is felt more sharply when new senior appointments brings their own particular interests (sometimes rejecting all that has gone before) and wish to lead innovations that enhance reputations. Unfortunately, sustainability, environmental management, or ESD, may be insufficiently 'sexy' to maintain the support of senior champions; leaving a legacy of campus expansion, or a higher position in a league table might hold greater allure for personal career development. It is therefore critically important that change agents do not depend on the support of just one or two senior staff; endorsement from the entire senior team must be visible and authentic. It is also important that when the corporate strategy and subsequent policies are drafted, that change agents go into over-drive to influence, and to ensure that commitment is set in stone. This requires change agents to be fully engaged with the process of strategy development (through appropriate committees), using the full range of influencing and negotiating skills to ensure that the right words appear in the vision, values, strategy and policies; finding diplomatic ways to suggest better alternatives is critical. It does not necessarily mean that what is implemented will be quite as anticipated, but at least it makes the agenda harder to ignore; some action will be taken forward. Our success at influencing at this level (getting the right words into documents), has been greater than our success with implementation, which has at times been blocked by university structures.

University structures have certainly tended to create a binary divide between 'academics' and 'professional services' staff; those who find themselves working in a 'third space' (Whitechurch 2012) face challenges and may experience dissonance. A particular frustration within the environmental management function has been the perceived failure of many academics to acknowledge professional services colleagues as practitioners and engage collaboratively with them. Whilst the shared endeavour of the two key change agents has had many benefits, it has not been possible to replicate this across the university and collaboration on sustainability across the academic and non-academic domains has been limited in scope as a result. In part this may be down to the instrumental orientation of many academics with an attitude towards professional services which is largely to ignore them, or involves 'you are there to give me what I need but don't tell me how to do my job'.

The 'gender issue' was considered earlier but its impact was felt at the organisational level, particularly with reference to the male dominated culture of estates but also in relation to an entirely male senior university executive team. The issue was not fully unpicked except to acknowledge that gendered ways of working might either enhance or detract from a broader approach to sustainable development; patriarchal approaches impede sustainability (Doppelt 2010).

Consideration of the culture of Estates then led to reflection on different academic and professional services cultures, expectations, resources and budget constraints. Although all employees within BU are on a single pay spine, there is a clear structural and cultural divide between academic and professional services staff. The academic champion has more freedom from bureaucratic demands but no resources; the environment manager is overwhelmed by committee attendance, reporting and day-to-day operational issues but has some budgetary control. The academic faced teaching and research demands and increasing pressure to secure funding; the environment manager faced greater pressure to secure KPIs and manage an expanded team (the latter had been a critical factor in promoting change, albeit primarily on the Estates-related agenda). The dis-investment on the academic side (the removal of the Centre as a 'hub') had served to under-mine the educational work, which although promised to continue, has not been picked up by other parts of the organisation. It has also (intentionally or not) given out a strong signal that sustainability is the remit of Estates. In a culture where professional services/administrative staff are perceived as secondary to the educational endeavour, there has been a tendency for academics to disengage with an agenda that is owned by practitioners as a consequence.

Other organisational factors have contributed to positive change and shared successes, most notably the change agents have made good use of their different opportunities to influence across the academic and professional services domains, particularly through committee structures. For example, the academic change agent is a member of a Faculty, has access to the Professoriate, and sits on groups such as the Education and Student Experience Committee and the Research and Knowledge Exchange Committee. It was this ability to work through the committee structures that ensured that the original vision was endorsed by Senate in 2000. The Environment and Energy Officer has access to the Board, can raise issues at a quarterly meeting with the Chief Operating Officer and is a member of Estates senior management team; her team has membership on every Estates Development project design team and sign off of plans. However, working separately in different organisational spheres has resulted in change but not always resulted in a joined up approach. As Sharp (2002, p. 130) notes, "the complexity of the organisation itself, compounded by the complexity of the environmental imperative, thwarts most attempts to gain organisational agreement on goals, alternatives and solution programs", particularly across traditional academic and non-academic boundaries. That said, the high profile of sustainability in the BU Vision and Values and Strategic Plan, coupled with genuine engagement at board level (albeit primarily on carbon), should be considered important 'wins' at a strategic level which have been jointly achieved.

There has also been a long history of collaborative project successes, including a HEFCE 'Leadership for Sustainable Development' project, the EcoCampus Gold Award, and Fairtrade University Status, as well as a comprehensive portfolio of projects within the environmental programme, some of which have been high profile. But as Sharp notes we should differentiate between project success and institutional transformation (Sharp 2002, p. 130). There is a risk that an unintended consequence of success at project level is that senior staff become complacent, believing change is 'in hand', and this in itself can become an organisational barrier to genuine transformational change.

This raises the question of how to overcome such complacency and how to challenge those organisational factors which impede progress. An integrative approach to sustainability would undoubtedly work better within an inclusive and appreciative culture which values professional services and academics as part of a team. The culture would need to embrace a collaborative ethos, with high levels of connectivity and communication; budget allocation and reward systems would need rethinking to inspire more integrative ways of working and to engender collegiality between the operational and academics parts of the system; structures would need to allow for accountability but be sufficiently flexible to facilitate systemic change.

7 External Factors

Finally reflection moved to consider external factors. At various times the external context has represented either a fair or foul wind to the progress of the institutional change agenda. Reflection began with a consideration of neo-liberalism and the corporatisation of HE, with the focus on action plans, competition, growth targets and KPIs (key performance indicators). Undoubtedly, there is a tendency for education for sustainability to "become an instrument of a managerialist culture" (Blewitt 2013, p. 52); the challenge is to contest the system (economic policies based on growth) and to ensure that an audit/commodity/KPI culture does not limit what might be achieved.

In relation to the sustainability agenda, between 2005 and 2010, the external context was a driver for change in many UK universities, including BU; and working in partnership, the change agents were able to capitalise on external developments to increase momentum internally.

The UK sustainable development strategy, 'Securing the Future: Delivering UK sustainable development strategy' (2005) set out the government's goals on sustainable development. In response to this, the Higher Education Funding Council for England (HEFCE) published its own strategy Sustainable development in Higher Education (HEFCE 2005b) following consultation with the sector (HEFCE 2005a), setting out HEFCE's approach. A further update was published in 2009 (HEFCE 2009) along with a consultation on challenging carbon reduction targets for the sector, sufficient to ensure satisfactory progress towards national government targets to reduce carbon emissions by 80 % by 2050 and at least 34 % by 2020 (against 1990 levels).

The publication of the Carbon reduction target and strategy for higher education in England (HEFCE 2010) was also an important trigger, confirming a sector level carbon reduction target in line with UK targets, a requirement for institutions to set their own targets for 'scope 1 and 2 emissions', and introduced a link between capital funding and carbon management performance through the Capital Investment Framework (second iteration).

External Funding opportunities followed, including those available through the Leadership Governance and Management Fund, and the HEFCE/Salix Revolving Green Fund. BU was successful in achieving funds through both and implementing change, with collaboration between the champions on the former, and an Estates lead, on the latter. External recognition, including awards, raised the profile of the collaborative work.

Legislative drivers included the introduction of the Carbon Reduction Commitment, which would require large energy consuming organisations to buy carbon allowances annually, based on consumption, to cover their carbon emissions. The revenue was to be recycled by government, based on organisational performance on carbon reduction during the year, benchmarked against other participating organisations, resulting in a potential financial and a reputational impact. Carbon management had become something that universities must do, rather than something they should do.

The People and Planet Green League, launched in 2007, has also been a critical driver in raising the profile of sustainable development with senior staff, with high profile rankings published in the Times Higher Education Supplement initially and later the Guardian. As McGowan noted when he presented People and Planet with a British Environment and Media Award for Best Campaign in 2007, "the green league succeeded in dragging environmental issues in from the fringes and making them a central concern for many Vice Chancellors" (McGowan 2007).

However, questions around education and learning were not included in the Green League until 2011.

Whilst the external context raised the visibility of sustainability across the higher education sector, it has also served to advantage campus greening as a high priority with senior staff. An unintended consequence of this may actually have been to disadvantage curriculum developments, as the focus gradually shifted from the holistic approach that collaboration had secured within BU, to a more compliance and reputation orientated focus. Even within campus greening activity, it has resulted in the subordination of some areas of the broader environmental programme, and led to a position where carbon has become the key focus of senior leaders (at Director level, and above).

Consideration was also given to the Decade for Education for Sustainable Development, Rio + 20, and the work of the UK Higher Education Academy (HEA) in sustainability. It was felt that while these things had been of impact for the change agents (and the academic champion had been to Rio and participated in the HEA's Green Academy), there had been limited impact on the organisational context. Unfortunately the 'decade' came and went; few within the organisation noticed its passing. The HEA's Green Academy had been very helpful for participating institutions but their wider sustainability work had had less impact. Individual researchers had engaged with Rio + 20, but the wider community were less concerned.

In May 2010, the Conservative Liberal Democrat coalition came to power in the UK, claiming that they would be the 'greenest government ever' (Cameron 2010), however, this has yet to be realised. Perhaps the first indication of changing government priorities was the decision to dissolve the Sustainable Development Commission in the same year. Changes to the Carbon Reduction Commitment Scheme aimed at simplification took away both the reputational (league table) and financial (recycling of funds) drivers for carbon reduction. An evaluation of progress towards the government's pledge in year one showed that real progress had been made in just six of 77 policy areas considered (Porritt 2011). Some of the initial momentum provided by HEFCE has also since diminished, particularly since the departure of the Senior Policy Advisor for Sustainability (who was not replaced) and the disbandment of the Sustainable Development steering group. HEFCE's engagement has become far less visible as a consequence. The link between performance on carbon reduction and capital funding has become less impactful, as the capital funding pot has diminished and left uncertainty over future capital funding.

This changing context, along with the organisational changes detailed above, resulted in the change agents being increasingly pulled in divergent directions, and defending ground that had been considered safe.

At the time of writing, there are signs of the external context shifting once more. In the 2013 grant funding letter to HEFCE (DBIS 2013), Government recognised the good progress higher education has made on sustainable development and called for further action "to build on the achievements of universities" and requested the development of a new sustainable development framework, which HEFCE is currently consulting on. It remains to be seen whether this will have sufficient teeth to effect real transformational change. At the same time, the Government announcement to remove the undergraduate student numbers cap from 2015, has resulted in plans to accelerate the most significant period of estates development the University has ever seen. The impacts remain to be seen.

8 Outcomes

Engaging in a more formal process of reflection yields some therapeutic benefits for change agents and champions of sustainability. It was interesting to note that during the process (and before the second formal meeting) both champions became more positive in outlook and had moved from an acceptance of the status quo (organisational barriers) towards developing approaches for challenging institutional hurdles. At the time of the second meeting, it was acknowledged that without actually being aware of a change, both had re-engaged with seeking to influence others and to reinvigorate the agenda through committee structures. Both had developed personal longer term plans and strategies for coping with a difficult organisational context.

In terms of developing collaborative actions (outcomes from reflection) it was acknowledged that while little can be done to shift the external context, internally it was important to reinforce for university leaders that the agenda had slowed pace: the Environment and Energy Manager felt that it was difficult to achieve a wider agenda and "true ecological citizenship" (Savelyeva and Park 2013, p. 190) from within Estates, where the organisational culture had placed too strong an emphasis on carbon and compliance; the academic champion felt that apart from her contribution to the Environment Strategy Group in particular, and other committees where she continued to raise the profile of sustainable development, it was more difficult to engage within the institution and easier to lead developments within the external community. As positive action for change, both agreed to seek opportunities to highlight to senior leaders:

- The importance of an inter-disciplinary unit, or function that brings together students and staff but which also serves as an interface between the academic and professional service divides. Such a unit would facilitate cross-institutional working, lead on staff development, but also play an instigative role in developing projects and campus-based activities. Further consideration needs to be given to 'third space' (Whitechurch 2012) working.
- To reflect back the detrimental effects of an over-emphasis on carbon which eclipsed other concerns.
- The challenges of leading on sustainable development from within the Estates function which impedes neutrality.
- The need for visible and authentic leadership, and in particular stronger senior leadership of the educative agenda.

- The need for mid-level leaders to follow through on implementation, to "ensure that graduates develop a global perspective and understand the need for sustainable development by seeking to embed sustainable development across the curriculum" (BU Vision and Values p. 19).

Both agreed to collaborate on organising further cross-university staff development sessions to encourage other champions. Both agreed to continue working within the system but to be more challenging of worldviews and ways of working, which compromise the radical potential of sustainable development.

9 Conclusion

This paper has presented the outcomes of reflection of two change agents (one academic, one practitioner), who have worked together for almost 10 years, to develop synergy through an integrative approach to sustainable development. The reflective account has been presented in a spirit of inquiry, with suggestions made as to where the tensions lie, in developing an integrative approach to sustainable development. Although a limited piece of research in the context of the experience in one institution, it is hoped that consideration of what has contributed to success and what has hindered, under the categories of personal, organisational and external factors, will resonate with others.

In terms of lessons to be drawn from the paper, the first would be to emphasise the importance of honest and critical reflection on those factors that impede integrative approaches to sustainability; sharing the difficulties is as important as showcasing success. The authors would also stress the value of more formal engagement with the process of reflection; reflective learning is at the heart of a change agenda and integral to institutionalising sustainability, where progress is challenging. There is always a danger that a focus on actions, drives out space for deeper engagement with reflection.

A good starting place for an integrative approach is to develop a positive relationship between the champion leading environmental management in estates and the academic champion leading sustainable development more broadly within an institution. The authors suggest that the personal relationship developed and shared vision and values, has been a significant factor in contributing to a joined up approach for sustainable development. The perspectives from the operational and academic domains (when brought together) offer an enriched and broader context for developing shared understandings; working collaboratively enables access to different forums, enhances the ability to tackle conflicting and challenging agendas as a team, and offers the opportunity to make a genuine impact. Such collaborations also provide mutual support which encourages perseverance. However, such relationships need to be established and maintained in contexts which place individuals in competition for status and resources; individuals need to be robust and devise ways to reduce anxieties and work around organisational constraints. Maintaining a collaborative front is critical to success; personal qualities and shared values are helpful at the outset but unless relationships are nurtured, collaboration may not be sustained. Another key lesson is the importance of finding ways to bridge the binary divide, created by University structures which contribute to 'silo' mentalities; such structures present a challenge to collaboration across the academic and professional services spheres.

The success of the approach at BU has been rewarded by external recognition but it is suggested that organisational factors are serving to inhibit further change, particularly where university leaders fail to fully grasp the implications and urgency of sustainable development; the synergy created by working across organisational boundaries is not always recognised by university leadership; university structures create barriers and sub-cultures which are difficult to transcend and which lead to competing agendas. A critical lesson is to ensure that external project success does not result in complacency; an acceptance of the status and too much tolerance of organisational hurdles will not serve the sustainability agenda in the long-term. Champions need to continually reinforce the agenda with senior leaders and highlight the impact of organisational barriers.

Finally, we concur with Sharp (2002, p. 133) who suggests: "that transformation will only come about when a large number of people set up different priorities in both the large and small arenas of the university, establishing new routines and structures, despite local conflicts and set-backs." This paper has emerged in a context of set-backs and local conflicts, however engaging formally with the process of critical reflection has enabled the change agents to become more optimistic; the determination to persevere and to champion better ways of working has increased. However, whether the subsequent efforts of champions will lead to "a reworking of the design and operation of institutional reward systems" and an "appropriate linkage between the operational and academic functions of the university" (Sharp 2002, p. 78) remains doubtful.

Looking to the future, champions for sustainability have to challenge the system; they need to be cautious of the bargains they strike to secure a place at the table (Selby and Kagawa 2011). Compromise may secure some change but a lack of criticality, and projects which deny the true values and radical roots behind the agenda, may not only lead to personal dissonance but may also mean that a university's contribution to sustainability remains unrealised. Future research needs to contribute to a more critical account of the barriers and how these were overcome, so that change agents may learn more fully from others' experiences.

References

Acevedo B, Bonner C, Johnson S, Malevicius R (2012) Students' [passionate] engagement with processes of greening the campus. In: Leal Filho W (ed) Sustainable development at universities: New Horizons. Peter Lang Scientific Publishers, Frankfurt, pp 374–383

Ballard D (2005) Using learning processes to promote change for sustainable development. Action Res 3(2):135–156

Bartlett P, Chase G (2004) Sustainability on campus: stories and strategies for change. MIT Press, Cambridge
Blewitt J (2013) EfS : contesting the market model of higher education. In: Sterling S, Maxey L, Luna H (eds) The sustainable university. Routledge, Abingdon, pp 71–86
Boud D, Keogh R, Walker D (1985) Reflection : turning experience into learning. Kogan Page, London
Boyd EM, Fayles AW (1983) Reflective learning: key to learning from experience. J Humanistic Psychol 23(2):99–117
BU (2018) Bournemouth university strategic plan. http://strategicplan.bournemouth.ac.uk/. Accessed Feb 2014
Cameron D (2010) Speech given by the Prime Minister at the department of energy and climate change, 14 May 2010. https://www.gov.uk/government/speeches/pms-speech-at-decc. Accessed March 2014, pp 19–53
DBIS (Department for Business, Innovation & Skills) (2013) Higher education funding 2013–2014 letter. http://www.hefce.ac.uk/media/hefce/content/whatwedo/fundingandinvestment/fundinginstitutions/annualfundingallocations/grantletter/HEFCE%20Grant%20Letter%202013%20-%20Tim%20Melville-Ross.pdf. Accessed March 2014
Dewey J (1933) How we think. D C Heath and Co, Boston
Doppelt B (2010) Leading change towards sustainability. Greenleaf Publishing, Sheffield
Eagly AH, Johnson BT (1990) Gender and leadership style: a meta-analysis. Psychol Bull 108 (2):233–256
Eagly AH, Johannesen-Schmidt MC (2001) The leadership styles of women and men. J Soc Issues 57(4):781–797
Haigh M (2005) Greening the university curriculum: appraising an international movement. J Geogr High Educ 29(1):31–48
HEFCE (2005a) Sustainable development in higher education: consultation on a support strategy and action plan. Higher Education Funding Council for England, Bristol
HEFCE (2005b) Sustainable development in higher education: strategic statement and action plan. Higher Education Funding Council for England, Bristol
HEFCE (2009) Sustainable development in higher education: 2008 update to strategic statement and action plan. Higher Education Funding Council for England, Bristol
HEFCE (2010) Carbon reduction target & strategy for higher education in England. Higher Education Funding Council for England, Bristol
Holley KA (2009) Understanding interdisciplinary challenges and opportunities in higher education. ASHE High Educ Rep 2009(35):1–131
Jones P, Selby D, Sterling S (eds) (2010) Sustainability education: perspectives and practice across higher education. Earthscan, London
Leal Filho W (2010) Teaching sustainable development at university level: current trends and future needs. J Baltic Sci Edu 9(4):273–284
Marshall J, Coleman G, Reason P (eds) (2011) Leadership for sustainability: an action research approach. Greenleaf Publishing, Sheffield
McGowan A (2007) Presentation at British Environment & Media Awards. http://peopleandplanet.org/bema_award_2007. Accessed March 2014
Moon JA (2000) Reflection in learning & professional development: theory and practice. Kogan Page, London
Moore J, Pagani F, Quayle M, Robinson J, Sawada B, Spiegelman G, VanWynsberghe R (2005) Recreating the university from within: collaborative reflections on the University of British Columbia's engagement with sustainability. Int J Sustain High Educ 6(1):65–80
Petford N, Shiel C (2008) A holistic approach to the 'globalisation' agenda at Bournemouth University: the role of senior management. In: Shiel C, Mckenzie A (eds) The Global University: the role of senior managers. BU/DEA, London, pp 20–25
Porritt J (2011) The greenest government ever: one year on, friends of the Earth. http://www.foe.co.uk/sites/default/files/downloads/greenest_gvt_ever.pdf. Accessed March 2014

Reason P, Bradbury H (eds) (2001) Handbook of action research: participative inquiry and practice. Sage, London
Reason P, Rowan J (eds) (1981) Human inquiry: a sourcebook of new paradigm research. Wiley, Chichester
Richter DM, Paretti MC (2009) Identifying barriers to and outcomes of interdisciplinarity in the engineering classroom. Eur J Eng Educ 34(1):29–45
Rosenthal R, Jacobson L (1968) Pygmalion in the classroom: teacher expectations and pupil's intellectual development. Holt, Rinehart and Winston, New York
Savelyeva T, Park J (2013) Complexity of Campus Sustainability Discourse. In: Filho Leal (ed) Sustainable development at universities: new horizons. Peter Lang, Frankfurt, pp 183–192
Schön D (1983) The reflective practitioner: how professionals think in action. Basic Books, New York
Schön D (1987) Educating the reflective practitioner: toward a new design for teaching and learning in the professions. Jossey Bass, San Francisco
Selby D, Kagawa F (2011) Development education and education for sustainable development: are they striking a faustian bargain? policy and practice: a development education review, vol 12. Spring, pp 15–31
Sharp L (2002) Green campuses: the road from little victories to systemic transformation. Int J Sustain High Educ 3(2):128–145
Shiel C (2007) Developing and embedding global perspectives across the university. In: Marshall S (ed) Strategic leadership of change in higher education. Routledge, London and New York, pp 158–173
Shiel C (2011) Are we there yet? In: Shiel C (ed) Global vision, local action: education for sustainable development and global citizenship. In: Proceedings 4th international conference. Bournemouth University, Bournemouth, pp 13–41
Shiel C (2013) Developing global perspectives: global citizenship and sustainable development within higher education. Ph.D. thesis (PhD). Bournemouth University, School of Applied Sciences, Bournemouth
Shiel C, Bunney A (2002) Global responsibility and sustainable development: tailoring the concepts to business programmes. In: Paper presented at the 1st annual teaching and learning conference, Bristol
Shiel C, Mann S (2005) A global perspective at Bournemouth University: education for global citizens and sustainable development. Internal strategic report. Bournemouth University, Bournemouth
Shiel C, Williams A, Mann S (2005) Global perspectives and sustainable development in the curriculum: enhanced employability, more thoughtful society? Paper presented at the enhancing graduate employability: the roles of learning, teaching, research and knowledge transfer. In: Proceedings of the Bournemouth University learning and teaching conference. Bournemouth University, Bournemouth
Steinaker N, Bell R (1979) The experiential taxonomy: a new approach to teaching and learning. Academic Press, New York
Sterling S, Maxey L, Luna H (eds) (2013) The sustainable university: progress and prospects. Earthscan/Routledge, London
Sterling S, Scott W (2008) Higher education and ESD in England: a critical commentary on recent initiatives. Environ Educ Res 14(4):386–398
Velazquez L, Munguia N, Sanchez M (2005) Deterring sustainability in higher education institutions: an appraisal of the factors which influence sustainability in higher education institutions. Int J Sustain High Educ 6(4):383–391
Wade BH, Stone JH (2010) Overcoming disciplinary and institutional barriers: an interdisciplinary course in economic and sociological perspectives on health issues. J Econ Educ 41:71–84
Wals AEJ, Blewitt J (2010) Third-wave sustainability in higher education: some (inter)national trends and developments. In: Jones P, Selby D, Sterling S (eds) Sustainability: perspectives and practice across higher education. Earthscan, London, pp 55–74

Wals A, Corcoran PB (2006) Sustainability as an outcome of transformative learning. In: Holmberg J, Samuelsson BE(eds) Drivers and barriers for implementing sustainable development in higher education. UNESCO, Paris, pp 103–108

Whitchurch C (2012) Reconstructing identities in higher education: the rise of third space professionals. Routledge, New York

White RM (2013) Sustainability research: a novel mode of knowledge generation to explore alternative ways for people and planet. In: Sterling S, Maxey L, Luna H (eds) The Sustainable University: Progress and prospects. Routledge/Earthscan, London, pp 168–191

Whitfield J (2008) An indifference to boundaries. Nature 451:872–873. http://www.nature.com/news/2008/080220/full/451872a.html. Accessed 27 Jan 2014

Author Biography

Chris Shiel is an Associate Professor in the Faculty of Applied Sciences, the former Director of the Centre for Global Perspectives and a Leadership Foundation Fellow. Her doctorate 'Developing global perspectives and sustainable development within Higher Education' represents a summary of her work as a champion for institutional change. She has led ESD within the UK for 15 years.

Amanda Williams is Environment and Energy Manager at Bournemouth University, with a background in geography, she has worked in environmental management for 10 years in higher education and is a full member of the Institute of Environmental Management and Assessment. Her previous role was Research Assistant—Education for Sustainable Development.

Part II
Convergent Approaches

Advanced Training for Sustainability Change Agents—Insights and Experiences from a Seminar Series Using the Method of Systemic Constellations

Georg Müller-Christ and Anna Katharina Liebscher

Abstract

The implementation of sustainability in universities is a task that is increasingly assigned to sustainability coordinators or sustainability change agents. This is a pleasing development, however, the people that hold these jobs hardly know where to "place" sustainability issues in their universities. This paper presents the insights and experiences from a seminar series offered for sustainability coordinators in 2013 aiming at the implementation of sustainability in research, teaching and institutional management. Using the method of systemic constellation, the participants can easily take a closer look at the place sustainability is given in their institution. When trying out several alternatives for developments it is possible to not only visualize problems and barriers, but also to identify drivers and feasible solutions. There are specific patterns that exist in different specific contexts at all universities. The aim of the paper is to present these patterns and ways to deal with the problems sustainability change agents face in their institutions. These include barriers build up by persons, making support available, learning processes and the monitoring of changes and possible tensions and conflicts.

Keywords

Education · Sustainability · Sustainability coordination · Systemic constellation · Sustainable change

G. Müller-Christ (✉) · A.K. Liebscher
Department of Business Studies and Economics, University of Bremen,
Wilhelm-Herbst-Str. 12, 28359 Bremen, Germany
e-mail: gmc@uni-bremen.de

A.K. Liebscher
e-mail: liebscher@uni-bremen.de

W. Leal Filho et al. (eds.), *Integrative Approaches to Sustainable Development at University Level*, World Sustainability Series, DOI 10.1007/978-3-319-10690-8_31

1 Introduction: Sustainability Coordination at German Universities

A remarkable development is taking place in Germany: Following the UN Decade of Education for Sustainable Development more and more universities create new jobs for sustainability coordinators. These people shall give the issue of sustainability a good place in their respective universities. The assigned tasks are very different at the universities: at some of them the coordinators shall develop a sustainability strategy, at some of them they are asked to create new course offers, some universities want to install an environmental management system, some want to establish new research foci. The diversity of these tasks gives an insight into the problem that there is no common understanding of how sustainability can be integrated into university structures yet. In our seminar series with several sustainability coordinators we learned that all of them are concerned with one particular question: How much importance does the university management attribute to the issue of sustainability?

The authors tried to answer this question via the method of systemic constellations. This method applies a kind of transverbal language using space, which makes it possible to constellate the actual relationship of university management and sustainability. For this, persons act as representatives of the rectorate, of executive committees and of sustainability and are positioned in a room in order to visualize the relationships. Usually, this procedure reveals that sustainability coordinators want sustainability to stand directly next to university management claiming that sustainability must be the most important topic within the university. However, this picture turned out to be incongruous. Not surprisingly, university management was very busy managing the existing leadership problems within their institutions. They only recognized the task of managing sustainability from the corner of their eye. So the search for a good place for sustainability in the university system had to go on. This paper presents the possibilities the method of systemic constellations gives and shows some results of the constellations conducted during the seminar series that reveal some alternatives for the integration of sustainability in the system of a university. The innovative character of the paper lies especially in the application of the method of systemic constellations. This method has hardly been applied when organizational problems in the university surrounding are addressed. The paper will show how the application of the method can contribute to innovative insights in the higher education sector. A systemic view on the development of the implementation of sustainability issues in the higher education system can be given when applying the method in the context of sustainability coordination.

2 A Map for Sustainability at Universities

There are two kinds of universities in Germany: regular universities and universities of applied sciences. They differ in their focus, which is more on research at universities and more on teaching at universities of applied sciences. However, both

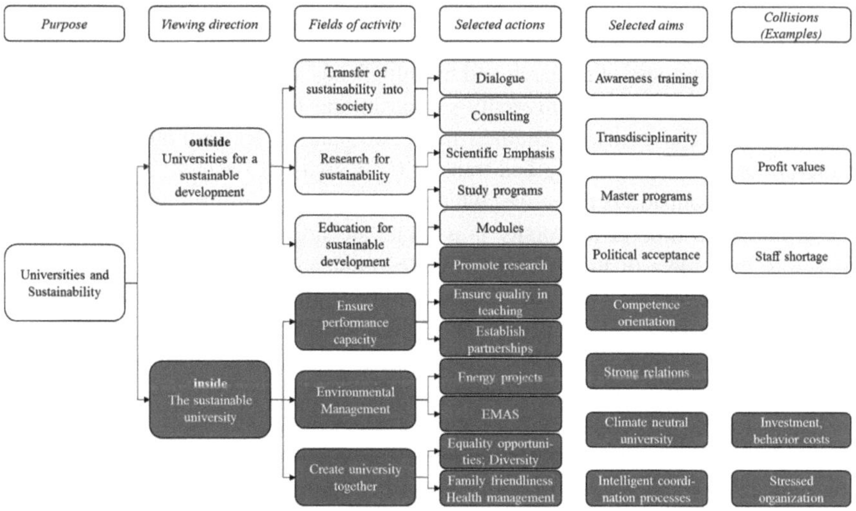

Fig. 1 Map for the connection of university and sustainability

kinds have the task to create and convey knowledge through research *and* teaching. Managing a university therefore means, first of all, to facilitate research and teaching processes with a high quality level. These processes require good infrastructure; so another main task is the management of operating services (i.e. buildings, research infrastructure, etc.). More and more, a close connection of universities and their regions is demanded, which especially means that issues and questions are conveyed to the social environment. These different challenges can be illustrated in a map serving as a frame of reference for universities as 'fields of activity', which is presented below.

When one wants to integrate sustainability into the university maps a scenery evolves that is difficult to overlook and that is based on three topographic elements:

- Universities are microeconomic institutions that need to be managed sustainably.
- Universities are social actors that need to contribute to a sustainable development of the world.
- These two perspectives can be differentiated from each other. The following statements refer to the figure explained below (see Fig. 1).

2.1 Two Perspectives on the Sustainable University

Universities need to fulfil many demands their environments make. Most of these demands can be handled on two levels: from an *inside* perspective and an *outside* perspective. Humanization of work, gender equity, environmental protection, democracy and now also sustainability require innovative research. In this connection, a university is the acting subject creating new knowledge for a livable society. The closeness of research and teaching leads to direct knowledge transfer to students which proves the university's contribution for society. The sustainability reports of the University of Bremen from 2005 and 2010 are both based on this structure of research, teaching and transfer (University of Bremen 2005, 2010) that represents the outside perspective, which is also mentioned on the map in Fig. 1.

At the same time, a university is a social subsidiary system of society and is therefore the object of its own insights. Recommendations for enhanced environmental protection, progress in gender equity, humanized work processes and democratic decision-making processes should always lead to changes in the institution itself. As a public employer, a university is in charge of following its own insights. This is the inside perspective.

The external perspective and the internal perspective also apply for the issue of sustainability. At the first fork shown in the figure it is necessary to decide whether the university wants to contribute to a more sustainable society via research and teaching and/or whether it wants to develop towards more sustainability itself. This differentiation is important because it leads to different decision-making processes and concerns different decision-makers. Researchers decide if they want to focus on the question of how to reduce the side effects destroying the sources for social and natural resources and to strengthen the sources' productive power. University policies as well as university management can accelerate such a realignment with incentives, framework structures and arguments.

The extent to which a university as a microeconomic actor faces up to its own insights on sustainable management is decided by university management and can be accelerated by the engagement of professors.

2.2 From the Perspective to the Fields of Activity of Sustainable Universities

The two perspectives of the linkage of sustainability and university logically lead to different fields of activity. These fields of activity clarify the content of the construct of sustainability.

2.2.1 Contributions of Universities to a Sustainable Development of Society

The purpose of universities is to create new knowledge through research and to convey this knowledge through teaching. Simultaneously, they have the purpose to transfer the knowledge to other social institutions and to encourage a critical debate on social development.

In the field of research, a social-scientific sustainability research emerges and takes its place next to the strong engineering and natural-scientific environmental technology research. The latter focuses on first-order knowledge concentrated on exploitation, the former focuses on second-order knowledge aiming at more orientation regarding the evaluation of main and side effects of economic and technical developments.

In the field of teaching, the existing environmental education is enhanced by education of sustainable development (de Haan et al. 2008). This approach is based on the insight that under modern conditions problems occur and need to be solved that are far more complex than usual. Competencies of organization for sustainable development comprise the competencies needed for the solution of complex consideration and decision-making problems, which become more intensive due to the requirement of a more sustainable development. Universities face the challenge to create forms of knowledge transfer with which they can reach the greatest possible number of students of different fields of study.

2.2.2 Sustainability of the Institution Itself

In the end, a sustainable development of society aims at all social institutions (being microeconomic social systems). Thus, universities are asked to reflect the main and side effects they cause for humankind and nature and to organize them in a more trouble-free way. Universities therefore need to focus more on the autonomies of the sources that produce the material and immaterial resources they are dependent on. This challenge could be based on the efforts taken for the installation of an environmental management system. Environmental management systems aim at the monitoring of material and energy flows, so that harmful impacts can be revealed and be regulated systematically. The University of Bremen, for example, has treated this field of activity by taking part in the Eco-Management and Audit Scheme process and by being validated for EMAS.

While environmental management usually means that with different measures material and energy flows are reduced, meaning that natural resources are applied more efficiently, sustainable management focuses rather on the resource base (Müller-Christ 2011). This is a serious differentiation: Even if resources are applied most efficiently, they will someday be used up if reproduction is not considered at the same time. This is the focus of the research field 'sustainable resource management': Which natural, social and economic resources is a university dependent on? And what can it do in order to secure the long-term supply with these resources?

The previous efforts for environmental management have revealed that institutions, which want to regulate their material and energy flows in a way that causes less side effects, need to initiate a development of their whole organization. This is increasingly so when the issue of sustainability shall also be taken into account. The field of activity 'organizational development' is extremely important because the running processes of organizational development now also need to be aligned to the decision-making premises of sustainability.

The previous efforts for the humanization of work, for gender equity, for democratization, and for the reconciliation of work and family life are not included when sustainability is focused solely on the resource relationships of a university.

Through this, the concept of sustainable development of universities is sharpened but the excluded needs to be re-included in the following steps of concretization. However, the humanization of work, gender equity, democratization and the reconciliation of work and family life are normative concepts that can claim relevance even without having to use the sustainability term.

3 The Method of Systemic Constellations

Systemic constellations facilitate emotional, affective and cognitive experience and learning in divergent groups. They work with scenic presentations of relationship structures of a system: A client places persons in the room (in the constellation space) as elements of a system, and the relationships of the elements are being visualized through the distance between persons and their respective viewing directions. Persons standing as representatives make use of their "representative perception" which allows them to empathize with the elements they represent and to serve as their mouthpiece. They can physically feel if the place they were assigned and the relationships they have with others are acceptable, pleasant, distracting, strengthening etc. and they gain implicit knowledge of the system through intuition (for this frequently verified phenomenon the definite explanation has not yet been provided, see Klein and Limberg-Strohmaier 2012). In this way, a certain system of origin can be reproduced realistically in a different place under the responsibility of a trained facilitator.

The process work comprises specific questioning of elements, the inclusion of new elements or exclusion of old ones, and precise demands concerning the representatives' condition. It aims at creating a "good" system and reveals interpretation offers for the original problem, which often lead to great progress regarding the insights of the clients. The following discussion between all participants and the audience is of particular importance. In this discussion, the representatives as well as the observers describe and compare the associations they had. Experience teaches that during these discussions researchers and practitioners can debate on the same level because they refer to commonly created pictures and experienced changes. The conclusions drawn from constellation work would, in most cases, not have been reached through document analyses, interviews or empirical surveys—at least not in such a short period of time (constellations usually take 1–2 h).

Empirical evidence is growing that in constellations a kind of sign language or sign system is used which is interpreted in similar ways by different people (Varga von Kibéd and Sparrer 2009). Schlötter proved in an often cited work that persons share supra-individually similar experiences regarding the meaning of the positions persons occupy in a certain space which is why they derive not only similar experiences, but also similar interpretations. When persons in constellations are exchanged or constellations are repeated somewhere else with different representatives, the statements the representatives make correspond significantly (Schlötter 2005; Kohlhauser and Assländer 2005). Baecker concludes that systemic constellations generate a self-commenting structure in space, which does not demand more than the copying of specific characteristics of the system of origin (Baecker 2005).

From constellations practice can be seen that people are able to take the role of representatives in constellations in a realistic way without extensive theoretical explanation, solely through professional guiding: Without having explicit causal explanation yet, the apprehension of interrelations and their verbal expression emerge from the simulation of a system of origin (Rosner 2007). The prevalent attempts to explain this phenomenon work with constructs such as "representative perception" (Varga von Kibéd 2005), "morphogenetic fields" (Sheldrake 2008) or "knowing fields" (Mahr 2003).

The method of systemic constellations has reached a maturity that makes it an interesting instrument for sustainability-related management research and can be viewed as part of action research. Giving recommendations for institutions' activities requires substantiated knowledge and understanding of the institution that is about to be changed. At an early stage the insight emerged that the people who want to develop an organization can cause more effects once they get into the institution and develop better understanding though participation. However, a participant automatically becomes an element of the system he or she wants to change. From the view of an intendedly neutral science great challenges arise for this kind of action research. Action research belongs to qualitative empirical social research whose methodological self-confidence continually grows with increasing quality of recommendations (Atteslander 2003; Krause 2006). With their solution-orientation system constellations can contribute to this self-confidence and can provide greater significance for the demands for analyses and insights the transdisciplinary action research makes.

In the end, action research builds a bridge between the observers' and representatives' self-experience and the scientifically distanced research (Rosner 2007). The latter can not only explore the method's causal links, but also cluster repeatedly occurring relationship patterns and condense them to new hypotheses on systemic relationships. In the process of integrating sustainability into predominant decision-making routines such new relationship patterns are essential.

Compared to conventional methods of qualitative empirical social research systemic constellations have the advantage that within a very short period of time even complex systems can be illustrated and be given a voice. Systemic constellations are an effective instrument for the visualization of the order of interactions between a system's elements, in the sense of a grammar of systemic action.

The term action research carries terminologically both the search for new insights and organizational will. Research insights and recommendations for organization and action develop in the same process. Every constellation helps understand the grammar of systemic action better. For this, every constellation recalls the principles of systemic order and at the same time makes the testing of different solutions possible on the basis of these principles. At the end, there may not be specific recommendations for action (however, no method can provide these with absolute certainty), but a clearer and deepened insight into the analyzed system which subsequently leads to more appropriate behavior.

4 Sustainability at Universities—Examples from the Seminar Series

4.1 Example No. 1: University Management and Sustainability

During the seminars with the sustainability coordinators it became clear that the coordinators have very different pictures in mind when thinking of the meaning sustainability should have for the rectorate or the university management. In systemic visualizations the coordinators made their "inner maps" visible, following the question: What should the relationship of university management, sustainability coordinator and sustainability itself look like?

4.1.1 Description of the Constellation

The university management or rectorate was assigned a fixed place. The sustainability coordinators then positioned the issue of sustainability and themselves in relation to the university management. Different pictures evolved (see Fig. 2): Constellation #1 showed that the coordinator placed herself next to the rectorate, placing sustainability right between them without leaving room for anyone else. This picture reveals that the coordinator sees herself in one line with the university management. Constellation #2 showed a different picture: this sustainability coordinator adopted a challenging attitude towards the university management by

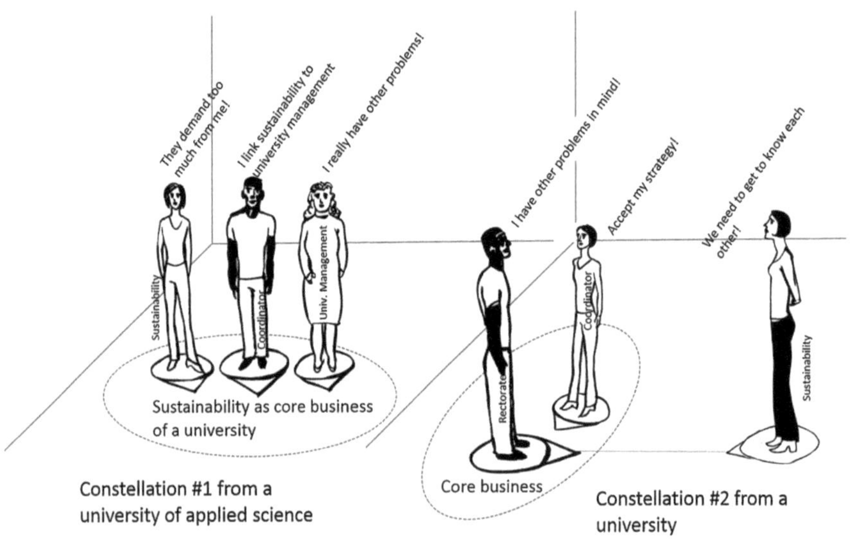

Fig. 2 Constellations regarding the role of university management

placing herself near the management, but diagonally staggered. She gave sustainability a central position by placing it in the management's direct field of vision.

4.1.2 Process Work

The process work revealed for both constellations that sustainability felt insecure and thought it was in the wrong place. It felt that the university had not yet accepted it and searched for a more inconspicuous position. The respective university managements were pleased and distanced themselves, too. However, they felt the need to keep sustainability in view. It turned out that the university management has to deal with too many different other problems. The most urgent problem entered into the system and immediately drew the management's attention on itself. This shift in perception then led to the sustainability coordinators' feeling of being weak and irritated. They had problems with finding a good place for themselves afterwards.

4.1.3 The Final Picture

The final picture revealed a phenomenon that the coordinators could hardly accept although it actually occurs very often: The distance to the problem made the university management feel comfortable and benevolent. They were annoyed when sustainability tried to move closer to them. In the concluding discussion after the systemic constellation the sustainability coordinators developed different ideas and strategies on how to prepare the issue of sustainability in a way that university

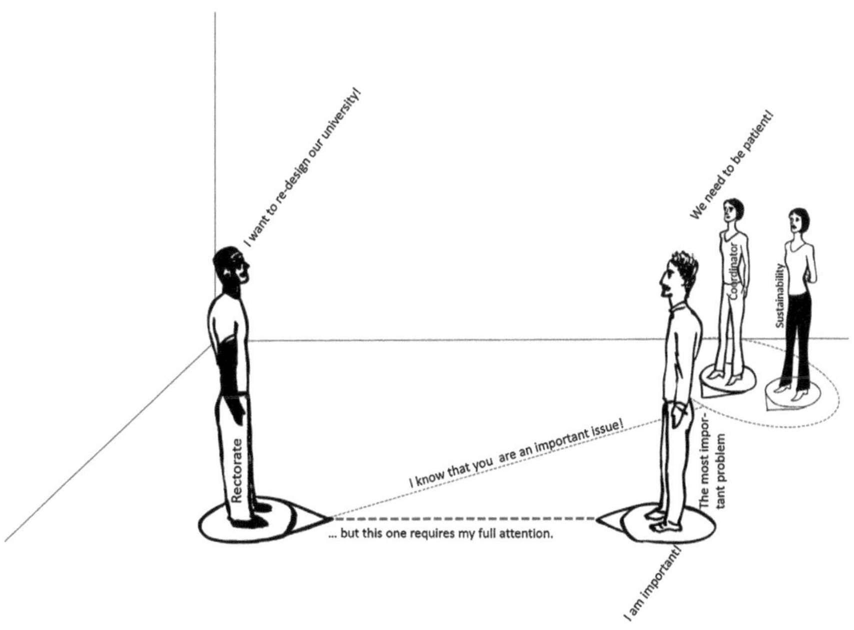

Fig. 3 Final picture of one constellation regarding the role of university management

management might not only want to look at sustainability and support it morally, but also make it part their strategy (Fig. 3).

4.1.4 Interpretation Offers

Usually, the jobs for sustainability coordinators are created by the university managements. Therefore, the jobholders find it extremely important to build a good relationship with the management. They know that they can fulfill their tasks better when the university management is committed to the issue of sustainability—and attaches importance to the coordinators by doing so. It is hard to accept that after the job is created sustainability is only one topic among many. One sustainability coordinator even linked arms with the university management and explained to them that they would walk together from now on. However, the assignment of so much meaning is rather uncommon. The risk is that sustainability coordinators have to spend most of the time during their time-limited jobs on finding a good place within the system. From a systemic perspective, new topics start on a very low level and have to show a lot of appreciation for those issues that had already been there when trying to reach a higher level with more importance. This process of overtaking requires a lot of empathy from the sustainability coordinators.

4.2 Example No. 2: New and Old Players in the Sustainability System at a University

The work during the seminars showed that there are often problems from the past that can complicate the work of sustainability coordinators. Coordinators need situationally suitable resources to support them in their work. With systemic constellations these resources were made clear and the major problems hampering the work were revealed.

4.2.1 Description of the Constellation

In this case, the sustainability coordinator had the task to implement a new, living strategy for sustainability. The coordinator placed the following five elements in the constellation space according to her inner picture of the constellation: the sustainability strategy, the rector, herself, potential alliances (as the resources she might need) and unsuccessful sustainability projects from the past whom she identified as the major barriers for her work (see Fig. 4). The picture shows that the unsuccessful projects from the past stand in the center of the system, occupying a crucial place within the constellation. From the outside the coordinator looks at the system and sees the sustainability strategy and potential alliances in the distance far away. Typically, the rector looks at the system from a point that provides a good overview.

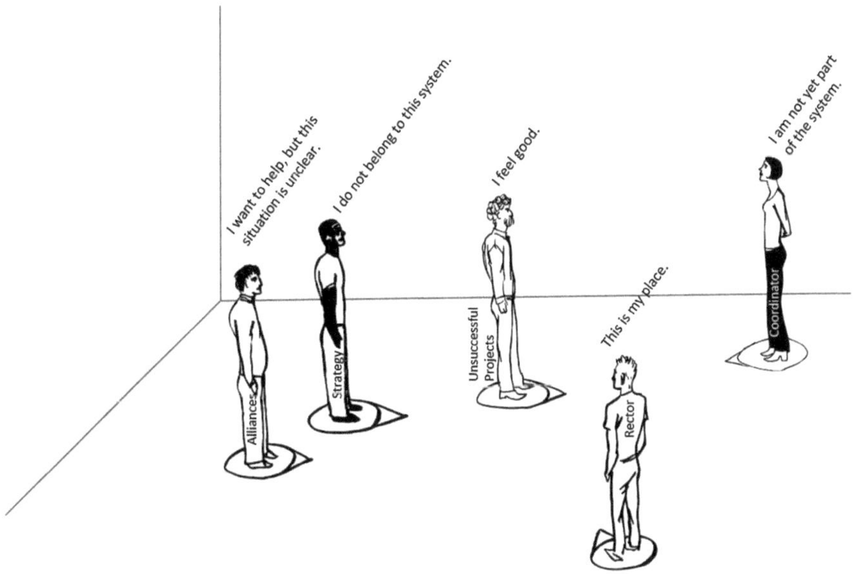

Fig. 4 Constellation regarding the implementation of a sustainability strategy (*initial picture*)

4.2.2 Process Work

The process work revealed that the strategy does not yet feel in sight. Too many other problems need to be discussed and solved first. Since the elements thought that something was missing, two more representatives were placed in the constellation: a research center dealing with sustainability issues at the university and the old ethos of the university. These two elements together with the unsuccessful projects dominated the system, so that the coordinator felt there was no place for her and concluded that it was not yet the time for a completely new strategy. Under the direction of the facilitator the elements searched a long time for their respective good place in the system aiming at a situation in which the coordinator can see her possibilities of developing and implementing a new strategy.

4.2.3 The Final Picture

The final picture revealed that there is a strong field of tension the sustainability coordinator has to pass through on her way to implementing a new sustainability strategy (see Fig. 5). The elements that already existed before the new strategy was discussed and before the sustainability coordinator started her job remained very strong and felt that they belonged to each other. They represented the old system: the university. The coordinator found out that she definitely needs to form alliances in order to finally reach the new strategy. These elements represent the new system: sustainability.

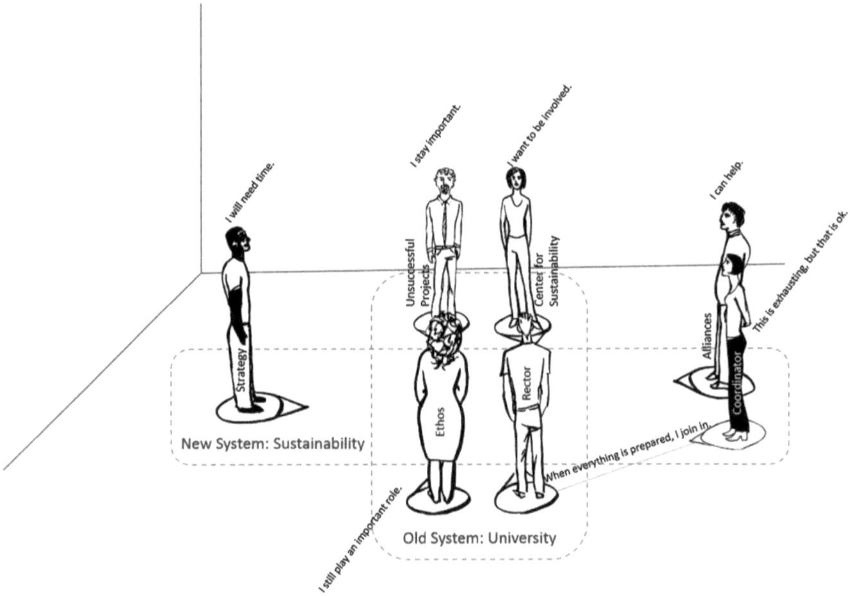

Fig. 5 Final picture of the constellation regarding a new sustainability strategy

4.2.4 Interpretation Offers

This constellation revealed an important, but often underestimated phenomenon: the role of older elements and their demands. From a systemic perspective, 'the old' has priority when compared to 'the new'. When sustainability and all the people, projects and structures that go along with it enter into a system, they have to be aware of the older structures that are being changed and the older elements that might lose importance. In this example, an advice to the sustainability coordinator could be to act carefully when in contact with the university system and to deal with old structures, with people that where there before she came in and with failures or disappointments in the past. Moreover, the problem mentioned above is valid for this example, too: the rector has the university development in mind and initially supported the coordinator. But in daily work life he has to keep in mind many different projects. And he is naturally primarily part of the old system—the university. The process of walking through the field of tension requires a lot of strength and much respect for 'the old' from the coordinator on the way to developing the new sustainability strategy.

4.3 Some Short Stories from the Constellations

These stories serve to rethink relationships and to point out the difficulties occurring during the process of integrating sustainability in universities. The constellations outlined took place during the seminar series, which the participants regarded as very helpful and realistic. Due to lack of time and space, only the central points of the constellations are described.

- Laughter about eco-efficiency:
  A constellation about a research institution was conducted. From the view of management, professors, senior researchers, doctoral students and administrative staff were placed in the room. In addition, as one central external challenge eco-efficiency was positioned in the system. Some other elements reacted with uncontrollable laughter. How could this be interpreted? The research institution is dependent from energy-intensive facilities for its natural scientific experiments. Energy saving is stressful for the institution because this requires a different organization of research work. This means that the researchers' self-autonomy is reduced when they cannot utilize the facilities at any time as they could before. Humor is a well-known measure to reduce inner tensions. The constellation revealed why it is not as easy to combine research and eco-efficiency as often expected.
- Who is stronger: sustainability of diversity?
  The cross-sectoral issues health, gender, diversity, operational safety and sustainability sort of bicker about importance within the institution. Often, they need to be realized with few human resources. What is the best way of dealing with this problem for the different issues: going alone, working together, forming one unit, taking weaker issues with them? The constellation showed that the gender issue wanted to exploit the sustainability issue for its own ends. The health issue felt closely related to the sustainability issue, but felt weak and unhelpful. Operational safety seemed to live in its own world. Diversity and sustainability fought for the leading position. The constellation revealed the warning for the different issues that quick fraternization of cross-sectoral issues could hold dangers.
- Do students actually want education for sustainable development?
  Creating a picture with the relationships of lecturers, students and departments often shows that students are completely focused on their lecturers and draw little attention to the departments or the university. When education for sustainability is added to the picture, only the lecturers react to this new element, attaching (some) importance to this new issue. Sometimes the students then moved closer to the lecturers. How could this be interpreted? Certainly, in every university there are some students that are interested in sustainability issues. However, most students probably concentrate on meeting the demands their lecturers make. If the lecturers teach sustainability, the students will learn these topics; if not, they will not.

5 Conclusion

Working on the development of universities towards more sustainability is a matter of personal importance for many sustainability coordinators. If they could, many of them would take the issue and put it immediately into every process, every course and every office at their universities. However, sustainability is still one issue among many. University management has to deal with many problems, any of them also connected with people for whom their projects are equally important. Demanding that sustainability needs to become the top issue of university management is therefore a too ambitious goal. Sustainability coordinators need to realize that the development and stabilization of sustainability at their universities is rather a problem of organizational development than a one-time-implementation-process.

The map for sustainability and universities can help to identify the paths sustainability coordinators could take. The decision to become a sustainable university (inside view) requires different approaches than the aim to contribute to a sustainable development of society (external view). Often sustainability coordinators try to combine both paths. A clear demarcation of the system that shall be made sustainable can help when trying to find the processes and people that need to participate if sustainability shall be reached.

The method of systemic constellations can visualize such insights. The method can not only make current problems visible, but also provides the opportunity to give different future scenarios a try. By this, one can get a feeling for recommendable or inadvisable objectives. For many sustainability coordinators that took part in the seminar series the pictures evolving were surprising and revealed new perspectives or showed clearly what they actually already knew, but never wanted to admit: they assigned the great issue of sustainability a too central place in the whole university system. In both cases the sustainability coordinators found the insights very helpful, supporting them to (re)define their ways of fulfilling their tasks as coordinators or to make clear in which direction their future foci could be aimed.

Summing up, the following insights were gained from the work with sustainability change agents in German universities: (1) The persons trying to integrate sustainability into research, teaching, transfer, administration and university management face some difficulties when trying to *find the proper place for themselves and for the sustainability topic in their universities*. The creation of such jobs has made the importance of sustainability visible; however, the sole creation of jobs that focus on the integration and organization of such cross-sectoral topics cannot automatically make the institutions sustainable. Sustainability coordination usually has to stand in line with other topics like quality, diversity or equity improvements. (2) Although one might wish for university management to *put sustainability on top of their agendas*, this is unlikely to happen. There are too many topics university management has to deal with in everyday decision-making. It seems that university management is more willing to adopt approved approaches than to develop these themselves. Sustainability coordinators (or sustainability change agents) need to accept that their work on changing the ways universities deal with sustainability issues will only slowly produce concrete results.

The great task to bring sustainability into universities, both into internal processes like the institutional management and into external processes like educating students in the matter of education for sustainable development, requires people that organize these processes. The creation of job descriptions like sustainability coordinators is a first step to organize the rearrangement of university issues towards sustainability. The revelation of the barriers and drivers, of potential strategies and concrete measures is the next step that needs to be assigned importance and attention. Systemic constellations can help to make these steps clearer. From the insights of the constellations conducted the following consequences can be deduced besides the need for sustainability change agents:

From the look at the role of sustainability change agents one can see that *the task of integrating sustainability into universities* is not only a structural one, but rather *a process-oriented one*. The process of system development is not yet organized in a way that could lead to the embedding of sustainability in the overall system of higher education.

The higher education system is a system that can only change slowly. Too many different aspects hamper a quick development. Integration processes initiated within the established and well-known structures might take too long. Therefore, one might have to think about the *constitution of new institutions* that do not have to follow the usual structures. One example from the University of Bremen in Germany is the Virtual Academy for Sustainability (www.va-bne.de) which provides online courses with credit points for students who want to take sustainability courses but find no offers at their home universities. Examples like this, which try to bypass the usual university structures, could help establish a sustainability culture among universities. Sustainability coordinators would then not have the task to change the university system on their own but to initiate and support projects and institutions that might themselves lead to a more sustainable higher education system in the long run.

References

Atteslander P (2003) Methoden der empirischen Sozialforschung, 10th edn. Berlin

Baecker D (2005) Therapie für Erwachsene. Zu Dramaturgie der Strukturaufstellung. In: Groth, T./ Stey, G. (ed): Systemaufstellung als Intervention in Organisationen—Von der Praxis zur Theorie und zurück. Heidelberg, pp 14–31

de Haan G, Kamp G, Lerch A, Martignon L, Müller-Christ G, Nutzinger H-G (2008) Nachhaltigkeit und Gerechtigkeit. Grundlagen und schulpraktische Konsequenzen, Berlin, Heidelberg

Klein P, Limberg-Strohmaier S (2012) Das Aufstellungsbuch. Familienaufstellung, Organisationsaufstellung und neueste Entwicklungen. Wien

Kohlhauser M, Assländer F (2005) Organisationsaufstellungen evaluiert. Studie zur Wirksamkeit von Systemaufstellungen in Management und Beratung. Heidelberg

Krause O (2006) Performance management. Eine Stakeholder-Nutzen-orientierte und Geschäfts-prozess-basierte Methode, Diss. Wiesbaden

Mahr A (2003) Konfliktfelder—wissende Felder. Systemaufstellungen in der Friedens- und Versöhnungsarbeit. Heidelberg

Müller-Christ G (2011) Sustainable management. Coping with the dilemmas of a resource-oriented management. Berlin

Rosner S (2007) Systemaufstellungen als Aktionsforschung. Grundlagen, Anwendungsfelder, Perspektiven. Schriftenreihe des Instituts für systemische Aktionsforschung, Band 1. München, Mering

Schlötter P (2005) Vertraute Sprache und ihre Entdeckung. Systemaufstellungen sind kein Zufallsprodukt—der empirische Nachweis. Heidelberg

Sheldrake R (2008) Das schöpferische Universum. Die Theorie der morphogenetischen Felder. Frankfurt

University of Bremen (2005) Sustainability Report of the University of Bremen 2005 (German only). http://www.ums.uni-bremen.de/pdf/UniHB_NHB_2005.pdf

University of Bremen (2010): Sustainability Report of the University of Bremen 2010 (German only). http://www.ums.uni-bremen.de/pdf/UniHB_NH_Bericht10_gesamt.pdf

Varga von Kibéd M (2005) Ein Metakommentar. In: Weber G, Schmidt G, Simon FB (eds) Aufstellungsarbeit revisited—nach Hellinger? Heidelberg, pp 200–250

Varga von Kibéd M, Sparrer I (2009) Ganz im Gegenteil. Tetralemmaarbeit und andere Grundformen systemischer Strukturaufstellungen, 6th edn. Heidelberg

Author Biographies

Prof. Dr. Georg Müller-Christ studied business studies at the University of Bayreuth where he also received his doctoral degree and venia legendi for business studies. Since the year 2001 he has been holding the Chair in Sustainable Management at the Department of Business Studies and Economics at the University of Bremen. He is speaker of the working group "Universities and Sustainability" of the Round Table of the UN-Decade "Education for Sustainable Development" and is a trained facilitator for systemic constellations.

Dr. Anna Katharina Liebscher studied Industrial Engineering at the University of Bremen where she also received her doctoral degree in business studies. Since 2010 she has been working at the Chair in Sustainable Management at the Department of Business Studies and Economics at the University of Bremen. She has been working with the method of systemic constellations for several years in the university context.

Together, Prof. Müller-Christ and Dr. Liebscher give university courses on sustainability and management issues using the method of systemic constellations. They also carry out seminars for sustainability coordinators and other interested parties that aim at finding a good place for sustainability in universities.

Business and Management Education for Sustainable Development: A Process for Creating Sustainable Public Value

Mirjam Minderman

Abstract

As business schools provide training and education to both future and current business leaders, they can be regarded as a very important influence on how businesses operate. As such, business schools could play an important role in the change process of the profit sector towards sustainability. An increasing number of business schools sees its responsibility in this field and explicitly claims to adapt curricula and research in such a way that corporate social responsibility and sustainable development become fully embedded. This paper forms the introduction to a research project that aims to describe and analyse the process that enables business schools to create sustainable public value through their education for sustainable development. The paper provides a brief overview of the developments in management and business education for sustainable development, leading to the definition of the research question and strategy. The use of a value-based theoretical framework and the focus on the process behind education for sustainable development provide a highly relevant alternative to the merely quantitative assessment tools that are currently used to demonstrate the value of education for sustainable development. The research offers a qualitative method and a continuous process—that includes stakeholders—for assessing the contribution of management and business education for sustainable development towards the common good.

M. Minderman (✉)
Sustainable Development Initiative, Postbus 19272, 3501DG Utrecht
The Netherlands
e-mail: info@opzoeknaarmeerwaarde.nl

W. Leal Filho et al. (eds.), *Integrative Approaches to Sustainable Development at University Level*, World Sustainability Series, DOI 10.1007/978-3-319-10690-8_32

Keywords
Business and management education · Sustainable development · Value · Process · Assessment

1 Introduction: Sustainability in Management and Business Education

Higher education is generally seen as a major potential catalyst towards sustainable development, in particular through its functions of education, research and outreach (Barth et al. 2011; Madeira et al. 2011; Waas et al. 2012; Yarime et al. 2012). Within this field of higher education for sustainable development, management and business education holds a special position. Management and business education has been and still is seriously criticised because of its contribution to the current unsustainable practices in the private sector. As business schools provide training and education to both future and current business leaders, they can be regarded as a very important influence on how businesses operate (Morsing and Rovira 2011, p. xviii; Slater and Dixon-Fowler 2010). The economic and financial crisis of the last years reinforced criticism of the sort of education that business schools provide; one of the main points of critique being that business schools form business leaders with a single-minded and short-term view on the pursuit of profits (Ghoshal 2005; Morsing and Rovira 2011, p. xx; Swaen et al. 2011, pp. 179–180). While businesses are being challenged to expand their purpose beyond mere profit-making to value-added contributions to society and the environment (Porter and Kramer 2011; Visser 2011; WBCSD 2010) and business leaders are being urged to lead responsibly and address societal and environmental needs of the world (Sachs 2013), business schools (and management education in general) are called upon to train such responsible leaders (Alcaraz and Thiruvattal 2010; Muff et al. 2013, p. xxiv; Slater and Dixon-Fowler 2010; Von der Heidt and Lamberton 2011, p. 676).

Over the last years, several initiatives have started with the goal of promoting management and business education for sustainable development. As early as 1991, a group of dedicated scholars within the Academy of Management joined hands and formed the 'Organizations and the Natural Environment (ONE) Division', with the aim of bringing the natural environment to the forefront of management research and education (only in 2007, ONE became an official division of the Academy) (Stead and Stead 2010).

While ONE has a focus on the 'green' or 'planet' aspect of sustainability, other groups use a broader scope of the concept, including 'people' and 'profit' (or 'prosperity'), and stress a more holistic approach towards management and business education for sustainable development. By 2010, three initiatives were simultaneously and to some extend similarly working on the broader transformation of business schools towards sustainability: the Globally Responsible Leadership Initiative (GRLI) had launched project SB21, which was about developing a blueprint for the business school of the twenty-first century. Some deans at a meeting of the

World Business School Council of Sustainable Business (WBSCSB) 'decided it was about time that business schools engaged in the public debate on sustainable development and reflected on what contribution management education could make to create a better world' (Muff et al. 2013, p. xvi) and thus initiated 50+20. The mission of the United Nations-supported Principles for Responsible Management Education (PRME) initiative is to inspire and champion responsible management education, research and thought leadership globally. In January 2011 the three organizations signed a joint agreement: project SB21 and 50+20 merged into one and continued under the name 50+20, with support of the three organizations GRLI, WBSCSB and PRME. There are many more initiatives with a regional scope and/or a focus on specific issues such as business ethics. Many are networks of academic institutions and companies, thus stimulating the interaction between academics and practitioners, across sectors.

The work of these initiatives and of an increasing number of engaged scholars and practitioners has contributed to a much wider acceptance of the need to include 'sustainability' in business and management education. This acceptance is both reflected and further stimulated by the fact that in early 2013 the two leading accreditation organisations for business schools, the AACSB and EFMD, included sustainability in their accreditation criteria.

Despite, or possibly because of, this wider acceptance of the need for business and management education for sustainability, there is a vivid discussion amongst both academics and practitioners about the contents and methods of such education (similar to the discussion concerning higher education for sustainability in general). Should it be added to existing programs (possibly as electives), or fully integrated in all courses? Is it about providing knowledge or rather about training specific competences? (Rusinko 2010; Stubbs and Cocklin 2008) Following the famous quote by Einstein 'We cannot solve our problems with the same level of thinking that created them' some claim that a drastic reorientation of pedagogy is needed in order to allow for the integration of holistic and systemic thinking that (management and business) education for sustainable development requires (Waas et al. 2012; Muff et al. 2013; Wals 2011). Especially with regard to management and business education, the need for different thinking, metaphors and paradigms is stressed (Audebrand 2010).

Not only are there many different approaches and ideas about what education for sustainability should be like, also the terminology itself differs amongst groups and scholars. Concerning sustainability and higher education in general, Wals and Jickling (2002) illustrate the different perspectives on 'sustainability', 'education for sustainable development', and 'education for sustainability' and stress that conflicting norms, values and interests can be underlying these terms and concepts. With regard to sustainability and management and business education, also the terms 'responsible business education' and 'responsible management education' are being used. While acknowledging that different views and values may underlie the various terms, this paper uses 'management and business education for sustainable development' as a comprehensive term, following the terminology and definition provided by the United Nations and promoted through the UN Decade of Education for Sustainable Development (2005–2014): 'Education for Sustainable Development allows every human being to acquire the knowledge, skills, attitudes and values necessary to shape a sustainable future' (UNESCO 2014).

Given the rapidly increasing number of business schools that explicitly claim to include sustainability in their curricula and research and the many different ways in which this is done, the question is to what extent the business schools' educational and research activities for sustainable development indeed contribute to 'a more sustainable world'. Answering this complex question would require a longitudinal study, but in order to be able to come to some (tentative) answers on the shorter term, this paper proposes a research project that uses an alternative approach that allows for statements about the value of management and business education for sustainable development.

2 Research Question

This paper forms the introduction to a research project that describes and analyses the contribution of management and business education for sustainable development to the change process towards sustainability of the profit sector. The description will be made from the perspective of business schools. The research will look into how a small group of selected business schools develop and implement their strategy with regard to education for sustainable development and how they (expect to) monitor and evaluate the results of this strategy. By looking at these cases through a specific theoretical lens, it will be possible to come to statements about the process that enables the selected schools to create sustainable public value. The central question of this research is:

> What process enables business schools to create sustainable public value through their education for sustainable development?

In order to be able to answer the central research question, a number of sub-questions need to be taken into account. First of all core concepts such as sustainability, (sustainable) value, education for sustainable development and business schools need to be defined. Because these are broad concepts of which many different interpretations exist, it is important to define clear meanings in the context of this research.

Furthermore a theoretical framework needs to be developed that can be used to look at the practice of education for sustainable development provided by business schools. As the research focuses on the value of this type of education, theories concerning value creation, sustainability and the role of business schools in society need to be taken into account. The starting point of the theoretical framework will be a theory about the creation of value for the common good: public value theory. Its main features and relevance for this research are explained in the next paragraph.

The theoretical framework will serve as a lens to look at the practice of a number of business schools who claim explicitly that they aim to provide management and business education for sustainable development. By means of case studies the research will look into how these schools (expect to) create sustainable value through their education for sustainable development and how they monitor this value. These case studies will not only describe, but also stimulate a dynamic process in which the business schools develop, ensure and monitor the creation of sustainable public value. The researcher serves both as a facilitator and an (engaged) observer in this process, that is supported by the theoretical framework. Thus, the case studies also contain elements of action research.

The combination of these concepts, theories and case studies will not only provide an answer to the question of what process enables business schools to create sustainable public value through their education for sustainable development, but it will also result in a coherent framework that business schools can use to demonstrate the sustainable public value of their education for sustainable development.

Before elaborating on the research strategy, more needs to be explained about the theory that underlies the research.

3 Theoretical Framework: Public Value Theory

As the research looks into the value of management and business education for sustainable development, the starting point for the theoretical framework is a theory that is about the creation of value for the common good: public value theory.

Since 1995, Moore's concept of *public value* has become an essential frame of reference for policy-makers and managers in the public sector around the world. Over the years, the theoretical framework has been further developed and applied in practice. Challenges related to sustainability have explicitly been included, especially by Benington (2011) and Swilling (2011), who even introduced a tentative conception of 'sustainable public value'.

As Benington (2011) points out, there are two perspectives when defining public value: the first is 'what does the public most value', which emphasizes dialogue and engagement with current users, citizens, and communities. The second is 'what adds value to the public sphere', which focuses on the wider public interest, not just on the needs of current users but also on the longer-term public good, including the needs of generations to come (Benington 2011, p. 43). As sustainability is precisely about the longer-term public good, including the needs of generations to come,[1] 'public value' is used here in its wider notion of 'what adds value to the public sphere'.

Key to the conceptualization and operationalization of public value is the *strategic triangle*.

Benington and Moore (2011, p. 4) describe the strategic triangle as 'a framework for aligning three distinct but inter-dependent processes which are seen to be necessary for the creation of public value:

[1] There are numerous definitions of sustainable development, but the most frequently quoted is from *Our Common Future*, also known as the Brundtland Report: *Sustainable development is development that meets the needs of the present without compromising the ability of future generations to meet their own needs* (WCED 1987).

- Defining public value—clarifying and specifying the strategic goals and public value outcomes which are aimed for in a given situation.
- Authorization—creating the 'authorizing environment' necessary to achieve the desired public value outcomes—building and sustaining a coalition of stakeholders from the public, private and third sectors (...) whose support is required to sustain the necessary strategic action.
- Building operational capacity—harnessing and mobilizing the operational resources (finance, staff, skills, technology), both inside and outside the organization, which are necessary to achieve the desired public value outcomes.'

The concept of public value highlights the importance of focusing on processes and outcomes ('what value is being added to the public sphere, by whom and how?') not just on inputs and outputs (Benington 2011, p. 47). Thus, public value is not measured by 'objective numbers' (if such numbers are possible at all) but established by the judgment of—in this case—the business school and its authorizing group of stakeholders: if they argue—jointly and convincingly—that achieving specific goals and effects has resulted in sustainable public value, such value has indeed been created.

The translation of public value theory and specifically the strategic triangle with regard to management and business education for sustainable development could be highly relevant. The focus on the creation of value for the common good and the important role for stakeholders fits in with generally acknowledged prerequisites for implementing change towards sustainability. Concerning management and business education for sustainable development, it does link especially to the abovementioned 50+20 vision, which is in short about 'the transformation of management education, in which the common tenet of being the best *in* the world is revised in favor of creating businesses that are designed to be the best *for* the world' (Muff et al. 2013, p. 55). Muff and colleagues argue that 'engaging in the transformation of business and the economy will require measures that assess a business school's presence in, contribution towards and impact on the greater public. Defining measures of quality and success relates to evaluating a school's level of engagement in society both in terms of intensity and perceived success. Such public engagement ranges from local to international, and needs to be evaluated according to its public relevance, taking into account competences and potential contributions of the school. (...) By working closely with its relevant stakeholders, the school is recognized and compensated for the value it contributes. This recognition may lead to a redefinition of sources for funding, opening potential new avenues of cooperation.'

The constant alignment between goals, resources and especially the authorizing environment of which the strategic triangle exists, provides a continuous and dynamic process that ensures focus and stakeholder involvement at various stages of management and business education for sustainability, ranging from development, implementation and monitoring to accountability. This process offers an alternative method to establish and legitimize the public value of management and business education for sustainable development. Such an alternative method is useful because current sustainability assessment tools for higher education tend to focus more on operational eco-efficiency instead of education and research, while they generally ignore broad stakeholder involvement (Shriberg 2002; Yarime and Tanaka 2012). The mechanism provided by the strategic triangle refrains from instrumental, quantitative indicators and allows for a qualitative, value based monitoring process that is based on stakeholder involvement and allows to take into account different perspectives, circumstances and needs.

Because a continuous stakeholder dialogue is inherent to public value theory, Benington stresses that it can be used 'not only as a conceptual tool for strategic planning, but also as a heuristic device to stimulate debate between competing interests and perspectives, and to generate dialogue about how to improve services, about who gains and who loses, and about relative benefits and costs' (Benington 2011, p. 49). Such a debate is relevant in a context where internal and external stakeholders need to be taken aboard and where resources are (increasingly) limited.

Over the last years, public value theory has been applied to education in general and as described above Benington (2011) and Swilling (2011) have linked public value theory to sustainability. This research will add a new realm to the existing theory on creating public value by widening its scope to business schools and by building a link between public value and education for sustainable development. In addition, the combination of public value theory and monitoring is new. So far, the theory has focussed on *creating* public value, not on monitoring. However, the same mechanism used through the strategic triangle for creating public value, also offers a relevant mechanism for monitoring the created value.

4 Research Strategy

The research looks at the strategies and actions of business schools concerning their educational and research activities (inside and outside the regular curricula) with which they intend to contribute to the change towards sustainability of the for-profit sector. The use of a theoretical framework that is based on the creation of public value allows to describe the process through which business schools create value through their education for sustainable development. The focus of the research on the concept of 'value' and on the underlying process, means that it goes beyond pedagogy and beyond the measuring of indicators.

After having *defined* the key concepts, the research aims to *develop* a theoretical framework based on public value theory that can be used to *describe* the efforts of business schools to create sustainable public value. Consequently these efforts are *analysed and compared* to find common characteristics and to refine the theoretical framework. Finally the main issues for further implementation and research will be *formulated*. This research strategy is explained below and subsequently summarized in Table 1.

Table 1 Research strategy

Steps	Key activity	Art of activity	Research method
1. Defining key concepts and relevant context	Define	Theoretical research	Desk research
2. Developing a framework of analysis for sustainable value creation	Develop	Theoretical research (adapted in the course of the research)	Desk research
3. Describing the practice of business schools (using the framework)	Describe	Field work	Case research by desk research, interviews, workshops, etc.
4. Analysing and comparing the practices	Analyse and compare	Finding common opportunities and threats, chances and dilemma's; adjusting the theoretical framework	Desk research
5. Concluding on the main outcomes and issues	Formulate	Answering research question; defining needs for further research	Desk research

1. *Defining key words and relevant context*
   Defining the key terms such as 'value' and 'sustainability' is essential in a social and academic environment where the same words may carry so many different meanings. Furthermore it is important to describe the specific context and background of business schools.
2. *Developing a framework of analysis for sustainable public value creation*
   Public value theory will be the starting point for the theoretical framework of the research. The application of public value theory and specifically the strategic triangle with regard to business schools serves to establish a process that enables them to create sustainable public value through their education for sustainable development. In addition, the alignment between goals, resources and especially the authorizing environment allows for a monitoring process. Precisely this combination of public value creation and monitoring makes the strategic triangle a highly relevant concept for this research. During this phase, (aspects of) other theories may be included in order to optimise the theoretical framework for the context of business schools and education for sustainable development. Based on the experiences with the case studies, the theoretical framework may still be further refined throughout the research.
3. *Describing the practice of business schools*
   As indicated, this research is in line with the tenet of vision 50+20, in which a drastic transformation of management and business education is promoted and in which business schools are recognized and compensated for the value they—together with their stakeholders—contribute towards the common good. Therefore, the research will concentrate on business schools that are part of the 50+20 initiative. An invitation to participate in the research as a case study will

be spread amongst the 50+20 community, which should lead to the participation of some 3 to 5 business schools.

In the case studies, the theoretical framework will be applied to the participating business schools. Each school will engage in a series of workshops (three workshops in some two years). In the first workshop, the school defines its targets concerning its education for sustainable development, its network of authorizing stakeholders and the organization and capacity concerning the realization of its targets. After this defining workshop, the second workshop will focus on monitoring progress and dilemmas. The third workshop will evaluate the process.

Parallel to the workshops, stakeholder meetings will be used to analyze the process of cooperation concerning the realization of targets and the monitoring of effects. This method has aspects of the 'Executive Sessions' of Moore c.s. and focuses on the alignment of goals and the level of cooperation between business schools and stakeholders in the creation of sustainable public value.

Through these workshops and meetings, data on the creation and monitoring of sustainable public value are gathered, including the process of stakeholder involvement and stakeholder management. The case studies consist of a report of the process of each participating business school.

4. *Analysing and comparing the practices*

   The data from the case studies will be compared and analysed, in order to find common opportunities, dilemmas and threats. As the case studies entail a continuous confrontation between theory and practice, the analysis will not only look at the practices of the business schools, but also reconsider the theoretical framework. If it has turned out that certain aspects of the theoretical framework are not useful in practice, or that other relevant aspects are missing, the framework needs to be further refined based on these experiences. The best methods for the comparison and analysis of data yet need to be determined.

5. *Concluding on the main outcomes and issues*

   All information and data gathered during the four previous steps will allow for conclusions in this last phase. The use of the theoretical framework in the practice of several business schools should result in an answer to the question of what process enables business schools to create sustainable public value through their education for sustainable development. Based on the adjusted version of the theoretical framework that resulted from the previous phase, bottlenecks and recommendations will be indicated concerning the creation and monitoring of sustainable public value in the context of education for sustainable development by business schools and areas for further research in this field will be identified.

5 Conclusion

Over the last years, business schools have been seriously criticized for their contribution to the current unsustainable practices in the for-profit sector. Consequently, there is now increasing pressure—amongst others by means of accreditation criteria, but also through peer pressure and demands from leading companies—to integrate issues of sustainable development in education and research programmes. Many business schools are responding to this pressure, but in many different ways. This leads to the question of what process these schools enables to indeed contribute to the common good, i.e. to create sustainable public value? Answering this question is not about ticking boxes on a checklist, but about engaging in a continuous process, together with business and society, in which the role of management and business education and of business in society is being (re) defined, translated into concrete actions and being assessed in terms of sustainable public value. This paper proposes a research project that aims to facilitate such a process with a small number of business schools that participate in case studies. Based on these cases, the research will result in a coherent framework that business schools in general can use to demonstrate the sustainable public value of their education for sustainable development. Thus, the outcome is replicable and offers business schools a qualitative and value-based alternative to current sustainability assessment tools that are merely quantitative and instrumental. In this way, the research intends to contribute both in theory and practice to the reorientation of business schools towards the creation of sustainable public value.

References

Audebrand LK (2010) Sustainability in strategic management education: the quest for new root metaphors. Acad Manag Learn Educ 9(3):413–428

Alcaraz JM, Thiruvattal E (2010) An interview with Manuel Escudero. The United Nations' principles for responsible management education: a global call for sustainability. Acad Manage Learn Educ 9(3):542–550

Barth M, Michelsen G, Sanusi ZA (2011) A review on higher education for sustainable development-looking back and moving forward. J Soc Sci 7(1):100–103

Benington J (2011) From private choice to public value? In: Benington J, Moore MH (eds) Public Value: Theory and Practice. Palgrave Macmillan, New York, p 314

Benington J, Moore MH (eds) (2011) Public value: theory and practice. Palgrave Macmillan, New York, p 314

Ghoshal S (2005) Bad management theories are destroying good management practices. Acad Manage Learn Educ 4(1):75–91

Madeira AC, Carravilla MA, Oliveira JF, Costa CA (2011) A methodology for sustainability evaluation and reporting in higher education institutions. High Educ Policy 24(4):459–479

Morsing M, Rovira AS (2011) Business schools and their contribution to society. Sage, London, p 254

Muff K, Dyllick T, Drewell M, North J, Shrivastava P, Härtle J (2013) Management education for the world: a vision for business schools serving people and the planet. Edward Elgar, Cheltenham, p 234

Porter ME, Kramer MR (2011) Creating shared value. Harvard Business Review, January, pp 62–77

Rusinko CA (2010) Integrating sustainability in management and business education: a matrix approach. Acad Manag Learn Educ 9(3):507–519

Sachs JD (2013) Werken aan de toekomst. http://www.project-syndicate.org/commentary/creating-a-sustainable-development-path-for-the-global-economy-by-jeffrey-d–sachs/dutch. Accessed 2 Feb 2014

Shriberg M (2002) Institutional assessment tools for sustainability in higher education: strengths, weaknesses, and implications for practice and theory. High Educ Policy 15(2):153–167
Slater DJ, Dixon-Fowler HR (2010) The future of the planet in the hands of MBAs: an examination of CEO MBA education and corporate environmental performance. Acad Manag Learn Educ 9(3):429–441
Stead JG, Stead WE (2010) Sustainability comes to management education and research: a story of coevolution. Acad Manag Learn Educ 9(3):488–498
Stubbs W, Cocklin C (2008) Teaching sustainability to business students: shifting mindsets. Int J Sustain High Educ 9(3):206–221
Swaen V, de Woot P, de Callataÿ D (2011) The business school of the twenty-first century: educating citizens to address the new world challenges. In: Morsing M, Rovira AS (eds) Business schools and their contribution to society (pp 175–192). Sage, London, p 254
Swilling M (2011) Greening public value: the sustainable challenge. In: Benington J, Moore MH (eds) Public Value: Theory And Practice. Palgrave Macmillan, New York, p 314
UNESCO (2014) Education for sustainable development (ESD). http://www.unesco.org/new/en/education/themes/leading-the-international-agenda/education-for-sustainable-development. Accessed 2 Dec 2014
Visser W (2011) The age of responsibility: csr 2.0 and the new dna of business. Wiley, Chichester, p 389
Von der Heidt T, Lamberton G (2011) Sustainability in the undergraduate and postgraduate business curriculum of a regional university: a critical perspective. J Manage Org 17 (5):670–690
Waas T, Hugé J, Ceulemans K, Lambrechts W, Vandenabeele J, Lozano R, Wright T (2012) Sustainable higher education—understanding and moving forward. Brussels, Belgium: Flemish Government—Environment, Nature and Energy Department
Wals AEJ (2011) Learning our way to sustainability. J Educ Sustainable Dev 5(2):177–186
Wals AEJ, Jickling B (2002) 'Sustainability' in higher education: from doublethink and newspeak to critical thinking and meaningful learning. Int J Sustain High Educ 3(3):221–232
WBCSD (World Business Council for Sustainable Development) (2010) Vision 2050: the new agenda for business. WBCSD, Geneva, Switzerland
WCED (1987). Our common future. World commission on environment and development. Oxford University Press, New York, p 400
Yarime M, Tanaka Y (2012) The issues and methodologies in sustainability assessment tools for higher education institutions a review of recent trends and future challenges. J Educ Sustain Dev 6(1):63–77
Yarime M, Trencher G, Mino T, Scholz RW, Olsson L, Ness B, Frantzeskaki N, Rotmans J (2012) Establishing sustainability science in higher education institutions: towards an integration of academic development, institutionalization, and stakeholder collaborations. Sustain Sci 7 (1):101–113

Author Biography

Mirjam Minderman holds a master's degree in International Relations with a specialization in development cooperation. When working in the field of microfinance, fair trade and responsible investing, she became convinced that development cooperation will not be effective unless the broader notion of sustainable development is taken into account. Thus, her focus shifted more and more towards sustainability and CSR and training and education on these issues. During the last 4 years, she worked at the Faculty of Management and Business of University of Applied Sciences

Leiden (The Netherlands), where she developed and implemented the faculty's policy on education for sustainable development and in addition developed and coordinated 'International Sustainable Development, post-graduate course for professionals in business and society'. Since 2014, she fully concentrates on starting her PhD research about the value of management and business education for sustainable development.

Aiming for Full Coverage—Integrating Sustainability Education into All Undergraduate Courses at La Trobe University, Australia: Achievements, Lessons Learnt and Barriers Addressed

Colin Hocking and Matthew Riddle

Abstract

In 2012, La Trobe University committed to ensuring that every undergraduate student, across all disciplines, will have significant and assessed experience of three 'Essentials' of learning: Sustainability Thinking; Global Citizenship; and Innovation and Entrepreneurship. These broadly align with the principles promoted through the Decade of Education for Sustainable Development (ESD). This initiative by La Trobe constitutes one of the first examples of whole-of-institution embedding of sustainability education into all undergraduate courses. La Trobe Essentials are more than content; they are designed to develop students' capacity to address our most pressing global challenges. To achieve this, action was needed so that there was alignment of policy, strategy, resourcing and regulation at all levels. Each Essential is tailored within each discipline, in partnerships between University-wide curriculum officers, Faculty education teams, and course and subject coordinators. This in turn has led to agreed approaches to the development of curriculum assessment and reporting for Sustainability Thinking and the other Essentials. Auditing methods have been developed to map the occurrence of subjects likely to contain the Essentials, and as a starting point for exploring which subjects can be developed as Essentials subjects, or be re-designed for this purpose. These approaches to curriculum development and auditing may be of assistance to other higher education institutions. Examples of how Sustainability Thinking has already been incorporated include a core Sustainability Thinking unit across all Business Management courses, a cross-disciplinary elective subject on Climate, Sustainability and

C. Hocking (✉) · M. Riddle
Educational Development, La Trobe Learning and Teaching (LTLT), La Trobe University, Melbourne Campus, Bundoora, VIC 3086, Australia
e-mail: c.hocking@latrobe.edu.au

W. Leal Filho et al. (eds.), *Integrative Approaches to Sustainable Development at University Level*, World Sustainability Series, DOI 10.1007/978-3-319-10690-8_33

Society, and a large, high enrolment Humanities-based elective entitled Food for Thought. The Essentials are widely supported by academic staff. Their enthusiasm, along with senior management endorsement and curriculum expertise centrally and in Faculties, have helped to overcome many of the significant barriers encountered in implementing Sustainability Thinking across all courses.

Keywords
Whole-of-Institution · Audit · Curriculum development · ESD · Efs

1 The Challenge: Integrating Sustainability Education into Every Discipline

> Sustainability becomes particularly meaningful in higher education when student and faculty find ways to infuse it with meaning in the own everyday future professional and personal lives…We need …vital coalitions between teachers, students, administrators and the wider university community (to allow joint) design learning processes that are consistent with … principles of good education and …the core ideas underlying sustainable development (Wals 2011).

The incorporation of sustainability education into courses and programs is by no means the only challenge facing Universities in the next decade and beyond. It is however, arguably one of the most important when put in the context of our rapidly degrading planet. Higher Education in general is at a major turning point worldwide (Fullan and Scott 2009), as access is widened, and costs are limited within an environment of increasing regulation and financial constraint. Student expectations are changing as they increasingly seek just-in-time support, real-world learning and value for money. The Information Technology revolution is reshaping the world of knowledge generation and sharing. The professions and employers are seeking graduates who can manage rapid change, in the midst of increasing globalization, and the impacts of rapidly unfolding global challenges such as climate change and the financial crisis (Scott et al. 2012). The increasing expectation that Higher Education institutions will help to create more sustainable futures, through education, research and community engagement (GUNI 2011) needs to be pursued in ways that interweave with these other challenges for change. The international GUNI (2011) study has identified several key challenges for integration of sustainability into higher education courses that highlight the need for institution-wide approaches.

The Decade for Education for Sustainable Development (DESD) 2005–2014 calls for

> a world where every human being has the opportunity to acquire the knowledge, skills, attitudes and values required for a sustainable future. The concept of ESD challenges the way we think about education today and encourages us to learn for change and learn to change
>
> (UNESCO Bangkok 2014).

Universities Australia (UA) has recently endorsed the 'Commitment of Sustainable Practices of Higher Education' in preparation for the United Nations Conference on Sustainable Development in Rio in 2012 (Rio +20). This included a commitment to "Teach sustainable development concepts, ensuring they form part of the core curriculum across all disciplines" (Sawahel 2012). Many universities in Australia now have aspects of Education for Sustainability (EfS) spread across some courses and disciplines. The Australian Government (2009) National Action Plan for Education for Sustainability goal for Universities is that: "2.2 Education for sustainability is integrated into all university courses/subject areas, and campuses are managed in a sustainable way."

However, in practice the incorporation of EfS into tertiary courses in Australia has been ad hoc and patchy. A national stocktake of sustainability in the curriculum in 2010, funded by the Australian Government (reported in Scott et al. 2012), found that most universities had given some attention to aspects of sustainability in their courses across a wide range of disciplines. However, little consistency, systemic approaches or linkages within each individual institution, or across institutions, was evident.

Expert analysis of Education for Sustainable Development for the DESD has highlighted the importance of attending to processes that engage the whole system of education within which specific EfS initiatives are embedded:

> ESD (needs to give) not only specific learning approaches and techniques used within education but also to the professional and management processes adopted across educational systems themselves. This holistic approach means that ESD is brought to life not only in the curriculum or educational programmes, but also in institutions and organisations which facilitate these learning processes (Tilbury 2011).

This paper describes a new approach by La Trobe University to put in place institution-wide policies, systems and processes to ensure that sustainability education and other inter-related areas of broad learning are embedded across all undergraduate programs. The paper reports on the success of this approach to date, and reflects on the challenges that have arisen, and how these have been addressed.

2 La Trobe University: Sustainability Thinking for All Students

The La Trobe University response to the challenges of developing a whole-of-institution approach to EfS has been enacted through its Future Ready Strategy (La Trobe University 2012a). This plan mandates that, from 2015 on, all undergraduates will have the opportunity and encouragement to experience each of three broad areas for learning at least once in their course of study: Sustainability Thinking, Global Citizenship, and Innovation and Entrepreneurship. Together named the *La Trobe Essentials*, each area has been defined in terms of broad student learning outcomes (La Trobe University 2013a). University-level definitions of the Essentials are shown in Table 1.

Table 1 University level descriptors of the La Trobe Essentials

Essential	Descriptor
Sustainability Thinking	Sustainability Thinking is a capacity to engage effectively with social, environmental and economic change and challenges in the contemporary world. These include, for example, climate change, food and water security and human and labour rights, ...(so) that all La Trobe University students reflect on:
	• the complex interactions between natural, economic, social, political and cultural systems;
	• our obligations to future generations;
	• how the choices we make will affect the public good and the well being of future generations.
Global Citizenship	During their degrees all La Trobe University students will reflect on the opportunities and obligations of their citizenship in a globalising world. This entails:
	• recognising the broader global context in which their studies exist;
	• understanding the diversity of values and perspectives across the globe;
	• developing personal and technical skills to interact effectively with a diverse range of people in a world of complex interdependence;
	• actively engaging with a wide range of communities and institutions to meet the demands of global challenges and obligations.
Innovation and Entrepreneur-ship	Innovation relates to the introduction of the new; renewal, and change. Entrepreneurship at La Trobe refers to seizing opportunities and managing them effectively to realise their potential ... (so that) students will develop skills and knowledge ... to include:
	• understanding and experiencing the dynamics of change;
	• the generation and effective management of new ideas;
	• the ability to network, forge partnerships, collaborate and effectively communicate;
	• flexibility and creativity in developing ideas, and in resolving complex problems;
	• the capacity to apply theory, locate and synthesise knowledge resources, prioritise action and manage associated risks.

The overall purpose of the three La Trobe Essentials is: "to foster the skills and knowledge needed [for graduates] to become *future ready* in the workplace and the wider world. This includes being able to:

- engage with the major economic, technological, political and social issues, and understand how these are often inter-woven;
- demonstrate flexibility to adapt to rapidly changing environments in the work-place and beyond; and
- apply their knowledge and skills effectively and responsibly to help find solutions to these major challenges."

  – from La Trobe University (2012b) Teaching and Learning Framework.

In addition to developing a set of University level definitions of the Essentials, a set of descriptors for students has also been written—see Essentials for Students—La Trobe University (2012b). The Essentials are somewhat different, but complementary to, La Trobe's Graduate Capabilities, which are defined across four broad domains: (1) Literacies and Communication Skills; (2) Inquiry and Analytical Skills; (3) Personal and Professional Skills; and (4) Discipline-specific Knowledge and Skills (La Trobe 2013b).

Details of what the Essentials look like in individual subjects, across curriculum areas, and how these subjects are to be organized in relation to students' programs of studies, are covered in more detail below. An outline of what the Essentials are, and how they are being implemented, can be found at La Trobe University (2013a).

When the Essentials, including Sustainability Thinking, are fully implemented, La Trobe will be the first University in Australia to have embedded education for sustainability (EfS) across all undergraduate courses and programs. It will be the first instance of fully implementing the Australian Government (2009) National Action Plan goal for EfS in Universities of being "integrated into all … courses/subject areas", and one of the few instances internationally of institution-wide incorporation of sustainability education in the higher education sector.

The policy decision to incorporate sustainability into all undergraduate courses, and a clear, institution-wide approach to what Sustainability Thinking means, is only one aspect of what has been put in place to achieve full EfS coverage for all students at La Trobe. The University also has developed a suite of policies, procedures, resources and other support and regulatory mechanisms that form the 'machinery' of educational renewal to drive the change process.

3 Institutional Alignment and Scaffolding Required for Full Coverage

The underpinnings of the Essentials, including Sustainability Thinking, are fully expressed at the *University level*. Under the La Trobe University (2012a) Future Ready: Strategic Plan 2013–2017, the distinctive La Trobe Framework for teaching and learning will (p. (3): "best equip (students) to understand and engage with the global issues of today and … be *work ready*, *world ready* and *future ready*". This approach is fully consistent with La Trobe's ethos and history of social justice and progressive outlook, and are consistent with the newly aligned, interdisciplinary and futures oriented Research Focus Areas around which research at La Trobe is organised (La Trobe University 2012a).

The Essentials and other related educational developments are overtly intended to contribute to the University goals (2013–2017) to: (1) Increase student numbers; (2) Increase student retention and … success …; (3) Improve student (course) satisfaction; and (4) Improve student employment outcomes …. These educational developments form part of La Trobe's approach to marketing and promotion.

At the *Learning and Teaching Plan Level* the three Essentials are fleshed out in more detail, in line with "the University … re-imagining how learning and teaching is conducted and how the experiences across all campuses reflect the University's standards, culture and values" (La Trobe University 2012b, p. 2). One of the La Trobe Learning and Teaching Plan aims, under Quality of Learning and Teaching, is "for students to leave La Trobe with capabilities that are highly relevant to the workforce and the needs of global communities (and) provide rich and engaging opportunities for students to grow and develop, in the realm of co-curricular and curricular activities." and "for academic staff … to engage in development to refresh approaches and to engage with innovation …".

At the *Delivery level*, implementation of these clearly articulated high level strategies and policies are scaffolded and supported through several new educational structures and processes now in place at La Trobe. These include:

- The re-organisation of educational development, technical educational support and co-curricular/student enrichment initiatives into a single centre—La Trobe Learning and Teaching (LTLT);
- Within LTLT, the appointment of two fully-time specialist Education Development Officers to assist with the implementation of the Essentials, including Sustainability Thinking;
- Integrated processes of educational development with each of the Associate Deans Academic and with Curriculum Officers located in each Faculty;
- Coordinated design and development of the Essentials with other University-wide education development initiatives, including Flexible and On-Line delivery, Blended Learning, and the Radical Learning (New Technologies) Project.

As with the other educational development projects at La Trobe, the Essentials are being developed through a multi-level educational design process that coordinates actions in each of the following: subjects, courses, auditing, tracking, improvement, reporting, professional development and technical support. How these are brought to bear on the implementation of the Essentials is described below.

4 How the Essentials Are Organised in Subjects

Each of the Essentials, including the Sustainability Thinking Essential, is integrated into appropriate subjects located within and across undergraduate courses, in ways that either align with each discipline area, or in subjects that are cross-disciplinary in nature (La Trobe University 2013a). These Essentials subjects have significant content and assessment that cover the Essentials descriptors. Every La Trobe undergraduate student will be required to complete at least one of each type of Essentials subjects during their undergraduate course. Many students will focus on one Essential each year, but in some courses, subjects may contain significant content on more than one Essential. There is no preferred order of completion.

Some students will complete more than one of each of the Essentials subjects. So the key focus for delivery of the Essentials is not by embedding them in specific courses. In some learning areas this would be problematic, as what constitutes a clearly identifiable 'course' can be difficult to define, especially where there are a wide range of electives or optional subjects from which students can choose. Rather, the focus is on ensuring that, whatever their program of study, students will have the opportunity and encouragement to complete at least one of each of the Essentials subjects, and course/program area coordinators are charged with the task of ensuring that this happens.

Within each of the designated Essentials subjects, there must be one major assessment task (or equivalent) based on the learning requirements of the Essential, often integrated with other learning requirements for the subject. Students who pass the subject will automatically meet the requirement for the Essential(s) also.

A statement about La Trobe University's Essentials will be included on the Australian Higher Education Graduate Statement (AHEGS), which accompanies the student transcript record. This will be accompanied by a list of the Essentials subjects completed by the student. The AHEGS and academic transcript will be available to employers when graduates applying for employment, along with statements about the Essentials and their value to graduates and the broader community.

5 How the Essentials Are Organised in Courses

Each of the La Trobe Faculties have already assessed current course and subject offerings for significant coverage of each of the Essentials and a gap analysis completed. Faculties have also identified subjects that can be offered as electives to students in other faculties, and subjects are being designed or re-designed as needed so that all La Trobe undergraduate students have the opportunity to complete all three Essentials. All majors must either include the Essentials, or space made available within the degree for students to take relevant electives from other areas. Subject learning guides and the University Handbook will be updated to indicate which subjects meet the Essentials requirements, and in what ways.

6 Auditing and Tracking the Essentials

Initial audits of the Essentials across all courses/programs were conducted across all five Faculties of La Trobe University during 2013, using methods designed by teaching and learning staff in each Faculty, in line with the university goal of conducting a gap analysis of the La Trobe Essentials across all degrees. The Essentials audit was designed to focus on the requirement that all students will experience each Essential at least once in their degree, in appropriate disciplinary contexts via core subjects, or via electives containing the Essentials. It was also recognised that subjects may well contain more than one Essential, and for the purpose of this initial audit, that all elements of the Essential need not be fully present in every case.

The audit was carried out by subject co-ordinators, in a process of self-evaluation, using qualitative methods, to make an estimate of the extent to which the attributes of the Essentials were present in their subjects. The overall process of auditing was coordinated through the Associate Dean Academic of each Faculty, with assistance from Faculty curriculum staff. This audit also included a requirement to assess the structure of the undergraduate degree programs and their rules, as well as individual core subjects and opportunities for taking electives. This approach to auditing across each student learning pathway was primarily directed towards gauging the opportunities that existed in the undergraduate programs for the students to experience each Essential at least once.

How this process was carried out varied between the Faculties, and the process of how best to audit the Essentials at the Faculty and course level is still being refined. For some Faculties, the development of fully articulated Faculty-level descriptors for the Essentials was helpful in carrying out the Faculty audit—these descriptors included three or four elements for each Essential, matched with an indicator consisting of a short descriptive statement. For example, in the Faculty of Business, Economics and Law these descriptors were informed by the discipline descriptors and Faculty Standards for graduate capabilities developed under the C-Ren Project (Spencer et al. 2012) as part of the university's *Design for Learning* project. The resulting set of descriptors was used in the audit analysis, via a survey deployed electronically to subject co-ordinators as a self-assessment instrument.

The audit process established a baseline from which development of fully formed Essentials subjects could proceed. While the rapid qualitative analysis of subjects for the audit by subject coordinators was only cursory, it formed the basis for more extensive engagement at the Faculty, course and individual subject level around what the descriptors of the Essentials means, what assessment for these looks like, and how they should be incorporated into student learning activities and subject's Intended Learning Outcomes (ILO's). This follow-up work has relied on the work of extensive support networks within the University to assist in this development process—as outlined above.

7 Outcomes of the Audit Process

Outcomes of the audits were variable between Faculties, depending on the in-Faculty processes used and the extent to which the Faculties and individual course coordinators understood the purposes and descriptors of the Essentials. Overall the audit was a valuable process because it sketched out the extent to which aspects of each of the Essentials, including Sustainability Thinking, were already being addressed in subjects, and in what course/programs areas they were substantially present, or lacking. Out of this process arose the realization that aspects of each of the Essentials were covered in a significant number of subjects, but often in a more background or cursory way. This led to the question as to whether the Essential could be more foregrounded in the re-development of the identified subjects, or whether new subjects needed to be developed to meet the descriptor requirements of the Essential.

Assessment of the extent to which 'Essentials-capable' subjects were present in major courses and programs revealed some valuable patterns of occurrence. These formed the basis for follow up engagement with subject and course coordinators around what needs they had for Faculty-level and central Education Development Unit-level support. For example, it was revealed that many of the Health Science courses were low on Sustainability Thinking. In other areas, even in various science-based courses, where aspects of Sustainability Thinking might be assumed to be strong, the occurrence of this Essential was more patchy than anticipated. By contrast, there were courses/programs and large-scale elective subjects across Faculties, where Sustainability Thinking was strongly represented, sometimes in unexpected ways. For example, Sustainability Thinking and the other Essentials were all strongly represented in Accounting (Fig. 1). There appeared to be a pattern where the way in which the discipline was interpreted by the particular discipline group affected the extent to which individual Essentials were present or not. For the Accounting group, the high level of Sustainability Thinking resulted in part from their engagement with triple bottom line reporting and Corporate Social Responsibility.

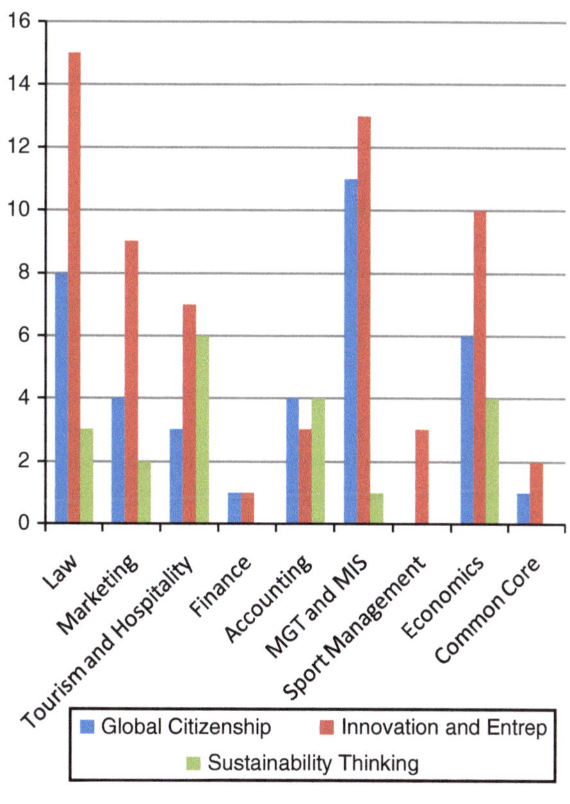

Fig. 1 Number of Essentials-containing subjects by discipline in the Faculty of Business, Economics and Law. From Ambrose, McCormack and Riddle (2012)—used with permission

Fig. 2 Number of Essentials-containing subjects by discipline the first year of the Faculty of Business, Economics & Law. From (Ambrose et al. 2012)—used with permission

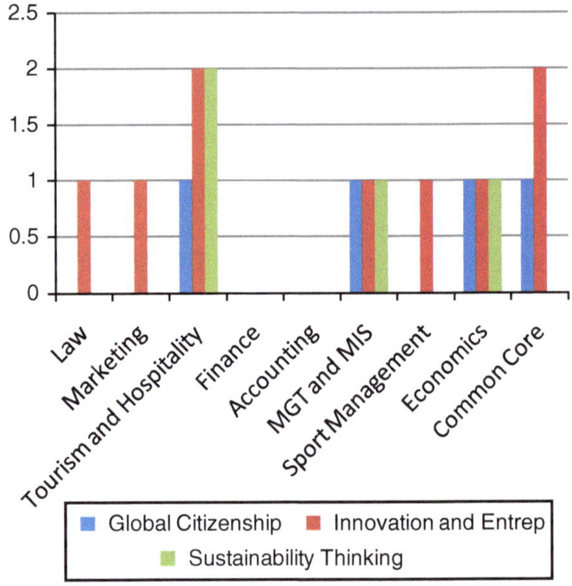

Arguably the most fully developed outcomes of the Auditing process came from the Faculty of Business, Economics and Law—FBEL (Ambrose et al. 2012). The following are two examples of the outcomes of the audit, expressed as comparisons of the occurrence of Essentials subjects between discipline areas. In Fig. 1, several trends are evident in this analysis. Firstly, the Innovation and Entrepreneurship Essential is well represented in all courses—as would be expected for courses focused around the disciplines of Business, Economics and Law. Secondly, Accounting has all three Essentials well represented, whereas Sustainability Thinking currently appears to be missing from Sports Management and Finance, and is under-represented in many discipline areas.

If each student at La Trobe is going to be given the opportunity of taking at least one of each of the Essentials subjects during their course of study, then it will be important to have these Essentials subjects well represented across year levels. Figure 2 shows an analysis of the Essentials audit data for the first year for each discipline area in FBEL, as a result of the audit. This reveals that many discipline areas are lacking in both Sustainability Thinking and Global Citizenship Essentials subjects. This outcome has led to a major coordinated effort to introduce these Essentials more substantially into either core subjects or large enrolment electives in FBEL, in 2014 and beyond.

8 Processes for Subject Re-design and Development

Following on from the audit process, some curriculum design intensive workshops and other equivalent processes have already been undertaken across the university with groups of staff on subjects identified in the audit, and these are ongoing.

One outcome of these has been the development of a series of procedural steps and curriculum tools to assist staff to successfully incorporate the Essentials into chosen subjects. These procedures have now been tested in several rounds of curriculum development, and they are still evolving. Approaches to curriculum design and re-design of subjects to include the Essentials have included the following:

(1) Beginning with subjects identified in the audit, a more detailed analysis is undertaken of the extent to which the subject already incorporates elements of the Essential, and the areas of weakness. For this, a diagnostic checklist is used incorporating the University-level and Faculty-level descriptors.
(2) This analysis is used to promote discussion about how the subject under consideration, including learning resources and student learning activities, might be re-configured to more effectively deliver the requirements of the Essentials. In this process, the specialist Essentials staff (e.g. Sustainability Education Officer for Sustainability Thinking), Faculty based curriculum staff, and other Faculty-based colleagues can be of assistance. The overall purpose of this process is to align and integrate the Sustainability Thinking (or other) Essential with the overall discipline focus of the subject, and with any other required educational purposes (e.g. delivery of some of the graduate capabilities).
(3) Coming out of this process, but not necessarily in a lock-step fashion, consideration is given to the assessment tasks that will enable the subject coordinator to determine the extent to which students have achieved the learning requirements of the Essential. These assessment tasks are also crafted in ways that integrate with the other assessable learning requirements of the subject.
(4) As it becomes clearer how the Essentials element of the subject can be best integrated, the Intended Learning Outcomes (ILOs) for the subject can be re-written to take account of the ways in which the subject content, learning activities and assessment have been adjusted. Sometimes words or phrases from the University-level or Faculty-level Essentials descriptors are useful in assisting the re-write of the ILOs.

Where there is a broader interest across a Faculty or sub-Faculty group to incorporate Essentials across a range of subjects, a more comprehensive Course Design Intensive (CDI) process has been deployed, adapted from the Oxford Brookes model (Oxford Brookes 2014; and see also Dempster et al. 2012).

As of early 2014, the Essentials development and tracking process is being coordinated with other major educational developments at La Trobe. For example, La Trobe has a target of lifting delivery of subjects by Flexible and On-line delivery to 60 % by 2017. Teams of educational developers and education technicians are beginning to work with Faculty staff to re-design subjects so they have a mixture of at least 25 % on-line delivery, and at the same time improve quality of student experience and retention rates. The targets for these developments will be large-scale core and limited-option subjects, especially in year 1. Where there are opportunities to incorporate Essentials into these subjects, the Essentials team at La Trobe will be involved as part of this overall education development process.

9 Examples of Sustainability Thinking in Practice at La Trobe

Exemplars of subjects which had been successfully adapted to incorporate the Essentials are in the process of being mounted on the La Trobe website and elsewhere. Some examples of subjects that have been recognized as containing the Sustainability Thinking Essential have been posted on the national Learning and Teaching Sustainability website at www.sustainability.edu.au—under La Trobe University. Three current examples are described below.

Climate, Sustainability and Society: This is a cross-disciplinary elective subject developed and taught jointly by staff in the Faculty of Science, Faculty of Health Sciences and Faculty of Business, Economics and Law, to over 150 students. The subject is delivered in Flexible and On-line mode to students at three campuses. The subject aims to help students to put together the bigger picture of how climate is changing, and how this is affecting the economy and society … and develop students understanding of the need for appropriate public policy, business approaches and responses to the underlying science, to effectively address climate change.

Food for Thought: Discovering the world through commodities: This is a limited-options breadth subject, currently under development to be delivered by staff in the Faculty of Humanities and Social Sciences in 2015. The subject will be in Flexible and On-line Delivery Mode, and will be made available to all other Faculties, across several campuses. The subject introduces students to the essential nature of food and drink to human existence, as well as being revealing of the nature of human society and culture. In this exploration, students engage actively with key historical and contemporary problems such as colonialism, fair trade, famine, and sustainability.

Essentials in Core Subjects at 3 Year Levels in the Business Management Course: In the re-organisation of the Business Management degree program, involving the rationalization of a range of smaller electives, and a shift to Flexible and On-line Delivery mode, design is currently under way for three of the core subjects in the new Business Management degree program to include the three Essentials: Global Citizenship in year 1; Sustainability Thinking in year 2; and Innovation and Entrepreneurship in year 3—with scaffolding of the concepts in each year level into the Essentials subjects in the following year.

10 EfS Interwoven as a Strand of Educational Development

In pursuing a whole-of-institution approach to Education for Sustainability (EfS) at La Trobe University, it has been vital for this endeavour to be located in, and coordinated through, the educational infrastructure and processes of the University. This highlights a shift in approach, from EfS being pursued either through Operations and Infrastructure, or through a specialist sustainability unit, to one where development of EfS is identified primarily as an educational issue (Hocking et al. 2011). Ryan and Tilbury (2013) have described this process as follows:

"(In universities there is a) need to access deeper currents of teaching and learning to make ESD a viable education proposition, as well as the potential transfer to other parts of the education and skills sector. This not only provides access to the processes of educational development of the institution, but at a more fundamental level it normalises EfS as being part of a forward-thinking, innovative approach to Higher education, one that assumes and promotes EfS as part of best educational practice."

This shift in approach is additionally important because EfS calls for a shift in the pedagogy of teaching and learning, away from knowledge transmission and toward collaborative exchange of ideas, which is in line with recent research and understanding of effective tertiary education (Barth et al. 2007; Jones et al. 2010). EfS also needs to move with other major shifts in tertiary education development, and be incorporated into broader changes in curriculum and learning towards more blended modes of delivery. The supporters and practitioners of EfS need to avoid being stranded with out-dated modes of tertiary learning and teaching, lamenting the times when staff-student ratios and face-to-face teaching allowed a more intimate form of education, assuming that this is required to explore the complex issues that are essential to EfS. It has been shown that this requirement for high student-to-staff ratios are not needed in other areas of learning that require complex, interdisciplinary and values-related approaches (for example Global Citizenship and Internationalisation of the Curriculum—see Leask 2012) and nor should it be for EfS.

11 Challenges and Opportunities for EfS at La Trobe

Sustainability Thinking, Global Citizenship and Innovation and Entrepreneurship are in the process of being incorporated into a wide range of subjects, across diverse discipline areas in all five faculties at La Trobe University. These will be promoted to commencing undergraduate students in 2015 as subjects that they can complete to meet the requirements of the Essentials.

For each discipline area, incorporation of Essentials into undergraduate subjects has presented significant challenges for La Trobe University staff, and these challenges are particular for each discipline or grouping of courses. For example, in the Faculty of Business and Law, the initial audit revealed that staff teaching in each discipline area have particular ways of interpreting and developing their discipline that affect how they perceive EfS, which in turn affects the ease with which this might be strengthened. Similar patterns were observed in other discipline groups in the University, with differing strengths and weaknesses.

The *La Trobe Essentials* initiative is still in the early stages of practical development and implementation. We are still unclear about the extent to which these will be embraced and further developed across disciplines—however we have made a start. The challenges encountered by the university are likely to be present for other institutions attempting to follow similar pathways of curriculum development. There is the ongoing challenge for how best to map and track the development of

Sustainability Thinking and the other Essentials across courses, and the processes that emerge may be useful elsewhere. The signs are that many teaching staff are engaging positively with the need to bring issues of sustainability, global citizenship and innovation into their learning and teaching, even if historically one or more of these has not been a part of their discipline outlook. This is grounds for quiet hope. But considering that this is the *first* instance of a University-wide incorporation of sustainability across all undergraduate courses anywhere in Australia, in light of the sustainability challenges that are emerging or already evident, we will need to learn from one another quickly and act in future with a greater level of urgency. In the words of David Orr over two decades ago (1994):

> The truth is that many of the things on which our future health and prosperity depend are in dire jeopardy: climate stability, the resilience and productivity of natural systems, the beauty of the natural world and biological diversity ... It is worth noting that this is not the work of ignorant people. Rather it is largely the results of work by people with B.As, B.Scs, LLBs, MBAs and Ph.Ds...

References

Ambrose K, McCormack S, Riddle M (2012) Faculty of Business, Economics and Law. Report on the La Trobe Essentials Audit. Internal La Trobe University paper available from the authors

Australian Government (2009) Living Sustainably—The Australian Government's National Action Plan for Education for Sustainability. Department of Environment, Water, Heritage and Arts—http://www.environment.gov.au/education/publications/pubs/national-action-plan.pdf

Barth M, Godemann J, Rieckmann M, Stoltenberg U (2007) Developing key competencies for sustainable development in higher education. Int J Sustain High Educ 8(16):416–430

Dempster JA, Benfield G,. Francis R (2012) An academic development model for fostering innovation and sharing in curriculum design. Innovations Edu Teach Int 49(2):135–147

Fullan M, Scott G (2009) Turnaround leadership for higher education. Jossey Bass/Wiley, San Francisco

GUNI (2011) Global University Network for Innovation. Higher Education in the world 4: Higher education's commitment to sustainability: from understanding to action. Palgrave, McMillan

Hocking C, Daddow A, Ford R (2011) building sustainability into the core business of teaching and learning—Victoria University as a case study. In: Paper and presentation to australasian campuses towards sustainability (ACTS) annual conference, Adelaide, September

Jones P, Selby D, Sterling S (2010) More than the sum of their parts? Interdisciplinarity and sustainability. In: Jones P, Selby D, Sterling S (2010) Sustainability education. Perspectives and Practices across higher education. Earthscan, London, Washington

Leask B (2012) Internationalisation of the Curriculum. www.ioc.net.au. Accessed 25 Feb 14

La Trobe University (2012a) Future ready: strategic plan 2012–2017. www.latrobe.edu.au/futureready/. Accessed 30 Mar 2014

La Trobe University (2012b) Teaching and learning framework—Essentials. http://www.latrobe.edu.au/teaching/la-trobe-framework/essentials. Accessed 30 Feb 14

La Trobe University (2013a) Definitions of essentials. http://www.latrobe.edu.au/teaching/la-trobe-framework/essentials/definitions. Accessed 30 Feb 2014

La Trobe University (2013b) La trobe graduate capabilities. http://www.latrobe.edu.au/teaching/la-trobe-framework/graduate-capabilities. Accessed 24 Feb 2014

Orr D (1994) The earth in mind: on education, environment and the human prospect. Island Press, p 7

Oxford Brooks (2014) The curriculum design intensive process. https://wiki.brookes.ac.uk/display/CDIs/Home. Accessed 30 Mar 2014

Ryan A, Tilbury D (2013) Uncharted waters: voyages for education for sustainable development in the higher education curriculum. Curriculum J 24(2):272–294

Sawahel W (2012) University leaders worldwide sign sustainability declaration. University World News 25 May 2012 Issue No 223 www.universityworldnews.com/index.php?page=UW_Main. Accessed 25 Jan 2014

Scott G, Tilbury D, Sharp L, Deane E (2012) Turnaround leadership for sustainability in higher education. Learning and Teaching Excellence Division, DEEWR, Australian Government, Canberra

Spencer D, Riddle M, Knewstubb B (2012) Curriculum mapping to embed graduate capabilities. High Edu Res Dev 31(2). http://dx.doi.org/10.1080/07294360.2011.554387

Tilbury D (2011) Education for sustainable development. An expert review of the processes and learning. Report to Section for Education for Sustainable Development, Division of Education for Peace and Sustainable Development. UNESCO, United Nations (Paris)

UNESCO Bangkok (2014) Education for sustainable development—about US 'Vision'. http://www.unescobkk.org/index.php?id=976. Accessed 30 Mar 2014

Wals A (2011) The "E" in ESD: from green washing the ivory tower to deep learning for sustainability. In: Barth M, Rieckmann M, Sanusi Z (eds) 2011. Higher education for sustainable development: looking backward, moving forward. VAS-Verlag fur Akademishe Schriften (Hamburg)

Authors Biography

Dr. Colin Hocking Prior to La Trobe University, Dr. Colin Hocking was Education for Sustainability (EfS) Officer at Victoria University, helping to shape EfS into the University Strategic Plan. He has experience in EfS program design and evaluation, including development of interactive 3-dimensional simulations of energy efficiency for engineering education. Colin has a background in facilitation and design of sustainability education and change programs with communities and businesses over 15 years, and a parallel research background in ecology and management of natural ecosystems.

Associate Professor Matt Riddle Associate Professor Matt Riddle joined La Trobe University in 2008 as an Educational Designer and then as Acting Associate Dean Academic in the Faculty of Business, Economics and Law, after several years as Research Associate at CARET at the University of Cambridge. Matt has a background in multimedia development and e-Learning, contributing to a range of education-related projects over 13 years at the University of Melbourne. Matt has a strong interest in applied educational research, including student engagement and retention, active and collaborative learning and virtual learning space design.

Green Roofs and Urban Campus Greening: Learning About Sustainability Through Doing

Karolina Barkowicz and Judy Rogers

Abstract

This paper reports on an on-going RMIT University project that aims to link campus greening initiatives with learning and teaching outcomes with a particular emphasis on sustainability—what it means and how it can provide a useful framework for evaluation. The focus is on education for and through sustainability rather than education about sustainability. Stage 1 of the project involved designing, developing and evaluating green roofs to consider what is best practice sustainable design. It began with a series of comparative precedent and practice based case studies and background research on various typologies, growth mediums and plant materials. From these initial studies, pilot green roof modules were developed. The modules tested suitable vegetation and alternative growth mediums. The project was then taken into a living laboratory context, where third-year students built, maintained, and monitored the performance of the green-roof modules. Stage two of the project aims to build on these outcomes through further student-led environmental monitoring including the development and testing of techniques to evaluate biodiversity and productive landscape benefits as well as microclimatic condition. The final stage of the project will focus on the development and testing of a Post Occupancy Evaluation Framework for the roof garden pilot project to inform future design development and curriculum innovation.

Keywords

Education · Green infrastructure · Sustainability · Curriculum · Green roofs

K. Barkowicz · J. Rogers (✉)
School of Architecture and Design, RMIT University, Swanston Street, Melbourne 3000, Australia
e-mail: judy.rogers@rmit.edu.au

W. Leal Filho et al. (eds.), *Integrative Approaches to Sustainable Development at University Level*, World Sustainability Series, DOI 10.1007/978-3-319-10690-8_34

1 Introduction

As the imperative to deliver sustainability education gathers pace at the close of the UN decade of education for sustainable development it is timely to step back and consider what that education looks like and how it reflects the breadth and depth of sustainability principles and practices. Universities have responded to the challenge of sustainability through a range of measures including campus greening and sustainability education initiatives or Education for Sustainability (EfS). These measures are rarely linked. As Savelyeva and McKenna (2011) argue:

> Greening campuses implies infusing environmentally friendly practices in all dimensions of university operations and infrastructure. EfS refers to the set of ethical standards for academia to become sustainable. Both approaches have different origins and are well supported at the national and international levels. However, the idea of greening campuses is often interpreted as the main understanding of sustainability in higher education. In turn, the concept of EfS has found its applications within a narrow set of disciplinary fields, specifically, environmental education and the natural sciences.

This paper discusses a project delivered at RMIT University that attempts to link these two dimensions of higher education sustainability, offering opportunities for landscape architecture students to be involved in campus greening initiatives, effectively learning through doing. The focus of the project—The Matter of Landscape: Sustainable Design Strategies for RMIT City Campus—is on designing, building and then testing green roof modules. However, rather than simply accept that green roofs deliver 'sustainability' outcomes students are required to interrogate and question the benefits of green roofs using sustainability as a framework for evaluation. Using this framework they are asked to look beyond the 'easy' and often unsubstantiated assumptions that all green roofs and green infrastructure are 'sustainable' and to consider implementation across the three poles of sustainability—economic, environmental and social.

The paper begins with a discussion of green roofs—what they are, what their claimed benefits are and why implementation presents an interesting way to educate design students in the theory and practice of sustainability. The paper then provides an overview of sustainability and sustainability education and the way in which these concepts have been deployed here. A distinction is made between education about, for or through sustainability before the discussion moves to an overview of the curriculum developed as part of the broader project and outcomes to date. As the project is on-going the paper ends with an overview of future directions.

2 Why Green Roofs?

Along with green walls, rain gardens, street trees and permeable paving, green roofs are part of an emerging global discourse around greener cities. 'Green roof' is an umbrella terms used to describe a number of systems for green rooftops of buildings. The list of benefits attributed to green roofs is broad and widely shared. They include reduction in the heat island effect, noise reduction, energy conservation, amenity, replacement of lost green space, increased property values, storm water management along with what has been described as 'green relief' in highly dense cities. Green roofs it is further claimed 'transform cities from urban grey to urban green' (Weiler and Scholz-Barth 2009) while at the same time mitigating the effects of climate change.

Green roofs can be either extensive or intensive, depending on the load bearing potential of the building, depth of soil, maintenance required, access and the aims of the project. Extensive green roofs it has been argued present a:

> …tremendous opportunity to retrofit and green existing structures in cities, thereby realizing multiple benefits of green roofs, including: reduced runoff and flood management; reducing energy demands in heating and cooling houses; improving the longevity and replacement costs of roofs; and benefiting city climate all while also providing for aesthetics and wildlife benefits in downtown areas (Tourbier 2011).

An intensive green roof in contrast has deeper soil to support a larger variety of plants. Intensive green roofs can also provide accessible open space and often include larger plants, food growing areas and trees as well as walkways, water features and irrigation systems. Because of this difference specifications differ dramatically from extensive green roofs. Intensive green roofs are feasible only on flat roofed buildings and the added weight of soil and water require a substantial building structure. They are therefore more costly but are more energy efficient than extensive green roofs and have the capacity for greater storm water retention. The potentials for retrofitting existing buildings with intensive green roofs are limited.

Use of the terms intensive or extensive to describe green roofs has however been criticized because it is 'terminology' driven rather than use driven and can lead to confusion and inaccuracy in the design documentation and client expectations (Hopper 2007). The problem with the use of blanket terms, according to Weiler and Scholz-Barth (2009) is that:

> …neither clearly reflects the system's expected purpose or use nor adequately conveys design or maintenance requirements. Furthermore, a terminology-driven, rather than use-driven, approach to the design and construction of green roofs can lead to additional confusion and inaccuracy in design, documentation, and client expectations.

What the authors are referring to is that rather than being simply 'green', green roofs serve a range of different purposes depending on design and client intent. Depending on intent green roofs do not necessarily tick all of the sustainability boxes, nor of the claimed benefits. What is significant is the way in which the term 'green' dominates discussions around rooftop plantings and as a result assumptions are made about the benefits of ALL green roofs.

One needs to ask, therefore, why the focus on the 'greenness' of green roofs? 'Green' as a leitmotif or symbol for green roof raises a whole series of questions about what a green roof actually is and does. Does 'green' in fact denote 'sustainability'? As Latz (2007) has argued:

> The topic of green roofs is closely connected to "ecological correctness", and is part of the indemnification vocabulary of sins committed in bad conscious like landscape consumption and the displacement of nature.

What Latz is suggesting is that much of the rhetoric around green roofs is about language-in-use. Contained within the rhetoric is a future vision of what a city should look like. In this future 'green', 'livable', resilient dense city public open space on the horizontal plane is at a premium and so, as the argument goes, elevating that space onto walls and rooftops of buildings will allow for increased densities to create a system of 3 dimensional green spaces in cities, without the loss of open space.

The unspoken question in all of this is of course, access. Defined as Landscape over structure or the integration of landscape and architecture green roofs are often seen as a replacement for 'green' space at ground level a claim that is often couched in terms of adding to green spaces in cities. However, who lives or works in the buildings above the ground and who can occupy space at the ground level needs careful consideration. Or put another way if green space at ground level is replaced by elevated green space what are the socio-political consequences of this move.

Implementation of green roofs and green infrastructure therefore, in and of itself does not necessarily lead to 'sustainability'. This simple observation became the starting point for the project outlined in this paper. The focus is not on educating students about green roofs as sustainable infrastructure—but rather asking students to question, interrogate, and challenge current understandings and practices through active involvement in designing, constructing and testing—or learning through doing. This is education not about, but for and through sustainability.

3 Sustainability Education: About, for or Through?

Sustainability education, it has been argued, is often delivered in a University setting as content mapped onto the curriculum of various disciplines and fields (Alvarez and Rogers 2006; Savelyeva and McKenna 2011). The dominant approach in such education is prescriptive: environmental targets, audits, energy and water efficiency, sustainable design form the focus of an education about sustainability but not necessarily for or through sustainability (Rogers 2013a, b). There is a consequent need to distinguish between sustainability education delivered as content added onto curriculum and sustainability education that engages students in real world problem based learning.

According to Huckle and Sterling (1996) sustainability education 'is essentially process-driven, is participative and empowering, is liberatory and continuous and that it is necessitated by the possibilities and dangers presented by an emerging 'postmodern' world (Huckle and Sterling 1996, p. 14). Open dialogue, discussion, debate and exchange of ideas are foregrounded rather than the transmission of content or the 'how to' of sustainability.

Similarly, in their study of sustainability education Wals and Jickling (2002) suggest because decisions about sustainability ultimately rest on different interests and values the concept needs to be openly challenged, negotiated and discussed rather than masking its complexity under a seemingly 'shallow consensus'. They contrast two approaches to education for sustainability. The first adopts an instrumental view of sustainability where 'sustainability is fixed, pre-and expert determined (i.e. academics) and to be reproduced by novices (i.e. students)' (Wals and Jickling (2002). They argue that such an approach is not necessarily educational. For Wals and Jickling an emancipatory approach to education for sustainability must necessarily be participatory, open and respectful of different perspectives and attitudes and provide 'a means to become self-actualized members of society, looking for meaning, developing their own potential and jointly creating solutions' Wals and Jickling (2002). This approach to education is based on a process of seeking rather than setting the sustainability agenda.

A similar distinction is made by Warburton who argues that sustainability education requires 'deep' approaches to learning:

> Deep learning involves paying attention to underlying meaning. It is associated with the use of analytic skills, cross referencing, imaginative reconstruction and independent thinking. In contrast, surface learning strategies typically place more emphasis on rote-learning and simple description (Warburton 2003).

The distinction between shallow and deep learning is an important one. It suggests that shallow learning is reached based on a pre-prescribed idea of what sustainability is and that the desire for consensus leads to the imposition of a particular moral and ethical agenda based on defining 'right' and 'wrong' behavior (s) and ideas. Values, life experiences and concerns about sustainability are lost in the quest for a common, shared vision that potentially has more to do with the teacher imposing a particular viewpoint than with the student. Deep learning, on the other hand, can only occur within a learning environment that is respectful of differing perspectives and ideas.

In the project outlined here the focus was on encouraging students to think beyond prescriptive and fixed ideas about sustainability, to grapple with the possibilities of what it could mean, to effectively turn the concept back upon itself and open it up for critique. Applying the key dimensions of sustainability to a real world example, students learn through doing.

4 The Matter of Landscape: Learning Through Doing

In 2013 staff members and students in the RMIT Landscape Architecture Program in collaboration with RMIT Property Services designed and built a green roof pilot and demonstration project as stage 1 of an on-going project titled *The Matter of Landscape*. The project began with a series of comparative precedent and best practice case studies in Melbourne, Australia of existing green roofs, background research on various typologies of green roofs, soil or growth mediums and appropriate plant materials. From these initial studies the pilot green roof modules were developed and suitable vegetation (testing biodiversity, edible landscapes, and carpeting species for thermal insulation and fast growth) as well as alternative growth mediums (testing light weight, soil mixtures) were tested. The project was then taken into a living laboratory context, where 3rd year Landscape Architecture

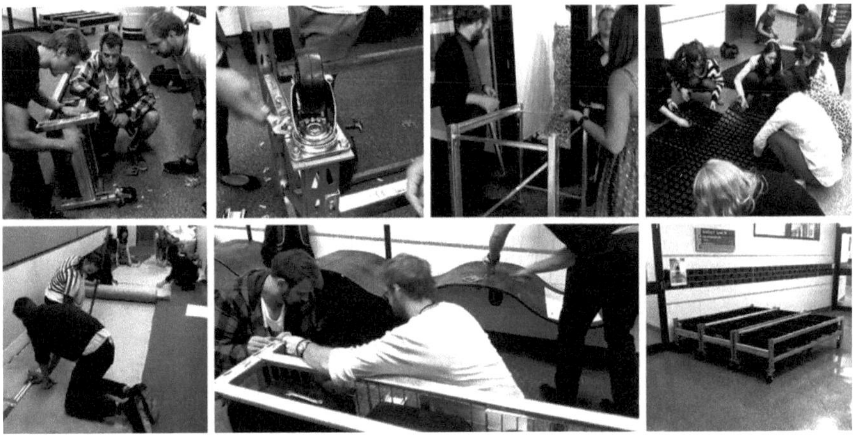

Fig. 1 Module construction by landscape architecture students

students built, maintained, and monitored the performance of a series of green roof mobile modules (Bartkowicz et al. 2013).

The initial green roof experimental system is located on a south-westerly balcony on the tenth floor of RMIT University City Campus. This particular site had been chosen for its predominant harsh wind conditions and minimal exposure to sunlight. The individual test plots consist of 16 elevated mobile planters (modules) that have been assembled together to accommodate the trials of extensive and intensive planting modes. These comprise of nine deep planters (600 mm in depth) and six shallow planters (180 mm in depth) providing a range of growing conditions according to their positioning. One deep planter was included as a control without any plant material (see Fig. 1).

Students took part in a work integrated learning program where core sustainability metrics and research methodologies were embedded into the green roof concept. Landscape Architects often concentrate on the design of spaces utilizing vegetation without understanding how to quantify and qualify the affects of their design work. This project sought to combine learning to design and construct a roof garden and to measure its tangible affects. These goals further promoted interdisciplinary learning and external industry collaboration. Students with the advice of industry proposed planting schemes and growth medium mixes. The plantings included: biodiversity plots, edible landscapes plots, and carpeting species plots. Students collected data on plant growth, vegetation coverage, soil performance (porosity, drainage, hydration, electrical conductivity) as well as albedo effect (reflective surface temperatures) and thermal insulation performance (see Fig. 2).

Performance-based data collected in regard to plants including survival rates, plant proliferation rates, physical damage, sun scorch and evidence of reaction to the watering regime (see Fig. 3). Experimental cladding systems applied to the green roof modules were also installed and offered a range of benefits. The cladding provided an insulation barrier to the growth medium containers, opportunities for

Fig. 2 Environmental monitoring by landscape architecture students

Fig. 3 Plant performance data collection

seating and catchment for storm water and self-watering opportunities. Some cladding systems also encompassed lighting to provide visibility during night hours.

In addition, to quantitative environmental metrics the students investigated various qualitative or more socially sustainable ideas. Because the roof top space can be viewed from offices above, it was important to design it as a place for visual respite. In addition, students designed and tested various bench and seating prototypes (see Fig. 4).

Fig. 4 Constructing bench and seating prototypes

They installed light-weight and recycled material seating modules, so visitors could spend time on the balcony. This had an added effect of providing further insulation to module growing mediums. Students also designed removable cladding or siding panels. Other environmental aspects such as water collection, drainage, wind protection and enhancement of natural sunlight were integrated within module cladding systems.

The momentum of the project also allowed for wider university involvement including interdisciplinary collaborations between RMIT's Environmental Science Program and Property Construction and Management Program. The Student Union's organic café is also now involved in the project and hopes in the future to supply their produce directly from the green roof modules.

Beyond the student learning aspects of the project, a final research report was developed for RMIT Property Services as a briefing document, case study comparison, and initial design guidelines for the commissioning future roof gardens.

Concurrent to this project, the City of Melbourne was developing a series of green roof guidelines and scoping potential project partnerships for green roof opportunities. The City of Melbourne became actively involved with the RMIT pilot project and it has informed the framework for draft policy and guidelines (State Government of Victoria 2014).

5 The Matter of Landscape: Stage 2

Stage two of the project began at the beginning of 2014 with the aim of building on outcomes from stage 1 through further student-led environmental monitoring including the development and testing of techniques to evaluate biodiversity and productive landscape benefits as well as microclimatic condition. The final stage of the project will focus on the development and testing of a Post Occupancy Evaluation Framework for the roof garden pilot project to inform future design development and curriculum innovation.

A new site has been selected for the experiment that allows for increased access for University staff and student in order to investigate how such a space can or is used. The parameters of this new experiment will respond to the data collected in 2013 and 2014 and will include a post-occupancy evaluation. In this way the project will begin to address social sustainability criteria in greater detail.

The educational focus in stage 2 of the project is also to encourage students to look beyond the 'easy' and often unsubstantiated assumptions that all green roofs and green infrastructure are 'sustainable' through investigation on a broad scale through to the 1:1 construction of a pilot project green roof. The class began, therefore, with a provocation where students were asked—Green Roofs and Green Infrastructure—are they sustainable? Responding to this question involved an evaluation of an existing green roof using sustainability principles as a framework (see Table 1). The evaluation framework reflects claimed sustainability benefits drawn from broader literature and students were asked to seek out available evidence that supported the claimed benefit.

The students were also asked in the initial stages of the semester to divide into groups of five to develop a master plan for RMIT city plan identifying potential sites for green roofs. Each group was given a policy document from the city of Melbourne—and they were asked to draw out key ambitions from that document that reflected social, economic or environmental dimensions of sustainability.

While students were divided into groups to examine one key policy that reflected either social, economic or environmental sustainability in small group discussions by week 2 what they all very soon realized is that it is not possible to separate each of these dimensions out so easily (see Fig. 5).

For instance, students working on environmental sustainability quickly realized that they needed to engage with the social and economic dimensions of sustainability in order to realize their ambition for RMIT city campus. What is interesting here is that student could have been told that each of the sustainability dimensions needed to be integrated but rather they learnt this though the process of negotiating their master plan. This was reflected in the process and methodology the students used to generate their group master plans, combining and integrating various elements of the three sustainability pillars to realise their key ambitions through development of specific roof-top typologies applied to the RMIT City Campus. In addition realizing the need to engage with all three dimensions of sustainability all of the groups identified that in order to engage with the social realm the question of access needed to be foregrounded as a key challenge.

In the coming weeks students will begin constructing green roof modules for the new site. This will involve building and then testing planting modules at a larger scale with tree species, further testing of light-weight soil/growth mediums, and identification of species appropriate for supporting biodiversity and edible

Table 1 Evaluating green roofs

Environmental	
Sustainability benefit	Evaluation method
a. Reduce heat island effect (reduce radiated heat from buildings)	• Testing the internal building temperature prior to green roof installation versus temperature with a green roof installed
	• Substrate type used (weight and depth)
	• Type of plants used to insulate (their positioning to one another and their evaportranspiration ability)
	• Positioning of the Green Roof (aspect and pitch)
	• Input resources (hydro, nutrients, lights)
	• Measuring the albedo effect and solar radiation
	• Measuring heat radiation from roof prior to green roof installation and post green roof installation—infrared technology
	• Measuring the R—Value of the roof insulation
b. Noise reduction	• Measuring internal building decibels prior to green roof installation versus internal building decibels post green roof installation
	• Plant species used (size, root structure, density of foliage) and coverage ability (initial planting positions and growth rate/direction)
	• Substrate ingredients, depth and weight used
	• Local climate affecting growth rate
c. Increase Biodiversity:	• Microclimate proposed by green roof
• plants	
• invertebrates	• Current climate affects
• birds	
• mammals	• Green roof type applied (substrate, accessibility, maintenance, input resources etc.)
	• Green roof components used—modular/insitu
	• Watering regime
	• Plant species and communities proposed—e.g. Natives
	• Plant density and coverage ability
	• Animal refuges present (wooden platforms/burrows)
d. Reduce storm water run-off contributing to:-	• Testing stormwater runoff amount, quality (pH, dissolved oxygen, turbidity, ammonia, nitrite, alkalinity, hardness) and travel rate from similar environmental conditions versus stormwater runoff amount, quality and travel rate collected from a green roof.
• water quality	• Type of drainage system applied (SUDS/type of cells)
• flood mitigation	• Type of plant species used—Wetland planters? Rain harvesting strategies
	• Type of substrate used (ingredients and depth)—WHC/AFP
	• Roof slope

(continued)

Table 1 (continued)

Environmental	
e. Improve air quality	• Compare air quality surrounding the green roof against the Ambient Air National Environment Protection Measure: carbon monoxide, nitrogen dioxide, ozone, sulfur dioxide, lead and particles (measured as PM10, a measurement that relates to the size of the particle)
	• Height of the green roof above ground and proximity to the pollution generators
	• Types of plants species used which capture pollen, dust, filter noxious gases and volatile organic compounds
	• Amount of sunlight for photosynthesis to occur—Is the roof North facing?
	• Water supply and amount accessible to the green roof as the CO_2 comes from water rather than air pollution
f. Carbon absorption	• Plant species absorption levels/abilities (amount of stomata)
	• Substrate absorption levels: AFP/WHC—pores
	• Plant species evoptranspiration ability (exposure to sun and water)
	• Drainage systems used (cell/gravel)
Economic	
Sustainability benefit	Evaluation method
a. Reduce energy use within a building	• Building costs (energy bills, maintenance etc.) prior to green roof installation and post green roof installation
• heating	• Green roof insulation properties (type: substrate, vegetation, drainage cells, aspect, area, pitch)
• Cooling	• Plant species used and their evoptranspiration ability (exposure to sun and water)
	• Efficiency of the existing heating/cooing equipment
	• Local weather-climate variation
b. Increase roof longevity	• Type of waterproofing membrane used and its initial lifespan compared to an installed green roof lifespan (inc. cost of maintenance)
	• Green roof coverage—plot sizes, component depths, orientation (protection from sun and rain)
	• Leakage amount—(anti leakage systems employed)
	• Type of maintenance and regularity specified for the green roof (service lifecycle)
	• Green roof type and purpose
c. Increase worker productivity	• Number of employee sick days
	• Quality of worker output
	• Number of worker 'errors'

(continued)

Table 1 (continued)

Environmental	
d. Increase property values	• Additional floor area/garden space (usability of the green roof)
	• Amount and height of other properties overlooking the green roof (who can see it?)
	• Seasonal appearance of the green roof
	• Further economic opportunities: cafes etc
e. Job creation	• Maintenance work for the upkeep of the green roof
	• Amount of green roof specialized industry experts
	• Food production possibilities
Social	
Sustainability benefit	Evaluation method
a. Increase recreational use of space	• Number of people who use/visit the green roof
	• Green roof amenities (seating/bins)
	• Seasonal appearance and positioning (protection from sun)
	• Levels of noise and surveillance
b. Green relief	• Visual/physical distance to a green roof from inside a building
	• Time spent viewing the 'GREEN'
	• Seasonal appearance of the green roof
	• Distraction factor—measurable?
	• Type of plant species and their aesthetical qualities (colours, sizes etc.)
c. Increase food production in city	• Type of food production crops available and needs in particular climates
	• Space/weight of substrate required to grow crops
	• Buildings appropriate for heavy loads
	• Accessibility and requirements by maintenance/ harvesting team
d. Increased amenity	• Type of design elements proposed (seating etc.)
	• Amount of time spent by people using amenities
	• Levels of addressed needs of the building occupants (more shade etc.)
e. Worker health	• Number of sick days of employees recorded (decline?)
	• Decrease in stress levels…measurable?
	• Amount of money spent on medicine per employee per year
	• Productivity… measurable?

Fig. 5 Negotiating and discussing a master plan for RMIT city campus

landscapes. Students will also be involved in testing recycled materials and a more robust cladding system and develop ongoing maintenance regimes.

Each of the 4 tutorial groups has been divided into 2 with one team responsible for the hard landscape component of the green roof modules. This includes: working to an allocated budget, designing, documenting and constructing a cladding system to be fixed to one or more of the allocated green roof modules. The other team will be responsible for the soft landscape component and will involve developing a planting plan, documenting and then implementing a planting scheme within the allocated green roof modules. These activities will be followed in semester 2, again with Landscape Architecture Students, with further environmental monitoring and the development of, and testing of a Post occupancy Evaluation with a particular focus on how the site is used.

6 Conclusion and Future Directions

The paper reported on outcomes of an on-going project that links campus greening initiatives with curriculum innovation at RMIT University, Australia. The project aimed from the outset to move away from generalised claims about the sustainability benefits of green roofs towards a focus on performance-based criteria for evaluating green roofs. The focus is therefore not on educating students about green roofs as sustainable infrastructure—but rather asking students to question, interrogate, and challenge current understandings and practices through active involvement in designing, constructing and testing—or learning through doing. This is education not about, but for and through sustainability.

In the future the potential exists to build on the master plans developed in stage 2 of the project allowing implementation of green roof modules on other building on city campus. Data gathered by students during the testing and monitoring phases of the project, including the Post Occupancy Evaluation, will be used to drive further developments and campus greening initiatives. There is also enormous potential to further build on industry and interdisciplinary collaborations as the project continues to gain momentum.

References

Alvarez A, Rogers J (2006) Going 'out there': learning about sustainability in place. Int J Sustain High Educ 7(2):176–188

Bartkowicz K, Howard M, Rogers J (2013) Green skins reconsidering green roofs as sustainable infrastructure. Paper presented at 6th International Urban Design Conference, Sydney

Hopper LJ (2007) Landscape architectural graphic standards: Student Edition. Wiley, New Jersey

Huckle J, Sterling S (1996) Introduction. In: Huckle J, Sterling S (eds) Education for sustainability. Earthscan, London

Latz P (2007) Introduction. In: Werthmann C (ed) Green roof-a case study: michael van valkeenburgh associates' design for the headquarters of the american society of landscape architects. Princeton Architectural Press; New York, pp 14–15

Rogers J (2013) Education for and through sustainability. In: Caeiro S, Leal Filho W, Jabbour C, Azeiteiro UM (eds) Toward interdisciplinary dialogue in sustainability assessment tools in higher education institutions mapping trends and good practices around the world. Springer, Berlin

Rogers J (2013) Green, grey or brown: green roofs as sustainable infrastructure. In Brebbia CA (ed) Sustainable development and planning V1, WIT Press, UK

Savelyeva T, McKenna JR (2011) Campus sustainability: emerging curricula models in higher education. Int J Sustain High Educ 12(1):55–66

State Government of Victoria (2014) Growing green guide: a guide to green roofs and walls and facades in Melbourne and Victoria Australia. http://www.growinggreenguide.org/

Tourbier JT (2011) Green Roofs, urban vegetation and urban runoff. In: Douglas I, Goode D, Houck MC, Wang R (eds) The routledge handbook of urban ecology. Routledge, Oxon, pp 572–582

Wals A, Jickling B (2002) Sustainability in higher education: from doublethink and newspeak to critical thinking and meaningful learning'. Int J Sustain High Educ 3(3):221–224

Warburton K (2003) Deep learning and education for sustainability. Int J Sustain High Educ 4 (1):44–56

Weiler S, Scholz-Barth K (2009) Green roof systems: a guide to the planning, design, and construction of landscapes over structure. Wiley, Hoboken, New Jersey

Author Biography

Karolina Bartkowicz and Judy Rogers teach in the Landscape Architecture Program at RMIT University, Australia. Karolina Bartkowicz also works as a research assistant on the project *The Matter of Landscape*. Judy Rogers is Program director of the Bachelor of Design (Landscape Architecture)/Bachelor of Urban and Regional Planning. Her Ph.D. was concerned with examining and critiquing sustainable city discourse, particularly within the Australian context. She has taught a wide range of courses that are concerned with sustainability and sustainable development, including a fully on-line course titled Challenges in Sustainability

Enhancing ESD Through the Master of Clean Energy Engineering Co-op Program: A Canadian Case-Study

Jenny Reilly and Tatiana Teslenko

Abstract

Like many other universities worldwide, the University of British Columbia in Canada promotes teaching and research on sustainability and has a history of pursuing strong operational sustainability goals. UBC demonstrates innovation by going a step further to integrate efforts with the help of its University Sustainability Initiative (USI). By fostering partnerships and collaborations that extend beyond traditional disciplinary boundaries, the initiative intersects two important themes: using the campus as a living laboratory and empowering students as agents of change. In addition to 480 sustainability-related courses, UBC has 41 sustainability-related programs. Our paper will describe one of them—the Master of Engineering in Clean Energy Engineering with a co-op option, offered by the Faculty of Applied Science and the Clean Energy Research Centre. This program is the only one of its kind in Canada. It provides advanced training in energy efficiency and conservation, including technologies that will help to meet the global need for energy while reducing electricity consumption, as well as greenhouse gases and other emissions. The co-op option provides a deeper and richer form of learning by presenting an authentic learning space and adding a significant value to education for sustainable development. Students complete co-op work terms within various industry sectors including consulting, municipal government, and in the pulp and paper industry, in positions focused on demand-side energy efficiency and conservation work. The paper describes the Engineering Co-op Program and two significant partnerships with the BC Hydro

J. Reilly (✉) · T. Teslenko
The University of British Columbia, Vancouver, Canada
e-mail: jenny.reilly@ubc.ca

T. Teslenko
e-mail: tteslenko@gmail.com

W. Leal Filho et al. (eds.), *Integrative Approaches to Sustainable Development at University Level*, World Sustainability Series, DOI 10.1007/978-3-319-10690-8_35

Power Smart Initiative and FortisBC, electricity and natural gas corporations in British Columbia, Canada, that support the program by providing expertise and co-funding of co-op work terms. Knowledge of this successful initiative could be valuable for many universities that are engaged in promoting university-industry partnerships within the global engineering curriculum.

Keywords
Clean energy engineering · Energy efficiency and conservation · Transformative learning · Co-operative education · Industry partnership

1 Introduction

In the years since the Bruntl and Report, sustainability has become a global priority (Anderberg et al. 2009; Barth et al. 2007; Leal Filho 2009), and the global community engaged in developing strategies to fight climate change, poverty and environmental degradation (Cebrián et al. 2012). The Declaration of the United Nations Decade of Education for Sustainable Development (abbr. DESD 2005–2014) brought forth a socio-political understanding that nothing less than a paradigm shift was on the agenda—an epistemological change towards sustainability. A consensus was reached that "it is imperative that society transforms its relationship with the use of resources" (Desha and Hargroves 2012, p. 29). Sustainability "implies a change of fundamental epistemology in our culture" and, consequently, in our education, research and practice (Sterling 2004). Instead of simply reproducing the existing social structure, a paradigm shift enables social transformation for a better, more sustainable world. Education for sustainable development (ESD), therefore, includes education for a sustainable social transformation (Cebrián et al. 2012). Such a social transformation stimulates critical reflection, clarification and reassessment of values (Tappeser and Meyer 2012, p. 395).

ESD plays a key role in fostering "a more sustainable, equitable and socially just society" (UNESCO 2005). Universities around the globe are engaged in "transforming students to become future leaders, empowered and motivated to make decisions for sustainability" (Anderberg et al. 2009; Barth et al. 2007; Shriberg 2012). Many universities in North America made a conscious effort to promote education and research for SD and to pursue strong operational sustainability goals. In Canada, the University of British Columbia, a Tier One research university, started a curriculum and organizational change using a "whole-system" approach described in the literature (see, for example, Desha and Hargroves 2012).

This paper describes the journey undertaken by the University of British Columbia (UBC) in order to encourage education and research for SD, to offer a transformative learning experience, and to develop learning communities for sustainability. The authors present the milestones reached along this path and discuss significant initiatives, among them "UBC as a living lab" and "UBC as an agent of change". The authors further introduce one of UBC's 41 sustainability-related programs– the Master of Engineering in Clean Energy Engineering degree program, focusing on its co-operative education option, offered by the Engineering Co-op Program at the Faculty of Applied Science. Because engineering students will be required to play a leading role in the enactment of sustainable development, they should learn how to deliver solutions that are not only technically and commercially viable, but also environmentally and socially sustainable. In order to develop appropriate competencies and skills, partnerships between the university and industry are instrumental. The authors discuss the role of industry partnerships in the development of the co-operative education option and the resulting change in the students' and employers' appreciation of core competencies and skills.

2 Theoretical Framework and Methodology

Due to the need for a paradigm change, the universities around the globe were required to transform yet again, thanks to the new understanding of sustainability, pedagogy and methodology. The University of British Columbia included in its Strategic Plan the goal to provide every student the opportunity for transformative student learning through outstanding teaching and research, enriched educational experiences, and rewarding student life (Place and Promise 2014). Learning takes place "within the social and emotional context of each student's life" (Place and Promise 2014). Applying new pedagogical research, UBC is expanding learning opportunities outside of the classroom, e.g., through community service learning, international opportunities, internship and mentorship, experiential learning and co-op experiences (Place and Promise 2014).

Transformative learning is "the capacity to change existing patterns and worldviews, to construct new knowledge collectively, to challenge and improve practice, and to critique and examine sustainability issues" (Sterling 2004). Transformative learning theory (Mezirow 2009) is particularly useful for this study because it emphasises critical reflection, dialogue and holistic learning. In particular, transformative learning theory and action research are instrumental for examining the value of experiential learning and its impact on the change of students' understanding of core competencies and skills.

Action research is a useful method for our study that can be combined with observation, interviewing, and pedagogical reflection. It is also important in projects involving industry partners because it can contribute to the transformation of professional practice and generate new knowledge (Cohen et al. 2000; Cebrián et al. 2012; Somekh 2006). Action research promotes emancipatory rather than technical change, so it is can help faculty and staff to embed ESD in the curriculum. It can be undertaken by a group of instructors working at the same university, researchers within the same institution, students and advisers (Cebrián et al. 2012; Somekh 2006). Our research team, comprised of faculty and staff members, chose emancipatory action research (EAR) due to its potential to engage participants in a learn-by-doing process accompanied by critical reflection, clarification of essential values, and exploration of contrasting viewpoints (Cohen et al. 2000; Somekh 2006).

Transformative learning theory and EAR were used to develop flexible learner-centered pedagogy for the needs of co-operative education. Data about students' and employers' evaluation of competencies and skills were collected through surveys, focus group discussions, observation, and analysis of institutional and program resources.

3 UBC's Sustainability Milestones

In 1990 the University of British Columbia signed the Talloires Declaration and started an important transformation of its ethos. ESD became an agenda of growing significance that resulted in changes in curriculum, pedagogy, structure, and organization. Bridging the gap between knowledge and action, UBC became the first university in Canada to adopt a sustainable development policy and to open a campus sustainability office. UBC Strategic Plan, *Place and Promise*, contains nine commitments with goals and actions designed to realize the university's vision for the future. Sustainability is a key commitment, recognizing that, "in order to meet society's needs without compromising those of future generations, the best efforts of the brightest minds are required" (Place and Promise 2014).

The University has marked its progress with numerous milestones summarized in Table 1. Its Vancouver campus has been transformed into a vibrant, sustainable community where students, faculty, staff and local residents live, work and learn. In response to the challenge of creating a more sustainable society, in 2011 UBC opened the Centre for Interactive Research on Sustainability (CIRS). CIRS has become an internationally recognized research facility that accelerates the adoption of sustainable building technologies and sustainable urban development practices (UBC Sustainability 2014).

4 UBC as a Living Lab and an Agent of Change

The University Sustainability Initiative (USI) was established in 2010 to integrate academic and operational sustainability efforts and to act as the University's agent in ESD innovation. It fosters partnerships and collaborations that extend beyond traditional boundaries of disciplines and sectors and works in two related areas: establishing the campus as a living laboratory and transforming the whole University as an agent of change (UBC Sustainability Initiative 2014).

As Shriberg (2012) notes, "leading institutions of higher education are increasingly utilizing the campus as a laboratory to move beyond organizational "greening" into developing the skills set in students to lead deep institutional changes for sustainability in academia and beyond" (p. 19). The entire UBC campus is regarded as a living laboratory, as a "teaching tool" (Savanick Hansen and Wells 2012), providing the community with the freedom to explore the technological, environmental, economic and social aspects of sustainability. The four parts of the "living laboratory" initiative are:

Table 1 UBC sustainability milestones

Period	Initiative
1990–2000	1990—UBC signs the international talloires declaration
	1996—Faculty members of UBC's School of Community and Regional Planning originate the award-winning ecological footprint concept
	1997—UBC is Canada's first university to adopt a sustainable development policy
	1998—UBC is Canada's first university to open a campus sustainability office
2001–2005	2001—UBC launches EcoTrek (2001–2008), the largest energy and water retrofit at a Canadian campus at the time
	2003—UBC pioneers the U-Pass program
2006–2010	2006—UBC publishes *Canada's first campus-wide sustainability strategy*
	2007—UBC reaches its kyoto targets, reducing GHG emissions from academic buildings to six per cent below 1990 levels; faculty member Dr. John Robinson shares the 2007 Nobel Prize with former US Vice President Al Gore as a member of the intergovernmental panel on climate change
	2008—UBC is the first to sign the *climate change statement of action for canada* and *achieves an A-rating in the sustainable endowment institute's green report card*, among the top in North America
	2009—UBC develops the Sustainability Academic Strategy, which recommended establishing and expanding shared infrastructure and programs to connect sustainability across academics and operations; UBC integrates sustainability as one of nine core commitments in "Place and Promise", further institutionalizing sustainability in UBC's highest-level strategic plan; UBC publishes a series of case studies on sustainability
	2010—UBC commits to reducing GHG emissions 33 % by 2015, 67 % by 2020, and 100 % by 2050, compared to 2007 levels; UBC launches the Sustainability Initiative (USI) that promotes and unites sustainability efforts in teaching and learning, research and campus operations
2011–2012	2011—UBC opens the *Centre for Interactive research on sustainability (CIRS)*, the most innovative and high performing building in North America; UBC's Vancouver campus is designated *Canada's first Fair Trade campus* by FairTrade Canada;
	UBC breaks ground on one of the largest *steam-to-hot water conversion projects* in North America, which will reduce GHG emissions by 22 %;
	UBC receives *Canada's first gold rating in STARS* (*Sustainability Tracking, Assessment and Rating System*), a comprehensive university sustainability rating system
	2012—UBC opens the *bioenergy research and demonstration facility* which will reduce campus GHG emissions by 9 % and introduce clean energy technologies at a community scale

Source Our Story (2014)

1. The integration of UBC's core academic mandate (research and teaching) with the University's operations.
2. Partnerships between the University and private sector, public sector or NGO organizations.

3. Sound financial use of UBC's resources and infrastructure.
4. The potential to transfer the knowledge UBC gains into practical, positive action applicable to the greater community.

UBC is uniquely suited to act as a living laboratory, as a community of over 49,000 students, 13,000 faculty and staff, and residents, with over 50 % of campus households occupied by someone who studies or works at UBC. The university has about 500 buildings covering 402 hectares of land, owns and operates its own utilities including electrical, heating, water and waste, and is responsible for its roads and infrastructure (UBC Sustainability Initiative 2014).

UBC is representative of many communities; therefore, many communities can implement sustainability solutions developed here. For example, while transforming the campus into a completely sustainable community, UBC researchers are studying new approaches to UBC's energy generation and distribution systems (UBC Sustainability Initiative 2014).

As an agent of change, the University is training the future "green leaders" who will take ESD principles beyond the gates of campus and effect positive change. As noted by Shriberg (2012), "leadership skills and practices to guide society on a sustainable path are and will continue to be in high demand" (p. 19). UBC scholars and students conduct important research, while the University's private, public and NGO partners contribute expertise, human capital and resources to projects that might not otherwise be available. Private sector partners help to commercialize and take to market technologies and innovations developed on campus; public sector partners help to develop policies and regulations and share them with other communities; NGO partners help with social licence and community engagement efforts.

To support students in reaching their potential, UBC has established sustainability "pathways" that allow students, regardless of their disciplines, to integrate sustainability into their studies. UBC offers more than 480 sustainability courses and 41 programs that range in scope from the highly specialized to the multidisciplinary. A number of non-credit options are also available, such as UBC Reads Sustainability events with authors, volunteer opportunities at the UBC Farm, participation in the SEEDS Program, etc. (UBC Sustainability Initiative 2014).

5 UBC's Master's Program in Clean Energy Engineering

One of the University's important initiatives is the graduate program "Master of Engineering in Clean Energy Engineering" offered through the UBC Clean Energy Research Centre[1] in the Faculty of Applied Science. This is the only program of its kind in Canada and one of only a handful in the world. This program enables the University to stay at the forefront by providing advanced training in energy

[1] The UBC Clean Energy Research Centre (CERC) is "dedicated to developing knowledge and solutions aimed at reducing the environmental impact of energy consumption and ensuring the sustainability of our energy supply" (CERC 2013, 2014a, b, c).

Table 2 Scheduling of the master of engineering in clean energy engineering co-op program

Academic calendar	Fall (Sept–Dec)	Winter (Jan–April)	Summer (May–Aug)
Year 1	Study	Study	Work-term 1
Year 2	Work-term 2	Study	Graduation

The North American academic year commences in September
Source Engineering Co-op Program (2013b)

efficiency and conservation. Offered by the Faculty of Applied Science and the Clean Energy Research Centre, the program is focused on the supply-and-demand side of energy. This includes energy conservation, social change concepts, efficient use of electricity and natural gas, energy supply, and methods for comparing and evaluating alternative energy scenarios (CERC 2014c). Students study sustainable energy sources such as biomass, solar, wind and small-scale hydro, in addition to having a concentration on business, management, leadership and other aspects of energy efficiency and conservation (CERC 2014c).

This program has a co-operative education option,[2] allowing the students to complete a four- or eight-month work experience (Engineering Co-op Program 2013a). The option is facilitated through the Engineering Co-op Program, the largest co-operative engineering co-op program in Western Canada. The Engineering Co-op Program is offered at both the Vancouver and Okanagan (in the Interior region of British Columbia) campuses. The Engineering Co-op Program at the Vancouver campus commenced in 1978; since its inception, the program has secured over 22,000 co-op work term opportunities for students with employers locally, nationally and internationally. The co-op option is a non-mandatory program and available in all 11 undergraduate disciplines and Master of Engineering, Master of Software Systems (the only mandatory co-op option) and Master of Applied Science degree programs. The University of British Columbia Okanagan campus opened its doors in 2005 and has co-op options in all three undergraduate disciplines in addition to Master of Engineering degree programs.

The UBC Engineering Co-op Program combines classroom learning with relevant, technical, paid, monitored and evaluated co-op work experiences. Students participating in the Master of Engineering in Clean Energy Engineering co-op option alternate between three academic terms and one- or two- engineering co-op work terms. Table 2 represents the academic scheduling of the Master of Engineering in Clean Energy Engineering Co-op Program.

The Master of Engineering in Clean Energy Engineering, non-mandatory co-op option, that is now in its 5th year, adds a significant value to ESD by presenting an authentic learning space that provides students advanced training that will help

[2] As outlined in the UBC academic calendar, "Co-operative Education is a partnership between students, employer and the University of British Columbia. As an educational process, co-operative education formally integrates a student's academic studies with paid, approved, career-related work experience in participating employer organizations" (UBC Vancouver Academic Calendar 2014/15 2014).

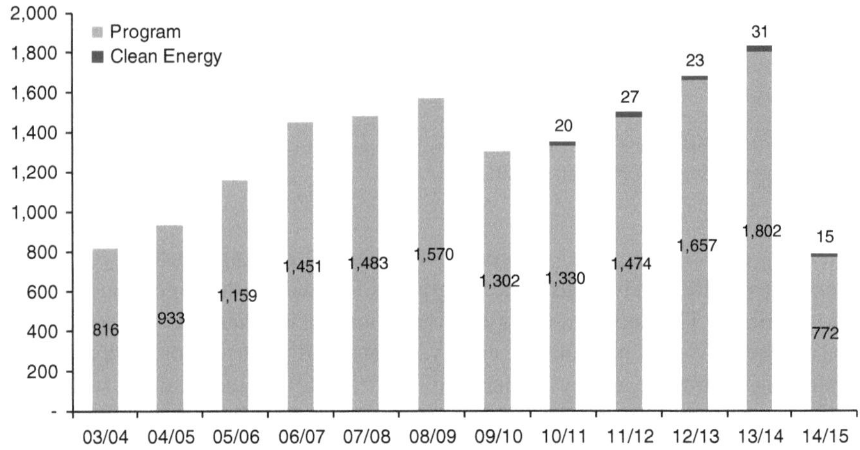

Fig. 1 Co-op Work Term Statistics for the master of engineering in clean energy program. *Source* Engineering Co-op Program (2013b)

meet the global need for energy while reducing energy consumption, greenhouse gases and other emissions. Since the inception of the program, Master of Engineering in Clean Energy Engineering co-op students have secured 116 co-op work terms. Co-op work term statistics are provided in Fig. 1.

The value of authentic learning spaces has been widely discussed in literature (Pretorius 2012). The advantages of the co-op option are significant and multiple: UBC students gain up to eight-months of paid, relevant and invaluable work experience, while businesses gain access to bright, motivated students for short-term employment. Many co-op students, in fact, return to employers with whom they have had their co-op experience for employment after graduation as energy professionals and have a significant impact on the engineering community both domestically and internationally (Engineering Co-op Program 2013a).

From the commencement of the program, there was great interest and support from BC Hydro, a provincial Crown corporation in British Columbia, Canada, that generates, purchases, distributes and sells electricity. BC Hydro was instrumental in providing the first co-op salary subsidy offered in the program, called the 'BC Hydro Power Smart Salary Subsidy for Co-op Work Terms'. The objective of BC Hydro Power Smart co-funding was to provide opportunities for students to gain experience in the application of energy efficiency, conservation skills and concepts in the commercial and industrial sectors of British Columbia (Engineering Co-op Program 2013a). More recently, another company, FortisBC (a natural gas, electricity and alternative energy retail company in British Columbia, Canada) followed with support through the 'FortisBC Salary Subsidy for Co-op Work Terms' (UBC's Master of Clean Energy 2014). In Canada BC Hydro and FortisBC are leaders in sustainable energy and in producing and delivering electricity and natural gas in environmentally and socially responsible ways. Through successful partnerships involving the UBC

Sustainability Office, the UBC Faculty of Applied Science, BC Hydro Power Smart and FortisBC, two types of co-op salary subsidies are now available. Both subsidies were solely created for provincial employers hiring Master of Clean Energy Engineering students in approved industry co-op work terms in the areas of electrical and natural gas energy efficiency and conservation (Engineering Co-op Program 2013a) and have been instrumental in the steady growth of the program.

Salary subsidy funding is provided to companies offering approved co-op industry experiences in the form of 50 % or up to $5,500 of a co-op student salary over a four-month work term. Approved industry experiences are full-time positions of 35–40 h per week for a minimum of 12 consecutive weeks to a maximum of 32 consecutive weeks (Engineering Co-op Program 2013a).

The provision of co-op salary subsidies offered through public and private partners in the Province demonstrates the University's strong support and commitment to sustainability and transformational learning. Master of Engineering in Clean Energy Engineering Co-op students that are employed by industry partners who have utilized either the BC Hydro Power Smart or FortisBC salary subsidy have worked in a number of positions. The focus of their work term has been on the application of energy efficiency and conservation skills on the demand side of the energy system in either electrical or natural gas energy savings. Students apply ESD concepts during their co-op industry work terms in one or more of the following areas (Engineering Co-op Program 2013a):

- Research and development of new, demand-side technologies that save electrical energy or natural gas energy
- Analysis, energy modelling, or design of new or existing homes, commercial buildings and industrial facility energy use
- Testing, energy studies or measurement of energy performance technologies and facilities
- Monitoring, targeting and reporting; measurement and verification methods
- Energy management or energy coaching including operation and maintenance in support of efficiency
- Energy economics, decision-making and behaviour in homes or businesses
- Policy support of energy efficiency initiatives such as building codes equipment standards and government or utility incentive programs
- Local government programs that include the efficient use of natural gas
- Conservation programs for natural gas
- Natural gas energy efficiency business case development and financial analysis.

As part of the ongoing efforts to evaluate and improve the UBC Engineering Co-op Program, during each work term students and employers are given the opportunity to complete a brief on-line survey concerning their experiences. Results are compiled per discipline annually. Numerical data supplied by students are compiled on the following topics: students' perception of job information, job quality, technical competencies and transferrable sills, job recommendations, goals and reflection, administration review, co-op coordinator review, and co-op program review. Numerical data from employers are compiled on student technical competence and transferable skills, overall performance, and co-op program feedback. In 2012/2013 the following results were obtained from participating students and employers (Engineering Co-op Program 2013b):

- Hundred percent of students agreed that they were satisfied with the work term experience, that the work term experience was influential in developing their career and academic goals, and that their academic program had prepared them well for the work term.
- Employers ranked the students on a list of technical competencies and transferable skills that have been defined by the Canadian Engineering Accreditation Board (2012). The following competencies and skills were ranked the highest with an Excellent-Good rating: 'Knowledge Base for Engineering', 'Problem Analysis', 'Investigation', 'Teamwork', 'Individual Work', 'Professionalism', 'Impact of Engineering on Society and the Environment', 'Ethics and Equity', and 'Life-long Learning'. Ranked second was 'Use of Engineering Tools' (93.3 %), and third 'Communication Skills' (92.8 %).
- Employers judged 100 % of students to be good or outstanding in their overall work term performance.

The high percentage of stakeholder satisfaction levels demonstrates the value perceived, impact of the experience, and value of authentic learning spaces that the co-op option provides.

6 Conclusion

In 1990, along with hundreds of other leading educational institutions who signed the Talloires Declaration, the University of British Columbia pledged to make sustainability the foundation for campus operations, research and teaching (UBC Sustainability 2014). Since that time the whole university community has engaged in participatory action research in order to enable the paradigm shift towards embracing sustainability. This was undertaken by encouraging research, partnership, curriculum and organizational change. A deep integration of operational and academic efforts in sustainability challenged the UBC community to reach across traditional boundaries of disciplines and sectors. Over the years, by using the "whole-system" approach, the University has engaged its community in ESD in the following ways:

1. At the organizational level UBC has committed to a deep integration of its operational and academic efforts.
2. UBC research teams reported on the devastating consequences of unsustainable development, as well as studying people's behaviours and advancing sustainability scholarship inside and outside UBC.
3. On the curriculum level, hundreds of ESD-related courses, programs and educational events were developed for faculty, staff and students.

Importantly, UBC has established partnerships with private companies, the public sector, and NGOs. The partnerships formed with the BC Hydro Power Smart Initiative and FortisBC support the Master of Engineering in Clean Energy Engineering program by providing expertise and co-funding of co-op work terms. This training includes technologies that will help to meet the global need for energy while reducing electricity consumption, as well as greenhouse gases and other emissions. The results discussed in the paper could be valuable for many universities that are engaged in promoting university-industry partnerships within the global engineering curriculum.

References

Anderberg E, Norden B, Hansson B (2009) Global learning for sustainable development in higher education: recent trends and a critique. Int J Sustain High Educ 10(4):368–378

Barth M, Godemann J, Rieckmann M, Stoltenberg U (2007) Developing key competencies for sustainable development in higher education. Int J Sustain High Educ 8(4):416–430

Canadian Engineering Accreditation Board (2012) Accreditation Criteria and Procedures Report. http://www.engineerscanada.ca/files/w_Accreditation_Criteria_Procedures_2012.pdf. Accessed 27 Mar 2014

Cebrián G, Grace M, Humphris D (2012) Developing people and transforming the curriculum: action research as a method to foster professional and curriculum development in education for sustainable development in higher education. In: Leal Filho W (ed) Sustainable development at universities: new horizons. Peter Lang, Frankfurt, pp 273–284

Clean Energy Research Centre (CERC) (2013) FortisBC Annual Impact Report 2012–2013. Faculty of Applied Science, the University of British Columbia, Vancouver

Clean Energy Research Centre (CERC) (2014a) BC Hydro Annual Impact Report 2012–2013. Faculty of Applied Science, the University of British Columbia, Vancouver

Clean Energy Research Centre (CERC) (2014b) BC hydro power smart program. http://www.cerc.ubc.ca/prospective_students/powersmartposter.pdf. Accessed 27 Mar 2014

Clean Energy Research Centre (CERC) (2014c) Master of engineering in clean energy engineering. http://www.cerc.ubc.ca/prospective_students/cleanenergy.php. Accessed 27 Mar 2014

Cohen L, Manion L, Morrison K (2000) Research methods in education. RoutledgeFalmer, Abingdon

Desha CJ, Hargroves K (2012) Fostering rapid transitions to education for sustainable development through a whole-system approach to curriculum and organizational change. In: Leal Filho W (ed) Sustainable Development at Universities: New Horizons, Peter Lang, Frankfurt, pp. 29–46

Engineering Co-op Program (2013a) Report on Masters of Clean Energy Student and Employer Evaluations (2012/2013). The University of British Columbia, Vancouver

Engineering Co-op Program (2013b) Engineering Co-op Program 2012/2013 Year End Report. The University of British Columbia, Vancouver

Leal Filho W (ed) (2009) Sustainability at universities—opportunities, challenges and trends, vol 31. Peter Lang, Frankfurt

Mezirow J (2009) Transformative learning theory. In: Mezirow J, Taylor EW, (eds) Transformative learning in practice: insights from community, workplace and higher education. Jossey-Bass, San Francisco, CA, pp 18–32

Our Story (2014) UBC sustainability. http://sustain.ubc.ca/our-commitment/our-story. Accessed 27 Mar 2014

Place and Promise: The UBC Plan (2014) http://strategicplan.ubc.ca/the-plan/sustainability/. Accessed 27 Mar 2014
Pretorius R (2012) The value of authentic learning spaces to facilitate education for sustainable development in open and distant learning. In: Leal Filho W (ed) Sustainable development at universities: new horizons. Peter Lang, Frankfurt, pp 169–182
Savanick Hansen S, Wells CW (2012) The campus as a teaching tool: a case study of Macalester College's EcoHouse. In: Leal Filho W (ed) Sustainable Development at Universities: New Horizons. Peter Lang, Frankfurt, pp 479–492
Shriberg M (2012) Building sustainability leaders: a framework to prepare students to thrive on complexity and lead transformative changes. In: Leal Filho W (ed) Sustainable Development at Universities: New Horizons. Peter Lang, Frankfurt, pp 273–284
Somekh B (2006) Action research. A methodology for change and development. Open University Press, Milton Keyes
Sterling S (2004) Higher education, sustainability and the role of systemic learning. In: Corcoran PB, Wals AEJ (eds) Higher education and the challenge of sustainability: problematics, promise and practice. Kluwer Academic Publishers, Dordrecht, pp 49–70
Tappeser V, Meyer A (2012) Change-agents in sustainability governance: institutional transformation at three institutions of higher education. In: Leal Filho W (ed) Sustainable Development at Universities: New Horizons. Peter Lang, Frankfurt, pp 395–404
UBC's Master of Clean Energy Engineering program receives boost of support from Fortis BC (2014) In: UBC clean energy research centre news. http://blogs.apsc.ubc.ca/cercnews/2012/11/19/ubc%e2%80%99s-master-of-clean-energy-engineering-program-receives-boost-of-support-from-fortisbc/. Accessed 27 Mar 2014
UBC Sustainability (2014) http://sustain.ubc.ca/. Accessed 27 Mar 2014
UBC Sustainability Initiative (2014) http://sustain.ubc.ca/our-commitment/about-sustainability-initiative. Accessed 27 Mar 2014
UBC Vancouver Academic Calendar 2014/15 (2014) Co-operative education program. http://www.calendar.ubc.ca/vancouver/index.cfm?tree=12,195,272,31. Accessed 30 Mar 2014
UNESCO (2005) United Nations decade of education for sustainable development (2005–2014): draft international implementation scheme. http://www.env-edu.gr/Documents/files/Basika%20Keimena/DESD.pdf. Accessed 27 Mar 2014

Author Biographies

Jenny Reilly holds a Diploma of Human Resources from Monash University, Australia, and is currently enrolled in a Master of Business Administration program at the Edinburgh Business School, Heriot-Watt University. She is Director of Co-op and Professional Development in the Faculty of Applied Science at the University of British Columbia. She is also Director of the Canada-Japan Co-op Program which is a national consortium made up of 17 universities and colleges across the country. She is the Chair of the UBC Co-op Council and has held past positions as President of the Association for Co-operative Education, British Columbia and the Yukon; Chair, Canadian Association of Co-operative Education Accreditation Council; UBC Delegate of the Accountability Council of Co-operative Education.

Dr. Tatiana Teslenko holds a Kandydat of Philological Sciences degree from Odessa University, Ukraine (1989) and a Ph.D. from Simon Fraser University in Canada (2000). She is Professor of Teaching at the Department of Mechanical Engineering in the University of British Columbia. Her research interests include education for sustainable development, engineering communication, and experiential learning. Dr. Teslenko authored and co-authored four books and dozens of papers in communication studies and writing pedagogy. She designed and launched several programs for international students (such as ASSIST UBC and the Graduate Teaching Assistants training program). She was the founder and inaugural Director of the Centre for Professional Skills Development at the Faculty of Applied Science.

Termite Tales: Organisational Change —A Personal View of Sustainable Development in a University—As Seen from the 'Tunnels'

Jack Christian and Liz Walley

Abstract

This paper tells the story of an ongoing greening process in a HE Institution from the perspective of an individual on the 'shop floor'. It adds to the work of Walley and Stubbs (1999, 2000) on organisational greening at the level of the individual in context. Adopting Morgan's (1993) metaphor of how termite mounds—organizations—are shaped, the 2000 paper investigated the role of the environmental champions in large organisations exploring the notion of strategic termites as inspiration for would-be environmental change agents. Whilst Walley and Stubbs (2000) focused on formally appointed champions, this paper records the journey of an informal agent, taking place against a backdrop of organisational change in a large university. Although much of this change had little to do with greening it presented opportunities for an industrious termite to tint the mound a little greener. The journey notes successes and failures from the 'termite's' perspective and how his thinking gradually shifted from purely opportunistic to something more strategic. The strategic termite metaphor aids the interpretation of this journey. It is hoped this paper will inspire would-be sustainability change agents. Like termites we all move along seemingly predetermined pathways but if we remain sensitive to our surroundings, opportunities to reshape the future will present themselves.

J. Christian (✉)
Department of Accounting, Finance and Economics, Manchester Metropolitan University Business School, All Saints Campus, Oxford Road, Manchester M15 6BH, UK
e-mail: j.christian@mmu.ac.uk

L. Walley
Department of Management, Manchester Metropolitan University Business School, All Saints Campus, Oxford Road, Manchester M15 6BH, UK
e-mail: l.walley@mmu.ac.uk

W. Leal Filho et al. (eds.), *Integrative Approaches to Sustainable Development at University Level*, World Sustainability Series, DOI 10.1007/978-3-319-10690-8_36

Keywords
Metaphor · Greening · Sensemaking · Inspiration · Change agent

1 Introduction

This paper aims to presents a story of (part of) a university's greening process from the perspective of an individual academic in the Business School. The story is interpreted through the medium of metaphor—specifically Morgan's strategic termite metaphor (1993)—and structuration theory (Giddens 1984). Recent literature on 'greening' or sustainable development in higher education reveals a dearth of perspectives from the grassroots of the organisation—exploring, for example, how change might or might not be accepted by staff or students. Following Rasanen et al. (1994) this paper offers a richer, deeper, interpretivist approach, casting light on sensemaking and sensegiving at the 'bottom' of an HE organisation. This innovative approach adds to the body of knowledge on organisational greening within HE.

The story is presented first as a 'termite's tale'. Our termite's recollections of his behaviours and actions are then interpreted in the light of Walley and Stubbs (1999, 2000) interpretation of Morgan's strategic termite metaphor to reveal insights and lessons learnt. We anticipate these insights will be useful as a source of inspiration to would-be sustainability change agents in HE and maybe beyond.

2 Organisational Change and Termites

Morgan, together with Burrell, (Burrell and Morgan 1979) is credited by many with the introduction of a new multi-paradigm approach to organisation science (McCourt 1997). He and Burrell suggested that several paradigms were possible at any one time, each of them giving rise to different sets of theories. What then occupied Morgan was how these paradigms gave rise to different theories.

In due course he came to the conclusion that the different paradigms are based on the different perspectives with which we can view an organisation (Clegg et al. 2006; Cunliffe 2008; Brooks 2009). These perspectives can be linked to "insights associated to different metaphors for the study of organisations" (Morgan 1980, p. 611). In Images of Organisation (Morgan 1986) he presents eight such metaphors including the organization as an organism. In Imagination: The art of creative management (Morgan 1993) he takes this metaphor one step further to introduce the organisation as a termite mound. It was this metaphor that inspired the work of Walley and Stubbs (2000) who describe his ideas thus:

"The imagery Morgan seeks to invoke by the termite metaphor is of a "self-organizing process where order emerges 'out of chaos'" (Morgan 1986, p. 44). He describes how distinct piles of earth emerge from apparently random behaviour. These structures somehow attract the attention of other termites and become the

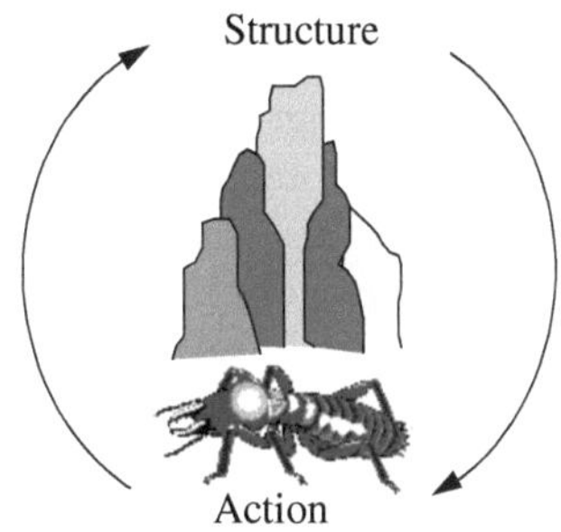

Fig. 1 Termite mounds (reproduced from Walley and Stubbs 2000, pp. 44–45)

focus for sustained building activity, leading to the formation of columns. Columns that form sufficiently close together can be joined by arches as the mound structure gradually grows in complexity. In this self-organising process, the mound structure both emerges from the termites' actions and shapes the way in which those actions are taken (Fig. 1)."

In utilising the 'structure and action' framing for the termite metaphor, Walley and Stubbs (2000) also draw on the work of social theorist Giddens (1984). They suggested that environmental initiatives cannot be understood solely in terms of the organisational structures that surround them. Nor is it appropriate to focus exclusively on the actions taken by individuals. Rather they argue that environmental initiatives should be seen in terms of order emerging from the mutually producing relationship between action and organisational/social structure. In other words, structure shapes action and action shapes structure.

The image of an evolving, infinitely variable structure, continually being made and remade through the actions of many individual agents, offers a mental frame with which to approach the complexity of individuals' actions in the context of organisations.

Walley and Stubbs (2000, p. 41) explore "how inspirational lessons for organizational greening might be drawn at the level of the individual in context." Building on their earlier paper (Walley and Stubbs 1999) they look at the activities and processes undertaken by environmental champions in large organisations, exploring in particular the notion of strategic termites as a source of inspiration for would-be environmental change agents. Notably however the paper focused for the most part on formal champions, i.e. those with a specific role within their organisation, and presented a structuralist or organizational view of the changes that took place. Whilst this would of course enable a would-be environmental change agent to see what changes had taken place, and to that extent act as an incentive, it did not have anything to say on how the agents involved had themselves made sense of the situations they were in and why they acted in the way they did. One of the aims of this paper is to add to the work of Walley and Stubbs in this respect.

There is no doubt that the termite mound is a powerful metaphor. This is evidenced by Wheatley (1996) in her article The Unplanned Organization: Learning from Nature's Emergent Creativity where she describes a termite tower in the Australian savannah as a life-changing image. The tower is twenty feet high and the interior is a complex mixture of tunnels and arches that move heat and moisture about to provide cool living places and appropriate growing conditions for the termite's fungi food plant. The tower is built without any apparent design or planning beforehand and there are no obvious leaders, it is seemingly the product of all the termites within the group.

Such imagery is powerful and must be handled with care. Morgan himself is always careful to warn us of the limitations of metaphorical analysis. Despite that, as McCourt (1997) reminds us, his work has been subject to a number of critiques. Writers of a more realist bent, for example Bourgeois and Pinder (1983) see a more objective world where literal and precise descriptions and explanations are both possible and appropriate. Others have noted its failure to address issues of power (Tinker 1986) and the difficulty in transferring metaphoric insight into action (Boje and Summers 1994).

Writing in 'Cultural Perspectives on Organizations' Alvesson (1993) is much more positive about the use of metaphors, calling on work by Ricoeur (1978), Brown (1976) and Lakoff and Johnson (1980) he describes metaphors as "illustrative devices" and "crucial elements in how people relate to society" (ibid p. 10). Critiquing Bougeois and Pinder he claims it is impossible to let the objective data speak for themselves, echoing perhaps Lyotard (1984) who points out that scientists must always revert back to language and metaphor to explain their findings.

Nevertheless Alvesson acknowledges there are problems with the use of metaphor. These include the extent to which a metaphor can truly describe the subject under consideration and both under-use and over-use of particular metaphors. Alvesson calls for a self-critical and reflective approach to choosing metaphors constantly "reminding oneself and the reader that they do not tell the whole story" (ibid p. 13).

3 Changing the Colour of Higher Education

Two papers exemplify the green tide flowing through higher education in the United Kingdom; Integrating Sustainability into Business Schools—analysis of 100 UN PRME Sharing Information on Progress (SIP) Reports by Godemann et al. (2011), and Facilitating organizational change for embedding sustainability into academia: a case study by Exter et al. (2013). The first is exactly as the title describes it, the authors analysing and reporting on the first 100 reports submitted to the UN Principles of Responsible Management Education initiative. This initiative germinates in the UN Global Compact, a collaboration between the United Nations and several thousand trans-national enterprises, who called upon business schools around the world to build the concept of responsible management—which includes ethical and sustainable elements—into their curriculum. So far approximately 450 business schools have signed up to the initiative including circa 25 in the UK.

The second paper is a description of sustainability being embedded into the curriculum at Cranfield. This paper is primarily a description of the individuals involved in and the processes undertaken to implement PRME at Cranfield together with an analysis of the changes achieved using the model developed by Godemann et al. (2011) as part of their research.

Both papers are essentially functionalist, positivist and built around a realist ontology. They portray the data as out there to be discovered and, in the case of the Cranfield paper, there are summary recommendations on how to embed sustainability into business schools. There is no real discussion on how change might or might not be accepted by the staff or students in the business school. If indeed a business school can be compared to a termite mound there is nothing heard from the termites in these papers.

More interest in the grassroots can be found in the work of Adrianna Kezar who has been writing about change in higher education institutions in the USA for over a decade. In recent work she has researched the formations, functions and challenges of grassroots leadership teams and the processes of sense-making and sense-giving in transformational change processes *from the bottom up*. In her work there is some sense of what the termites might be thinking.

A major institutional change in US higher education in recent times that can be compared to the greening of, or introduction of sustainability into, the curriculum is the call for Interdisciplinary Learning. This call is noted in, for example, Interdisciplinary Strategies on Transformative Change in Higher Education (Holley 2009) and Understanding sense-making/sense-giving in transformational change processes from the bottom up (Kezar 2013). It is in the latter paper that Kezar notes "In recent articles most scholars of sense-making, note understanding sense-making/sense-giving from the bottom up and over a period of time as the key new areas for research" (ibid p. 762).

Arguably however even Kezar's papers are too reductionist to convey how sense-making and sense-giving happens. In her 2013 paper she tries to condense data from 28 institutions and inevitably the reader sees little of what is actually happening at 'the bottom', we just see Kezar's own interpretation. It seems what is missing is a rich, deep interpretivist approach giving voice to those at the bottom.

This paper then seeks to achieve two things. Firstly by employing a quasi-ethnological approach, i.e. by entering the mound and describing the life of a green termite, it hopes to cast light on sense-making and sense-giving at the 'bottom' of the Higher Education establishment and in doing so add to the body of knowledge concerning organizational change within Higher Education. Secondly and possibly more important from the termite's point of view, it hopes to encourage other green termites who are endeavouring to green their organization.

4 Methodology

Essentially one of the authors—a termite—relates his experiences since taking up employment in Higher Education—arriving in the mound. He focuses on green issues which for one reason or another are important to him and tells how he found other termites with similar interests and how these meetings and other chance circumstance allowed him to contribute to the greening of the mound. An interpretation of this narrative is then offered as seen through the lens of Walley and Stubbs (1999, 2000). However given the primary aims of this paper are to understand termite sense-making and to motivate would-be termites the narrative is given in full to allow the reader to make their own interpretation.

4.1 Caveat

Following Alvesson (1993) above, we must of course remember that we are not dealing with a real termite. In reality our termite belongs to several mounds, e.g. social and professional relationships, and can import new ideas and resources from these mounds. Also in his world some termites have (much) more power than others, e.g. vice-chancellors and deans, something missing from Margaret Wheatley's (1996) image described above. Finally and most importantly our termite is self-aware and can think ahead so as to try and influence other termites and the future in general.

5 Termite Recollections

I have always exuded Green, it's in my DNA. I have studied natural history for as long as I can remember though until the last decade it was always something I did outside my career as a professional accountant. I have taken part in numerous national recording schemes covering birds, insects and reptiles over the last 20 years.

In 2007 I applied for a job at the university, I had been teaching quantitative methods and accounting part-time for a couple of years and a full time post became available. A few weeks prior to the interview I had read a book review and one of the books contained several chapters on 'environmental management accounting'. It had stuck in my mind because in 30 years of professional praxis I had never heard of this! I had read the professional journals and occasionally attended regional meetings of the Chartered Institute of Management Accountants but environmental management accounting had never come to my notice.

At the job interview I was asked if I would like to undertake research. I wasn't sure what that would mean but I said "Yes" as that seemed appropriate.

Then came the follow up question, "What would your research?"

"Environmental Management Accounting," sounded like a good answer so I gave it and got the job.

On arrival the department's head of research put me in touch with LW, a lecturer in another department who was seen as a business school sustainability expert. I was to start studying for a Ph.D. and LW was to be one of my supervisors. I was in the mound and I was running in the Green tunnels. Chance and chance alone had put me there and so it was to continue over the next few years. Openings would appear in the tunnel walls, often caused inadvertently by the activities of other termites, and I would move through to build new tunnels or follow existing tunnels all the while exuding Green.

Initially of course I had to find my way around a new organisation as well as learn more about environmental management accounting. Probably the first door to open was in another faculty. LW introduced me to RH in the Science faculty. RH was a keen proponent of the Sustainability agenda in the University (by now I had come to recognise that Green was an integral part of Sustainability) and had set up an Education for Sustainable Living (E4SL) group. One of her first tasks was to determine the extent to which Sustainability was included in teaching activities throughout the University.

LW and I joined the Group and I was motivated to carry out an audit of courses delivered in the Business School looking for signs of Sustainability in the course specifications. At the same time LW and I discussed what we were doing with a number of colleagues. Soon a number of colleagues had joined us to form a Curriculum for Sustainable Learning (C4SL) group in the Business School. This happened in 2009 and perhaps our major achievements were the setting up of a C4SL website and blog, and the delivery of a workshop to encourage and support the explicit inclusion of Sustainability elements in business school courses.

C4SL was eventually subsumed in the University's Sustainable and Ethical Enterprise Group that came into being at the end of 2011. In its short existence C4SL raised the profile of Sustainability. Further it opened two more doors for me; first I was asked to take over a postgraduate course—Accounting, Society and the Environment (ASE), second I was asked to write a top-up course for part-qualified accounting professionals. I called this unit Ethics and Sustainability Accounting (ESA). ESA was particularly exciting as it was to be internet based and would have a truly international reach. In fact the first delivery in 2011/2012 went out to over 200 students across 40 countries!

ASE commenced delivery in 2010 which turned out to be a busy year with two more opportunities to push the Sustainability agenda. First the Business School was undertaking a major revision of its curricula in a process known as EQAL and second it began to seek AACSB accreditation. As part of the latter process the Dean asked LW and I if we knew anything about PRME (The Principles of Responsible Management Education), a United Nations sponsored initiative. It so happened that I knew the UK leader of the PRME movement and was able to introduce him to the Dean. The Dean became convinced that PRME was a concept that reflected the School's values and subsequently LW and I helped her convince the faculty that it would be in the School's interest to sign up to the PRME agenda.

To support the School's commitment to Responsible Management I worked with other colleagues as part of the EQAL process to introduce "Responsible" units or courses into (nearly) all our undergraduate programmes. Over 90 % of our undergraduates take one of these courses: there are three of them—'the Responsible Accountant', 'the Responsible Marketer' and 'Responsible Business'—which allows them to be tailored to the programme within which they are delivered, and they are primarily focussed on ethics and sustainability.

2010 had been frenetic. The vast majority of the work carried out had been "extra-curricula". Some of it carried on into 2011 and frankly I needed to consolidate these new initiatives so I undertook very little new work choosing to by-pass some "open doors". For example I chose not to get involved in Green Impact, a programme which encourages staff and students to live and work more sustainably. I decided that I would focus on sustainability in the taught curriculum.

This hiatus in activity carried on into 2012, due in large part to some major changes in my personal life, but also because the Business School moved into a new building and the new geography did not support old relationships. However late in the year circumstances once again combined to ignite my Green impulses. Every 2 years PRME signatories have to produce a Sharing Information on Progress Report. I volunteered to work with the School's administration and marketing teams to collate material for this report which we submitted in December. At the same time leading exponents of PRME in the UK (PRME by now having the support of some 25 Business Schools in the UK) were advancing the idea of a UK Chapter of PRME. They called one or two meetings which I attended and, returning inspired, I began to wonder how we could further advance the progress we had made as a Business School.

Eventually I had an idea, this was that all new undergraduates should be made aware of the School's commitment to PRME and the associated concepts of ethical and sustainable business. I devised four lectures which I could deliver to all new students, probably as a guest lecturer in suitable slots in their programme. I put this idea to the Dean this time asking if I could have a time allowance in my workload to prepare and carry out the lectures. I copied my suggestion to my head of department and rather pleasingly he and the Dean agreed it was a worthwhile idea and I was given a time allowance.

Not everything goes to plan however and it was well into 2013 by the time I was able to start organising the lectures. I had to find suitable lecture slots in the numerous programmes that we run at the Business School and this meant speaking to departmental and programme heads as well as colleagues who ran the actual courses. Anyone who has been involved in Higher Education will appreciate how time consuming this turned out to be. Some doors opened quickly, others less so and then there were some that were kept closed! Somewhat disappointingly to-date I have only reached out to circa 60 % of our first year students.

However as one door closes another often opens and in this instance that door was the University's quinquennial review. Every 5 years the University reviews all its programmes and courses and at this point in time, early 2014, lecturers are rewriting their first year courses for 2014/2015. This has given me a second opportunity to approach them and ask them to write in a space for my lectures. An opportunity I intend to take.

6 Termite Reflections

The first and most obvious realization is that there was no plan. A combination of opportunity and 'something inside me' made me take the actions I did. By and large this has not changed, I do not plan ahead I simply have ideas and I look for opportunities (doors opening) to put them into effect. I think this opportunist approach arises out of a feeling that I have no way of influencing events at university, or mound, level and I must work within whatever scope of influence that I have. What has changed over time is an increased awareness of the mound, the structure of the university, and its tunnels, i.e. the rules, regulations and hierarchies.

As to what the 'something inside me' is I can only assume it an aggregation of the discourses that have shaped my life and my understanding of reality. Essentially I love 'nature' and I believe we have a moral obligation to the other life forms we share the planet with, now and in the future. My beliefs have has been enhanced by regular forays beyond the mound, to conferences and presentations, notably to the Congress for Social and Environmental Research (CSAER).

Motivation then has not usually been a problem. However there have been knock-backs. As well as the closed doors referred to above two of the three 'Responsible' units were almost lost in the machinations of the quinquennial review process and it took a little bit of negotiating to reinstate them. What this made me realise is that my own tunnels can fall in and I need to lookout for this, other termites are tunnelling all around me each with their own DNA; and so it should be, in diversity we have strength.

Finally I note termites are non-dominant and do not seek control and that was a source of inspiration to me. What is required is resilience and hope. Hope that we are working to make a better world and resilience to enable to us to get up again when it looks like the battle is being lost.

7 Interpreting Termite Tales

Key to Walley and Stubbs (2000) findings is Morgan's image of a 'strategic termite'. They précis Morgan as suggesting "this could open up new horizons for those who have been struggling without success to mobilise action across their organisation" (ibid p. 51). Further they cite him directly as explaining strategic as "while their activity is open to the influence of random opportunity, decisions and actions are always informed and guided by a strong sense of what they are ultimately trying to achieve" (Morgan 1993, p. 45 in Walley and Stubbs 2000, p. 51). Finally they describe how a 'strategic termite' "might look to create, foster and nurture 'mounds' of activity consistent with the direction in which they hoped things might evolve" (ibid p. 51).

In this section we aim to make the connections between our termite's recollections and reflections and the metaphorical concepts and other relevant theory referred to earlier in this paper. In particular it will be interesting to note which elements of the strategic termite metaphor (Morgan 1993) have been pertinent to this would-be environmental change agent. We start by offering a pictorial representation of the termite mounds described in our termite's story—see Fig. 2.

From the chronology of events it is evident that JC—our termite—started working at the Business School in the same year as a commitment was made by the university to the environmental sustainability agenda; most notably through the appointment of an Environmental Sustainability Manager. But the termite story

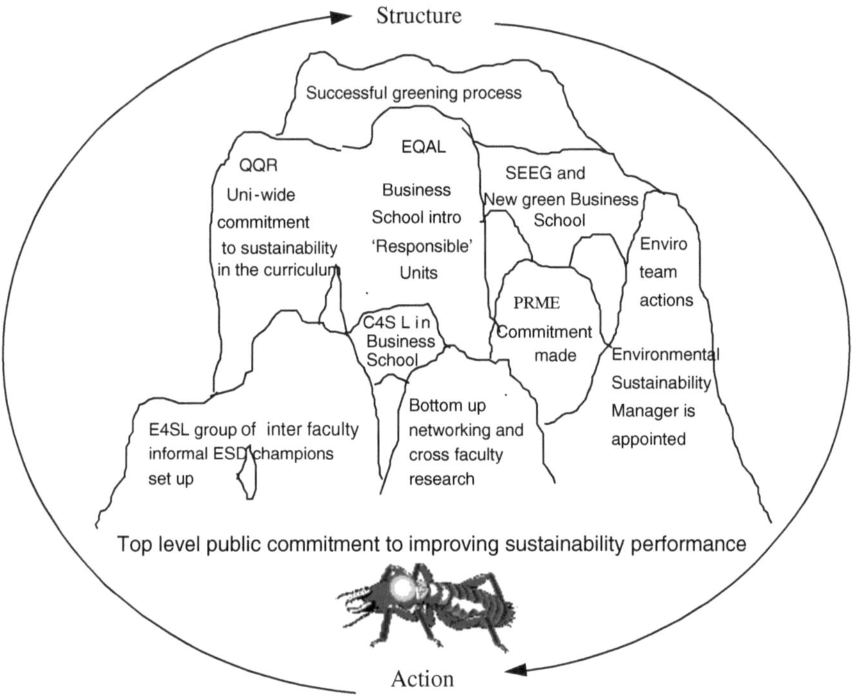

Fig. 2 The organization from a termite's viewpoint

reveals that this top-level commitment was not evident to JC until much later in his career at the university. Nevertheless, from the outset, JC's decision to research environmental management accounting was informed and guided by—in what we might recognise as a strategic termite manner (Morgan 1993)—a strong sense of what he felt was the right thing to do and he took advantage of random opportunity to start building his first mound. In response to questions asked subsequent to JC writing his recollections and reflections JC recalled reading the Walley and Stubbs (2000) paper and being uplifted by the concept of the lone termite pursuing its cause. It seems likely therefore that, consciously or unconsciously, the termite metaphor was a motivating image to his subsequent, apparently random and opportunistic actions.

In earlier research on environmental champions Walley and Stubbs (1999, p. 5) identified "'networking', 'a sense of audience' and 'agenda translation'" as significant aspects of the greening process. In this termite's tale the E4SL group was identified as a cross faculty networking group which shared ideas and achievements and 'built on each others' mounds'. For example, RH organised the first E4SL Workshop in the Science faculty. Inspired by that experience and benefiting from those ideas the business school members of E4SL, including JC, began to build a similar mound, organising a business school curriculum workshop and a C4SL group in the Business School the following year. It is clear from the termite story that JC also benefited from external networking at CSAER conferences and later, with external PRME networks.

From the C4SL mound tunnels ran into the rest of the Business School leading to the development of two new units. One of these formed a link or arch that reached across to that part of the Business School mound representing strategy. It was an external course aimed at part-qualified professionals, the strategy in this case being to develop links with professional bodies and offer more external courses. JC's links to the PRME network also connected with a Business School strategy, i.e. gaining AACSB accreditation, and this laid the foundations of a new PRME mound. Recognising the 'green' content of the PRME initiative JC translated, in part at least, the search for a valuable marketing tool into a green agenda, a process sometimes referred to as 'greenjacking' (Walley and Stubbs 1999).

The PRME mound has continued to grow as a result of a number of opportunist interactions. For example at one point it came together with part of the University mound which was undertaking a fundamental reorganisation of the University's degree structure known as EQAL. This top-down, structural, process change gave JC and two other termites the opportunity to develop the three new Responsible Management units referred to in JC's story. Notably in the course of building the PRME mound JC has negotiated a time allowance and is currently working with colleagues on PRME-related courses and lectures for 2014/2015. Whilst his recollections do not explicitly point to a 'sense of audience' (defined by Walley and Stubbs (1999, p. 5) as "about sending different messages to different audiences, according to their priorities, needs, etc.") it is implicit in his description of his negotiations with the Dean, his Head of Department and the other staff.

Throughout JC's tale, there are examples of the 'structure and action' (Giddens 1984; Walley and Stubbs 2000) reinforcing, iterative processes at work. For example as heard in the termite story, the university's curriculum processes (structure) provide opportunities for termites to take action in terms of proposing new sustainability units. Termite action was also enabled by top-down structural change, e.g. the executive decision to aim for AACSB accreditation. Equally the work of termites (action) has facilitated 'structural' change, for example the growth of sustainability content across the curriculum facilitated Aim 1 of the Dean's restated business school mission: to develop socially and environmentally responsible early career professionals for successful careers.

At the current time the PRME mound continues to support and build onto the Business School strategy and indeed the University Environmental Management strategy both of which seek to increase the visibility of ethics and sustainability in the curriculum. It is taking advantage of the University's quinquennial review of its programmes (structure) to once more push this agenda (action) to help to maintain the University's successful transition to a Green university.

7.1 Notable Absences

Researchers such as Post and Altman (1994) have suggested that successful greening processes seem to be associated with 'using all the levers' of organisational change, e.g. strategic direction, working the corporate culture, communication, rewards, performance appraisal systems, etc.; however in this instance, this was not the case. Formal rewards and appraisals appear to have had very little to do with these changes although there were doubtless intrinsic awards for the individuals involved.

Also absent in JC's tale is any reference to exemplar imagery as identified by Walley and Stubbs (2000) in their work. When asked about the latter subsequent to writing his recollections JC suggested that the intrinsic rewards were sufficiently motivating for him and he had never really concerned himself with imagery. However on reflection he could see how others had been influenced by the new business school building and the success of the university in the 'Green League' and this had helped him in his endeavours.

8 Conclusions

Following McCourt's (1997) analysis of Morgan's (1993) work on the model of change we see that metaphoric thinking $\Longrightarrow$ new understanding $\Longrightarrow$ creative action. The authors believe the use of the termite metaphor to describe organisational change has illustrated how an individual can contribute to organisational change. Specifically we have seen how an individual working at the operational level has contributed to the greening of a large HE organisation. Further the authors believe the metaphor shows how smaller structures can contribute to the formation of larger structures in such a way as to illustrate Gidden's (1984) structuration theory whereby individual actions contribute to the shaping and reshaping of structures.

Obviously it has not been the aim of this paper to make generalised conclusions. In truth the primary aim of the paper is to "motivate and inspire individuals that it is worthwhile to act individually when faced with the enormous scale and complexity of the environmental agenda" (Walley and Stubbs 2000, p. 52). The authors have sought to achieve this by offering a "more process oriented and interactive view of greening" (Rasanen et al. 1994, p. 15) and sharing their bottom up attempts at sense-making in the context of greening in HE sector (as suggested by Kezar (2013)).

However there are perhaps lessons to be drawn from the case that may be applicable in similar contexts. Firstly we note that our termite is a professionally qualified individual working at the operational level of a very large organisation who was able to find the space to adapt his work in a way that suited his ideas on education for sustainable development. Further by making use of a variety of internal and external networks he was able to develop these ideas. This was of course made possible by working 'with the grain' of the organisation, i.e. by watching organisational developments and opportunistically using these to advance his own goals. For the individual in a large organisation who wishes to make changes the key lessons therefore seem to be 'be flexible', 'network' and 'look for and take advantage of appropriate opportunities'.

There are also perhaps lessons for 'senior termites', i.e. policy makers in large organisations. Firstly they need to remain alert to the 'modus operandi' within their organisation. This seems obvious, it is a critical aspect of good governance, but it is much easier said than done in a large organisation. At the very least there must be good two-way communication channels reaching to every level of the organisation. Secondly the policy makers must be open to new ideas and look for opportunities to improve operational policies or even organisational strategy by 'bringing in' new concepts and ideas germinating at lower levels of the organisation.

For the 'transformational' change that has and is still occurring in the case outlined in this paper, 'top-down' commitment has depended upon 'bottom-up' commitment and action. Whilst the Dean's support and that of the University's environmental management team has been fundamental to the changes that have taken place at the Business School, the curriculum has been shaped and the sustainability discourse carried by independent termites acting sometimes alone and sometimes together. The result appears to be a success story as being 'Green' is now seen as one of the Business School's and indeed the University's major selling points.

References

Alvesson M (1993) Cultural perspectives on organizations. Cambridge University Press, Cambridge

Boje DM, Summers DJ (1994) Review of 'Imaginazation: the art of creative management'. Adm Sci Q 39(4):688–690

Bourgeois VW, Pinder CC (1983) Contrasting philosophical perspectives in administrative science: a reply to Morgan. Adm Sci Q 28:608–613

Brooks I (2009) Organisational behaviour individuals, groups and organisation. Prentice Hall, Harlow

Brown RH (1976) Social theory as metaphor. Theory Soc 3:169–197

Burrell G, Morgan G (1979) Sociological paradigms and organizational analysis. Heinemann, London

Clegg SR, Courpassen D, Phillips N (2006) Power and organisations. Sage Publications, London

Cunliffe AL (2008) Organization theory. Sage Publications, London

Exter N, Grayson D, Maher R (2013) Facilitating organizational change for embedding sustainability into academia: a case study. J Manag Dev 32(3):319–332

Giddens A (1984) The constitution of society: outline of the theory of structuration. Polity Press, Cambridge

Godemann, J., Herzig, C., Moon, J., and Powell, A., (2011), *Integrating sustainability into business schools – analysis of 100 UN PRME Sharing Information on Progress (SIP) reports*, Nottingham University Business School, ICCSE Research Paper Series

Holley KA (2009) Interdisciplinary strategies as transformative change in higher education. Innov High Educ 34:331–344

Kezar A (2013) Understanding sensemaking/sensegiving in transformational change processes from the bottom up. High Educ 65:761–780

Lakoff G, Johnsson M (1980) Metaphors we live by. University of Chicago Press, Chicago

Lyotard J-F (1984) The postmodern condition a report on knowledge. University of Minnesota Press, Minneapolis

McCourt W (1997) Discussion note: using metaphors to understand and to change organizations: a critique of Gareth Morgan's approach. Organ Stud 18(3):511–522

Morgan G (1980) Paradigms, metaphors and puzzle solving in organization theory. Adm Sci Q 25:605–622

Morgan G (1986) Images of organization. Sage, London

Morgan G (1993) Imaginization: the art of creative management. Sage, London

Post J, Altman B (1994) Managing the environmental change process: barriers and opportunities. J Organ Change Manag 7(4):64–81

Rasanen K, Merilainen S, Lovio R (1994) Pioneering descriptions of corporate greening: notes and doubts on the emerging discussion. Bus Strategy Environ 3(4):9–16

Ricoeur P (1978) Metaphor and the main problem of hermeneutics. In: Regan CE, Stewart D (eds) The philosophy of Paul Ricoeur. Beacon Press, Boston

Tinker T (1986) Metaphor or reification: are radical humanists really libertarian anarchists? J Manag Stud 23(4):363–384

Walley EE, Stubbs M (1999) 'Greenjacking'—a tactic for the toolbag of environmental champions? Reflections on an SME success story. Eco-Manag Auditing 6(1):26–33

Walley EE, Stubbs M (2000) Termites and champions case comparisons by metaphor. Green Manag Int 29:41–54

Wheatley MJ (1996) The unplanned organization: learning from nature's emergent creativity. Noetic Sci Rev 37. http://www.margaretwheatley.com/articles/unplannedorganization.html. Accessed 13 Jan 2014

Authors Biography

Jack Christian is a Chartered Management Accountant and former Financial Director and European Financial Controller; since joining academia he has specialised in Ethics and Sustainability Accounting. He is the Business School's PRME link and champions the concept of educating responsible managers. His primary research interests are Environmental Accounting and Accounting for Biodiversity.

Liz Walley has been developing and teaching Sustainable Business units at undergraduate and postgraduate level in MMU Business School for the last 15 years. In her role as 'Co-convenor, Curriculum' in the Sustainable and Ethical Enterprise Group (SEEG), and elsewhere, she champions sustainability across the Business School curriculum. Liz has published in the areas of environmental change agents and green entrepreneurs. Prior to MMU, Liz worked in consultancy, banking and industry in various roles.

Integrating Sustainable Development Within Teaching Fashion Education

Amanda Langdown and Valeria Ruiz Vargas

Abstract

It is strongly assumed that the commodification of services and, increasingly, experience itself is killing off positive incentives towards the pursuit of the common good. For instance, Fashion design is guilty of some of the most wasteful practices, both in terms of the speed at which it demands change and the resources that it consumes. This paper examines practices that could influence and drive fashion curricula and programmes of study towards an entirely more sustainable future. Integrating sustainable development into common practice and integrating them within teaching fashion education, in order to motivate individuals and organisations to follow a more thoughtful and critical approach to design. There are designers and academics who already advocate a more sustainable paradigm of fashion education. Therefore, it is imperative that strategies are implemented internationally to encourage designers, new and old, to adopt adaptive rather than transgressive forms of creativity. The research identifies meaningful teaching, learning and engagement methods that support slowing down the metabolism of fashion and the exploration of alternative approaches. This is with a view to create practices that appreciate and integrate both the novel and the sustainable aspects of fashion, ensuring that they become commonplace in education as an essential consideration in design for the common good.

A. Langdown (✉)
Department of Design (Fashion), Manchester School of Art, Manchester Metropolitan University, Benzie Building, Cavendish Street, Manchester M15 6BR, UK
e-mail: a.langdown@mmu.ac.uk

V.R. Vargas
Centre for Excellence in Learning and Teaching Environment Team, Manchester Metropolitan University, All Saints Building, Oxford Road, Manchester M15 6BH, UK
e-mail: v.vargas@mmu.ac.uk

W. Leal Filho et al. (eds.), *Integrative Approaches to Sustainable Development at University Level*, World Sustainability Series, DOI 10.1007/978-3-319-10690-8_37

Keywords
Fashion · Appreciative inquiry · Maker culture · Slow fashion · Education

1 Introduction

According to evidence gathered by (Circular Economy and Resource Efficiency Experts), a third of all clothes bought in the UK end up in landfill (WRAP 2012).[1] This is an astonishing statistic, and yet the fashion system continues to move at break neck speed. We are now presented with new ideas and new fashion identities nearly every 6 weeks. We barely have time to wear the clothes before they become obsolete, and therefore unwanted. We also continue to see fashion design programmes recruiting large numbers of students both from home and abroad. Fashion design courses have increased in popularly over the last 30 years and offer huge marketing potential and revenue for universities in this fee paying era. As fashion becomes ever more connected to every aspect of our lives, an increasing number of teenagers dream of careers as fashion designers.

Clearly, there is a dilemma facing universities. Popular programmes can recruit well but It could be argued that the commodification of higher education could result in more programmes providing course content based on what people want rather than what they need. The human default setting to experience the novel and new (characteristics that fashion has in abundance) could, according to Sandel (2009) "crowd out the good incentive"[2] to develop programmes of study that equip design students for a 21st century design system; a system that must acknowledge the enormous amount of waste and unethical work practices that keep the industry running.

---

[1] Waste and Resources Action Programme All data and references are included in 'Valuing our clothes: the evidence base', WRAP (2012), available from www.wrap.org.uk/clothing. Waste data are based on the analysis in 'Benefits of re-use case study: clothing', WRAP (2012). *Data limitations* This report seeks to identify overall patterns of environmental impact, based on estimates of the quantities and impacts of clothing at each stage of the life-cycle. Due to the complexity of the supply chain, consumer use and disposal routes, and limited availability of data specific to the UK, values are approximate.

[2] Michael Sandel at the 2009 Reith lecture told the following anecdote to illustrate the way in which the market place can kill off good incentives. A study of some Israeli childcare centres offers a good real world example of how market incentives can crowd out non-market norms. The centres faced a familiar problem—parents sometimes came late to pick up their children, and so a teacher had to stay with the children until the tardy parents arrived. To solve this problem, the childcare centres imposed a fine for late pick-ups. What do you suppose happened? Late pick-ups actually increased. Now if you assume that people respond to incentives, this is puzzling. You would expect, wouldn't you, the fine to reduce, not increase the incidence of late pick-ups? So what happened? Introducing the fine changed the norms. Before, parents who came late felt guilty; they were imposing an inconvenience on the teachers. Now parents considered a late arrival a service for which they were willing to pay. Rather than imposing on the teacher, they were simply paying her to stay longer.

2 Background

Certainly fashion as a cultural phenomenon is in part responsible for perpetuating a desire to consume more than we need. It is not in itself sinister, but has been so commoditised that it now appears to be part of a "predatory culture" as expressed by McLaren (1995). He suggests that cultural identity is formed almost exclusively around extreme levels of marketing and consumption.

Recent interviews with first year fashion design students confirm that many are driven by a need to stand out from the crowd. They are a highly individualised generation and see fashion design as a means of expressing themselves with total freedom and a way of gaining recognition. At interviews some applicants report that their main ambition is to take part in Graduate Fashion Week and then hopefully to be spotted by and prepared for a career in what they perceive as a glamorous or cool industry.

Von Busch has used the metaphor of religion to help us understand the far-reaching effect of the fashion phenomenon. He describes designers as 'agents of legitimation' who perform 'acts of consecration'. Consequently, the process of making becomes an act of 'transubstantiation'. To see this metaphor in action, one need only watch students preparing for a fashion show resembling a holy ritual.

This godlike treatment of designers is said to have originated from the ideals of the Romantic Movement and Woods (2007, p. 104) suggests that the current popular view of creativity is still characterised as "theatrical, confident and self-absorbed". During the periods of "market triumphalism" (Sandel 2009) and conspicuous consumption through the 1980s and 1990s, the creative maverick gained the reputation as being a messianic genius producing the unique and unprecedented (Woods 2007, p. 105).

However interviews with first year fashion students' reveal their uncertainty and insecurity about understanding creativity. They can often compare themselves unfavourably with peers whose work they categorise as crazy or *colourful.* It is this self-absorbed transgressive creativity (Woods 2007) that most students aspire to and it is also this resource of imagination that needs to be nurtured and channelled towards the acquisition of the knowledge, skills and values necessary to develop sustainable literacy.

During the interviews, that will be discussed below one final year fashion design student had negotiated a project that took local production as the context for the designs. However, it was the images that she produced that were commented on. The sustainable aspect was marginalised.

Transforming students into talented mavericks who can use fashion's unique elements for achieving the spectacular in social and environmental terms (Thorpe 2012) is surely the dream of most design academics.

It is clear that the fashion system is far reaching and in a state of flux. Developments in digital technology are bringing about unprecedented change including breaking down existing hierarchies. If the 20th century was about large scale, grand gestures and mass markets, the 21st century is it's polar opposite. Small scale, niche marketing and customised manufacturing are more fitting for the digital age.

It is time to encourage and teach adaptive creativity to fashion students. We need to equip designers with the skill to be agents for intervention and to understand their role in relation to engaged forms of consumer participation (Von Busch 2008) and particularly their roles as global citizens.

3 Valuing Well-Being

We can see that the green agenda in HE is being driven by the NUS, HEFCE and HEA. Economics students at Manchester University became very vocal recently about the way in which their curriculum still focused on a system that had brought about the virtual collapse of the banking system. Thorpe argues for the pursuit of market reform for corporations, as well as banks, 'so that they are legally allowed to pursue goals that are broader than simply maximising profits' (Thorpe 2012). Values should no longer be attached solely to the shareholder model but should take into account a range of metrics based around ecological and social consumption, with an emphasis on wellbeing for all.

Porritt (2014) insists that governments have the power to reform capitalism and acknowledge the importance of social and ecological metrics by setting prices correctly, for example, levying higher taxes on carbon and other natural resources. But Thorpe (2012) argues that these reforms will not necessarily address some of the problems of individualism, and the pace at which we demand novelty and stimulation. For this we must strive to encourage and educate our students towards a more sustainable and socially responsible position.

Clark (2008), Research chair of Fashion at Parsons new School for Design, New York, questioned fashion's role in perpetuating this emphasis on newness, novelty and image. She suggests a conceptual framework that could help raise the collective ethical consciousness based on principles derived from the Slow + Design Manifesto (2006) may be the answer.

1. Valuing local economies and distributed economies.

The Global identity is made from a network of local alternative initiatives that share common characteristics.

"A network system in which whatever is locally available is used to the best advantage and whatever cannot be produced locally is exchanged and shared, giving rise to a society and an economy that is at the same time both local and cosmopolitan." (Slow + Design Manifesto 2006).

2. Transparent production systems, with less intermediation between consumer and producer.

There are opportunities here to break down the traditional design hierarchies of designer, producer, and consumer. New systems and services are required and solutions can be found through peer-to-peer relationships for example. Fashion starts to be an opportunity for social engagement, a choice and not a mandate.

3. Production of sustainable and sensorial products that have a longer life and more highly valued than typical consumables.

This is a way of understanding a product from the knowledge of how it is made, from the raw material to the end product, rather than just through the gratification that comes with consuming it.

Clark is justifiably concerned by the speed at which the large commercial fashion brands update their offer. They tap into our need for immediate rewards rather than taking a long view. We are more inclined towards the things that we want than the things we may need. We see the shiny, new and novel idea and feel compelled to have it. We forget the future and are happy for momentary, instant gratification (Offer 2006).

Offer (2006) also tells us that this problem of short-term thinking is the subject of much research in psychology, animal behaviour, and increasingly now in economics. As affluent societies place so much importance on the superficial aspects of materialism, defining themselves through their earnings, possessions, appearance and celebrity, that they have lost contact with fundamental values. When we have so much of everything we cannot even appreciate what we do have, we experience less pleasure and search for the next hit of the new. James (2008) agrees that constantly seeking the things that we desire prevents us meeting the basic needs for well-being, and this can result in a lack of contentment for many.

The New Economics Foundation (NEF) (Thorpe 2012, p. 66) was commissioned by the UK government to find out why materialism wasn't working for so many so called affluent citizens. Thorpe feels that growth leaves out social and environmental values and she is developing a range of social metrics in order to capture those values in a meaningful way.

Her in depth research suggested 5 approaches to measuring well-being (Thorpe 2012, p. 66).

- Social connection
- Physical activity
- Reflection and awareness
- Continuous learning
- Cooperation and giving

It appears that the time is right to follow the trend for marrying up social business education with design programmes in order to ensure social innovation (Thorpe 2012).

4 Preparing for a New Business Paradigm

Design is undergoing a revolution. Technology is empowering more people to create and disseminate designs, and professionals and enthusiastic amateurs are using it to share their practice around the globe. Open design is changing everything including how designers make a living.

The British Design council has suggested that more open and innovative business models could be adopted as the way forward (Jankel 2013). Reconfiguring existing boundaries of production could help reorganise the existing systems and networks that link farming, food production with apparel and textiles for a more connected and sustainable design future.

Planning the curriculum for a network generation caught up in fashion's religious fervour presents us with certain challenges. For many young people their fashion awareness is informed by the speed of change and the dictatorship of fashion trends.

However, at the same time we can see how in the process of social change that new values are gaining greater importance. Castells (2012, p. 230) suggests this as a period of change from individuation to autonomy. He describes individuation as 'the cultural trend that emphasises the projects of the individual as the paramount principle orienting her/his behaviour'. This is not the same as individualism which may see many individuals collaborating towards achieving a common goal. Autonomy however, allows individuals and groups to work independently of the existing institutions of society, and Castells (2012) contends that the Internet, formed around open source protocols, provides the commons for activities that benefit freedom of practice.

We must be mindful in our role as educators that we do not miss this paradigm shift we are reminded by this quote by Hoff:

> In times of change learners inherit the earth while the learned find themselves beautifully equipped to deal with a world that no longer exists.

Harvard Business School (HBS) have recently changed their approach to teaching their world renowned MBA. They have reduced their reliance on case study teaching and included real life problem based learning. HBS reflected that their graduates, who go on to become some of the most highly paid leaders of business and industry, were being held somewhat responsible for causing the recent financial crisis. Often these graduates were perceived as reckless and over confident in their decision making. Educational leaders at HBS hoped to change their students' values through experiential learning and on the spot encounters, which could result in them developing more humanistic values (In Business 2012).

There are similar examples of this more engaged model of learning and teaching in fashion Education. University of the Arts London (UAL) was part of the second green academy group with 'Lightening the Load': Creating Change for sustainability through Fashion Education. Alongside HEA mentors, they are developing a 10-year plan for institutional change in terms of ESD. The textile Environment

Design (TED) project was established in 1996 and UAL currently offers BA (Hons) and MA Textile Design (Chelsea College of Art and Design) with key researchers and lecturers such as Rebeca Farley, textile up-cycling expert and Lorna Bircham, low-impact dye expert. UAL also offers MA Textile Futures (Central Saint Martin College) and their academic staff include Kate Goldsworthy, working on technologies for recycling fabrics and Carole Collet, resilient textiles expert. UAL's aim is to embed sustainability at all levels of the curriculum and "Develop processes, projects and benchmarks that articulate and embody this vision for teaching and learning, sharing across subject areas".[3] Higher Education Academy (HEA) in association with the Environmental Association for Universities and Colleges (EAUC) and the National Union of Students (NUS) created Green Academy and Green Academy 2, which funded projects in eight and nine universities respectively. Diverse and substantial programmes were initiated to embed ESD in the curriculum. However, it is in important to acknowledge that initiatives such as the above mentioned are only the starting point and ESD must be consistently embedded throughout curricula.

UNESCO (2010)'s programme: *Teaching and Learning for a Sustainable Future* endorses the HBS model by suggesting a field immersion experience as an invaluable method of learning about real life issues.

Approaches to embed sustainability into fashion curriculum may include learning as part of a community, a practical and multi-disciplinary approach, and an innovative curriculum for the future. These are widely known teaching practices, however, the following discussion aims to clarify the reasons for applying them into fashion curricula as means to embed ESD throughout teaching and learning in this area.

Firstly, learning as part of a community gives a wider context in order to understand global and local consequences of current and future practice. In this scenario self-importance moves to a different level. Individualism and its expression through fashion shows may have a more balanced value in relationship with sustainability and design service. Community in these terms should be interpreted as future and past generations as well as a global and local system in fashion industry. This leads to understanding fashion consumption and production from the point of view of consumers but also from the point of view of producers, which is often neglected in the industry. The needs of the consumer and producer, in these terms, must also include present and future needs in terms of climate change. Clothes must be designed in an innovative way as the weather is bringing new challenges and our capability to adapt will be decisive in terms of the future of humanity. For instance, energy usage must have a place in learning about fashion design. Fashion as a consequence of climate change needs to be flexible, have excellent quality and be adaptive. For instance, it is important to understand the multiple layers of the relationship between the different parts of the chain in fashion industry. Deciding

---

[3] 'Lightening the Load': Creating Change for sustainability through Fashion Education. http://www.heacademy.ac.uk/projects/detail/changeprogrammes11-12/CPs_12_13/GA-ESD/10_university_of_the_arts.

the type of fabric or the techniques that will be used to make a piece of garment may be seeing as detached from the other parts of the industry but they are crucial in the whole chain of events. Fashion as any other industry works as a natural ecosystem where each part of the system is crucial for the whole. In consequence, a better understanding of the whole system can help individuals in problem solving and decision making for a more sustainable future in the fashion industry.

5 Maker Culture and the Craft of Use

One of the important themes emerging from the interviews with students was their desire to make clothes. Maker culture is part of the trend towards valuing autonomy. The Maker Culture advocates, see the sewing machine as an 'instrument for liberation and skills are a path to freedom' (Von Bosch 2008). It is a reaction against fashion as a phenomenon of dictatorship with all the anxieties that that can induce. In many respects, Maker Culture is a form of transgressive creativity as it allows for individuation, however it fits well with a more sustainable approach to fashion production. As Twigger-Holroyd (2012) suggests, by addressing issues around personal well-being through making, it combines both personal and environmental benefits.

There are moves to reintroduce the skills to enable a renaissance of the British clothing industry, which sounds very much like a business as usual approach to manufacture.

Recent statistics from WRAP's Love your clothes initiative (WRAP 2012) reveal that £140 m of used clothing goes into landfill and £30 billion worth of clothing, hangs in wardrobes unworn. Fletcher (2014) suggests that the majority of people can only see fashion in terms of shopping for clothes but that her personal journey, has taken her away from these traditional encounters with fashion through materials, to a position where she is connecting to people and new fashion systems.

Twigger-Holroyd (2012) uses the metaphor of fashion as a land that has been taken over by global industries. There is so much emphasis on trends and image, that an imbalance is created, by affording more importance and value to the unknown fashion viewer rather than to the wearer.

In addition, students value the making process and are often inspired from their knowledge in this area. Most of them think it is important to understand how garments are made in order to make informed decisions towards a more sustainable practice.

6 Appreciative Inquiry as a Tool for Change

In order to discover our student's attitudes towards sustainable fashion we conducted interviews with small focus groups using appreciative inquiry. This provided us with a double dividend. It revealed both the level of their existing knowledge and attitudes towards sustainability, while at the same time provided us with a forum to ask positive questions about sustainability. It was interesting to note that a marked change in attitude occurred almost as soon as the questions were asked. These questions invited students to shift their perspectives from the individuated mode to the autonomous mode, enabling them to express their visions for the common good.

The problem based teaching and problem-solving model, using an authentic project as the context for learning, has long been the traditional method for teaching and learning in design education. This can however act as a constraint on the imagination as project briefs often retain so many of the conventional codes and promote traditional systems and values. Appreciative inquiry can foster a spirit of discovery and affect change.

Discursive tutorials with small groups of students have also revealed that they recognise the dilemmas facing them but are unsure about how to break free, or find other ways. Working with students as learning partners, using appreciative inquiry could help form a new curriculum with an emphasis on discovering alternatives. Final year students describe how they feel their values have changed over the years of study and that it is important for them that the companies they eventually work for share the same values.

Fletcher reminded us at the Women of the World conference (2014) that in order to become better consumers and producers we have to design future visions through grounding ourselves in the here and now. She also suggested that the business as usual model needs revising, including its unethical and wasteful practices and the anxiety that this type of model induces in consumers.

It was to this end, that in 2012, a cohort of final year fashion students were asked for their thoughts on the big issues of waste surrounding the fashion industry. They were asked to consider some of the guiding principles of the slow fashion movement, described earlier, at the deductive phase of their final design projects. Several students started to show a new level of awareness and this was evidenced in the project briefs that they wrote for themselves.

6.1 Case Study Paula

Paula was an overseas student from Cyprus. She had been discussing the economic situation in her country and it was depressing her, but through appreciative inquiry and story telling, she started to see a project whereby she could work with the local crafts people in her area. She worked with them to co-produce designs that showcased local skills. For this student, fieldwork resulted in her developing greater levels of confidence and consciousness of the role she had in being part of a solution and not the problem. The experience resulted in a deep level of learning and she felt empowered by working collaboratively through this authentic project.

6.2 Case Study Kim

Kim had taken part in an open discussion about the benefits of critical thinking when planning a self-initiated design brief. The learning facilitator told the story of her own research, described her own process of reflection and exploration of slow fashion practice. Kim identified strongly with the ideas of examining where you come from to identify local materials and process that could be re valued. She designed garments using local wool and worked with company who were promoting local yarns. They introduced her to groups of local knitters, who worked with her on producing the garments.

6.3 Case Study Emma

Emma explored the theme of up cycling. She had struggled to understand the transgressive form of creativity that some of her peer group aspired to. However she had very good making skills and she employed these to produce beautifully reconfigured formal garments. Her confidence grew as she realised that she was not obliged to produce garments of shock and awe.

7 Conclusion

In conclusion, universities must be at the forefront of two aspects of the future employers and employees in the fashion industry; the preparation for a new business model that takes into account sustainable practices and the value of well-being. It's clear that design students would appreciate a curriculum that would equip them with understanding more about sustainable enterprise and CSR beyond notions of the bottom line.

The student interviews and focus groups show that through the process of studying fashion their perceptions about the role of the designer have changed. However, it is clear that when students discover the wider picture of fashion industry in a global context, including unethical issues, and wasteful practices the mismatch between their newfound awareness with the lack of enthusiasm from tutors leads to disillusionment. Also, the wide spread teaching and learning practices hardly ever are supported by a progressive curriculum that enables students to imagine new paradigms for sustainable fashion practice. The issues mentioned above create barriers to implement ESD into fashion curricula. However, tools to overcome them and develop a model for the future, include an imaginative approach to learning and teaching, appreciative inquiry, knowledge about production including maker culture, and learning as part of the community.

Despite some tensions that may occur between tutor's roles as academics and learning facilitators, the experiences discussed through the paper suggest that by teaching by example, and encouraging students to collaborate with staff through immersive projects, ESD can be achieved both formally and informally.

References

Castells M (2012) Networks of outrage and hope: social movements in the internet age. Polity, Cambridge

Clark H (2008) Slow fashion—an oxymoron—or a promise of the future? J Fash Theory 12 (4):427–446

Fletcher K (2014) Wow conference; loubutin and landfill: how to be a sustainable fashionista, Royal Festival Hall. https://www.youtube.com/watch?v=zovDGuTqjiA. Last Accessed 27 Mar 2014

In Business (2012) Battle of the business schools BBC 4 http://www.bbc.co.uk/programmes/b01sd2n0. Last Accessed 27 Mar 2014

James O (2008) The selfish capitalist: origins of afluenzza. Vermilion, London

Jankel J (2013) The design council. http://www.designcouncil.org.uk/our-work/insight/the-big-rethink/designs-role-in-innovation/transcript/. Last Accessed 26 Nov 2013

McLaren P (1995) Critical pedagogy and predatory culture: oppositional politic in a postmodern era. Routledge, New York

Offer A (2006) The challenge of affluence: self control and well being in the United States and Britain since 1950. Oxford University Press, Oxford

Porritt J (2014) Sustainable and ethical enterprise group (SEEG) launch. Manchester Metropolitan University, Manchester

Sandel M (2009) BBC reith lecture: 'A new citizenship'. http://www.bbc.co.uk/programmes/b00kt7rg. Last Accessed 30 Mar 2014

Slow + Design Manifesto (2006). http://www.experientia.com/blog/uploads/2006/10/slow_design_background.pdf. Last Accessed 30 Mar 2014

Thorpe A (2012) Architecture & design versus consumerism: how design activism confronts growth. Routledge, London

Twigger-Holroyd A (2012) Fashion diggers: transgressive making for personal benefit. Mak Futur 2. http://makingfutures.plymouthart.ac.uk/journalvol2/pdf/Twigger_Holroyd_Amy.pdf. Last Accessed 15 Mar 2014

UNESCO (2010) UNESCO Teaching and learning for a sustainable future; a multimedia teacher education programme. http://www.unesco.org/education/tlsf/. Last Accessed 22 Mar 2014

Von Busch O (2008) FASHION-able. Hacktivism and engaged fashion design, Dissertation in Design. http://www.konst.gu.se/english/ArtMonitor/dissertations/otto_von_busch/. Last Accessed 30 Mar 2014

Woods J (2007) Relative abundance in designers, visionaries and other stories: a collection of sustainable design essays. In: Chapman J, Gant N (eds) Earthscan, London

WRAP (2012) Waste & resources action programme. http://www.wrap.org.uk/SCAP2020targets traderelease. Last Accessed 30 Mar 2014

Authors Biography

Amanda Langdown has spent 23 years teaching fashion and since 2009 she became particularly interested in sustainable fashion. Amanda actively encourages a sustainable approach to design through her teaching practice. Recently, she wrote a business commentary on Slow fashion as an alternative to mass production—a fashion practitioners journey for the Journal of Social Business.

Valeria Ruiz Vargas was working in Manchester Art School reviewing Education for Sustainable Development issues in the curriculum and engaging staff and students in this area. Valeria is now Education for Sustainable Development Co-ordinator working alongside CELT and the Environment Team. This role is allowing her to expand her previous work in the Art School supporting MMU's commitment to Embedding sustainability into the curriculum. Her job will also contribute to the new Environmental Strategy led by the Environment Team at MMU and the Global Citizenship Initiative led by CELT.

Her research interests focus on ethical and ecological issues related to cultural shifts, well-being, and indigenous thinking in her home country Colombia.

Amanda and Valeria are collaborating since they met in the Sustainability Forum, one of the Art School initiatives to embed ESD into the curriculum in Manchester Metropolitan University.

Global Consensus Is a Dream, but Twitter Is Real: Simulating a Sustainable Development Goals Summit Through Interdisciplinary Classroom Politics and Negotiation by Social Media

Simon Kemp, Julia Kendal, Adam Warren, Laurence Wright, John Canning, Marcus Grace and Clare Saunders

Abstract

Rio+20 saw commitment from the international community to develop Sustainable Development Goals (SDGs) to shape the global efforts towards sustainable development. As part of an interdisciplinary curriculum innovation module at the University of Southampton, students take part in a 'SDGs Summit'. Interdisciplinary student groups represent nation 'blocs' attempting to reach consensus on six priority SDGs, from a starting set of sixteen, through personal and social media negotiation. The exercise requires students to inhabit the perspectives of different 'blocs', challenging them to extend their understanding and application of sustainable development beyond their own discipline. This paper shares best practice on this innovative vehicle for teaching students the complexities surrounding international political negotiation and agreement where the outcomes

S. Kemp (✉)
Centre for Environmental Sciences, University of Southampton, Building 7, Engineering and the Environment, University Road, Southampton SO17 1BJ, UK
e-mail: sk2@soton.ac.uk

J. Kendal
University of Southampton, Building 7, University Road, Southampton SO17 1BJ, UK
e-mail: J.Kendal@soton.ac.uk

A. Warren
Centre for Innovation in Technologies and Education, University of Southampton, Building 32, University Road, Southampton SO17 1BJ, UK
e-mail: ajdw@soton.ac.uk

L. Wright
Maritime and Technology Faculty, Southampton Solent University, Room 306, East Park Terrace, Southampton SO14 0YN, UK
e-mail: Laurie.wright@solent.ac.uk

W. Leal Filho et al. (eds.), *Integrative Approaches to Sustainable Development at University Level*, World Sustainability Series, DOI 10.1007/978-3-319-10690-8_38

will have differing social, economic and environmental consequences for the 'blocs', depending on their economic prosperity, natural resources, states of development, and political ideologies. The paper also explores how the summit incorporates the challenges around inequality of access and influence on the global negotiating platform. The adaptation of existing simulation and negotiation pedagogies to address current global political concerns and the use of Twitter in the classroom is considered to be a suitable approach to address the complex interdisciplinary subject area of sustainable development. Although initially focused at ESD practitioners, the interdisciplinary, social media and international focus of the SDGs Summit renders this approach relevant for pedagogical innovators and students across the globe.

Keywords
Education · Sustainable development · Curriculum · Sustainable development goals · Simulation · Gaming · Twitter

1 Introduction

The commitment of the international community to agree Sustainable Development Goals (SDGs) as one of the key outcomes of Rio+20 offers Education for Sustainable Development (ESD) practitioners opportunities to create innovative multidisciplinary pedagogies. One suggested approach is the delivery of a 'SDGs Summit' ('the summit') providing a new multidisciplinary global sustainability focus to simulation pedagogies.

The University of Southampton is a research-intensive university in Southern England. Part of the prestigious 'Russell Group' of research-led Higher Education (HE) Institutions, the University offers a 'Curriculum Innovation Programme' (CIP) of interdisciplinary modules for students to develop skills and understanding outside their traditional curriculum (University of Southampton 2014).

J. Canning
Centre for Learning and Teaching, University of Brighton, M104 Mayfield House, Falmer, Brighton BN1 9PH, UK
e-mail: j.canning@brighton.ac.uk

M. Grace
Southampton Education School, University of Southampton, Building 32 m University Road, Southampton SO17 1BJ, UK
e-mail: M.M.Grace@soton.ac.uk

C. Saunders
Environment and Sustainability Institute, University of Exeter, 2.33 Penryn Campus, Penryn, Cornwall TR10 9FE, UK
e-mail: C.Saunders@exeter.ac.uk

The value of interdisciplinarity in ESD is well recognised (Parker 2010). Over two-thirds of UK undergraduate students consistently state a desire for sustainable development (SD) to form part of their studies (Drayson et al. 2013). CIP is an appropriate vehicle for a cross-university interdisciplinary module on sustainability open to students who usually lack formal curriculum options to study this 'wicked problem' (Rittel and Webber 1973). The CIP module 'Sustainability in the Local and Global Environment' (module code 'UOSM2015') has been offered to students across the University since 2012. The module has an interdisciplinary and collaborative focus, with taught content delivered by academics from different disciplines across various faculties.

It may be argued that single sustainability modules work against the current trend towards embedding ESD across HE curriculum (Winter and Cotton 2012; Kemp et al. 2012), but the interdisciplinary approach of the UOSM2015 module ensures a balanced approach to social and economic and environmental SD content. Single modules also introduce sustainability to large numbers of students (Hegarty et al. 2011). The greatest value of single interdisciplinary modules is the ability to explore sustainability in the necessary depth for such a contested and complex subject. This is particularly important for students pursuing a career in this field, and for developing sustainability-literate graduates within a 'greener economy'.

Simulations are an established pedagogical approach for students learning about the complexities of international agreements and summits (Lantis 1996; Starkey and Blake 2001). Krain and Lantis (2006) developed a Global Problems Summit to introduce international relations students to the complexities of diplomacy and negotiations, demonstrating that simulations and traditional lecture environments both promote learning but result in significantly different impacts on the learning outcomes gained. Students benefit from recreating complex political negotiations that encourage participants to consider the motives and interactions between groupings (Smith and Boyer 1996). McIntosh (2001) advocates the use of a 'Model United Nations' as a useful framework integrating elements of an undergraduate international studies curriculum.

Simulation pedagogies can be interpreted as a variation of the Kolb (1984) experiential learning cycle. Dieleman and Huisingh (2006) advocate experiential learning in an ESD context to address systems thinking, collaboration between teachers and learners, and active learning for SD. ESD should enable students to move beyond critique to creating change towards a desired future based on particular values; experiential learning fosters the necessary emotional engagement with the subject to achieve this (Dieleman and Huisingh 2006). Students move from a passive position as outsiders from the process to active participants of a specific actor group (Dieleman and Huisingh 2006), in this case political negotiators working for global consensus on SD.

Technology is a recognised tool for enhancing student learning (Jisc and The HEA 2009), particularly social media as it facilitates information sharing, participation and collaboration (Evans 2013). The UOSM2015 module aspires to innovate in pursuit of the most effective pedagogies; using Twitter as a learning and communication tool is an important component of the overall delivery, and especially the summit.

Twitter is a social networking site, where individuals microblog and interact with other users. Twitter is currently the second most popular social media site (eBizMBA 2013), with an estimated 996.5 million accounts worldwide (Twopcharts 2014), and over 15 million UK users (Socialmediatoday 2014). Integrating social media into HE is still relatively new, with empirical evidence of its effectiveness in facilitating learning still emerging (Evans 2013). Education practitioners have focused more on incorporating Twitter into the learning process than other social media sites (Evans 2013). Twitter is better suited to ongoing discussion between learners (Ebner et al. 2010), and enables students to be co-producers in current public debate (Blair 2013). Twitter has advantages over Virtual Learning Environments such as Blackboard which have their own norms and procedures (Blair 2013). Students may be more familiar with Twitter and benefit from developing their digital literacy (Hargittai 2007) and therefore employability within their discipline-specific studies.

Integrating Twitter into the UOSM2015 module provides a voice for students lacking the confidence or skills to contribute verbally (Junco et al. 2011). This is vital to ensure all students can meaningfully participate in the fast-moving, vocally-driven political summit simulation exercise. Throughout the module staff posted resources and commentary from a dedicated account '@UOSM2015' and their own accounts using #UOSM2015. Students were encouraged to use and monitor this hashtag.

The use of Twitter as a pedagogical tool to complement simulation and gaming is insufficiently researched to date, rendering this a valuable area for exploration. The majority of work in summit simulations has taken place through single disciplines, typically in international studies. This paper is innovative through demonstrating how the adaptation of existing simulation and negotiation pedagogies to address current global political concerns and the use of social media in the classroom addresses the complex interdisciplinary subject area of sustainable development.

2 The Sustainable Development Goals Summit

The 2012 UN Conference on Sustainable Development (Rio+20) saw international commitment to agree SDGs (United Nations 2014). These will build on the success of the Millennium Development Goals (MDGs), set at the 2000 Millennium Summit (United Nations 2000) to meet the needs of the world's poorest communities, whist minimising harmful environmental impacts.

The challenges inherent within attempting to reach global consensus on sustainability targets form an important component of the UOSM2015 module. The established pedagogical approach of simulation, combined with emerging pedagogies of gaming and use of Twitter have been utilised to tackle this challenge through the 'SDGs Summit'. This is a potentially effective approach to understand the complexity behind reaching global consensus on international SD priorities, and enhance student understanding of global sustainability issues.

Table 1 Aims of the SDGs summit

1.	Enhance student understanding of SD
2.	Enable students to improve understanding of SD through experiential learning
3.	Introduce students to the process of negotiation for international agreement on SD priorities and targets
4.	Assess the pedagogical value of using Twitter in and beyond the classroom
5.	Assess the pedagogical value of simulation and gaming as a technique for the delivery of ESD

Interdisciplinary student groups represent selected international 'blocs' attempting to reach consensus on SDGs through face-to-face and social media Twitter negotiation. The summit ends with the global community agreeing six priority SDGs from a starting set of sixteen. Students have to inhabit the perspectives of different 'blocs', challenging them to extend their application of SD beyond their own discipline or personal views. The pedagogy of active ESD through simulation, strategy setting, negotiation, and use of social media in the classroom is considered a suitable vehicle for the delivery of the aims (see Table 1) of this particular challenge.

3 Methodology

Simulation and gaming within a credit-bearing module require precise planning, simplifying a potentially complex process where misinterpretations by students may lead to failure of the exercise and non-achievement of module aims. Equally, execution of the simulation should consider the construct of the wider module to avoid the 'game' dominating over intended learning outcomes (McIntosh 2001).

The summit was developed by an interdisciplinary team of seven academics representing politics and international relations, education, modern languages, environmental science and engineering, and a learning technologist specialising in gaming design. The team divided the international political community into nine 'blocs' (see Table 2), representing some of the differing ideological, financial, cultural and resource perspectives within the international community. The blocs have identifiably different priorities, enabling students to engage with the difficulties in achieving global consensus from differing standpoints.

The blocs represented not just geographical units but also the major powers on the international stage. Each bloc is allocated different voting power on a sliding scale to represent the theoretical levels of influence held by the groups of nations. For example, a 'North America' influence card is worth +10 votes, whilst an 'Africa' card is only worth +2 points. This engages students with the unequal platforms from which different nations currently negotiate whilst attempting to reconcile global community interests and, more importantly, the needs of their own populations.

Table 2 Nation blocs and the number of votes allocated to each voting card

Bloc	Influence card	Top priority card	Blocking move card
North America	+10	+30	−30
China	+8	+24	−24
Europe	+6	+18	−18
Russia	+4	+12	−12
Pacific Rim	+4	+12	−12
India	+3	+9	−9
Middle East	+3	+9	−9
South America	+3	+9	−9
Africa	+2	+6	−6

Each bloc has six voting cards (see Table 2):

- Four standard Influence cards
- One Top Priority card, which can be used to try and push through a key goal One Blocking Move card, which can be used to try and eliminate a goal

The number of blocs was limited to nine to ensure the game was manageable for the student cohort size (53 in year one, 56 in year two). Students are randomly and equally allocated to blocs. The module team act as the 'United Nations', managing the summit, monitoring negotiations (including Twitter), and reporting results.

4 The Proposed Sustainable Development Goals

The proposed goals (see Table 3) build upon the MDGs and reflect current global priorities. Aims and likely targets were developed for each goal to assist in identifying which align with bloc priorities, for example,

Controlling Population Growth (E):
This goal aims to help developing nations accelerate their demographic transition to lower birth rates by improving maternal health, reducing child mortality and providing effective education and resources for family planning. It also aims to control accelerating human population growth globally in recognition of the pressure placed on resources and biodiversity.

If adopted, this goal might lead to targets such as:

Universal access to maternal health care by 2020

Universal access to family planning education and resources by 2020

Reducing the maternal mortality rate by 75 % by 2025 based on 2014 levels

Reducing child mortality by 75 % by 2025 based on 2014 levels

Table 3 The sixteen proposed goals, five in each of the three key areas of SD: social, environmental and economic issues

Code	Sustainable development area	Goal name
A	Social	Improving health
B	Social	Promoting equality and diversity
C	Social	Universal education
D	Social	Reducing violence
E	Social	Controlling population growth
F	Environmental	Protecting biodiversity
G	Environmental	Minimising climate change
H	Environmental	Sustainable land management
I	Environmental	Sustainable water management
J	Environmental	Reducing pollution and waste
K	Economic	Eradicating extreme poverty
L	Economic	Sharing knowledge and technologies
M	Economic	Enabling access to resources
N	Economic	Promoting free and fair trade
O	Economic	Supporting economic growth
P	Multi	Ecosystems services

Goal P cuts across all three areas

The ideal outcome would be a balanced approach to SD, with equal numbers of social, environmental and economic goals. Offering equal numbers in each area highlights any bias towards particular concerns in the final goals.

5 Preparation for the Game

One week before playing round one, the students were allocated to their group and provided with game instructions and the proposed goals, giving time to research together the likely priorities of their bloc. Dedicated Twitter accounts are provided for each bloc and the UN (@SDGSUNM), with the internationally used #SDGs hashtag adopted for the duration. These serve the dual purpose of an alternative negotiating platform and engaging the wider sector with the summit.

The game is played over a two-day period. The opening round allows the bloc members to work together to agree upon their targets, followed by two rounds of negotiation and voting, with elimination of five goals at the end of the first vote. The use of Twitter for negotiation is encouraged throughout the two-day period, extending beyond formal, timetabled sessions. The six goals with the most votes in the final vote form the SDGs for the global community for the next 10 years (Table 4).

Table 4 The SDGs summit rounds of activities, aims and outcomes

Round	Activity	Aim	Outcome(s)
1	Preliminary bloc discussions	Agreement of bloc 6 priority goals	• Group decision on priorities
			• Group decision on goals to be blocked
			• Priorities submitted to the UN only
2	Negotiation	Agreement between blocs to support common priority goals	• Bloc members negotiate with other blocs in person and via Twitter
3	First vote	Identify 5 goals with least votes	• Blocs vote, seeking to ensure priority goals are not in the bottom 5 goals
4	Outcome of negotiations	Announcement of 5 eliminated goals	• Eliminated goals only announced through UN Twitter account
			• Votes for 11 remaining goals are not carried over to future rounds
5	Final negotiation	Agreement between blocs to support common priority goals	• Blocs re-evaluate and re-negotiate priorities based on remaining SDGs
6	Final vote	Identify 6 goals with most votes	• Blocs agree and cast final votes
7	Summit close	Announcement of 6 SDGs	• UN identifies and announces 6 goals with the most votes, which become the SDGs for the next 10 years
			• Discussions about summit and learning outcomes
			• Comparison of SDGs with blocs' submitted priority goals

6 Results

The results of the voting by the student cohorts over 2 years were collated, including first votes, final votes and the agreed six SDGs resulting from each summit.

7 SDGs Summit Year 1 (2012/2013)

Six goals were eliminated after the first round of voting: B, E, F, J, L, N (Table 5). Usually only five goals would be eliminated, but B, E, F, J, and N were all equal on zero points (Table 6). The six goals with the most votes are shown in Table 7.

8 SDGs Summit Year 2 (2013/2014)

The five goals eliminated after the first round of voting were: C, D, F, M, P (Table 8). The six goals with the most votes in the final round are shown in Table 11. In 2013/2014 the option to make UN agreements was introduced.

Table 5 Results of the first round of voting in the SDGs Summit 2012/2013

	A	B	C	D	E	F	G	H	I	J	K	L	M	N	O	P
North America			10				−30	10	10				30		10	
China			8				16		8			−24	24			
Europe							42						−18			
Russia	8			−12					8						12	
Pacific Rim							8		12				−12			8
India	6		3				−9		3						9	
Middle East				3			−9		3			3	3		9	
South America				12			−9						9			
Africa	6						−6				6		2			
Total	20	0	21	3	0	0	3	10	44	0	6	−21	38	0	40	8

Table 6 Results of the final round of voting in the SDGs summit 2012/2013

	A	C	D	G	H	I	K	M	O	P
North America		10		−30	10	10		10	30	
China		8		24		8		8	8	
Europe	6		6	24			6	−18		
Russia	8			−12		12			8	
Pacific Rim			4	20		4			−12	
India	3			−6		9	6			
Middle East			6	−9		3		3	9	
South America			21	−9						
Africa	10			−6		2	2			
Total	27	18	37	−4	10	48	14	3	43	0

Goals B, E, F, J, L, N were eliminated in the first round

Blocs would agree to vote for a particular goal. For each bloc included in the agreement, the UN added two votes to the goal. This provided additional incentive to negotiate. (Table 9).

A balance across social, environmental and economic goals was not achieved in either year (Tables 7 and 10). Combining the votes from both years shows an overall bias towards social concerns (Table 11). Supporting Economic Goals (O) received the most votes over 2 years.

Table 7 The six goals with the most votes at the end of voting in 2012/2013

UOSM2015 2012/2013	
Final 6 SDGs	Area
I—Sustainable water management	Environmental
O—Supporting economic growth	Economic
D—Reducing violence	Social
A—Improving health	Social
C—Universal education	Social
K—Eradicating extreme poverty	Economic

Table 8 Results of the first round of voting in the SDGs summit 2013/2014

	A	B	C	D	E	F	G	H	I	J	K	L	M	N	O	P
North America								10					−30		60	
China		8			24				8	8			−24		8	
Europe				−18			24		6	6	6					
Russia		4	0	4				4							4	
Pacific Rim	4						4		−12	16					4	
India	3		3		9		3			3			−9			
Middle East				3			−9	3	6			9				
South America				3	−9		3	3			9				3	
Africa							2		2	−6	6			4		
UN					4		12	8	6	4			−4		6	
Total	7	12	3	−8	28	0	39	28	16	31	21	9	−67	4	85	0

9 Discussion

The overarching aim of the summit was to enhance student understanding of SD, particularly in its context as a 'wicked problem' (Rittel and Webber 1973). Throughout early module discussion sessions students commonly criticised the lack of sustainability action on the international stage, questioning why international conventions and policies falter or set apparently unambitious targets. The summit was developed as a formative teaching method to address these concerns, helping students develop an empathetic appreciation for the complexities surrounding political agreement and the restrictions faced by different political actors. Students learn that the range of presented outcomes will have differing social, economic and environmental consequences for countries (through the use of blocs), depending on

Table 9 Results of the final round of voting in the SDGs summit 2013/2014

	A	B	E	G	H	I	J	K	L	N	O
North America					10	20	−30				40
China		8	24	−24		8			8		8
Europe				30		6	6			−18	
Russia		4		−12	20						4
Pacific Rim				4		−12	4	4		16	
India	6	3	9	−6							
Middle East					6	6	−9		9		
South America				−6	3			12		3	
Africa	2			−4	4	6					
UN	4	4	4	8	12	14	−2	4	4	4	6
Total	12	19	37	−10	55	48	−31	20	21	5	58

Goals C, D, F, M, P, were eliminated in the first round

Table 10 The six goals with the most votes at the end of voting in 2013/2014

UOSM2015 2013/2014	
Top 6 SDGs	Area
O—Supporting economic growth	Economic
H—Sustainable land management	Environmental
I—Sustainable water management	Environmental
E—Controlling population growth	Social
L—Sharing knowledge and technologies	Economic
K—Eradicating extreme poverty	Economic

Table 11 The six proposed goals with the highest combined totals of votes across the 2 years

UOSM2015 2012/2013 and 2013/2014	
Combined top 6 SDGs	Area
O—Supporting economic growth	Economic
I—Sustainable water management	Environmental
H—Sustainable land management	Environmental
A—Improving health	Social
D—Reducing violence	Social
E—Controlling population growth	Social

their economic prosperity, natural resources, states of development, and political ideologies, enhancing their overall understanding of the challenge of achieving global sustainability.

Contextualisation is critical to simulation exercises as they should ultimately seek to represent reality (Dieleman and Huisingh 2006). The summit attempts to replicate the political agreement process, whilst equally focusing on the use of negotiation to achieve SD. This is not only for the successful operation of the game, but also for the development of sustainability skills, understandings and personal attributes, the achievements that assist graduates in employment and career development; otherwise known as student employability (Yorke 2006). The value of this is highlighted by a student quote from anonymous module evaluation questionnaires:

> The summit highlighted some of the difficulties that stem from conflicting global interests and is but one example where the team strives to parallel university learning experiences as close as possible to real world scenarios. It is because of this that I think I will be better prepared for my professional life after university.

The development of sustainability negotiation skills is central to the summit, with students altering their personal perspectives on global sustainability as they attempt to achieve their target goals. The need to negotiate is driven by the transparent disparity in levels of influence from the blocs involved in the summit. Further to this, blocs are also required to identify six priority goals, but only have five positive voting cards, so must negotiate with other blocs to achieve all six priority goals.

In both years to date, each of the final six SDGs only gained the necessary number of votes through collaboration. Negotiation was successfully used by blocs with less influence to ensure their priority goal was included in the final SDGs despite large numbers of votes cast by more powerful blocs for other goals. For example, in 2012/2013 North America cast ten votes for 'Sustainable Land Management (H)' but this goal was excluded from the final six SDGs by collaborative votes for 'Eradicating Extreme Poverty (K)' by Europe (six votes), India (six votes) and Africa (two votes). Negotiation was also successfully used to exclude goals from the final six, for example in both years 'Minimising Climate Change (G)' received enough positive votes to be included in the top six SDGs, but these were vastly exceeded by the negative votes received. UN agreements were introduced to the game in 2013/2014, with 17 agreements made over both voting rounds. Students recognised the value of negotiation in achieving their nation's goals on an international stage.

The final SDGs highlight how students inhabited the values of their blocs. The overwhelming majority of participants identified the environment as their main area of personal interest upon commencement of the module, yet were prepared to eliminate a goal to minimise climate change if it reflected the concerns of their bloc. Their personal focus on the environment has not led to a bias towards these goals in the summit. To the contrary, in year one, social goals were dominant (Table 7), whilst in year two economic goals held priority (Table 10). Even though students were less interested in or even personally adverse to 'Supporting Economic Growth (O)', this goal achieved most combined votes across 2 years of UOSM2015. Through simulation, students develop an emotional understanding of others' motivations and consequent actions (Dieleman and Huisingh 2006).

As one student stated:

> (I now) appreciate where other perspectives of sustainability come from.

The voting patterns highlight that students were developing an understanding and empathy for the focus of national goals on economic growth, and why this can inhibit global progress towards SD.

The use of Twitter as a summit negotiation tool was considered a success. All groups participated in communications with other blocs during the summit, and students continued to use twitter as an information retrieval tool for their studies post-summit. One student has stated how they now use Twitter to keep updated with UN Conferences, supplementing their studies. Blair (2013) explores how Twitter can be particularly useful educationally in engaging students in politics by providing another platform to not just actively engage with current practitioners and debates but also have the opportunity to be co-producers of knowledge. By integrating Twitter and use of the hashtag #SDGs, which has been adopted by the UN for the global public SDGs discussion, students contribute to the current discussion about the international community's global community for sustainability.

Learning through enjoyment is an important element of any rounded pedagogy. There remains, however, a risk that whilst students might be enjoying the summit, they might not be enhancing their understanding of the complexities of SD. Student comments regarding their overall experience of the summit and the wider module sheds some light on this:

> I really enjoyed this (summit). And I urge to keep it non-assessed, as if it was an assessment I feel everyone would be on edge and the summit would not have played out as it did.

> The structure of the module was genuinely engaging and covered a much wider scope of sustainability issues than I imagined. The global summit was a fantastic exercise.

> We were encouraged to communicate with staff and fellow students via twitter in order to keep our debates ongoing outside of teaching time. Such innovative ideas made the teaching incredibly academically rewarding and perhaps most importantly, thoroughly enjoyable!

These comments demonstrate that students were enthused by the summit, the teaching, and the use of social media. The comment requesting it remains a formative exercise is particularly pertinent, highlighting the value of non-assessed learning experiences for students in reaching the final intended learning outcomes for a module. A post-summit reflection discussion was held; this is critical to ensure the simulation is focused on student learning for their personal and professional life, not on the gamification itself (McCarthy and Anderson 2000).

10 Conclusions

The SDGs summit sought to adapt and build on the existing successes in experiential learning and simulation pedagogies by applying and adapting international summit simulations through a student-led interdisciplinary ESD lens, and by introducing social media (specifically Twitter) as a negotiation and communication tool. Through integrating these approaches into an interdisciplinary module, it is evident students develop their sustainability literacy and political understanding essential to developing globally responsible citizens for the future workplace.

Through the interdisciplinary, international focus of a SDGs Summit this paper has demonstrated that ESD is a valuable tool for pedagogical innovators to teach students the complexities surrounding international political negotiation and agreement. The use of simulation pedagogies teaches students that political decisions will have differing social, economic and environmental consequences for nations or 'blocs', depending on their economic prosperity, natural resources, states of development, and political ideologies. This serves to also enhance students' understanding of the complex systems nature of sustainable development. The SDGs summit tackles the challenges around inequality of access and influence on the global negotiating platform, and is valuable learning tool for students across the globe.

References

Blair A (2013) Democratising the learning process: the use of twitter in the teaching of politics and international relations. Politics 33(2):135–145

Dieleman H, Huisingh D (2006) Games by which to learn and teach about sustainable development: exploring the relevance of games and experiential learning for sustainability. J Clean Prod 14:837–847

Drayson R, Bone E, Agombar J, Kemp S (2013) Student attitudes towards and skillsin sustainable development. Higher Education Academy, York. http://www.heacademy.ac.uk/assets/documents/sustainability/ESD_student_attitudes_2013_v4.pdf. Last Accessed 30 Mar 2014

eBizMBA (2013) Top 15 most popular social networking sites. The eBusiness Guide. http://www.ebizmba.com/articles/social-networking-websites. Last Accessed 28 Apr 2014

Ebner M, Lienhardt C, Rohs M, Meyer I (2010) Microblogs in higher education—a chance to facilitate informal and process-oriented learning. Comput Educ 55:92–100

Evans C (2013) Twitter for teaching: can social media be used to enhance the process of learning? Br J Educ Technol. doi:10.1111/bjet.12099

Hargittai E (2007) Whose space? Differences among users and non-users of social network sites. J Comput-Mediat Commun 13:276–297

Hegarty K, Thomas I, Kriewaldt C, Holdsworth S, Bekessy S (2011) Insights into the value of a 'stand-alone' course for sustainability education. Environ Educ Res 17(4):451–469

Jisc and The HEA (2009) Enhancing learning through technology: a guide to resources provided by the academy and JISC collaboration network. http://www.jisc.ac.uk/media/documents/publications/enhancinglearningthroughtechnology.pdf. Last Accessed 28 Apr 2014

Junco R, Heiberger G, Loken E (2011) The effect of Twitter on college student engagement and grades. J Comput Assist Learn 27:119–132

Kemp S, Scoffham S, Rands P, Robertson A, Robinson Z, Speight S, Raghubansie A, Luna H (2012) A national programme to support education for sustainable development in the United Kingdom: university experiences of the HEA green academy programme. In: Leal Filho W (ed) Sustainable Development at Universities. Peter Lang, New Horizons

Kolb DA (1984) Experiential learning experience as a source of learning and development. Prentice Hall, New Jersey, p 36

Krain M, Lantis JS (2006) Building knowledge? Evaluating the effectiveness of the global problems summit simulation. Int Stud Perspect 7:395–407

Lantis JS (1996) Simulations as teaching tools: designing the global problems summit. Int Stud Notes 21(1):30–38

McIntosh D (2001) The uses and limits of the model united nations in an international relations classroom. Int Stud Perspect 2:269–280

McCarthy JP, Anderson L (2000) Active learning techniques versus traditional teaching style: two experiments from history and political science. Innov High Educ 24(4):279–294

Parker J (2010) Competencies for interdisciplinarity in higher education. Int J Sustain High Educ 11(4):325–338

Rittel H, Webber M (1973) Dilemmas in a GENERAL THEORY OF PLANNING. Policy Sci 4:155–169

Smith ET, Boyer MA (1996) Designing in-class simulations. PS: Polit Sci Polit 29(4):690–694

Socialmediatoday (2014) UK social media statistics for 2014. Social Media Today. http://socialmediatoday.com/kate-rose-mcgrory/2040906/uk-social-media-statistics-2014. Last Accessed 28 Apr 2014

Starkey BA, Blake EL (2001) Simulation in international relations education. Simul Gaming 32 (4):537–551. doi:10.1177/104687810103200409

Twopcharts (2014) Twitter account monitor. Twopcharts. http://twopcharts.com/twitter activitymonitor. Last Accessed 28 Apr 2014

United Nations (2000) 55/2 United Nations millennium declaration. United Nations general assembly, New York NY, 18 September, 2000. http://www.un.org/millennium/declaration/ares552e.pdf. Last Accessed 28 Apr 14

United Nations (2014) Sustainable development goals. United Nations sustainable development knowledge platform. http://sustainabledevelopment.un.org/?menu=1300. Last Accessed 29 Apr 2014

University of Southampton (2014) Flexible learning at Southampton, University of Southampton. http://www.southampton.ac.uk/cip. Last Accessed 24 Apr 14

Winter J, Cotton D (2012) Making the hidden curriculum visible: sustainability literacy in higher education. Environ Educ Res 18(6):783–796

Yorke M (2006) Employability in higher education: what it is—what it is not, HE Academy/ESECT, learning & employability series. http://www.employability.ed.ac.uk/documents/Staff/HEA-Employability_in_HE(Is,IsNot).pdf. Last Accessed 29 Apr 2014

Authors Biography

Simon Kemp Simon Kemp is a Principal Teaching Fellow in the Centre of Environmental Sciences, University of Southampton, and was nationally recognised in 2013 as Times Higher Education (THE) Most Innovative Teacher of the Year, and as a National Teaching Fellow in 2010.

Julia Kendal Julia Kendal is the Sustainability Action Officer at the University of Southampton, a cross-faculty role working to embed sustainability into the University staff and student experience.

Adam Warren Adam Warren is a Senior Learning Designer in the Centre for Innovation in Technologies and Education, University of Southampton.

Laurence Wright Laurence Wright is a Lecturer of environmental sciences in the Maritime and Technology Faculty at Southampton Solent University and Director of SOSA Ltd, a management support consultancy service.

John Canning John Canning is a Senior Lecturer in the Centre for Learning and Teaching at the University of Brighton, and a Fellow of the Higher Education Academy and of the Staff and Educational Development Association.

Marcus Grace Marcus Grace is a Senior Lecturer in Science Education in the Southampton Education School, University of Southampton, and a Member of the Academic Committee of European Research in Didactics of Biology, and the Society of Biology's Education, Training and Policy Committee.

Clare Saunders Clare Saunders is a Senior Lecturer in Environmental Politics in the Environment and Sustainability Institute, University of Exeter.

Attitudes, Barriers and Motivators as Factors for Sustainability of Higher Education e-Learning Programmes at Universidade Aberta, Portugal

Paula Bacelar-Nicolau, Sandra Caeiro, Ana Paula Martinho and Ulisses Miranda Azeiteiro

Abstract

In higher education distance learning institutions, where studies are carried on through a variety of distance learning regimes, from the correspondence and earlier forms of distance education to fully online e-learning programmes, and where planners and administrators seize the effective adoption and deployment of technology-enabled education, student attitude and motivation assume considerable significance. Attitudinal pre-dispositions, institutional and allied barriers (including appropriate policy initiatives), besides the scientific and pedagogical quality of degree programmes are assumed to play a crucial role in

P. Bacelar-Nicolau (✉) · S. Caeiro · A.P. Martinho · U.M. Azeiteiro
Department of Sciences and Technology, Universidade Aberta, Lisbon, Portugal
e-mail: Paula.Nicolau@uab.pt; pnicolau@uab.pt

S. Caeiro
e-mail: scaeiro@uab.pt

A.P. Martinho
e-mail: aptm@uab.pt

U.M. Azeiteiro
e-mail: ulisses.azeiteiro@uab.pt

P. Bacelar-Nicolau · U.M. Azeiteiro
Centre for Functional Ecology, Department of Life Sciences, University of Coimbra, Coimbra, Portugal

S. Caeiro
Center for Environmental and Sustainability Research, Faculdade de Ciências e Tecnologia, Universidade Nova de Lisboa, Lisbon, Portugal

A.P. Martinho
Laboratory of Distance Education and E-learning, Universidade Aberta, Lisbon, Portugal

W. Leal Filho et al. (eds.), *Integrative Approaches to Sustainable Development at University Level*, World Sustainability Series, DOI 10.1007/978-3-319-10690-8_39

sustainability of higher education systems. This article reports the findings of a study conducted at Universidade Aberta, Portugal—the Portuguese Distance Learning Education University—to examine student attitudes towards e-learning and to identify barriers and motivators of e-learning adoption, which are key factors for decreasing dropout rates of a 2nd cycle degree e-learning programme, analysed as a case study.

Keywords
Higher education · Distance learning · Barriers and motivators · Dropout rates · Sustainability

1 Introduction

Distance learning education is a form of delivering education and instruction to students who are not physically present in a traditional setting such as a classroom. Distance learning thus provides access to learning when the source of information and the learners are separated by time and/or space. It is mediated by a diversity of instructional media ranging from paper books, radio and television, as in the correspondence courses to internet as in online e-learning courses (Martinho et al. 2014; Moore et al. 2011).

The distance learning Universidade Aberta (UAb), Portugal, has offered all of its programmes as online e-learning regimes since 2006, covering everything from Lifelong learning courses to, 1st cycle (Bachelor degree), 2nd cycle (Master degree) and Doctoral programmes. All programmes follow an online teaching and learning system, based on UAb's pedagogical model (Pereira et al. 2008) and using the learning management system (LMS) Moodle. This aims to provide students with equal access to learning resources and communication channels with their teachers, fellow students, and academic and administrative support services. The learning communities interact with secretarial/administrative staff, coordinating team of the programme and teachers, in individual virtual spaces. A virtual "Café" space is also available for social interaction. Prior to every programme there is a short (2 week) initiation course to the platform and the pedagogical model. All programmes at UAb are directed to the public over 21 years old or working students.

E-learning programmes at UAb, as reported by other higher education institutions (HEIs), had an increasing acceptance and success rate in recent years (Bacelar-Nicolau et al. 2009, 2012; Martinho et al. 2014; Wagner et al. 2008). However, failure by students to complete the degree is a factual and real problem, both for the student who has invested his/her time, intellect and economical resources, and for the education institution which needs the student in order to be sustainable. This is often reported as dropout.

Several studies have analysed dropouts from higher education institutions, both distance learning and on-campus, although the definition of dropout has not always been clear. In this study the authors adopt the definition used by Levy (2007) and define *dropout students* (or non-completers) as students who voluntarily withdraw from e-learning degrees while acquiring financial penalties. The authors also define *early dropout* as those who, having enrolled in a degree, opt to drop out before the start of the courses, without financial penalties.

Dropout rates have been reported to be greater in distance learning HEIs than in conventional universities and vary depending on the mode of distance learning adopted by the institution, as well as on the degree study area (Eisenberg and Dowsett 1990). Dropout rates were reported to range between 20 and 30 % in European countries and up to 50 % in Asian countries (Schnepf 2014).

Although greater, dropout causes in e-learning programmes do not appear to differ much from those of conventional universities (Araque et al. 2009; Cingano and Coppoli 2003; Sittichai 2012). Many of the dropout causes, and their relative importance, reflect particular realities and contexts such as social, economic, religious, ethnic, cultural or study areas (Paechter et al. 2010; Panda and Mishra 2007; Sun et al. 2008; Wan et al. 2008).

Student dropout studies in higher education e-learning programmes (Ariwa 2002; Dirkx and Jha 1994; Parker 2003; Xenos 2004; Xenos et al. 2002) point out several main causes for dropping out of the system: informatics' literacy (e.g. Xenos et al. 2002), internet access and lack of training in e-learning (Panda and Mishra 2007), administrative issues, satisfaction with e-learning (Levy 2007), social interaction (Muilenberg and Berge 2005) and weak educational strategies, low academic performance (Araque et al. 2009) and cost to access the internet (Muilenberg and Berge 2005).

The aim of this study was to examine student's attitude towards e-learning in a 2nd cycle degree programme at Universidade Aberta, as well as to identify barriers and motivators of e-learning adoption which are key factors for decreasing dropout rates in this 2nd cycle degree e-learning programme—Master degree in Environmental Citizenship and Participation.

2 Master Degree in Environmental Citizenship and Participation, Universidade Aberta, Portugal

The Master degree in Environmental Citizenship and Participation has been on offer at the Universidade Aberta (UAb, Portugal) since 2006. It is a formal course, organized according to the European Credit Transfer and Accumulation System (ECTS), and taught in a blended-learning system (b-learning). Curricular units (CU) are delivered online, on e-learning mode, except for one CU which comprises of a 1 day face-to-face module, although this may be attended online via direct transmission by the students who cannot be present. The number of students per class is fixed between 25 and 30. The programme has the duration of three semesters, the first two semesters being dedicated to the curricular units (total of 60 ECTS) and the third semester being dedicated to the planning, developing, writing and defending of the Dissertation/Project (40 ECTS). The semester is defined as a period of

20 weeks, the 4 final weeks being dedicated to the final evaluation. The open source Moodle software is used as the LMS.

This programme was designed to prepare individuals who complete it for working on environmental policy making, as well as for improving their environmental citizenship, participation and planning abilities. The programme was also designed to be aimed at the public, specifically those working within government, as public and private environmental advisors, members of Environmental NGOs, teachers, researchers and individuals involved in environmental practices, policies, planning, training, participation and citizenship. Subjects, methods and the case studies used in this programme are drawn from the Environmental, Sustainable and Social Sciences. The pedagogical model underlying the learning process was developed specifically for e-learning 2nd cycle degrees at the UAb (Pereira et al. 2008).

Each CU has a virtual classroom in the LMS, which serves a similar purpose to the face-to-face lectures, being a privileged space where the learning process takes place, and where a series of attitudes and problems can be detected and solved.

Virtual classrooms of the CUs are organized into topics, each topic being developed for periods of one to 3 weeks, depending on the subject area. Each topic is generally associated with one learning activity. Learning activities (e-activities) are mostly based on the 'Moodle activities' and include lessons, mini-tests, glossaries, discussion *fora* (mostly involving group work), short written assessments, and blogs. Communication is mostly asynchronous, although some synchronous moments are used, e.g. to evaluate the learning progress or for evaluation. Support materials include books, e-books, research papers, internet sites, power point presentations and short videos.

3 Methods and Research Instruments

3.1 Student Data Analysis

Data was collected from the Academic Office of UAb relating to the candidate's enrolment and to the students progression in the Master degree in Environmental Citizenship and Participation for the scholar years 2007/2008 until 2013/2014. This included age, gender, study regime (full time and part time), nationality, enrolment year, and status (dropout, student, graduate). Statistical univariate and bivariate analyses were performed on these data (Excell, Windows).

3.2 Students' Questionnaire Surveys

The Master degree in Environmental Citizenship and Participation was evaluated, every year at two distinct moments throughout the course (end of first semester and end of second semester) using two confidential questionnaire surveys (published in

Bacelar-Nicolau et al. 2009). The first evaluation moment aimed to perceive the motivation and satisfaction level of students. The second survey aimed to evaluate student's knowledge acquisition on key concepts of environmental sciences, student's sensing of personal attitudinal and behavioural changes in specific environmental areas and student's global achievements relating to their expectations (e.g. acquisition of professional competences, public participation). In the context of this study the definition of attitude according to Myers et al. (2010) was used, i.e. a favourable or unfavourable evaluative reaction towards something or someone, rooted in ones beliefs, and exhibited in ones feelings and inclinations to act. This study was conducted between June 2007 and January 2010, during the four editions of the Master degree (n = 78 students) (published in Bacelar-Nicolau et al. 2012). A third new questionnaire survey was later conducted in June 2010 with the students who had completed the Master programme, 1–2 years after the defence of their dissertation thesis (n = 10). This 3rd questionnaire survey was similar to the 2nd evaluation moment and aimed to confirm individual's perception of knowledge acquisition, as well as their attitudinal and behavioural changes (earlier perceived), 1–2 years after the completion of the Master programme. Appropriate criteria of clarity and objectivity were used when preparing the questionnaire surveys (Quivy and Campenhoudt 2003) and pre-test was also performed. Response rate to the third questionnaire survey was 100 %.

3.3 Semi Structured Interviews to Teachers

The semi structured qualitative interview to teachers (Fontana and Frey 1994) is an instrument of interpretative nature (Denzin and Lincoln 1994). In order to accomplish its objective, an interview guide for 4 teachers, who taught in this master since the beginning, was designed and applied in April 2014, taking into account the following dimensions: (i) The collaborative pedagogical model/teacher-student interaction; (ii) Online Peer Community/peer interaction dimensions/learning community; and (iii) Barriers and motivators of e□learning adoption, which are key factors for decreasing dropout rates of a 2nd cycle degree e-learning programme. The method used for the qualitative approach included, as procedure, the content analysis of Bardin (1977).

4 Results and Discussion

The analysis of data from the Academic Office for 3 academic years, from 2010/2011 to 2012/2013, registered a total of 140 students enrolled in the Master degree, of which 49.6 % were female and 50.4 % were male. Most enrolled students were in the age group of 40 to 49 years old (35.6 %) and 50–59 (32.6 %), less were between 30–39 years old (9.2 %) or over 60 years old (17 %); a minority was aged between 21 and 29 years old (2.8 %). Most students, 59 %, were registered as full time and

Table 1 Bivariate statistical analysis of student status versus characterisation variables

	Dropout freq.	%	Student freq.	%	Graduate freq.	%
Age						
20–29	7	14	5	10.9	1	2.3
30–39	20	40	16	34.8	14	31.8
40–49	15	30	16	34.8	15	34.1
50–59	0	0	8	17.4	11	25.0
>60	5	10	1	2.2	3	6.8
No response	4	8	0	0	0	0
Student regime						
Full time	49	98	16	34.8	17	38.6
Part time	1	2	30	65.2	27	61.4
Gender						
Female	20	40	21	45.7	20	45.5
Male	30	60	25	54.3	24	54.5
Nationality						
Portuguese	39	78	35	76.1	34	77.3
Angola	1	2	0	0	2	4.5
Cape verde	0	0	4	8.7	4	9.1
Mozambique	1	2	1	2.2	1	2.3
Brazil	8	16	5	10.9	0	0.0
Guinea	1	2	0	0	0	0.0
Spain	0	0	0	0	1	2.3
Germany	0	0	0	0	1	2.3
France	0	0	1	2.2	0	0.0
Italy	0	0	0	0	1	2.3

41 % were part time students. Their nationalities included Portuguese (77 %), Portuguese speaking countries (Angola, Cape Verde, Mozambique, Brazil, Guinea; 19 %) and European (Spain, Germany, France, Italy; 5 %). From these, 36 % dropped out, 31 % graduated and 33 % are still students.

The bivariate statistical analysis of the students' status versus the other studied variables (Table 1) indicated that the major difference between dropout students and others is the study regime in which they enrol; full time study. Dropouts were also related to the age. There was a higher proportion of younger as well as older students in the dropout group among the 20–29 and 30–39 age groups, and the above 60s, when compared to the graduate or students group. This can be partly explained for the younger students (21–40 years old), by the fact that they may have greater work and family constraints (personal communication by dropout students

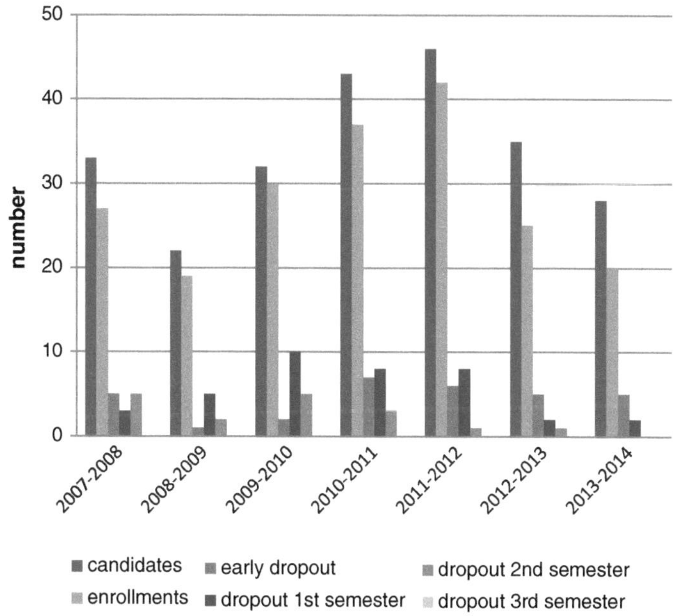

Fig. 1 Numbers of candidate, enrolment and dropout per academic year, from 2007/2008 until 2013/2014

to Coordination team, unpublished) than individuals aged 40–60, which was also observed by Xenos et al. (2002). Gender, which has been related to student dropout by the same authors, was not associated to dropout in our Master degree. Data relating to student nationality indicated that Brazilian students had higher dropouts than others.

The analysis of dropout through the 7 editions of the Master degree from 2007/2008 onwards showed that there was an increase of candidates and enrolments from 2007/2008 until 2011/2012 and a decrease afterwards (Fig. 1). This decrease in candidates and enrolment can be related to the current European economic crisis. The analysis also showed that the global dropouts in the first 3 editions was higher (average 15 %, Fig. 2) than in the latter (10 %). This corresponded to the year (2011) in which the part time regime study was introduced as an option for students' enrolment.

The "global" dropouts can be divided in 4 sequential phases: "early dropouts", dropouts during 1st semester (mostly in the first two months), dropouts in the 2nd semester and 3rd semester. The average of "global" dropouts in these 4 phases is respectively 16, 18, 9 and 0 %. The sequential decrease of dropouts towards the end of the programme has been also reported by Levy (2007) and explained by an increased motivation by the nearing of graduating semester. In fact, most dropouts occur during the first 2 months of the 1st semester of the programme, which coincides with the period in which students are suddenly faced with the Maters'

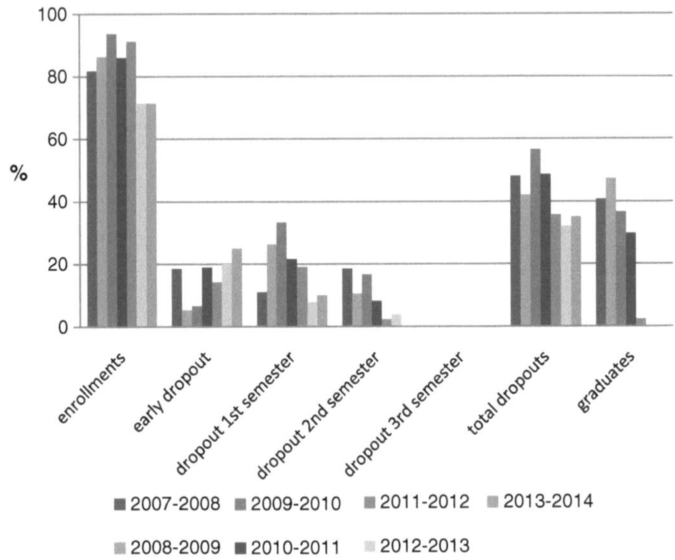

Fig. 2 Evolution of candidate, enrolment and dropout (as percentage) between 2007/2008 and 2013/2014

workload and either adapt to that reality, or they do not. "Early dropouts", prior to the start of the programme, also represent a loss of potential students (OECD 2010, 2011, 2013).

The questionnaire surveys applied to students during the programme (end of 1st and 2nd semesters) indicated that students felt a very high level of motivation and satisfaction with the Master degree (Bacelar-Nicolau et al. 2009).

In global terms, the curricular plan was considered good and very good (93 %) for the acquisition of sustainable environmental behaviour. Main contributors to the high satisfaction level were good didactic materials, good learning activities, very good competence and support of the teaching staff, and good interaction among students. At the end of the curricular year (2nd questionnaire survey), students felt an increase in their knowledge acquisition (93–100 %), as well as a modification of their attitudes (80–100 %) and behaviour (73–93 %) in key concepts/matters for a sustainable environment. Most students (93–100 %) also felt that they had improved their competences at the professional level, and that the developed tools for the acquisition of an active environmental citizenship were very good. Also, at the end of the second semester, all students felt that the programme had increased their intention to alter the attitudes and behaviour of others and 75 % had started active collaboration in public participation for environmental sustainability.

The 3rd questionnaire survey, conducted 1–2 years after the Master's conclusion, indicated a persistent high level of motivation and satisfaction with the Master degree. Namely, individuals still felt an increase in the acquired knowledge (89–100 %), as well as a change of their attitude (63–90 %) and behaviour

(75–100 %) towards key concepts/matters for a sustainable environment. Most students (80–100 %) also maintained that they had improved their competences at the professional level, and that the developed tools for the acquisition of an active environmental citizenship were very good. These results are in accordance with those reported by other authors (Crompton et al. 2002; Roy and Potter 2008) in which students enrolled in environmental studies e-learning courses maintained to have changed their behaviours and attitudes.

The 2nd and 3rd questionnaire surveys thus indicated similar results, although there were some interesting differences in the latter, namely; the drop in the perception of the individual's attitudinal change (80–100 to 63–90 %), the increased perception of the improved competences at the professional level and finally the increased perception of developed tools for the acquisition of an active environmental citizenship (80–100 to 93–100 %). Behaviour changes should only be evaluated at long and not short term, as behaviour tends to regress to its original status. Therefore, the results from the latter survey should be more realist and indicative of changes than the second one. Moreover, in the 3rd questionnaire survey all students maintained that the programme had increased their intention to alter the attitude and behaviour of others and 90 % were actively collaborating in public participation for environmental sustainability (compared to 75 % in the 2nd questionnaire survey, conducted before the development of the Master's dissertation). In addition to this, most students perceived that they had reached an affective learning outcome of values, attitudes and behaviours, which, as suggested by Sheppard (2008), is a central element of education for sustainability.

Teachers intervening in the eight editions of the Master degree were interviewed about the collaborative pedagogical model, student-student interaction and teacher-student interaction. When asked to describe the teaching and learning asynchronous tools and collaborative pedagogical model, they stated their evolution dealing with the platform and adaptation to the pedagogical model:

> (…) I'm much more confident using this pedagogical model, that works very well at master level (very collaborative) Teacher #1

Teachers reinforced that collaborative activities promoted *the collaborative work (…) and interactions* (Teacher #2) and they also reported on having adapted to students' response capabilities as a co-evolution process:

> In the early years of this course I realised that the students had too much work load in my curricular unit and in the whole course, consequently I reduced the number of activities Teacher #3 Due to excessive workload in my own curricular unit, the number of activities was reduced. Teacher #3

Teachers also felt that students evolved in the use of the platform and in their interaction with the model:

> The evolution has stabilised, besides small yearly improvements in the courses materials Teacher #1

> The students are well adapted to this methodology and they work in a collaborative way. Teacher #2

All the interviewed teachers stated that this collaborative Pedagogical Model enhanced students' motivation, activities engagement and competences acquisition. However the teachers' role was also important:

> (…) Yes, but teacher role is very important in promoting all these interactions and motivations. Teacher #2

> I have learned that being present, and giving continuous feedback to their learning process is fundamental for their motivation, activities engagement and competences acquisition. An absent on-*line teacher is the worst thing the students can have.* Teacher #1

Thus, the "construction" of an online peer community and peer learning community was perceived has promoting students' motivation, activities engagement and competences acquisition. In this peer learning community, the presence of the teacher is fundamental. In fact, e-learning curricula should be guided by pedagogy, rather than led by technological possibilities, if they are to deliver the intended learning outcomes (Csete and Evans 2013). This is patent in the importance also given to the "*teaching pedagogical skills*" (Teacher #2) in the teaching-learning process.

Also, experience as an e-learning teacher, together with the established collaborative e-learning model, was perceived as influencing the peer interaction/learning community processes (e.g. conflicts and complicity):

> With experience, and like any regular face-to-face teacher, we learn how to deal with students' conflicts (e.g. different personalities and cultures) and constraints (much related with their time limitation due to having full time jobs), and advise them how to overcome their problems/difficulties. Being flexible is a very important issue that asynchronous learning also allows more easily. Teacher #1

> teachers' presence is important to moderate discussions in order to eliminate possible conflicts, to calm down overly participative students, to motivate the least participative ones, to answer any questions they have, (…) Teacher #3

In fact, some of the requirements for best practice in learner-centred environments include 'frequent and ongoing feedback' to students (Habron et al. 2012) and, in this way, teachers' presence and patterns of regular participation and assessment are important motivators and impact on students' satisfaction and performance. The teachers' perception of the importance of their frequent "presence" and their role as moderators, facilitators and scientific counsellors concurred with students perceptions (Bacelar-Nicolau et al. 2012).

All the interviewed teachers stated that this conflict absence and peer complicity contributed to the students' motivation, activities engagement and competences acquisition, and that this sensing of a learning community was very important for the success of the e-learning process.

Transformative learning involves collaborating with others and constructing new knowledge for sustainability (Cebrian et al. 2013). Interaction is a central construct in online learning (Moura et al. 2010) and interaction among peers, and with teachers, is privileged by online students (Swan 2003; Moura et al. 2010). Teaching presence (Swan 2004) has a central role to promote not only a conceptual learning but also a learning community (Peltier et al. 2007; Moura et al. 2010).

The sense of belonging to a peer group, and the changes in values and attitudes, involve the affective domain that is very important in higher education for sustainability (Shephard 2008).

When asked to identify the barriers and motivators of e-learning adoption which were the key factors for decreasing dropout rates of the Master degree e-learning programme, teachers indicated:

- for *barriers*: students' lack of time, students' different ways of interacting and/or working (particularly the absent students that can impair/disrupt a team and mismatch activity schedules) and family and employment issues affecting availability to learning activities. Informatics literacy, internet access and lack of training in e-learning were also mentioned by one of the teachers.
- for *motivators*: the flexibility (in time and space) that e-learning allows and collaborative environment engagement that students largely appreciate; the sense of a community, with change of experiences in environmental subjects; good interaction in small group activities and peer complicity. The new competences acquired and the consequent possibility of social and professional evolution is also a strong motivator for these students that can combine their studies and their professional activities (only possible in the e-learning classes).

Teachers were also asked to identify factors for decreasing dropout rates for this Master degree programme. Some key points are: the importance of the enrolment phase, the construction of a learning community/peer community, the part-time study regime, and the teacher's role and presence:

> Difficult question … but maybe enhancing, at the beginning of programme, or even at the enrolment phase, how important their commitment is, and the time that they need to spend to be able to fulfil all the learning tasks and courses. Also, convincing them that who is a full-time worker should enrol the programme not in a full time basis, but in "partial" regime! Also reducing the fees and facilitating their payments. Another important issue is that the teacher has an fundamental role of continuous motivator and being always "present" on their courses, giving feedback on the learning process along the semester and not only at the end. This is very important in my point of view. Teacher #1

Also relating to decreasing dropouts, another teacher mentioned the importance of introducing synchronous activities:

> (…) with presentation of works by the student and promoting discussions open to all students, in a synchronous way Teacher #3

Other teachers mentioned communication issues.

Better coordination and communication between students and supervisors. (…) Better communication strategies, namely an online journal for the students in the second year (master thesis year) Teacher #4

Therefore, the study indicates that students felt highly motivated and satisfied with the e-learning Master programme on Environmental Citizenship and Participation. The main motivators were the good didactic materials, good learning activities, very good competence and support of the teaching staff, good interaction among students during the curricular year, and later the perception of the improved competences at the professional level and developed tools for the acquisition of an active environmental citizenship. However, completer students corresponded to 60 % of the total of students enrolled in the Master degree. The dropout students appeared to find the following as main barriers: TIC literacy limitations and lack of time for study and e-activities (mostly due to family and work constraints).

5 Conclusion

In this study the authors aimed to examine student attitudes towards e-learning and to identify barriers and motivators of e□learning adoption, which are key factors for decreasing dropout rates of a 2nd cycle degree e-learning programme, used as a case study.

Universidade Aberta has been offering online e□learning programmes for almost a decade. The pedagogical model of UAb for 2nd cycle programmes was designed to meet and overcome the most reported barriers in e-learning. For example, the exclusion of TIC/lack of e-training and social interaction were particularly targeted during the initiation module to degrees at UAb, and through the dedicated coordination team, secretarial virtual spaces and virtual "Café" this has been resolved. Strongly collaborative study and teacher's "presence"—both as scientific counselor and motivator—are key factors for student motivation and satisfaction within the Master degree. However, too much collaborative study and e-activities result in excessive workload (both for students and teachers) and some teachers have resolved this by adapting their activities whilst ensuring they maintain the bond within the e-community.

Herein the authors reported that students of the Master degree in Environmental Citizenship and Participation, at UAb (as in Bacelar-Nicolau et al. 2009, 2012) as well as its graduates, after 1–2 years, were still highly satisfied and motivated both by the scientific and the pedagogical quality of the degree.

Through its eight editions the coordinators have aimed to improve the institutional procedures, including policy initiatives, which helped overcome some perceived existing barriers, namely the inclusion of the part-time regime study, which appears to be lowering dropouts in recent years.

The most significant barriers perceived, although to some extent dependent of the student context (such as age or nationality), mainly included limitations to internet access, excess of workload, and family and/or work constraints. Taking into consideration the fact that these are working students the coordinators have, in recent years, implemented the part-time study regime option, which has resulted in an increase of student persistence and a decrease of dropout rates in the Master degree. In line with this, in 2013, the possibility of enrollment in single curricular units has also been implemented, which has resulted in an increased persistence of those students.

Exclusion in TIC issues is still a reality among some candidates and students in our Master degree (mainly from some Portuguese speaking countries) which may contribute to a dissatisfaction with e-learning, social exclusion from the e-community and consequent dropout. The early dropout, prior to the start of the programme, may be reduced by administratively introducing an earlier fee, which is not applied. These are some aspects which need addressing in the future in order to increase sustainability within HEIs.

Both the academic and social systems of a university are regarded as important frameworks from which the dropout process must be examined. However the path to sustainability of higher education systems also includes the need for high quality both in science and pedagogy and the need for effective institutional policies and technological initiatives that makes it possible to overcome barriers and enlarge the number and quality of the motivators.

Institutional processes have evolved, and they will continue to do so, seeking to adapt to novel public, cultural and/or social and economic circumstances in order to guarantee sustainability. In this context, a fuller characterisation of the tertiary dropout students at UAb is being prepared, aiming to better understand their profile.

References

Araque F, Roldan C, Salguero A (2009) Factors influencing university dropout rates. Comput Educ 53:563–574

Ariwa E (2002) Evaluation of the information, communication and technology capabilities and online learning. USDLA J 16(11):59–63

Bacelar-Nicolau P, Martinho AP, Amador F, Caeiro S,' Azeiteiro UM (2012) Online learning for sustainability: the student perception in an environmental science post-graduation. In: Gonçalves F, Pereira R, Leal-Filho W, Azeiteiro UM (eds) Contributions to the N decade of education for sustainable development in the series Umweltbildung, Umweltkommunikation und Nachhaltigkeit—Environmental Education, communication and sustainability, Peter Lang, Frankfurt, 33, 281–294

Bacelar-Nicolau P, Caeiro S, Martinho AP, Azeiteiro UM (2009) E-learning for the environment. The Universidade Aberta (Portuguese Open Distance University) experience in the environmental sciences post-graduate courses. Int J Sustain High Educ 10(4):354–367

Bardin L (1977) Análise de Conteúdo, Edicções 70, Lisbon (in Portuguese)

Cebrian G, Grace M, Humphris D (2013) Organisational learning towards sustainability in higher education. Sustain Account Manag Policy J 4(3):285–306

Cingano F, Coppoli P (2003) Determinants of university drop-out probability in Italy. Bank of Italy—Research Department, Rome http://www.econ.upf.edu/~cingano/papers/dropout_EALE.pdf. Accessed 23 Apr 2014

Crompton S, Roy R, Caird S (2002) Household ecological footprinting for active distance learning and challenge of personal lifestyles. Int J Sustain High Educ 3(4):313–323

Csete J, Evans J (2013) Strategies for impact: enabling e-learning project initiatives. Campus-Wide Inf Syst 30(3):165–173

Denzin NK, Lincoln YS (1994) Introduction the discipline and practice of qualitative research. In: Denzin NK, Lincoln YS (eds) Handbook of qualitative research. Sage Publications, London, pp 1–20

Dirkx JM, Jha LR (1994) Predicting rapid completion, persistence, and attrition in Adult Basic Education. Adult Educ Quart 45(1):269–285

Eisenberg E, Dowsett T (1990) Student dropout from a distance education project course: A new method analysis. Distance Educ 11(2):231–253

Fontana A, Frey JH (1994) Interviewing the art of science. In: Denzin NK, Lincoln YS (eds) Handbook of qualitative research. Sage Publications, London, pp 361–376

Habron G, Goralnik L, Thorp L (2012) Embracing the learning paradigm to foster systems thinking. Int J Sustain High Educ 13(4):378–393

Levy Y (2007) Comparing dropout and persistence in e-learning courses. Comput Educ 48: 185–204

Martinho AP, Caeiro S, Caetano F, Azeiteiro UM, Bacelar-Nicolau P (2014) Training and employability, competences from an e-learning undergraduate programme in environmental sciences. In: Azeiteiro UM, Leal Filho W, Caeiro S (eds) E-learning and sustainability in the series Umweltbildung, Umweltkommunikation und Nachhaltigkeit—Environmental Education, Communication and Sustainability. Peter Lang, Frankfurt (in press)

Myers D, Abell J, Kolstad A, Sani F (2010) Social psychology. McGraw-Hill Education, United States 685 p

Moore JL, Dickson-Deane C, Galyenb K (2011) E-learning, online learning, and distance learning environments: are they the same? Internet High Educ 14:129–135

Moura APM, Cunha LM, Azeiteiro UM, Aires L, de Almeida MDV (2010) Food consumer science post-graduate courses: comparison of face-to-face versus online delivery systems. Br Food J 112(5):544–556

Muilenburg L, Berge Z (2005) Student barriers to online learning: A factor analytic study. Distance Educ 26(1):29–48

OECD (2011) Education at a glance 2011: OECD Indicators, OECD Publishing. http://dx.doi.org/10.1787/eag-2011-en. Accessed 23 Apr 2014

OECD (2013) Education at a glance 2013: OECD indicators, OECD Publishing. http://dx.doi.org/10.1787/eag-2013-en. Accessed 23 Apr 2014

OECD (2010) Highlights from education at a glance 2010, OECD Publishing. http://dx.doi.org/10.1787/eag_highlights-2010-en. Accessed 23 Apr 2014

Paechter M, Maier B, Macher D (2010) Students' expectations of and experiences in e-learning: Their relation to learning achievements and course satisfaction. Comput Educ 54:222–229

Panda S, Mishra S (2007) E-learning in a mega open university: Faculty attitude, barriers and motivators. Educ Media Inter 44(4):323–338

Parker A (2003) Identifying predictors of academic persistence in distance education. USDLA J 17 (1):55–62

Peltier JW, Schibrowsky JA, Drago W (2007) The interdependence of the factors influencing the perceived quality of the online learning experience: a causal model. J Mark Educ 29(2): 140–153

Pereira A, Mendes AQ, Morgado L, Amante L, Bidarra J (2008) Universidade Aberta's pedagogical model for distance education. Universidade Aberta, Lisbon. https://repositorioaberto.uab.pt/bitstream/10400.2/2388/1/MPV_uaberta_english.pdf. Accessed 20 Dec 13

Quivy R, Campenhoudt L (2003) Manual de investigação em Ciências Sociais. Gradiva, Lisboa, p 290

Roy R, Potter S (2008) Designing low carbon higher education systems. Environmental impacts of campus and distance learning systems. Int J Sustain High Educ 9(2):116–130

Schnepf SV (2014) Do tertiary dropout students really not succeed in European labour markets?, Discussion Paper No. 8015, http://ftp.iza.org/dp8015.pdf Accessed 23 Apr 2014

Shephard K (2008) Higher education for sustainability: seeking affective learning outcomes. Int J Sustain High Educ 9(1):87–98

Sittichai R (2012) Why are there dropouts among university students? Experiences in a Thai University. Int J Educ Dev 32:283–289

Sun P, Tsai RJ, Finger G, Chen Y, Yeh D (2008) What drives a successful e-Learning? An empirical investigation of the critical factors influencing learner satisfaction. Comput Educ 50:1183–1202

Swan K (2003) Learning effectiveness: what the research tells us. In: Bourne J, Moore JC (eds) Elements of quality online education: practice and direction. The Sloan Consortium, Needham, MA, pp 13–450

Swan K (2004) Learning online: a review of current research on issues of interface, teaching presence and learner characteristics. In: Bourne J, Moore JC (eds) Elements of quality online education: into the mainstream. The Sloan Consortium, Needham, MA, pp 63–80

Wagner N, Hassanein K, Head M (2008) Who is responsible for E-learning success in higher education? a stakeholders' analysis. Educ Technol Soc 11(3):26–36

Wan Z, Wang Y, Haggerty N (2008) Why people benefit from e-learning differently: The effects of psychological processes on e-learning outcomes. Inform Manag 45:513–521

Xenos M (2004) Prediction and assessment of student behaviour in open and distance education in computers using Bayesian networks. Comput Educ 43(4):345–359

Xenos M, Pierrakeas C, Pintelas P (2002) A survey on student dropout rates and dropout causes concerning the students in the Course of Informatics of the Hellenic Open University. Comput Educ 39:361–377

Author Biographies

Paula Bacelar Nicolau Ph.D. in Environmental Microbiology from University of North Wales, Bangor, UK, is Assistant Professor at Universidade Aberta, Portugal, where she teaches Biology, Microbiology, Conservation Biology, and Food Biochemistry. She is Senior Researcher at the Centre for Functional Ecology of University of Coimbra. Her main research interests include e-learning in biological and environmental sciences, and environmental microbiology. She is the coordinator of the MSc in Environmental Citizenship and Participation at Universidade Aberta.

Sandra Sofia Caeiro holds an undergraduate degree in Environmental Engineering from Universidade Nova de Lisboa (1992), a Masters in Science of Coastal Zones from Universidade de Aveiro (1997) and a Doctorate on Environmental Engineering from the Universidade Nova de Lisboa (2004), Portugal. She is currently Assistant Professor in the Department of Science and Technology at Universidade Aberta (UAb) and a senior researcher at CENSE—Center for Sustainability and Environmental Research from Universidade Nova de Lisboa, Portugal. Her main research and teaching areas include environmental risk assessment, environmental management and spatial planning support tools, education for sustainable development. She is the coordinator of the B.Sc. programme on Environmental Sciences, and the institutional coordinator of the European Virtual Seminar in Sustainable Development. She is on the editorial board of the international journals of Ocean and Coastal Management, Elsevier, Latin American Journal of Management for Sustainable Development, Inderscience, and BioMed Research International, Hindawi Publishing Corporation and reviewer of several international scientific journals and books. She has mentored several post-graduate students and postdoctoral researchers, published papers in peer-review ISI journals, chapter books and international conference proceedings, and coordinated and participated in several research projects.

Ana Paula Teixeira Martinho holds a Ph.D. on Environmental Engineering (2003), a M.Sc. in Sanitary Engineering (1999) and a B.Sc in Environmental Engineering (1993) from Universidade Nova de Lisboa, Portugal. She is currently Assistant Professor in the Department of Science and Technology at Universidade Aberta and a and a senior researcher at LEAD, Laboratory of distance education and e-learning, Universidade Aberta, Portugal. Her main research and teaching areas include environmental impact assessment, integrated waste management, environmental ethics and environmental citizenship and participation, and e-learning in science education. She is now the vice-coordinator of B.Sc in Environmental Sciences.

Ulisses Miranda Azeiteiro was educated at the Universidade de Aveiro, Portugal, from which he received his Master Science Degree in 1994, Universidade de Coimbra, Portugal, where he received his Doctorate (Ph.D.) in Ecology in 1999 and Universidade Aberta, Portugal, where he received his Aggregation in Biology in 2006. He holds an Assistant Professor teaching position at the Universidade Aberta (Assistant Professor with Aggregation), Portugal, in the Department of Sciences and Technology, where he teaches Environmental Education, Education for Sustainable Development, Biology, Biodiversity and Research Methodologies. He is Senior Researcher at the Centre for Functional Ecology of Coimbra University where he is the Coordinator of the Group Ecology and Society.

Fostering Community Outreach Activities for Environmental Sustainability Through a Cross-Border Academic Research Partnership

Mihaela Sima, Ines Grigorescu, Dan Balteanu and Georgi Zhelezov

Abstract

The focal aim of this study is to provide a general framework of the role of an academic research partnership in fostering community outreach activities to improve environmental sustainability through developing tailored science-society interfaces. The paper seeks to present an overview of the EU Romania—Bulgaria Cross Border Cooperation Programme project entitled "Romanian—Bulgarian cross-border joint natural and technological hazards assessment in the Danube floodplain. The Calafat-Vidin—Turnu Măgurele-Nikopole sector (ROBUHAZ-DUN)" and, particularly, the research collaboration, scientific outcomes and dissemination activities carried out during the 18 month project. The aim of the paper is to offer an example of how to promote environmental sustainability to community members in a rural transboundary area in order to respond to their needs in terms of hazard assessment and mitigation, but also to increase knowledge and awareness of disaster risk reduction, climate change and environment sustainability. A special attention was paid to the main scientific and informative products (promotional materials, maps, posters, guidebooks, university course, reports) which were used as support materials for the dissemination activities undertaken throughout the project in terms of raising awareness and informative campaigns in schools, meetings with local authorities, joint round tables, media events, summer school etc. These activities were

M. Sima (✉) · I. Grigorescu · D. Balteanu
Romanian Academy, Institute of Geography, 12 Dimitrie Racovita, sector 2, 023993 Bucharest, Romania
e-mail: simamik@yahoo.com

G. Zhelezov
Bulgarian Academy of Sciences, National Institute of Geophysics, Geodesy and Geography, Acad. Georgi Bonchev str., bl.3, 1113 Sofia, Bulgaria

W. Leal Filho et al. (eds.), *Integrative Approaches to Sustainable Development at University Level*, World Sustainability Series, DOI 10.1007/978-3-319-10690-8_40

aimed at bridging the gap between the academic research and local communities in an area prone to natural and human-induced hazards in order to support environmental sustainability through disaster risk reduction education.

Keywords
Community outreach · Environmental sustainability · Research partnership · Cross-border cooperation · Dissemination activities

1 Introduction

Universities and research institutions, through the activities they develop, play an important role in shaping communities' awareness on environmental sustainability issues. Accordingly, scientists are more and more required to carry out tailored research that can be effectively communicated to stakeholders and communities (Hanson et al. 2006).

Global environmental concerns related to land use/land cover change, biodiversity loss, climate change-related extreme events, atmospheric pollution, biogeochemical cycle becomes negative irreversible trends as acknowledged once more by the Rio+20 United Nations Conference on Sustainable Development which was held in Rio de Janeiro in June 2012. In line with that, the final document, 'The Future We Want' recognizes that a close collaboration between researchers and stakeholders, such as business, governments and civil society is highly required (Leemans and Solecki 2013).

Moreover, developing reliable partnerships between science and society able to define relevant integrated research problems and explore successful pathways towards global sustainability is a key objective of the Future Earth framework, the new 10-year international initiative on global sustainability research launched during the Rio+20 United Nations Conference (Mauser et al. 2013).

The recent societal, policy and scientific evolutions are aimed at fostering the scientific utility and applicability by providing effective linkages with the international conventions not only through the official science–policy assessments (e.g. IPCC) but also through new active dialogues between researchers and decision makers (Leemans and Solecki 2013).

The focal aim of the recent studies undertaken worldwide is to encourage higher education institutions to commit their intellectual, economic and human resources to community change, through knowledge transfer, research and outreach activities (Vidal et al. 2002).

Regional scale approaches seems to be more appropriate for successfully implementing sustainable development actions as they are small enough to be of direct interest to citizens and large enough to involve key stakeholders in the planning and decision-making processes (Adomssent 2011).

The Strategy of the United Nations Economic Commission for Europe (UNECE) for ESD, designed as a contribution to and in line with UN Decade of Education for Sustainable Development implementation scheme, admitted that special emphasis should be given to South-Eastern and Eastern European countries in solving their ESD-related problems in terms of "a lack of adequate instruction materials, the inefficient use of the capacity of higher education and research institutions, a shortage of skilled educators and insufficient awareness-raising, as well as a lack of interdepartmental and multi-stakeholder cooperation on ESD" as well as "the poor quality of education for children living in rural areas and the lack of financial and human resources to develop ESD in those areas" (UNECE 2005). Under the given circumstances, it becomes central to enhance the involvement of higher education institutions and research networks through "teaching, research and community engagement to strengthen global and local knowledge of Education for Sustainable Development" as stated in the Bonn Declaration (UNESCO 2009a).

Moreover, the need for enhancing ESD through community-science partnerships is also acknowledged by the European Strategy for Sustainable Development in terms of promoting high quality education for sustainable development at all levels of education through life-long learning, sharing knowledge and good practice and providing key competences to achieve sustainability (European Council 2009; Adomssent et al. 2014), but also through a multi-stakeholder engagement (Storni et al. 2012).

It is recognized the role of universities and academic research institutions play as societal actors especially when cooperating and networking with regional and local actors in order to take up society-driven challenges relying on an integrative development model (Mader et al. 2013). It is also acknowledged the critical leadership role in sustainability by offering solutions adapted to nowadays challenges through applied sustainability research (Bilodeau et al. 2014). As an example, the Model for Integrative Development (Graz Model) proposes five main principles in approaching sustainable regional development through higher education and communities partnerships (Mader 2009): Leadership and Vision; Social Network (people and stakeholders concerned); Participation (people of the social network involvement in the process); Education and Learning (knowledge building activities and the ability to learn from one's experiences, the ESD having an important role) and Research Integration (seeking new solutions for existing and upcoming challenges), each principle having three hierarchical levels (Trummler et al. 2011) (Fig. 1).

Interchanging higher education institutions with regional networks, exchange of knowledge and expertise, as well as sharing and collecting information about innovative projects promoting sustainable development in Europe was also the aim of the Database on Learning for Sustainable Development, which was developed in the framework of the 3-LENSUS project (Trummler and Mader 2011).

In line with this, the core idea of this cross-border cooperation project between Romania and Bulgaria entitled "Romanian—Bulgarian cross-border joint natural and technological hazards assessment in the Danube floodplain. The Calafat-Vidin —Turnu Măgurele-Nikopole sector (ROBUHAZ-DUN)", was to develop a

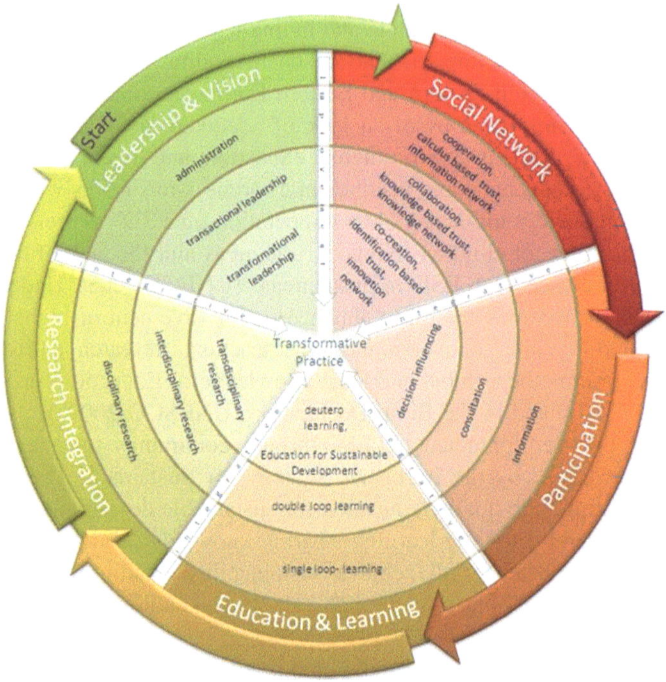

Fig. 1 The Graz Model for Integrative Development (Mader 2009)

network partnership of higher education and research institutions at regional and local level in order to tackle specific needs in terms of environment sustainability and hazard assessment and mitigation.

The aim of the paper is to offer an example of how to promote environmental sustainability to community members in a rural transboundary area in order to respond to their needs in terms of hazard assessment and mitigation, but also to increase knowledge and awareness of disaster risk reduction, climate change and environment sustainability. A wide variety of local stakeholders have been approached (local municipalities, schools, civil protection, environmental protection, population), the methods, tools and dissemination materials used being targeted to address their needs, interest and knowledge.

2 The Showcase of Including Environmental Sustainability Education in a Cross-Border Cooperation Project

The study-area of the project lays in the south-eastern part of Europe, overlapping the Lower Danube Valley in Carpathian-Balkan region and covering southern Romania and northern Bulgaria (Fig. 2). The main objective of the ROBUHAZ-DUN project, aimed at producing hazard and vulnerability maps and rise local

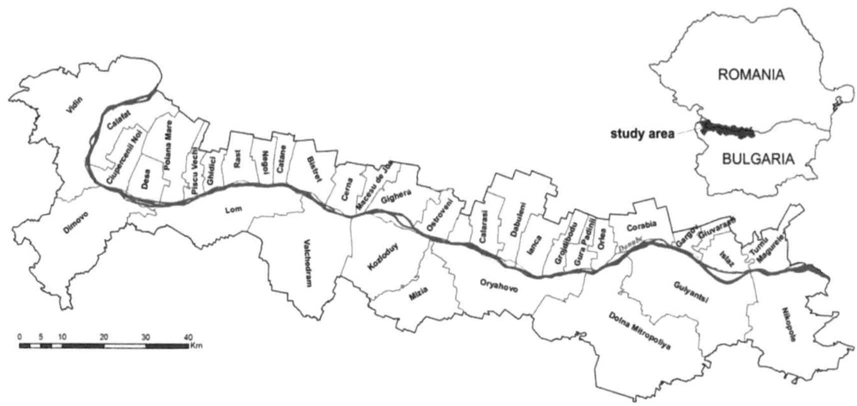

Fig. 2 The Calafat-Vidin—Turnu Măgurele-Nikopol Danube floodplain sector in the Romanian and Bulgarian territories (Grigorescu et al. 2013)

stakeholders awareness and education on the main environmental concerns the area is exposed to is a primary focus of education for disaster risk reduction programmes, as a component of education for sustainable development.

Generally, the major environmental issues the study-area deals with the intensification of extreme weather events, expansion of areas with sand dunes, land degradation, decrease of groundwater level, thus triggering an increase of nitrogenous components concentration that impact the quality of water etc. In addition to the natural-driven environmental issues there are a series of man-made drivers that have led to an increase in societal vulnerability of the Danube Floodplain to various types of hazards. In this respect, the socio-demographic and economic characteristics of the Danube settlements, as well as the institutional frame and political stability are key drivers that change the resilience and vulnerability of local communities to different forms of disturbances (Bălteanu and Dogaru 2011; Bălteanu et al. 2013).

Consequently, a key factor in undertaking community outreach activities on environmental issues in a hazard-prone area is its population. The entire study-area enfolds 36 municipalities (26 in Romania and 10 in Bulgaria) and covers a total surface of 6,259 sq.km out of which 2,216 sq.km in Romanian side and 4,043 sq. km in the Bulgarian side, thus gathering a total population of 355,379 inh. (TEMPO online 2010; NSI 2011).

The main socio-economic features of the study-area refers to the increased employment share of the occupied persons in agriculture, the reduced number of industrial activities, the high rate of ageing population, the weak occupation degree of working-age people etc. All of these are elements that describe the profound rural character of the Danube settlements (Geografia Romaniei 2005) and, ultimately their high dependence to natural and human-induced environmental disturbances.

The assessment of natural and technological hazards was carried out by a joint interdisciplinary partnership spanned over 18 months which included Romanian and Bulgarian researchers and teachers (geographers, chemists, geologists, hydrologists, climatologists, social and environmental protection specialists): *Institute of Geography, Romanian Academy—lead partner*; *National Institute of Geophysics, Geodesy and Geography, Bulgarian Academy of Sciences*; *Research Institute for Analytical Instrumentation (ICIA), Cluj-Napoca*; *Geological Institute, Bulgarian Academy of Sciences*; *University of Craiova, Geography Department*. The ROBUHAZ-DUN project involved several activities which can be grouped into two main categories: scientific and community-oriented (Fig. 3). In view of that, the crucial step in undertaking project's scientific-oriented objectives was the elaboration of a joint, comprehensive and homogenous Romanian-Bulgarian data base on natural and human-induced hazards assessment for the analysed Danube Floodplain sector and the use a common methodology for creating hazard maps. The community-oriented activities on increasing environmental sustainability in terms of hazard mitigation had in view a close cooperation with local stakeholders to identify their needs and practices in order to jointly develop solutions and actions. In this respect, meetings and campaigns with local authorities, experts, school children, university students have been performed, based mainly on a participatory approach, more than 20,000 persons attending people-to-people meetings.

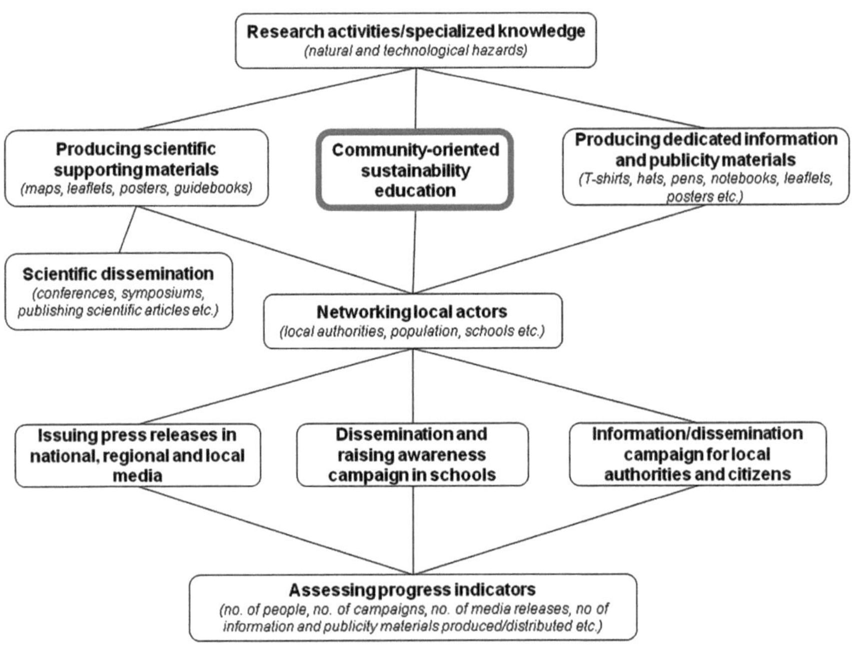

Fig. 3 The ROBUHAZ-DUN project workflow targeting community outreach activities

A major target group of the project, besides the university students, was the primary and secondary schools' children. The project aimed to enhance the capacity of primary and secondary schools for interdisciplinary teaching on sustainability issues, with a focus on climate change and disaster risk reduction issues, by establishing partnerships with universities and research institutes. In this framework, an example for presenting the sustainability issues using tailored information and dissemination materials has been offered.

In Romania and Bulgaria, as in other Central and Eastern European countries, it is recognised a lack of competence in ESD in the education sector as well as a poor quality of education for children living in rural areas due to a lack of financial and human resources (UNESCO 2009b). A review study on the presence of sustainable development education in the Romania's national curriculum in the primary and secondary school indicates that this is not a specific educational objective and sustainable development education elements does not appear systematic and integrated throughout the curriculum (Dumitru and Stoenescu 2011). In the higher education, the situation is more or less the same, with only one higher education institution signatory of the Talloires Declaration and very few examples reported on embedding sustainability elements throughout the curriculum, but being far from a radical institutional change.

3 Project's Approach and Main Results

The consortium placed an important emphasis on experts/specialists involvement as well as on local communities' potential needs in order to be able to produce useful and easy-to-use scientific products.

Based on the specialised knowledge developed within the project, the joint Romanian-Bulgarian team prepared several scientific and informative products to be used in the unfolding of the community-oriented activities of the project: leaflets, posters, informative panels, guidebooks, university course, scientific papers etc.

The community-oriented approach of the project was traded on ever since the first scientific products of the project were ready to be put forward to the end-users in the Calafat-Vidin—Turnu Măgurele-Nikopole Sector by both Romanian and Bulgarian partners. Therefore, the consortium made use of several communication/raising awareness tools in order to transfer knowledge-based information to stakeholders, local authorities, schools, local communities etc. in terms of raising awareness campaigns in schools, meetings with administrative leadership of localities, media releases, information and publicity materials to be distributed, dedicated website etc.

Generally, four key approaches on public awareness and education for disaster risk reduction and sustainability education are recognised, in terms of organising campaigns, participatory learning, informal education as well as formal school-based interventions (Rose et al. 2009). In the current project, the dedicated campaigns approach targeting local communities has been largely used due to the

Table 1 Main community outreach activities in the project

What?	Whom?	How?	When?
Joint Round-tables on "Hazard maps assessment and mapping", Bucharest and Sofia	Experts in hazard assessment and mapping	Workshop discussion	At the beginning of the project, once the methodology was ready to be discussed and adapted based on experts' analysis
Joint field campaigns for assessing soil and water vulnerability to pollution	Population, local authorities	Small meetings, reports on water and soil quality	4 campaigns in different seasons throughout the 18 months
Raising awareness and informative campaigns in schools	School children, teachers	Slide presentations, games, flyers, audio-video products	In the second half of the project, once the tailored promotional materials where prepared
Meetings with local authorities	Local authorities	Informal meetings	In the second half of the project, once the tailored promotional materials where prepared
Elaboration of the *Technical Guide* on "Hazard assessment and mitigation in the Danube Floodplain (Calafat-Vidin—Turnu Măgurele-Nikopol sector)" Romanian-English and Bulgarian-English versions	Professional and non-professional stakeholders	Bilingual books	End of the project, one of project's main results with a focus on methodology
Summer school on *Sustainable development and natural and technological hazards in the Danube Floodplain* organised in the study-area	University students from Romania and Bulgaria	Slide presentations, group exercises, field trips, promotional materials	End of the project
University Course entitled *"Natural and technological hazards assessment and mitigation"*, elaborated by the partner university in the project	Master students	Book	End of the project, the course is now included in the university's curriculum
Media releases	Public	News at local and national newspapers, magazines, journals, websites, radio and television	After every organised meeting and event meeting was organised in the study-area

(continued)

Table 1 (continued)

What?	Whom?	How?	When?
Project's web-site	Public	Updates of the project's activities, scientific papers, reports, products	Periodic updates
Scientific publications	Scientific community	Papers, reports, presentations at conferences	During/after the project

advantage of reaching a large number of people and attracting media attention, being also preferred since one of the measurable indicators of the project was focused on this quantitative aim. Informal education initiatives targeting school children and participatory techniques during workshops and focus groups were also included.

There are several reported initiatives on developing community based partnerships lead by universities (Leal Filho 2012; Kawabe et al. 2013; Cole and Fieselman 2013), the basic methods being used referring to participatory workshops, group interviews, face-to-face interviews or classes with school children, campaigns etc.

A participatory-based approach to interact with local communities have been used also by this cross-border project, a synthesis of the main types of community outreach activities in terms of *what* is disseminated, the target groups addressed (*whom?*), the appropriate way to do it (*how*?) and the calendar—the moment of their implementation throughout the project (*when?*) providing a general overview of project's roadmap (Table 1).

The undertaken field campaigns involved permanent contact with local authorities, schools and population which help the research team to identify specific needs in terms of awareness, preparedness and adaptation in case of natural and technological hazards (Bălteanu and Sima 2013b).

The resulted scientific products of the project (joint hazard maps, environmental quality maps, graphs, reports) were turned into tailored information for local authorities, citizens, pupils, teachers, university students and, further on disseminated by means of information and raising awareness activities which were organized throughout the entire project period in both countries.

Due to the dominant rural character of the region with lower access to the e-world, the web-based dissemination materials and products have been less used print materials and audio-video products were largely employed. Thus, the main promotional products developed within the project were: bilingual information panels, bilingual posters, bilingual leaflets, T-shirts, mugs, notebooks, folders, memory sticks, hats, paper bags, pens.

Among them, an important promotional tool able to provide tailor-made dissemination and raising awareness products were the bilingual posters which displayed significant information produced within the project on hazard assessment and mitigation in the study-area (Fig. 4). A particular attention was given to the depiction of the most relevant natural and human-induced phenomena the area is exposed to into comprehensible and easy-to-use hazard maps tailored for local stakeholders. Additional customized information on soil and water quality was also displayed in order to raise local people's awareness on the environmental concerns their community is dealing with.

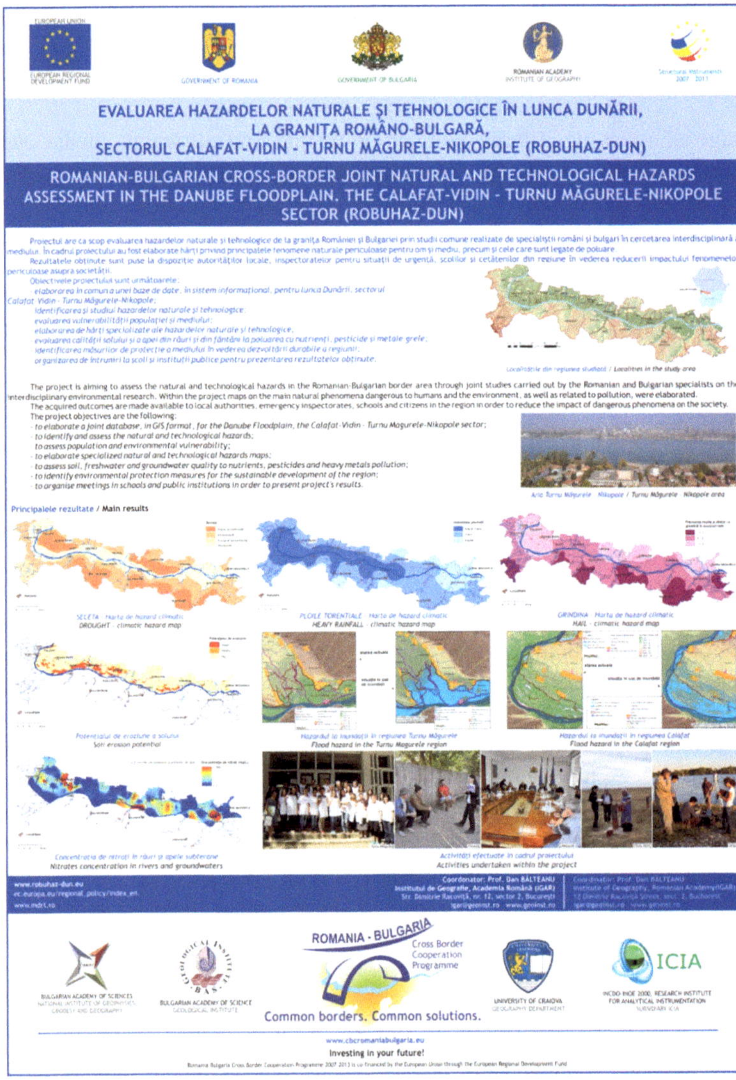

Fig. 4 Promotional materials used in the project e.g. bilingual poster

Table 2 Community outreach-oriented indicators

Result indicators		Indicator value
Designation	Unit	Quantity
Number of raising awareness campaigns in schools	Campaign	12
Number of media events with local authorities and citizens	Campaign	23
Number of promotional materials developed and distributed in Romania and Bulgaria	Copy	8,245
Number of people benefiting from awareness raising activities on environmental protection	People	23,280
Number of people graduating cross-border summer school *Romanian-Bulgarian Summer School on "Sustainable development and natural and technological hazards in the Danube Floodplain"*	People	23
Number of people participating in people to people actions	People	23,280
Number of targeted schools in Romania and Bulgaria	School	62
Number of targeted local authorities (city halls, city councils, local administration) in Romania and Bulgaria	Local community	36
Number of scientific conferences attended (national and international)	Paper	17
Number of scientific papers published by the team members (national and international peer-reviewed journals)	Article	17
Number of media releases	Article	18

The results of the project can be measured in specific indicators aimed at weighing the extent to which the community outreach activities covered the project study-area, and furthermore, at EU level the eligible joint Romanian-Bulgarian territory (Table 2).

Overall, the number of people in the project benefitting from raising awareness campaigns and attending people to people actions rose up to more than 14,000 in Romania and over 9,000 in Bulgaria through a 12 raising awareness campaigns in schools and 23 media events with local authorities, 18 media releases etc. During this project, a total number of 62 local communities and 36 schools were targeted in Romania and Bulgaria (Table 2).

In addition to the communication and raising awareness tools used within the project, three of the outcomes can be considered as part of the science-society interfaces that the consortium tried to develop and promote: *a Technical Guide, a University Course and a Cross-Border Summer School.*

The *Technical Guide* on "Hazard assessment and mitigation in the Danube Floodplain (Calafat-Vidin—Turnu Măgurele-Nikopol sector)" was published in Romania (Romanian-English version) and Bulgaria (Romanian-English version) aiming to provide a progress on Romanian and Bulgarian joint assessment of natural and technological hazards at regional level. The Technical Guide hereby reveals the added value of cooperation between research institutes and universities from both countries, by using common methodologies and joint field campaigns in approaching environmental hazards (Bălteanu and Sima 2013a).

The *University Course "Natural and technological hazards assessment and mitigation"*, published in Romanian included essential theoretical information on natural and technological hazards assessment. The course is mainly targeting university students, Ph.D students and master students, university teachers, in the field of environmental sciences.

The *Cross-Border Summer School* entitled *"Romanian-Bulgarian Summer School on Sustainable development and natural and technological hazards in the Danube Floodplain"* was an important way to share specialised knowledge on hazard assessment to master and PhD students in Romania and Bulgaria. The lectures were given by both Romanian and Bulgarian project scientists and experts in the field of climate hazards, Geographical Information System and spatial data analysis, soil and water quality assessment, social vulnerability to natural and technological hazards etc. Also, field surveys in the flood-affected and flood-prone areas, the sectors affected by geomorphological processes etc. were organised, as well as meetings with the local authorities (Calafat and Vidin City Hall) and stakeholders (Oltenia Regional Meteorological Center; "Romanian Waters" National Administration, Jiu Water Branch; Craiova Wastewater Treatment Plant). Hereby, the summer school contributed to the strengthening of local multi- stakeholders' partnerships on environmental sustainability education and the engagement of cross-border students towards sustainability targets.

4 Conclusions

The essential view of the project was not only to produce scientific reliable information for the scientific communities but to be able to build effective partnerships on sustainability education with local authorities and local communities in order to provide tailored comprehensible scientific information. The topic of the current research frames into one of the key priority areas of the EU Strategy for the Danube Region, an important macro-regional development strategy for the Danube regions that targets development and cooperation in view of reaching environmental sustainability through the protection of its natural areas, landscapes and cultural heritage.

Although the interest in such initiatives proved to be high, especially in terms of school-related activities, such singular initiatives would not likely change significantly the environmental and sustainability literacy of target groups. There is a need for long-term partnerships, continuous and systematic programmes, coupled with a change in the curriculum at national level both for schools and higher education institutions. However, the informative dissemination materials, slide presentations, the technical guide and the university course can be further used by the community in future related activities.

The university partner in the project directly benefited from the activities developed within the project by establishing longer-term partnerships with the schools and municipalities in the region to bring their students for field trips and teaching activities in the area, but also by including in the curriculum dedicated courses on natural and technological hazards, vulnerability assessment and sustainable development.

For the research institutes partners from Romania and Bulgaria, the main benefit of the stakeholder oriented approach of the project was to adapt the research methodology for hazard assessment and products to a particular non-scientific language and to present the results in a way that can be actively used by the community, but also to develop locally relevant sustainable solutions for hazard management in the area. The main challenge for having institutions from two countries working on a cross-border area was to define a common methodological approach of hazard assessment considering the existing data in both countries, but also to jointly organise the activities in the regions overcoming the language, tradition, or cultural barriers. The successful implementation of the project was to create a model that to be easily replicated and adapted to other cross-border regions.

Acknowledgments The results presented in this chapter were obtained in the framework of the project "Romanian—Bulgarian cross-border joint natural and technological hazards assessment in the Danube floodplain. The Calafat-Vidin—Turnu Măgurele-Nikopole sector (ROBUHAZ-DUN)", funded by the Cross-border Cooperation Programme Romania-Bulgaria, MIS-ETC code 350, coordinated by the Institute of Geography, Romanian Academy.

References

Adomssent M (2011) In search of the knowledge triangle for regional sustainable development: the role of universities. In: Barton A, Dlouha J (eds) Multi-actor learning for sustainable regional development in Europe: a handbook of best practice. Grosvenor House Publishing Ltd., UK, pp 5–18

Adomssent M, Fischer D, Godemann J, Herzig C, Otte I, Rieckmann M, Timm J (2014) Emerging areas in research on higher education for sustainable development management education. Sustainable consumption and perspectives from central and eastern Europe. J Clean Prod 62:1–7

Bălteanu D, Dogaru D (2011) Geographical perspectives on human-environment relationships and anthropic pressure indicators. Rev Roum Géogr Rom J Geogr 55(2):69–80

Bălteanu D, Dogaru D, Dragotă C, Grigorescu I, Dumitraşcu M, Corbu A, Sima M, Kucsicsa Gh, Călin D (2013) Dryness and drought phenomena in the Danube Floodplain: the Calafat-Vidin—Turnu Măgurele-Nikopole Sector. A local community-based participatory approach. In: The 13th international multidisciplinary scientific geoconference SGEM, 16–22 iunie, 2013, Albena, Bulgaria, Ecology, Economics, Education and Legislation Conference Proceedings, vol I, 291–298

Bălteanu D, Sima M (2013a) Natural and technological hazards—theoretical aspects. In: Bălteanu D, Sima M (eds) Hazard assessment and mitigation in the Danube Floodplain (Calafat-Vidin—Turnu Măgurele-Nikopol sector), Technical Guide. Editura Universitaria, Craiova, pp 20–24

Bălteanu D, Sima M (2013b), Introduction. In: Bălteanu D, Sima M (eds) Hazard assessment and mitigation in the Danube Floodplain (Calafat-Vidin—Turnu Măgurele-Nikopol sector), Technical Guide. Editura Universitaria, Craiova, p 11

Bilodeau L, Carter K, Gage S (2014) Advancing campus and community sustainability: strategic alliances in action. Int J Sustain High Educ 15(2):157–168

Cole E, Fieselman L (2013) A community-based social marketing campaign at Pacific University Oregon: recycling, paper reduction, and environmentally preferable purchasing. Int J Sustain High Educ 14(2):176–195

Dumitru D, Stoenescu C (2011) Education for sustainable development. Romania. Research Report. Center Education 2000 + , Bucharest, 44 p

European Council (2009) Mainstreaming sustainable development into EU policies: 2009 review of the European Union strategy for sustainable development. http://eur-lex.europa.eu/LexUriServ/LexUriServ.do?uri=COM:2009:0400:FIN:EN:PDF. Accessed 14 Mar 2014

Geografia R (2005) Campia Romana, Dunarea, Podisul Dobrogei, Litoralul romanesc al Marii Negre si Platforma Continentala, vol V. Romanian Academy Publishing House, Bucureşti 968 p

Grigorescu I, Dragotă C, Kucsicsa Gh, Nikolova M (2013) Joint assessment of climate hazards in the Danube Floodplain: the Calafat-Vidin—Turnu Măgurele-Nikopole Sector. A preliminary approach. In: The 13th international multidisciplinary scientific geoconference SGEM, 16–22 iunie, 2013, Albena, Bulgaria, Energy and Clean Technologies Conference Proceedings, Air Pollution and Climate Change, pp 689–695

Hanson CE, Palutikof JP, Dlugolecki A, Giannakopoulos C (2006) Bridging the gap between science and the stakeholder: the case of climate change research. Clim Res 31:121–133

Kawabe M, Kohno H, Ikeda R, Ishimaru T, Baba O, Horimoto N, Kanda J, Matsuyam M, Moteki M, Oshima Y, Sasaki T, Yap M (2013) Developing partnerships with the community for coastal ESD. Int J Sustain High Educ 14(2):122–132

Leal Filho W (2012) Putting concepts of sustainable development into practice at higher education institutions: Experiences from the SKINT project. In: Leal Filho W (ed) Sustainable development at universities: new horizons. Peter Lang, Frankfurt, pp 771–778

Leemans R, Solecki W (2013) Redefining environmental sustainability, Editorial overview. Curr Opin Environ Sustain 5(3–4):1–6

Mader C (2009) Principles for integrative development processes towards sustainability in regions. Doctoral thesis, University of Graz, 141 p

Mader M, Mader C, Zimmermann F, Gorsdorf-Lechevin E, Diethart M (2013) Monitoring networking between higher education institutions and regional actors. J Clean Prod 49:105–113

Mauser W, Klepper G, Rice M, Susanne SB, Hackmann H, Leemans R, Moore H (2013) Transdisciplinary global change research: the co-creation of knowledge for sustainability. Curr Opin Environ Sustain 5(3–4):420–431

NSI—National Statistical Institute of Bulgaria (2011). http://www.nsi.bg/census2011/index.php. Accessed Mar 2013

Rose C, Rouhban B, Tovmasyan K, Schick O (2009) Workshop 4: education for sustainable development and disaster risk reduction: building disaster-resilient societies. In: UNESCO world conference on education for sustainable development 31 March—2 April 2009, Bonn. http://www.esd-world-conference-2009.org/fileadmin/download/workshops/ESD2009WS4_Report.pdf. Accessed 10 Mar 2014

Storni A, Castellani V, Sala S (2012) Working with stakeholders for a successful research on sustainable development. In: Leal Filho W (ed) Sustainable development at universities: new horizons. Peter Lang, Frankfurt, pp 669–680

TEMPO online (2010) Statistical databases of Romania. https://statistici.insse.ro/shop/. Accessed April 2013

Trummler M, Mader C (2011) Database on learning for sustainable development—analysis of projects. In: Barton A, Dlouha J (eds) Multi-actor learning for sustainable regional development in Europe: a handbook of best practice. Grosvenor House Publishing Ltd., UK, pp 149–156

Trummler M, Mader C, Zimmermann F, Gorsdorf E, Diethart M (2011) Networking and interaction between regions and higher education institutions. In: Barton A, Dlouha J (eds) Multi-actor learning for sustainable regional development in Europe: a handbook of best practice. Grosvenor House Publishing Ltd., UK, pp 106–123

UNECE—United Nations Economic Commission for Europe (2005) UNECE Strategy for education for sustainable development. Vilnius: CEP/AC.13/2005/3/Rev.1. http://www.unece.org/fileadmin/DAM/env/documents/2005/cep/ac.13/cep.ac.13.2005.3.rev.1.e.pdf. Accessed 21 Feb 2014

UNESCO (2009a) Bonn declaration. In: UNESCO World conference on education for sustainable development held in Bonn, Germany on 31 March to 2 April 2009, Bonn. http://www.unesco.or.kr/upload/data_center/2009_ESD_BonnDeclaration_eng.pdf. Accessed 18 Mar 2014

UNESCO (2009b) Review of contexts and structures for ESD (by Wals, A.E.J.) Paris, UNESCO. http://unesdoc.unesco.org/images/0018/001849/184944e.pdf

United Nations (2012) The future we want. In: Conference outcome A/CONF.216/L.1. New York

Vidal A, Nye N, Walker Ch, Manjarrez C, Romanik C (2002) Lessons from the community outreach partnership center program, Final Report, U.S. Department of Housing and Urban Development Office of Policy Development and Research Washington, DC. http://www.cpn.org/topics/youth/highered/pdfs/COPC_Program.pdf. Accessed 16 Feb 2014

Author Biographies

Dr. Mihaela Sima is a senior researcher at the Institute of Geography and the coordinator of the Environment and GIS Department. Her main interests are in the field of environmental sustainability and quality with a special emphasis on evaluating the impact of human activities on the environment as well as climate change impacts on different societal fields. Dr. Sima has been involved in several research projects on coordinating the stakeholders involvement and interaction on topics as climate change impacts, environmental sustainability, and hazard assessment.

Dr. Ines Grigorescu is a senior researcher at the Institute of Geography, Romanian Academy, Environment and GIS Department. Her research activity deals mainly with the study of man-environment relationships, integrated environmental assessment and change, community outreach activities and GIS. She attended several national and international conferences and symposia and published over 70 papers on environmental-related issues in Romania. She was involved in different national and international research projects (e.g. FP7, South Eastern Europe, Cross-Border Cooperation).

Prof. Dan Bălteanu is the Director of the Institute of Geography, Member of the Romanian Academy, and Chair of the Romanian National Committee on Global Environmental Change. He is senior researcher at the Institute of Geography, Ph.D supervisor and professor at University of Bucharest. His academic competences are in the field of natural hazards, land use/land cover change, environmental and climate change. He is the author and editor of over 15 volumes and more than 150 scientific paper published in national and international journals and coordinated several national and international projects in the field of environmental sciences.

Dr. Georgi Zhelezov is an associate professor and head of Section “Physical Geography” in Department of Geography of the National Institute of Geophysics, Geodesy and Geography, Bulgarian Academy of Sciences. His research area is landscape ecology, wetland studies, ecosystem services and spatial modeling of environmental systems. His current work is related on evaluation of the potential and dynamic of the landscapes of the wetlands and mountainous regions, implementing GIS and spatial modeling. He was involved in participation and coordination of different national and international research projects related with sustainable development, of the nature systems, in the mountain regions of South Eastern Europe, comparative studies and spatial transformation of the wetlands, risk assessment of the cross-border regions.

A Journey in Sustainable Development in an Urban Campus

Darren Reidy, Maria J. Kirrane, Barrie Curley, Denis Brosnan, Stephan Koch, Paul Bolger, Niall Dunphy, Michelle McCarthy, Mark Poland, Yvonne Ryan Fogarty and John O'Halloran

Abstract

University College Cork is located in an urban setting in the heart of Cork city. The university was the world's first Green-Campus awarded by the Foundation for Environmental Education (FEE), Copenhagen, in 2010 (renewed 2013) and the world's first university to achieve ISO50001 standard certification for Energy Management Systems: we are student led, research informed and practice focussed on matters of sustainability and this ethos is embedded across the entire university. During the period campus recycling rate increased from 21 to 75 %, total energy consumption decreased by 9 %, the number of staff choosing to cycle to work- increased from 6 to 12 % across campus. €1,000,000 on waste costs was saved and a saving of over 7,50,000 m^3 of water. We are also committed to engaging in training our staff in environmental awareness in biodiversity and environmental management. Student led initiatives range from food production in on-campus poly-tunnels to collaboration with our estates to maintain biodiversity. This paper describes the journey for the university staff and its students in advancing our Green-Campus, with a wider impact on our stakeholders and service providers.

D. Reidy (✉) · M.J. Kirrane (✉) · B. Curley · D. Brosnan · S. Koch · P. Bolger · N. Dunphy · M. McCarthy · M. Poland · J. O'Halloran
University College Cork, Western Road, Cork, Ireland
e-mail: d.t.reidy@umail.ucc.ie

M.J. Kirrane
e-mail: m.kirrane@umail.ucc.ie

Y.R. Fogarty
Unit 5a, An Taisce, Swift's Alley, Francis St, Dublin 8, Ireland

W. Leal Filho et al. (eds.), *Integrative Approaches to Sustainable Development at University Level*, World Sustainability Series, DOI 10.1007/978-3-319-10690-8_41

Keywords
Voluntary environmental initiatives • Green education • Education for sustainable development • Eco-Universities

1 Introduction

The need for change in social, political and economic agendas to include environmental responsibility has repeatedly been highlighted as essential if we are to protect the interest of future generations without adverse environmental effects. Universities have long been charged with being agents of social change (Lozano 2010) and it has been said that it is an "ethical imperative" that universities act as role models and educate the next generation in sustainable development (Nicolaides 2006). In recent years many universities have attempted to institutionalize the concept of sustainable development by incorporating it into the curriculum, research agendas, infrastructure, administration and operations. However, it is rare that it simultaneously becomes established in all disciplines (Sharp 2002; Lozano 2010) "The challenge for educational institutions is not in the teaching of facts" but in the creation of an "active, transformative process of learning that allows values to be lived out and debated, and permits a unification of theory and practice" (Warbuton 2003).

Founded in 1845, University College Cork (UCC) is one of Ireland's premier universities. It boasts a vibrant, modern city centre campus encompassing historic buildings located on a site of over 30 ha. The university is home to some 20,000 full-time students and employs c.2800 people. The University is highly committed to sustainability and the green agenda. It was the world's first Green-Campus university awarded by the Foundation for Environmental Education (FEE 2014), Copenhagen, Denmark and is currently ranked second globally in the UI Green Metric World University Rankings (Green Metric 2014). The university took an innovative approach in transforming to a sustainable campus through a student-led, research informed and practice focused approach. The methodologies employed are framed around our adaptation of the existing seven step Green-Schools programme. UCC researches, teaches and practices environmental sustainability and this ethos is embedded across the entire university. This paper outlines the origins of UCC's Green-Campus programme and how it has been implemented in all aspects of university life. The paper further describes how the UCC model can, and has been, used as a model for sustainable development across a variety of institutions.

2 Origins of the Green-Campus Programme

Green-Schools (known internationally as Eco-Schools) is an international education programme run by the Foundation for Environmental Education and active in over 23,000 schools across 43 different countries. It is operated in Ireland by An Taisce, the National Trust for Ireland. It promotes environmental sustainability and action through a seven step programme, awarding a Green-Flag to successful applicants. The steps begin with the *establishment of a committee* representative of all major stakeholders. This committee must then undertake an *environmental review* and develop an *action plan* which is implemented, *monitored and evaluated* continuously. The programme must be *linked to the curriculum*/learning on campus and the committee must *inform and involve* the wider community. Establishing a *green charter* marks the final step towards the awarding of a Green-Flag (Madison and Ryan 2012). Central to the Green-Flag ethos is the involvement of all stakeholders and the embedding of environmental sustainability and awareness throughout all aspects of the institution.

Currently, in UCC, there are over 120 student clubs and societies representing a range of interests. In 2006, the Environmental Society was formed and issues with environmental management of the campus became their focus. With many of their members having come from Green-Schools, and aware of the positive impacts that come with the adoption of a Green-Flag ethos, they decided that it should be possible to adapt the programme to suit a university model. In continuing Green-Schools to third level it would ensure students continued to learn and contribute as valuable working citizens in an environmentally healthy and equitable society. Before Green-Campus, student involvement in the environmental management of their university was non-existent. It was felt that challenging this could be the catalyst required to make UCC a more sustainable place to study, work, live and socialise. The students believed that the programme could provide a framework and a common goal that all aspects of university life could work toward. An Taisce agreed that UCC would be a suitable candidate to roll out a pilot Green-Campus initiative. The students contacted the relevant stakeholders, including other students, academics, administrative staff, and the Buildings and Estates office. This office is responsible for the management, services and maintenance of the campus buildings and grounds. A committee was formed (Fig. 1) and the first meeting was held in December 2007.

The Green-Campus initiative "encourages a partnership approach to environmental education, management and action in third level institutions" (An Taisce 2014). It requires the institute to complete the programme under seven different environmental themes; Litter and Waste, Energy, Water, Travel, Biodiversity, Climate Change and Global Citizenship. An Taisce and FEE recognise that universities operate on an extensive scale and that fulfilling these steps under these themes will happen on a rolling basis. "The programme primarily aims to ensure that members of a campus community can engage in a meaningful way to enhance sustainability on campus" (An Taisce 2014). It "rewards long term commitment to continuous improvement from the campus community" (Madison and Ryan 2012).

Fig. 1 Committee structure and function (An Taisce 2011)

3 A Student Led Initiative

Researchers agree that lack of knowledge is a barrier to behavioural change, however it has also been shown that knowledge alone will not motivate someone to adopt a new behaviour (Schultz 2002; Stern 2000; DeYoung 2000). Student leadership in activism has been shown to set precedent for behavioural change (Altbach and Cohen 1990), it is thus critical that universities adopt a student led approach to informing and involving the entire university community as a means of complimenting the knowledge based curricular activity. Sharp (2002) noted that when it came to the greening of campuses "students have much more potential power to catalyse and drive organizational transformation than they believe is possible".

One primary objective of the Eco-Schools and Green-Campus programmes is to provide an educational platform through which students will become environmentally and socially responsible, traits they will carry with them and disseminate into their future communities and workplaces (Krnel and Naglič 2009; Madison and Ryan 2012). Linking sustainability to the curriculum is thus imperative in order for such knowledge to permeate all disciplines. Multiple studies have found that most students identify the term sustainability with environmental issues and less so with economic and social ones (Azapagic et al. 2005; Summers et al. 2004). Universities must find a way of integrating all three strands of sustainability into the education system in order to ensure a well-rounded knowledge base in graduates (Kagawa 2007).

The constitution of the UCC Green-Campus Committee therefore makes active student representation a requirement. Provisions are made each year for the engagement of incoming students to ensure continuity. The deputy president of the Student's Union is a mandated member of the committee; thus despite the transient nature of the student body they have consistent representation. The committee has been acknowledged as a fully functioning administrative body by the university president. It is open for any member of the university to join and meets on a regular basis to discuss current issues in environmental management of the university. Every suggestion is considered, investigated and its feasibility explored.

The Green-Campus programme was championed by a small number of students who were interested in environment and sustainability. A major aim of the programme therefore is to inform and involve students from all academic disciplines and backgrounds. Every year the student members of the Green-Campus Committee together with the Environmental Society organise a Green-Week on campus, complimented by standalone events across the year. Events take the form of social, political, creative and charitable platforms and provide opportunity for discussion, debate and engagement from all students. The committee liaises with societies from a range of disciplines in order to reach as wide an audience as possible. Societies that have actively engaged with the programme include the Photography, Fashion, Art, English Literature and International Development societies and clubs such as the Kayaking and Snow Sports clubs.

In 2013 UCC Envirosoc and Green-Campus Committee held a "Feed the 500" event to highlight the problem of excessive food waste. Food normally destined for landfill was donated by restaurants and wholesalers across Cork city. Just under 700 students gathered on campus to be fed a warm nutritious meal on a cold February day. The event attracted much publicity across the university as well as in the press. It was awarded best event at the National Board of Irish College Society Awards 2013. Following this success, Cork City Council in collaboration with UCC have established the Cork Food Policy Council and subsequently held a "Feed the City" event which successfully fed over 5000 people in March 2014 (CFPC 2014).

As a direct result of the Green-Campus programme new University Wide Modules on environment and sustainability have been developed (http://www.ucc.ie/modules/). These are available to all students and cover topics such as ecology, environmental law, environmental health, philosophy and the psychology of climate change. Students earn academic credit upon completion of these modules. Over thirty other courses related to environment and sustainability are also offered to students.

4 Research Informed

The importance of linking high quality research to the objectives of sustainable development is acknowledged by the academic community and policy makers. This is emphasised by calls for the European Research Area to be "firmly rooted in society and responsive to its needs and ambitions in pursuit of sustainable development" (Georghiou et al. 2009). Innovation which stems from research is an essential tool in safeguarding the social, economic and environmental interests of future generations (Georghiou et al. 2009). Many fields of science, engineering, arts and humanities are related to the study of sustainable development, however some authorities question whether there is coherent collaboration between the various sectors, and whether sustainable development should be a central research objective of institutions.

Historically sustainable development and environment was not a core research objective of the university. UCC has responded and has established environmental research as an integral part of the university's strategy to develop inter-disciplinary research institutes that have a critical mass of world-class researchers in prioritised research areas. The report of the Research Prioritisation Steering Group (UCC 2012a) stated that continued "research is required to understand better how Ireland will enable a transition to a low emissions, sustainable and climate resilient economy and society".

The Environmental Research Institute (ERI), is an interdisciplinary research institute which facilitates and supports environmental sustainability in UCC. The institute aims to foster collaborative, multidisciplinary research through key research thematic areas and to facilitate the transfer of research to industry. The institute brings together expertise in biological, chemical and environmental sciences as well as environmental engineering, energy, humanities and environmental law. In 2011 an independent report into the commercial and economic impact of research investment in Ireland found that the ERI had a high impact in all six evaluation categories (commercial, human capital, capability, reputation, national policy, wider impacts) (UCC 2012a). UCC intends a €14 m expansion of the ERI with a particular focus on marine resources and renewable energy by 2015. Students and staff from the ERI have a significant role in the Green-Campus Committee and measures are in place to ensure continued involvement.

A 2009 European Commission report recommended that sustainable development needed to be imbedded in practise and how research is conducted. The report suggested that in order for research outputs to have their greatest impact sustainability must be practised in the research process (Georgiou et al. 2009). The ERI is an excellent of example of these recommendations in practise. The facility has been built to the highest standards of contemporary sustainable design with features such as solar panels, geothermal heat pumps and heat recovery systems. Natural lighting and ventilation were a priority in the building design. The design criteria for the building were explicit; design that demonstrated a best practice approach to environmental design as well as one that stood out as a low-energy laboratory building to complement the work undertaken inside. The passive building serves as a full scale test bed investigating how buildings perform in relation to their design intentions. (O'Sullivan et al. 2004; Ahmed et al. 2009).

5 Practise Focused Implementation

Upon establishment, the Green-Campus Committee conducted a review of the environmental status of UCC. Baseline data were gathered for each theme. Many review questions became part of research projects conducted by students of science, engineering and humanities. The environmental reviews inform actions which are implemented, the success of each action is monitored and continuously evaluated. We set out below highlights of a practise focused implementation of the seven step Green-Campus programme for each theme.

5.1 Waste

In 2008 a student initiative to audit waste being disposed of at UCC chose random waste disposal skips throughout campus, itemizing their contents. This quantified for the first time exactly what was being disposed and how. This resulted in the provision of improved recycling facilities, changing of waste collection contractors, and the separation of biodegradable waste at source from catering facilities and landscaping activities. This change in operations was complemented by campaigns to inform the campus community of changes undertaken.

Implementation of such actions has seen an overall increase in recycling from 21 to 75 % and over 1015 tonnes diverted from landfill. The university has saved over €1million in waste costs. In an example of the cooperation between stakeholders central to the Green-Campus ethos, compost produced from university waste is now used for landscaping while biodiesel produced from waste oils collected in campus restaurants runs two university vehicles. The committee ensures sustained improvement through continual monitoring. D'Arcy (2012) identified shortfalls in the introduction of recycling facilities (Table 1) and surveys were undertaken by students on attitudes towards recycling. Findings of such reports are reviewed periodically to insure continued improvement toward a sustainable campus.

Table 1 Assessment of general waste facilities in UCC (D'Arcy 2012)

S	University system in place	Rating system:	
F/N	facilities on-site / not	(green)	Good
A/NA/~	accessible /not / semi	(yellow)	OK
E/NE/~	effective / not / semi	(red)	Needs work

4a: General waste		Recycling												
Location	General waste	Dry mixed recycling	cardboard/ paper	Newspaper &Magazine	Glass & Cans	Poly-styrene	Batteries	Print Cartridges	Computers /parts	WEEE	Other Equipment	Pallets	Food/ Compost	Catering waste
Tyndall	S/F/A/E	S/F/A/E	S/F/A/E	S/F/A/E	S/F/A/E	N/NA/NE	S/F/A/E	S/F/A/E	S/F/A/E	S/F/A/E	S/F/A/E	S/F/A/E	N/NA/NE	S/F/A/E (K)
Main campus														
Kane/Windle	S/F/A/E	S/F/A/E	S/F/A/E	S/F/A/E	S/F/A/E	N/NA/NE	S/F/A/E	S/F/A/E	S/N/A/~E	S/N/A/~E	N/NA/NE	N/NA/NE	N/NA/NE	S/F/A/E (K)
Biosciences	S/F/A/E	N/NA/NE	N/NA/NE	N/NA/NE	S/F/A/E	N/NA/NE	S/F/A/E	S/F/A/E	S/N/A/~E	S/N/A/~E	N/NA/NE	N/NA/NE	N/NA/NE	none
Student Centre	S/F/A/E	S/F/A/E	S/F/~A/~E	N/NA/NE	S/F/A/E	N/NA/NE	S/F/A/E	S/F/A/E	S/N/A/~E	S/N/A/~E	N/NA/NE	N/~A/~E	N/NA/NE	S/F/A/E
Western campus														
Brookfield	S/F/A/E	S/F/A/E	S/F/A/E	S/F/A/E	S/F/A/E	N/NA/NE	S/F/A/E	S/F/A/E	S/N/A/~E	S/N/A/~E	N/NA/NE	F/A/~E	F/NA/NE	S/F/A/NE (K)
Gateway	S/F/A/E	S/F/A/E	S/F/A/E	N/NA/NE	S/F/A/E	N/NA/NE	S/F/A/E	S/F/A/E	S/N/A/~E	S/F/A/~E	N/NA/NE	F/A/~E	F/NA/NE	S/F/A/NE (K)
North Mall	S/F/A/E	S/F*/A/~E	N/NA/NE	N/NA/NE	S/F/A/~E	N/NA/NE	S/F/A/E	S/F/A/E	S/N/A/~E	S/N/A/~E	N/NA/NE	N/NA/NE	N/NA/NE	S/F/A/NE (K)
ERI	S/F*/A/~E	S/F*/A/~E	N/NA/NE	N/NA/NE	N/NA/NE	N/NA/NE	S/F/A/E	S/F/A/E	S/N/A/~E	S/N/A/~E	N/NA/NE	N/NA/NE	N/NA/NE	none
Dental school	S/F/A/E	N/NA/NE	S/F/A/~E*	N/NA/NE	N/NA/NE	N/NA/NE	N/NA/NE	S/F/A/E	S/N/A/~E	S/N/A/~E	N/NA/NE	N/~A/~E	N/NA/NE	none

5.2 Energy

A campus energy review in the academic year of 06/07 revealed that overall energy consumption was 53.6 MWh. Average energy usage was 332 kWh/m^2 allowing the identification of buildings that were the largest consumers. Student projects, undertaken with the assistance of Buildings and Estates staff, have identified opportunities to improve the energy status of buildings and infrastructures throughout campus.

UCC was the first university in the world, and the first Irish public sector body, to be certified for ISO50001. This is an international standard that sets out conditions for the establishment, implementation, maintenance and improvement of an organisation's energy management system. It enables a systematic approach for continued improvement of energy performance, including energy efficiency, security, use and consumption (ISO 2011). ISO50001 was released in 2011 and UCC immediately began working towards achieving the standard.

An integrated energy management programme was initiated creating a structured approach by the following methods; strategic energy management initiatives, technical efficiency, behavioural changes, awareness, reduction projects and effective energy procurement. This has resulted in a number of key achievements. Overall energy consumption fell by almost 20 % despite the campus increasing in size during the period. In 2013 53 MWh were used, representing 267 kWh/m^2; a reduction of 18 % per full time equivalent (FTE) student was achieved. This meets a 2020 target of a 33 % reduction from the baseline.

UCC now employs a number of on-site renewable energy sources including solar, wind and geothermal systems. Over 20 % of UCC's purchased energy comes from renewable sources. Energy efficiency is a design priority for all new university developments, as is the continued upgrading and retrofitting of historic campus properties. UCC's example of successfully implementing ISO50001 was adopted as a case study by other higher education institutions, including Massachusetts Institute of Technology (Ledewitz 2012).

5.3 Water

Average water use from 2002–2005 was 2,23,287 m^3 which was higher than expected. Student-led studies found that large quantities of water were being lost to leaks in dated plumbing networks, some dating back to 1845.

A process of upgrading to a more efficient plumbing network ensued and is still in progress. A campaign to change user's attitudes toward water consumption was also initiated. To date the actions taken have resulted in a cumulative saving of over 7,50,000 m^3 of water since 2007; consumption trends becoming progressively lower annually. Consumption in 2013 was 1,25,414 m^3 less than the baseline (Fig. 2).

5.4 Transport

A survey of staff and student travel patterns was conducted in 2007, updating previous 2001 data. The survey showed that in 2007 less than half of the UCC population walked to campus. Only 6 % cycled while 3 % staff and 11 % of

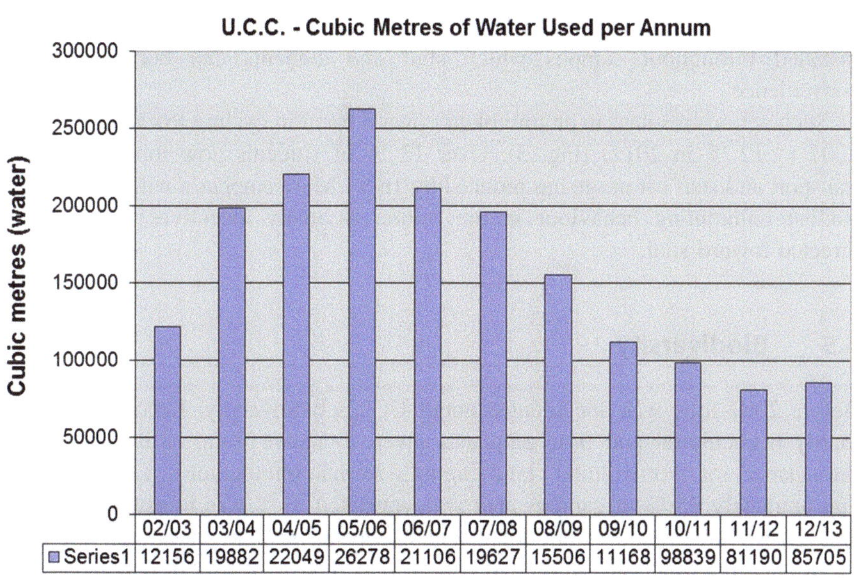

Fig. 2 Annual water consumption 2002–2013

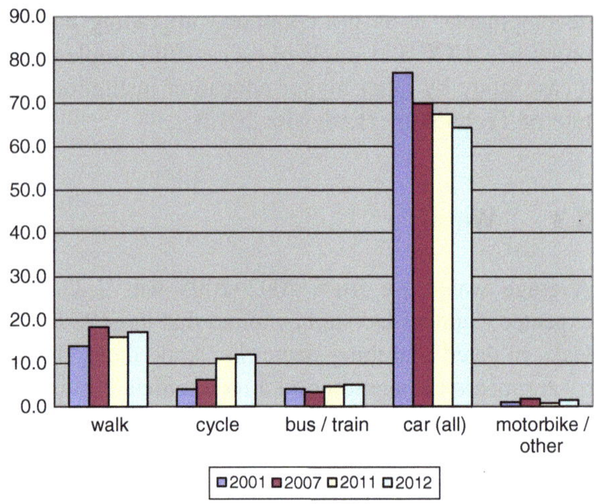

Fig. 3 Staff (%) travel modes to and from the university

students used public transport. 36 % of students and 70 % of staff arrived on campus by car.

A Sustainable Commuter Plan was developed and implemented. UCC 'Park and Ride' services were expanded, particularly during peak times. Carpooling is actively encouraged through provision of reserved parking spaces and enterprising 'shared transport' smartphone applications were piloted by UCC in 2008. A "Cycle to Work" bicycle purchase scheme, tax-incentivised by the Irish government, was adopted and bicycle parking and changing facilities improved. In association with Cork City Council cycle lanes were upgraded on all commuter routes making cycling to UCC more convenient and safe. A fleet of campus bicycles have been installed throughout campus which staff and students can borrow at their convenience.

Such actions resulted in an unbroken upward trend in cycling to campus (6 % in 2007 to 12 % in 2012) (Fig. 3). Over 12 % of students now travel by public transport and staff car usage has reduced by 16 %. More emphasis will be placed on student commuting behaviour in the future, as many incentives are currently directed toward staff.

5.5 Biodiversity

Before 2007 little was documented about UCC's biodiversity. Records were primarily horticultural and little emphasis given to native flora, or the origin and provenance of horticultural landscaping. Aerial photography identified that approximately 65 % of campus area was vegetated, 6 % was forested. Extensive areas of invasive species were documented. Importantly, the review identified that the River Lee, which flows through campus, becomes a legally designated Natura 2000 habitat under the EU Birds Directives (NPWS 2011) downstream.

A number of actions were implemented to enhance biodiversity on campus: A database of trees on campus has been generated, including information on species, age, condition and origin. A tree planting schedule for the next 5 years is established giving native species priority. As part of their Green-Week activities students organised a native tree planting day with local schoolchildren. Students have also established an Invasive Species Action group, utilising academic expertise to inform management of invasive species. A riverbank restoration scheme has subsequently removed large stands of invasive species and re-established native communities. Habitats for a number of protected species have been improved through the provision of roosting and nesting boxes for bats and birds. Swift boxes have been fitted with cameras which broadcast to the web.

The Green-Campus Committee is committed to the training of staff and students in more sustainable and environmentally friendly types of biodiversity management. All horticultural, landscaping and grounds staff have been trained in biodiversity conservation. An organic community garden has been established on campus. It is operated by students and facilitates the sustainable production of food for the community who actively use it. The garden also functions as a teaching resource for sustainable food production.

Monitoring and evaluation of biodiversity in UCC will be significantly increased in 2014 by undertaking a complete ecological assessment and habitat map. This will be complimented by a student led initiative to record campus biodiversity. This "Bioblitz" will be the first of its kind to be held on a university campus and will see students and staff record campus flora and fauna over 24 h, with support from the National Biodiversity Data Centre (NBDC 2014). Initiated by UCC students this BioBlitz will be replicated on campuses nationwide.

5.6 Climate Change

UCC prepared an inventory of greenhouse gas emissions or 'carbon footprint' associated with its activities for the 2008–2009 academic-year, with a follow-up study for 2011–2012. These studies were performed in accordance with WRI/WBCSD Greenhouse Gas Protocol Corporate Accounting and Reporting Standard, (GHG Protocol Standard), the most widely used standard and guidelines (Dunphy et al. 2013).

Emissions are categorised into three 'scopes'. Scope 1 are the emissions from activities under the direct control of the organisation e.g. onsite combustion; Scope 2 are offsite emissions arising from the purchase of electricity, heat or steam; and Scope 3 are any other indirect emissions from sources not directly controlled by the organisation e.g. arising within the organisation's supply chain. UCC adopted an inclusion approach to selection of scope 3 emissions and included amongst others GHG emissions associated with travel, water consumption, and waste management.

UCC's Carbon Footprint scope 1 and 2 reduced between the study periods on both an absolute basis (−8 %) and normalised for floor area (−13 %) or student numbers (−13 %) reflecting ongoing efforts of the university to improve its energy performance. When including the scope 3 emissions, there was an increase in absolute terms (+4 %) although a decrease was evident when normalised for floor area (−2 %) and student numbers (−2 %). The increase in scope 3 emissions was chiefly due to a significant increase in student commuting. This may be due to wider economic and societal issues—anecdotal evidence suggests that many students are choosing to commute rather than incur the cost of accommodation away from home (Dunphy et al. 2013). As mentioned, student commuting behaviour is a priority for the Green-Campus committee over the coming years.

6 UCC's Green Charter

The steps outlined above demonstrate UCC's commitment to environmental sustainability through an all-encompassing approach. UCC's Strategic Plan (UCC 2012b) states that the university is "committed to the intellectual growth, social formation, and welfare" of its students. It recognises that in order to fulfil this commitment it must continue to "further develop itself as a 'Green University". UCC has thus put in place an Environmental Policy which firmly commits to the principals of sustainability. The policy recognises that university activities impact upon the environment through education, research and operations. The policy states that the university is thus "cognisant of its environmental responsibilities in local, national and global terms", acknowledging a responsibility for, and commitment to, the protection of the environment throughout activities and endeavouring to continuously improve its environmental performance by the implementation of a documented Environmental Management System (UCC 2012b).

The implementation of the University Environmental Policy is overseen by the Green-Campus Committee and the overall University Green Forum. The forum has adopted a code of practice which highlights that a true Green-Campus should be "Student Lead, Research Informed and Practise Focused". As a symbol for all students the Green-Flag hangs over the main quadrangle on UCC campus and other off campus sites. It shows that there is a framework in place for all stakeholders in the university, including the students, to have their input in the environmental management of the campus. It is an internationally recognised symbol of the commitment that the university shows to environmental sustainability and demonstrates that on UCC campus it is a priority.

7 Conclusions

UCC's successful implementation of a pilot Green-Campus programme is evidenced by the fact that to date 22 Campuses in Ireland are formally registered on the programme: Six Universities, six Institutes of Technology, four Colleges of Education, four Colleges of Further Education and two specialist training institutes, including a hospital. The benefits of the programme are seen in both the preparation of students as environmentally conscious citizens and activists as well as the sustainable development of an institution whose scale equates to that of a small town. Indeed, it has been demonstrated through Green-Campus that this stakeholder driven, bottom up approach to environmental management is also of benefit to the local and wider communities (Madison and Ryan 2012). In these testing economic times it is prudent that graduates possess wider skillsets that can help deal with the challenges presented by environmental degradation and resource depletion. From an educational viewpoint it is imperative that these skills are taught to all students, not just those with an environmental leaning. Graduates with these competencies are essential in building the smart and green economies of the future. The continuation of the Green-Flag ethos from pre-school to primary, secondary and tertiary levels provides a framework for "action-oriented continuous education" focusing on a more sustainable future, not just in Ireland but globally (Madison and Ryan 2012).

By embracing the student perspective and enthusiasm, UCC has established itself as a global leader in sustainable development in universities. UCC continues to review and improve its performance in accordance with the Green-Campus approach to sustainability. Having developed a successful template the UCC Green-Campus Committee openly share their experiences, both positive and negative, with other institutions. These links ensure that best practise guidelines are made available to as wide an audience as possible.

References

Ahmed A, Ploennigs J, Gao Y, Menzel K (2009) Analyze building performance data for energy-efficient building operation. http://www.academia.edu/4086186/. Accessed 25 Mar 2014

Altbach PG, Cohen R (1990) American student activism. J Higher Educ 61(1):32–49

An Taisce (2014) http://www.greencampusireland.org/. Accessed 25 Mar 2014

Azapagic A, Perdan S, Shallcross D (2005) How much do engineering students know about sustainable development? The findings of an international survey and possible implications for the engineering curriculum. Eur J Eng Educ 30(1):1–19

CFPC (2014) http://corkfoodpolicycouncil.com/. Accessed 29 Mar 2014

D'Arcy A (2012) Waste management in UCC, scoping study and environmental action plan. Draft for Consultation, Cork

DeYoung R (2000) Expanding and evaluating motives for environmentally responsible behavior. J Soc Issues 56(3):509–526

Dunphy NP, Ryan MB, Morrissey J, Poland M (2013) Tracking carbon footprints—experience of an Irish university. In: 16th conference of the European roundtable on sustainable consumption and production (ERSCP), 7–9 June 2014. Istanbul, Turkey

FEE (2014) http://www.fee-international.org/en. Accessed 28 Mar 2014

Georghiou L, Esterle L, Hubert C Kuhlmann S. (2009) Gearing European research towards sustainability. European commission RD4SD exercise. http://ec.europa.eu/research/sd/pdf/rd4sd/rd4sd_final_report.pdf. Accessed 29 Mar 2014

Green Metric (2014) http://greenmetric.ui.ac.id/. Accessed 28 Mar 2014

ISO (2011) ISO50001 energy management systems—requirements with guidance for use. Energy review, international organization for standardization, Geneva, Switzerland
Kagawa F (2007) Dissonance in students' perceptions of sustainable development and sustainability: implications for curriculum change. Int J Sustain High Educ 8(3):317–338
Krnel D, Naglič S (2009) Environmental literacy comparison between eco-schools and ordinary schools in Slovenia. Sci Educ Int 20:5–24
Ledewitz J (2012) Early lessons learned from building an ISO 50001: conformant energy management system for MIT. ACEEE Summer Study Energ Effi Buildings 3:271–283
Lozano R (2010) Diffusion of sustainable development at universities' curricula: an empirical example from Cardiff University. J Clean Prod 18:637–644
Madison O, Ryan Y (2012), Best practises for Eco-schools international programme implementation for tertiary level in Ireland and Russia. Lifelong learning, continuous education for sustainable development. In: Proceedings of the 10th anniversary international cooperation, vol 10, pp 112–114
NBDC (2014) http://www.biodiversityireland.ie/. Accessed 29 Mar 2014
Nicolaides A (2006) The implementation of environmental management towards sustainable universities and education for sustainable development as an ethical imperative. Int J Sustain High Educ 7(4):414–424
NPWS (2011) Conservation objectives for Cork Harbour SPA [004030]. Generic Version 4.0. Department of arts, heritage and the gaeltacht, Government of Ireland
O'Sullivan D, Keane M, Kelliher D, Hitchcock R (2004) Improving building operation by tracking performance metrics throughout the building lifecycle (BLC). Energ Build 36:1075–1090
Schultz PW (2002) Knowledge, information, and household recycling: Examining the knowledge-deficit model of behaviour change. In: Dietz T, Stern PC (eds) New tools for environmental protection: education, information, and voluntary measures. National Academy Press, Washington DC, pp 67–82
Sharp L (2002) Green-Campuses: the road from little victories to systematic transformation. Int J Sustain High Educ 3(2):128–145
Stern PC (2000) New environmental theories: Toward a coherent theory of environmentally significant behaviour. J Soc Issues 56(3):407–424
Summers M, Corney G, Ghilds A (2004) Student teachers' conceptions of sustainable development: the starting-points of geographers and scientists. Educ Res 46(2):163–182
Taisce An (2011) The green campus committee. Environmental Education Unit, Dublin
University College Cork (2012a) Environmental research institute strategic plan 2012–2016. UCC, Cork
University College Cork (2012b) UCC strategic plan, sustaining excellence 2013–2017. http://www.ucc.ie/en/strategicplanning/2013/. Accessed 29 Mar 2014
Warburton K (2003) Deep learning and education for sustainability. Int J Sustain High Educ 4 (1):44–56

Author Biographies

Darren Reidy holds a B.Sc. and M.Sc. from UCC and was part of the original student-led group which formed the first UCC Green-Campus Committee. He has worked in Science and Environmental Education for state/semi-state agencies for a number of years. He is currently pursuing a Ph.D. in the School of BEES at UCC.

Maria Kirrane is completing her Ph.D. on honey bee behaviour in the School of BEES at UCC. As a founding member of the university Environmental Society she was one of the main instigators of the Green-Campus programme and has remained involved in the initiative since its inception.

Barrie Curley HDip SHAWW, HDip BFIS, BComm, is Estates Administrator in University College Cork. A Graduate of UCC he oversees all Grounds related matters, a large part of which involves ensuring the Estate is run in a sustainable manner while promoting UCC as a Green Campus.

Dennis Brosnan is the Energy Manager in UCC and represents the UCC energy committee in Green-Campus. This committee has responsibility over the consumption of energy, gas and water and has led the implementation of numerous energy initiative such as ISO 50001.

Stephan Koch is Commuter Plan Manager within the UCC Office of Buildings and Estates, in charge of all daily travel matters to and from the Campus and the promotion of sustainable transport. He has a degree in Transport Planning and Operation (Technische Universität Berlin) and joined UCC in 2006.

Paul Bolger is Manager of UCC's Environmental Research Institute, one of the leading national institutes in Ireland in the area of Environmental, Marine and Energy research. He holds a PhD in Chemistry and Masters in Business Administration. He is currently Chair of Environmental Sciences Association of Ireland.

Niall Dunphy is the Director of the Cleaner Production Promotion Unit (CPPU) in the School of Engineering at UCC. CPPU conducts funded research and works closely with businesses on commissioned research and consultancy to promote sustainable consumption and enable cleaner production. Niall has been Principal Investigator of numerous National and EU funded research, demonstration and training projects.

Michelle McCarthy has worked in the Buildings and Estates Office since 2008 has been a member of the Green Campus Committee since 2009. She has a Bachelor of Business Studies (Marketing) from Cork Institute of Technology.

Mark Poland is a chartered civil engineer and Fellow of Engineers Ireland. He is a graduate of UCC and has experience in Ireland and the UK as a consulting engineer and project manager on commercial and industrial projects. He is currently Director of Buildings and Estates at UCC and leads a team of professionals involved in the management of the University's estate.

Yvonne Ryan Fogarty received her doctorate in environmental science from the University of Limerick where she researched the application of industrial ecology principles on small to medium sized enterprises and business clusters. She is currently the Green-Campus Coordinator at An Taisce.

John O'Halloran currently holds the Chair in Zoology and is Head of School of Biological, Earth and Environmental Sciences (BEES) at UCC. He is a student led, research informed and practice focussed Professor and chairs the UCC Green Forum.

The Experience of Training of Experts in Sustainable Development at the University of Perugia

Antonio Boggia, Cinzia Buratti, Adriano Ciani, Stefano Casadei and Stella Giordano

Abstract

This paper presents the experience of the doctorate in "Rural Sustainable Development, land and environment" of the Faculty of Agriculture at the University of Perugia (Italy). The entire period of activity of the doctorate, founded in the academic year 2000/2001, including the ongoing cycles, covers a time span of 15 years with 10 cycles completed. Its principal aim was the training of researchers and managers with good skills to address the integration and sustainability of interventions within rural and environmental systems. All the doctoral students who have completed the course until now, six months after the acquisition of the Ph.D., had an employment contract or a job. At the present time 17 % of them are uin permanent employment at Italian Universities, 29 %

A. Boggia (✉) · A. Ciani · S. Giordano
Department of Agricultural Food and Environmental Sciences, Perugia University,
Borgo XX Giugno 74, 06121 Perugia, Italy
e-mail: boggia@unipg.it

A. Ciani
e-mail: adriano.ciani@unipg.it

S. Giordano
e-mail: stellagiordano@libero.it

C. Buratti
Biomass Research Center, University of Perugia, Via G. Duranti,
Road St. Lucia Sn, 06125 Perugia, Italy
e-mail: cinzia.buratti@unipg.it

S. Casadei
Department of Civil and Environmental Engineering, Perugia University,
Borgo XX Giugno 74, 06121 Perugia, Italy
e-mail: stefano.casadei@unipg.it

W. Leal Filho et al. (eds.), *Integrative Approaches to Sustainable Development at University Level*, World Sustainability Series, DOI 10.1007/978-3-319-10690-8_42

have a fixed-term contract at the Italian University, 27 % are in Public-authority, 10 % are in Private authority and 17 % are Professional Agronomists. After the last reform of the Italian University, the doctorate has been incorporated into the new doctorate "Energy and Sustainable Development" of the Biomass Research Centre. This will be a single training course in which all major issues related to sustainable development will be handled: engineering, energy, environment, land-use planning, economy and medicine.

Keywords
Education · Rural sustainable development · Environment · Human capital · Energy

1 Introduction: The History and the Role of Higher Education for Sustainable Development

Since the beginning of the debate on sustainability, education was regarded as one of the key instruments for the realization of sustainable development. In principle n. 19 of the Stockholm Declaration (1972), it is stated that: "Education in environmental matters, for the younger generation as well as adults, giving due consideration to the underprivileged, is essential in order to broaden the basis for an enlightened opinion and responsible conduct by individuals, enterprises and communities in protecting and improving the environment in its full human dimension." (United Nations 1972). In 1977, following the work of the Stockholm Conference, an international conference on environmental education took place in Tbilisi. The Conference produced the Tbilisi Declaration, the first international declaration concerning the role of education in preserving the environment. In this document, for the first time were defined goals, objectives, and guiding principles for education as a strategy for the changing of lifestyles and improving the protection of the environment. According to the Tbilisi Declaration, environmental education should consider the environment in its totality; be a continuous lifelong process; have an interdisciplinary approach and examine major environmental issues from local, national, regional, and international point of view.

Also the goal of environmental education is the realization of:

- "Awareness: to help social groups and individuals to acquire an awareness and sensitivity to the total environment and its allied problems;
- Knowledge: to help social groups and individuals to gain a variety of experience in, and acquire a basic understanding of, the environment and its associated problems;
- Attitudes: to help social groups and individuals to acquire a set of values and feelings of concern for the environment and the motivation for actively participating in environmental improvement and protection;

- Skills: to help social groups and individuals to acquire the skills for identifying and solving environmental problems;
- Participation: to provide social groups and individuals with an opportunity to be actively involved at all levels in working toward resolution of environmental problems." (United Nations 1977).

In the history of HESD the contribution of the Tbilisi Conference was very important. The pedagogical principles developed during the conference represented, in fact, the basis for the creation of all the disciplines relating to what now is defined as education for sustainable development (ESD). Furthermore, in the Declaration, for the first time the importance of higher education for environment preservation was affirmed: "The conference (…) recommends to Member States (…) that they integrate environmental education into their general policy and, in accordance with their national structure, take appropriate measures in order, more especially to: (…) expand environmental studies at the higher education level." (United Nations 1977). Moreover in the years following the conference, thanks to the innovative vision offered, all over the world training within school systems have been set up, at various levels, regarding environmental education, which gave rise to new curricula and teaching materials. At the end of 1980s and at the beginning of 1990s, in correspondence with the Brundtland Commission (1986) and the United Nation Conference on Environment and Development (UNCED), the Rio Earth Summit (1992), a second important period for the development of ESD, took place. During the Rio Summit for the first time the term "education for sustainable development" was used. Furthermore in the document Agenda 21 education is recognized as a strategic tool to achieve sustainable development. In Agenda 21, in Chap. 36, entitled "Education, Training and Public Awareness", indeed, are recovered and expanded concepts of the Tbilisi Declaration in the perspective of sustainable development: "education is critical for promoting sustainable development and improving the capacity of the people to address environment and development issues." (United Nations 1993). In the same chapter there are also specific references to Higher Education and its role in Sustainable Development: "Countries, assisted by international organizations, non- governmental organizations and other sectors, could strengthen or establish national or regional centres of excellence in interdisciplinary research and education in environmental and developmental sciences, law, and the management of specific environmental problems. Such centres could be universities or existing networks in each country or region, promoting cooperative research and information sharing and dissemination." and "Countries could support university and other tertiary activities and networks for environmental and development education. Existing regional networks and activities and national university actions which promote research and common teaching approaches on sustainable development should be built upon, and new partnerships and bridges created with the business and other independent sectors, as well as with all countries for technology, know-how, and knowledge exchange." (United Nations 1993). In these statements we can find the most important topics about ESD, that are still at the center of international debate: the importance of scientific research on issues relating to sustainable development, the need to involve in the ESD all the stakeholders of the society and the need of an interdisciplinary approach in the creation of academic curricula on sustainable development. After the Rio Conference and after Agenda 21, work to shape a common strategy in relation to HESD have been numerous. The United Nations Educational, Scientific and Cultural Organization (UNESCO), and the United Nation Commission on Sustainable Development (CSD) had a very important role in this process. In 1996, in particular, during the fourth session of CSD, the two organizations started together an International Work Programme on Education, Public Awareness, and Training for Sustainability, with the aim of developing and implementing the strategic program on education presented in Chap. 36 of Agenda 21. Futhermore, in 1998 the UNESCO organized the "World Conference on Higher Education" (WCHE). In the Declaration adopted after the WCHE, the "World Declaration on Higher Education for the Twenty-First Century: Vision and Action" it is stated that education is a "fundamental pillar of human rights, democracy, sustainable development and peace" and that "the core missions and values of higher education, in particular the mission to contribute to the sustainable development and improvement of society as a whole, should be preserved, reinforced and further expanded" (UNESCO 1998). After the WCHE, another fundamental moment, promoted by UN, in the development of HESD was the Johannesburg World Summit on Sustainable Development (2002). During the conference it was proposed that the Decade of Education for sustainable Development (DESD) would be throughout the years 2005–2014. The implementation of DESD was entrusted to UNESCO which developed an International Implementation Scheme for the Decade. In this programmatic document, UNESCO affirms that the goals of DESD are to:

- "Provide an opportunity for refining and promoting the vision and transition to sustainable development through all forms of education, public awareness and training;—Give an enhanced profile to the important role of education and learning in sustainable development." The objectives for the DESD are to: "—facilitate networking, linkages, exchange, and interaction among stakeholders in ESD;—foster an increased quality of teaching and learning in education for sustainable development;—help countries to make progress towards and to attain the millennium development goals through ESD efforts;—provide countries with new opportunities to incorporate ESD into education reform efforts." (UNESCO 2005). In 2003 and in 2009 there were other two World Conferences on Higher Education organized by UNESCO, in which the global higher education community could work to put into action HESD all over the world as a driver for sustainable development. After these conferences there was the third Earth Summit in Johannesburg, Rio + 20, in 2012. This was a further opportunity for participating countries to take stock of the situation on HESD, to open working groups and to promote DESD and its principles. Indeed in the declaration "The future we want. Our common vision" it is stated: "We resolve to promote education for sustainable development and to integrate sustainable development more actively into education beyond the United Nations Decade of Education for Sustainable Development." (United Nations 2012). One of the outcomes of this conference was the activation of the "Higher Education Sustainability Initiative" (HESI), which is a program that aims to create a network of educational institutions actively engaged in spreading HESD, according to the principles of the DESD and the Sustainable Development Goals (SDGs). In particular, in order to join this program they have to sign a declaration in which they agreed to: teach sustainable development concepts, ensuring that they form a part of the core curriculum across all disciplines; encourage research on sustainable development issues, improving scientific understanding through exchanges of scientific and technological knowledge; greening their campuses by reducing the environmental footprint, adopting sustainable procurement practices, adopting effective programmes for waste minimization, recycling and reuse, and encourage more sustainable lifestyles; supporting sustainability efforts in the communities in which they reside. The creation of this network could be a great opportunity. Through the active involvement of the institutions that produce education, creating a strong international movement for the dissemination of ESD can actually use education as a tool to achieve sustainable development. For these reasons HESI partners are organizing a preparatory meeting for the World Conference on Education for Sustainable Development, organized by UNESCO in Aichi-Nagoya, Japan, on 10–12th November 2014. Their aim is to take the conference as an occasion to analyze the results obtained, to create new partnerships, and to set new objectives.

2 The Experience of the Doctorate in "Rural Sustainable Development, Land and Environment" of the Faculty of Agriculture of the University of Perugia: Goals, Motivations and Structure of the Educational Path

The doctoral program in "Rural Sustainable Development, Land and Environment" (RSDLE) was founded in 2000 at the Faculty of Agriculture of Perugia. The entire period of activity of the doctorate, including the ongoing cycles, covers a time span of 15 years with 10 cycles activated. The motivations that led to the birth of RSDLE were the will to activate a path of ESD within the academic curricula following stimulation of the international debate, and the need to respond to the territorial design issues through the lens of sustainable development. The influence of globalization on the economic and environmental processes in the rural landscape caused a strong pressure on local systems creating the need for new professional competencies able to act in the territory and planning development in a sustainable way (Van Der Ploeg 2010). The choice of the European Union to give more attention to the policies of integrated rural development, compared to traditional agricultural ones, expressed in the Agenda 2000 document confirmed such a need, also with the affirmation of the multifunctional model for agriculture. In order to implement this strategy of development in rural territories, it was essential to rethink in an innovative way the methods of systems analysis, to be carried out in multidisciplinary form, and the decision-making procedures and management, to be carried out using the criteria of efficiency and effectiveness and making use of apropriate tools for the selection and evaluation. In this particular historical and social period, the Ph.D. in "Sustainable Rural Development, Land, and Environment" was meant to represent a useful tool for the training of researchers and managers, who have good skills to address methodological issues related to the management of the integration and sustainability of interventions in rural and environmental systems. The main goal was to get the participants to acquire scientific knowledge about sustainable development, with particular reference to the analysis, planning, and management of socio-economic space, to the environment and its resources, to soil, air, and water. Another specific objective was to form a professional role that was able to address in an autonomous way the complex and variable topics of study, and to understand and manage the processes related to sustainable development of rural systems, as well as the practices of transformation of land and environment. For these reasons in the RSDLE Ph.D., all topics were thought to be studied in depth and highly multidisciplinary, under the powers of the various scientific disciplines involved in the teaching staff (economists, agronomists, environmental engineers, hydraulic engineers, planners).

The main scientific topics of the RSDLE Ph.D. were: social-economical aspects of sustainable rural development; techniques of analysis, evaluation, and management of sustainable rural development paths; and Land use planning and environment. The topic "Social-economical aspects of sustainable rural development" was considered as fundamental because the possibility of achieving the goal of economic and environmental sustainability of a territorial system is linked to the way the relationship between the social-economical system and the natural resources present in it is defined. Regarding this, the topics studied were:

- Economic development, with particular reference to the development of local systems;
- Welfare Economics;
- Economics of the environment, with reference to the micro and the macro-economic environment;—Economic planning of land and environment;
- Management and control of land use plans and environment;
- Policies for Sustainable Rural Development.

Another essential requirement for the preparation of rural development programs is the knowledge of local systems and related resources which is fundamental for the development of physical-mathematical models. Therefore, the collection and computerization of data from different sources and their integration with other information (both qualitative and quantitative) was becoming essential. In this regard, in the macro topic "Techniques of analysis, evaluation, and management of the paths of sustainable rural development" the following were studied during the Ph.D:

- Techniques of information acquisition and return of information for the creation of databases;
- Innovative methods for assessment of land resources and environmental and sustainable land use planning, through new criteria for the analysis of land resources based on the use of Geographic Information Systems (GIS) also in the Web (WebGIS);
- Methods of spatial analysis and geospatial analysis applied to operational research;
- Decision support system on natural resources management (air, water, soil).

The development of the theme "Land use planning and environment" refers to the elaboration of strategies in the processes and practices of land transformation. The educational-scientific path, which was proposed in this regard inside the Ph.D., was based on the analysis of highly significant real situations, texts, and experiences. Beyond that, examined in depth, were the subjects:

- The design of plans, actions, and territorial programs and environment;
- The decision-making process in planning; Management of environmental quality;
- The integration of regulatory instruments, economic, and negotiating for the construction of sustainable policies;
- The sustainability of the changes in rural areas;
- Assessing the visual quality of the landscape.

3 Analysis of the Effectiveness of RSDLE Doctorate in the Creation of Human Capital for Sustainable Development

The experience RSDLE doctorate had important results in multiple dimensions: the training of experts in sustainable rural development, in the production of teaching materials and expertise about ESD, in the production of scientific research, and the self-training of professors in a perspective of life-long learning.

The data reported below were collected by a direct interview of the Doctoral Board, which is tasked with gathering information concerning the results of the doctorate.

There were 21 professors involved, 19 of which belonging to the Departments of Man and Territory, Civil and Environmental Engineering (section of agricultural hydraulics and forestry), Agricultural Economics, Farm Appraisal, and Food Sciences of the University of Perugia, and the remaining 2 belonging to the Italian Ministry of the Environment. There were 52 students enrolled in the doctorate during the 10 cycles, of which 41 completed the course, 5 abandoned it and 6 are ongoing. The 41 doctoral theses' produced focused on various topics: analysis of sustainable development projects at an international and local level, the multi-functionality of agriculture, the preservation of the rural landscape, the protection of soil and water resources, the role of organic agriculture in sustainable rural development, the supply chain of sustainable agro-energy, the multi-criteria analysis applied to rural development, the construction of models of land analysis using GIS, just to give some examples. In the same subjects in the last 5 cycles of the Ph. D., 45 scientific papers were produced and published by graduate students. Whereas in these cycles, 18 people have graduated which means that each student during the doctoral has published an average of 2.5 scientific works, which is 0.83 each year. In addition, as doctoral students are unlikely to publish this data during their 1 year, this was considered a good result by the Doctoral Board.

All the doctoral students who had completed the course until now (until the 9th cycle), 6 months after the acquisition of the Ph.D., were in an employment contract or a job. At the present time, 17 % of them are employed permanently in Italian Universities, 29 % have a fixed-term contract at the Italian University (departments, spin-offs, university bodies), 27 % in Public-authority (FAO, European Union, Region, Municipality, Forestry authority, Authority for the environment), 10 % in Private authority (Firm, NGOs) and 17 % are Professional Agronomists (Fig. 1).

In addition to these results, also thanks to the experience of RSDLE, in September 2011 an International Summer School entitled "The Sustainable Management and Promotion of the Territory" (SMPT) was organized. The Summer School was promoted by the Department of Agricultural Economics, Farm Appraisal, and Food Sciences (DSEEA), one of the departments involved was also in RSDLE, with the partnership of the State Technical Agricultural College A. Ciuffelli of Todi, The Ciuffelli Alumni Association, and the Todi Municipality. Now in its fourth edition, the SMPT Summer School represents an important success in HESD, for its capacity to create a link between the local and the international level, with participants from over 20 countries all over the world. Another positive externality of

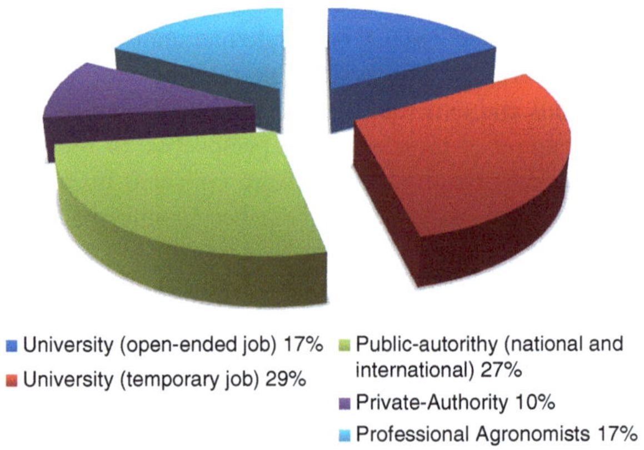

Fig. 1 Workplaces of the total employed at present time of the Ph.D. student (until the 9th cycle)

RSDLE was also the activation, in 2011, at DSSEA of the spin-off CARE (by the acronym in Italian of Environmental Preservation Economic Reinforcement), which is specialized in environmental subjects. It is composed by university professors and technical staff, professionals, graduated experts, and a nonprofit association, the Centre for the rural and agricultural development (Ce.S.A.R.). Its principal aim is to support private firms and public authority in achieving sustainable development. In particular, CARE helps firms to comply with environmental laws and promoting the culture of respect of the environment as an opportunity rather than as a constraint. Moreover, CARE supports public bodies (ministries, regional, provincial, and local administrations) in handling territorial policies for sustainable development. Some of the services offered to firms are: environmental certification; environmental management systems; environmental impacts evaluation; integrated environmental authorization; environmental monitoring plans, and legal assistance for any kind of controversy in environmental subjects. Some of the specific services for public bodies are: support for normative and planning fulfillments, concerning environment and territory; support in implementing environmental and sustainable development politics; assistance in implementing environmental certification systems for public bodies, and territorial environmental certification; assistance in implementing and managing local Agenda 21 programs; strategic environmental evaluation of politics and programs, and planning and development of models for sustainable development of territories. In its 3 years of activity, CARE also created an original tool, the environmental network "EnviNet" (http://envinet.ning.com), which allows dissemination and information activities. With almost 750 users, "EnviNet" is becoming an important virtual space where events, experiences, and solutions for sustainability are shared. Its potential in the dissemination of sustainable development has been recognized and awarded in 2012, the year in which "EnviNet" won the 2012 National Prize "Seven Green Awards", organized by the "Corriere della Sera" newspaper in the category of Culture.

4 Looking Forward: Doctorate in "Energy and Sustainable Development"

In 2010, Italy was made a reform, which came into force in 2013 and has affected the entire University system. This reform led to a spending review, and as a result many doctorates were combined. The University of Perugia has tried to take this reform an opportunity to build interdisciplinary doctoral which is able to respond to current social, economic and environmental issues with a broader vision. In this regard it was decided to merge the RDSLE Ph.D. and the Ph.D. in "Energetic Engineering", giving rise to the Ph.D. "Energy and Sustainable Development" (ESD Ph.D.). The first cycle started in November 2013. The aim was to create a multi-disciplinary approach able to deal with Sustainable Development more in line with the holistic vision that has always distinguished HESD on international debate. The ESD Ph.D. is oriented on sustainable development of territory and energy systems and is based on the integration of engineering, agricultural, and medical skills. The areas of education and research are: the engineering expertise for the development of technologies in the service of traditional and renewable energy sectors, the energy saving technologies in the agri-food and agricultural buildings, land management and monitoring and analysis tools of the same, and the environmental pollution analysis in terms of health hazard assessment in connection with the energy technologies studied. The doctoral program aims to train researchers and specialists in energy planning and the government and the promotion of the territory in all fields (energy, economic, social, cultural, environmental, services) and in terms of fertile combination between tradition and technological innovation of process and product. The Ph.D. program is organized in two curricula: curriculum in "Energy and its effects on human health" and curriculum in "Energy, agricultural and territorial resources". The curriculum in "Energy and its effects on human health" aims at training professionals geared up for the challenge of sustainable development, and the experts on the impact of energy systems on health and environment in the field of energy efficiency and of technologies for the use of conventional and renewable sources. The curriculum in "Energy, agricultural and territorial resources" has the goal to create professional figures for the sustainable management and promotion of territory, in compliance with the objectives expressed in the documents "The Future we want", the Strategy of Sustainable development Goals 2015–2030, and the Framework Programme Horizon 2020, with particular attention to strategic and sustainable energy planning. The Ph.D. students admitted to the 1styear of Doctorate Course (Cycle XXIX) were 12:6 in the curriculum "Energy and its effects on human health" and 6 in "Energy, agricultural and territorial resources".

5 Conclusions

The experience of the Ph.D. RDSLE was significant for the spread of HESD at the University of Perugia and consequently in all Umbrian territory. In particular, it had numerous positive impacts on multiple levels: it initiated, for the first time in Umbria, a Ph.D. degree based on the HESD principles, creating educational materials and research opportunities; formed a considerable number of experts in sustainable development of the territory, which subsequently applied the knowledge acquired in a local and international level; it matured the knowledge and the awareness among professors, encouraging them to integrate sustainability into their teaching programs and in their fields of research; it was fundamental in the creation of the new Ph.D. "Energy and Sustainable Development" in which the agricultural, engineering, and medical areas of research were merged; it was the basis for the establishment of an international summer school and enabled the creation of a spin-off which deals with environment and sustainability. The relevance of the result confirms the ability of HESD in creating positive dynamics in multiple dimensions, which can have a strong influence in the effective implementation of sustainable development both on a regional and international level.

References

Ciani A (2012) The sustainable management and promotion of territory: a strategic operative education plan and training as a result of collaboration between Perugia University, Todi's State Technical Agricultural College and Local Municipality, in "Sustainable Development at Universities: New Horizonns" Peter Lang Internationaler Verlang der Wissenschaften, pp 235–246

UNESCO (1998) World declaration on higher education for the twenty-first century: vision and action and framework for priority action for change and development in higher education. United Nations educational, scientific and cultural organization education sector, Parigi, p 138

UNESCO (2005) United Nations decade of education for sustainable development (2005–2014): international implementation scheme. United Nations educational, scientific and cultural organization education sector, Paris, ED/DESD/2005/PI/01, p 31

United Nations (1972) The Stockolm declaration on human environment. Harward Int Law J 14:423–515

United Nations (1977) The Tblisi declaration. United Nations educational, scientific and cultural organization education sector, Paris, ED/MD/49, p 96

United Nations (1993) In: Daniel S (ed) Agenda 21: the earth summit strategy to save our planet. EarthPress, Boulder, p 351

United Nations (2012) Report of the United Nation conference on sustainable development, resolution 1: the future we want. United Nation, New York, pp 1–53

Van Der Ploeg JD (2010) The peasantries of the twenty-first century: the commoditisation debate revisited. J Peasant Stud 37(1):1–30

Author Biographies

Antonio Boggia is Associate Professor of Environmental Appraisal at the University of Perugia. His scientific activity is focused on environmental evaluation and assessment and on economics and management of natural resources. He is specialized in: *environmental assessment and management in agricultural development* and in *GIS and computer-aided support systems in environmental assessment and management*, Wye College, University of London; *Euro Course in Environmental Impact Assessment*, EU, JRC Ispra (VA). He is member of research groups both at national and international level. Author of more than 100 scientific publications, mostly on the above mentioned topics. He was aVisiting professor at the University of Minnesota (USA). He is member of: SIDEA—Italian Society of Agricultural Economics; CeSET—Appraisal and Territorial Economy Research Center; MCDM—'International Society on Multiple Criteria Decision Making, IAIA—International Association for Impact Assessment.

Cinzia Buratti is Associate Professor of Applied Thermodynamics and Heat Transfer and HVAC Plants, from 2005, at the Department of Engineering, University of Perugia (Italy), Director of Biomass Research Centre, Coordinator of Ph.D. Course in Energy and Sustainable Development, and Delegate for Umbria Section of AIDI (Italian Illuminating Association). Her current research is in: Transparent insulation materials (aerogel) foe energy saving in buildings; Indoor thermal comfort: instrumental and adaptive approach; Biomass and bioenergy; Environmental Impact and Life Cycle Assessment (LCA) of energy chains; Heat Transfer (thermal resistance of building elements, thermal expansion of materials); Acoustics (Environmental acoustics, measurements of sound insulation in buildings and of buildings elements); Natural lighting; Thermo igrometrical conditions in museums and in presence of works of art. She has written some books and over 150 publications about thermal comfort, fluids for refrigerating absorption machines, energy, biomass and bioenergy, acoustics, thermophysical properties of fluids, heat transfer, indoor lighting. Activity of reviewer for several international Journals and Conferences. Applied Research activities for: ANPA (National Agency for Environment Protection), Italian Ministry of Environment, Italian Ministry of Agriculture, ENEL (Italian Society for Electrical Energy), IKEA Italy, Società Autostrade S.p.A., Municipality of Rome; Municipality of Perugia; Municipality of Terni; Superintendence for Archaeology of Rome; Superintendence for Works of Art of Umbria; Biennale of Venice.

Adriano Ciani is Full Professor of Agricultural Economics and Rural Appraisal at Department of Agricultural , Environment and Food Sciences-DSA3at Perugia University. He is also Professor of Environmental Economics and Farm Management. He speak 6 foreign languages more Italian mother language. His current research is in: Sustainable Management of Territory, Common Goods, Water Resources, ICT and Sustainable Rural Development. He has written several books and 235 other publications. He is Director of the SIS-SMPT International Summer School in Sustainable Management and Promotion of Territory. He has taken part in collaborative research, teaching, and technical and dissemination activities in over 50 countries. He founded the BIOSPHERA-Scientific and Cultural Association for Sustainable Development in 1997. In 2000, organized the First World Forum on Rural Tourism and Agro-tourism. He also founded the International Association of Experts in Rural Tourism and Agro-tourism (IAERT). From 2001 to 2009 he was commissioned by the MoFA as Scientific Attaché at the Italian Embassy Tirana, where he promoted the wide scientific program "A Sustainable Development for Albania". In the years 1980–1990 he was Minister to the Provincial Government of Perugia, to SASE Lt. Airport (Perugia), Vice-President of the Umbria-Tuscany Irrigation Authority, and Auditor of ERI New Editions. He is an honorary member of the Academic Senates of the seven universities of Romania. He has been appointed of "Symbol of the City of Durres" and "Honour of the City" in the Girocastro Region in Albania. In October 2014 he received by the Agricultural University and Veterinary Medicine the "Laurea Honoris Causa" in Economic Engineering for Agriculture.

Stefano Casadei is an Associate Professor of Agricultural Hydraulics at the University of Perugia. He has taught in the following academic courses: Water management and irrigation systems, advanced degree of Science in Sustainable Agriculture; Technology Design, advanced degree of Science in Sustainable Rural Development. He is the Scientific Coordinator at Ph.D. School Sustainable Rural Development, Land and Environment and the chief executive of the Academic Spin-Off T4E Srl (One Technology Four Elements).

His current research activity is on the following subjects: Water Resources Management, Decision Support System Model, management of spatial data with particular reference to the field of surface water resources. He is the Author and co-author of about 90 scientific papers on these topics. During his career was the Coordinator in the following collaborative research projects: Integrated Special Fund for Research, "Integrated model for the evaluation of climate change impact in natural and agricultural ecosystems in the Mediterranean area (MICENA)"; Por-Fesr Project—Umbria Region, "Drought and Climate Change (SECLI); Tiber River Basin Authority, "Convention for the development of a simulation software of the surface river network"; Tiber River Basin Authority, "Development of scenarios for the water management in the Umbro-Tuscan system, Water Balance of the system".

Giordano Stella is a Ph.D Student in "Energy and Sustainable Development" of the Biomass Research Centre of the University of Perugia. His principal interests are: food security, sustainable rural development, organic farming, Ecological Economics, alternative social and economic models. He is particularly specialized in food security and about this topic he was the co-author of the article "Revisiting the food component of the ecological footprint indicator for autonomous rural settlement models in Central Italy" published in the international journal "Ecological Indicator". Actually he is developing a Decision Support System for Food Security.

The Student Voice: Experiences of Student Engagement in Education for Sustainable Development

Rachel Drayson and Charlotte Taylor

Abstract

Significant policy level action aimed at driving implementation of education for sustainable development (ESD) across Higher Education Institutions is currently taking place in the UK. Taking a student focused perspective; this paper looks at student attitudes, skills and experiences of sustainable development 'on the ground'. The paper presents findings from longitudinal research (2010–2014) conducted by the National Union of Students (NUS) on behalf of the Higher Education Academy designed to ascertain student understanding of sustainable development and their experiences of ESD in teaching and learning. The research also assessed the link between skills for sustainable development, employability and employment. The longitudinal nature of the research provides an insight into the changes that occur throughout students' university careers and how attitudes have changed over time. These findings are coupled with an insight into the approaches adopted by NUS and the wider student movement in this field. Approaches across the formal and informal curriculum are presented as case studies in student engagement with sustainable development. This paper will be useful to anyone interested in understanding more about the student voice in relation to sustainable development, and learning about some of the practical steps that have been taken to engage the student body.

R. Drayson · C. Taylor (✉)
National Union of Students, Macadam House, 275 Gray's Inn Road, London, WC1X 8QB, UK
e-mail: charlotte.taylor@nus.org.uk

R. Drayson
e-mail: Rachel.drayson@nus.org.uk

W. Leal Filho et al. (eds.), *Integrative Approaches to Sustainable Development at University Level*, World Sustainability Series, DOI 10.1007/978-3-319-10690-8_43

Keywords
Education · Sustainable development · Research · Students · Skills · Employability

1 Introduction and Context to the Research

The UK continues to face a continued period of economic difficulty and uncertainty. Associated with this have been high levels of unemployment, particularly amongst 16–24 year olds in the UK with numbers topping 1 million in November 2011 (The Guardian 2011). Looking in more detail at graduate unemployment, research has found that graduate full-time (FT) employment has fallen continuously between 2002 and 2010, from 57 % of graduates in 2002 to 51 % of graduates in 2010 (Mehta and Rutt 2012).

Therefore a need exists to ensure UK graduates are fully equipped with the skills desired and valued by their future employers. Moving towards a 'green economy' has frequently been put forward as a major part of the solution to the current situation by the UK government meaning that the field of education for sustainable development bears increasing relevance. Recognition that all businesses will need to develop resilience to climate change and to use natural resources efficiently, and that a workforce with appropriate skills is required is increasing (BIS 2011a). This is matched by an expectation amongst the vast majority of recent graduates (96 %, Sky 2011) that they will be involved in sustainability in some way during their careers.

NUS and the HEA have conducted research over three academic years with the aim of improving the limited base of research on ESD beyond the setting of further education.[1] The research aimed to better understand student attitudes to and aspirations towards developing skills in SD both within and beyond higher education as well as understanding the policy context within which higher education institutions (HEIs) are operating. The longitudinal element to the research aims both to further understanding, and build a clear picture of how attitudes may change throughout university careers as well as tacking any changes in attitudes and experiences of SD amongst students entering their first year of study.

Auxiliary influences on the research have included the vote by the UK government, in December 2010, to increase student tuition fees to a maximum of £9,000 per annum and to increase the basic threshold fees to £6,000 up from £3,290 per annum. The increase in fees came into force during the 2012–2013 academic year and is hypothesised to have the potential to become a shaping influence on the "student voice". Therefore a clear and detailed understanding of student

[1] Research reports on each year of research are available at: http://www.heacademy.ac.uk/assets/documents/sustainability/firstyearattitiudes_finalreport.pdf and http://www.heacademy.ac.uk/assets/documents/esd/Student_attitudes_towards_and_skills_for_sustainable_development.pdf.

expectations is necessary to ensure graduates leave university with optimal skills from the perspective of students, employers and policy-makers.

Taking place throughout the UN Decade of Education for Sustainable Development, this research has presented a timely opportunity to take stock of current practices and identify where attention might best be focused going forward. This, in turn, has focused the work of NUS and its member students' unions on the area of Education for Sustainable Development, and some of these techniques and approaches will be touched upon in this paper.

The overarching objectives of the research can therefore be summarised as follows:

- Current attitudes towards and understanding of sustainable development (SD) amongst students
- First year student attitudes towards and skills in SD over time
- Attitudes towards and skills in SD as students progress their university career
- The impact of changes in the HE landscape in the UK with the introduction of increased tuition fees for 2012 first year students by tracking first year responses from 2010 to 2012

2 Research Methodology

A two-phase methodology was conducted and repeated over all 3 years of the research; a desk-based research period reviewed existing policy-based research and existing research into student attitudes and skills feeding into empirical research, which took the form of a national online survey. Further information on the research methodology can be found in the full research report.

Skills were identified within the desk research to be key indicators of the core skills in sustainable development: living within environmental limits; ensuring a strong, healthy and just society; achieving a sustainable economy; promoting good governance; and using sound science responsibly. The skills were defined as:

- Understanding people's relationship to nature
- Analysing using many subjects
- Act as a responsible citizen globally and locally
- Plan for the long term as well as the short term
- Use resources efficiently
- Think of the whole system and the links when considering new ideas
- Adapt to new situations
- Consider the ethical issues of your subject
- Problem solving using many subjects (introduced in 2011/12 and 2012/13)

Table 1 Guide to graphical representations

Year of research[2]	Survey audience and sample
2010–2011	Survey data from:
	1st years (n = 5,763)
2011–2012[3]	Survey data from:
	1st years (n = 1,552)
	2nd years (n = 1,641)
2012–2013	Survey data from:
	1st years (n = 4,099)
	3rd years (n = 2,657)

Table 1 outlines the audiences for the research across the 3 years of research and the samples achieved in each case. In each year, respondents were sent an invitation to complete the survey in subsequent years.

This paper will look in detail at the following areas of the research results, mentioning some of the work that NUS and members of the student movement have begun as a result:

- Current understanding of SD amongst HE students in the UK
- Student desire for ESD
- Use of and attitudes towards the skills of sustainable development
- The links between employability, future graduate employment and SD

3 Current Understanding of SD Amongst HE Students in the UK

Respondents were asked to define sustainable development in their own words, without prompting. They were not provided with a definition of sustainability until the end of the survey in order to avoid influencing respondents and ensure responses were based on their own understanding of the environmental and social skills associated with sustainable development. Figure 1 shows a word cloud of responses.

All 3 years of research found the focus of definitions provided by respondents to be in line with the concept of sustainability as conveyed within the Brundtland definition. The environmental dimension of this definition was found to be more prominent in respondent understanding, as shown by Fig. 4. This suggests a potential need to review the teaching provided to ensure the full capacity of SD is being covered. Here it is worth elaborating on the methodology of the research—within the eight skills used to describe sustainable behaviours, one specifically focussed on the environment and another on ethics. The remaining six were deliberately broad in order to reduce the risk of leading respondents into associating

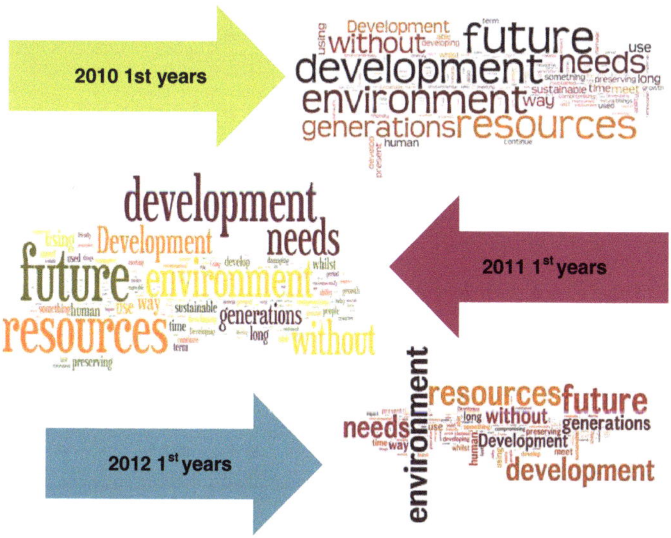

Fig. 1 What do you understand the term 'sustainable development' to mean

the survey with the environment. Only two standalone questions within the survey directly addressed the environment. In order not to bias the bulk of the survey, the definition of sustainability was taken towards the end of the survey to allow analysis of the different understandings of the term. With student respondents placing weight upon understanding sustainability within a narrow spectrum, associated with the environmental needs of today rather than the wider global needs of tomorrow, standalone research of the definition of sustainability is needed to ensure no positioning effect is in place (Fig. 2).

Further in depth exploration into student understanding and definition of sustainable development is needed to unpack the focus on the environmental elements uncovered in this research.

4 Student Desire for ESD

Despite these positive responses, the responses to the following question demonstrate the further scope for incorporation of sustainability within university practices and course leadership as three fifths of respondents in 2012 express a desire to learn more about sustainable development (59.6 % n = 2,362 2012 1st years, and 60.7 %, n = 1,519 2012 3rd years). There continues to be no significant change in desire to learn more about sustainable development between year groups which suggests that the capacity for increased coverage remains despite entering their final year at university. Student understanding of sustainability is key to comprehending where coverage can be improved.

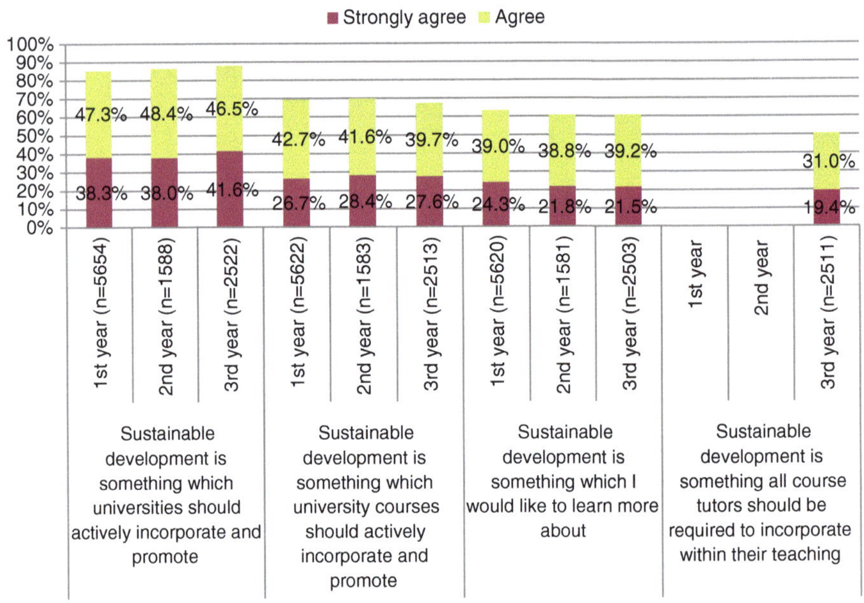

Fig. 2 To what extent, if at all, do you agree with the following statements

In terms of 1st year responses, 2012 1st years are significantly less likely to agree that sustainable development is something they would like to learn more about than their 2010 counterparts (59.6 %, n = 2,362 2012 1st years compared to 63.3 %, n = 3,557 2010 1st years). As before, further investigation would be required to assess whether this decrease is due to improved teaching and learning, or a reduction in engagement amongst 2012 1st years.

Within the same question, respondents confirmed their desire for sustainable development to be actively promoted and incorporated by their university with 2012 3rd years significantly more likely to agree with the statement than 2010 1st years (88.1 %, n = 2,222 compared to 85.6 %, n = 4,840). This finding corroborates the expressed views on relevance and importance for their time in higher education earlier within this report.

An additional statement introduced to the 2012 questioning also revealed a desire for an obligation to be set for course tutors to provide teaching and learning on sustainable development, with approximately 50 % of respondents agreeing with the proposition (49.4 %, n = 1,972 1st years 2012 and 50.4 %, n = 1,266 3rd years 2012 strongly agree or agree).

5 Use of, and Attitudes Towards, the Skills for Sustainable Development

With the majority of respondents across all years of research reporting at least partial coverage of SD skills during FE it calls into question the extent to which this learning is both useful and transferred to the HE setting.

According to the research, relatively low numbers of respondents cite conducting sustainability skills 'all the time' and the difference between year groups for each skill is relatively minor although there is evidence of an increase in the practicing of skills as respondents progress through their university careers (see Fig. 3). For example, 2nd years in 2011 report a significant increase in 'analysing using many subjects' 'all of the time' or 'most of the time' on 2010 1st years. The same significant increase applies between 2nd and 3rd years (61.5 %, n = 3,489

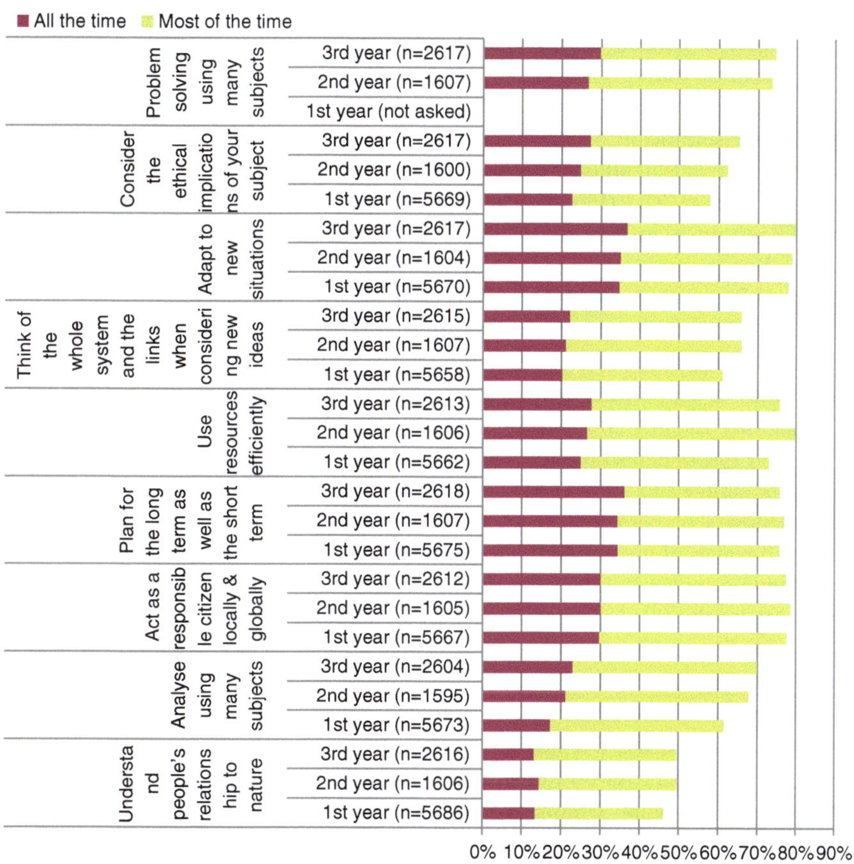

Fig. 3 To what extent, if at all, do you think that you personally carry out the following skills? (2010 cohort tracker—*1st, 2nd and 3rd years*)

2010 1st years, 67.7 %, n = 1,080 2011 2nd years and 69.9 %, n = 1,820 2012 3rd years). One skill that appears to defer from the trend is the ability to 'use resources efficiently'. The practicing of this skill to some extent showed a significant increase between 2010 1st year and 2011 2nd year (72.9 %, n = 4,128 to 79.8 %, n = 1,282) however this was reversed in 2012 with a significant decrease in the number of 3rd year respondents reporting carrying out this skill (75.7 %, n = 1,978).

In terms of ranking SD skills by the importance of inclusion in their courses, both overtly SD skills (according to respondents definitions of SD) and more generic skills have seen a significant decrease from 2010 1st year responses to 2011 1st years, and again from 2011 to 2012 1st years. For example, 26.6 % of 1st years in 2010 ranked 'understanding people's relationship to nature' as 1 or 2 in terms of importance (n = 1,510), falling to 23.5 % (n = 334) in 2011 and finally to 20.6 % (n = 822) in 2012. The ability to 'act as a responsible citizen globally and locally' has also seen decreases amongst the first year intakes with just 30.5 % (n = 1,223) ranking this skill as very important compared to 33 % (n = 483) in 2011 and 43.4 % (n = 2,446) in 2010. Respondents placed high value on many aspects of sustainable development in terms of the associated skills increasing their ability to perform well in their courses during HE.

As respondents have progressed throughout their university careers there has also been a significant decrease in the ranking of skills as 'most important', for example there is a significant decrease between 1st year responses in 2010 and 2nd year responses in 2011 (Fig. 4). The decrease is also repeated between the latter and 3rd year responses in 2012 for the ability to act as a responsible citizen globally and locally. For example only 27.9 % of 3rd year respondents in 2012 compared to 31.1 % of 2nd years in 2011 (n = 474) and 43.4 % of 1st years in 2010 (n = 2,446).

Despite university choice being influenced more greatly by the traditional indicators of quality, such as reputation and teaching, there is a clear demand for universities to support the development of the wider suite of skills associated with sustainability literacy and thereby improving employability. Approximately 70 % of 1st year respondents continue to agree (strongly agree or agree) that universities should be obliged to include SD skills development as part of their courses.

Perceptions of performance of clubs and societies remains consistently lower across all waves of the research (the option was included in 2011 and 2012 surveys. This finding suggests that the informal learning opportunities, identified as important by the 2011 desk review, may be currently under-used and has implications for the preference for extra-curricular learning activities described later on in this report.

In the past 8 years NUS' work to provide informal learning opportunities for SD on campus' across the UK has grown exponentially, with NUS programmes currently being run on over 80 further and higher education campuses across the UK.[2] The work of Students' Unions has also grown within this time; notably seeing a

[2] For more information on NUS programmes: http://www.nus.org.uk/greener.

[3] For more information on Students' Green Fund: http://www.studentsgreenfund.org.uk.

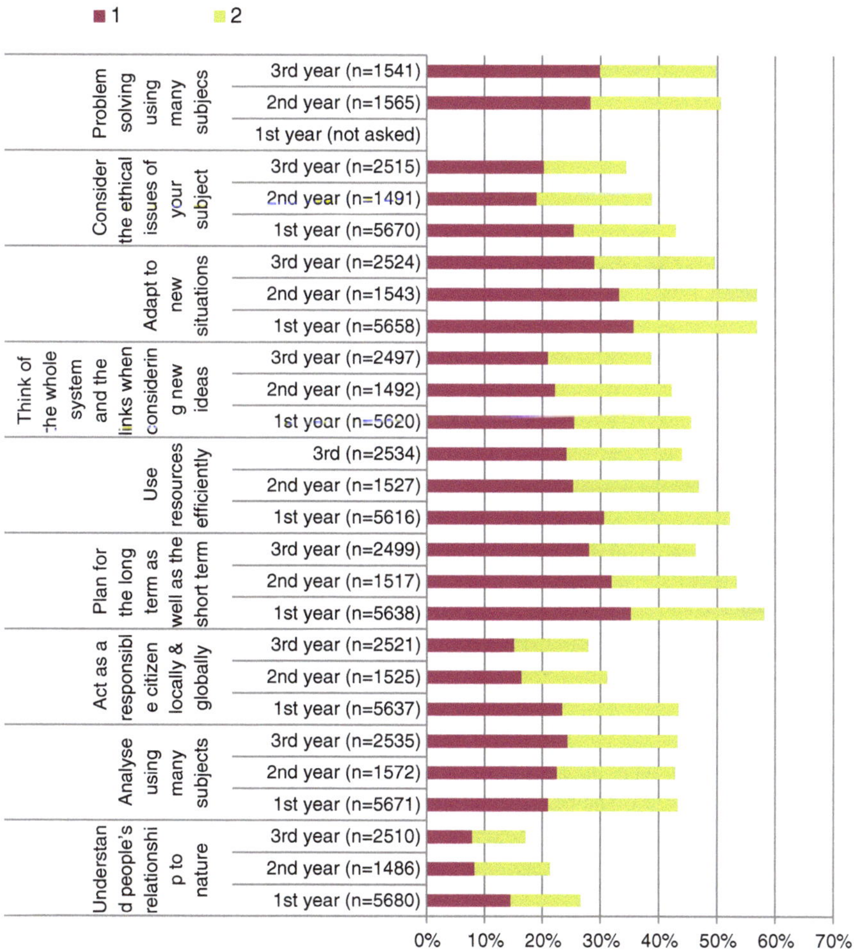

Fig. 4 Please rank the following skills in terms of their importance in being included in your course for a graduate in your field, where *1* is the most important and 8 is the least important (2010 cohort tracker—*1st, 2nd and 3rd years*)

step change in 2013 with NUS' Students' Green Fund.[3] This was a £5.2million fund given to NUS by the Higher Education Funding Council for England (HEFCE) to distribute to 25 students' unions so that they can run two-year sustainability projects with a budget in the region of £300,000 with four key focuses: impact, partnership, legacy and engagement. One key benefit resulting from the funding is that over 60 new jobs have been created in students' unions to deliver projects that provide informal learning opportunities for SD on 25 campuses.

[3] For more information on Students' Green Fund: http://www.studentsgreenfund.org.uk.

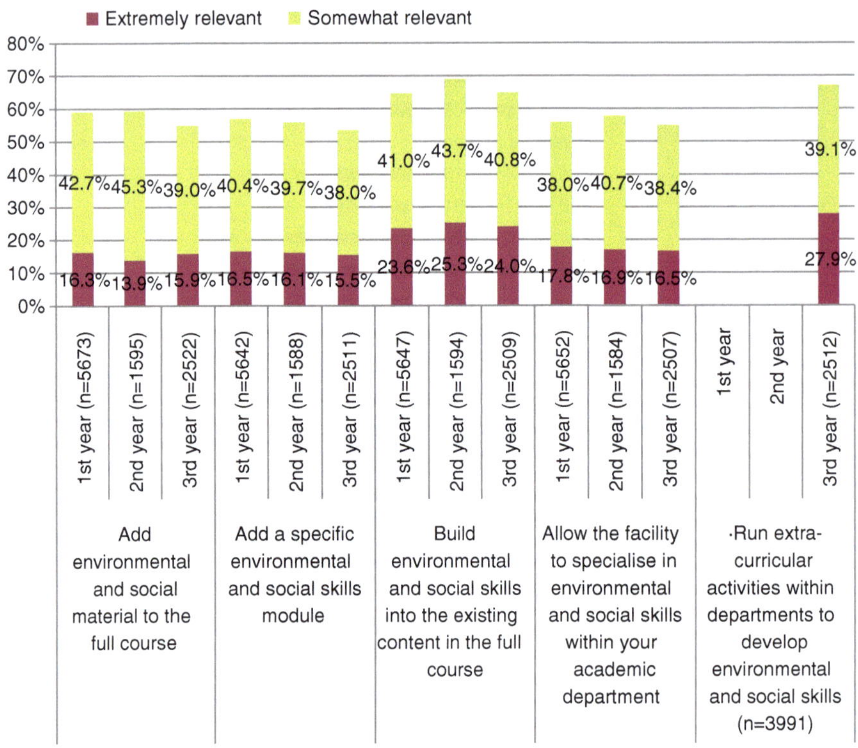

Fig. 5 What do you think the most relevant way of including social and environmental skills within your own course would be? (2010 cohort tracker—*1st, 2nd and 3rd years*)

Indirect results include students' unions that applied for funding but were unsuccessful having their ideas match-funded, or sustainability staff employed to increase their potential to receive funding in the future, and students' unions increasing their sustainability work to match the experiences offered by other unions.

In terms of the formal curriculum, respondents have indicated across all 3 years of research that they are amenable to the inclusion of skills for sustainable development into their courses, with a preference for sustainability content to be built into existing content rather than creating additional or specialised sustainability modules or courses (Fig. 5).

This preference for building environmental and social skills into existing content changes as respondents' progress through their academic careers. There is a significant increase in preference for this method between first and second year, and then a decrease between second and third year (64.6 %, n = 3,648 2010 1st year, 69 %, n = 1,100 2011 2nd year and 64.8 %, n = 1,626 2012 3rd year). There is also a significant decrease in preference for adding environmental and social material to the full course amongst 3rd years in 2012, when compared to 1st year responses in 2010 and 2nd year responses in 2011 (54.9 %, n = 1,385, compared to 59 %, n = 3,347 and 59.2 %, n = 944 respectively). One hypothesis for this change, which requires further investigation, is the potential influence of existing course commitments for third year students. Many may feel they don't have time within their schedules to fit in extra teaching and learning on sustainable development. This may also explain the preference for sustainability skills to be developed as extra-curricular activities, allowing students to participate as and when they are able (for example 69 %, n = 2,754 2012 1st years and 67 %, n = 1,683 2012 3rd years indicate extra-curricular activities within departments as 'extremely' or 'somewhat' relevant).

When unpacking the preferences for SD skills in teaching and learning further and linking to employability, there is a clear preference amongst 2012 1st and 3rd years for employability skills to be targeted at chosen subject areas rather than developing transferable skills, such as environmental and social skills (for example 71.1 %, n = 1,804 2012 3rd years would prefer an module developing skills in their chosen subject area). This shows the importance of understanding student definitions of sustainability (environmental and social skills) and also of enhancing precise understanding of the skills needs and desires of employers by sector to ensure graduates with the necessary attributes.

6 The Links Between Employability, Future Graduate Employment and SD

Skills in sustainable development are expected to be overwhelmingly important for employment in the future: between 80 % and 90 % of 2012 3rd years rank the majority of skills as important or very important in terms of employment (Fig. 6). Again the exception is the ability to 'understand people's relationship to nature' which is seen as less important by 3rd years, continuing a trend seen in their 2nd year (2011) and 1st year (2010) responses (55.9 %, n = 1,459, compared to 60.1 %, n = 960 and 61.1 %, n = 3,459 respectively).

Analysis of changes throughout their time spent in higher education highlighted that the perceived importance for employers of the ability to 'analyse using many subjects' has significantly increased between 1st and 2nd year responses (86.3 %, n = 4,874 to 90.4 %, n = 1,440). The explicit SD skills however, such as 'understanding people's relationship with nature' decrease significantly in terms of importance for employers between 1st year in 2010 (61.1 %, n = 3,459) and 3rd years in 2012 (55.9 %, n = 1,459).

Comparing the longitudinal first year responses, only 'acting as a responsible citizen locally and globally' showed any significant difference between 2010 and 2011 first year cohorts with 2010 respondents seeing this as more important (85 % in 2010 (n = 5,612) compared with 82.3 % (n = 1,517) in 2011).

Respondents in 2012 continue the beginnings of a trend uncovered by the 2011 research in that the ability to use social and environmental skills is ranked relatively low in terms of importance. Figure 7 shows little change as students progress

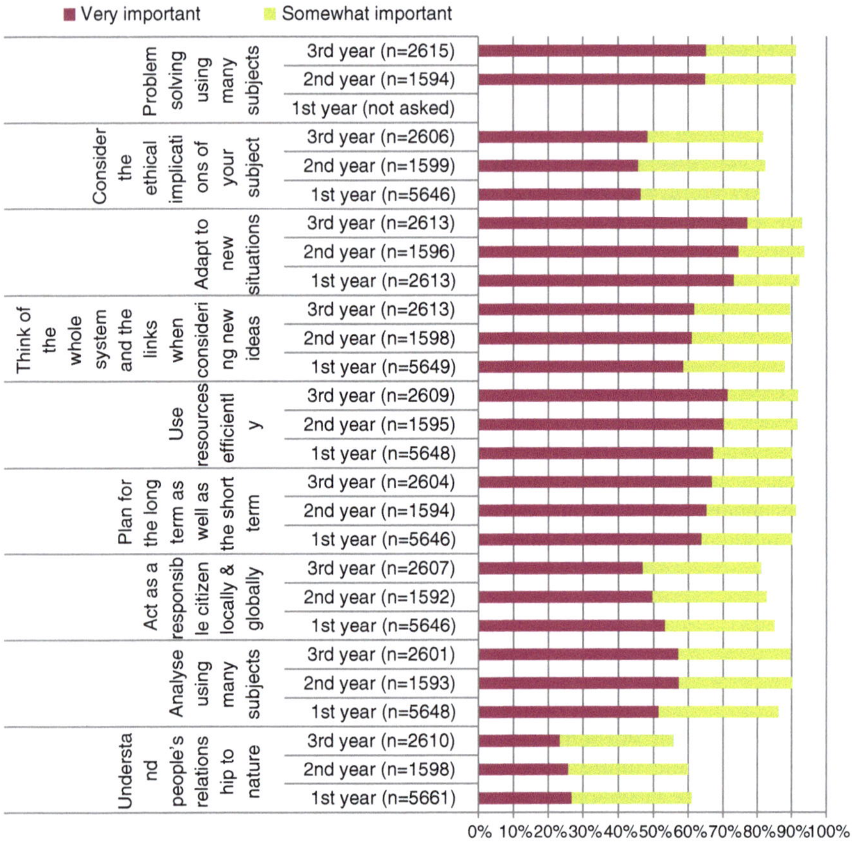

Fig. 6 How important do you think the following skills are to your future employers? (2010 cohort tracker—*1st, 2nd and 3rd years*)

through higher education with 35.1 % (n = 561) 2nd years 2011 and 34.8 % (n = 901) 3rd year 2012 respondents ranking this skillset as 1 or 2/most important.

Other key skills such as application of information technology (39.7 %, n = 637 2011 2nd years and 41.1 %, n = 1,076 2012 3rd years ranked as 1 or 2), application of numeracy (34.4 % 2011 2nd years and 2012 3rd years (n = 549 and 896 respectively) ranked as 1 or 2) and business and customer awareness (47.8 % 2011 2nd years and 2012 3rd years (n = 762 and 1,245 respectively) ranked as 1 or 2) receive relatively low ranking by students again suggesting that further work is needed to ensure student perceptions match with the reality of employer preferences and demands.

The 2012 research included a greater focus on employability, including respondent perceptions of employer demands and expectations. The majority of 3rd year respondents (89.4 % (n = 2,265) in 2012 are confident that they will be able to put into action the skills that employers are looking for upon entering the

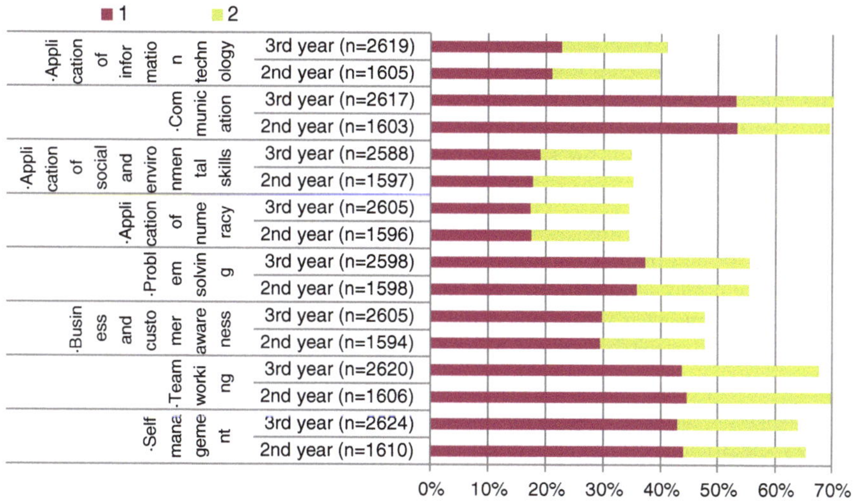

Fig. 7 How important do you think the following skills are to your future employers when compared against each other? (*1* is the most important) [2010 cohort tracker—*1st, 2nd and 3rd years*]

workplace. Whilst on the surface this is a positive result, ensuring students and HEIs have a correct understanding of the skills needed and valued by employers is key given previous reviews have highlighted a disparity between graduate evaluations of their own skills and employers. In terms of improving skills, those respondents who currently did not feel confident and able to put into action the

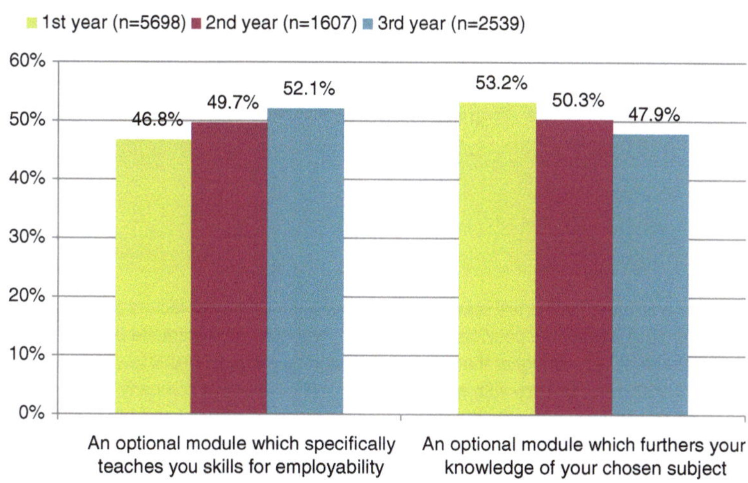

Fig. 8 We are interested in your prioritisation of social and environmental aspects in the future. Please select which option you would choose (2010 cohort tracker—*1st, 2nd and 3rd years*)

skills desired by employers believed the two main options to be completing an internship or work placement (69.1 % 2012 3rd years, n = 166) or undertaking further academic study (38.8 %, n = 93). This finding again reinforces the recommendation made by this research and elsewhere to engage with employers in skills identification and prioritisation.

Continuing the theme of learning for employability a clear trend has emerged over the three waves of research which demonstrates a shift in focus amongst respondents as they progress through their university career. Figure 8 shows that preference for an optional module which teaches specific skills for employability is significantly higher amongst 3rd year respondents in 2012 (52.1 %, n = 1,323) than amongst the same year group when surveyed in 2011 (2nd years, 49.7 % n = 799) and 2010 (1st years 46.8 %, n = 2,667).

When comparing responses from 1st year students, preference for an optional module which furthers subject knowledge of their chose subject above that which focuses specifically on employability continues from 2010 into 2011 and 2012. In each wave of the research, just over half of 1st year respondents indicated a prioritisation of furthering subject knowledge (2010 53.2 %, n = 3,030, 2011 53.1 %, n = 816, 2012 54 %, n = 2,184).

Please note that in 2010 this question was asked prior to the vote on fees in England, and the 2011 data was captured prior to the first year group to pay increased tuition fees. The 2012 data would suggest that so far, the increased tuition fees have not impacted on 1st years learning preferences however further tailored research in the future will be needed to assess the long term impacts.

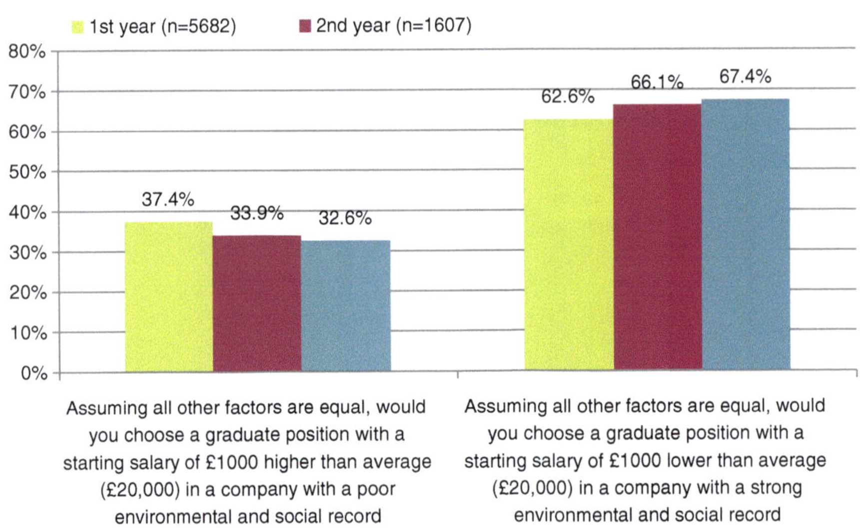

Fig. 9 We are interested in your prioritisation of social and environmental aspects in the future. Please select which option you would choose (2010 cohort tracker—*1st, 2nd and 3rd years*)

Approximately two thirds of respondents say they would sacrifice £1,000 from salary to work in a responsible company in alignment with findings from 2010 and 2011 (Fig. 9). The proportion of 2nd year students (2011 66.1 %, n = 1,062) stating they would be willing to make a salary sacrifice is significantly higher than 1st year respondents (2010 62.6 %, n = 2,557) supporting the hypothesis that employment becomes an increasing focus for students as they progress through their university career. This is continued into the 3rd year responses in 2012 with 67.4 % (n = 1,711).

In 2012 the choice of a starting salary £1000 lower than average (£20,000) in a company with a strong environmental and social record was also increasingly frequent amongst 1st year respondents (68.2 %, n = 2,761) with as significant increase on the responses from 2011 1st years (64 %, n = 983). The 2011 report put forward the hypothesis that the increase in tuition fees would have a negative influence on the number of 1st year respondents willing to accept a salary sacrifice due to increased concern over load repayments and debt. This does not seem to be apparent in the responses collected in 2012, although further investigation would be required to verify this. An alternative hypothesis for the responses seen might be the effect of continued levels of unemployment following graduation, with a third remaining unemployed within 6 months of graduation.[4]

The trend is however less pronounced when the sacrifice is increased to £3,000. At this salary point, approximately two-fifths of respondents report that they would still sacrifice £3,000 of their salary to work in an environmentally and socially responsible company. These findings add further depth to the research on 'Future Business Leaders' which revealed that whilst 41 % of respondents cited an impressive sustainability record as an important factor when looking at potential employers, 88 % would be influenced by the pay and benefits package.[5] The 1st year intake in 2012 also exhibited an increased willingness to accept a position in company with strong environmental and social performance with nearing a half of respondents opting for a salary sacrifice of £3,000 (45.8 %, n = 1,849), which continues to reflect the trend seen in the previous research with 2011 1st years showing a significant increase on 2010 1st years (42 % n = 1,847 compared to 38.7 % n = 2,198). It is possible that these results are a reflection of an increasingly altruistic population however it is also possible to hypothesise that these results reflect a concern over the possibility of securing a job upon leaving university over and above concerns about salaries. Further research would be required to unpack the finer detail of these findings.

---

[4] http://www.hecsu.ac.uk/current_projects_what_do_graduates_do.htm.

[5] Sky (2011), The Sustainable Generation: The Sky Future Leaders Study. Results based on research with 751 graduate trainees and recently graduated MBA students and potential middle managers earmarked for leadership positions.

7 General Recommendations

The findings discussed in the research cited in this paper resulted in the following recommendations for actors in the field of Education for Sustainable Development across the UK and internationally.

Educators, both of the formal and informal curriculum in HE need to address the mismatch between student definitions of SD and the definitions used by policy makers.

Resources produced to support teaching and learning for ESD should consider the full definition of sustainable development to ensure that those subjects that are not as immediately linked to sustainable development are able to integrate the subject. Again, linking with employability presents a strong potential to ensure the relevance of teaching. Previous waves of this research have also advocated the use of case studies as a means of demystifying content for those subjects who are distant from the more obvious elements of ESD.

The results presented here suggest a demand for employability, and sustainability skills, to be developed in a way that is relevant to their chosen subject area, reinforcing a need for teaching and learning on SD to be targeted rather than generalised. Nevertheless, in practise NUS notices that institutions in the UK that have been most successful in embedding ESD have done so through a variety of techniques, including through the informal curriculum, embedding within a chosen subject area, and creating a module available to the whole student body on sustainability.

Just as employers have an obligation to communicate clearly to HEIs and students the skills they are looking for, there is also a role for HEIs to assist with interpretation of business skills needed.

Student organisations have a role in vocalising demands and expectations from higher education in particular surrounding skills development. Respondents in 2012 highlighted the role extra-curricular activities could play in developing skills for sustainability. Student organisations could play a crucial role in delivering these activities in partnership with their parent institution.

Policymakers must ensure that skills for sustainable development are de-coupled from the ‘environmental’ or ‘sustainability’ sector. Skills for a Green Economy (BIS 2011b) notes that all businesses will need to develop generic or softer light green skills and there is a need to ensure that business is able to identify these skills and is able to communicate their needs to the future workforce.

Given the strong link between sustainability and employability it is important that resources and programmes for teaching and learning developed by education bodies in the arena of employability also include an element of sustainable development, and vice versa.

Employers of all sizes and sectors need to continue, and increase their engagement with HEIs to improve the communication (and identification) of skills needs expected amongst graduates. Work placements and internships were identified by survey respondents as being the most appropriate. Alongside this, professional organisations can play a role in identifying skills needs across their membership, and engaging fully with HEIs to ensure these needs are translated to students and graduates.

8 Recommendations for Further Research

This research looks at student attitudes towards and skills for sustainable development within the UK. Although it is hoped that the results can still give useful guidance internationally, it is recommended that more research is undertaken in other countries to map SD in HE. As well as giving a global picture of what can be done to embed ESD in HE, this could also aid the sharing of good practise and form even stronger international links in this area.

In the UK, evidence of the contribution of delivering education for sustainable development in further education continues, demonstrated by respondents' recall of coverage of a range of SD skills, however recall of the schemes delivering SD in FE has fallen consistently since 2010. Further research is needed to understand whether the reduction in recall of SD schemes in FE translates to a reduction in understanding and practicing of skills for SD, as well as a review of the coverage of SD within FE to ensure the full definition is being covered and not just the environmental elements reflected in respondent definitions of the term.

In continuing turbulent times for higher education and graduate employment, longitudinal work into the student demands, aspirations and considerations will be key in understanding the impacts of increased financial pressures and ensuring higher education is able to meet expectations of both students and their future employers.

References

BIS (2010) (Department for Business, Innovation and Skills) Higher ambitions: the future of universities in a knowledge economy. http://www.bis.gov.uk/assets/BISCore/corporate/docs/H/09-1447-higher-ambitions

BIS (2010) Sustainable development action plan. http://www.bis.gov.uk/assets/biscore/business-sectors/docs/09-p59-bis-sustainable-development-actionplan-2009-11

BIS (2011a) Skills for a green economy: a report on the evidence. http://www.bis.gov.uk/assets/biscore/further-education-skills/docs/s/11-1315-skills-for-a-green-economy

BIS (2011b) Higher education: students at the heart of the system. https://www.gov.uk/government/uploads/system/uploads/attachment_data/file/32409/11-944-higher-education-students-at-heart-of-system.pdf

BRASS (2010) Sustainability tool for auditing curricula in higher education (STAUNCH). http://www.brass.cf.ac.uk/projects/Rethinking_the_Future_for_Sustainability/rethinking-thefuturefor-sustainability–STAUNCH.html

Cade A (2008) Employable graduates for responsible employers. Research on the links between sustainability and employability in the graduate job market in relation to higher education teaching and learning. http://www.heacademy.ac.uk/assets/York/documents/ourwork/sustainability/EmployableGraduates2008.pdf

DCSF (2010) (Department for Children, Schools and Families) Sustainable development action plan for 2010–2012. http://www.dcsf.gov.uk/aboutus/sd/docs/SDAPthe-best-place-to-grow-up.pdf
Department for Education (2013) http://www.education.gov.uk/schools/teachingandlearning/curriculum/nationalcurriculum2014/nationalcurriculum
ECOTEC/LSC (2007) The implementation of an internal baseline survey of sustainable development across the learning and skills council (LSC). A draft report to the learning and skills council (LSC)
Forum for the future (2008) The future leaders survey 07/08. http://www.forumforthefuture.org/files/FutureLeaders0708_0.pdf
Higher Education Academy (2009) Education for sustainable development and global citizenship (ESDGC). Review of a curriculum audit in Wales. http://www.heacademy.ac.uk/assets/York/documents/aboutus/wales/ESDGC_Wales_June_2009.pdf
Martin S, Martin M, Cohen J, Correo R (2006) (for the Learning and Skills Network) Sustainable development in the learning and skills sector: a national baseline survey. https://crm.lsnlearning.org.uk/user/login.aspx?code=062427
Oakleigh Consulting Ltd, CRAC for HEFCE (2011) Increasing opportunities for high quality higher education work experience. http://www.agcas.org.uk/assets/download?file=2606&parent=1028
Institute for learning (2011) Green economy survey: a survey of teachers and trainers, members of the Institute for Learning. http://www.ifl.ac.uk/__data/assets/pdf_file/0003/24483/IfL-Green-Economy-Survey-report.pdf
Mehta P, Rutt, S (2012) (for Local Government Association) Hidden talents: a statistical review of destinations of young graduates (LGA research report), Slough: NFER. http://www.local.gov.uk/c/document_library/get_file?uuid=0eb95341-f5d8-4ee1-b09a-b988201e9513&groupId=10171
LSC and ESD Consulting Ltd (2009) Embedding sustainable development in the curriculum: guidance for staff within learning institutions on how to embed sustainability into what and how they teach. http://www.eauc.org.uk/sorted/files/embedding_sustainabilyt_in_the_curriculum_guide.pdf
Selby D, Jones P, Kagawa F (2009) Sustainability promotion and branding: messaging challenges and possibilities for higher education facilities. Sustainability 1(3):537–555. http://www.mdpi.com/2071-1050/1/3/537/pdf
Sky (2011) The sustainable generation: the sky future leaders study. http://efsandquality.glos.ac.uk/toolkit/sky_future_leaders_study.pdf
The Guardian (2011) http://www.guardian.co.uk/news/datablog/interactive/2011/nov/16/youth-unemployment-map

Authors Biography

Rachel Drayson is the co-author of NUS' sector-leading research 'student attitudes towards and skills for sustainable development.' She leads on the environmental research for all areas of NUS' work.

Previously to her current role, Rachel has 5 years' experience of working in environmental and sustainability research, having previously worked for Brook Lyndhurst, a specialist sustainability research consultancy. Rachel works in areas of both qualitative and quantitative research through a range of different projects including community action on environmental issues such as climate change, evaluating behaviour change and communicating environmental issues with SMEs.

Charlotte Taylor leads on NUS' work in education for sustainable development, helping to develop their strategy in this area. She chairs the Greener Jobs Alliance, and has developed schemes for increasing student involvement in NUS programmes.

In addition to this, Charlotte's principal role is to expand and develop opportunities for students and organisations off-campus by adapting NUS' behaviour change programmes for organisations such as the NHS, local government, private sector and third sector organisations.

A Complex Framework: Expanding the Understanding of the Human Being and Organizations to Integrate Sustainability in Education and Promote Transformative Learning

Lisiane Celia Palma and Eugênio Ávila Pedrozo

Abstract

Sustainability is a multidimensional issue and consequently, it requires multiple alternatives to be viewed through a complex lens. Thus, this paper presents a complex conceptual framework that relates levels of individual and organizational learning to different understandings that people have of sustainability and different ways of understanding the human being and organizations. The results of this study may help educational institutions, coordinators and professors to integrate sustainability in education and to promote transformative learning, presenting some key elements to this end. This study also presents ways to relate different approaches in order to deal with the complexity and opens space for the emergence of new solutions that are required to deal with issues related to sustainability. It brings questions related to the role of individuals and their relationships with others and with nature, and the role of educational institutions in society.

Keywords

Sustainability · Complexity · Learning

L.C. Palma (✉)
Federal Institute of Education, Science and Technology of Rio Grande do Sul (IFRS), Rua Dra Maria Zélia Carneiro de Figueiredo, 870 A, Canoas 92412-240, Brazil
e-mail: lisianepalma@yahoo.com.br

L.C. Palma · E.Á. Pedrozo
Federal University of Rio Grande do Sul (UFRGS), Washington Luiz, 855, Porto Alegre 90010-460, Brazil
e-mail: eapedrozo@ea.ufrgs.br

W. Leal Filho et al. (eds.), *Integrative Approaches to Sustainable Development at University Level*, World Sustainability Series, DOI 10.1007/978-3-319-10690-8_44

1 Introduction

The problems related to sustainability are complex, requiring new responses and ways of thinking that are different from those that generated such problems. Several actors and interests that are simultaneous competitive, complementary and opposing are involved. Hence, it is possible to say that sustainability is a multidimensional issue and consequently it requires multiple alternatives to be viewed through a complex lens.

However, the conceptual framework of neoclassical theory, which dominates our society, has led to a limited understanding of human beings (Soderbaum 1999, 2009) and organizations (Stubbs e Cocklin 2008; Soderbaum 2009) and consequently of sustainability.Nevertheless, seeing sustainability beyond the status quo requires a paradigm shift, which is related to transformation. Thus some authors have indicated transformative learning as an alternative to bring effective changes to address sustainability (e.g.: Sterling 2010–2011; Thomas 2009; Sipos et al. 2008;). This is related to changes that go beyond behavioral change, representing a challenge to existing beliefs and ideas, and promoting the reconstruction of meanings. It suggests a more radical change, a paradigm shift, rather than a change within the existing paradigm.

Thus, this paper presents a complex conceptual framework that relates levels of individual and organizational learning to different understandings that people have on sustainability and different ways of understanding the human being and organizations. Given the complex nature of the issue of sustainability, this paper proposes a shift from the simplification paradigm toward a complexity paradigm (Morin 2008, 2011). This framework may help educational institutions, coordinators and professors to integrate sustainability in education and to promote transformative learning, presenting some key elements to this end.

2 Sustainability and Its Different Interpretations

Because sustainability is a complex issue, it can be understood in different ways. In this sense, some people argue that sustainability can be achieved within the present structures, maintaining the status quo and business as usual. Others argue that a fundamental reform is necessary, but without a complete breakdown of the current system. In other words, that decisions and behaviors must be slightly altered to meet the new environmental and social challenges. A third position argues that economic and power structures of society itself are the root of the problems, and therefore, a more radical change or transformation is necessary (Hoopwood et al. 2005; Soderbaum 2009).

To provide a generalized view of the trends within the sustainable development debate, Hopwood et al. (2005) suggest a mapping methodology based on a combination of environmental and socioeconomic issues on two separate axes (Fig. 1). The socio-economic axis covers the level of importance given to human well-being and equality and the environment axis covers the priority of the environment from low environmental concern through technocentred to ecocentred. The central shaded area of the map indicates the range of views within the sustainable development debate, combining socio-economic and environmental issues. There are views outside of this area, concerned with either environmental or socio-economic issues but not both.

It is possible to relate these different understandings of sustainability with different conceptions of economic actors and organizations. The first discussion is related to a critique of the neoclassical limited view of the economic actor, which is presented by many actors who propose a broader understanding of being related with sustainability (Becker 2006; Siebenhüner 2000; Soderbaum 1999). In this sense, Bina and Vaz (2011) summarize the key dimensions shaping the different conceptions of the self, in an attempt to compensate for the limitations of the concept of *homo economicus,* to address resulting distortions of the self and what it means to be human (Table 1).

The focus on self-relation and egoistical behaviour is in conflict with the need to promote altruism and responsibility for others and future generations, as well as for the environment and the global commons, at the heart of sustainability. The idea of utility maximisers is also linked to equally narrow conceptions of justice, focused on rights to property and autonomous action, with limited regard for responsibility in relationships between people, with future generations, and with nature—and thus with notions of distributive justice (Bina and Vaz 2011).

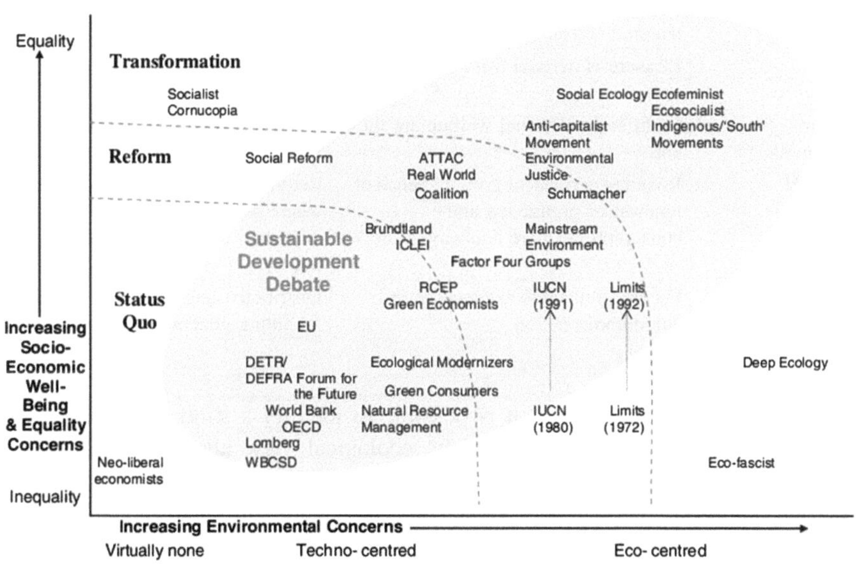

Fig. 1 Mapping of views on sustainable development. *Source* Hopwood et al. 2005

Table 1 Conceptions of economic actor: beyond reductionist and distorted 'selves'

Dimensions shaping the economic actor	Homo economicus: the narrow self	Homo sustiens/politicus/ecologicus: the wider self
Economic tradition	Neoclassical economic theory	Unorthodox economics, including ecological economics
Human being	Individual	Social being
Idea of the self	Narrow, reductionist	More holistic and balanced
	Narcissistic, individualistic	
Human behaviour	Egoistical behaviour, self-interest	Egoistical and altruistic behaviour
	Rational utility maximiser	Capable of sympathy, cooperation
	Competitive	Reciprocator
Human being and'others'	Self-regarding	Self and other-regarding, emphasising the ethical dimensions of community relations (active participation in the polis)
	Denial of links to the 'community' and of ethical dimensions therein	Moral responsibility towards others
Human being and nature	Nature is a mere input to the economic system	Also acknowledges non-instrumental bonds with Nature: sympathy, respect and emotional relationship with Nature; a source of inspiration and creativity
	The relationship with Nature is one of instrumental	
	self-interest	
	Responsibility limited to its efficient exploitation	Moral responsibility towards Nature, including its protection for non-human interests
Human wellbeing	Depends on increasing levels of material consumption	Depends on a wider range of material and non-material sources, including beauty, spirituality
	Pleasure is derived from goods and service	
Socio-economic context	Profit and individual welfare are the aim	Wellbeing and social welfare are the aim. Profit is a means to an end
	Requires permanent growth, hence of renewal of production and consumption (more is always better)	Requires a steady state economy, where quality becomes more important than quantity (less can be better)
Concept of justice	Focused on rights to property and autonomous action	Distributive justice. Responsibility for future generations and for nature

Source Bina and Vaz 2011

In relation to organizations, it is possible to identify a similar discussion that goes from the neoclassical view to the ecological view about business. These different views have influenced organizational strategy, which can be competitive—focused manly on shareholders' value—or sustainable—focused on stakeholders' value (Martinet and Reynaud 2004; Barin-Cruz et al. 2006).

In this sense, Stubbs and Cooklin (2008) have developed a sustainability framework, based on a review of the corporate sustainability literature, which outlines the characteristics of a typical business under each sustainability worldview. It summarizes the key organizational assumptions underlying each paradigm (Table 2).

Each of these paradigms, or worldviews (ecocentric, ecological modernization and neoclassical), is based on sets of shared fundamental assumptions about sustainability. Yet, they are not closed or static. Individuals and organizations may draw assumptions from different worldviews in a variety of ways (Hopwood et al. 2005) and organizations' behaviors and practices may reflect aspects of more than one worldview (Stubbs and Coklin 2008).

Different understandings of individuals, organizations and sustainability are co-related. At one end of the spectrum is the idea that it can be achieved within the present structures, maintaining the status quo and business as usual. This idea is related to the neoclassical worldview which has a narrow definition of the self as a self-interested, utility maximiser, the *homo economicus* view, which understands that the purpose of business is to increase revenue/profit and maximize shareholder value.

On the other hand, there is an ecocentric worldview, which is related to a view that economic and power structures of society itself are the root of the problems and, therefore, a more radical change or transformation is needed. It is related with a more holistic understanding of the human being, and hence the sense of (human) life that is proposed by ecological economists, along with other un-orthodox economists—referred to as *homo sustiens/politicus/ecologicus*. It also proposes that business must increase the quality of life and enhance social equity.

Between these two extremes, there is an understanding that a fundamental reform is necessary to achieve sustainability, but without a complete breakdown with the status quo. Proposing that decisions and behaviors must be slightly altered to meet the new environmental and social challenges. This is related with the ecological modernization worldview.

It is important to understand that these positions are not dichotomous, instead they are situated on a continuum. All of these conceptions are related to different levels of learning as well. This is the topic of the next discussion.

3 From Single Learning to Transformative Learning

To address the issue learning and sustainability, several authors suggest transformative learning as an alternative (e.g.: Sterling 2010–2011; Thomas 2009; Sipos et al. 2008). According to Sterling (2010–2011), in the current context, where there is a call for the re-examination of assumptions and values, critical thinking and new creativity, the concept of transformative learning is coming more to the forefront.

Table 2 Characteristics of business within alternative worldviews

	Ecocentric	Ecological modernization	Neoclassical
Purpose of business (business objectives)	Increase quality of life and enhance social equity (human and non-human species)	Pursue social, economic and environmental goals. Firms need to make a profit to exist but do not just exist to make a profit	Increase revenue/profit ("sustainable profit growth") and maximize shareholder value
Strategic approach	Organizational commitment to nature: minimize the use of virgin materials and non-renewable energy; eliminate emissions and effluent; minimize the life-cycle costs of products and services; andrenew natural resources.	Profitability through minimizing environmental impact and balancing stakeholder claims.	Pursuit of technical efficiency—use of technology to decrease labor costs and increase productivity and profits. Sustainability is typically viewed as a cost but some firms use it to gain competitive advantage
Structure	Non-hierarchical Decentralized authority Industrial ecosystems—each organization is part of a whole system	Systems approach (not maximizing the "parts") Horizontal integration across organization (cross-functional teams) Network structure. Intra- and inter-organizational cooperation	Hierarchical (functional, divisional or matrix) Command and control Maximizing individual parts (silos) Competition between departments
Approach to nature and the environment	Harmony with nature. Resources are finite	Technological innovation to minimize resource usage. Use renewable/human-made resources instead of non-renewable. Stewardship	Above nature Resources are free, plentiful and to be exploited now, not saved for future generation
Attitude to technical progress (to solve resource constraints and pollution)	Technological skepticism	Prudent skepticism	Technological optimism

(continued)

Table 2 (continued)

	Ecocentric	Ecological modernization	Neoclassical
Timeframe focus	Long	Medium-long. Concern for current generation and future generations	Short. Typically quarterly reporting and 1–4 year planning cycles
The stakeholders	Nature is the dominant stakeholder. Place nature at the centre of organizational concerns	Address needs of all (human) stakeholders. Acknowledges interconnections between humans and rest of nature—nature is a stakeholder	Shareholder is dominant. Place shareholder concerns at the centre of organizational concerns
Treatment of externalities	Fully internalized—pollution and waste are eliminated	Supports internalizing costs. Focus on reducing pollution and waste through technology and redesigned processes	Pollution, waste and costs of social degradation are externalized (unless legislated)
Product life-cycle	Focus on selling services. Responsible for complete life of product/service: cradle-to-cradle.	Sell low-impact products and services. Responsible for disposal/reuse of products at end of life	Not responsible for product end-of-life (disposal/reuse) nor concerned with designing low-impact products unless it increases profit/ revenue
Production systems approach	Circular (closed loop) production systems such as biomimicry and industrial ecosystems	Employs recycling. Moving towards closed loop systems	Linear production systems. Energy-and resource-intensive. Limited downcyclin
Cooperation or competition?	Cooperation	Coopetition	Competition
Performance measures Triple bottom line (TBL)	Emphasis on environmental and social impacts such as waste and pollution, use of resources and energy, biodiversity, and wellbeing and welfare of stakeholders and local communities	TBL—more equal weighting of economic, environmental and social performance	Financial and market measures such as revenue, profit, shareholder value, market share, return on equity, return on assets and return on investment

Source adapted from Stubbs and Coklin 2008

This concept arose from the work of the adult educationalist Mezirow (1978). It refers essentially to a qualitative shift in perception and meaning making on the part of the learner in a particular learning experience such that the learner questions or reframes his/her assumptions or habits of thought. It is the process of effecting change in a frame of reference.

Frames of reference are the structures of assumptions through which we understand our experiences. They selectively shape and delimit expectations, perceptions, cognition, and feelings. They set our "line of action." We transform our frames of reference through critical reflection on the assumptions upon which our interpretations, beliefs, and habits of mind or points of view are based (Mezirow 1997).

Paulo Freire (1987), when addressing transformative learning, emphasizes awareness. He argues that it allows people to insert themselves in the historical process, as a subject, questioning the status quo.

Sterling (2003) introduced the concept of "levels of knowing" in reference to transformative learning and sustainability (Fig. 2). This helps illuminate that learning can involve and affect different levels of consciousness. Deeper perceptions and conceptions inform, influence and help manifest more immediate ideas and they, in turn, affect more everyday thoughts and actions.

Sterling (2010–2011) states that this model raises an important and often missed dimension: that we can learn at different levels of knowing and meaning. Transformative learning is normally understood as learning which touches our deeper levels of knowing and meaning, and, thus influences our more immediate and concrete levels of knowing, perception, and action.

As Sterling (2010–2011) suggests, the work of Gregory Bateson is useful to clarify what transformative learning may entail. Bateson (1972) distinguished three kinds of learning and changes (in addition to "zero learning"), which corresponds to

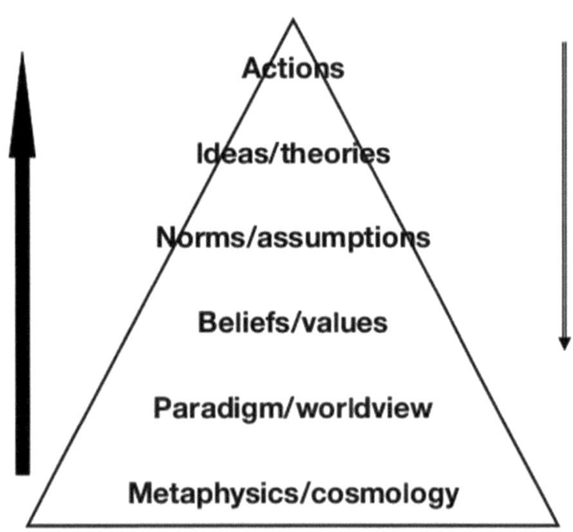

Fig. 2 Levels of knowing
Source Sterling 2010–2011

increased learning ability. First-order change refers to doing "more of the same", that is, change within particular boundaries and without examining or changing the assumptions or values that inform what you are doing or thinking. In this sort of learning, meaning is assumed or given and relates primarily to the external objective world. Second-order change refers to a significant change in thinking or in what you are doing as a result of examining assumptions and values. It is about understanding the inner or subjective world. In this sort of learning, meaning is recognized and negotiated amongst those involved.

Third-order change is an epistemic learning. It involves a shift of epistemology or operative way of knowing and thinking that frames people's perception of, and interaction with, the world. It means, an expansion of consciousness and that a more relational or ecological way of seeing arises, inspiring different sets of values and practices. It is the experience of seeing our worldview rather than seeing with our worldview so that we can be more open to and draw upon other views and possibilities. The case for transformative learning is that learning within a paradigm does not change the paradigm, whereas learning that facilitates a fundamental recognition of a paradigm and enables paradigmatic reconstruction is by definition transformative (Sterling 2010–2011).

To better understand organizational learning, the organizational change process is explored below to contribute to the present discussion.

4 From Marginal to Deeper Change

Wollin (1999) proposes that the deep structure of organizational systems is organized as nested, branching, multi-level hierarchies. As such, the organization of deep structure consists of a nested hierarchy where elements or sub-systems at less fundamental levels are sub-sets, and therefore dependent on elements or sub-sets at more fundamental levels on the same branch. Therefore, organizational change can be marginal or deep. According to Wollin (1999), organizational change results from the interaction and interdependences of the different levels of the organization's durable deep structure, its external environment and the purposeful action of actors internal and external to the organization.

In this sense, Waddell (2007) presents three types of change organizations presented in Table 3.

As revealed above, the changes as well as the learning process may occur at different levels. It is possible to relate organizational change with levels of knowledge and learning presented in Fig. 2. It can be said that changes only in the actions characterize marginal (Wollin 1999) or first order (Waddell 2007) changes as well as first order learning. At the other extreme are deep (Wollin 1999) or third order (Waddell 2007) changes that only happen with a paradigm shift. In this case it has to be transformative learning. In between these two is second order change and learning.

Table 3 Types of change in problem-solving initiatives

Criteria	First order change	Second order change	Third order change
Desired outcome	"More (or less) of the same."	Reform	Transformation
Purpose	To improve the performance of the established system	To change the system to address shortcomings and respond to the needs of stakeholders	To address problems from a whole-system perspective
Participation	Replicates the established decision making group and power relationships	Brings relevant stakeholders into the problem solving conversation in ways that enable them to influence the decision making process	Creates a microcosm of the problem system, with all participants coming in on an equal footing as issue owners and decision makers
Process	Confirms existing rules. Preserves the established power structure and relationships among actors in the system	Opens existing rules to revision. Suspends established power relationships; promotes authentic interactions; creates a space for genuine reform of the system	Opens issue to creation of entirely new ways of thinking about the issue. Promotes transformation of relationships toward whole-system awareness and identity; promotes examination of the deep structures that sustain the system; creates a space for fundamental system change

Source Waddell 2007

5 A Complex Framework

Summarizing what has been discussed up to this point, it is possible to say that there are different interpretations related to sustainability. These are related to understandings about individuals and organizations.

The neoclassical view of both individuals (*homo economicus*) and organizations (profit-maximizing firms and the interests of shareholders), is dominant, because it is consistent with the dominant paradigm (simplification). It can be related with the first level of learning, with a more limited and simplified view of the world.

However, if the objective is a real change towards sustainability, it must go beyond the status quo, seeking the transformation. It is related to the third level of learning, which involves a wider and more complex view. It suggests a paradigm shift. Thus, it is possible relate these levels of learning with a shift from the simplification paradigm towards the complexity paradigm.

The complexity paradigm is related to a theoretical opening, it enables emergency insertion of the world and the subject. It proposes to address the multidimensionality of things, the link, the bond, leading to the challenge of recognizing the complexity of the real (Morin 2011).

Therefore, some principles and concepts of the complexity theory were noticed in the framework construction (Morin 2003):

- the systemic or organizational principle: connects the knowledge of the parties to the knowledge of the whole;
- the recursive circle principle: the products and the effects are producers and creators of what they produce;
- dialogical principle: links two principles or notions where one must exclude the other, but they are inseparable in the same reality;
- the principle of reintroduction of knowledge in all knowledge: all knowledge is a reconstruction /translation by a spirit/brain in a certain culture and a certain time.

Based on Complexity Theory, it can be stated that to move towards sustainability a reconnection between the individual, society and species is required. Furthermore, we must relearn how to learn. This involves transformative learning, which is a process of constant order/disorder/organization (Morin 2008). In this process it is important to understand and analyze individual, organizational and environmental aspects in a recursive manner, which means they influence and produce each other.

Transformative learning searches for a change that includes our understanding of ourselves, our self-location and our relationships with other humans and with the natural world. It involves the expansion of individuals' the mental models at the individual level and shared mental models at the organizational level. In the framework (Fig. 3), to enhance the difference between the learning of individuals and organizations' process, we chose to use the term learning to refer to individuals, and change to refer to organizations. Emphasizing that there is change from one level to another only if there is learning.

In Fig. 3 the box "zero sustainability" represents mental models whose sustainability aspects are not considered, it means a level where there is no concern for the issue. First-order learning and change, named respectively as conformative (Sterling 2010–2011) and marginal (Wollin 1999) refer to the stage where sustainability issues are beginning to emerge, but are reflected only in actions do not involve the expansion of the mental model. The individual observes changes in the environment, accesses, designs and implements new actions, without however, changing their mindset. Then they make adjustments or adaptations to keep things stable in the face of change—seeking maintenance. This stage can therefore be connected to the status quo view in relation to sustainability. For individuals, meaning is taken as given and changes occur only in the actions and behavior, without questioning what is behind them. In organizations, the goal is to improve the performance of the already established system, replicating the decision-making group and power relationships, confirming the existing rules. Thus, the characteristics of *homo ecologicus* and neoclassical organization are maintained.

Second-order learning already implies an expansion of mental models and a significant change in the understanding of what is being done as a result of the examination of assumptions and values. In this type of learning, the significance is

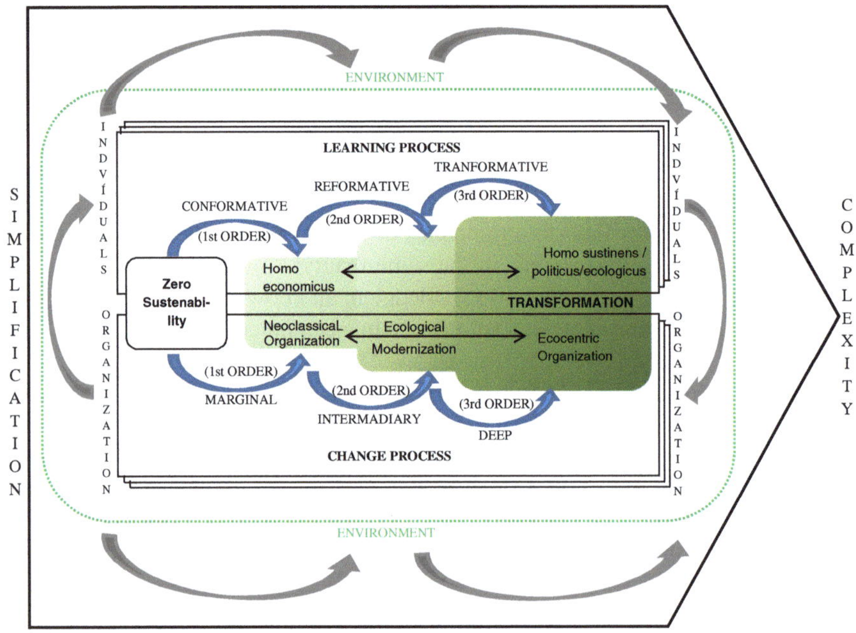

Fig. 3 A complex framework: expanding the understanding of the human being and organizations to integrate sustainability in education and promote transformative learning. *Source* the authors

recognized and negotiated among stakeholders. Individuals who go through this learning level—called reformers (Sterling 2010–2011)—start to question the values behind their actions. Organizations that undergo a second-order change (Waddell 2007)—called intermediate to signal it is between the marginal and profound change (Wollin 1999)—seek to insert concerns with social and environmental issues, beyond the economic, shifting the system to treat deficiencies and respond to stakeholder's needs. At this level there is the understanding that a fundamental reform is necessary for sustainability, but not a complete break with the existing order.

Finally, third-order learning and change—nominated transformative (Sterling 2010–2011) and deep (Wollin 1999) respectively—are more radical and imply an even greater expansion of frameworks, signifying a paradigm shift. It means, an effective expansion of understanding in relation to individuals and organizations. At this level, the social role of organizations is understood and the goals of the stakeholders are consider beyond the shareholders' goals. Organizations incorporate ecocentric characteristics. They seek to proactively address problems and seize opportunities from a whole-system perspective, allowing all participants to come in on an equal footing as issue owners and decision-makers. They also open issues for the creation of entirely new ways of thinking, promoting the examination of the deep structures that sustain the system, creating a space for fundamental systemic change. The individual, in turn, considers not only their interests (selfish behavior/

self-interest) but also their responsibility towards others and nature, expanding the reductionist view of "I".

Regarding the framework, the external arrows connecting individual and organization reinforce the idea of recursion, in which they both influence each other. The environment is also represented and influences individuals and organizations recursively. Multiple boxes represent that there are many individuals and organizations with different mental models in the same environment.

Based on the discussions presented and the framework, it is possible to highlight some key elements that can help educational institutions, coordinators and professors to integrate sustainability in education and to promote transformative learning. For this it is important to promote:

- a broader understanding of the human being and organizations;
- a broader understanding of sustainability (beyond the status quo);
- discussions related to the role of individuals and their relationships with others and with nature;
- working with multiple approaches, going beyond the neoclassical view;
- a more complex understanding of the world and its connections;
- the acceptance of different ways of thinking;
- spaces for dialogue and reflection, encouraging dialogic process;
- critical thinking and creativity;
- re-examining assumptions and values;
- interdisciplinary and transdisciplitanarity;
- a broader understanding of the process of learning, going beyond the transmission process;
- a learning experience that leads to critical reflection regarding the assumptions upon which our interpretations, beliefs, and habits of mind or points of view are based;
- the insertion of all individuals that compose the organization (professors, students, etc.) in the historical process, as subjects questioning the status quo and promoting changes.

These key elements open space for the emergence of new solutions that are required to deal with issues related to sustainability.

6 Conclusion

The discussion and the framework presented in this paper, which connects different elements related to sustainability, individuals, organizations and learning can help to illustrate that there is a paradigm behind our view of the world. This understanding is fundamental to promote the transformative learning that is related to paradigm change.

Some ways to promote this is by focusing on interactive, experiential and participatory learning and offering the practical skills and strategic thinking required to face the ecological, economic and social challenges. Some authors suggest the dialogic processes/pluralism (e.g. Wals and Schwarzin 2012; Waddell 2007) and the development of sustainability competences like systems-thinking, anticipatory, normative, strategic and interpersonal competencies (Wals and Schwarzin 2012; Wiek et al. 2011) as elements to achieve a transition towards sustainability in people, organizations and society as a whole.

In this sense, it is important to ask about the role of educational institutions in society. If they exist to promote the status quo, they can continue being content-led and focused on transmissive pedagogies that promote mainly the first level of learning. Yet this will not be enough to promote a transformation, because it searches for solutions to sustainability issues within the existing standards, replicating the same way of thinking that have resulted in our unsustainable world.

However, if educational institutions seek to lead a transformation to sustainability, it is important to expand their understanding of learning and to change into places that promote discussion, critical and complex thinking. In other words, they should become places that promote new responses and other ways of thinking that are different from those that generated the sustainability problems. Thus, educational institutions, as organizations that learn, should relearn how to learn. They should go through a third-order change themselves, to promote individual transformative learning, deep changes in other organizations and to transform the society.

It isn't an easy process, considering the conservative nature of the educational institutions, mainly, universities. But some movements in this direction are happening around the world, as revealed by the map constructed by Reevo project that aims to build a virtual platform for the dissemination, visibility, meeting, connection and collective action of educational experiences, organizations and people around the world. They have constructed the Alternative Education Collective Mapping, which groups those practices, theories, philosophies, different approaches to understanding the traditional hegemonic education established. 400 schools are listed on this map, including some famous schools around the world, as Schumacher College and Escola da Ponte (Reevo 2014). Thus, it shows that some changes are happening and are possible. However, it challenges our current ways of understanding the world, thinking and acting, and requires empathy, resilience and learning.

For future studies, we suggest some possible empirical researches. One suggestion is to develop empirical research to identify drivers and barriers to this transformation. Another possibility is investigate what is being developed in these alternative schools that can be used as a model in traditional educational institutions.

References

Barin-Cruz L, Pedrozo E, Estivalete V, De FB (2006) Towards sustainable development strategies: a complex view following the contribution of Edgar Morin. Manag Decis 44(7):871–891

Bateson G (1972) Steps to an ecology of mind. Chandler, San Francisco

Becker C (2006) The human actor in ecological economics: Philosophical approach and research perspectives. Ecol Econ 60:17–23

Bina O, Vaz SG (2011) Humans, environment and economies: From vicious relationships to virtuous responsibility. Ecol Econ 72:170–178

Freire P (1987) Pedagogia do Oprimido. Rio de Janeiro: Paz e Terra, 17ª ed

Hopwood B, Mellor M, O'brien G (2005) Sustainable development: mapping different approaches. Sustain Dev 13(1):38–52

Martinet AC, Reynaud E (2004) Entreprise durable finance et stratégie. Revue Française de Gestion 30(152):121–136

Mezirow J (1997) Transformative learning: theory to practice. New Directions for Adult and Continuing Education 74:5–12

Mezirow J (1978) Perspective transformation. *Adult.* Education 28(2):100–110

Morin E (2008) *O Método 1*: a natureza da natureza. 2. ed. Porto Alegre: Sulina

Morin E (2011) *O Método 2*: a vida da vida. Porto Alegre: Sulina

Morin E (2003) Para Navegar no século XXI—Tecnologias do Imaginário e Cibercultura. 3ª Ed. Porto Alegre—Edipucrs/Sulina

Reevo (2014) Mapeo Colectivo de La Educacion Alternativa. www.map.reevo.org. Accessed 30 Mar 2014

Siebenhüner B (2000) Homo sustinens—towards a new conception of humans for the science of sustainability. Ecol Econ 32:15–25

Sipos Y, Battisti B, Grimm K (2008) Achieving transformative sustainability learning: engaging head, hands and heart. Int J Sustain High Educ 9(1):68–86

Soderbaum P (1999) Values ideology and politics in ecological economics. Ecol Econ 28:161–170

Soderbaum P (2009) Making actors, paradigms and ideologies visible in governance for sustainability. Sustain Dev 17:70–81

Sterling S (2010–2011) Transformative learning and sustainability: sketching the conceptual ground. Learn Teach High Educ 5:17–32

Sterling S (2003) Whole systems thinking as a basis for paradigm change in education: explorations in the context of sustainability. 476 f. Tese (Ph.D. in education, systems thinking and sustainability)—centre for research in education and the environment, University of Bath, Bath

Stubbs W, Cocklin C (2008) Teaching sustainability to business students: shifting mindsets. Int J Sustain High Educ 9(3):206–221

Thomas I (2009) Critical thinking, transformative learning, sustainable education, and problem-based learning in universities. J Transforma Educ 7(3):245–264

Waddell S (2007). Realising global change: developing the tools; building the infrastructure. J Corp Citizensh 26:69–84

Wals AEJ, Schwarzin L (2012) Fostering organizational sustainability through dialogic interaction. Learn Organ 19(1):11–27

Wiek A, Withycombe L, Redman CL (2011) Key competencies in sustainability: a reference framework for academic program development. Sustain Sci 6:203–218

Wollin A (1999) Punctuated equilibrium: reconciling theory of revolutionary and incremental change. Syst Res Behav Sci 16:359–367

Author Biographies

Lisiane Celia Palma is a professor and researcher at the Federal Institute of Education, Science and Technology of Rio Grande do Sul (IFRS), in the city of Canoas, Brazil; a PhD candidate to the Post Graduate Program in Management (PPGA) at the Federal University of Rio Grande do Sul (UFRGS) and a researcher member of the Technological Innovation and Education Group at the IFRS and at the Organizations Study Group (GESTOR) at UFRGS, in the city of Porto Alegre, Brazil. Recently, she spent three months as a visiting doctoral student at the Centre for Sustainable Futures (CSF) in Plymouth University in England sponsored by the Coordination for Higher Education Personnel Improvement (CAPES). Her research interests encompass sustainability, education, organizations and management. She has worked in projects related with these topics in Brazil, Italy and Colombia and has worked with secondary and higher education and also as a professor in MBA courses. She holds a Master's Degree from the Center for Studies and Research in Agribusiness (CEPAN) at UFRGS sponsored by CAPES.

Dr. Eugenio Avila Pedrozo is Professor of Strategy, Interorganizational Relationships, Sustainable Innovation and Complexity (morinian approach), in the Group ITS "Innovation, Technology and Sustainability" at Federal University of Rio Grande do Sul, Brazil (UFRGS). Eugenio received his Ph.D. from INPL, Institut National Polytechnique de Lorraine, Nancy, France. He is ex-director of Cepan (Center for Studies and Research in Agribusiness) and was Chief of Department of the Department of Administrative Sciences in the Management School at UFRGS. His research interests include sustainable development, BOP, strategy, innovation, interorganizational relationships, agribusiness, interdisciplinarity, system analysis and complexity. He serves on the board of many Brazilian journals. He was director of thesis of more than 90 master of science and Ph.D. students. He has more than 80 publications in journals and more than 200 publications in congress.

Civic Action for Sustainable Futures: What Role for Adult Environmental Education?

Vincent Caruana

Abstract

The paper draws on research conducted by the author in part fulfilment of his PhD research, submitted in December 2013. The main aim of this research was to identify salient situations that enable participatory action as well as create barriers to the transition towards sustainability and to identify key conditions for Adult Environmental Education that would sensitise and mobilise sustained civic action for sustainable development. The research methodology adopted a case study approach—using four case studies across the Mediterranean geographic territory—using multiple sources of data gathering as appropriate to the different contexts of the case studies chosen and as permitted by practical considerations. The study evidences the power of the case studies to reframe and critically challenge hegemonic core beliefs. With inspiring leadership and a real commitment to a mental mode that embraces citizen mobilisation and new spaces for conversations, both CSOs and local institutions can be a positive force towards sustainable solutions at a local level. Adult Environmental Educators and Education for Sustainable Development practitioners need to re-define their roles—in particular their mentorship role and leadership training in addressing the sustainability transition.

Keywords

Case study approach · Environmental education · Adult education · Education for sustainable development · Emerging paradigm · Participation

V. Caruana (✉)
Centre for Environmental Education and Research, University of Malta, Msida, Malta
e-mail: censu.caruana@um.edu.mt

W. Leal Filho et al. (eds.), *Integrative Approaches to Sustainable Development at University Level*, World Sustainability Series, DOI 10.1007/978-3-319-10690-8_45

1 Introduction

The main aim of the author's doctoral thesis was to identify salient situations that enable participatory action as well as create barriers to the transition towards sustainability and to identify key conditions for Adult Environmental Education that would sensitise and mobilise sustained civic action for sustainable development. The research questions were classified under three major headings: Context as pertaining to an ever changing glocal world, mindsets and paradigms; Participation issues; and Educational processes.

The research methodology adopted a case study approach using multiple sources of data gathering as appropriate to the different contexts of the case studies chosen and as permitted by practical considerations. Four case studies were chosen through convenience within the Mediterranean geographic territory: an Intentional Community in Malta; a Fair Trade Network in Egypt; and two Local Agenda 21 processes: one in Modena (Italy) and one in Bethlehem (Palestine). The results were evaluated and analysed against the objectives of this research and narrated in a non-technical language that highlighted the specificity of each case study and identified themes, patterns and commonalities.

As Kjorven (2003) remarks, change requires bold political action both in the North and in the South, while education can bring us closer to realizing our grand hopes for the future by preparing the ground for such political action. Yet what is effective and not still remains somewhat elusive. The author's endeavour to engage in the four case studies, considered by peers as being examples of good practice, was to get closer to the elusive search for effectiveness. In drawing up his conclusions he freed himself from the above mentioned major headings used throughout all chapters to outline more all-encompassing strands that emanated from the analysis of the data gathered. This paper draws heavily on these concluding reflections in an attempt to connect some of the dots that can help accelerate the process towards a better common future, while contextualising the research into a wider framework of transformative learning.

These reflections capitalise on the learning from the case studies examined and help charter the unknown terrain referred to by Jickling and Wals (2012). They define a number of options for taking the sustainability transition forward, in particular how more effective non-formal Environmental Education may help overcome some of the key obstacles to greater citizen engagement in sustainable development and promote local choices that carry a societal as well as a global message. They are intended to provide some guidelines to assist educators in addressing the sustainability transition together with interacting groups of envisioning citizens in a locally appropriate manner and to guide the re-imagining of education that asks of us (as Environmental Educators and Educators for Sustainable development) to explore unknown terrain.

2 Infinite Solutions

The variety of the case studies explored reflect a variety of responses available in the transition towards sustainability, which is why in Education for Sustainable Development the talk is often about infinite solutions rather than searching for a solution. However, as a girl participating in the Maastricht 2002 European Global Education Congress remarked, the world is too large to be moved from one point—we need a network to make changes (O'Loughline and Wegimont 2003). Perhaps this is the result of remnants of the dominant paradigm, so well captured by McTaggart (2011), in the foreword to her book, *The Bond*:

> For centuries, Western science and many Western cultures have taught us to think of ourselves as individuals. But today, a revolutionary new understanding is emerging from the laboratories of the most cutting-edge physicists, biologists, and psychologists: What matters is not the isolated entity, **but the space between things**, the relationship of things.

This research adds evidence to what McTaggart is discovering through her extensive research, this time from a completely different angle: a case study approach within the social sciences. The implications are many and *ecological identity*, pioneered so beautifully by Thomashow (1995), takes a new and forefront significance in the efforts and endeavours of adult environmental educators. He directs environmental educators to "*connect their inner voices with understanding of ecology, community, and citizenship*" (p.xiii).

Freire emphasized revolution not as a starting point or an end point, but rather as a process in our daily life (Darder 2013). The author's doctoral research was a peek within the daily life of such revolutionary processes. In each case there was a good dose of touching the lives of those that were the most vulnerable—whether through modern technologies such as Geographical Information Systems, through fairer trade, through a forum on the alternative economy or through permaculture principles and pesticide free farming.

Each case study illustrated a different way of standing against injustice as a long term process. Even where supported by a short term process, such as the three-year funded project which kick started the Local Agenda 21 called Bethlehem 21, the action plan was still over 20 years. These projects served as a reminder that short term interventions, either by do-gooders or by outside *experts* with no roots in the territory, lack the continuity necessary for what is essentially a long term process. Of course short term interventions can be contextualised within long term programmes, but here the emphasis was on the process of getting people together, of owning their destiny and of engaging in the process, the emancipatory process of the emerging paradigm, still with no clear form, but with the *double-challenge* of social justice and the preservation of the environment at its core.

3 Country Contexts

One important strand that emerged from the analysis of the country contexts and the case studies is the role of history in conditioning the various paradigms of development and the need for adult educators to engage at this level and to produce local culturally appropriate guidelines for good practice. Specific skills to work effectively with others, such as those to do with conflict resolution, decision-making, community leadership and problem resolution, emerged as important components in any structured inputs in community processes for a better environment. Another emerging issue was the provision of open spaces for reflective thinking, participatory democracy, formulation of proposals, free exchange of experiences and networking for effective action (World Social Forum 2001).

Students often tend to misunderstand case studies, in that sometimes they are too eager to transfer the benefits emerging from one case study to other contexts, perhaps forgetting that the case study was a success because it was the result of the creation of knowledge and intelligence from a process that belongs to the people involved and the context in question. Yet there is another pitfall in this reasoning: the rendering of passiveness, i.e. being presented with readymade solutions to implement without questioning. This in itself is anti-pedagogic and potentially dangerous, especially if those presenting the solution have their own hidden agendas. Since Tbilisi (UNESCO 1977), Environmental Education was defined as an active learning process—this can never be rendered passive—but rather take life with all its complexities and ambiguities.

4 Impact

Globally there seems to be some concern on the impacts of the various adjectival educations on individuals and the wider community. In some areas, such as Environmental Education and Global/Development Education, most impact assessments have been conducted in schools. Further research in the area is needed, in particular in the adult, youth and community sectors. According to Pedrazzini (2012) individuals exposed to Global Education, develop an outward-looking and inclusive attitude and a more positive and sensitive approach to daily life. Pedrazzini (2012) adds that at the local scale it contributes to shaping more sustainable and diversity-respectful societies.

However, little focus has been given overall to which impacts one needs to study. For example, the Fair Trade Egypt case study indicated conclusively the positive benefits and therefore impact of Fair Trade on the livelihoods of disadvantaged producers. Such a qualitative impact assessment does not assess the impact of any Development Education programme, but does provide a tool for Development Educators. By focusing on the impact of examples of good practice as another possibility to the educational processes of adjectival educations, educators will automatically be equipped with new tools to enhance their work.

Each case study demonstrated the various transdisciplinary efforts of silent stakeholders in creating a better world—a world which increasingly requires culturally and ecologically sensitive responses. In trying to extrapolate the implications of observing such change agents on the ground for the initial and on-going formation of educators, the author cannot but feel humbled. Each situation presented the amazing and pioneering work of colleagues and fellow change agents in bringing about social justice and peace and in advancing Education for Sustainable Development in a mostly informal or non-formal setting. This in itself is motivating.

5 A Reasonable Revolution

In an Open Letter to the green movements, Secrett (2011) asks what more can NGOs do to help humanity out of the mess we are in? He surmises that tactics based on protest and outrages are not working and suggests going from the defensive to the offensive, sparking a reasonable revolution. The case studies documented are a chronicle of leaders, perhaps not famous but respected, who were at the forefront of a reasonable revolution in their daily lives and communities. Again what are the implications for Education for Sustainable Development courses?

Viewing the four cases studies from a distance, with perhaps the exception of the Malta case study, there was too much of the cognitive and too little of the other spaces—namely mind, heart, and spirit in the explicit discourse of the interviewees. Is this is in itself strengthening the problem in that the inherent imbalance is not challenged?

A core strand of Environmental Education is what we call Values Clarification, which essentially is about challenging ourselves so that our behaviour is more coherent with our stated values. Of course each and every person is himself or herself responsible for his or her behaviour and we cannot force political engagement on anyone. Yet what Modena and Bethlehem were able to do was to start the right conversations in official processes. The same can be said regarding Fair Trade Egypt, which initiated a new discourse that jars with the mainstream discourse of trade. Regarding the Malta case study, perhaps one can only appreciate its radicalism when one takes into consideration the conservativeness of Maltese society and the way not only were they were able to start a new conversation—too new for the Church structures to actually grasp and too innovative for the planning authority to comprehend—and also manage to move through and implement it.

6 Beyond Project Cycle Management

On a global level, Agenda 21 has had limited impact and spread. Projects did try to make up for this failure, as in the case of Bethlehem 21. However this reflected more the good will of the then current mayors and the foresight of the coordinating NGO rather than a broad commitment to the idea. What happens when the project cycle ends? How can we promote less dependency on funding and short term projects? What about long term programmes as opposed to short term projects? What about the sustainability of the projects beyond their life cycle? An issue to deal with, that requires further reflection, is the day to day reality the educator is faced with: the quality-quantity nexus within a fast-slow world nexus. The issue of time, of wanting results fast, often within project established time frames, has overall been ignored in most discourses that took place under the auspices of the Decade. Perhaps this has been considered as requiring too much of a paradigm shift to put as a central concern. Yet clearly it comes as a stumbling block to progress—even within the most successful of initiatives.

7 The Power of Community

In each case-study, but in particular in the Malta and Egypt one, which were less hindered by official protocols, the author could clearly discern the community empowerment as a dialectical process involving the individuals and the community, based on the responsibilities individuals have towards themselves and those they come in contact with. Yet even in the Bethlehem and Modena case study, the author could share the concerns of the interviewees in dedicating so much energy to get people together, to understand face to face that the concerns of the individual are also to be taken into account in a dialectical process of the concerns of those they come in contact with, the frustrations expressed when people still consider the municipality as the door to knock if one needs a job. Of course this is no criticism of knocking the doors as when one needs a job, it is often about survival and one tries out every door to keep up the pun. But rather perhaps that due to the hegemony of the dominant paradigm it is still often hard to understand the power of community solutions. The two case studies in Bethlehem and Modena do bring out the other dialectical relationship between authority and freedom and the way—at a municipality level—the two communities tried to live this process within the tensions of an Agenda 21, in particular the re-definition of the relationships between citizens and authorities.

8 Leadership

The author's journey over the four case studies highlighted something which was not sought out explicitly during the field-work conducted, but which came out forcefully in each and every case study. Each case study had its own successes due to having leaders dedicated to socially just and community empowerment. The data collection chronicled how the leaders in each case study waded their project through various obstacles to arrive where they arrived, drawing lessons not to be rigid and to adapt according to need rather than stick religiously to original plans.

Furthermore, their discourse, though often not explicit, was a discourse of love. It was the great Paolo Freire, who in an education hegemonic system built on valuing the cognitive over the heart, had the guts to talk about *pedagogy of love*. Interestingly enough the author now considers these leaders to be his friends. Perhaps *friendship studies* is a missing link of Education for Sustainable Development, and Education for Sustainable Development is actually an Education in Human Relations. Perhaps it is too political. It was Derrida (2005) who exclaimed that the future of the political becomes the future of friends, "*which, furthermore, bind together, in their very essence, friendship and democracy* (p.198)". The author's itinerary over the years in his research and other experiences has led him to conclude that the human element of relations and lovability came out too often to be a coincidence. It is science, yet not scientific in the sense that the language of love is left outside of the realm of science. In fact, the key people and main contact points within each case study earned the author's admiration for their moral commitment in their leadership roles, aware of the huge tasks faced working in a dominant paradigm which creates inequalities and the destruction of the environment. As Freire (1998) claims, "*human existence is, in fact, a radical and profound tension between good and evil, between dignity and indignity, between decency and indecency, between the beauty and ugliness of the world.*" (p.53).

9 Mentorship

Darder (2013) so ably highlighted the role of critical leadership: "*to have an uncompromising commitment to the empowerment and self-determination of the most disenfranchised*" (p.16). The author witnessed this most strongly in Palestine where it became evident to him that it takes guts to take leadership roles in such situations; to stay close to the people and together with them ask "Why?"—when that is exactly the question which an oppressing order will do all to stop you from asking. This led the author to strengthen his belief in mentorship as a necessary role for the adult environmental educator, a mentorship that goes beyond a neo-managerial paradigm and that offers a ministry of accompaniment to the leaders involved. At the end, the role of Education for Sustainable Development is to support the learner in finding his or her path, the path of passion, the path that leaves you with a reserve of energy to engage in the double challenge. Freire himself had to leave his official role, as he understood the force of hegemony, and he left to remain in the struggle. Here again mentoring can be of an enormous support to leaders and potential leaders in defining their role, in defining their uniqueness and what they can do best for themselves in relation to others.

The hegemonic forces are strong and dominant and a critical friend (a mentor) can be of immense support in the process of finding one's calling and remaining true to it. Furthermore, while motivation is a huge subject in education in the formal sector, more needs to be invested in trainings on motivating adults.

During the various interviews and engagements during this inquiry, the author witnessed his interviewees sharing their doubts—doubts which are themselves pedagogic in that it is not about claiming the knowledge that exists or about managing to find it in a haystack. Rather, as Freire (1970) reminds us, it is about problem posing education. It is about creating the knowledge together in the local context and spaces where the people gather together. Every time we deal with human society we are dealing with change and complexity. That is why the case studies stood out as a mere snapshot of some highlights of a complex ever changing reality. The interviews themselves were creating new knowledge in that the author often set the interviewees thinking in directions they had not considered before. This poses challenges to what type/s of leadership do educators need to support and facilitate to bring to the table complexity as a starting point and real context of emancipatory change? As Darder (2013) posited: "*the vision here is to learn together as a way of life by which we transform the world as community*" (p.14). It is crucial to consider critical leadership as a pedagogy in formal Education for Sustainable Development training courses and within an emerging paradigm (which is more cooperative as against competitive) re-valorise communal forms of leadership and further research and praxis in this area.

10 On Violence

At Bethlehem, the author observed the intentionality of violence with the daily humiliations at check points, sympathising with the Freiren differentiation between justice as intentionality and violence as emanating from communities that struggle (Darder 2013). Calculated daily intentional violence, which degrades and dehumanises the person, needs to be taken into account in any community educational project for empowerment. From the case studies under examination, the violence in Bethlehem was more visible, as expected in an occupation that knows that it can get away with violating 67 UN resolutions with no impunity. Under the violence of the occupation, the subject of what are our sources of hope, often discussed during my initial Environmental Education training, took a completely new meaning.

Yet the violence of poverty, as expressed in Cairo, where the state offers no welfare for vulnerable people is also another form of violence—structural, visible, and hegemonic—and the emancipatory educational efforts of Fair Trade Egypt are as necessary as in Bethlehem. Again Conflict Resolution, in particular the Non-Violent resolution of conflict, does not appear as a core component of most Education for Sustainable Development courses. This needs to be addressed, in particular from an emerging paradigm point of view, which observes violence as creating mass poverty in the south and ever increasing poverty in the North. Social learning also reminds us that we tend to model the behaviours we see around us, which explains the need to make non-violent examples of good practice more known and compassion central to Education for Sustainable Development courses. This is just an extension of the pedagogy of love that Freire was bold enough to passionately believe in and promote.

11 On Revisiting the Research Questions

As adult environmental educators we stand at an important time as every choice we make is done in the context of the challenges, mishaps and chaos of our current reality. The author is aware that work of adult environmental educators and ESD practitioners in the midst of crises, as related to financial issues, to environmental issues (such as climate, energy, transport, food, water), to political dishonesty and to horrific conflicts, has a huge potential impact on our generation and future generations. As the struggles and stories of the leaders and core stakeholders in the case studies analysed indicate, the beauty of an emerging paradigm lies in the reality that no permission needs to be sought in grasping the opportunity to change outmoded mind-sets that no longer serve us and our well-being. Passion and motivation seem to be crucial.

The power of the case studies examined lies in their ability to reframe and critically challenge the taken for granted hegemonic core beliefs of how a municipality, a Church organisation and a trade organisation ought to act. The case studies provide a snapshot of a genuine attempt in moving forward from the competitive technocentric mind-set built on an *I win at the expense of you losing*, which is often the mind-set behind the current challenges and chaos. Winning or success takes the form of keeping a hope alive under a brutal occupation, of privileging the alternative economy that puts people before profit and of rediscovering a sense of community. Winning takes the form of becoming an active participant in fixing the crises that beset us by ripping up the no longer serving hegemonic rules and norms of the dominant paradigm and starting afresh by figuring out what one can best do, from his or her vantage point, to make change towards sustainable development happen.

Today's glocal reality signifies that CSOs and local institutions are constantly faced with a choice to position themselves between a dominant neo-managerial organisational model (mainly based on log frames and project deliverables and largely inspired by the corporate world) and a value-based model (privileging empowerment, capacity building and systems thinking). In positioning themselves CSOs can opt for any form of hybrid that emphasises one end or the other of the spectrum in different measures. With inspiring leadership and a real, as against a token, commitment to a mental mode that embraces citizen mobilisation and new spaces for conversations, both CSOs and local institutions can be a positive force towards sustainable solutions at a local level.

12 Participation

The political space for participation is not a constant, but an ever changing space negotiated through the struggles, doubts and hopes of the relevant stakeholders and conditioned by the socio-political reality of the context in which they operate. The research whose conclusions are presented here focused on three completely different models of creating such political spaces: through local agenda 21 processes; through networking marginalised producers in the South amongst themselves and with consumers in the North; and through recreating a welcoming community. Whatever the tools used, whether traditional training courses, the setting up of thematic working groups, Open Space Technology or product development, each can be contextualised successfully into a wider framework of expanding the political space for civil society's voice and actions. Clearly, a commitment from the part of the institutions to frame the participation of civil society as a positive factor can be an important process in opening up new spaces for important conversations on sustainable futures.

While resource mobilisation and funding remains a crucial issue for any successful programme, the quality of the links needs to be given more attention, both in research and in the work of adult educators. Whether it is Fair Trade consumers in the North, Palestine solidarity groups, support for product development and entrepreneurial development through a long term presence or the sharing of participatory budgeting experiences, the emerging solidarity is a noteworthy response. It goes beyond the aid paradigm, and in itself is a re-positioning based on joint learning, both in the North and in the South, in face of the changing relationships of globalization and localization.

The vulnerability of local processes was evident through this research. The limited capacity, the difficulties in resource mobilisation, the dependence of the life-cycle of a process on the project time-frame, the dependence of the life-cycle of a process on national (and local) elections, the lack of role models, and the hindering political and social welfare contexts are just some of the vulnerabilities identified. The focus on the immediate results can be a distraction from the wider longer term goals of sustainability. The case studies analysed illustrated positive systemic changes in the *modus operandi* of both institutions and CSOs, yet a longer term longitudinal study is necessary to understand the longer term outcomes of such changes. Furthermore, more study and further research is required to understand the different expressions of success in different contexts and groups. To what extent is it more important to implement all of the goals and targets identified in the local Agenda 21 processes or rather to provide by default inclusive and accountable institutions and new opportunities for citizen expression, empowerment and participation in what effects most in their daily life?

13 Educational Processes

Adult Environmental Educators and Education for Sustainable Development practitioners need to re-define their roles. The eagerness of the author's interviewees to narrate their stories had another message behind the stories told. Their stories were asking for a renewed attention to citizen efforts and conscientisation. One important role of the educator is to support the process of CSOs in articulating their mission and vision and identity beyond the trappings of neo-managerialism and service delivery. This needs to be sustained by the necessary resources—in proportion to their energy and capacity—to maintain their core mission and relevance to the context in which they operate.

The situations presented had their own doubts, ambiguities, difficulties, at times oppression; and one role of the educator is to accompany CSOs in this uncertainty. An active accompaniment that includes the enhancement of capacities; the building of new, stronger and more effective networks; the mobilisation of those groups normally excluded or hard to include such as youth and migrants; the use of effective tools in communication as per context; the *critical friend* support towards leaders and change makers; and the encouragement to CSOs and institutions of good will in their experimentation in co-creating sustainable communities and futures.

The two local Agenda 21 case studies indicate recognition on the part of the authorities of the role and contribution of civil society. A process with its own tensions, yet a process that can be mediated and further supported by the transformative educator, in particular with regards to processes of critical reflection, analysis, evaluation, systemisation of experiences and social learning, and the capitalisation of experiences. In seeking to open up a space for enabling the increased participation of civil society, authorities also define their own space as catalysts of collaboration, and as taking a leading role in opening up new spaces for diverse perspectives to be debated, confronted, challenged and channelled towards positive solutions for more well-being and a better environment.

14 Conclusion

In documenting the processes, the Educator-Researcher takes on another role—that of bringing to light and sharing the social learning taking place at grassroots and local level. This can be done strategically to scale up social learning, the systemisation and capitalisation of experiences, active solidarity, and participatory and active citizenship. In the face of social, environmental, cultural and economic challenges presented by an ever-changing glocal world, investing in leadership, social learning and in the enhancement of the capacity of both institutions and CSOs, to *be* and *become* innovative co-creators of a new emerging paradigm, is a challenge for educators to embrace with passion and urgency!

References

Darder A (2013) Leadership for social justice and community empowerment: a Frierian approach. In: Conference presentation, University of Malta, October 2013

Derrida J (2005) The politics of friendship. Verso, London

Freire P (1970) Pedagogy of the oppressed. Herder and Herder, New York

Freire P (1998) Pedagogy of freedom: Ethics, democracy and civic courage. Rowman and Littlefield Publishers Inc, Lanham

Jickling B, Wals A (2012) Debating education for sustainable development 20 years after Rio: a conservation between Bob Jickling and Arjen Wals. J Educ Sustain Dev 6(2):49–57

Kjorven O (2003) The millennium development goals: a challenge to global education to 2015. In: O'Loughline E, Wegimont L (eds) Global education in Europe to 2015: strategy, policies and perspectives: outcomes and papers of the Europe-wide Global Education Congress Maastricht, The Netherlands. North-South Centre of the Council of Europe, Lisbon, 15–17 Nov 2002, pp 29–33

McTaggart L (2011) The bond: connecting through the space between us. Hay House, London

O'Loughline E, Wegimont L (2003) Global education in Europe to 2015, strategy, policies and perspectives: Outcomes and papers of the Europe-wide global education congress Maastricht, The Netherlands. North-South Centre of the Council of Europe, Lisbon, 15–17 Nov 2002

Pedrazzini E, obo North South Centre Secretariat (2012) The impact of global/development education. In: North-South Centre, 2nd European Congress on Global Education: education, interdependence and solidarity in a changing world. North-South Centre, Lisbon, pp 56–62

Secrett C (2011) An open letter to the green movement. The Guardian. http://www.guardian.co.uk/environment/2011/jun/21/charles-secrett-open-letter-activists

Thomashow M (1995) Ecological identity: becoming a reflective environmentalist. Massachusetts Institute of Technology Press, Cambridge

UNESCO (1977) Tbilisi declaration (1977). http://www.gdrc.org/uem/ee/tbilisi.html

World Social Forum (2001) Charter of principles. http://www2.portoalegre.rs.gov.br/fsm2013_ing/default.php?p_secao=5. Accessed Nov 10, 2013

Author Biography

Vince Caruana was born on the 10 April 1967. In 1991 he obtained an honours degree in Education from the University of Malta, specialising in Physics and Mathematics. Three years later he obtained a Diploma in Political Studies from the same University. Subsequently between 1995 and 1998 he studied Environmental and Development Education, obtaining an M.Sc. through London South Bank University. Vince is currently a full-time lecturer at the Centre for Environmental Education and Research (University of Malta), where he has just obtained his Ph.D., focusing on education for sustainability, in particular among adults, youth and the community. Vince has been for the past 25 years very active in the Development NGO scene, both locally and at a European level.

Who Am I? The Role(s) of an Academic at a 'Sustainable University'

Rehema M. White

Abstract

The contemporary university dances to a neoliberal tune, waltzing through agendas such as globalisation, the commodification of education, the audit culture and (in UK) research impact. The roles of the academic have changed with increasing and contradictory pressures imposed. The notion of a sustainable university offers an alternative mode for academic performance. The aim of this paper was to explore the current role(s) of the academic in a 'sustainable university', aided by critical reflection on the Transition: University of St Andrews initiative. In the context of the sustainable university, the possibility of academics to act as change agents is reinstated through teaching, research and activism; research expectations may be reconciled partly through a different realisation of research impact; relationships and self fulfilment can be embraced; and individualism and competition shifted to collective action and collaborative working. Whilst a sustainability focus offers hope for academia, the roles of an academic cannot be fully resolved until a paradigm shift in higher education institutions is achieved.

Keywords

Sustainable University · Academic · Sustainability research · Education for sustainability · Higher education corporatisation

R.M. White (✉)
Department of Geography and Sustainable Development, University of St Andrews, North Street, St Andrews, Fife KY16 9AL, Scotland
e-mail: rehema.white@st-andrews.ac.uk

W. Leal Filho et al. (eds.), *Integrative Approaches to Sustainable Development at University Level*, World Sustainability Series, DOI 10.1007/978-3-319-10690-8_46

1 Introduction

The pursuit of scholarship as an academic is a complex venture these days. As individuals, we seek intellectual satisfaction, try to inspire a new generation of students, hope to 'make a difference' through our research. Simultaneously, many of us precariously balance a family life, raising children, supporting a partner, perhaps caring for other kinfolk. We also live within a local community and may even dare to have other, non-academic interests. (How) is it possible to reconcile these different roles within the current tensions of the academic system and wider societal pressures? *What if* we re-framed our roles as academics within the context of a 'sustainable university'? What would this mean for our roles, and how might expectations and support alter to enable this new way of being?

Universities are expected to function as leaders and mentors in society; to enable deeper understanding of ecological crises and social ills and to offer technological and policy solutions (e.g. Zilahy and Huisingh 2009). Universities are supposed to deepen the schooling of the next generation. And, more recently, universities are clearly expected to manage their campuses in an environmentally friendly way (Rowe 2007). A more rigorous interrogation of what a 'sustainable university' might be is now questioning how higher education functions, and how universities might best facilitate a transition to sustainability (Sterling et al. 2013).

Academics have long been instrumental in conceptualising and (sometimes) facilitating sustainable development, and so their place in this new configuration is a critical piece of the puzzle. In this paper I will examine the current context of higher education institutions, drawing from the analysis of Morrissey (2013) and others on the performance of academic practice. I then build on the work of Sterling et al. (2013) on the sustainable university to re-frame the challenges for academics in "enacting alternative academic subjectivities" (Morrissey 2013) within this context. I use a reflexive process (see Finlay and Gough 2003) to develop insights from personal experiences, drawing especially on Transition: University of St Andrews. The aim of this paper is thus to explore the role(s) of an academic in a 'sustainable university' and to offer this exploration within the current debate on the changing nature of academia.

2 The Current Context in the Higher Education System

Universities currently navigate contemporary agendas including the positive and negative aspects of globalisation (Young et al. 2006). A more corporate face of higher education has emerged such that pervasive managerialism has altered the way that universities work (Martin 2009; Blewitt 2013). With the commodification of education, there is a risk that business targets supersede goals for excellent teaching (Doring 2002; Blewitt 2013). Defunding (in USA at least) has turned universities from functioning for public to private good, to the pursuit of patents and an uneasy entry into the world of commerce (Martin 2009). Productivity and the bottom line are emphasised at the expense of meaning, causing a crisis of reality for many academics (Sparkes 2007).

Within this increasingly neo-liberal context, Morrissey (2013) "consider[s] academics as self-governing subjects operating within a university governmental architecture that has been increasingly inflected in recent years by neoliberal designs to affect a performing, optimal individual in and for a performing institution." He suggests that Davies' (2006) analysis of each student being considered an "economic unit of use" should also apply to academics.

Research assessment exercises also form part of the audit culture. In UK, some believe that the research assessment process has diminished teaching effort, has had some perverse effects on research and innovation and has caused significant stress for academics (Sparkes 2007). An attempt to counter the favouring of theoretical, monodisciplinary research led to the development in UK of the impact agenda, in which each department is expected to demonstrate key impact case studies. However, there has been little recognition of how impact should influence student and other learning (Meagher et al. 2008). In addition, not all impact will be measurable (Walter et al. 2004); rather impact can be serendipitous, cumulative and can be maintained after cessation of a research project (Meagher et al. 2008; O'Brien et al. 2013).

This rather grim general context is offset by positive instances in higher education, in which some interdisciplinary courses and many innovative research centres address real world problems in a critically rigorous way (see Sterling et al. 2013 for examples). However, it also demonstrates the need for the development and performance of "alternative subjectivities" by academics within contemporary universities (Morrissey 2013).

3 Contemporary Academic Roles

Academics have in the past been influential through their activities in teaching and research, acting as potential change agents for students and for society (Doring 2002). Houston et al. (2006) define service as a separate activity. However, the contemporary university context described above threatens the ability of academics to contribute critical dialogue and significantly change the wider setting (Doring 2002). Higher workloads and stress (Houston et al. 2006) impede the ability of academics to both meet these new expectations and retain more traditional value based approaches to academia (Meyer and Evans 2005).

Research is a process by which we attempt to derive information, systematically and through the collection, analysis and interpretation of data; it can thus be the answer to a question, the resolution of a problem or lead to greater understanding of a phenomenon (Leedy 1997). Yet the erosion of classic distinctions between 'basic' and 'applied' forms of scientific research is obscuring the socio-political debate over what form of research we now desire or promote (Roll-Hansen 2009). Whilst universities increasingly seek funding through applied research projects (Doring 2002), and teaching is valuable income (Morrissey 2013), academics are incentivised through theoretical research outputs, leaving them struggling to reconcile the demands of audited accountability and intellectual recognition (Houston et al. 2006).

Education is an "open ended process that helps people to make sense of an increasingly complex world" (Sarkissian et al. 2009). However, as teaching is being made more 'efficient', the ability of academics to empower students is eroded (Doring 2002). Even education for sustainability, in collaborating with this managerial culture, has compromised its radical potential (Sterling and Maxey 2013; Blewitt 2013).

Academics now spend an inordinate amount of time fulfilling administrative requirements (Sparkes 2007; Sterling and Maxey 2013; Martin 2009). Managing is emphasised over thinking. And finally, many academics in contemporary universities pay little attention to real world imperatives. Barriers to academics currently engaging in practice, particularly in regional sustainability initiatives, include a lack of incentivisation (focus on research and teaching), external drivers reinforcing mono-disciplinary silos and the time and money required to engage in interdisciplinary research and in stakeholder engagement (Zilahy and Huisingh 2009).

In the past, academic activism, both through writing and direct action, was more prevalent (Martin 2009). Martin (2009) cites an increasing focus on professional expertise, a decline in regard for public service, self appointed right wing watchdogs for liberal activity and other factors as reasons for this change.

This analysis demonstrates that academics currently struggle to fulfil their roles within the changing nature and contradictory context of contemporary universities. I will now introduce the notion of another current factor: 'the sustainable university'.

4 What Is a 'Sustainable University'?

The notion of a 'sustainable university' has recently been explored (Sterling et al. 2013). Whilst the concept is contested, Sterling defines it as,

> one that through its guiding ethos, outlook and aspirations, governance, research, curriculum, community links, campus management, monitoring and *modus operandi* seeks explicitly to explore, develop, contribute to, embody and manifest—critically and reflexively—the kinds of values, concepts and ideas, challenges and approaches that are emerging from the growing global sustainability discourse (Sterling 2013: 23).

The sustainable university thus embodies, lives and creates sustainability. It moves beyond the notion of a sustainable university as one that emphasises operational sustainability (Rowe 2007). Whilst recognising the validity of basic, specialised scholarship, it sees most sustainability research as being both focused on the real world challenges facing us, and comprising research modes such as interdisciplinarity, participation and reflexivity (White 2013). It acknowledges the need for both curricular focus on sustainability issues and pedagogic support for the transformative process (Ryan and Cotton 2013). Critically, the sustainable university expects academics to act out the normative impulse of sustainability. It traverses the professional and personal, through demanding reflexive practice, critical engagement and transformative teaching. It offers a framework by which academics can consciously adapt process as well as output and perform an integrated form of scholarship. It highlights the "serious mismatch between the purposive and operational norms of higher education" and proposes that addressing sustainability challenges is a mechanism to reshape the university (Sterling and Maxey 2013). In other words, it might offer an alternative valuation and performance for academics today, as requested by Morrissey (2013).

5 Reflection on the Lived Experience

In order to further explore the possibility of the notion of sustainability to assist academics in navigating contemporary trends in higher education institutions, I offer a reflexive analysis (Finlay and Gough 2003) of experiences at the University of St Andrews.

5.1 Case Study Transition: University of St Andrews

The University of St Andrews is a small, research intensive, university in Scotland, with approximately 7,000 (many overseas) students. The University has world class humanities and arts, and excellent science. Whilst the students rarely engage in active protest, a focus on sustainability has been nurtured by undergraduate and postgraduate Sustainable Development (SD) Programmes, an active Estates department and St Andrews Sustainability Institute (SASI). In 2009 a group of students and staff launched Transition: University of St Andrews (TUSA), a grassroots initiative that aims to promote and enable sustainability in the university and wider communities. With some external funding, we have established many projects, including community gardens, Cook Smarter, Carbon Conversations, a Local Economic Trading Scheme and home energy audits (in partnership with the community run StAnEn). It also actively attempts to link sustainability action to curriculum and research activities across the University, for example, encouraging student dissertations on University operational topics. Whilst student engagement has been largely successful, the initiative has struggled to engage many academic staff.

As a founding and steering group member of TUSA, past Director of both SD Programmes, an active participant in SASI and a member of the University's SD Working Group, I have been immersed in sustainability initiatives at the University. A desire to more fully understand the imperatives of sustainability action in the context of academia prompted me to begin a reflective diary guided by questions on process, motivations, roles and reflection on my community engagement.

Whilst I make no claims that this institution has achieved the status of 'sustainable university', it has stated intent and made significant progress in that direction (see White and Harder 2013).

6 Potential Roles of an Academic in a Sustainable University

6.1 Reinstating the Role of Change Agent

Sterling and Maxey (2013) propose that universities should act as "agents of change", but acknowledge that to do so will require changes in the institutions themselves. We can thus see that a shift to sustainable universities could help to reinstate the lost role of change agent that Doring (2002) laments. Certainly in TUSA my colleagues and I are motivated by values to 'make a difference' both locally and globally.

How then might this role be achieved? Firstly, a sustainable university is expected to lead by example through sustainable operations (e.g. Sterling et al. 2013; Rowe 2007). Academics can contribute specialist knowledge. For example, at the University of St Andrews academics have played roles in engaging the community, advising on energy strategies, changing behaviour and assisting with design of biodiversity strategy. As well as through expertise, White and Harder (2013) suggest that academics need to play a role as active members of the university community to permit sustainable action.

Secondly, Doring (2002) proclaims the importance of academics as change agents through their interactions with students. Indeed, Sterling (2001) suggests that, "The key to creating a more sustainable and peaceful world is learning. It is the change of mind on which change towards sustainability depends; the difference of thinking that stands between a sustainable or chaotic future." (Sterling 2001: 12). In contrast to the drive towards efficiencies in teaching (Doring 2002) and the privileging of research over teaching (Sparkes 2007) in contemporary universities, the role of teaching in the sustainable university is thus given significant recognition. However, as Orr (2004) points out, many of our current unsustainable practices are initiated and maintained by highly educated university graduates. Teaching in a sustainable university will thus have to enable transformative learning that also prepares students to be moral and responsible citizens (Ryan and Cotton 2013). Students have reported profound changes in thinking through their engagement in discursive and practical action in TUSA, through fieldtrips and through novel curricular assignments. Curriculum can also directly address institutional or local community needs (Cortese 2003). At the University of St Andrews in the SD Programme students have undertaken dissertations, for example, to reduce food waste in university residences, to develop a carbon calculator for student use and to assess influences of attendance at a local outdoor nursery on children.

Thirdly, the notion of change agent can be imposed through research and research outputs. Sustainability research may offer one approach by which the impact agenda (as described above for UK) and the need for meaningful work (Sparkes 2007) can be at least partly reconciled. The normative imperative of sustainability research permits, indeed embraces, value driven scholarship (White 2013). Sustainability research could support a form of impact in which measurement should be triangulated; it is related to connectivity; and it is possible to set in place processes and activities but not guarantee outcomes (Meagher et al. 2008).

Finally, academics can (re)develop a role as change agents in wider society. My TUSA experiences have simulated my engagement in community and in NGOs outside of the university. I do not always use specific academic knowledge, but may also contribute through transferable skills: fund raising, minute taking, strategic development and report writing. For example, my role in TUSA has been partly to facilitate process, strategic visioning and continuity, in contrast to the more action based role played by many of the students involved. Such activities can enable a return to the critical activism lamented by Martin (2009) and Blewitt (2013).

6.2 Research Expectations

The role of the academic as researcher in a sustainable university could offer some relief from contemporary conflicting research expectations. There is some alignment between the attributes of sustainability research (White 2013) and Mode 2 forms of research that promote a wider framing of problems, with stakeholder engagement and a shift towards more collaborative knowledge production (Gibbons et al. 1994). Indeed, it is suggested that science can be perceived as a consilience of knowledge formats, working with porous disciplinary boundaries and across different scales, leading to the emergence of shared visions and more democratic forms of knowledge production (Costanza 2003; Carolan 2006). Taking this view one step further, we can define knowledge production as a "process of social negotiation involving multiple actors and complex power relations" (Nygren 1999: 267). Because of these complex power relations, we need to maintain open communication channels and help create opportunities for mutual learning for those engaged in research processes (Van Kerkhoff and Lebel 2006). Our role as an academic is often thus not that of an isolated individual but rather that of a knowledge broker, facilitating input and interpretation from others. Different roles might emerge within this more complex context (Brand and Karnoven 2007). In fact, stakeholders can develop as much benefit from other stakeholders and from the collaborative processes co-developed in a large participatory interdisciplinary project as they can from the research outputs themselves (O'Brien et al. 2013). As an academic we may promote not only factual learning but also social learning, which includes cognitive (new knowledge) and normative (new ways of acting) aspects (Webler et al. 1995). Impact can thus occur throughout research processes, not just from research outputs (Pain et al. 2011).

Real world 'wicked problems' are not framed within academic disciplines and so often require interdisciplinary approaches and engagement with stakeholders

(Rittel and Webber 1973). Interdisciplinarity can promote a triangulation of depth, breadth and synthesis (Klein 1996). However, such approaches can be challenging and time consuming, requiring relationships and trust to be established across academics and other participants (O'Brien et al. 2013). Research assessment exercises in UK and associated academic incentivisation and funding opportunities limit interdisciplinarity (see also O'Brien et al. 2013, Sparkes 2007) but they are promoted via sustainability research (White 2013).

6.3 Values, Relationality and Self Fulfilment

The dissatisfaction, disillusionment and stress of academics in the neoliberal business management model of universities (Houston et al. 2006; Sparkes 2007; Blewitt 2013; Sayce et al. 2013) illustrate the need for a more fulfilling and enabling environment within universities. Academics with meaningful roles who experience personal development and who have some autonomy are more highly motivated and less subject to burnout (Bilge 2006). A sustainable university would facilitate a more meaningful, value based scholarship and recognition of the relational nature of knowledge generation and mobilisation. The way in which academics perform their roles will influence this potential. For example, trust and relationships built between academics facilitate interdisciplinary research (O'Brien et al. 2013). Engagement with the university community enhances the potential for sustainability action (White and Harder 2013). The facilitation of the dialogic learning that is considered to be appropriate for education for sustainable development requires personal engagement, a reflexive approach and rather than the transmission of facts to be memorised, as in banking education, the presentation of opportunities for learning and a mentoring through this experience (see Sterling 2013; Freire 2000). Co-production of knowledge with student learners is a key role, and one that may have more impact than our research activities. The personal engagement and critical reflexivity brought to learning processes also demands that we revisit the core of our knowledge, permitting us to interrogate our ontology and understanding of fundamental concepts; to make new linkages; to articulate our passion for our subject. Impact is not impersonal either; policy influence is affected by the relationships and trust built between academics and policy makers over time. It has been noted that "…the production, interpretation, dissemination and implementation of evidence is a social process subject to the forces and vagaries of social life." (Upshur 2000: 96).

The embodied aspects of the sustainable university are partially fulfilled through the people who work there and there are emotion filled aspects to the roles played by academics; they are not neutral or theoretical. Normative understanding of sustainable development reflect the values of an individual. This is where an apparent paradox emerges. Whilst academics in a sustainable university need to be trusted to offer scientific 'evidence' and 'sound policy advice', values influence their views and beliefs. Recognition of this demands a moral stance from academics, but also acknowledgement that the scientific process has always been influenced by human interactions (Escobar 1998). Through sustainability research academics need to permit civil society to ask not just 'how' but 'why' (Brand and Karnoven 2007).

Self fulfilment will also increase when academics can develop effective work/life balances (Sayce et al. 2013) which means recognition of the external roles of an academic: as mother/father, daughter/son, brother/sister, friend or colleague, mentor or mentee. This is particularly important to encourage contributions from ethnic minorities and women. The proportion of women in higher education institutions declines with seniority (Bordons et al. 2003). Underlying reasons for this disparity include the greater responsibility for family commitments often held by women (Shen 2013), educational localism because mobility is constrained and poorer professional networks (Paige Miller et al. 2006).

6.4 Individual and Collective; Collaboration and Competition

The contemporary university model promotes individualism and competition, which the premise that such processes will be linked to inter-institutional competition (Morrissey 2013). However, sustainability principles promote collective and collaborative approaches. Collective, community based action can significantly support sustainability action (White and Harder 2013). Ironically, the latter approaches are more suited to a Mode 2 form of research, as reported above, in which interdisciplinary and participatory teams together tackle research issues. In a sustainable university, collaborative working might enable academics to achieve their multiple and complex roles within a framework that permitted individual flourishing and a life outside of the university. This vision of collective action for the common good contrasts with the notion of academic as individual economic unit (Morrissey 2013).

It will take radical changes to persuade higher education leaders to change their vision, mission and faculty performance evaluation criteria and to support interdisciplinary and partnership working (Zilahy and Huisingh 2009). In addition, each university is unique. A 'sustainable university' should be interpreted within the place and position of a particular institution. I have focused in this paper on the experiences of those in the global north. Likewise, the roles of an academic will differ with the focus of their university, and with the disciplinary and topic based focus of the academic.

7 Conclusions: Reconciliation or Contradiction?

In this paper I have attempted to examine how the roles of an academic in a sustainable university (as explored by Sterling et al. 2013) could offer an alternative mode of performance within the contemporary university, as sought by Morrissey (2013). We have seen that a sustainable university offers new routes by which to traverse the often hostile terrain of the neoliberal model. In particular, the role of change agent may be reinstated through active teaching modes, sustainability research modes and activism. Whilst some of the current paradoxical research expectations may be reconciled, including the drive for impact in its widest sense, others remain contradictory. Enhanced self fulfilment would improve wellbeing and hence productivity, thereby ironically satisfying performance indicators. Acknowledgement of the academic as a *person* with external roles would bring a rich plurality of perspectives to scholarship; and acknowledgement of *personality* would allow acceptance of relational aspects of knowledge generation and mobilisation. Collective and collaborative ways of working could thrive. Yet—sustainability studies tell us that we need a paradigm shift in higher education, not a tweaking of the current system (see Sterling et al. 2013). The notion of the sustainable university offers a tantalising glimpse of what the roles of the academic might become, but we must not delude ourselves. In reality, we will need to to re-value, re-inspire and re-incentivise academics. We need to work with the critical pedagogy that is now emerging from sources outside of academia (Blewitt 2013) to enable the paradigm shift required. Only then will we, as academics, be able to overcome the current contradictions of universities and reconcile our roles with the needs of people and planet.

References

Bilge F (2006) Examining the burnout of academics in relation to job satisfaction and other factors. Soc Behav Personal 34:1151–1160

Blewitt J (2013) EfS: contesting the market model of higher education. In: Sterling S, Maxey L, Luna H (eds) The sustainable university: progress and prospects. Routledge, Abingdon

Bordons M, Morillo F, Fernandez M, Gomez I (2003) One step further in the production of bibliometric indicators at the micro level: differences by gender and professional category of scientists. Scientometrics 57:159–173

Brand R, Karnoven A (2007) The ecosystem of expertise: complementary knowledges for sustainable development. Sustain Sci Pract Policy 3:21–31

Carolan MS (2006) Science, expertise, and the democratization of the decision-making process. Soc Nat Resour 19:661–668

Cortese AD (2003) The critical role of higher education in creating a sustainable future. Plan High Educ 31:15–22

Costanza R (2003) A vision of the future of science: reintegrating the study of humans and the rest of nature. Futures 35:651–671

Davies B (2006) Subjectification: the relevance of Butler's analysis for education. Brit J Sociol Educat 27:425–438

Doring A (2002) Challenges to the academic role of change agent. J Further High Educ 26:139–148

Escobar A (1998) Whose knowledge? Whose nature? Biodiversity, conservation, and the political ecology of social movements. J Polit Ecol 5:54–82

Finlay L, Gough B (2003) Reflexivity: a practical guide for researchers in health and social sciences. Wiley-Blackwell, US

Freire P (2000) Chapter 2 Pedagogy of the oppressed. Continuum, New York

Gibbons M, Limoges C, Nowotny H, Schwartzman S, Scott P, Trow M (1994) The new production of knowledge: the dynamics of science and research in contemporary societies. Sage, London

Houston D, Meyera LH, Paewai S (2006) Academic staff workloads and job satisfaction: expectations and values in academe. J High Educ Policy Manage 28:17–30
Klein JT (1996) Crossing boundaries: knowledge, disciplinarities and interdisciplinarities. The University Press of Virginia, Virginia
Leedy PD (1997) Practical research: planning and design Upper Saddle River. Prentice Hall, NJ
Martin R (2009) Academic activism. PMLA 124:838–846
Meagher L, Lyall C, Nutley SM (2008) Flows of knowledge, expertise and influence: a method for assessing policy and practice impacts from social science research. Res Eval 17:163–173
Meyer LH, Evans IM (2005) Supporting academic staff: Meeting new expectations in higher education without compromising traditional faculty values. High Educ Policy 18:243–255
Morrissey J (2013) Regimes of performance: practices of the normalised self in the neoliberal university. British J Soc Educ. http://dx.doi.org/10.1080/01425692.2013.838515
Nygren A (1999) Local knowledge in the environment-development discourse: from dichotomies to situated knowledges. Critique Anthropol 19:267–288
O'Brien L, Marzano M, White RM (2013) 'Participatory interdisciplinarity': Towards the integration ofdisciplinary diversity with stakeholder engagement for new models of knowledge production. Sci Pub Policy 40:51 61
Orr DW (2004) Earth in mind: on education environment and the human prospect. Island Press, Washington
Paige Miller B, Soortamoorthy R, Anderson M, Palackal A, Shrum W (2006) Gender and science in developing areas: has the internet reduced inequality? Soc Sci Q 87:679–689
Pain R, Kesby M, Askins K (2011) Geographies of impact: power, participation and potential. Area 43:183–188
Rittel HWJ, Webber MM (1973) Dilemmas in a general theory of planning. Policy Sci 4:155–169
Roll-Hansen N (2009) Why the distinction between basic (theoretical) and applied (practical) research is important in the politics of science. In: Fennel D (ed) Contingency and dissent in science. Centre for the Philosophy of Natural and Social Science, The London School of Economics and Political Science, London
Rowe D (2007) Education for a sustainable future. Science 317:323–324
Ryan A, Cotton D (2013) Times of change: shifting pedagogy and curricula for future sustainability. In: Sterling S, Maxey L, Luna H (eds) The sustainable university: progress and prospects. Routledge, Abingdon
Sarkissian W, Hofer N, Shore Y, Vajda S, Wilkinson C (2009) Kitchen table sustainability: practical recipes for community engagement with sustainability. Earthscan, London
Sayce S, Bradley JF, Ritson J, Quinn F (2013) Well-being: what does it mean for the sustainable university? In: Sterling S, Maxey L, Luna H (eds) The sustainable university: progress and prospects. Routledge, Abingdon
Shen H (2013) Mind the gender gap. Nature 495:22–24
Sparkes AC (2007) Embodiment, academics, and the audit culture: a story seeking consideration. Q Res 7:521–550
Sterling S (2001) Sustainable education: revisioning learning and change. Green Books, London
Sterling S (2013) The sustainable university: challenge and response. In: Sterling S, Maxey L, Luna H (eds) The sustainable university: progress and prospects. Routledge, Abingdon
Sterling S, Maxey L (2013) Introduction. In: Sterling S, Maxey L, Luna H (eds) The sustainable university: progress and prospects. Routledge, Abingdon
Sterling S, Maxey L, Luna H (eds) (2013) The sustainable university: progress and prospects. Routledge, Abingdon
Upshur REG (2000) Seven characteristics of medical evidence. J Eval Clin Pract 6:93–97
Van Kerkhoff L, Lebel L (2006) Linking knowledge and action for sustainable development. Annu Rev Environ Resour 31:445–477
Walter I, Nutley SM, Davies HTO (2004) Assessing research impact. Research Unit for Research Utilisation, University of St Andrews, St Andrews

Webler T, Kastenholz H, Renn O (1995) Public participation in impact assessment: a social leanring perspective. Environ Impact Assess Rev 15:443–463
White RM (2013) Sustainability research: a novel mode of knowledge generation to explore alternative ways for people and planet. In: Sterling S, Maxey L, Luna H (eds) The sustainable university: progress and prospects. Routledge, Abingdon
White RM, Harder M (2013) The Journey towards Sustainability via Community: Lessons from two UK universities. In: Sterling S, Maxey L, Luna H (eds) The sustainable university: progress and prospects. Routledge, Abingdon
Young OR, Berkhout F, Gallopin GC, Janssen MA, Ostrom E, Van Der Leeuw S (2006) The globalization of socio-ecological systems: an agenda for scientific research. Glob Environ Change 16:304–316
Zilahy G, Huisingh D (2009) The roles of academia in regional sustainability initiatives. J Clean Prod 17:1057–1066

Author Biography

Dr Rehema White is a sustainability generalist with interests and experience in the production, mobilisation and implementation of knowledge for sustainable development. Her research, teaching and practical experience over the past 25 years have enabled her to employ interdisciplinary, holistic, co-productive methodologies that recognise the complexity of systems and the difficulties in linking theory and practice. She has explored learning for sustainability through, for example, community development in Africa; the support of outdoor learning for early years children in Scotland; and whole university approaches to sustainability. She strives for synthesis and integrative analysis across different fields, drawing from her epistemological journey across the natural and social sciences and her international experience. She has thus published and held grants across a wide range of areas. Transition: University of St Andrews has taught her much about the personal and political aspects of sustainability.

Trans-disciplinarity in Sustainability Science and Education

Markus Will and Lars Rydén

Abstract

The paper presents a characterization of trans-disciplinary research and sustainability science and education, and argues for its necessity in addition to traditional forms of knowledge production. A short analysis on the obstacles met within the academic community for those working in trans-disciplinary research and education in an often rather conservative environment is also made. Finally, some examples are given, and conclusions on the implementation of integrative, trans-disciplinary research and sustainability science against the background of freedom of teaching and research will be drawn.

Keywords

Trans-disciplinary research and education · Sustainability science

1 Background

A shift towards a more sustainable world requires a strong support from science. However science, as it is most traditionally conducted within the academic disciplines, is not able to fulfil the needs required by a transition to sustainable world. It is not merely a question of technological innovations. These changes are part of a

M. Will (✉)
Zittau/Goerlitz University of Applied Sciences, Theodor-Körner Allee 16, 02763 Zittau, Germany
e-mail: m.will@hszg.de

L. Rydén
Baltic University Programme, Uppsala Centre for Sustainable Development,
Uppsala University, Villavägen 16, SE 752 36 Uppsala, Sweden
e-mail: lars.ryden@csduppsala.uu.se

W. Leal Filho et al. (eds.), *Integrative Approaches to Sustainable Development at University Level*, World Sustainability Series, DOI 10.1007/978-3-319-10690-8_47

social-economic transformation process, which requires that science be conducted in a trans-disciplinary mode, beyond the disciplines. Accordingly, the role of science and scientists is changing. Parts of the academic system face a crisis of confidence, legitimacy and influence in society. This can clearly be seen in scarce real-life (or political) impacts of academic reports on global warming, decreasing biodiversity, a looming resource crisis, but also on societal inclusion and coherence. This is neither only a crisis of systems knowledge, i.e. on the origin and development of sustainability problems, nor of target knowledge. It is rather a crisis of transformative knowledge, that is, how science can make a change in a pluralistic society, and a crisis of normative positioning, as individual scientists have difficulties in assuming and defending an ethics of sustainability.

2 Sustainable Development as a "Wicked" Problem

Sustainable development, the transition to a sustainable world, is an immense challenge. It requires no less than a complete reorganization of our society and how it is run. The complex challenges we are confronted with can be characterized as "wicked problems". Presently a profound and comprehensive definition of sustainability and sustainable development is lacking. Despite of the large body of literature available, defining and contextualizing sustainability is still subject of ongoing discussions (cp. Leal 2011:428). Obviously, there are different aspects of sustainability: (i) sustainability is described from a spiritual or religious perspective, calling for living with or within nature. Ecosystem services are considered as an essential foundation for anthropogenic activities ("nested systems approach"). Other understandings of sustainability focus on the question of (ii) resource management and efficiency. At this point, renewable resources should be used respecting regeneration times, while non-renewable resources may only be used in a most efficient and frugal way or not at all. The main paradigm here is efficiency, i.e. achieving more output with less input. Of course, opponents argue that efficiency is an insufficient condition in a limited system due to rebound effects, caused by (unlimited) growth of demands, production and economies as a whole (Polimeni et al. 2008; Jackson 2009). The "doomsday"-group with prominent members such as Thomas R. Malthus, Kenneth Boulding or Dennis Meadows and the Club or Rome—to name a few—has been concerned with the carrying capacity of the planet, referring to (iii) limits of growth in resource consumption and production in a finite global system. In the late 1980s the focus of the discussion was adapted again, now focusing on issues of ethics and equity, taking questions of (iv) intra-generational and intergenerational justice and fairness between industrialized and developing countries into account. A common denominator was found in the standpoint, that current and future generations both have the right to fulfil their basic needs. However, the intra- and intergenerational time perspective makes it difficult to assess whether something is sustainable or not, as major impacts may occur in the remote future. There is nothing like a 'hidden hand' ensuring that egocentric individual actions automatically contribute for the sake of social and ecological systems. Indeed, value choices and ethics cannot be avoided (cp. Funtowicz and Ravets 2003:2).

This short overview on different approaches to sustainable development illustrates, that there is no complete and exclusive description of the term sustainability itself and therefore no profound understanding of the different problems related to the desideratum of sustainability, which makes the issue even more "wicked" (see Box 1). Multiple actors, different and not always complementary understanding of the concept, different values and normative backgrounds, as well as uncertainties in cause-impact chains and theoretical models and inadequate fore- and back casting makes it difficult for decision makers and governance systems to react in a relevant way. Instant political reaction and common decisions for collective action is, however, desperately needed and demanded in order to mitigate the most serious consequences. As in most societal challenges, the precise task is ill-defined (Rittel and Webber 1973:160) and planning decisions are relying on value judgements for resolution. As there is no exclusive problem definition, there is no exclusive solution and there is no definite 'right' or 'wrong', only a gradual improvement and permanent reformulation and re-solution seems to be possible.

Box 1: A characterization of wicked problems (own elaboration based on Rittel and Weber 1973; Funtowicz and Ravetz 1993; Sterrman 2012)

Sustainability problems are typically "wicked", i.e.:

1. *No final, i.e. complete and exclusive problem definition:* In the times of cybernetic management, there was the hope to have perfect information, i.e. all characteristics, elements and their linkages are known. Optimization models treated problems like in a chess play: as if the rules of the game are clear, all actions of involved players are known and all situations and system conditions could be predicted in a computational model with help of probabilities. As we know from assessment of emerging technologies and from behavioral psychology, there is no much hope for final prediction of reactions of players or side effects, or for comprehensive problem definition. Instead, when defining a problem or developing an exhaustive inventory of characteristics, we have the idea of the solution already in mind: "*Problem understanding and problem resolution are concomitant to each other. Therefore, in order to anticipate all questions (in order to anticipate all information required for resolution ahead of time), knowledge of all conceivable solutions is required.*" (cit. Rittel and Webber 1973:161)
2. *No final, i.e. complete and exclusive solution:* If there is an ill-defined problem, there are also ill-definable solutions. Though, a variety of different approaches can be expected. However, a judgment about the feasibility of plans relies on realistic judgments, depending on normative backgrounds and a capability for appraisal based on trust and credibility.

There is the danger to expulse "out-of-the-box" innovations at early stages when openness and willingness for trials is not present.

3. *There is no inherent stopping rule:* If defining the problem is close to solve it, but there are no criteria saying when a complex problem is understood sufficiently, a never-ending problem-solving-cycle is induced. Every additional effort in analysing the problem, may enhance the chance to finding a better solution, so the hope. There is a risk to get stuck in "analysis-paralysis", which is dysfunctional for a managed problem solving process. Therefore, external considerations are often applied to stop the continuous cyclic process of problem-description-solution, i.e. budget constraints.
4. *No "right-or-wrong"*: Due to the fact, that problem description is challenging and ex ante assessment about future sustainability impacts is uncertain due to lack of objective criteria, there are no "true-or-false". Instead, only a gradual assessment which is dependent on the perspective taken and values and mindsets involved, and which is participatory can lead to generally acceptable appraisals. Therefore expressions like "better/worse" are more appropriate also in case of ambiguous results.
5. *Flexible, collective actions needed:* Due to high levels of complexity, uncertainty and self-organizing properties of ecosystems, understanding and management of wicked problems needs to involve actors from different public authorities and from the private sectors at different territorial levels, i.e. multinational, national, regional and local ("multi-level-governance"). This is, obviously, ridden with prerequisites, regarding consensus-finding mechanisms etc. Furthermore, permanent critical reflection, updating and adjusting of problem solving strategies is necessary as coined in the concept of adaptive management (Folke et al. 2005).
6. *No trial-and-error:* In puzzle-solving exercises (such as mechanical engineering), it is possible to try various solutions in experimental runs without penalties in case of failure. With wicked problems, every implemented solution is consequential and will cause effects and side-effect that are not necessarily beneficial. Every attempt to solve a problem leaves a trace—so to say. There is not much room for trial-and-error strategies and ultimate tests ensuring success is not possible, for instance because impacts are occurring with a time lag in the far distant future and are not predictable. It is impossible to asses ex ante, if a measure to solve a problem will be accomplished successfully. Directly related to this high uncertainty, decisions are needed in a short term and there are high stakes for politicians and decision makers. False decisions will take revenge from the public. However, actions to be taken should be considered as learning process.
7. *Uniqueness of problems:* Despite of similarities between problems—as described with the syndrome-approach by the WBGU (1997)—there are often new, unrecognized distinguishing properties that make a transfer of

solutions from a previous problem to another improbable. There is nothing like a cookbook-recipe.

8. *Every problem is a symptom of another problem:* Only in a systemic or holistic perspective, when linkages of elements and between problem spheres are understood, failure in problem solving may be avoided. Non-systemic policy responses, i.e. removing or influencing single systems elements is likely to pose new problems or side-effects (e.g. in the case of biofuels, or risks related to CFCs, asbestos, mad cow disease, genetically modified organisms etc. (Dörner 1997).

3 Limitations of the Conventional Way of Knowledge Production

Science and the academic system are considered as important actors in producing knowledge and therefore influencing sound decision making and contributing to progress of societies. The traditional way of knowledge production found in hard sciences such as natural sciences or engineering was and still is more or less straightforward. It is the task of a mathematician to solve an equation. The chemist is analysing the molecular structure of unknown substances or the faith or chemicals in the environment. In these examples of traditional sciences, the task is clear. The result is tangible: whether it is right or wrong, demonstrated by a philosophical argument or by a mathematical proof, whether the problem is solved or not (Rittel and Webber 1973:160). In the hard sciences, a central quality criterion is objectivity and scientists try to present reliable and robust evidence. Let's call this way of scientific reasoning the "Newtonian"–after Isaac Newton—or reductive sciences. Newtonian sciences are occupied with explanation and description of subjects by application of general methods and theoretical models. New theories are elaborated by testing of variables and hypothesis in empirical and experimental research, contributing to falsification of theories. Research questions are concerning inner-academic for the sake of explanation and generalisation.

The way of reasoning is reductionist while generalization and idealisation is strived for. The central aim of this 'normal science' or 'mode 1-knowledge production', as it is also referred as (Kuhn 1970, Gibbons et al. 1994, Funtowicz and Ravets 1993), is to advance the knowledge base predominantly within a certain academic discipline, also when the results may become of high relevance and use in society. The major quality criteria are reliability and validity rather than the applicability and usability of knowledge in a broader societal or practical context. Normal basic research is curiosity-driven (Funtowicz and Ravets 1993:740) as in the case of particle physics, which is done 'for its own sake' without primarily considering direct application and usability. Of course, results of pure and

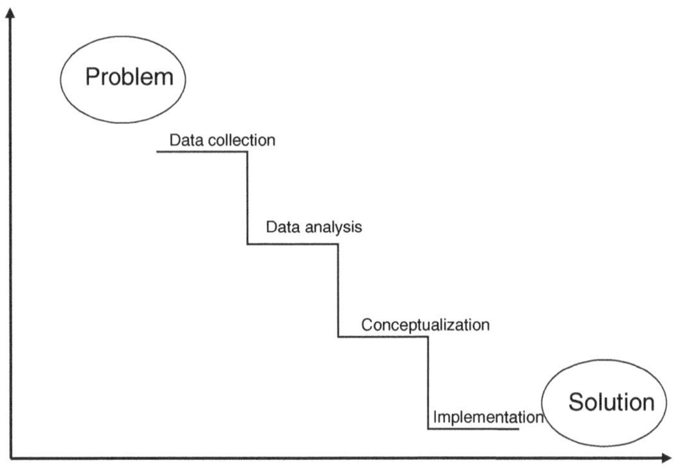

Fig. 1 The waterfall-model of problem solving (after Conklin 2009, modified)

fundamental normal research should and can be transferred into real life contexts (as in the case of Pasteur), but the academic and non-academics spheres and actors are mostly kept separated. Applied scientists (Thomas Edison) or consulting experts are often using results of fundamental research to solve real-life problems, such as in engineering and design (cp. Kates 2012, referring to Stokes 1997). Usually they follow a more or less linear and top-down routine. One starts with the problem formulation to get a comprehensive understanding (which is not possible when tackling with complex sustainability issues). This requires to study requirements and to collect data that can be analysed. Eventually one will be able to derive a problem solution, when enough and sufficient data was treated with theoretical models and all parts are puzzled together (cp. Funtowicz and Ravets 1993:740). Most people and organisations are quite familiar with this logic or pattern of thinking. It is widespread in most management systems, manuals and guidelines. However, this "waterfall logic" (see Fig. 1), as a flow down from problem definition to solution does not work with in wicked environments of any political relevance. Rather, different types of knowledge are needed and the process of reasoning is not as straightforward as in the waterfall model.

4 Types of Knowledge Needed for Sustainability Transitions

In order to deal with complex, wicked sustainability challenges three types of knowledge should be considered: systems, target and transformation knowledge (cp. Hirsch-Hadorn et al. 2008:30ff.):

- *Systems knowledge* is knowledge of the status quo, of the current status ("How is it?"). It involves empirical data and understanding of the characterisation, genesis and development of factors influencing a specific problem. System knowledge also refers to the problem interpretation, depending on the perception of the goal and options for change. It is necessary to provide information about uncertainty of data and models as well as philosophical and normative premises and assumptions which often remain unexpressed in conventional research (Funtowicz and Ravets 1993:2).
- *Target knowledge* is knowledge about desired goals about how it should be or not be. It is needed to ensure clarification, for priority setting amongst stakeholders and actors in relation to the common good. As it is obviously influenced by a plurality of norms, values and attitudes, it is also dependent on different world views and the interpretation of (policy) options for change.
- *Transformational knowledge* is needed to identify better technical, social, legal or cultural practices and how to realise them. It is related to change and innovation management. It is about learning, how to make existing procedures, structures and regulation more flexible and about processes to balance incompatible dimensions and decision making regimes, as well to navigate between technical inadequacy and political irrelevance (Guston and Sarewitz 2002) Furthermore, one needs to know how to respond to *Non-Knowledge*:
- *Uncertainty and Non-Knowledge:* When, for example, technology risks should be assessed, decision makers are confronted with limits of scientific knowledge. Scientists do not only generate knowledge, but also at the same time ignorance concerning possible side effects (Böschen et al. 2006). Decision makers and civil society need to "*Acknowledge and respond to ignorance, as well as uncertainty and risk, in technology appraisal and public policy making*" (cit. EEA 2001, p. 168), as the European Environment Agency (EEA) stated and still endorses (EEA 2013), when it comes to risks related to the precautionary principle and technologies.

Research begins with a politically and socially relevant problematique, requires big projects and is expected to lead to applied knowledge. As sustainability problems are characterised by uncertainty, complexity, values in dispute, high stake and urgent decisions, science under this conditions is anything but normal. No wonder then that there is a request for a new kind of science, post normal science, to accomplish this: "*By structure, method, and content, sustainability science must differ fundamentally from most science as we know it.*" (cit. Bolin et al. 2000)

5 Trans-disciplinary in Sustainability Research

Sustainable development has deep roots but it is not until the last decades that its consequences for how we conduct science have been expressly stated. Origins of a new kind of science, whether it is called trans-disciplinary research, post normal science, mode 2 or sustainability science (see Table 1 for definitions), trace back to the early 1990s.

Table 1 Overview on definitions

Sustainability science (Kates 2012, in: Weinstein/Turner 2012:3ff.)	The science and technology needed to develop sustainability is essentially integrative of the natural, social, and engineering sciences; seeks to bridge the communities engaged in promoting environmental conservation, human health, and economic development; and brings together the worlds of knowledge and action. […] sustainability science addresses fundamental questions of scale, nonlinear processes and complexity, and the unity of nature and society
Sustainability science (PNAS 2010 in: Weinstein/Turner 2012:7)	…an emerging field of research dealing with the interactions between natural and social systems, and with how those interactions affect the challenge of sustainability: meeting the needs of present and future generations while substantially reducing poverty and conserving the planet's life support systems
Action research (Hult/Lennung 2007:247)	Action research simultaneously assists in practical problem solving and expands scientific knowledge, as well as enhances the competencies of the respective actors, being performed collaboratively in an immediate situation using data feedback in a cyclical process aiming at an increased understanding of a given social situation, primarily applicable for the understanding of change processes in social systems and undertaken within a mutually acceptable ethical framework
Mode 2 knowledge production Gibbons et al. 1994, pg. 167	Knowledge production carried out in the context of application and marked by its: trans-disciplinarity; heterogeneity; organisational heterarchy and transience; social accountability and reflexivity; and quality control which emphasises context and use-dependence. Results from the parallel expansion of knowledge producers and users in society
Post-normal science (Funtowicz/Ravetz 2003:1)	Post-Normal Science (PNS) is a new conception of the management of complex science-related issues. It focuses on aspects of problem solving that tend to be neglected in traditional accounts of scientific practice: uncertainty, value loading, and a plurality of legitimate perspectives. PNS considers these elements as integral to science. By their inclusion in the framing of complex issues, PNS is able to provide a coherent framework for an extended participation in decision-making, based on the new tasks of quality assurance. The shift to a post-normal mode is a critical change

(continued)

Transdisciplinary science Burger/ Kamber 2003, pg. 44, Klein et al. 2001:4	We characterize 'trans-disciplinary science' as (1) cognitive and social co-operation across disciplinary boundaries, (2) an intention towards the direct application of scientific knowledge in both political decision making and societal problem-solving, and (3) the participation of non-scientific stakeholders within research processes
	The core idea of trans-disciplinarity is different academic disciplines working jointly with practitioners to solve a real-world problem. It can be applied in a great variety of fields

Meeting the challenge of sustainability science requires new styles of institutional organization to foster and support inter-disciplinary research over the long term. A permanent deliberate discussion about key questions, appropriate methodologies and institutional settings is needed in connection to policy agenda setting. Research itself must be focused on the character of nature-society interactions, on our ability to guide those interactions along sustainable trajectories. Trans-disciplinarity is, however, a science beyond disciplines (interdisciplinary), integrating non academic actors in research projects that are aligned to master sustainability issues (Decker 2007). The background for the request for trans-disciplinarity is that the urgent real-world problems to be addressed for sustainable development do not easily fit into the research process of normal science, as described by e.g. Kuhn (1970). Instead it is characterized by Hirsch-Hadorn et al. (2008:33) and Pohl (2011:620):

- Related to real-life problems and specific problem solving
- Transcending of disciplinary and paradigmatic boundaries
- Participation of (non-academic) stakeholders
- Acceptance of diverse perspectives, problem framings and interpretations
- Holistic, i.e. non-reductionist approach, accepting the complexity of the issue
- Linkage of abstract and case-specific
- Explicitly normative: knowledge should solve specific problems (transformative knowledge) based on transparent ethical values to promote what is widely seen as common good

It is clear that the results from sustainability science should be relevant for problems solving and thus being close to management. Trans-disciplinary sustainability science requires novel schemes, structures, instruments, methods and methodologies in order to address challenges ahead. However, there is a variety of methodological challenges to be considered in research (see Pohl and Hirsch-Hadorn 2007:92ff.; Kates et al. 2001:641; Swart et al. 2002:1994 and others), i.e.:

- Spanning the spatial scales between diverse issues, such as globalized economies and local faming
- Spanning the temporal scales related to intergenerational justice between generations and accounting of long-range effects in ecological systems
- Addressing cross-cutting issues
- Combination of qualitative and quantitative research methods
- Considering the urgency of problem solutions needed and inertia towards transformation and change in social system systems (psychological barriers, motivation and human behaviour, technological and institutional path dependencies)
- Dealing with complexity and multiple stressors in impact-chains and interconnectedness of themes and issues (i.e. ecosystem functions—poverty—climate change; bio based products/fuels—malnutrition and hunger;—land competition and extensive organic agriculture—land use for animal feedstuff; scarce resources and land grabbing etc.)
- Integration of different perspectives and mind-sets of actors involved
- Understanding and handling of unpredictability and uncertainty, esp. regarding critical emission loads, thresholds and systems tipping points,
- Avoiding analysis-paralysis and entering a problem-solving mode
- Participation of non academic stakeholders, such as representatives from business and industry as well as civil society, politicians, other decision makers and persons concerned, at early research stages and linking of policy development and action
- Respecting the unity of nature and society

A post normal research process Wicked sustainability problems imply not only the involvement of different disciplines and non-academic actors, but also an adjusted research process, which is to be considered as recursive or iterative and which consists of overlapping stages (Jahn 2012:3; Pohl and Hirsch-Hadorn 2007:42):

- Problem identification, structuring and transformation:
- Identification of sustainability problems and related societal actors and academic disciplines
- Transformation into epistemic objects by applying theories in order to derive a theoretical framework to determine adequate research questions or problem solving strategies
- Determination which and how actors should be involved in research, taking into account pluralistic interests, values and circumstances and heterogeneous problem perceptions
- Production of new knowledge by applying (disciplinary) qualitative or quantitative research methods or engineering and conceptualisation instruments
- Embedding and implementation of results in social situations, i.e. everyday-life and scientific contexts (i.e. "pilot intervention")
- Peer-Review procedures to test and evaluate the expected impacts

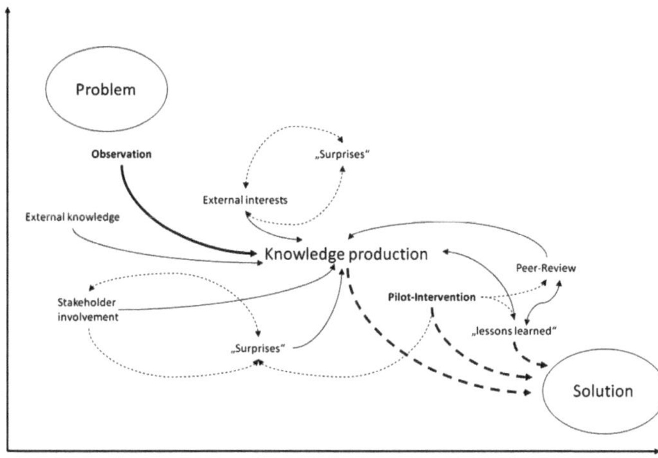

Fig. 2 A recursive trans-disciplinary research process (after Gross et al. 2005:275, modified)

The research process is less straightforward as idealised (see Fig. 2), e.g. in empirical research or engineering. Instead it needs to be considered as a recursive learning process, where kick-backs due to "surprises", i.e. by conflicting interests of stakeholders or trade-off dilemma should be expected. However, being recursive may considered as an advantage: intermediary results can be critically reflected and adjusted against diverse perspectives (single-and double loop learning). Although new research questions or information demand may arise, a recursive research design is also a mean of reduction of complexity by problem identification and structuring (cp. Pohl and Hirsch-Hadorn 2007:43).

6 Quality Criteria

From the characteristics and requirements of a trans-disciplinary and sustainability oriented research, quality criteria may be derived, which count in addition to reliability and validity, both considered as traditional criteria in quantitative empirical research. Some of them will be summarized here:

- Relevance of research results for academia, i.e. validity of results, orientation at state-of-the art knowledge related to research findings and methodology, transparency about own and research from other colleagues, honestly, correct citation
- Practical relevance for stakeholders and decision makers, i.e. methodological benefits of concept, instruments and tools in real-life applications, adaptability, practicability, usability and adequateness
- Comprehensible and reproducible documentation of all results during the research, e.g. of workshops, focus groups or case studies

- Justification in case of non-conformities related to the planned research design in order to ensure a analytical and systematic—as opposed to arbitrarily—procedures that follow an theoretical framework
- Communicative validation by extended peer-reviews, i.e. critical and constructive discussion with experts from different scientific disciplines, layman and actors involved in the research process
- Application of "mix-method"-approaches, i.e. triangulation, to avoid bias and to reduce challenges in admission of research subjects and to contribute to generalization of results.

7 Obstacles for Trans-disciplinarity

Sustainable Development has for a long time had its focus on the environment and natural sciences. Other academic disciplines have felt excluded; aspects of economics, human welfare, the political systems etc. have been underdeveloped in the early academic discourse. Later the discourse was "high jacked" and changed towards the consent-oriented "three-dimension" approach, which causes difficult trade-off dilemmas and green-washing, as everything and nothing became labeled "sustainable". The nested systems approaches, i.e. environment in focus, was repressed in the course of academic in political discussions. The un-deliberated and non-selective usage of the word "sustainability" made it regrettably a buzzword, rather than a serious concept regarding the state of the world. (e.g. Goldie et al. 2005; Parr 2009; Platje 2006; Leal 2011).

In academia there seem to be different academic cultures, say tribes. Universities are structured along disciplines: "*The world has problems, universities have departments*" (cit. Bozemann 1990). Faculties often are quite independent, like small universities; many students never leave their faculty! Beside of the social organization this leads likewise to an enrolment in epistemic cultures, including learning-styles and procedures how knowledge is produced and what is considered as legitimate knowledge (Nerland et al. 2010; Knorr Cetina 1999). The prevailing research culture is very reductionist. In particular PhD students have great difficulties to think outside their narrow topic. There is often a quest from both academia and politicians for reliable and objective evidence as result of research studies, which is inadequate for wicked sustainability problems (Kastenhofer 2011). Different knowledge cultures came to different, sometimes incompatible results and thus public controversies e.g. in the evaluation of emerging technologies like biotechnology (Böschen et al. 2006). At the same time, as general education ("bildung") expects students to be familiar with literature, the existing political system, what money and banks are, technical "hard-knowledge" (kWh, phosphate or eutrophication) are, however, most often accepted as mysteries.

8 Cases of Trans-disciplinary Education for Sustainable Development (ESD)

Transition to a more sustainable world requires not only experts and practitioners, but also that large strata of society are informed and know about the key issues of sustainable development. This is the core position in the efforts to develop and implement Education for Sustainable Development (ESD) on all levels, from pre-school to universities. This mission needs to reach not only decision makers, but everyone in society. The United Nations for this reason opened a Decade for ESD, 2005–2014, under the leadership of UNESCO. In the documents we find very similar wordings as in the background to sustainability science: trans-disciplinarity, applicability, normative. Students should ideally become "change agents" and contribute to a future more sustainable world. Teaching and education in a trans-disciplinary way requires competences, capabilities and skills to solve sustainability related problems (Dusseldorp 2012:226), which demand situational approaches not routines or mechanical guidebooks.

As one example of how to deal with trans-disciplinarity in teaching and research activities by the Baltic University Programme (BUP), a network and platform for cooperation among universities in the Baltic Sea Region, will be presented. More than 200 institutions of higher education are members of the network, which is coordinated by a secretariat at Uppsala University. During more than 20 years, a variety of teaching and learning materials have been developed and integrated in the university courses in different ways (Table 2). ICT was used from the beginning of the network (i.e. by lectures broadcasted via satellites to Eastern European Countries, as well as audio and video conferences) in order to reach mass scale distance audiences. In 2012 more than 7,500 students have been enrolled in BUP courses at universities in 15 European countries.

Additionally, regular teachers and students conferences have been organised. The Baltic University Programme focuses its activities on sustainable development, in an academic, political and pragmatic sense (cp. BUP 2013). BUP develops a competence not provided by the normal curriculum. It addresses regional development, use a systems approach (systems analysis and mapping) and promote regional understanding. The courses are often connected to applied projects in cooperation with other actors in society. They ask for problem solving and practical results. A pragmatic perspective have been taken under integration of various non academic stakeholders, decision makers and civil society (i.e. within the Baltic University Urban Forum, see BUP 2007). Publications have been interdisciplinary (multidisciplinary) and regional in scope. Conferences have gathered teachers and students from many disciplines and a wide range of countries.

Another case of good practice for trans-disciplinarity can be presented from the University of Applied Sciences Zittau/Goerlitz, which is offering a Bachelor program specifically on "Ecology and Environmental Protection" since 1994. Although the title of the program may let us expect something else, the philosophy is to provide both theory and applied science that is not reduced to classical biotic

Table 2 The study courses offered by the Baltic University Programme (BUP)

	Name of the course	ECTS	Decription
Undergraduate level	Environmental science (ES)	15	The ES courses give multidisciplinary overviews of the situation of the environment in the Baltic Sea region and on how to manage and protect it. The courses provide an excellent background for studying the Baltic University courses on sustainable development
	English for environmental science		This is an intermediate level English language course which develops comprehension, vocabulary and skills of expression. Terminology used in environmental studies is explained and practised
	The Baltic Sea region area studies (AS)	15	The AS courses treat the Baltic Sea region and its societies under eight headings: history, culture, language, democracy, multiculturality, social conditions, economics and security. The courses deal mostly with the conditions in the region after the systems change in 1989–1991. Regional development is in focus with emphasis on democracy, human rights and economics
	A sustainable baltic region (SBR)	7.5	The course deals with sustainable use and management of natural resources and long-term protection of the environment. Important issues in the course are energy and energy use, material flows, economy and ethics, industry, agriculture, transport and community development
	Sustainability applied in international learning (SAIL)	7.5	A course at Uppsala University on sustainability organised by the Baltic University Programme. The education consists of workshops, seminars and group discussions on board a sailing ship
Master level	Sustainable water management (SWM)	7.5	The SWM courses are master's level courses focusing on issues of sustainable use of water and water resources in the Baltic Sea region. Students should preferably have a background in e.g., hydrology, geology, physical geography, agronomy, soil science, forestry or environmental engineering
	Community development (CD)	15	The CD courses are master's level courses focusing on sustainable community development and urban planning with an emphasis on the Baltic Sea region. The courses are

(continued)

Table 2 (continued)

	Name of the course	ECTS	Decription
			interdisciplinary, problem oriented and preparatory for a professional career
	Environmental management (EM)	30	The EM courses address management in all kinds of organizations, although industries are in focus. They are well adapted for competence development for professionals. The course focuses on environmental policy, cleaner production technologies, life cycle assessment and environmental management systems and certification
	Ecosystem health and sustainable agriculture (EHSA)	30	The EHSA courses address the field of rural development, sustainable agriculture and animal health pertaining to the Baltic Sea Region and to some degree also the Great Lakes region. The topics of the course are sustainable agriculture, ecology and animal health and rural development and land use

ecology and natural conservation. Instead, a holistic approach is taken, integrating a variety of additional technical and economically oriented measures and sustainability principles. This follows the basic idea, that specialization, i.e. the detailed investigation of phenomena from a discipline or sub-discipline, is necessary but not sufficient (Schaltegger et al. 2013). Capacity building and specialization in the first semesters is considered as productive, but needs to be re-integrated in a "bigger picture" in a problem-oriented way. Lectures in the program are committed to this understanding which is reflected in the program design. Still, the program is quite unique in Germany. Students from all over the country apply for the course and make it one of the most successful programs of the university.

9 Conclusion and Afterthoughts

In this paper we have argued that trans-disciplinarity is a necessary methodological starting point for research in order to tackle sustainability problems, if science and academia are expected to make a relevant contribution to problem solving. This is especially true, when societal relevant problems are complex, knowledge is uncertain and plural values and mindsets, stakeholders and actors need to be integrated in a process, where different and sometimes complementary academic paradigms are involved. A model for a research process was presented, appropriate to generate different kinds of knowledge: (a) systems knowledge, i.e. state-of-the-art, understanding of a status-quo real-life), (b) target knowledge (about what is desirable, needed, demanded by whom and when, (c) transformative knowledge (about how to make change and which political, economic, social, legal and cultural aspects need to be considered). It was, furthermore, argued that in the face of unsustainable development also education should embrace the notion of trans-disciplinarity and introduce it on a broad scale.

Trans-disciplinary research and education is, however, not only challenging as the inherent complexity need to be grasped, but because also values such as the liberty and freedom of research—as e.g. statutory in the German constitution—are touched. With Enlightment and the promotion of natural sciences, rationality, i.e. scientific reasoning instead of superstition or ideology was guiding academia and was a basis for independency of scientist from political, economic or societal interest (Mulder 2010; Scott 2002). This leads to the question if freedom of research is a higher value as the purpose to become effective in cultural and societal shaping by support of research findings. It is therefore necessary to reflect about values and ethical justification of value-loaded and value-driven research in order to protect against particular fashions or ideologies. Education at Universities is surely not about inculcation of norms (Jickling 1994), but development of skills for reflection and critique in order to support students in finding their own way. Researchers committed to trans-disciplinarity and sustainability need to answer themselves and others, if normative statements are meant as being imperative or conditional (Kopfmüller et al. 2001, Grunwald 2002). It is itself an ethical prerequisite that science contributes to a deliberate problem solving process.

References

Baltic University Programme www.balticuniv.uu.se

Bolin B, Clark W, Corell R, Dickson N, Faucheux S, Gallopan G, Gruebler A, Hall M, Huntley B, Jaeger J, Jodha N, Kasperson R, Kates R, Lowe I, Mabogunje A, Matson P, McCarthy J, Mooney H, Moore B, O'Riordan T, Schellnhuber J, Svedin U (2000) Sustainability science: statement of the Friibergh Workshop on sustainability science. http://sustainabilityscience.org/content.html?contentid=774. Last accessed 14 Apr 2014

Böschen S, Kastenhofer K, Marschall L, Rust I, Soentgen J, Wehling P (2006) Scientific cultures of non-knowledge in the controversy over Genetically Modified Organisms (GMO). Cases Molecular Biology and Ecology GAIA 15(4):294–301

Bozeman MC (1990) The environments of United-States R-and-D laboratories—political and market influences. Policy Sci 23:25–56

BUP (2007) Baltic University Urban Forum. http://www.balticuniv.uu.se/buuf/. Last accessed 14 Apr 2014

BUP (2013) Annual Report 2012. Baltic University Programme Uppsala Centre for Sustainable Development, Uppsala University

Burger P, Kamber R (2003) Cognitive integration in transdisciplinary science: Knowledge as a key notion. Issues in Integrative Studies 21:43–73

Conklin J (2009) Wicked problems and social complexity. White Paper, CogNexus Institute

Decker M (2007) Angewandte interdisziplinäre Forschung in der Technikfolgenabschätzung. Graue Reihe der Europäischen Akademie zur Erforschung von Folgen wissenschaftlich-technischer Entwicklungen, Bad-Neuenahr-Ahrweiler GmbH

Dörner D (1997) The logic of failure: recognizing and avoiding error in complex situations. Perseus Press, Cambridge

Dusseldorp M (2012) Bildungspotenziale der Simulation von TA-Institutionen: Das Beispiel des TAB. In: Dusseldorp, M. and Beecroft, R. (2012). Technikfolgen abschätzen lehren. Bildungspotenziale transdisziplinärer Methoden. Springer VS

EEA (European Envionment Agency) (2001) Late lessons from early warnings: the precautionary principle 1896 –2000. In: Harremoës P et al. (eds) Office for Official Publications of the European Communities, Luxembourg

EEA (European Envionment Agency) (2013) Late lessons from early warnings: science, precaution, innovation. In: Harremoës P et al. (eds) Office for Official Publications of the European Communities, Luxembourg

Folke C, Hahn T, Olsson P, Norberg J (2005) Adaptive governance of social- ecological systems. Annu Rev Environ Resour 2005(30):441–473

Funtowicz S, Ravets J (1993) Science for the post-normal age. Futures 25:739–755, Butterworth-Heinemann Ltd

Funtowicz S, Ravets J (2003) Post-normal science. Prepared for the international society for ecological economics—Internet encyclopedia of ecological economics. http://isecoeco.org/pdf/pstnormsc.pdf. Last accessed 12 Aug 2014

Gibbons M, Nowotny H, Limoges C (1994) The new production of knowledge: the dynamics of science and research in contemporary societies. Sage Pubn Inc, London

Goldie J, Douglas B, Furnass B (2005) In search of sustainability. CSIRO Publishing, Melbourne

Gross M, Hoffmann-Riem H, Krohn W (2005) Realexperimente. Robustheit und Dynamik ökologischer Gestaltungen in der Wissensgesellschaft. Soziale Welt 54:241–258

Grunwald A (2002) Normativ, aber nicht normativistisch. Bemerkungen zur Grenze zwischen Wissen, Werten und Bekehrung als Anwort auf Christian Berg. Technikfolgenabschätzung – Theorie und Praxis Nr. 2, 11. Jg., Juli 2002

Guston DH, Sarewitz C (2002) Real-time technology assessment. Tech Soc 24:93109

Hirsch-Hadorn G, Hoffmann-Riem H, Biber-Klemm S, Grossenbacher-Mansuy W, Joye D, Pohl C, Wiesmann U, Zemp E (eds) (2008) Handbook of transdisciplinary research. Springer Science + Business Media B.V, Dordrecht

Hult M, Lennung SA (2007) Towards a defintions of action research: a note and bibliography. J Manage Stud 17(2):241–250

Jackson T (2009) Prosperity without Growth? The transition to a sustainable economy. Earthscan, London

John T (2012) Transdisciplinarity as a research practice to approach sustainability challenges: a social-ecological perspective. John Benjamins Publishing, Amsterdam

Jickling B (1994) Why I don't want my children to be educated for sustainable development. Trumpeter 11(3). http://trumpeter.athabascau.ca/index.php/trumpet/article/view/325/497. Last accessed 14 Apr 2012

Kastenhofer K (2011) Risk assessment of emerging technologies and post-normal science. Sci Technol Human Values 36(3):307–333

Kates RW (2012) From the Unity of Nature to Sustainability Science: Ideas and Practice. In: Weinstein MP, Turner RE (eds) Sustainability science: the emerging paradigm and the urban environment. Springer Science + Business Media, LLC, New York

Kates RW, Clark WC, Corell R, Hall MJ, Jaeger CC, Lowe I, McCarthy JJ, Schellnhuber HJ, Bolin B, Dickson NM, Faucheux S, Gallopin GC, Grübler A, Huntley B, Jäger J, Jodha NS, Kasperson RE, Mabogunje A, Matson P, Mooney H, Moore B, O'Riordan T, Svedin U (2001) Sustainability science. Science 292:641–642, American Association for the Advancement of Science

Klein JT, Grossenbacher-Mansuy W, Häberli R, Bill A, Scholz RW, Welti M (eds) (2001) Transdisciplinarity: joint problem solving among science, technology, and society. Synthesebücher. Birkhäuser, Basel

Knorr Cetina K (1999) Epistemic cultures: how the sciences make knowledge. Harvard University Press, Cambridge

Kopfmüller J, Brandl V, Jörissen J, Paetau M, Banse G, Coenen R, Grunwald A (2001) Nachhaltige Entwicklung integrativ betrachtet. Konstitutive Elemente, Regeln, Indikatoren. Edition Sigma: Berlin

Kuhn TS (1970) The structure of scientific revolutions. International encyclopedia of Unified Science, vol 2/2, 1962. The University of Chicago, Chicago

Leal W (2011) About the role of Universities and their contribution to sustainable development. High Educ Policy 24:427–438

Mulder K (2010) Don't preach! Practice! Value laden statements in academic sustainability education. Int J Sustain High Educ 11(1):74–85

Nerland M, Jensen K, Bekele TA (2010) Changing cultures of knowledge production and learning in higher education: a literature review. University of Oslo Department of Educational Research, Oslo

Parr A (2009) Hijacking sustainability. The MIT Press, Cambridge

Platje J (2006) Who is interested in what kind of sustainable development? Time- horizons and stakeholder interests. Econ Environ Stud 8:13–20

PNAS (2010) Sustainability science. http://sustainability.pnas.org/. In: Proceedings of the National Academy of Sciences of the United States of America

Pohl C (2011) What is progress in transdisciplinary research? Futures 43:618–626

Pohl C, Hirsch-Hadorn. (2007) Principles for designing transdisciplinary research. Oekom, Munich

Polimeni, J.M., Mayumi, K., Giampietro, M., Alcott, B. (eds.) (2008). The Jevons Paradox and the Myth of Resource Efficiency Improvements. Earthscan

Rittel WJH, Webber MM (1973) Dilemmas in a general theory of planning. Policy Sci 4:155–169, Elsevier Scientific Publishing Company

Schaltegger S, Beckmann M, Hansen Erik G (2013) Transdisciplinarity in corporate sustainability: mapping the field. Bus Strategy Environ 22:219–230

Scott W (2002) Education and sustainable development: challenges, responsibilities, and frames of mind. Trumpeter 18(1):1–6

Sterrmann JD (2012) Sustaining sustainability: creating a systems science in a fragmented academy and polarized word. In: Weinstein MP, Turner RE (eds) Sustainability science: the emerging paradigm and the Urban environment. Springer Science + Business Media, LLC

Stokes DE (1997) Pasteur's quadrant: basic science and technological innovation. Brookings Institution Press, Washington

Swart R, Raskin P, Robinson J, Kates RW, Clarc WC (2002) Critical challenges for sustainability science. Science 297:1994–1995

WBGU (1997) World in transition: the research challenge. German advisory council on global change, Alfred-Wegener-Institute for Polar and marine research. Springer, Berlin

Authosr Biography

Markus Will Dipl-.Ing (FH), works as lecturer at the University of Applied Sciences Zittau/ Goerlitz. His academic interests lie in life cycle assessment, technology assessment and innovation theory. Recently, he deals with methodological questions of greenhouse gas accounting for regions. He regularly works as facilitator and consultant, trying to support the development of sustainability strategies in organizations and municipalities. Despite his deep involvement with the topic, his environmental footprint is still larger than would be adequate in a fair and sustainable world.

Lars Rydén PhD, is emeritus professor at the Centre for Sustainable Development at Uppsala University, Sweden. With a background in biochemistry he started in 1991 the Baltic University Programme as a response to the large-scale political changes at the end of the Cold War and new possibilities to cooperate between East and West over the old Iron Curtain. The Programme (www.balticuniv.uu.se) has grown ever since and now includes 232 member universities. From a start dominated by education on environmental issues it has developed a strong content of sustainable development with courses, applications and efforts to introduce changes by cooperation with authorities. He attempts to improve sustainability in everyday life, e.g. by biking and going by train when possible.

An Action Research Project for Embedding Education for Sustainable Development in a University Curriculum: Processes and Prospects

Gisela Cebrián, Marcus Grace and Debra Humphris

Abstract

Research on sustainable development in higher education has tended to focus on environmental management of university estates and operations, and case studies and examples of good practice, without presenting the coherent theoretical or methodological approaches required to look at the change processes of universities seeking to embed sustainable development. Although the value and contribution of university initiatives has been articulated, little holistic and structural transformation of universities has been achieved so far. This paper presents an action research project undertaken at the University of Southampton to develop an organisational learning model to embed Education for Sustainable Development (ESD) within the undergraduate curriculum. An action research approach guided by participatory and emancipatory approaches was used. The researcher aimed to learn from real practice through acting as a facilitator for curriculum development in ESD within an interdisciplinary group of academic staff members. A critical friend position was acquired within a community of practice to implement a programme that attempted to embed sustainable development within the student experience. This paper presents the qualitative methodology used in this project, providing a detailed overview of its research

G. Cebrián (✉) · M. Grace
Southampton Education School, University of Southampton,
Building 32, Southampton SO17 1BJ, UK
e-mail: gcb1d10@soton.ac.uk

M. Grace
e-mail: M.M.Grace@soton.ac.uk

D. Humphris
Imperial College London, South Kensington Campus, London SW7 2AZ, UK
e-mail: d.humphris@imperial.ac.uk

W. Leal Filho et al. (eds.), *Integrative Approaches to Sustainable Development at University Level*, World Sustainability Series, DOI 10.1007/978-3-319-10690-8_48

methods, processes and future prospects, which can inform other higher education institutions in their journey to embed ESD holistically.

Keywords
Action research · Curriculum · Organisational learning · Education for sustainable development

1 Introduction

Participatory and action-orientated approaches have been reported as critical to foster organisational learning and change towards sustainable development in private and public organisations (Doppelt 2010; Dunphy et al. 2007). Several publications have reported on case studies and the successful contribution of action research to foster organisational learning and change towards sustainable development (Molnar and Mulvihill 2003; Siebenhüner and Arnold 2007). Similarly, Marsh (2011) and Barasa Atiti (2008) used action research for this purpose in their doctoral studies. This evidences an increasing academic research practice and awareness in the use of innovative, participatory, collaborative and action-orientated research approaches to explore and create organisational learning and change towards sustainable development through the research process itself (Cebrián et al. 2012). In the area of Education for Sustainable Development (ESD), participatory action research and participatory research approaches have been used to work on professional development and curriculum development on ESD with school and university teachers (Gayford 2003; Tilbury et al. 2004). Action and reflection are critical to build the awareness, agency and association needed to deal with sustainable development (Ballard 2005). However little research has reported on the research methods and the change process of universities seeking to embed sustainability development in its activities holistically (Velazquez et al. 2006).

This paper presents an action research project undertaken at the University of Southampton to develop an organisational learning model to embed ESD within the undergraduate curriculum. First, the different dimensions of action research are discussed. Second, the action research project, the research stages and the roles adopted by the researcher are presented. Third, the research methods are outlined. Finally a discussion of the research processes and future prospects is provided to inform other higher education institutions in their journey to embed ESD holistically.

2 Framing the Different Dimensions of Action Research

The methodological principles for action research are: integration of research and action; conducted through collaborative partnerships; high level of reflexivity; stimulates transformative learning amongst participants; embraces broader political and historical contexts; and starts from aspiration and a vision of social transformation and social justice (Somekh 2006). These principles provide an open sight of action research and represent an umbrella for the different types and approaches such as participatory action research, critical action research, collaborative inquiry and action learning (Kember Associates 2000; Reason and Bradbury 2008). This study was conceived as a critical action research because it aimed at challenging current thinking and practice in ESD through critical reflection and action (Carr and Kemmis 2009; Kemmis 2010). Action research is a cyclical process of reflection, planning, acting and observing (McNiff and Whitehead 2002).

Some of the essential features of action research are: the vision that all the participants in the research have something to contribute and in turn have something to learn; participants are seen as co-researchers and co-learners, including the researcher; knowledge and theory are inseparable from practice; the main purpose is the improvement of a real situation or problem; reflection and action are two core elements; and the whole learning-by-doing process is what counts (McNiff and Whitehead 2002; Noffke 2009).

Five main characteristics of action research are identified as: human flourishing; participation; knowledge-in-action; practical issues; and emergent (Reason and Bradbury 2008). In this action research, academic staff members of the organisation were involved in developing their own solutions with the support and action of a facilitator for curriculum development towards ESD and as a critical friend for a community of practice focused on embedding sustainable development within the organisation (participation). Participants worked on a practical issue that mattered to people (practical issues). The research purpose and topic were meaningful to individuals and aimed at improving social well-being and sustainable development (human flourishing). The knowledge gained from the research emerged from real practice and interactions with the research participants (knowledge-in-action). In this project action research was envisioned as the means for developing new practices towards embedding ESD within the curriculum and for fostering the reflection that contributed to the improvement of a community of practice's work (emergent).

For Noffke (2009) there are three dimensions of the action research approach: the personal, the professional and the political. These dimensions can be differentiated according to the purpose of the action research work in educational settings, where different assumptions and practices can be recognised. The personal dimension makes reference to the practitioner as researcher and the process of self-reflection, planning and introducing changes to improve self-practice (McNiff and Whitehead 2002). The professional dimension corresponds to professional development purposes; usually in education (concerning teachers); with the aim of enhancing the teaching profession and developing teacher practice (Noffke 2009). The political dimension is usually embedded in the previous dimensions, but the purpose is mainly to generate democratic processes to empower groups often without voice, such as lower socio-economic groups and underprivileged communities (Carr and Kemmis 2009). It is linked to power and political issues, structures, participation and the decision-making of the community, and ideally leads to educational and social change (Kemmis 2010). For Noffke (2009) these three dimensions (personal, professional and political) are interconnected, however researchers tend to focus on a single aspect. By contrast Carr and Kemmis (2009) argue that it is not possible to give different emphasis to an action research project since it will inevitably involve:

- The self-transformation of participants as it develops personal understandings.
- The professional development of the community of practitioners.
- The empowerment of participants due to a democratic process that potentially can change the social system, the institution and the world.

The integration of these dimensions is what makes action research extremely relevant and transformational for the exploration of sustainable development in higher education, because of the different interpretations and complexity of the term (Filho 2000, 2011). The emancipatory or critical approach consciously explores the relationship between these three faces (Carr and Kemmis 1986). Acquiring a dialogue and a self-critical position with real problems of practice can lead to new insights and ways for embedding sustainable development holistically in the curriculum, which can potentially lead to educational and social change (Sterling 2004).

Moreover, a useful differentiation is made between first, second and third person (or order) action research (Reason and Bradbury 2008). An action research project that engages in these three modes of practice is more compelling and sustainable in the long-term. According to Reason and Bradbury (2008, p. 6), '*second person inquiry starts with interpersonal dialogue and includes the development of communities of inquiry and learning organizations*'. This research was carried out with an interdisciplinary group of academics and an existing community of practice. The researcher engaged them in collaborative, supportive and reflective discussions that aimed to develop new understandings and professional practice in ESD, fostering organisational learning towards sustainable development.

3 The Action Research Project and the Roles Adopted by the Researcher

3.1 The Action Research Project

The research design adopted comprised three stages of data collection:

- **Stage I:** Identify the factors influencing academic staff members' engagement in ESD and gain a deeper understanding of the University of Southampton context and organisation. Fourteen academic staff members from different disciplines were interviewed as part of an exploratory study and reconnaissance phase of a typical action research cycle.

- **Stage II:** Work with an interdisciplinary group of five academic staff members to critically reflect and act towards embedding ESD in their teaching practice through the creation of an action learning. The role as researcher was to be a facilitator for curriculum development in ESD through providing staff with support and space for interdisciplinary and critical reflection.
- **Stage III:** Fulfil the role of a critical friend for the Sustainability Programme team. The role as a researcher in this stage was to provide feedback to the group, assisting the process of reflection and articulating experiences by asking critical and sometimes provocative questions (Table 1).

Figure 1 summarises the overall research design and of the action research cycles.

Table 1 The research stages, its research aims, participants and the research methods used to collect data

Research stage	Research aims	Participants	Research method
Stage I Exploratory phase	Collect baseline data and gain a deeper understanding of the university Identify the factors influencing academic staff engagement in ESD Inform the research design and identify possible participants for subsequent stages	Fourteen academic staff members from different university departments and disciplines Roles included: director of programme, professor, teaching fellow, lecturer, senior manager and reader	Semi-structured interviews Participant observation
Stage II Facilitator of curriculum development in ESD	Foster critical reflection and action, and transform thinking and action towards embedding ESD within their teaching practice Gain learning from real practices and from participants' journeys on embedding ESD within their teaching practice	Five academic staff members from different roles and departments, and a range of teaching and research responsibilities Roles included: director of programme, lecturer, professor, senior manager and teaching fellow	Semi-structured interviews Action learning conversations Participant observation
Stage III Critical friend for the Sustainability Programme	Contribute to the self-reflection and collective reflection of the group Promote a critical group reflection which could facilitate acquiring new perspectives and discourses	The group was formed of six people of differing placements, responsibilities and groupings within the university Roles included: staff member, lecturer, teaching fellow, undergraduate student and senior manager	Individual interviews Reflective sessions Participant observation

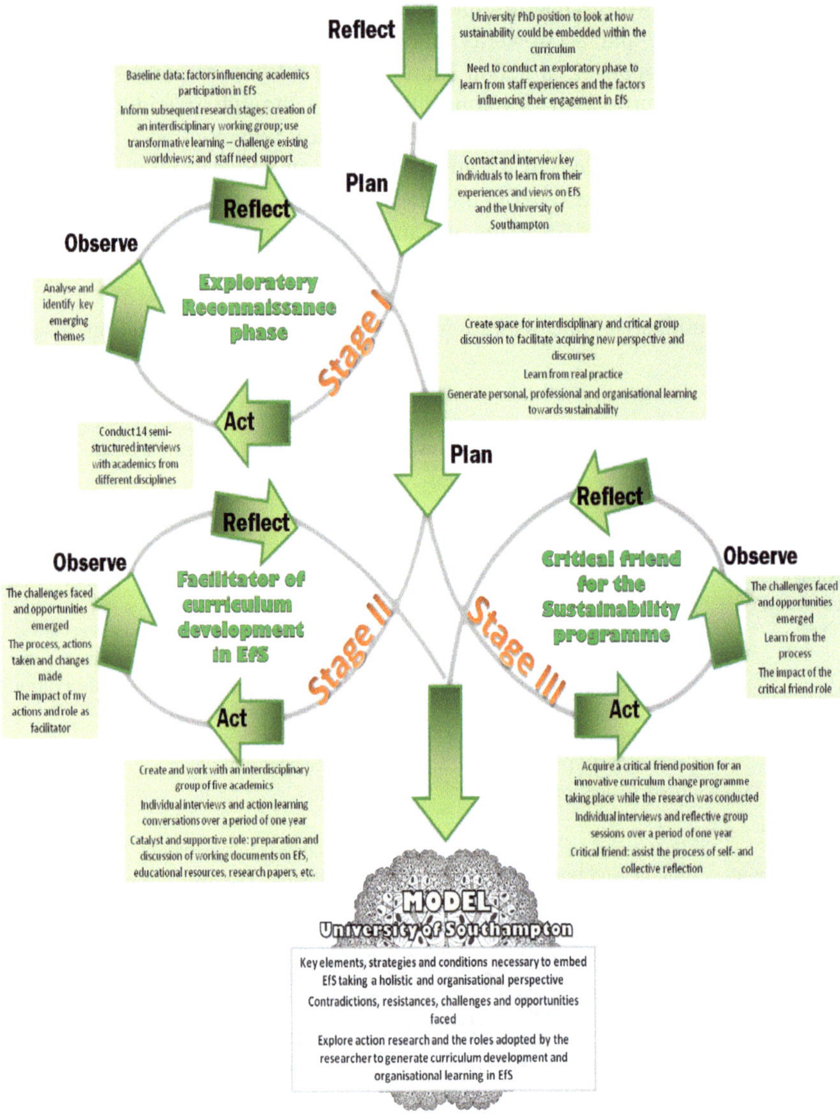

Fig. 1 Action research cycles of the research

3.2 The Facilitator and Critical Friend Role

A critical friend, as the name indicates, is a person who assists reflective processes in a supportive and helpful way. Costa and Kallick (1993, p. 50) defined a critical friend as:

> A trusted person who asks provocative questions, provides data to be examined through another lens, and offers critique of a person's work as a friend. A critical friend takes the time to fully understand the context or the work presented and the outcomes that the person or group is working towards.

Critical friends are facilitators of learning. They are able to listen, step back from the process, and assist through providing another perspective (Kember et al. 1997). Critical friends allow for time to reflect on processes and actions. Elliot (1991) further elaborated the role of critical friends in education by employing the phrase '*second order action research*'. The insider in the self-reflective and action research process leads to improving one's own teaching practice. Also, the outsider assumes a proactive role to promote both the creation of a reflective and collaborative learning community to aid the teachers' ability to learn and reflect on their practice. The outsider is defined as a '*second order action researcher*' and adopts a supportive role for the insider. In this sense, action research is about the process of learning in both, the insider-teacher and the outsider-supportive role. The outsider-supportive role, named critical friend or facilitator, can adopt different forms and roles, being more proactive or passive, depending on the project and the needs of the group (Kember et al. 1997).

Some examples of second-order action research projects also exist in the higher education context (Donche and Van Petegem 2004; Nascimento Botelho et al. 2010). Critical friends, being external advisers (Kember et al. 1997) or peers (Fulcher and Paull 2010), have been used in higher education to facilitate action research projects to improve professional practice of academic staff members. Critical friends are key agents in organisations that envision themselves as learning organisations (Senge 2006) for their ability to foster reflection and improvement of educational situations. However according to Fulcher and Paull (2010), little research exists on the real practices and processes related to the critical friend role.

The role of the critical friend or facilitator in an action research project is complex and controversial. Critical friends face a number of challenges, such as effective communication, ownership of the project, power relations and time constraints (Baskerville and Goldblatt 2009; Kember et al. 1997). In this sense, the challenges as facilitators of an interdisciplinary working group using action research, according to Tilbury et al. (2004), include: the type of support required by lecturers; the complex interactions between people from different disciplinary backgrounds; the lack of lecturers' time available; and the maintenance of motivation within the participants for the duration of the study. Thus, flexibility in the approach and in the different roles adopted is an intrinsic quality of facilitators of action research projects (Kember et al. 1997).

Moreover, critical friends and facilitators of action research need to be able to interact with people with different interests and who hold different worldviews and values. Establishing a personal and professional relationship with the participants and agree on the facilitator and critical friend role as an action researcher is critical. In this research a differentiation is made between the facilitator role (Stage II) and the critical friend role (Stage III). The term facilitator for curriculum development in

ESD is used to make reference to the role adopted for the interdisciplinary team of academic staff members at the University of Southampton (Stage II). This is because this was a leadership role, being more pro-active and supportive to the group and project development. In Stage III, the role of critical friend for the Sustainability Programme team was adopted. The term critical friend was used for this stage because the researcher became more of an outsider to an on-going initiative seeking to embed sustainability holistically within the student experience.

4 Research Methods

4.1 Semi-structured Interviews

A semi-structured interview approach was chosen because it allowed gaining a deep understanding of the research problem and world through unfolding participants' perspectives and experiences (Kvale and Brinkmann 2009; Minichiello et al. 1990). This allowed more flexibility and the possibility to generate a more comfortable environment, where rapport and trust could be developed between the researcher and the participants, and where participants' viewpoints and invisible factors related to the research topics, and understanding the world from the participants' perspectives could emerge (Minichiello et al. 1990). The interviews were designed with main topics for discussion, using open-ended questions and prompts, and allowing the participants to talk in-depth about other ESD-related issues that were important and relevant to them. The interview questions explored participants' experiences, views and understanding of ESD. Interview method is seen as a more comprehensive way to explore ESD, because of the complexities and controversies associated with the term (Cotton et al. 2009).

4.2 Action Learning Conversations

Action learning is an incessant process of reflection and learning that takes place with the support of a 'set' of peers with the aim to take action to improve practice (McGill and Brockbank 2004). In this research action learning was both a teaching and a learning approach to ESD that guided the research and a research technique. The creation of an action learning set was chosen as a research method for Stage II, to promote critical reflection, learning, and the development of new insights and practices in ESD amongst academic staff members (Revans 2011). Reflection and action are at the heart of the action learning process, given that it promotes reflection on past experiences, rethinking current action, and developing new insights in the members of the action learning sets (McGill and Brockbank 2004). Participants were asked to bring a challenge in each set, focusing on personal past experiences and ESD framework in the first set, on curriculum design and deliver in the second set and on personal, professional and organisational conditions in the last set. In this research each action learning set followed the structure of action learning conversations, which are based on a process of a repeated cycle for each challenge brought by participants (Marsick and Maltbia 2009, p. 162):

- Framing of the challenge as a question
- Unpacking meaning through sharing information about the context and prior action
- Peer questioning (to which the problem holder does not immediately respond) to unlock mental models that make one blind to other points of view
- Identifying assumptions that underlie current ways of framing the challenge
- Reframing one's understanding of the situation
- Making more informed decisions and taking informed action to address the challenge

Action learning conversations differ from other action learning approaches because they put emphasis on the critical reflection on possible assumptions made by the set members when framing the challenge, or when thinking about new insights and future action. This approach was chosen due to the interest of the researchers in promoting critical reflection and the questioning of current assumptions, values and beliefs in ESD amongst the research participants (Carr and Kemmis 2009).

4.3 Reflective Sessions

The group sessions conducted with Stage III participants, to provide them with my feedback and insights consisted on reflective sessions. Different techniques such as the Diamond 9, a tool for prioritising on different themes that were relevant to the group progress were used to help the group reflect and to foster strategic thinking and planning. In these sessions oral and written feedback was given on a number of themes identified through individual interviews and participant observations with the group. Other tools for reflection such as the learning organisation ideal (Senge 2006) were used. The following techniques were used to give effective feedback: to be helpful; to focus on behaviour and not on individual action; to use 'I' and not 'you'; to restrict feedback to elements about which I am certain; to focus on descriptions rather than judgements; to choose the appropriate time and place; to be precise and succinct; to make sure members understood the feedback; and to appeal for future action (Costa and Kallick 1993). These sessions were designed to provide the group with time for critical group reflection and for the sharing of insights and ways forward (Bambino 2002).

4.4 Participant Observation

Participant observation, ethnography and fieldwork are used interchangeably, where the researcher observes participants in their everyday life and experiences (Silverman 2006). It has been argued that participant observation is intrinsic to all social research, because one cannot study society without being part of it (Denzin and Lincoln 2011). In participant observation the researcher establishes a direct relationship with the social actors by staying in their natural environment. The purpose is to observe and describe their social actions by interacting with them, participating in their everyday ceremonials and rituals, and learning their code (or at least parts of it) in order to understand the meaning of their actions (Gobo 2011). Participant observation was used throughout the whole research process where the researcher engaged in informal sustainability events held at the University of Southampton and in interactions with the research participants in different research situations and informal meetings and events.

4.5 Discussion of Research Process and Future Prospects

Research for sustainability in HE has, over the last decade, become more focussed: on interdisciplinary; in transforming rather than informing; on having social impact; on social and structural change; on the researcher as a partner; and on research with people rather than on people (Tilbury 2012). So for example McMillin and Dyball (2009) used an action research approach to look at what education and operations' benefits are gained from campus sustainability projects, with the purpose of enhancing participation through forums, discussions and surveys, and envision sustainability creating organisational change. Moore (2005) used a participatory action research approach to examine ESD programmes' implementation at the University of British Columbia. Action research is also emerging as an increasingly used methodology in leadership and in training programmes for leaders to create sustainability processes in the context of their organisations or companies (Marshall et al. 2011).

Action research allows researchers to become active participants in different sustainability projects and activities taking place at the university, engaging key individuals in participatory process, and challenging existing organisational practices. Action research allows improving a real problems and situations, generating knowledge and action through the research process itself. However there are a number strengths and challenges for critical friends such as (Bambino 2002; Costa and Kallick 1993): build rapport and trust amongst the group; be flexible to adopt different roles when engaging with groups in different projects or work; reflect on the process and learning from the personal action, the role as a critical friend; provide support and effective feedback to the group without preaching. These elements have emerged as critical to conduct action-research projects towards embedding ESD in HE.

The framework of critical action research for ESD was suggested as the method to address academics' needs for developing strategy and action in their teaching practice. This method has enabled reflection and action, alongside the identification of specific needs of academics and the factors influencing their engagement and action in ESD. The potential of using these approaches to rethink current practice and to lead to new practices and actions of academics is evident. The use of an critical action research approach contributes to empowering academics to take ownership of sustainability and to connect it to their discipline. More empirical studies using action research are needed because they empower academics and motivate action. Action research contributes to better decision-making in terms of sustainability because it questions practice, and current assumptions and worldviews.

Participatory and emancipatory approaches such as action research an foster transformation and learning in ESD, catalyse critical reflection, collaboration and learning-by-doing from real practice. Action research and the methods discussed permitted to work with practitioners on a real situation, and to reflect and promote changes in the curriculum and organisation towards sustainability. Action research methodology is in line with ESD processes such as learning-by-doing, empowerment, dialogue and collaboration, innovation, and active and participatory learning. Through action research, academics develop new ways of understanding and new practices for embedding ESD within their teaching practice and within the organisation. Via this type of research the learning of individuals can lead to wider organisational learning and change. Due to the fact that they are contextualised, ESD interventions that are based on action-orientated processes, participation and collaboration are likely to be more culturally appropriate, and therefore more sustainable.

References

Ballard D (2005) Using learning processes to promote change for sustainable development. Action Res 3(2):135–156

Bambino D (2002) Critical friends. Redes Prof Dev 59(6):25–27

Barasa Atiti A (2008) Critical action research: exploring organisational learning and sustainability in a kenyan context. Macquarie University, Sydney

Baskerville D, Goldblatt H (2009) Learning to be a critical friend: from professional indifference through challenge to unguarded conversations. Camb J Educ 39(2):205–221

Carr W, Kemmis S (1986) Becoming critical: education, knowledge and action research. Deakin University Press, Geelong

Carr W, Kemmis S (2009) Educational action research: a critical approach. In: Noffke S, Somekh B (eds) The SAGE handbook of educational action research. SAGE Publications Ltd, London, pp 74–84

Cebrián G, Grace M, Humphris D (2012) Developing people and transforming the curriculum: action research as a method to foster professional and curriculum development in education for sustainable development in higher education. In: Filho WL (ed) Sustainable development at universities: New Horizons. Peter Lang Scientific Publishers, Frankfurt, pp 273–284

Costa AL, Kallick B (1993) Through the lens of a critical friend. Educ Leadersh 51:49–51

Cotton D, Bailey I, Warren M, Bissell S (2009) Revolutions and second-best solutions: education for sustainable development in higher education. Stud High Educ 34(7):719–733

Denzin NK, Lincoln YS (2011) Introduction: the discipline and practice of qualitative research. In: Denzin NK, Lincoln YS (eds) The SAGE handbook of qualitative research, 4th edn. SAGE Publications Inc, Thousand Oaks, pp 1–19

Donche V, Van Petegem P (2004) Action research and open learning: in search of an effective research strategy for educational change. Educ Action Res 12(3):413–431
Doppelt B (2010) Leading change toward sustainability: a change-management guide for business, government and civil society, 2nd edn. Greenleaf Publishing Limited, Sheffield
Dunphy D, Griffiths A, Benn S (2007) Organizational change for corporate sustainability, 2nd edn. Routledge, London
Elliot J (1991) Action research for educational change. Open University Press, Buckingham
Filho WL (2000) Dealing with misconceptions on the concept of sustainability. Int J Sustain High Educ 1(1):9–19
Filho WL (2011) About the role of universities and their contribution to sustainable development. High Educ Policy 24:427–438
Fulcher P, Paull M (2010) Critical friends: reflections on peer review of teaching. Paper presented at the ANZMAC 2010—conference theme: 'Doing more with less', University of Canterbury, Christchurch, 29 Nov–1 Dec 2010
Gayford C (2003) Participatory methods and reflective practice applied to research in education for sustainability. Can J Environ Educ 8(1):129–142
Gobo G (2011) Ethnography. In: Silverman D (ed) Qualitative research, 3rd edn. SAGE Publications Ltd, London, pp 15–34
Kember D, Ha T-S, Lam B-H, Lee A, NG S, Yan L, Yum JCK (1997) The diverse role of the critical friend in supporting educational action research projects. Educ Action Res 5 (3):463–481
Kember D Associates (2000) Action learning and action research: improving the quality of teaching and learning. Kogan Page, London
Kemmis S (2010) What is to be done? The place of action research. Educ Action Res 18 (4):417–427
Kvale S, Brinkmann S (2009) Interviews: learning the craft of qualitative research interviewing, 2nd edn. SAGE Publications Inc, Thousand Oaks
Marsh CL (2011) Action research for sustainability in the UK National health service (NHS): linking theory and practice in organizational strategy for sustainable development. The University of Leeds, Leeds
Marshall J, Coleman G, Reason P (eds) (2011) Leadership for sustainability: an action research approach. Greenleaf Publishing Limited, Sheffield
Marsick VJ, Maltbia TE (2009) The transformative potential of action learning conversations. In: Mezirow J, Taylor EW Associates (eds) Transformative learning in practice: insights from community, workplace and higher education. Jossey-Bass, San Francisco, pp 160–171
McGill I, Brockbank A (2004) The action learning handbook: powerful techniques for education, professional development and training. RoutledgeFalmer, Abingdon
McMillin J, Dyball R (2009) Developing a whole-of-university approach to educating for sustainability: linking curriculum, research and sustainable campus operations. J Educ Sustain Dev 3(1):55–64
McNiff J, Whitehead J (2002) Action research: principles and practice, 2nd edn. RoutledgeFalmer, London
Minichiello V, Aroni R, Timewell E, Alexander L (1990) In-depth interviewing: researching people. Longman Cheshire, Melbourne
Molnar E, Mulvihill PR (2003) Sustainability-focused organizational learning: recent experiences and new challenges. J Environ Plan Manage 46(2):167–176
Moore J (2005) Barriers and pathways to creating sustainability education programs: policy, rhetoric and reality. Environ Educ Res 11(5):537–555
Nascimento Botelho Md, Kowalski R, Bartlett S (2010) Buttercups and daisies: building a community of practice amongst teachers in a Brazilian university. Educ Action Res 18 (2):183–196

Noffke SE (2009) Revisiting the professional, personal and political dimensions of action research. In: Noffke S, Somekh B (eds) The SAGE handbook of educational action research. SAGE Publications Ltd, London, pp 6–24
Reason P, Bradbury H (2008) Introduction. In: Reason P, Bradbury H (eds) The SAGE handbook of action research: participative inquiry and practice, 2nd edn. SAGE Publications Ltd, London, pp 1–10
Revans R (2011) ABC of action learning. Gower Publishing Limited, Farnham
Senge P (2006) The fifth discipline: the art and practice of the learning organisation, 2nd edn. Random House Business Books, London
Siebenhüner B, Arnold M (2007) Organizational learning to manage sustainable development. Bus Strategy Environ 16:339–353
Silverman D (2006) Interpreting qualitative data, 3rd edn. SAGE Publications Ltd, London
Somekh B (2006) Action research. a methodology for change and development. Open University Press, Buckingham
Sterling S (2004) Higher education, sustainability, and the role of systemic learning. In: Corcoran PB, Wals AEJ (eds) Higher education and the challenge of sustainability: problematics, promise and practice. Kluwer Academic Publishers, Dordrecht, pp 49–70
Tilbury D (2012) Higher education for sustainability: a global overview of commitment and progress. In: GUNI (ed) Higher education in the world 4. higher education's commitment to sustainability: from understanding to action. Global University Network for Innovation, Barcelona, pp 18–28
Tilbury D, Podger D, Reid A (2004) Action research for change towards sustainability: change in curricula and graduate skills towards sustainability: final report prepared for the Australian government department of the environment and heritage and Macquire University. Macquire University, Sydney
Velazquez L, Munguia N, Platt A, Taddei J (2006) Sustainable university: what can be the matter? J Clean Prod 14:810–819

Authors Biography

Gisela Cebrián has recently been awarded her PhD in Education at the Southampton Education School, UK. Her doctoral research focused on the development of an organisational learning model to embed Education for Sustainable Development within the undergraduate curriculum at the University of Southampton. Gisela holds a bachelor's degree in environmental sciences from Universitat de Girona and a Master's degree in Mathematics and Science Education research from Universtiat Autònoma de Barcelona (UAB). Gisela is member of the Edusost Research Network on Education for Sustainability and the Research Group Complex at UAB. She has been involved in several research projects on Catalan universities and climate change, professional competencies in sustainability in higher education and student-centred and problem based learning approaches. Gisela's research interests encompass education for sustainability and innovative teaching and learning approaches within the higher education context.

Marcus Grace is Head of Science Education Research in the Education School at the University of Southampton, UK. His research interests centre around teaching about socio-scientific issues, i.e. issues (including sustainability issues) which have a basis in science and a potentially large impact on society. His current work focuses on developing realistic teaching approaches to improving health-related attitudes and behaviour among young people. Marcus taught science at secondary schools in London for many years. He is now Chair-elect of the Academic Committee of ERIDOB (European Research in Didactics of Biology) and member of the UK's Society of Biology Education, Training and Policy Committee.

Debra Humphris was appointed Pro Rector (Education) in October 2012 and in August 2013 became Vice Provost (Education). Prior to her appointment at Imperial College, Debra was the Pro Vice Chancellor (Education and Student Experience) at the University of Southampton since 2008. Professor Humphris was responsible for the creation, leadership and delivery of the University's education strategic plan, which emphasised enhancing the student experience and the importance of graduate employability. In 2003 she was appointed Director, Health Care Innovation Unit at the University of Southampton, which focused on education innovation in health and social care. She first joined the university in 2000 as Director of the New Generation Project, which explored inter-professional education in health care programmes. Originally qualifying as a Registered Nurse, she later held positions as Senior Research Fellow at the South Thames Regional Health Authority, and at St George's Hospital Medical School. Professor Humphris achieved her PhD in 1999, which considered the implementation of policy into clinical practice.

A New Venture into Sustainability—Establishing a Multi-disciplinary Sustainability Project in a Higher Education Institution and Engaging the University Population

Scott Reid and Victor Pepple

Abstract

Establishing a student union led university wide project focussed on environmental sustainability in a higher education institution comes with challenges and success stories. This paper outlines the ongoing development of the UK National Union of Students funded Bedfordshire Green Hub project at the University of Bedfordshire which is managed by The University of Bedfordshire Students' Union. Launched in October 2013, the project is a new venture into sustainability for the Students' Union. The focus of this paper is to present the progress so far, highlighting some of the areas that are important to project success, this paper will focus particularly on student engagement. This paper will be useful to institutions already running a multi-disciplinary sustainability project, and those who are thinking about starting their own venture into sustainability. Sustainability encompasses many areas. The university wide nature of the project means that there are diverse project initiatives to manage as the project spans many departments and requires support from different groups of people. Early findings show the importance of understanding different approaches to best engage with a wide range of interests. Sustainability means different things to different people. Individuals have their own priorities, but these can form part of a wider, collaborative goal. This paper will illustrate the importance of developing a shared understanding, goal and language with multiple

S. Reid (✉)
University of Bedfordshire Students' Union, University of Bedfordshire, Park Square, Luton LU1 3JU, UK
e-mail: scott.reid@beds.ac.uk

V. Pepple
Engineering Business Management Postgraduate Unit, University of Bedfordshire, Park Square, Luton LU1 3JU, UK
e-mail: victor.pepple@study.beds.ac.uk

W. Leal Filho et al. (eds.), *Integrative Approaches to Sustainable Development at University Level*, World Sustainability Series, DOI 10.1007/978-3-319-10690-8_49

partners. The project aims to raise awareness in the office, classroom and the wider community. A final aim of this paper is to share progress from each project environment so as to develop a pool of best practice and learning resource for integrating sustainability as an organizational culture within the higher education sector.

Keywords
Sustainability · Higher education · Students' union · Engagement

1 Introduction: A New Venture into Sustainability

One of the major challenges of this century is the environmental impact of carbon footprints from human activities. As the quest for a more sustainable planet increases, higher education institutions are well placed to champion the modelling of good sustainability practice, equipping future employees with the skills and knowledge required to make a positive impact. There are a set of skills associated with sustainability and sustainable development that are broadly transferrable also known as 'skills to last' (UNEVOC 2003). Skills important to sustainability include creativity, innovation, futures thinking and critical thinking. There are also broad skills such as responsibility and the importance of ethics and values. By launching a sustainability project in a higher education institution, it provides an opportunity to start broadening a student's skill set, offering extra-curricular experiences and practical opportunities to innovate.

The University of Bedfordshire Students Union (Beds SU) in the United Kingdom champions a number of ongoing sustainability projects through the National Union of Students (NUS) 'Green Fund'(National Union of Students 2014) funded Bedfordshire Green Hub. The Bedfordshire Green Hub is university wide, engaging with many stakeholders from students to academic staff in a range of departments.

The NUS funding secured in July 2013 is used to fund a number of university wide sustainability initiatives such as Student Eats projects, a student led green project fund, work on education for sustainable development and an innovative pop up green pod. The Bedfordshire Green Hub also supports existing NUS energy saving projects and actively supports and promotes Fairtrade and Recycling. These are student focused, university wide projects that aim to both raise environmental awareness and foster a sense of responsibility among the student population. A list of project initiatives is included below in Fig. 1.

The Bedfordshire Green Hub is very much a new venture into sustainability for Beds SU, the Bedfordshire Green Hub has diverse project initiatives, this requires not only support from different groups of people, but also needs to facilitate individual priorities working towards a common goal.

Initiative	Description
Student Led Green Project Fund	A project fund that can provide funding to current students sustainability project ideas.
Snap It Off	An NUS campaign where students are encouraged to take photos of lighting that is left on unnecessarily and submit them online.
Student Switch Off	An NUS not-for-profit campaign that encourages students to save energy when living in university halls of residence
Student Eats	Another NUS campaign focused on students growing their own fruit and vegetables. The Bedfordshire Green Hub has a garden containing five raised beds in an enclosed space.
Fairtrade	The Bedfordshire Green Hub actively supports and promotes Fairtrade.
Greening Curriculum	Working towards Education for Sustainable Development and making links with academics at the University of Bedfordshire
Recycling	Actively supporting and promoting recycling both in the Beds SU offices and to the wider student population
Pop Up Green Pod	A unique inflatable promotional pod that will tour the University of Bedfordshire campuses and beyond – running workshops and acting as a promotional hub.

Fig. 1 Bedfordshire Green Hub project initiatives

However successful the green projects have been, securing the involvement of all stakeholders across the university and beyond, has offered key insights that must be embraced and managed in order to make progress. Consequently the paper is structured to share lessons learnt from the experience of the Bedfordshire Green Hub, for this reason this paper may be of interest to University Student Unions' who are about to start their very own venture into sustainability.

This paper broadly aims to do the following:

- To provide useful insights into launching a sustainability project from a Students' Union.
- To share engagement insight into running a diverse, university-wide sustainability project.

2 Methodology

The nature of this paper contributes to it being a more practical paper, based on observation and experience. Data will refer to observations from informal meetings and events, feedback, engagement levels and achievements.

The aims and general comment will be backed up by data where possible, this is mainly qualitative data rather than quantitative.

This paper aims to offer useful, practical insights.

3 Findings

Since launching in October 2013 the Bedfordshire Green Hub has increased engagement with sustainability at the University of Bedfordshire, specifically in the student population but also more broadly within University departments. There are many different project initiatives to manage, a list can be seen in Fig. 1. These diverse project initiatives bring with them the challenge of different forms of engagement. This is an issue of communication.

3.1 Moving Past 'Green'

In the early period of the Bedfordshire Green Hub there was a perceived perception that the project was very 'green' focused, namely recycling and waste. As the project has progressed this general perception appears to have changed. Both staff and students have started to engage with the idea that the Bedfordshire Green Hub is about more widespread than typical 'green' and embraces a broader spectrum of initiatives.

Effective engagement produces a learning curve. Focus groups have discovered that students in the first instance would struggle to define 'sustainability' or 'environmental sustainability' beyond waste and recycling issues or the narrow spectrum of sustainability that is relevant to their course of study. A simple timeline exercise was carried out (Fig. 2). Students after completing the timeline exercise were more able to see the interconnectedness of sustainability and to realise the broad range of events that contribute towards the environment that we have today.

A timeline exercise was carried out during two focus groups with students from specific departments. This timeline aimed to illustrate global events that are relevant to sustainability during an unspecified period of time.

At first, students struggled to identify events that were relevant, thinking perhaps that they didn't know enough about sustainability. As the session progressed and interaction expanded, numerous items were placed on the timeline.

The timeline closely related to the subject that the students studied – broadly Life Sciences. More recent detail (the last decade) was more in depth. Students focused on waste and recycling policies locally, recent extreme weather events and even the Fukushima Nuclear Disaster. As students went further back in time they illustrated China's One Child Policy, World War Two and even as far back as the Agricultural Revolution, Population Expansions and the Big Bang.

This exercise stimulated thinking and linked events. By the end of the exercise students acknowledged that they had more knowledge about sustainability than first thought, and felt more interested and more intrigued.

Fig. 2 Timeline exercise

Students left the focus group commenting that they had a better understanding of sustainability, and that actually they could get more involved with it than they think. Several students registered their interest with volunteering for the Bedfordshire Green Hub.

3.2 Communication

Communication of project initiatives is one of the most important processes that the Bedfordshire Green Hub undertakes. As sustainability is very new to Beds SU and on this scale, to the University of Bedfordshire, it was important to have a range of project initiatives that can appeal to a wide range of student interests. Communication is very much key to engagement.

At the University of Bedfordshire there aren't any specific environmental undergraduate courses. There are environmental aspects in a range of courses, but these are not dedicated or stand-alone. Having no stand-alone undergraduate environmental course offers the challenge of not having one particular course or department to focus engagement (Earth Sciences for example), but also offers the opportunity of reaching a wider number of students.

Particular findings suggest that a wide range of students are involved with the project initiatives, not just students from a particular year group or discipline. This was highlighted in the recruitment of Student Staff, with the current team of student staff studying different subjects, in different year groups and with different levels of expertise and experience. This is also true for the volunteers who are registered with the project, there is no direct correlation between courses studied and year groups. As this finding extends into the day to day engagement with project initiatives it creates a challenge. How do you create a one size fits all campaign for specific initiatives such as the Student Led Project Fund, which naturally appeals to the more innovation and business minded students? One solution has been to learn by experience, to realise that project initiatives can evolve to facilitate student feedback.

It is important to appeal to a wide range of students but to also target specific student groups, to make engagement more specific and more detailed—this is a challenge. It is noted that as a project progresses it becomes easier to target specific student groups. One of the biggest challenges with engagement is the task of 'hooking' people in. There is a difference between one off engagement and continued support. Feedback from a sustainability workshop has shown that students ask for sustainability engagement that is fun and varied. Events that are fun are generally harder to convey a deeper message that continues to engage students. By combining project initiatives and engagement techniques then it is possible to create events that can hook people in. To launch the Student Eats aspect of the Bedfordshire Green Hub, the project was coupled with Fairtrade for a one-off 'Be a producer' event. This combined the Student Eats project with Fairtrade fortnight. Here the message of Fairtrade could be communicated in a practical environment.

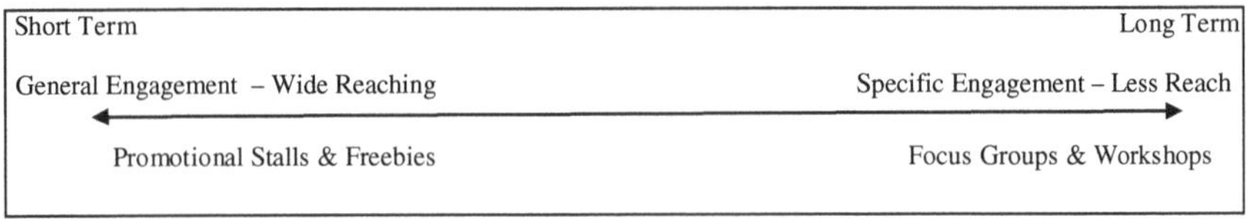

Fig. 3 Spectrum of engagement

This event both raised awareness and educated with regards to Fairtrade, but also engaged students with the Student Eats garden.

Engagement can to an extent be viewed as a spectrum (Fig. 3). At one side of the spectrum there is general engagement, this is wide reaching and more suited to promoting the Bedfordshire Green Hub brand as a whole. At the other side of the spectrum there is specific engagement, this has a reduced reach but is better for targeting specific groups of students (such as a focus group for Life Sciences) or for specific project initiatives (a workshop for student led projects). More general engagement often needs to be followed up with specific engagement, such as a follow up email. Therefore, general engagement (left of the scale) is better at short-term engagement whereas specific engagement (right of the scale) is better placed for providing a 'hook' and consequently more appropriate for long term followership.

Through observation, it has become clear that students are better engaged by fellow students, rather than (non-student) staff. The addition of student staff to the Bedfordshire Green Hub project has had a positive impact in this area. When running promotional stalls, students appear to be more engaged when talking to fellow students. There is an aspect of role models and students following students.

Students however aren't the only group that the project is communicated to, staff are also of high importance. More specifically this is Beds SU staff but generally, this is University wide for certain project initiatives. Early findings with Beds SU show the importance of not communicating too many channels, too quickly. Staff engagement has proved to be best if awareness and understanding is the priority, not what the project demands and needs. As the project has progressed, more and more changes have been asked of Beds SU staff. It has been highly important to respect the workload of fellow colleagues and also to be aware that sustainability is not always the top of an individuals list. Broadly speaking, sustainability often has to fight for attention, successful and well timed engagement can make this less of a fight and a smoother process. The Bedfordshire Green Hub has been delighted by the levels of Beds SU staff engagement.

Staff also have a role to play in advertising a project to students. Academic staff can engage with specific courses and have a somewhat captive audience in lecture theatres. Whilst holding focus groups attendance was secured by academic staff reminding students during lectures. There have also been experiences of students asking about the Bedfordshire Green Hub as a result of their lecturers informing them. This produces additional dimensions, the need to communicate to academic staff.

It is not just academic staff who are important to project success. There have been numerous examples of key relationships with other University support staff. Such as with the Sustainability Manager for the University, it has been important to collaborate on aspects of our work as there is a degree of overlap. Another example could be with staff at the campus of the Student Eats project, a strong relationship has been built with staff at that particular campus, without this managing the Student Eats project would be difficult to co-ordinate. Engagement with non-academic staff requires a different approach that understands the needs of different departments and the priorities.

A departing point on communication is that there is a 'snow ball' effect that takes place. Engagement doesn't just happen overnight. It requires the full spectrum of engagement in order to fully 'hook' people.

3.3 Timing

Closely linked to communication is timing. The various project initiatives have needed careful consideration with regards to launch dates and event dates. There is the very big divide between office working hours and student working hours. Various factors must be considered such as exam dates, holiday dates and even what days are generally busy in the University.

Further strengthening the link to keeping communications constant, it has been important to keep a constant message when launching events and project initiatives, but importantly not to lose the message. One particular example being during National Go Green Week 2014. This was a very successful week for the Bedfordshire Green Hub and a large number of both students, and staff were engaged with. During the week many events took place including a cooking demonstration, a smoothie bike, various promotions and campaigns and an acoustic night. The acoustic night provides an interesting example. An acoustic night was planned for the latter part of Green Week, during the evening. This event was well attended due to a final promotional push on the evening of the event but awareness leading up to the event wasn't high. The event was a success and many of those that attended expressed an interest to attend further acoustic themed events. It appears that the message was somewhat lost during the widespread promotion of Go Green Week 2014. This is a learning point to note, holding the event that corresponded to Green Week was very positive, but in future it could be advisable to hold such events separate to what are already communications saturated campaigns.

Something to also consider is the lead in time for events. Some events are simple and can be planned at short notice. Others need longer to plan and facilitate. For example, if items are required to be ordered in, if volunteers need to be sought or if materials need be created by a communications team. Early planning is essential for a successful event.

3.4 Common Goals

Sustainability has many definitions and is widely written about and debated about. Sustainable development has even more definitions and in the eyes of many is in fact an oxymoron. Can development ever be sustainable? The definition of Sustainable Development has caused much debate for many years (Redclift 2002). It is a confusing arena that both divides people and brings people together to work towards a common cause. Sustainability is one area that requires collective action, and individuals working towards a common goal, but it is also an area where individuals have their own priorities.

Sustainability means different things, to different people. This is both a beauty and a challenge. The beauty of this being that it remains open and wide reaching. This does present a challenge of engagement. One conflict that can be introduced is that of the individual and that of the wider, common goals of sustainability. Many individuals have their own individual priorities. Combining these individual priorities to work towards the common goals of sustainability can be a challenge. The Bedfordshire Green Hub is a university-wide project that requires engagement from many individuals. It has been interesting to observe the different priorities of different groups. Students generally have a focus on employability and grades (an individual priority). As tuition fees have increased and thus the financial burden on students, gaining a degree has become increasingly important, increasing the focus on employability (Yorke 2004). On the other hand, an academics focus could be on research and their teaching obligation.

Competing priorities of education, research and service among students and staff within the university framework have some dilution on the concentration of focus on sustainability programmes (Krizek et al. 2012). Furthermore, whereas teaching staff might be more interested in developing curricula through seeking out ways of integrating sustainability into it, students who use halls of residence may find recycling and 'switch it off' initiatives rewarding.

Finding common ground between these factors is important. Sustainability has the opportunity to be that common ground, and a higher education sustainability project can be the facilitator. An example here is with the Student Eats aspect of the Bedfordshire Green Hub. This project strand has its focus on student volunteers who volunteer at the allotment site and take ownership of the planted seeds and continued development of the crops. The boundaries of this project don't stop with students as academic staff can just as easily have input. One collaboration that has emerged has been with an academic member of staff who expressed an interest to run a research project relating to the breakdown of timber in a controlled compost bin. This is evidence of academics working alongside students, working on a common ground (the Student Eats garden) towards a common goal (a more sustainable future).

Having a wide range of project initiatives enables greater levels of engagement as different initiatives appeal to different individual priorities. This in turn creates opportunities for students and staff to make links with other issues that are not always anticipated. Links have been voiced to health and wellbeing aspects such as smoking, local foods and nutrition. This creates a greater common goal, able to bring together more people.

The Bedfordshire Green Hub has acted as a platform for multi-disciplinary, university-wide engagement. It is specifically project initiatives that become this platform such as Student Eats or Greening the Curriculum.

4 Impact

Ultimately, the project has been able to secure the collaboration of student and staff engagement in working towards a more sustainable university community. Early findings suggested that not all students had heard of the Bedfordshire Green Hub since its launch. However, since promotional campaigns during Go Green Week, recent feedback has suggested that a growing number of students are aware of the Bedfordshire Green Hub.

At the point of writing the Bedfordshire Green Hub has been established for 6 months. In this time the brand has been well established. With website, social media, promotional materials, hoodies and more, it is perceived through informal conversations with students that more and more people are aware of the Bedfordshire Green Hub. This has enabled the more specific engagement opportunities discussed, which can act as a hook. Go Green Week acted as a catalyst for engagement. The combination of events, promotional stalls and a marketing campaign resulted in over 200 students and staff registering their interest in the Bedfordshire Green Hub. This highlights the importance of visible, face to face engagement.

The Bedfordshire Green Hub has had an impact over different project environments. From the office, to the class-room to extra-curricular activities. Within the office environment the Bedfordshire Green Hub has already made a positive impact. One example being the addition of reusable paper boxes that fill up regularly or increased awareness about leaving computers switched on. The Beds SU offices are now more sustainable as a result of the Bedfordshire Green Hub project.

In the classroom there has also been a positive impact. Largely due to collaboration with academic staff. Focus groups have taken place as has a sustainability workshop where ensuring a curriculum that is ‘fit for the future’ was discussed and ways of embedding sustainability issues into the curriculum were explored.

The most visible project environment is involving extra-curricular activities. The impact here is more measureable. With increasing numbers of registered volunteers, visible promotional stalls and continued participation at the Student Eats garden. It is the extra-curricular environments that can offer students and staff the chance to learn skills that can’t always be taught in the classroom. Theory suggests that the greater the involvement of a student in university life and extra curricula activities, then the amount of student learning and personal development will be greater (Astin 1999).

5 Conclusions

To conclude, the progress so far of the Bedfordshire Green Hub has offered many lessons and shared learning opportunities. From the early days of the project where the project needed to move past the 'green' stereotype, to the ability to communicate and engage effectively. The spectrum of engagement that has been introduced highlights the differences between short term, widespread engagement that is somewhat shallow and the longer term, more focused and in depth engagement. This paper has aimed to offer learning examples that move on from communication towards other aspects such as timing and working towards a common goal. Sustainability can create a conflict between individual priorities and common goals. A sustainability project such as the Bedfordshire Green Hub can be that common goal. Finally the project has had a positive impact in different environments, from the classroom to the office and then on towards extra-curricular activities. There is still much progress to be made, but by continuing to learn and to understand the impact will surely be a positive one.

As the project progresses into year two, the lessons that have been learned with regards to engagement, communication and common goals will be used as wider community engagement is developed. Further broadening the project reach.

Launching a wide reaching, higher education sustainability project is both a unique opportunity and unique experience. It has the opportunity to bring people together and to make a positive impact.

References

Astin AW (1999) Student involvement: a developmental theory for higher education. J Coll Stud Dev 40(5):518–529. http://kvccdocs.com/KVCC/2013-Spring/FY125-OLA/content/L-17/Student%20Involvement%20Article.pdf. Accessed 25 Mar 2014

Krizek KJ, Newport D, White J, Townsend AR (2012) Higher education's sustainability imperative: how to practically respond? Int J Sustain High Educ 13(1):19–33. http://kevinjkrizek.org/wp-content/uploads/2012/04/SustainabilityCU.pdf. Accessed 16 Mar 2012

National Union of Students (2014) Students Green Fund. http://www.studentsgreenfund.org.uk/ Accessed 19 Aug 2014

Redclift M (2002) Sustainable development exploring the contradictions. Routledge, London

UNEVOC (2003) Skills to last—broadly transferable sustainable development skills for the Canadian workforce. UNEVOC, Winnipeg

Yorke M (2004) Encouraging the development of employability. ESECT, York

Authors Biography

Scott Reid is the Bedfordshire Green Hub Co-ordinator at the University of Bedfordshire Students' Union. Having studied Environment and Sustainability BSc at Keele University and then a work-based MA Leadership for Sustainable Development at Forum for the Future, he has a broad spectrum of awareness about sustainability issues. His current interest is in Higher Education sustainability and the communications associated with this.

Victor Pepple received his Bachelors in Petroleum and Gas Engineering (Gas Engineering Option) from the University of Port Harcourt where he studied from 2005 to 2011. He is currently studying at the University of Bedfordshire for a Master of Science degree in Engineering Business Management. His interest spans across integrating sustainability concerns into Engineering processes and operations.

MIX
Papier aus verantwortungsvollen Quellen
Paper from responsible sources
FSC® C105338

If you have any concerns about our products,
you can contact us on
ProductSafety@springernature.com

In case Publisher is established outside the EU,
the EU authorized representative is:
Springer Nature Customer Service Center GmbH
Europaplatz 3, 69115 Heidelberg, Germany

Printed by Libri Plureos GmbH
in Hamburg, Germany